W0018656

Transgenics in Endocrinology

CONTEMPORARY ENDOCRINOLOGY

P. Michael Conn, SERIES EDITOR

Difficult Cases in Endocrinology, edited by *MARK E. MOLITCH,* 2002
Selective Estrogen Receptor Modulators: *Research and Clinical Applications,* edited by *ANDREA MANNI AND MICHAEL F. VERDERAME,* 2002
Transgenics in Endocrinology, edited by *MARTIN MATZUK, CHESTER W. BROWN, AND T. RAJENDRA KUMAR,* 2001
Assisted Fertilization and Nuclear Transfer in Mammals, edited by *DON P. WOLF AND MARY ZELINSKI-WOOTEN,* 2001
Adrenal Disorders, edited by *ANDREW N. MARGIORIS AND GEORGE P. CHROUSOS,* 2001
Endocrine Oncology, edited by *STEPHEN P. ETHIER,* 2000
Endocrinology of the Lung: *Development and Surfactant Synthesis*, edited by *CAROLE R. MENDELSON,* 2000
Sports Endocrinology, edited by *MICHELLE P. WARREN AND NAAMA W. CONSTANTINI,* 2000
Gene Engineering in Endocrinology, edited by *MARGARET A. SHUPNIK,* 2000
Endocrinology of Aging, edited by *JOHN E. MORLEY AND LUCRETIA VAN DEN BERG,* 2000
Human Growth Hormone: *Research and Clinical Practice,* edited by *ROY G. SMITH AND MICHAEL O. THORNER,* 2000
Hormones and the Heart in Health and Disease, edited by *LEONARD SHARE,* 1999
Menopause: *Endocrinology and Management,* edited by *DAVID B. SEIFER AND ELIZABETH A. KENNARD,* 1999
The IGF System: *Molecular Biology, Physiology, and Clinical Applications,* edited by *RON G. ROSENFELD AND CHARLES T. ROBERTS, JR.,* 1999
Neurosteroids: *A New Regulatory Function in the Nervous System,* edited by *ETIENNE-EMILE BAULIEU, MICHAEL SCHUMACHER, AND PAUL ROBEL,* 1999
Autoimmune Endocrinopathies, edited by *ROBERT VOLPÉ,* 1999
Hormone Resistance Syndromes, edited by *J. LARRY JAMESON,* 1999
Hormone Replacement Therapy, edited by *A. WAYNE MEIKLE,* 1999
Insulin Resistance: *The Metabolic Syndrome X,* edited by *GERALD M. REAVEN AND AMI LAWS,* 1999
Endocrinology of Breast Cancer, edited by *ANDREA MANNI,* 1999
Molecular and Cellular Pediatric Endocrinology, edited by *STUART HANDWERGER,* 1999
Gastrointestinal Endocrinology, edited by *GEORGE H. GREELEY, JR.,* 1999
The Endocrinology of Pregnancy, edited by *FULLER W. BAZER,* 1998
Clinical Management of Diabetic Neuropathy, edited by *ARISTIDIS VEVES,* 1998
G Proteins, Receptors, and Disease, edited by *ALLEN M. SPIEGEL,* 1998
Natriuretic Peptides in Health and Disease, edited by *WILLIS K. SAMSON AND ELLIS R. LEVIN,* 1997
Endocrinology of Critical Disease, edited by *K. PATRICK OBER,* 1997
Diseases of the Pituitary: *Diagnosis and Treatment,* edited by *MARGARET E. WIERMAN,* 1997
Diseases of the Thyroid, edited by *LEWIS E. BRAVERMAN,* 1997
Endocrinology of the Vasculature, edited by *JAMES R. SOWERS,* 1996

Transgenics in Endocrinology

Edited by

Martin M. Matzuk, MD, PhD

Chester W. Brown, MD, PhD

and

T. Rajendra Kumar, PhD

Baylor College of Medicine, Houston, TX

Humana Press
Totowa, New Jersey

999 Riverview Drive, Suite 208
Totowa, New Jersey 07512

www.humanapress.com

Cover design by Patricia F. Cleary.
Cover illustration: From Fig. 3E in Chapter 10 "Knockout and Transgenic Mouse Models that Have Contributed to the Understanding of Normal Mammary Gland Development," by Tiffany N. Seagroves and Jeffrey M. Rosen.
Production Editor: Kim Hoather-Potter.

This publication is printed on acid-free paper. ∞
ANSI Z39.48-1984 (American National Standards Institute)
Permanence of Paper for Printed Library Materials.

Printed in the United States of America. 10 9 8 7 6 5 4 3 2 1

Library of Congress Cataloging-in-Publication Data
Transgenics in endocrinology/edited by Martin M. Matzuk, Chester W. Brown, and T. Rajendra Kumar.
p. cm. — (Contemporary endocrinology)
Includes bibliographical references and index.
ISBN 0-89603-764-9 (alk. paper)
1. Molecular endocrinology. 2. Transgenic mice. 3. Transgenic organisms. 4. Endocrinology—Research—Methodology. I. Matzuk, Martin M. II. Brown, Chester W. III. Kumar, T. Rajendra. IV. Contemporary endocrinology (Totowa, N.J.)

QP187.3.M64 T73 2001
573.4'19—dc21

00-067291

Preface

"Man's mind stretched to a new idea, never goes back to its original dimensions."
Oliver Wendell Holmes

The latter part of the 20th century has seen an amazing change in how we view and synthesize endocrinology. Prior to the 1980s, we understood endocrine disorders and the field of endocrinology through patients with genetic mutations, protein purification, physiological experiments in humans and whole animals, tissue culture cells, and radio-immunoassays. Little did we know that the field of endocrinology (and all of genetics) would move by leaps and bounds because of a simple mammalian model—a mouse that grew twice as fast as its fellow littermates. As students at the time, we were fascinated by these mice, glorified by their appearance on the 1982 and 1983 covers of *Nature* and *Science*. These transgenic mice, created by Drs. Ralph Brinster and Richard Palmiter and colleagues, were the first endocrine models created by genetic manipulation—mice carrying a mouse metallothionein I promoter driving the expression of either rat growth hormone (1982) or human growth hormone (1983). The expression of the foreign growth hormone genes (transgenes) resulted in "gigantic mice" because of the growth hormone excess. Clearly, for our field and all of biology, the phrase "a picture is worth a thousand words" rang true on those autumn days in 1982 and 1983 and generated a movement that revolutionized our thinking. Little did we realize that a second revolution was already evolving that would take hold of the field in the decade to follow.

In the early 1980s, Dr. Martin Evans' laboratory first isolated cell lines from the inner cell mass of mouse blastocysts that could be propagated in culture, maintain their pluripotency, and contribute to the germline. These so-called embryonic stem (ES) cell lines, first used with retroviral infection in an attempt to model the human Lesch-Nyhan syndrome, became valuable genetic conduits to mimic and better understand endocrine disorders and systems. In parallel with the development of ES cells, Oliver Smithies and colleagues showed in 1985 that they could achieve homologous recombination to correct a mutation in the human β-globin locus in mammalian tissue culture cells. Although this was heralded as a major breakthrough for the possible correction of human genetic diseases, it more importantly suggested that germline mutations of endogenous mammalian genes could be created. Homologous recombination in ES cell lines along with the so-called positive–negative selection strategy developed in the laboratory of Dr. Mario Capecchi, laid the foundation for "knockout" technology with more far-reaching implications than were envisioned at the early stages. The first knockout models were subsequently created with great excitement in the early 1990s including mice lacking the endocrine factors insulin-like growth factor II (1990), transforming growth factor-β1 (1992), and inhibin (1992). As you will see in the following chapters, thousands of transgenics have been created to study and manipulate the endocrine system. Some of these models have given expected results, whereas analysis of the phenotype of others has revealed novel functions for these endocrine factors. Clearly, transgenesis has given endocrinologists a new tool for understanding structure/function relationships in vivo.

In closing, we graciously thank all of the authors of *Transgenics in Endocrinology* for accepting our challenge to write state-of-the-art chapters on their specific topics. Writing reviews of ever-changing fields is not an easy task, but we honestly believe each chapter to be a work of art. Because all of the authors are experts in their respective areas, it has been a pleasure to read these bodies of work and to be part of the editorial process. We would also like to thank Michael Conn for having enough confidence in us to edit this volume of diverse topics, the first of its kind for the three of us. A great deal of thanks also goes to Ms. Shirley Baker who coordinated the correspondences with the authors and incorporated all of our editorial scribbles. Lastly, to all of the readers of this book, enjoy the chapters and the immense body of literature that has been published in the field of transgenics and endocrinology over the last two decades. We hope that *Transgenics in Endocrinology* will instill much excitement and insight into your endocrine research endeavors in the 21st century.

***Martin M. Matzuk,* MD, PhD**
***Chester W. Brown,* MD, PhD**
***T. Rajendra Kumar,* PhD**

Contents

CONTRIBUTORS

DOMENICO ACCILI, MD, *Developmental Endocrinology Branch, NIH-NICHD, Bethesda, MD*
ANDRZEJ BARTKE, PhD, *Department of Physiology, Southern Illinois University School of Medicine, Carbondale, IL*
RICHARD R. BEHRINGER, PhD, *Department of Molecular Genetics, University of Texas MD Anderson Cancer Center, Houston, TX*
DANIEL J. BERNARD, PhD, *Department of Neurobiology and Physiology, Northwestern University, Evanston, IL*
CAROLYN A. BONDY, MD, *Developmental Endocrinology Branch, NIH-NICHD, Bethesda, MD*
EMILIANA BORRELLI, PhD, *Institut de Genetique et de Biologie Moleculaire et Cellulaire, Strasbourg, France*
CHESTER W. BROWN, MD, PhD, *Department of Molecular and Human Genetics, Baylor College of Medicine, Houston, TX*
KATHLEEN M. CARON, PhD, *Department of Human Genetics, University of North Carolina-Chapel Hill, Chapel Hill, NC*
THOMAS CASE, BS, *Vanderbilt University Medical Center, Nashville, TN*
E. CHENG CHAN, PhD, *University of New Castle, Australia*
ALBERT S. Y. CHANG, MD, *Department of Molecular and Cellular Biology, Baylor College of Medicine, Houston, TX*
FRANCESCO J. DEMAYO, PhD, *Department of Molecular and Cellular Biology, Baylor College of Medicine, Houston, TX*
MARIA P. DE MIGUEL, PhD, *Kimmel Cancer Center, Thomas Jefferson University, Philadelphia, PA*
MARY JO DOHERTY, PhD, *Department of Molecular and Human Genetics, Baylor College of Medicine, Houston, TX*
PETER J. DONOVAN, PhD, *Kimmel Cancer Center, Thomas Jefferson University, Philadelphia, PA*
JULIA A. ELVIN, MD, PhD, *Departments of Pathology, and Molecular and Human Genetics, Baylor College of Medicine, Houston, TX*
THOMAS GÜNTHER, PhD, *Department of Obstetrics and Gynecology, Freiburg University Medical Center, Freiburg, Germany*
TOMONOBU HASEGAWA, MD, PhD, *Department of Pediatrics, Keio University, Tokyo, Japan*
NELSON D. HORSEMAN, PhD, *Department of Molecular and Cellular Physiology, University of Cincinnati, Cincinnati, OH*
THOMAS R. INSEL, PhD, *Emory University, Atlanta, GA*
PREETI M. ISMAIL, PhD, *Department of Molecular and Cellular Biology, Baylor College of Medicine, Houston, TX*
GERARD KARSENTY, MD, PhD, *Department of Molecular and Human Genetics, Baylor College of Medicine, Houston, TX*
SUSAN KASPER, PhD, *Vanderbilt University Medical Center, Nashville, TN*
RUTH A. KERI, PhD, *Department of Pharmacology, Case Western Reserve Medical School, Cleveland, OH*
T. RAJENDRA KUMAR, PhD, *Departments of Pathology, and Molecular and Cellular Biology, Baylor College of Medicine, Houston, TX*

MALCOLM J. LOW, MD, PhD, *Vollum Institute, Oregon Health Sciences University, Portland, OR*
JOHN P. LYDON, PhD, *Department of Molecular and Cellular Biology, Baylor College of Medicine, Houston, TX*
GRANT R. MACGREGOR, DPhil, *Center for Molecular Medicine, Emory University School of Medicine, Atlanta, GA*
NAOYA MASUMORI, MD, PhD, *Sapporo Medical University School of Medicine, Sapporo, Japan*
ROBERT J. MATUSIK, PhD, *Department of Urologic Surgery,Vanderbilt University Medical Center, Nashville, TN*
MARTIN M. MATZUK, MD, PhD, *Departments of Pathology, Molecular and Cellular Biology, and Molecular and Human Genetics, Baylor College of Medicine, Houston, TX*
MICHAEL MICHALKIEWICZ, DVM, PhD, *West Virginia University Health Sciences Center, Morgantown, WV*
YUJI MISHINA, PhD, *National Institute of Environmental Health Sciences/NIH, Research Triangle Park, NC*
JOHN H. NILSON, PhD, *Department of Pharmacology, Case Western Reserve Medical School, Cleveland, OH*
BERT W. O'MALLEY, MD, *Department of Molecular and Cellular Biology, Baylor College of Medicine, Houston, TX*
KEITH L. PARKER, MD, PhD, *Department of Internal Medicine, University of Texas Southwestern Medical Center, Dallas, TX*
MANIK PAUL, BS, *Vanderbilt University Medical Center, Nashville, TN*
MICHAEL J. REARDON, MD, *Department of Molecular and Cellular Biology, Baylor College of Medicine, Houston, TX*
JEFFREY M. ROSEN, PhD, *Department of Molecular and Cellular Biology, Baylor College of Medicine, Houston, TX*
ANDREA J. ROSS, PhD, *Center for Molecular Medicine, Emory University School of Medicine, Atlanta, GA*
MARCELO RUBENSTEIN, PhD, *Universidad de Buenos Aires, Buenos Aires, Argentina*
PAOLO SASSONE-CORSI, PhD, *Institut de Genetique et de Biologie Moleculaire et Cellulaire, Strasbourg, France*
TIFFANY N. SEAGROVES, PhD, *Department of Molecular and Cellular Biology, University of California, San Diego, CA*
SCOTT B. SHAPPELL, MD, PhD, *Vanderbilt University Medical Center, Nashville, TN*
SELMA SOYAL, PhD, *Department of Molecular and Cellular Biology, Baylor College of Medicine, Houston, TX*
TANIA THOMAS, PhD, *Vanderbilt University Medical Center, Nashville, TN*
DEANNE J. WHITWORTH, PhD, *Department of Molecular Genetics, University of Texas MD Anderson Cancer Center, Houston, TX*
TERESA K. WOODRUFF, PhD, *Department of Neurobiology and Physiology, Northwestern University, Evanston, IL*
LARRY J. YOUNG, PhD, *Center for Behavioral Neuroscience, Emory University School of Medicine, Atlanta, GA*
MORAG YOUNG, PhD, *Baker Medical Research Institute, Melbourne, Australia*
LIPING ZHAO, PhD, *Departments of Internal Medicine and Pharmacology, UT Southwestern Medical Center, Dallas, TX*

1 Germline Genetic Engineering Techniques in Endocrinology

Albert S. Y. Chang, MD, *Michael J. Reardon,* MD, *and Francesco J. DeMayo,* PHD

CONTENTS

INTRODUCTION

The latter half of the twentieth century exhibited great progress in the ability to understand and control the mammalian reproductive cycle and endocrinology. A direct result of the advancements in reproductive endocrinology was the ability to manipulate the mammalian genome. The advancements in reproductive endocrinology allowed foreign genes to be introduced in a regulated manner into the genome. Genes can now be ablated or mutated in subtle and specific ways, and entire genomes have been replicated—a process termed "cloning." Advances in reproductive endocrinology, combined with developments in molecular biology, led to an explosion in biotechnology at the end of the twentieth century. Although the advances in molecular biology were at the core of this explosion, the true power of the molecular techniques could not have been realized without the ability to generate living mammalian organisms with the appropriate genetic manipulations. This allowed the manipulations to be investigated in a model that met the constraints of developmental and physiological regulation. Using the combination of the advances in the disciplines of endocrinology and molecular biology, new models for development, physiology, and disease could be generated. The goal of this chapter is to review the technologies involved in the introduction, manipulation, and replication of the mammalian genome. This chapter examines these technologies and addresses the nuances and limitations of these approaches. Because the mouse is the easiest mammalian model to manipulate, this chapter will focus on the mouse model.

From: *Contemporary Endocrinology: Transgenics in Endocrinology*
Edited by: M. Matzuk, C. W. Brown, and T. R. Kumar

INTRODUCTION OF GENETIC ELEMENTS INTO THE MURINE GENOME

Success in the introduction of foreign DNA into a genome depends upon the ability to ensure that the DNA integrates into the germline. This allows the stable integration of the foreign DNA into the genome, and the transmission of the manipulation to future generations, thus establishing a new and novel strain of the organism. Therefore, techniques to introduce foreign DNA into a genome have centered on introducing the DNA into either the preimplantation embryo or the gamete. The most successful of these approaches has been the introduction of DNA into the male pronucleus of the one-cell mouse embryo *(1)*. However, recent advances in sperm freeze-drying and intracellular sperm injection (ICSI) of mouse oocytes may lead to the development of this technology for future use *(2)*.

Introduction of DNA into the preimplantation embryo ensures a high probability that the foreign DNA will incorporate into as many of the developing cells as possible. With introduction into these earlier cleavage stages, the probability of the integration of the DNA into the germline is higher. Two techniques have been successfully utilized to accomplish this goal. One approach utilizes murine retroviruses to infect eight-cell mouse embryos *(3)*. This approach has been very effective in generating animals with incorporation of the viral DNA into the murine genome. Because it involves a very hearty stage of preimplantation development, the eight-cell mouse embryo is very easily manipulated in vitro, and embryo transfer of these manipulated embryos results in a high percentage of mice born per embryo transferred. Because retroviral infection is very efficient with this approach, it holds a great deal of promise. However, there are two major flaws in this approach. First, since infection occurs at the eight-cell stage, the animals developed from such manipulations have been mosaic—the virus integrates at different sites in the different blastomeres of the eight-cell mouse embryo. Therefore, the integration sites needed to be sorted out in the first few generations of the newly established strain *(4)*. However, the fatal flaw in this approach was that one could not achieve reproducible expression of the foreign DNA. This was a result of the constraints of packaging the foreign DNA construction into the virus, in addition to still undefined reasons why the viral genome may have inhibited transgene expression when integrated into the murine genome *(5)*. Because of the limitations of the retroviral approach and the success of introducing DNA directly into the murine genome by microinjection of the one-cell mouse embryo, the latter approach has dominated the field, and has become the standard process for the generation of transgenic mice.

The ability to introduce foreign DNA into the murine genome by microinjection of DNA into the one-celled embryo was first reported by Gordon and colleagues in 1981 *(6)*. This technique has changed little in the 20 yr since its first report. The approach requires the ability to manipulate the one-cell mouse embryo, and is demanding in terms of skill and equipment. However, it confers the ability to reproducibly generate transgenic mice that express the transgene of interest. This approach has expanded many lines of research. First, it has made possible the investigation of elements that are responsible for the appropriate spatial and temporal regulation of gene expression in vivo. This has allowed the identification of tissue-specific promoters that will direct gene expression in an appropriate manner. Second, investigators can utilize these promoters to direct the expression of regulatory genes to desired tissues, allowing the analysis of the role these

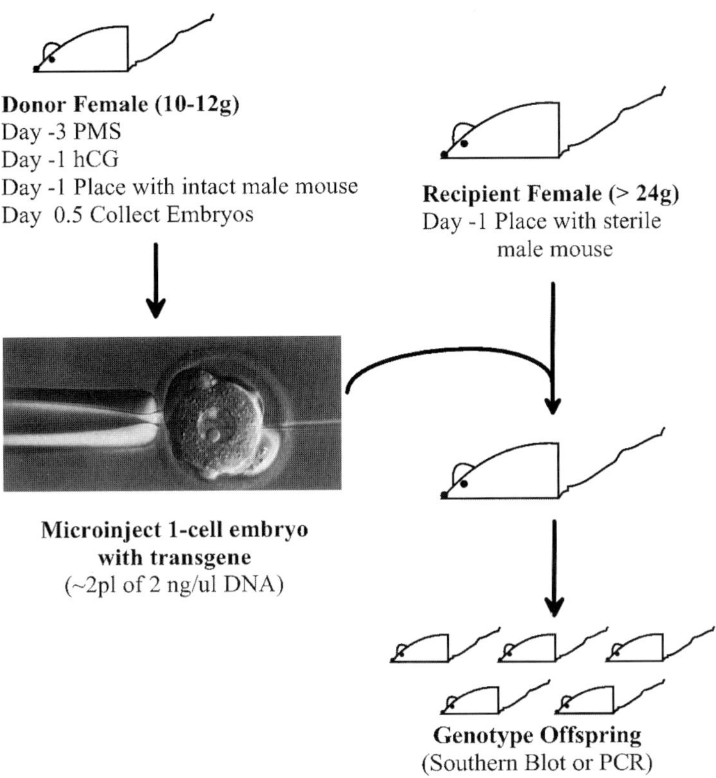

Fig. 1. Schematic for the generation of transgenic mice by DNA microinjection techniques. For embryo collection, female mice are treated with a superovulatory regimen of gonadotropins, consisting of pregnant mare serum gonadotropin (PMSG) followed by human chorionic gonadotropin (hCG) 48 h later. The donor females are then placed with a male mouse, and the embryos are collected the following morning. Embryos are harvested, processed, and readied for injection. Approximately 2 pl of DNA [2 ng/μL] is injected into the one-cell embryo and transferred to a synchronized, psuedopregnant female recipient mouse. The recipients are allowed to gestate to birth, and the offspring are screened for the presence of the transgene using Southern blot analysis or PCR.

proteins play in tissue development when their expression is disrupted. Finally, as a result of transgene integration, regulatory genes are sometimes disrupted. These insertional mutations have allowed the identification of genes involved in limb formation, which occurs at a low frequency, as well as the regulation of left/right asymmetry and other processes. Although the ability to generate transgenic mice by pronuclear microinjection is laborious, the power of the approach has proven the technology worthwhile *(1)*.

Hogan and colleagues have described the procedure for the generation of transgenic mice in detail *(7)*. Figure 1 describes the general schematic for the generation of transgenic mice by DNA microinjection, which consists of the preparation of the transgene, collection, injection, and transfer of the embryos, and identification of the transgenic offspring.

Preparation of Transgene DNA

In order to ensure maximum efficiency in the development of transgenic mice, the DNA fragment to be injected must be pure and accurately quantitated. The DNA to be injected must be digested with the appropriate restriction endonuclease(s) to free the transgene from prokaryotic vector sequences *(8)*. The incorporation of prokaryotic sequences inhibits expression of transgenes in the murine genome. The transgene is usually separated from the vector sequences by agarose-gel electrophoresis. The appropriate fragment is excised, purified, and quantitated by measuring the optical density (OD) of the sample at 260 nm. The purity of the fragment is estimated by measuring the OD at 280 nm and comparing the ratio of OD 260:OD 280. Once the isolated DNA is verified as the correct fragment and quantitated, the DNA is diluted to 2 ng/μL in TE buffer (10 m*M* Tris pH 7.5, 0.25 m*M* ethylenediaminetetraacetic acid [EDTA]). The DNA is then microinjected into the male pronucleus of one-cell mouse embryos. This buffer and concentration has been shown to be optimal for incorporation of the transgene into the murine genome *(8)*.

Embryo Collection, Microinjection, and Embryo Transfer

The strain of the mouse is important for the success of microinjection. Theoretically, transgenic mice can be made in any reproductively viable stain. However, to maximize the chance of obtaining transgenic mice, it is best to collect a large number of fertilized one-cell mouse embryos after treating female mice with a superovulatory regimen of gonadotropins and mating with an intact male mouse. The ability to collect these embryos is dependent upon strain, age, and dose of gonadotropins *(7)*. The majority of transgenic mice are made with either hybrid strains of mice, such as C57Bl/6J X SJL *(8)* or the inbred FvB/N strain *(9)*. Hybrid strains have the advantage of having females that respond well to the superovulatory regimen of gonadotropins and males that are sexually aggressive, mating at a high frequency with the induced female mice. However, the reproductive advantage of the hybrid males comes at the price of the lack of genetic uniformity that can be achieved by an inbred strain. The FvB strain of mice has the advantage of responding reasonably well to gonadotropins. Although inbred lines are not usually as virile and fertile as hybrids, the FvB strain produces suitable numbers of embryos for the generation of transgenic mice. This strain also offers the advantage of pronuclei that are slightly larger than other inbred strains of mice, giving the novice microinjector help with DNA injections *(9)*. These advantages have made the FvB strain of mice the major choice of mouse strain utilized in the generation of transgenic mice.

However, with the advent of knockout technology utilizing embryonic stem (ES) cells isolated from 129/Sv mice, the chimeras generated from this procedure have typically been bred to C57Bl/6J to rapidly analyze the contributions of the ES cells to the germline *(10)*. The mice initially generated from the chimeric cross are hybrid C57Bl/6JX129/Sv mice, since the two predominant strains for gene knockout analysis are 129/Sv and C57Bl/6J. Thus, the C57Bl/6J strain has re-emerged as a popular strain for transgenics.

Transgenic mice are produced as described by Hogan and colleagues *(7)*. Briefly, one-cell embryos are collected from mice receiving a superovulatory regimen of gonadotropins. The dose of gonadotropins and age of the donor female is critical. The response of female mice to the gonadotropins is maximal at prepuberty. Therefore, mice at 3 wk of age and 10–12 g of weight are optimal for the production of embryos. Also, the dose of gonadotropins may vary with the strain. Usually, 5 international units (IU) of pregnant mare serum gonadotropin (PMSG) or equine chorionic gonadotropin (eCG) is given by intraperitoneal injection, and 48 h later, the mice receive an intraperitoneal injection of human chorionic gonadotropins (hCG). The female is then placed with the male, and the embryos are collected the following morning. However, some inbred strains do not respond well to this regimen of gonadotropins, and usually respond to half the above dose to produce an acceptable number of embryos for microinjection.

The morning following mating, the females are examined for the presence of a post-coital vaginal copulatory plug. The mice are euthanized, and the oviducts are excised. The embryos are liberated and collected by flushing or mincing the oviducts. Once the embryos are collected, they are briefly incubated in medium with hyaluronidase to remove the cumulus mass. The embryos are washed, and the embryos containing visible pronuclei are separated from the unfertilized embryos, placed in a HEPES-buffered medium (M2), and readied for microinjection *(7)*. The embryos are microinjected with 2 pl of DNA at a concentration of 2 ng/μL *(8)*, using the aid of an inverted microscope equipped with differential contrast optics—usually Nomarski or Hoffman objective lenses—and micromanipulators.

Maximizing the production of transgenic mice is dependent upon the amount of DNA delivered to the male pronucleus, and also to the expansion of the pronucleus. It has been postulated that the hydrostatic pressure of the microinjection of DNA causes breaks in the chromatin. The DNA repair process then integrates the transgene into the chromatin. Therefore, it is important to deliver not only a specified amount of DNA, but also the appropriate volume of injection medium. The 2 pl of 2 ng/μL of DNA is the ideal compromise *(8)*. However, there is a fine line between the injection of the maximum volume of DNA and the volume tolerated by the embryo for survival. The usual survival after microinjection is 70–80%. In our studies, higher survival rates than this is associated with a low frequency of transgene incorporation.

The embryos surviving microinjection are transferred to the oviducts of synchronized, pseudopregnant ICR female mice. The recipients are allowed to gestate the embryos to birth, and to nurse the pups till weaning. Normally, 18–20 embryos are transferred to the oviducts of recipients, with only 20% of these transferred embryos surviving to birth because the survival of uninjected embryos is significantly higher, while the microinjected embryos suffer significant preimplantation loss. At weaning, the pups are identified by an ear tag, and screened for the presence of the transgene. The efficiency of the generation of transgenic mice is measured by the number of transgenic mice generated per number of embryos injected, and is generally at least 1–2% *(8)*.

Identification of Transgenic Mice

Once the mice born from the microinjection of DNA have been weaned, they are ready to be genotyped for the presence of the transgene. DNA is routinely isolated from tail biopsies and screened by either Southern blot analysis *(11)* or polymerase chain reaction (PCR) *(12)*. PCR screening of transgenic mice is by far the easiest and most rapid way to identify mice. However, limited information can be obtained by PCR screening of transgenic mice. Southern blot analysis, although slower and more laborious than PCR, is extremely useful to gain some initial information regarding the genetics of the founder transgenic mice (F_0). Southern blot analysis may be used to determine whether the integrated transgene is intact, as well as the number of copies of the transgene that have been integrated into the murine genome. The number of copies is important if the transgene integrates at more than one site in the murine genome, and helps to determine whether the transgene expression is proportional to the number of integrated transgenes—a rare phenomenon.

Transgenes usually integrate into the genome randomly, and usually in multiple, head-to-tail tandem copies *(13)*. The power of introducing DNA into the murine genome at the one-cell stage is that the incorporation of the transgene is early, and has the potential to be incorporated into every cell in the mouse. Therefore, when the founders are bred, they transmit the transgene to one-half of their offspring. However, in about 30% of cases, the transgene integrates at a later stage, and not every cell in the F_0 has incorporated the transgene. These animals are considered mosaics, and they transmit the transgene at a significantly lower frequency *(14)*. In some cases, the F_0 transmits the transgenes at a significantly higher frequency than 50%. Barring a phenotypic consequence of the transgene, this is usually a result of the transgene being integrated at two or more sites on separate chromosomes. It is important to monitor this phenomenon. F_0 mice with multiple integrations of the transgene must be bred to isolate each integration site. Each transgene-integration site should be treated as a separate strain of transgenic mouse, because the expression properties of each integration site may be significantly different. In this case, the use of Southern blot analysis of the first generation of offspring (F_1 mice) is important because it is necessary to design a strategy that can distinguish each of the transgene integration sites by either copy number or site of integration.

Transgene Expression

Once breeding lines are established for each F_0, expression analysis can safely be performed. The type of expression analysis conducted depends upon the specific question being addressed by the investigator. However, the generation of a specific phenotype from a transgene should not be used as the sole judgment in whether a transgene is expressed. RNA or protein analysis should be conducted to validate the expression of a transgene. Expression and phenotypic analysis should be confirmed with at least two independent lines of transgenic mice to ensure that the observed phenotype is caused by the expression of the transgene, and not the result of a change in the expression or disruption of an endogenous gene caused by the insertion of the transgene. In general, several rules should be followed regarding the expression of the transgene. First, the chromosomal environment in which the transgene is inserted usually influences its expression *(1)*. Therefore, without the incorporation of specific sequences to ensure that the transgene can be "insulated" from surrounding elements, the expression of the transgene will vary from F_0 to F_0. Second, expression of the transgene is rarely proportional to the number of copies of the transgene that have been inserted into the mouse genome. Only in rare instances do promoter fragments contain elements that allow the transgene expression to be proportional to the number of transgenes integrated into the chromatin *(15)*. Third, transgene expression is usually dependent upon the incorporation of genomic sequences in the transgene. Thus, transgenes containing cDNA sequences rather than genomic sequences are expressed at a significantly reduced frequency *(6)*. As a rule, genomic sequences—i.e., introns—should be incorporated into the transgene to maximize expression. First priority should be given to designing transgenes that contain the sequences corresponding to the gene to be expressed. If this is not possible, minigenes can be constructed from heterologous introns to enhance expression. However, no guarantee can be given that a prescribed minigene will be effective in conferring high levels of expression of the transgene *(17)*.

Bitransgenic Regulatory Systems

There are several disadvantages to the use of traditional transgenic technology in the generation of animal models. First, the initiation of transgene expression in mice is under the sole control of the elements regulating the developmental expression of these specific transgene promoters. Second, the level of expression of transgenes under the control of these promoters is dependent upon the relative strength of transcription of these promoters, as well as the site of integration of the transgene. Finally, the expression of the transgene is continuous and irreversible in the animal. This approach makes it difficult to differentiate the critical secondary events required for the generation of a specific phenotype. A system in which transgene expression could be regulated by the investigator would be a significant improvement over current models. Such a system could be used to investigate the changes resulting from the acute expression of the transgene. The system could also be used to examine the result of withdrawing transgene expression and identifying the events that follow. Over the last decade, three bitransgenic regulatory systems have been developed that allow for controlled transgene expression.

These bitransgenic systems consist of two lines of transgenic mice generated from two transgene constructions, as shown in Fig. 2. One transgene construction, the regulator, encodes a transcription factor with transcriptional regulatory activity controlled by the administration of an exogenous compound (ligand). The regulator transgene is placed under the control of DNA promoter elements needed to express the regulator in the tissue of interest. The second transgene construction, the target or responder, contains the gene of interest that has been placed under the transcriptional control of the DNA *cis*-acting elements that are responsive to the regulator. When the regulator line of transgenic mice and the target line of mice are crossed and the bitransgenic mouse is administered the exogenous compound, the target gene will be expressed. The bitransgenic regulatory system should fulfill the following criteria: 1) The target gene will be silent in the absence of the regulator; 2) The regulator will not impart a phenotype in the absence of the target gene; 3) In the bitransgenic animal, the target gene will be silent in the absence of the ligand; and 4) The ligand will not impart a phenotype. The ability to generate animals in which transgene expression can be regulated is now possible because of the development of three bitransgenic regulatory systems. These systems allow transgene expression to be regulated by the administration of either an insect hormone (ecdysone); an antibiotic (tetracycline); or an antiprogestin (RU486).

The most recent of the ligand-inducible bitransgenic regulatory systems is the ecdysone receptor-based regulatory system. This system exploits the insect steroid hormone receptor—the ecdysone receptor—as the regulator to control transgene expression in mice. Although this system fulfills the above criteria for a regulatory system, it is complicated by the fact that the transcriptional regulatory activity of the ecdysone receptor is dependent upon dimerization with members of the retinoic acid receptor family *(18,19)*. Therefore, in order for the bitransgenic system to be effective, the regulator transgene must contain both the ecdysone receptor and a member of the retinoic-acid receptor family under the control of a tissue-specific promoter. Despite this limitation, this system has been demonstrated to be effective in the regulation of the expression of the *Escherichia coli* β-galactosidase transgene in the T-cells of transgenic mice *(20)*. Recently, this system has been used to regulate transgene expression in the mammary gland *(58)*.

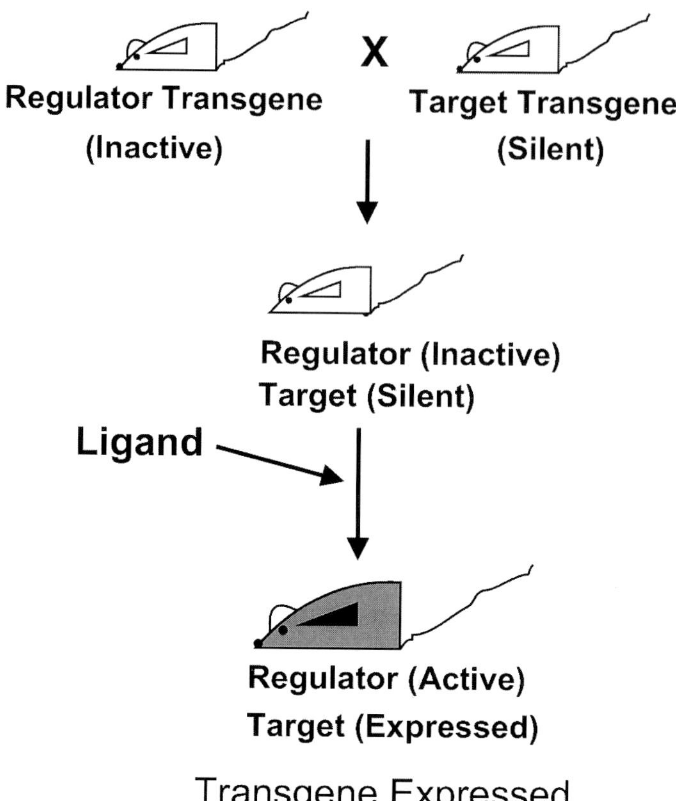

Fig. 2. Bitransgenic system. The regulator transgene encodes a transcription factor and remains inactive until the administration of an exogenous compound (ligand). The target transgene is the gene of interest, and is under the transcriptional control of the DNA *cis*-acting elements responsive to the regulator. The two lines of mice are crossed, and with the administration of the ligand, the regulator transgene is activated and the target transgene is expressed in the mouse (shaded mouse).

Currently, the most common ligand-dependent, bitransgenic regulatory system employed is the tetracycline-based regulatory system *(21)*. This system utilizes the bacterial tetracycline repressor gene (*tet*) to activate target genes that are under the control of promoter elements containing the tetracycline response element, the *tet* operon. Current engineering of the *tet* gene has generated two forms of this regulator. The first form, the *tet* transactivator (tTa) will activate transgenes in the absence of tetracycline. In this case, the administration of tetracycline causes the cessation of target-gene expression *(22)*. The second tetracycline-dependent regulator, the reverse-*tet* transactivator (rtTa), will activate transgene expression with the administration of tetracycline *(23)*. This system has been shown to regulate gene expression in a tissue-ubiquitous fashion *(24,25)*, and in a tissue-specific fashion in tissues such as the heart *(26)*, the beta cells of the pancreas *(27)*, the brain *(28)*, and the salivary glands *(29)*.

The third bitransgenic system used to regulate transgene expression in mice is one that utilizes a mutated human progesterone receptor that can activate gene transcription in response to antiprogestins *(30)*. This system is based on the observation that a deletion in the carboxy terminus of the human progesterone gene, which truncates the ligand-binding domain of this receptor, causes the receptor to stop binding the endogenous ligand progesterone. It was observed that this mutant could bind antiprogestins and activate gene expression *(31)*. This mutant ligand-binding domain was then used to create a chimeric receptor that consisted of the ligand-binding-domain mutant receptor, the DNA-binding domain of the yeast Gal4 transcription factor, and the activation domain of Herpes simplex virus protein 16, VP16 (VP16-AD). The chimeric receptor was called GLVP. In the presence of antiprogestins, the GLVP regulator transactivates regulatable target genes that contain the Gal4-binding site. This system has been shown to regulate gene expression in vitro using transient transfection, ex vivo after cellular transplantation *(30)*, and in vivo in transgenic mice *(32,33)*.

MODIFICATION OF ENDOGENOUS MURINE GENES

The ability to alter genes within the genome has become a powerful tool in the investigation of the genetic regulation of development. This has been made possible by advances in homologous recombination in embryonic stem cells *(34)*. Using this technology, gene expression can be ablated, or specific mutations as small as a single basepair change can be introduced into a desired gene. This technology has been extended to the point that entire genes can be substituted by inserting the coding region for one gene into the regulatory domain of another gene *(35)*. The power of this technology lies in the fact that the effects of these modifications can be observed in the context of a physiological and developmentally regulated system. The ability to modify endogenous murine genes requires the construction of the appropriate targeting vector for homologous recombination, the electroporation and selection of ES cell clones with the appropriate homologous event, the generation of chimeric mice with the targeted ES cells, and the establishment of mouse lines with the desired mutation.

The Targeting Vector

The targeting vector to be used in the introduction of mutations into the mouse genome by homologous recombination must be constructed from DNA that is homologous to the strain of mouse from which the ES cells have been derived. It has been shown that the frequency of homologous recombination is significantly increased by the use of syngeneic sequences *(36)*. Since most ES cells are derived from mice of the 129/Sv strain, most murine genes are now isolated from this strain of mice *(37)*. Once the gene to be mutated has been identified, a targeting construct can be generated. The targeting construct most commonly generated is that of the replacement vector type *(34)*. A schematic of a targeting vector is shown in Fig. 3. The replacement vector contains arms of the genomic sequences that flank a selectable marker. In some cases, the vector is designed to include both positive and negative selection *(38)*. If the goal is to ablate the desired gene, the targeting construct used in the homologous recombination must be designed to generate a deletion of the entire coding region of the gene or the functional domain of the gene. The targeting vector is constructed of homologous flanking DNA sequences ligated to the positive selection marker, usually the neomycin-resistance gene under the control of the phosphoglyceratekinase promoter (PGKneo) *(10)*. In order to maximize the recombination event, the targeting construction should contain at least 3 kb of total homologous DNA, although the more sequence included, the better the recombination frequency *(39)*. If the targeting vector incorporates a nega-

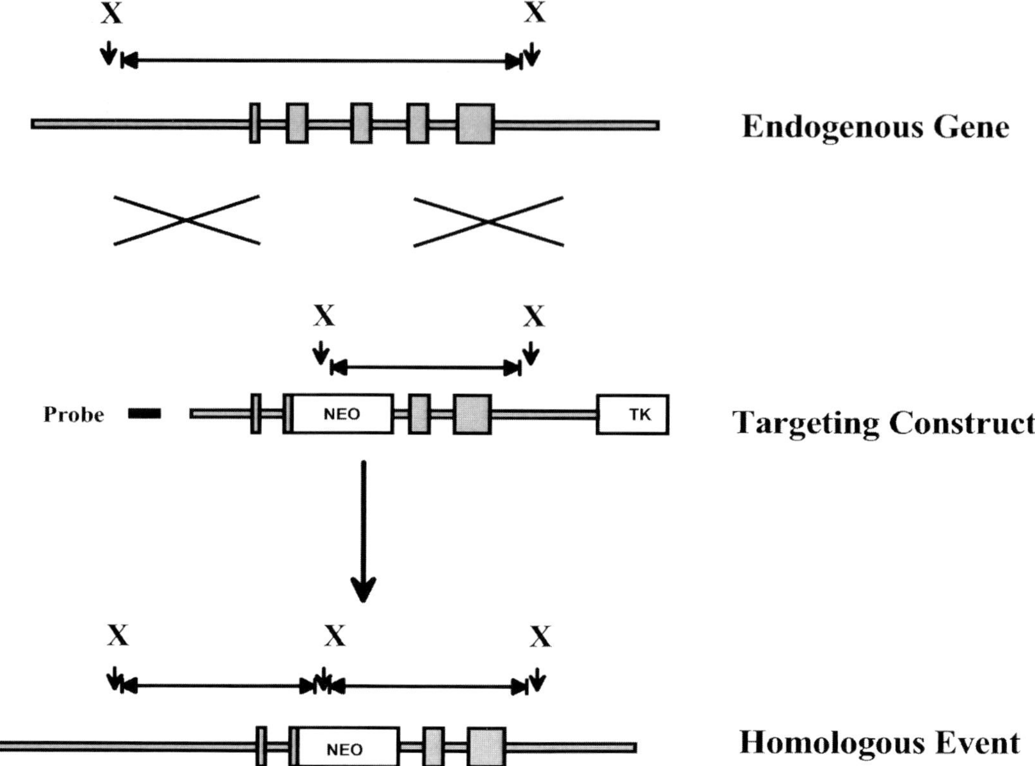

Fig. 3. Schematic for the construction of a targeting vector for targeted gene ablation. The gene of interest is identified and mapped (exons are illustrated as rectangles) with appropriate restriction endonuclease sites (*X*) noted. A targeting construct is made that consists of homologous flanking DNA sequences, a positive selection marker such as the neomycin (Neo) resistance gene, which disrupts exons in the coding sequence of the gene of interest, and a negative selection marker such as the thymidine kinase (TK) gene. A new restriction endonuclease site is introduced into the targeting construct by the NEO gene for screening by Southern blot analysis using a fragment of DNA (probe) outside the targeting construct. The homologous recombination event introduces the mutation into the mouse genome.

tive selection marker, this marker is removed upon correct homologous recombination. The negative selection marker usually consists of the thymidine kinase (TK) negative selection marker *(38)*. Cells cultured in the presence of gangcyclovir that express the TK gene will be selected against. Therefore, the TK negative selection marker is placed at the flanking end of the targeting vector. Upon homologous recombination, the TK is removed, allowing ES cells to survive in culture in the presence of gangcyclovir. However, the use of the negative selection marker only slightly enhances the selection of ES cells with the appropriate recombination event. Also, with the correct design of targeting vectors to insert genetic elements into the murine genome, it has been shown that the use of negative selection to achieve adequate recombination is not unnecessary *(40)*. Therefore, if the TK gene will be incorporated into the design of a construct, it is best to place this as a last step in the generation of the targeting vector. Then, it can be added only when the initial targeting frequency is unacceptably low.

Electroporation and Selection of ES Cell Clones

The success of the generation of mutant mice from ES cell manipulation results from the use of ES cells that have the potential to become part of the germline when aggregated with other embryos in the generation of chimeric mice *(41)*. The 129/Sv strain has been exploited to generate ES cells that are highly efficient in the generation of germline chimeric mice *(37)*. The culture requirements to maintain these ES cells in a pluripotent state depend upon the culture conditions used by the investigator to generate these ES cells. ES cells are usually derived using the procedures described by Robertson *(42,43)*. Blastocysts are collected from the uterus of postcoitum d 3.5 pregnant 129/Sv females. The blastocysts are cultured on feeder cells that are mitotically inactivated by treatment with mytomyocin *(10)* or gamma irradiation. It has been shown that leukemia-inhibitory factor (LIF) can be used to maintain ES cells in a pluripotent state in culture in lieu of the feeder cells *(44)*. However, the ES cells are usually cultured in the presence of feeders and the media supplemented with LIF to maintain the cells in the pluripotent state. The procedure for the culture of ES cells and the culture media has been previously described *(41)*, and consists of Dulbecco's modified Eagle's medium (DMEM) (high-glucose formulation) supplemented with 15% (v/v) fetal calf serum (ES-cell tested, Gibco-BRL, Grand Island, NY), 2 m*M* glutamine, 5×10^{-5} *M*, 2-mercaptoethanol (2-ME), 30 µg/mL penicillin, and 50 µg/mL streptomycin *(41)*. The serum used in the ES-cell culture must be tested from the vendor to ensure that the batch of serum used is optimal for ES-cell growth. Although ES cells are pluripotent, their ability to be maintained in pluripotency depends upon the culture conditions and the number of passages the cells have undergone. ES cells at a low passage number are very effective in developing into every cell of the mouse. However, this ability decreases with passage number *(45,46)*. Therefore, to ensure the success of the generation of germline chimeras, low-passage cells, usually passage 10–16, should be employed.

Once the targeting construct is generated, ES cells are electroporated with the linearized targeting vector. The ES cells are cultured on a monolayer of mitomycin C or gamma-irradiated feeder cells *(42)*. The ES cells are then placed under G418 selection with or without gangcyclovir selection, depending on whether negative selection is used. Ten days after electroporation, ES cell colonies can be visualized and are picked using a dissecting microscope. ES cell clones are replicated and cryopreserved, and duplicates are screened for homologous recombination using either Southern blot or PCR analysis. Because of the frequency of PCR contamination, Southern blot analysis is the preferred method of choice for screening clones for homologous recombination. The ES-cell genomic DNA is digested with the appropriate restriction endonucleases that will allow for a distinction of the targeted event from the endogenous gene. This endonuclease restriction site is usually a site that has been added or removed from within the targeting construction. The Southern blot is probed with a fragment of DNA from the endogenous gene that is located in a region outside of the targeting vector. This approach allows the identification of targeted clones without interference of random insertions of the targeting vector. Clones showing the appropriate insertion of the targeting vector are then expanded. Further, Southern blot analysis with a variety of restriction endonucleases and probes is usually recommended to verify the fidelity of the targeting event. Once the ES cells with the appropriate insertion are verified, they can be used to generate chimeras.

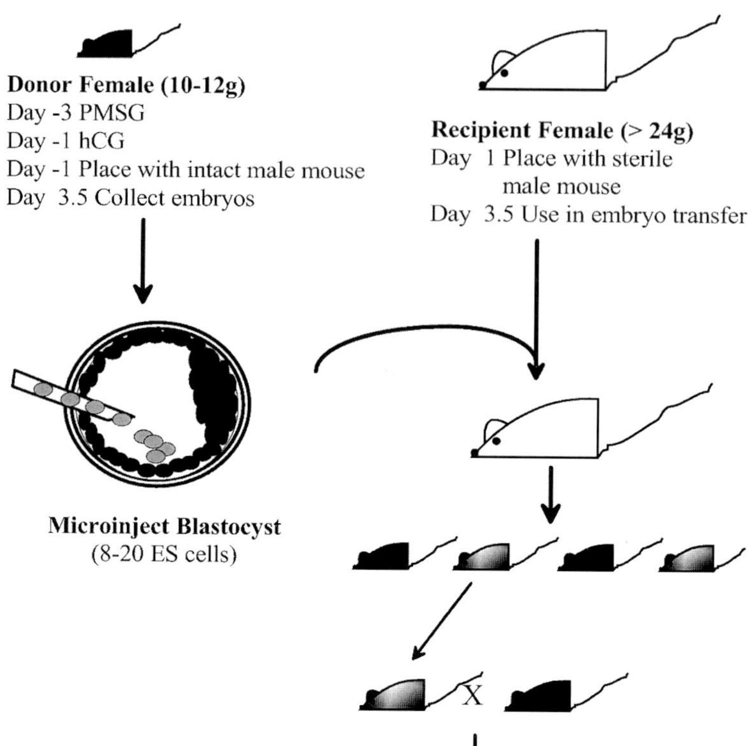

Fig. 4. Schematic for the generation of chimeric mice from embryonic stem cells. Day 3.5 embryos (blastocysts) are collected from females (illustrated here as a black mouse) treated with a superovulatory regimen of gonadotropins before mating with a male. The embryos (illustrated here as black cells) are microinjected with 8–20 mutant ES cells (illustrated here in color) and transferred to a d 2.5 pseudopregnant female mouse (illustrated here as a white mouse) and allowed to gestate to birth. Offspring are identified by coat color: 129/SV mice have agouti (illustrated here as a purple mouse) color, C57B1/6J have black color (illustrated here as a black mouse), and chimeric mice have mottled agouti and black (illustrated here as a mixed shaded mouse). Chimeric mice are bred with C57B1/6J mice to determine ES-cell lineage and germline incorporation, since the agouti phenotype is dominant over black. Agouti mice can be screened by Southern blot analysis or PCR for further analysis of insertion of targeting vector.

Generation of Chimeric Mice

The scheme for the generation of chimeric mice from ES cells is shown in Fig. 4. The procedure has been thoroughly described by Bradley *(41)*. Briefly, chimeric mice are generated by microinjection of 8–20 ES cells into the blastocoel of embryonic d 3.5 C57Bl/6J mouse embryos, depending on the source of the ES cells. The embryos are transferred to the uteri of postcoitum d 2.5 pseudopregnant ICR mice, and the embryos are allowed to gestate. Chimeric mice are identified by coat color. 129/Sv mice have an agouti coat, and C57Bl/6J mice have a black coat. The chimeric mice have a mottled black-and-agouti coat. The appearance of an agouti coat color indicates a high degree of chimerism in the mice. Gender can be used to provide an initial assessment of the success of the ES-cell transfer. Since most ES cells are derived from male embryos, male chimeras will have a higher probability of containing germ cells derived from the ES cells.

Male mice with a high degree of agouti coat color have a high probability of transmitting the mutation to the germline. These mice are bred to C57Bl/6J mice. Since the agouti phenotype is dominant over black, mice derived from the ES-cell lineage will be identified by the agouti coat color. These agouti mice can then be screened by PCR and/or Southern blot analysis to determine which of these have the desired mutation.

Phenotypic Analysis of Gene Mutation

Once the mutation has been incorporated into the murine germline, the phenotypic consequences of the mutation can be examined. If the heterozygous mice are viable, they can be intercrossed to generate mice homozygous for the mutation. The offspring from the heterozygous crosses will be genotyped to determine whether the frequency of wild-type, heterozygous, or homozygous for the specific mutation is the expected Mendelian 1:2:1. If this frequency is achieved, it can be concluded that the mutation in the gene had no impact on murine development. Any significant deviation from this frequency will most likely be seen as a decrease in the generation of homozygous mice.

In the generation of mice with a specific genetic ablation, several possible outcomes exist. First, the genetic ablation may be lethal and cause fetal, neonatal, or adult death. Second, the animals may be viable, and the gene ablation may impart a profound effect upon murine physiology. Finally, the mutation may have no obvious phenotypic effect upon murine development. If homozygous postnatal mice are not achieved, it can be concluded that the genetic ablation is lethal for either fetal or neonatal development. If this is the case, then a developmental analysis must be conducted to determine when and where the lethality occurred. If the mice produced from the gene mutation are viable, a detailed morphological, histological, and biochemical analysis must be conducted on these mice to first validate that the desired mutation has been generated and, second, to ensure that no subtle phenotype is observed.

Tissue-Specific Ablation

The approaches outlined here allow genes to be ablated, permit the generation of subtle mutations in genes, and even allow replacement of one gene with another (e.g. generation of a "knockin" of one gene into the locus of another). However, although this technology is powerful, it does have limitations. The genetic alteration introduced by homologous recombination in ES cells affects all cells in which the modified gene is expressed, and these modifications are irreversible. If the gene to be ablated is expressed during embryonic development and in a variety of tissues, the analysis of the effects of gene ablation may be limited. If the ablation of a gene effects embryo viability because of disruption of one particular tissue development, then the investigation of the function of that gene may be limited to one tissue and one particular time-point.

The problem encountered with ablating the expression of a gene that is expressed in a wide variety of tissues has been overcome by introducing specific DNA sequences, which are recognized by a specific recombinase, around a region of the gene to be removed. Currently, two recombinases have been used in mice, the Cre recombinase *(47)* and the FLP recombinase *(48)*. The specific recombinase can be expressed in a desired tissue using a chimeric gene introduced into the murine genome by transient transgenic techniques. The expressed recombinase then binds to the introduced DNA sequences and removes the DNA flanked by these recombinase recognition sequences. This system was first used to ablate the DNA polymerase β (pol β) gene specifically in T cells *(47)*.

In this original report, the Cre recombinase gene was expressed in the T cells of transgenic mice by placing Cre under the control of the *lck* promoter. Homologous recombination was used to flank the pol β gene with the recognition sites for Cre recombinase—*lox*P sites. The mice with the pol β gene "floxed" were bred to the mice expressing Cre recombinase. This procedure resulted in mice with T-cell-specific ablation of the pol β gene.

The limitation of the this approach is that gene ablation under the control of the Cre-*lox*P system is dependent upon the tissue-specific expression of the promoter used to express Cre. Also, the ablation of gene expression by this system is irreversible. Therefore, if the promoter used to express the Cre gene is expressed early in development, the investigation of the effects of gene ablation may be limited to fetal development. This approach can be further modified by fusing the Cre recombinase to the ligand-binding domain of the estrogen receptor *(49)* or the mutated progesterone receptor *(50)*. This has rendered the Cre recombinase active only when the specific ligand—tamoxifen or RU486, respectively—is given to the animal. Thus, a combination of these systems imparts spatiotemporal control of the ablation for the investigator.

NEW TECHNOLOGIES FOR GENETIC ALTERATION OF THE MURINE GENOME

The technologies of DNA microinjection and ES cell technology have created powerful tools for all aspects of developmental biology and reproductive physiology. However, there are limitations to these systems that relate to the strains of choice for the generation of transgenic and mutant mice. Although transgenic mice can be generated in theoretically any strain of mouse, these mice can only be generated efficiently in strains that respond well to superovulatory regimens of gonadotropins. This limits these mice to a very few strains. Mutant mice can only be generated in 129/Sv strains because of the lack of germline-quality ES cells in other strains of mice. Therefore, while the majority of transgenic mice made in inbred strains are of the FvB/N strain, mutant mice are made in the 129/Sv strain and 129/Sv X C57Bl/6J hybrid strain. Since the genetic background of the mouse may affect the observed phenotype, it is important to ensure the use of common genetic backgrounds when crossing transgenic and mutant mice *(51)*. Therefore, to combine transgenic experiments with mutant mouse experimentation requires working in a mixed genetic background, or breeding the transgenic or mutant mouse for several generations into the desired mouse strain. One solution to this problem may be found in the recent technology of sperm injection and nuclear transfer to mouse oocytes.

Sperm Injection

Sperm injection or ICSI has recently been shown to be effective with murine gametes. Although established with human gametes, the development of this technology has been slow in the mouse, largely because of the difficulty in microinjecting unfertilized mouse oocytes. Although the one-cell fertilized mouse oocyte is relatively amenable for the microinjection of DNA, the permeability of the unfertilized mouse oocyte is significantly more resistant to injection. Using conventional microinjection technology, the mouse oocyte either resists penetration by the microinjection needle or lyses soon after microinjection. However, it has recently been shown that the use of a Piezo micromanipulator can increase the efficiency of microinjecting mouse oocytes. Using Piezo manipulation, mouse oocytes have been injected with sperm, and efficient numbers of viable mice have been born from this technology *(52)*. The generation of mice using ICSI is not limited to freshly collected sperm, and has been successful using sperm that has been stored by freeze-drying *(53)*. For instance, rehydration of freeze-dried sperm containing transgenes has resulted in the generation of transgenic mice *(2)*. Although the efficiency of this technology is not significantly better than the traditional pronuclear injection protocol, this technology offers the potential for the generation of transgenic mice in strains of mice that were previously believed to be economically impossible. In most inbred strains, the difficulty in the generation of transgenic mice is not the inability of the male to produce sperm or the ability of the female to be stimulated to produce oocytes, but the ability of these strains to produce large numbers of fertilized eggs for the production of viable matings between the inbred males and the stimulated females. Therefore, transgenic mice utilizing ICSI may allow transgenes to be introduced into the 129/Sv strain and other inbred strains previously considered impossible or impractical.

Nuclear Transfer

Homologous recombination in mice is limited to strains of mice in which germline-quality ES cells can be isolated. However, with the advent of nuclear-transfer technology, (i.e. cloning), the ability to generate mutations may no longer be limited to the 129/Sv strain. Cloning of mammals was first established in domestic species—sheep *(54)* and cattle *(55)*. Again, the relative impermeability of the murine oocytes has made this technology virtually impossible. However, the same group that pioneered ICSI in mice has demonstrated that cloning can be accomplished with mice. Initially, cloning was first shown to be possible by the transfer of nuclei from ovarian cumulus cells *(56)*. However, other somatic cells have since been shown to be permissive to cloning technology *(57)*. The procedure for cloning mice has been described by Wakayama and colleagues *(56)*. In brief, murine cloning requires the collection of oocytes by superovulation. After oocytes are cultured in media containing cytochalasin B, the oocytes are enucleated using the Piezo manipulators. Following a period of recovery, the Piezo manipulator is used to introduce the somatic-cell nucleus into the oocytes. After the nuclear transfer, the oocytes are activated, using media containing cytochalasin B and strontium chloride. This activates the egg-plasma membrane without allowing the nucleus to be expelled. The egg is then cultured in traditional culture media and transferred to synchronized recipients. The efficiency of the generation of live mice from these procedures is very low. However, as this technology improves, the use of homologous recombination in somatic cells can be exploited to increase the efficiency and versatility of the generation of mutant mice.

Understanding the reproductive biology of the mouse has allowed the establishment of very powerful tools that has expanded the horizons of many scientific disciplines. The technologies for the manipulation of the murine genome are continually expanding. Incorporation of the new technologies will allow fields of research to progress at an even faster rate.

ACKNOWLEDGMENTS

This research was supported in part by NICHD/NIH through cooperative agreeement (U54[HD07495-28]) as part of the Specialized Cooperative Centers Program in Reproduction Research and in part by the NCI Special Program of Research Excellence in Prostate Cancer CA-91-35.

REFERENCES

1. Palmiter RD, Brinster RL. Germ-line transformation of mice. Ann Rev Genet 1986;20:465–499.
2. Perry ACF, Wakayama T, Kishikawa H, Kasai T, Okabe M, Toyoda Y, Yanagimachi R. Mammalian transgenesis by intracytoplasmic sperm injection. Science 1999;284:1180–1183.
3. Rubenstein JLR, Nicholas J-F, Jacob F. Introduction of genes into preimplantation mouse embryos by use of a defective recombinant retrovirus. Proc Natl Acad Sci USA 1986;83:366–368.
4. Soriano P, Jaenisch R. Retroviruses as probes for mammalian development: allocation of cells to the somatic and germ cell lineages. Cell 1986;46:19–29.
5. Jähner D, Jaenisch R. Retrovirus-induced de novo methylation of flanking host sequences correlates with gene inactivity. Nature 1985;315:594–597.
6. Gordon JW, Scangos DJ, Plotkin DJ, Barbosa J, Ruddle FM. Genetic transformation of mouse embryos by microinjection of purified DNA. Proc Natl Acad Sci USA 1980;77:7380–7384.
7. Hogan B, Beddington R, Costantini F, Lacy E. Manipulating the Mouse Embryo: A Laboratory Manual. Cold Spring Harbor, New York, 1994:151–204.
8. Brinster RL, Palmiter RD. Factors effecting the efficiency of introducing foreign DNA into mice by microinjecting eggs. Proc Natl Acad Sci USA 1985;82:4438–4442.
9. Taketo M, Schroeder AC, Mobraaten LE, Gunning KB, Hanten G, Fox RR, et al. FVB/N: an inbred mouse strain preferable for transgenic analyses. Proc Natl Acad Sci USA 199188:2065–2069.
10. Soriano P, Montgomery C, Geske R, Bradley A. Targeted disruption of the c-src proto-oncogene leads to osteopetrosis in mice. Cell 199164:693–702.
11. Southern PJ, Berg P. Transformation of mammalian cells to antibiotic resistance with a bacterial gene under control of the SV40 early region promoter. J Mol Appl Gen 1982;1:327–341.
12. Saiki RK. PCR Protocols. A Guide to Methods and Applications. Academic Press, San Diego, CA, 1990:13–27.
13. Lacy E, Roberts S, Evans EP, Burtenshaw MD, Costantini FD. A foreign beta-globin gene in tansgenic mice: integration at abnormal chromosomal positions and expression in inappropriate tissues. Cell 1983;34(2):343–358.
14. Wilkie TM, Brinster RL, Palmiter RD. Germline and somatic mosaicism in transgenic mice. Dev Biol 1986;118(1):9–18.
15. Grosveld F, van Assendelft GB, Greaves DR, Kollias G. Position-independent, high-level expression of the human (beta)-globin gene in transgenic mice. Cell 1987;51:975–985.
16. Brinster RL, Allen JM, Behringer RR, Gelinas RE, Palmiter RD. Introns increase the transcriptional efficiency in transgenic mice. Proc Natl Acad Sci USA 1988;85:836–840.
17. Palmiter RD, Sandgern EP, Avabock MR, Allen DD, Brinster RL. Heterologous introns can enhance expression of transgenes in mice. Proc Natl Acad Sci USA 1991;88:478–482.
18. Yao T-P, Forman BM, Jiang Z, Cherbas L, Chen J-D, McKeown M, et al. Functional ecdysone receptor is the product of EcR and Ultraspiracle genes. Nature (London) 1993;366:476–479.
19. Yao T-P, Segraves WA, Oro AE, McKeown M, Evans RM. Drosophila ultraspiracle modulates ecdysone receptor function via heterodimer formation. Cell 1992;71:63–72.
20. No D, Yao T-P, Evans RM. Ecdysone-inducible gene expression in mammalian cells and transgenic mice. Proc Natl Acad Sci USA (Genetics) 1996;93:3346–3351.
21. Shockett PE, Schatz DG. Diverse strategies for tetracycline-regulated inducible gene expression. Proc Natl Acad Sci USA 1996;93:5173–5176.
22. Gossen M, Bujard H. Tight control of gene expression in mammalian cells by tetracycline-responsive promoters. Proc Natl Acad Sci USA. 1992;89:5547–5551.
23. Gossen G, Freundlieb S, Bender G, Muller G, Hillen W, Bujard H. Transcriptional activation by tetracyclines in mammalian cells. Science 1995;268:1766–1769.
24. Furth PA, St. Onge L, Boger H, Gruss P, Gossen M, Kistner A, et al. Temporal control of gene expression in transgenic mice by a tetracycline-responsive promoter. Proc Natl Acad Sci USA 1994;91:9302–9306.
25. Shockett P, Difilippantonio M, Hellman N, Schatz DG. A modified tetracycline-regulated system provides autoregulatory, inducible gene expression in cultured cells and transgenic mice. Proc Natl Acad Sci USA 1995;92:6522–6526.
26. Yu Z, Redfern CS, Fishman GI. Conditional transgene expression in the heart. Circ Res 1996;79:691–697.
27. Efrat S, Fusco-DeMane D, Lemberg H, Emran OA, Wang X. Conditional transformation of a pancreatic beta-cell line derived from transgenic mice expressing a tetracycline-regulated oncogene. Proc Natl Acad Sci USA 1995;92:3576–3580.

28. Mayford M, Bach ME, Huang Y-Y, Wang L, Hawkins RD, Kandel ER. Control of memory formation through regulated expression of a CaMKII transgene. Science 1996;274:1678–1683.
29. Ewald D, Li M, Efrat S, Auer G, Wall RJ, Furth PA, Hennighausen L. Time-sensitive reversal of hyperplasia in transgenic mice expressing SV40 T antigen. Science 1996;273:1384–1386.
30. Wang Y, O'Malley BW, Jr, Tsai SY, O'Malley BW. A reulatory system for use in gene transfer. Proc Natl Acad Sci USA (Medical Sciences) 1994;91:8180–8184.
31. Vegeto E, Allan GF, Schrader WT, Tsai M-J, McDonnell DP, O'Malley BW. The mechanism of RU486 antagonism is dependent on the conformation of the carboxy-terminal tail of the human progesterone receptor. Cell 1992;69:703–713.
32. Wang Y, DeMayo FJ, Tsai SY, O'Malley, BW. Ligand-inducible and liver-specific target gene expression in transgenic mice. Nat Biotechnol 1997;15:239–243.
33. Pierson TM, DeMayo FJ, Matzuk MM, Tsai SY, O'Malley BW. Regulable expression of inhibin A in wild-type and inhibin a null mice. Mol Endocrinol 2000;14:1075–1085.
34. Capecchi MR. Altering the genome by homologous recombination. Science 1989;244:1288–1292.
35. Mansour SL, Thomas KR, Deng C, Capecchi MR. Introduction of a lacZ reporter gene into the mouse int-2 locus by homologous recombination. Proc Natl Acad Sci USA 1990;87:7688–7692.
36. Riele HT, Maandag ER, Berns A. Highly efficient gene targeting in embryonic stem cells through homologous recombination with isogenic DNA constructs. Proc Natl Acad Sci USA 1992;89:5128–5132.
37. Threadgill DW, Yee D, Matin A, Nadeau JH, Magnuson T. Genealogy of the 129 inbred strains: 129/SvJ is a contaminated inbred strain. Mamm Genome 1997;8(6):390–393.
38. Gu H, Marth JD, Orban PC, Mossmann H, Rajewsky K. Deletion of a DNA polymerase beta segment in T cells using cell type-specific gene targeting. Science 1995;265:103–106.
39. Hasty P, Rivera-Perez J, Bradley A. The length of homology required for gene targeting in embryonic stem cells. Mol Cell Biol 1991;11(11):5586–5591.
40. Abuin A, Bradley A. Recycling selectable markers in mouse embryonic stem cells. Mol Cell Biol 1996;16:1851–1856.
41. Bradley A. Production and analysis of chimaeric mice. In: Robertson EJ, ed. Teratocarcinomas and Embryonic Stem Cells, A Practical Approach. IRL Press, Oxford, UK, pp. 113–151.
42. Robertson EJ. Embryo-derived stem cells. In Robertson EJ, ed. Teratocarcinomas and Embryonic Stem Cells: A Practical Approach. Oxford, New York, pp. 71–112.
43. Robertson, EJ. Derivation and maintenance of embryonic stem cell cultures. In Robertson EJ, ed. Teratocarcinomas and Embryonic Stem Cells: A Practical Approach. IRL Press, Oxford, UK, pp. 223–236.
44. Williams RL, Hilton DJ, Pease S, Willson TA, Stewart CL, GearingDP, et al. Myeloid leukemia inhibitory factor maintains the developmental potential of embryonic stem cells. Nature 1988;336:684–687.
45. Nagy A, Gocza E, Diaz EM, Prideaux VR, Ivanyi E, Markkula M, Rossant J. Embryonic stem cells alone are able to support fetal development in the mouse. Development 1990;110:815–821.
46. Nagy A, Rossant J, Nagy R, Abramow-Newerly W, Roder JC. Derivation of completely cell culture-derived mice from early-passage embryonic stem cells. Proc Natl Acad Sci USA 1993;90:8424–8428.
47. Mansour SL, Thomas KR, Capecchi MR. Dispruption of the proto-oncogene int-2 in mouse embryo-derived stem cells: a geneal strategy for targeting mutations to non-selectable gene. Nature 1988;366:348–352.
48. Fiering S, Epner E, Robinson K, Zhuang Y, Telling A, Hu M, et al. Targeted deletion of 5'HS2 of the murine beta-globin LCR reveals that it is not essential for proper regulation of the beta-globin locus. Genes Dev 1995;9(18):2203–2213.
49. Feil R, Wagner J, Metzger D, Chambon P. Regulation of Cre recombinase activity by mutated estrogen receptor ligand-binding domains. Biochem Biophys Res Commun 1997;237(3):752–757.
50. Kellendonk C, Trouche F, Monaghan AP, Angrand PO, Stewart F, Schutz G. Regulation of Cre recombinase activity by the synthetic steroid RU 486. Nucleic Acids Res 1996;24(8):1404–1411.
51. Banbury Conference on Genetic Background in Mice. Mutant mice and neuroscience: recommendations concerning genetic background. Neuron 1997;19:755-759.
52. Kimura Y, Yanagimachi R. Intracytoplasmic sperm injection in the mouse. Biol Reprod 1995; 52(4):709–720.
53. Wakayama T, Yanagimachi R. Development of normal mice from oocytes injected with freeze-dried spermatozoa. Nat Biotechnol 1998;16(7):639–641.
54. Wilmut I, Schnieke AE, McWhir J, Kind AJ, Campbell KH. Viable offspring derived from fetal and adult mammalian cells. Nature 1997;385(6619):810–813.

55. Kato Y, Tani T, Sotomaru Y, Kurokawa K, Kato J, Doguchi H, et al. Eight calves cloned from somatic cells of a single adult. Science 1998;282(5396):2095–2098.
56. Wakayama T, Perry AC, Zuccotti M, Johnson KR, Yanagimachi R. Full-term development of mice from enucleated oocytes injected with cumulus cell nuclei. Nature 1998;394(6691):369–374.
57. Wakayama T, Yanagimachi R. Cloning of male mice from adult tail-tip cells. Nat Genet 1999; 22(2):127,128.
58. Albanese C, Reutens AT, Bouzahzah B, Fu M, D'Amico M, Link T, Nicholson R, Dephino RA, Pestell RG. Sustained mammary gland-directed, ponasterone A-inducible expression in transgenic mice. FASEB J 2000;14:877–884.

2 The Transgenic Mouse in Studies of Mammalian Sexual Differentiation

Deanne J. Whitworth, PHD
and *Richard R. Behringer*, PHD

CONTENTS

INTRODUCTION

Throughout history, the subject of sex has held an inherent fascination. The musings of Aristotle on the role of "an infinitesimally minute but essential organ" in determining whether "the animal will in one case turn to male (or) in the other to female" (Aristotle, *Historia Animalium*), offer an early insight into what has become one of the tenets in our understanding of sexual differentiation in mammals: that the sex of the gonad determines the sexual development of the individual.

Sex determination in mammals is a remarkable process that has its origin in an indifferent gonadal primordium, common to both sexes, which has the ability to differentiate into either a testis or an ovary. In the presence of a Y chromosome, the indifferent gonad develops as a testis; in the absence of a Y chromosome, and regardless of the number of X chromosomes present, the indifferent gonad develops as an ovary. Remarkably, until fairly recently, this was about all we knew with respect to the genetic events involved in testicular differentiation. With the discovery in 1990 of *SRY* (sex-determining-region, *Y* chromosome gene), the testis-determining gene on the Y chromosome, the field seemed set for the systematic isolation of other genes functioning in the testicular differentiation pathway. Eleven years have now passed since the discovery of *SRY*, and the pathway from indifferent gonad to testis appears ever more complex. Our understanding of the ovarian differentiation process is even more limited.

The use of transgenic mice has contributed enormously to our understanding of the mammalian sex determination and sexual differentiation pathways, and hence is the

From: *Contemporary Endocrinology: Transgenics in Endocrinology*
Edited by: M. Matzuk, C. W. Brown, and T. R. Kumar

focus of this chapter. While there is no denying the wealth of information gleaned from mouse models, it is also our belief that much can be learned from comparative studies of other mammalian systems. This chapter concludes with a brief introduction to the efforts that are underway to generate the first transgenic marsupial.

SEX DETERMINATION: THE FATE OF THE GONAD

The mammalian testis and ovary have a common origin in the indifferent gonad. In the mouse, the indifferent gonad is first discernible at 10.5 d postcoitum (dpc) as a thickening on the medial aspect of the mesonephros. At this stage, the gonad consists of four known cell types—the primordial germ cells, supporting cells, steroidogenic cells, and connective cell lineage—each capable of entering the testicular or ovarian differentiation pathway (Fig. 1).

The indifferent male gonad of the mouse becomes sexually dimorphic at around 12.5 dpc when the supporting cell lineage gives rise to the Sertoli cells, which organize themselves into seminiferous cords that enclose the primordial germ cells. At the time that Sertoli cells form seminiferous cords, they begin secreting the glycoprotein Müllerian-inhibiting substance (MIS; also known as anti-Müllerian hormone, AMH). MIS is responsible for inducing the regression of the Müllerian ducts, which would otherwise persist to form the female reproductive tract. The Sertoli cells in turn appear to direct the differentiation of Leydig cells from the steroidogenic cell lineage, and peritubular myoid cells from the connective tissue *(1)*. The production of testosterone by the Leydig cells stimulates the growth and virilization of the Wolffian ducts, which give rise to the epididymides, vasa deferentia, and seminal vesicles. At approx 13.5 dpc in the mouse testis, the germ cells become arrested in mitosis, marking their entry into the male pathway of development. Germ cells that fail to reach the gonad follow the female pathway and enter meiosis prematurely *(2)*.

In the female gonad of the mouse, the germ cells enter meiosis at approx 13.0 dpc, becoming arrested at the first prophase 2 d later. From approx 16.5 dpc, granulosa cells—derived from the supporting cell lineage—surround the oocytes to form primordial follicles. Granulosa cells do not produce MIS until after birth, thus creating a permissive environment for the differentiation of the Müllerian ducts into the oviducts, uterus, and upper one-third of the vagina. Similarly, the fetal ovary does not produce testosterone, and consequently the Wolffian ducts degenerate.

Studies of XX↔XY chimaeric mice provided the first clue that the differentiation of the male supporting cells into Sertoli, rather than granulosa, cells is under the cell-autonomous control of a Y chromosome-encoded testis-determining factor (TDF in humans, Tdy in mouse) *(1,3)*. In these mice, approx 95% of Sertoli cells are XY, and other testicular cell types are derived from XX and XY cells with equal probability. Thus, although the Y chromosome—and TDF/Tdy—is not essential for the differentiation of Leydig cells and peritubular myoid cells, it is required for Sertoli-cell differentiation.

The first example of a mutation to *Tdy* was described by Lovell-Badge and Robertson in 1990. Using XY embryonic stem (ES) cells that had been infected with a retroviral vector, they generated a male chimaera whose progeny included a small proportion of XY females. The retrovirally-induced mutation segregated with the Y chromosome. Further, during meiosis it was complemented by the Sxr and Sxr' translocations—both of which are translocations of the Y chromosome, including *Tdy*, onto the X chromo-

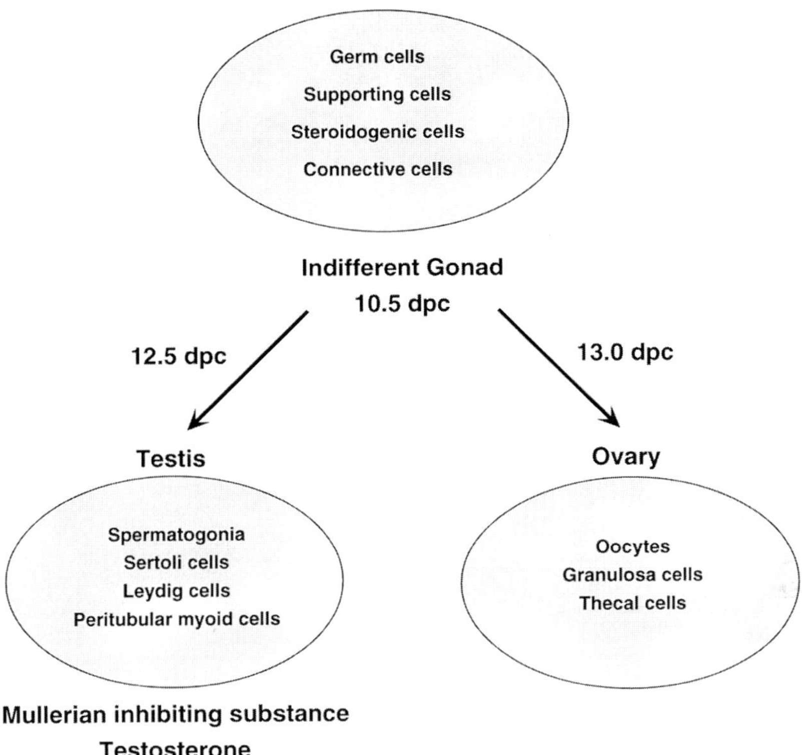

Fig. 1. Gonadal differentiation in the mouse. At 10.5 dpc in the mouse, the indifferent gonad contains four known cell types that are bipotential. The first stage of testicular differentiation occurs at 12.5 dpc, when the Sertoli cells differentiate from the supporting cell lineage to form cords that enclose the spermatogonia. Shortly after, Leydig cells differentiate from the steroidogenic cells and peritubular myoid cells arise from the connective cell lineage. Sertoli cells and Leydig cells produce MIS and testosterone, respectively, which are required for the development of the male internal reproductive structures. In the female at 13.0 dpc, the germ cells enter meiosis and are termed oocytes. Granulosa cells subsequently differentiate from the supporting cells, and thecal cells develop from the steroidogenic cells.

some—thus mapping the mutation to the testis-determining region of the Y. Consequently, this study provided the first case of XY sex reversal unequivocably caused by a mutation in *Tdy*.

EQUATING *SRY* WITH TDF

Major contributions to our understanding of sex determination and sexual differentiation in mammals have come from the study of mice and humans with primary sex reversal. The term "primary sex reversal" indicates XY individuals who develop as phenotypically female, despite the presence of a Y chromosome, and XX individuals who lack a Y chromosome but possess testicular tissue and consequently develop a male phenotype. Through analysis of the DNA of human patients with primary sex reversal, researchers have been able to progressively narrow down the portion of the Y chromo-

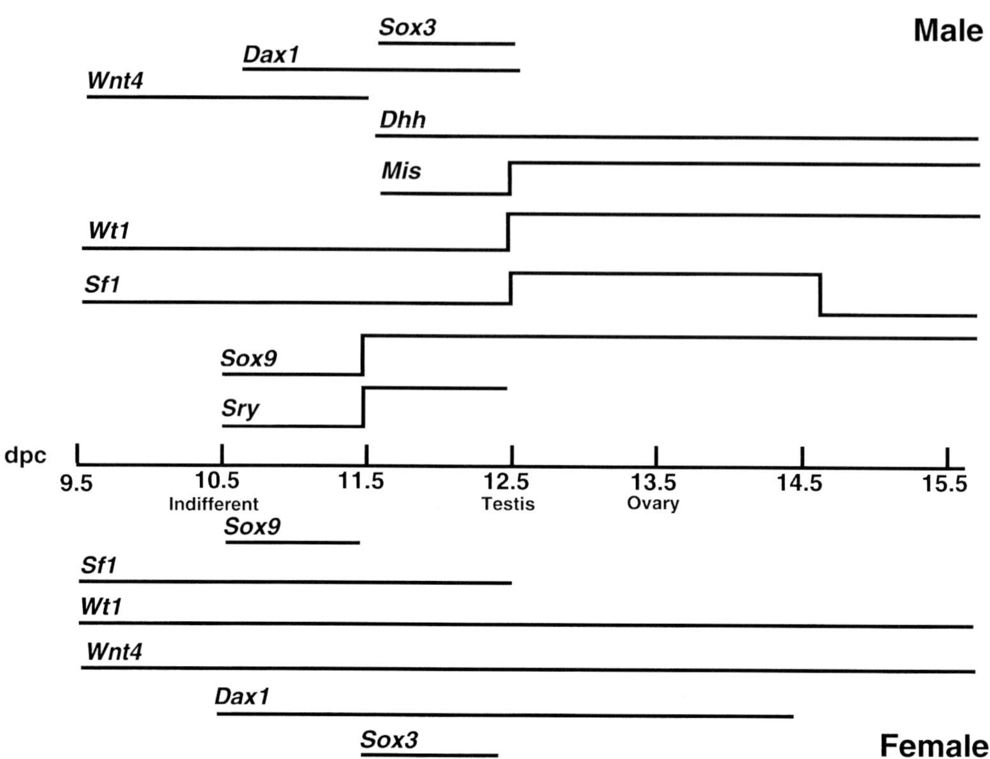

Fig. 2. Expression profiles of genes involved in sex determination and differentiation. Relative expression profiles of genes expressed in the supporting cell lineage and believed to be involved in sex determination and differentiation. Some genes, such as *Sf1* and *Dax1*, are also expressed in the steroidogenic cells; however, the expression profiles shown here are only representative of the supporting cell lineage and their descendents the Sertoli cells and granulosa cells.

some required to induce testicular differentiation in XX patients, which was missing in XY females *(5,6)*. In "walking" along this 35-kb "minimal" testis-determining region, Sinclair and colleagues identified a gene they named *SRY*, which is now generally accepted to be TDF *(6)*. In the mouse, *Sry* is absent from a strain in which XY individuals develop as females *(7)*, and conversely, is present in a small fragment of the Y chromosome known to cause sex reversal in the XX^{Sxrb} mouse *(7)*. *Sry* transcripts are detectable only within the male gonad, and during a very discrete period from 10.5–12.5 dpc, coincident with the onset of testicular differentiation at 12.5 dpc (Fig. 2) *(8,9)*. With a few unique exceptions *(10,11)*, *SRY* is conserved on the Y chromosome of all mammalian species examined *(6,7,12)*. The SRY protein encodes an amino acid motif called the high-mobility group (HMG) box which confers its ability to bind to and bend DNA, and thus act as a transcription factor—a function in keeping with its role as a "switch" in the sex-determining pathway *(7,8,13)*.

The most compelling evidence equating SRY with TDF came in the form of Randy, a transgenic mouse. When Koopman and colleagues microinjected into XX embryos a 14-kb fragment of the Y chromosome, containing only *Sry*, some of the resulting XX transgenics (including Randy) developed as phenotypic males complete with male mating behavior *(14)*. Thus, *Sry* is the only gene on the Y chromosome required for testis determination or, more specifically, for Sertoli-cell differentiation. In studies of $XY^{Sry+}\leftrightarrow XY^{Sry-}$ chimaeras, XY^{Sry-} cells very rarely contribute to the Sertoli-cell population (reviewed in *15*). Similarly, in the fetal gonads of XX^{Sxra} mice, in which the activity of the Sxra portion of the Y chromosome is influenced by the inactivation of the X chromosome, Sertoli cells display a preferentially active X^{Sxra} chromosome, while other testicular-cell types show a random inactivation of the X and X^{Sxra} chromosomes (*16*). Taken together these data reaffirm that *Sry* functions cell-autonomously within the pre-Sertoli cells to initiate Sertoli-cell differentiation.

SRY ALONE DOES NOT A TESTIS MAKE

With the exception of *SRY* on the Y chromosome, all other genes involved in testicular differentiation are on the X chromosome or the autosomes—a fact demonstrated by the sex-reversed *Sry* XX transgenic mice. Intriguingly, only 30% of these mice sex reversed, and this phenomenon appears to be independent of transgene homozygosity or copy number (reviewed in *15*). Over the course of breeding these mice, it became apparent that the background genotype of different strains of mice influences the propensity to sex-reverse (reviewed in *15*). More specifically, this indicates an incompatibility between the timing and/or level of *Sry* expression and that of other genes on the X chromosome, or autosomes involved in either the testicular or ovarian differentiation pathway. A similar scenario occurs in the $B6.Y^{Dom}$ mouse, in which the Y chromosome from *Mus poschiavinus* (Y^{Pos}) is placed onto an inbred *Mus musculus musculus* background (C57Bl/6). XY^{Pos} animals develop either as females with ovarian tissue or as hermaphrodites in which the gonads contain both ovarian and testicular tissue *(17–19)*. Sex-reversal in these animals appears to be caused by a misregulation in the timing of *Sry* expression *(20)* and a functional incompatibility of Sry^{Pos} with at least two autosomal alleles *(19)*.

While there is no disputing the role of *SRY* in initiating testicular differentiation, it is a sobering consideration that only approx 20% of XY sex-reversals in humans can be attributed to mutations or deletions of *SRY* or its flanking sequences. Similarly, not all cases (80%) of XX males can be explained by the presence of *SRY* in the genome. Sex-reversal in XY females with an intact *SRY*—and in the remaining 20% of XX males lacking *SRY*—may be explained by a loss or gain of function, respectively, of autosomal or X-linked genes acting downstream, or upstream, of *SRY* in the testicular-differentiation pathway. In the next section, we discuss a number of genes which are thought to play a direct, or in some cases indirect, role in testicular differentiation: some of these factors appear to function upstream of *SRY*, and others to function downstream. We also introduce what little is known about the genes thought to be involved in ovarian differentiation. Of central importance is the contribution that transgenic studies in mice have made in defining the roles of each of the putative sex-determining and -differentiating genes discussed.

Wt1

Heterozygous mutations of the Wilms' tumor-suppressor gene, *WT1*, are associated with Wilms' tumor, a childhood tumor of the kidney, and XY pseudohermaphroditism (reviewed in *21*). The urogenital abnormalities in individuals with *WT1* mutations suggests a role for *WT1* in sex determination. *WT1* encodes a nuclear protein with domains shared with transcription factors. In the mouse, *Wt1* expression is initially detected in the intermediate mesoderm that will give rise to the urogenital system, and later in the mesothelium and central nervous system (CNS) *(22,23)*. In the developing gonad, *Wt1* is expressed in Sertoli cells *(22)*. *Wt1* knockout mice are embryonic-lethal, and lack kidneys and gonads *(24)*. It appears that the initiation of genital-ridge formation occurs, but quickly fails. Thus, *Wt1* appears to be essential for the initial formation of the gonads acting in males upstream of *Sry*. WT1 may also have functions later in male sexual differentiation by regulating genes such as *Mis (25)*.

In the Denys-Drash syndrome, point mutations in the *WT1* locus that create missense, nonsense, or frameshift mutations result in more severe XY genital phenotypes in comparison to *WT1* null heterozygotes *(26)*. Recently, the Denys-Drash syndrome has been modeled in mice by the introduction of a translation-stop codon at codon 396 of *Wt1 (27)*. However, only one heterozygous mouse carrying the T396 mutation was obtained from chimaeras. This heterozygote had small, aspermic testes. Unfortunately, this heterozygote also had sex-chromosome aneuploidy (XXY), complicating phenotype interpretations. Chimaeras generated with the T396 heterozygous ES cells also had genital abnormalities. This too was complicated by sex-chromosome chimaerism (XX↔XY). The generation of mice carrying mutations that match those found in humans is a powerful approach for studying these mutations in vivo. However, it appears that these strategies may have to be modified to obtain interpretable information regarding sexual development.

M33

A more recently identified candidate for a role in gonadal development and testicular differentiation is the mouse *M33* gene—a homolog of the *Drosophila Polycomb* gene *(28)*. In *Drosophila, Polycomb* genes regulate the expression of homeotic genes that are required for segmental patterning in the embryo. Testicular differentiation in *M33* mutant mice is perturbed: gonadal phenotypes range from small ovaries to "indistinct" gonads, with some animals developing as hermaphrodites with one testis and one ovary *(28)*. To date, *M33* mutant mice are the only example of recessive true hermaphroditism. Gonadal development is retarded in both XX and XY embryos at 11.5 dpc, when the expression of *Sry* in the male would normally reach its peak. At this stage, the gonads appear to consist of little more than a thickening of mesenchymal tissue, with an absence of coelomic epithelium. The coelomic epithelium is thought to give rise to Sertoli cells *(29)*, and is thus likely to be the cell type in which *Sry* is expressed. Consequently, testicular differentiation in *M33* mutants may fail because of a paucity of coelomic epithelial cells, and subsequently, few or no Sertoli cells. Thus, the principal role of *M33* in the gonad appears to be in the process of gonadogenesis, with an indirect influence on testicular differentiation.

Sox9

Heterozygous mutations in *SOX9* are responsible for the human skeletal-malformation syndrome, ampomelic dysplasia (CD) *(30,31)*. Consistent with a role in testicular differentiation, approx 75% of XY patients with CD are sex-reversed *(30–33)*. Gonadal phenotypes range from partial testicular differentiation to ovarian development with fewer than normal follicles to the most extreme of cases, where the gonads are reduced to streaks of fibrous tissue *(34)*.

As with *SRY* and other members of the *SOX* (Sry-like HMG box) family of genes, *SOX9* encodes an HMG box. In addition, SOX9 has in its carboxy terminus a transactivation domain *(35)*. Patients with CD show a wide variety of mutations to *SOX9*, however in nearly all cases, the mutations result in a truncation of the transactivation domain *(35)*. This suggests that SOX9 is required for the transactivation of downstream genes involved in skeletal and testicular development. The disparate nature of the mutations, and the observation that all patients are heterozygous for the mutation, suggest that CD and its associated sex reversal are caused by a haploinsufficiency for *SOX9* rather than a dominant-negative effect.

In the mouse, low levels of *Sox9* expression are seen in both male and female urogenital ridges from 10.5 dpc, around the time the development of the indifferent gonad (Fig. 2) *(36,37)*. A sexually dimorphic pattern of expression begins at 11.5 dpc, when levels in the male gonad increase, coincident with the peak in *Sry* expression, while in the female, expression of *Sox9* is turned off. At 12.5 dpc, *Sox9* transcripts in the testis are localized to the Sertoli cells. The SOX9 protein is conserved among the vertebrate groups with nonmammalian sequences thus far identified for chicken, turtle, and alligator *(38–40)*. Expression of *SOX9* in these species is similarly correlated with testicular differentiation. However, in contrast to the pattern seen in mammals, *SOX9* expression in chickens and alligators occurs later in Sertoli-cell differentiation, and after the production of MIS *(38,40)*. These data suggest that while *SOX9* appears to play a role in instigating Sertoli- cell differentiation in mammals, in reptiles and birds it may function later in development—perhaps to maintain, rather than determine, Sertoli-cell fate.

We are addressing the role of *Sox9* in Sertoli-cell differentiation in the mouse, using both loss-of-function and gain-of-function approaches. Our loss-of-function study uses gene targeting in ES cells to generate mice with only one functional copy of *Sox9*. More specifically, we have deleted approx 450 bp of *Sox9* including the translation start site and the majority of the HMG box, thus generating a null allele *(41)*. Testicular development in mice heterozygous for the null allele is normal, while skeletal development is disrupted. This situation stands in contrast to the human condition, where the vast majority of XY patients with mutations of *SOX9* show both sex-reversal and skeletal malformations.

In a further attempt to understand the role of *Sox9* in Sertoli-cell differentiation, we have generated ES cells that are homozygous for the *Sox9* mutation. In chimaeras, *Sox9*-null ES cells give rise to Sertoli cells, indicating that *Sox9* is not required cell-autonomously for Sertoli-cell differentiation.

We have also generated fetuses that are homozygous for the *Sox9* mutation by injecting *Sox9*-null ES cells into blastocysts that are derived from tetraploid cells. In this instance, the embryo will be entirely ES-cell-derived, with tetraploid cells contributing only to the extraembryonic tissues. *Sox9*-null fetuses are grossly abnormal, and die shortly after 10.5 dpc. However, the gonad at this time appears to be normal, indicating that early expression of *Sox9* is not required for the development of the indifferent gonad.

More direct data implicating *Sox9* in Sertoli-cell differentiation has resulted from our gain-of-function study. In these mice, a tyrosinase minigene has fortuitously inserted within 1 centimorgan of *Sox9 (42)*. All XX mice carrying the transgene (XXtg) develop as phenotypically normal, although sterile, males. Significantly, all of these mice have a normal XX karyotype, and are devoid of known Y chromosome genes, including *Sry*. Gonadal development in the XXtg males follows the typical male pathway from the outset, with Sertoli cells and early seminiferous cords visible at 12.5 dpc. The differentiating Sertoli cells express *Sox9* and MIS. By 15.5 dpc, the XXtg testis contains well-organized seminiferous cords enclosed by peritubular myoid cells, interstitial tissue, and a tunica albuginea. In wild-type animals at 11.5 dpc, Sertoli cells have not yet differentiated; this state of morphological development corresponds with an upregulation in the male, and a downregulation in the female, of *Sox9* expression. In the XXtg gonad at 11.5 dpc, *Sox9* expression persists at high levels. These data indicate that the insertion of the transgene has potentially disrupted a repressor element upstream of *Sox9*, which would normally allow for the expression of *Sox9* to be downregulated in the female. Continued expression of *Sox9* in the XXtg gonad initiates testicular differentiation in the absence of *Sry*, implicating *Sox9* as a gene involved in the early stages of Sertoli-cell differentiation, perhaps immediately downstream of *Sry*.

If we extrapolate further, we can propose a model with *Sox9* as the critical testis-differentiating gene, and the function of *Sry* is to ensure that *Sox9* is upregulated in the XY gonad—perhaps by repressing a negative regulator of *Sox9* expression. In the XX gonad, this repression of *Sox9* expression would persist, preventing testicular differentiation in the presumptive ovary. In our XXtg mice, the insertion of the transgene may have disrupted the binding site for this repressor of *Sox9*, thereby permitting its continued expression. The role of *Sry* as a repressor of a repressor of testicular differentiation was originally proposed by McElreavey and colleagues in 1993, as an alternative to the more conservative view that *Sry* acts as a dominant testis inducer *(15)*. The reader will find that as we progress further into this chapter, the concept of *Sry* as an antagonist of an "anti-testis" factor becomes an increasingly attractive proposal.

Sf1 *and* Dax1*: Multifunctional Regulators of Gonadal Development and Function*

In addition to *Wt1* and *M33*, a strong candidate for a role in early gonadogenesis is the *Ftz-F1* gene, which encodes the orphan nuclear-receptor steroidogenic factor 1 (SF1). In the fetus, SF1 has multiple roles in early gonadal, adrenal, and brain development, and throughout life is involved in the regulation of steroidogenesis in multiple tissues (*see 43* and Chapter 8 for review). Mice homozygous for a null mutation in *Sf1* lack adrenals, show gonadal degeneration after 11.5 dpc, have impaired gonadotrophic function, and altered structural characteristics of the ventromedial hypothalamus *(44–47)*. Taken together, these studies demonstrate an essential role for *Sf1* at multiple levels of the hypothalamic-pituitary-gonadal axis.

At 9.0 dpc in the mouse, *Sf1* transcripts are present in the urogenital ridge, and expression continues in the indifferent gonad of both sexes (Fig. 2) *(48)*. A sexually dimorphic pattern of expression becomes apparent at 12.5 dpc, when *Sf1* transcription increases in the developing testis and is downregulated in the prospective ovary. At 14.5 dpc, *Sf1* transcripts in the testis are restricted to the Leydig cells. In the female, expression of *Sf1* is reinstated predominantly in the thecal cells of the postnatal ovary. The diminished expression of *Sf1* in the ovary at 12.5 dpc is coincident with the onset of testicular differentiation in the male, and suggests that SF1 may regulate the expression of testis-differentiating genes that are disruptive to ovarian development.

The expression profiles of *Sf1* overlap considerably—both temporally and spatially—with that of another orphan nuclear receptor, *Dax1*. Mutations in *DAX1* are responsible for the human disorder X-linked adrenal hypoplasia congenita (AHC), which is characterized by an absence of the adrenal cortex and hypogonadotropic hypogonadism *(49)*. In the mouse, *Dax1* is first expressed in the somatic cells of the indifferent gonad of both sexes at approx 10.5 dpc (Fig. 2) *(50,51)*. At 12.5 dpc, levels of *Dax1* expression in the developing testis decrease significantly to levels that are barely detectable. In the female gonad, *Dax1* continues to be expressed until approx 14.5 dpc. In the adult testis and ovary, *Dax1* transcripts are restricted to the steroidogenic Leydig and thecal cells, respectively. Similar relative expression profiles for *DAX1* and *SF1* are conserved in the pig *(52)*. The inverse expression profiles of *Dax1/DAX1* and *Sf1/SF1* during testicular differentiation, and their co-expression in the steroidogenic cells of the testis and ovary, imply that these genes interact during gonadal development and endocrine function.

SF1 and DAX1 have been shown to act antagonistically in the regulation of *Mis* expression in the developing testis. Previous studies have demonstrated that SF1 is required for the expression of *Mis* in vitro and in vivo *(53,54)*. More recently, Nachtigal and colleagues *(25)* have shown that DAX1 antagonizes SF1-mediated transactivation of *Mis*. Specifically, SF1 and the transcription factor WT1 are thought to associate and thereby synergistically promote *Mis* expression. DAX1, however, antagonizes this synergy, probably by interacting directly with SF1. Further, in vitro studies from several laboratories have indicated that DAX1 can inhibit the transcriptional activation by SF1 of genes involved in steroidogenesis *(55–57)*. It is suggested that DAX1 is able to bind to DNA hairpin-loop structures *(56)*. This DNA-binding ability is shared by SRY. Thus, a likely scenario might depict SRY and DAX1 influencing the ability of each other to either activate or repress SF1-mediated transcription of genes involved in steroidogenesis and testicular differentiation.

In addition to its antagonistic effects on SF1 transcriptional activity, *DAX1* has also been implicated as an ovary-determining gene, or "anti-testis" gene. *DAX1* maps to a 160-kb region of the X chromosome which, when duplicated, causes XY-dosage-sensitive sex reversal (DSS) *(49,58,59)*. In vivo confirmation that *DAX1* is indeed the gene responsible for DSS was obtained by generating transgenic mice expressing a *Dax1* cDNA under the regulation of a fragment of DNA taken upstream from the *Dax1* start of transcription: this transgene was designated Dax:Dax *(60)*. In one transgenic line, the levels of *Dax1* expression were five times that of normal levels, with a corresponding increase in protein levels. Surprisingly, all XY transgenics from this line developed testes, although development was initially retarded. However, when transgenic females were bred to males carrying a weak *Sry* allele (Y^{Pos}), 3 of 14 liveborns developed as hermaphrodites and eight developed as males, and at least four of these probably possessed ovotestes earlier in development. In fact, when fetal gonads were examined, 8 out of 11 XY^{Pos} transgenics had ovotestes. Similar results were obtained when the *Dax1* promoter was used to express *Sry* (Dax:Sry transgene) in XX animals carrying Dax:Dax.

Paradoxically, XY mice homozygous for the *Dax1* transgene, and expressing levels of DAX1 that are approx 11 times higher than endogenous levels, still develop normal testes. Yet in humans, a double dose of DAX1 is sufficient to induce a sex-reversal. Testicular development is only disrupted in those transgenic mice that possess a weak *Sry* allele (Y^{Pos} or Dax:Sry). The authors give a plausible account that takes into consideration two important observations. Firstly, testis development in both the Y^{Pos} and Dax:Sry mice lacking the Dax:Dax transgene is delayed. This becomes significant when one also takes into account that transcription of *Sry* in wild-type animals appears to pass along the gonad in such a way that each cell expresses high levels of *Sry* for only a brief period. Thus, most cells will have experienced the peak in *Sry* expression, and become committed to the Sertoli-cell fate, before experiencing high levels of *Dax1* expression. Consequently, *Sry* is able to outcompete *Dax1* for the activation of testis-differentiating genes. Secondly, in the case of Dax:Dax transgenics with the Y^{Pos} *Sry* allele, where levels of transcription are lower than normal, or XX Dax:Sry animals, where the onset of *Sry* expression appears to be delayed, *Dax1* expression increases before that of *Sry*, allowing *Dax1* to maintain its inhibition of testis-differentiating genes. Although the experiments are not optimal, these data suggest that DAX1 and SRY act antagonistically toward each other, competing to control the activation of genes in the testicular-differentiation pathway.

In apparent contradiction to the above hypotheses, mice deficient for *Dax1* develop normal ovaries and internal reproductive structures *(61)*. The only abnormality in females appears to be that some follicles contain more than one oocyte. These data do not support a role for *Dax1* as an ovary-determining gene. On the other hand, it is still plausible that DAX1 can act as an "anti-testis" factor by interfering with the expression of testis-differentiating genes, such as *Sox9*. Very low levels of *Sox9* expression are seen in both male and female urogenital ridges at 10.5 dpc, coincident with the expression of *Sf1*. It is possible that SF1 directs this very low level of *Sox9* expression. At 10.5 dpc, *Dax1* is expressed in male and female indifferent gonads, corresponding with the onset of *Sry* expression in the male. Critically, at 11.5 dpc, levels of *Sry* expression peak and *Sox9* levels are substantially elevated. In the female at this time, expression of *Sox9* is extinguished. If we assume that SF1 is responsible for maintaining low levels of *Sox9* expression, DAX1 may heterodimerize with SF1 and counter SF1-mediated transcriptional activation of *Sox9*. SRY may antagonize this effect of DAX1, by competing for the same binding site or altering the comformation of the DNA in such a way that DAX1 cannot bind or dimerize with SF1. Most importantly it may directly, or indirectly, cause the substantial increase in *Sox9* expression seen in the male at 11.5 dpc. This boost in *Sox9* expression is essential for testicular differentiation to occur. In the loss-of-function females lacking *Dax1*, expression of *Sox9* may continue at lower levels, but without the boost in *Sox9* expression seen in the male, the gonad continues along the ovarian-differentiation pathway. The expression of *Sox9* in the gonads of *Dax1*-deficient females is unknown.

Another possibility is that another gene, or genes, is also involved in the repression of *Sox9* expression in the female gonad, and that in the absence of *Dax1*, this other factor is sufficient to keep *Sox9* levels below the threshold for testicular differentiation. A possible candidate for this role is the *Sry*-related gene *Sox3*.

Sox3*: Ancestor of* Sry *and a Putative Ovary Determinant*

Of all the *SOX* genes, *SOX3* is most closely related to *SRY*, sharing 82% similarity in the HMG box *(62,63)*. Both SOX3 and SRY are encoded by one exon, and share homology outside of the HMG box—an intriguing finding, because no such homology exists between species for *SRY*. It has has been proposed that, in mammals, *SRY* evolved from *SOX3* during the evolution of the X and Y chromosomes *(64,65)*. *SOX3* has been highly conserved throughout vertebrate evolution, and is X-linked in both eutherian and marsupial mammals *(62–64,66)*.

The predominant site of *Sox3* expression in the mouse is in the developing central nervous system *(63)*. However, at 11.5 dpc, *Sox3* expression is observed in both male and female genital ridges (Fig. 2). Transcripts are localized to the somatic cells of the gonad and the expression level in the female appears to be twice that of the male, reaching levels equivalent to or greater than *Sry*. By 12.5 dpc, transcripts are no longer detectable in either male or female gonads (A. Hacker unpublished data, cited in *63*).

In vitro data indicate that SOX3 and SRY bind to the same DNA target sequence; however, SOX3 binds with a much lower affinity than SRY *(63)*. Thus, a likely scenario may be one in which SOX3 and SRY compete for the same target sequence involved in the regulation of ovarian- or testicular-differentiation genes. More specifically, Jennifer Graves *(65)* has proposed that SOX3 and SRY compete to regulate the expression of *SOX9*. In her model, SOX3 was once part of a dosage-regulated system of sex determination that involved the differential regulation of *SOX9*. With the evolution of *SRY* from *SOX3*, a more robust mechanism of sexual differentiation came into play. Thus, in the female, SOX3 would repress the expression of *SOX9* and consequently, other testis-differentiating genes. In the male, SRY would outcompete SOX3 for its binding site in the *SOX9* promoter region, and testicular differentiation would ensue. To date, nothing is known about the presence or absence of such a binding site in the *SOX9* promoter. XY human individuals deleted for *SOX3* develop small, but essentially normal, testes (*62*); XY individuals with a duplication of *SOX3*, or XX individuals deleted for *SOX3* have not been identified. However, loss-of-function and gain-of-function studies in mice will contribute substantially to defining the requirement of *Sox3* for ovarian differentiation.

Mis

MIS is one of the key hormones required in male development (*see 67* and Chapter 3 for review). MIS is produced by the Sertoli cells of the fetal testis, with the highest levels observed during the period in which the Müllerian duct regresses *(68,69)*. Expression of MIS continues after birth at reduced levels, and then declines sharply at puberty *(69–72)*. In the mouse, *Mis* transcripts are first detected in the differentiating Sertoli cells at 11.5 dpc, $1^1/_2$ d before the Müllerian ducts begin to regress (Fig. 2) *(69)*.

Initially, transgenic mice were used to explore the potential roles of MIS in vivo *(73)*. The mouse metallothionein promoter (MT) was used to direct widespread expression of human MIS (hMIS). Female MT-hMIS transgenic mice were born without a uterus or oviducts as expected. In addition, the ovaries became depleted of germ cells soon after birth, and eventually degenerated. These findings further confirmed the role of MIS in Müllerian-duct regression, and demonstrated that high levels of MIS were directly, or indirectly, toxic to female germ cells. Although most male MT-hMIS transgenic mice were overtly normal and fertile, some males from the highest expressing lines did not virilize and had undescended testes. Lyet et al. *(74)* determined that the high levels of hMIS caused a reduction in circulating testosterone. Presumably, those nonvirilized males from the highest expressing lines had severely reduced testosterone levels. More recently, Racine and colleagues *(75)* have shown that the overexpression of hMIS in male MT-hMIS transgenics blocks the differentiation of Leydig-cell precursors and decreases expression levels of the cytochrome p450 17α-hydroxylase gene, which is required for steroid synthesis. MIS appears to exert its effects on Leydig-cell differentiation and steroidogenesis directly via its receptor, which is now known to be expressed by Leydig cells in addition to Sertoli cells *(75)*.

Mis knockout mice have also been generated *(76)*. Although *Mis* is specifically expressed in postnatal granulosa cells of the ovary, *Mis*-mutant females are normal and fertile. However, it is possible that more subtle alterations in ovarian function are present.

Mis-mutant males developed as male pseudohermaphrodites, possessing testes and male internal organs as well as a uterus and oviducts. The majority of the *Mis*-mutant males proved to be infertile, not because of a germ-cell deficiency, but because of the physical abnormalities caused by the codevelopment of both male and female internal reproductive organs. In addition, a proportion of the older *Mis*-mutant males had Leydig-cell hyperplasia, indicating that *Mis* also influenced Leydig-cell proliferation. The finding that MIS receptors are expressed on Leydig cells suggests that this might be a direct interaction between Sertoli and Leydig cells *(75)*. By generating *Mis* mutant males that also carried the testicular feminization (*Tfm*) mutation, XY animals were produced that could not respond to the two primary male hormones: testosterone and MIS. The resulting animals overtly appeared as females, with a uterus and oviducts, but no Wolffian-duct derivatives. Testes were intra-abdominal at the position where ovaries would normally be located. These studies re-emphasized the hormonal control of sexual differentiation in mammals.

TGFβ signaling is mediated by transmembrane receptors with serine/threonine kinase activity *(77)*. Type I and type II receptors generate complexes upon ligand binding, and transduce their signals through Smad proteins to the nucleus. The type I receptor for MIS is unknown; however, the type II receptor for MIS has been isolated in the mouse *(78)*, rat *(79,80)*, rabbit *(81)*, human *(82)*, and tammar wallaby (D. Whitworth, unpublished data). We have generated MIS type II receptor knockout mice *(83)*. Interestingly, these mice are a phenocopy of the MIS ligand knockout mice, suggesting that the MIS signaling pathway involves a one ligand-one type II receptor pathway. The specificity of MIS signaling was established by generating female MT-hMIS mice that were also deficient for the MIS type II receptor *(84)*. In contrast to MT-hMIS females that lack a uterus, oviducts, and ovaries, the MT-hMIS/MIS type II receptor-mutant females were normal and fertile.

In the mouse, *Mis* expression is detected around 11.5 dpc, specifically in preSertoli cells and later in Sertoli cells of the fetal and postnatal testes *(9)*. The timing and cell specificity of *Mis* expression suggested that *Sry* may directly or indirectly regulate *Mis* transcription. However, tissue-culture and transgenic mouse experiments have indicated that steroidogenic factor 1 (SF1) is required to activate *Mis* transcription *(53,54)*. In contrast, targeted mutagenesis of the SF1 binding site in the endogenous mouse *Mis* locus does not block *Mis* transcription, but rather reduces *Mis* transcription levels *(85)*. This reduction is not sufficient to cause a persistence of Müllerian-duct-derived tissues. However, when a conserved SOX9 binding site is similarly mutated, no *Mis* transcription occurs and the mutant males develop as pseudohermaphrodites, similar to *Mis* mutant males. These manipulations of the endogenous promoter have revealed the distinct activities of these two transcription-factor binding sites in regulating *Mis* expression and have helped to piece together this important genetic pathway of male sexual differentiation.

Tfm

Mice carrying the X-linked *Tfm* mutation have female genitalia and lack Wolffian duct-derived structures, despite possessing testes and an XY karyotype *(86)*. *Tfm* is a naturally occurring mutation to the androgen receptor (AR) that results in complete androgen insensitivity (reviewed in *87* and *88*). In the absence of a functional AR, neither testosterone nor its more potent metabolite 5α-dihydrotestosterone can exert their effects on the Wolffian duct and the external genitalia, respectively. At the molecular level, *Tfm* is a single-base deletion in the N-terminal region of the AR gene. This leads to a premature termination of translation, which in turn gives rise to a truncated, unstable mRNA transcript *(89–92)*.

The external genitalia of *Tfm* male mice resembles that of wild-type females, although the vagina is shorter than normal and often blind-ending *(86)*. Testes of adults are smaller, and fail to descend beyond the internal inguinal ring into the scrotum (reviewed in *88*). Spermatogenesis proceeds in the testes of *Tfm* mutants; however, most germ-cells fail to progress beyond the spermatocyte stage *(86)*. Levels of the pituitary gonadotrophins luteinizing hormone (LH) and follicle-stimulating hormone (FSH) are elevated in *Tfm* males because of the loss of the AR-mediated negative feedback loop, which would normally regulate their production (reviewed in *88*). Leydig cells appear to be normal in number, but are hypertrophied *(86)*. In addition, the production of androgens is severely reduced, probably a result of the loss of the enzyme 17α-hydroxylase *(93)*.

By generating XX mice that are sex-reversed (XX^{Sxr}) and carrying the *Tfm* mutation, Ohno and colleagues were able to study the interactions during male development between cells that are wild-type for the AR, and those that are *Tfm*-mutated (reviewed in *87*). Because of X-inactivation, sex reversed mice that are heterozygous for *Tfm* are mosaics of cells which are sensitive to androgens (X^{WT}) and those that are insensitive (X^{Tfm}). The conclusion from these studies was that effects of testosterone could be mediated by local growth factors from wild-type cells to *Tfm* cells (reviewed in *87*). While cellular differentiation was seen to occur only in wild-type cells, embryonic induction, the morphogenesis of male reproductive structures, and the postnatal maintenance of these structures were all found to be mediated effects. The *Tfm* mouse provides an excellent model in which to study the breadth and mechanics of androgen function.

Wnt4

WNT4 belongs to the WNT family of secreted glycoproteins, with members that function during development as signaling molecules in a diverse range of cell types *(94)*. In the mouse, *Wnt4* expression is first detected at 9.5 dpc in the mesenchyme and coelomic epithelium of the mesonephros *(95)*. Shortly afterwards, expression of *Wnt4* extends into the mesenchyme of the indifferent gonad. Transcription of *Wnt4* in the gonads becomes sexually dimorphic at 11.5 dpc, when it is downregulated in the male gonad but maintained in the somatic cells of the female gonad.

In the female, *Wnt4* appears to be required for suppressing the differentiation of Leydig cells from the steroidogenic cell lineage. Ovaries collected from *Wnt4* homozygous mutant females at 14.5 dpc express the steroidogenic enzymes 3β-HSD and P450c17, both of which are involved in the synthesis of testosterone by Leydig cells and their precursors, and are also expressed by steroidogenic cells of the adult ovary. These ovaries also express type III 17β-HSD which, unlike 3β-HSD and P450c17, is a testis-specific enzyme. The Wolffian ducts of *Wnt4* mutant females persist and become coiled at their most cranial end, so that they resemble the epididymides of the male. This further suggests that the ovaries of these animals produce testosterone during fetal development. Lastly, *Wnt4* also has a role in the early formation of the Müllerian duct in both sexes. At 12.5 dpc, male and female *Wnt4* mutants lack an identifiable Müllerian duct and do not express the Müllerian duct markers *Pax8* and *Wnt7a*.

Wnt7a

Another member of the WNT family, WNT7a, also appears to be involved in Müllerian-duct development. In the fetus, *Wnt7a* is expressed in the epithelium of the Müllerian duct in both males and females from 12.5 dpc to 14.5 dpc *(96)*. In the male, expression of *Wnt7a* diminishes as the Müllerian duct regresses, while in the female expression persists into the adult where it is localized to the epithelial cells of the Müllerian-duct-derived oviducts and uterus *(96,97)*. Development of the Müllerian duct in *Wnt7a* homozygous mutant females is abnormal *(96)*. At birth, these mutant females lack coiled oviducts, and the wall of the uterus is thinner than that of wild-type littermates. Similarly, in adult females, the oviducts are still abnormal, the uterus is nearly devoid of uterine glands, and the stromal layer is thinner.

Male homozygous mutants have persistent Müllerian ducts resembling those seen in the MIS and MIS type II receptor mutants. Critically, *Wnt7a* mutant males (and females) fail to express the MIS type II receptor in the mesenchyme surrounding the Müllerian duct. This suggests that expression of *Wnt7a* in the epithelial cells of the Müllerian duct is required for regulating the expression of the MIS type II receptor in the periductal mesenchyme which, in turn, induces regression of the Müllerian duct epithelium.

Dhh

Hedgehogs are important signaling molecules that regulate diverse developmental processes *(98)*. Vertebrate hedgehog genes include *Sonic* (*Shh*), *Indian* (*Ihh*), and *Desert* (*Dhh*). It is *Desert hedgehog* that is most relevant to mammalian sex determination and differentiation. At the moment, *Dhh* has only been reported in mice, however *DHH* has also been isolated from a marsupial mammal, the tammar wallaby (C.-A. Mao, unpublished data). In the mouse, *Dhh* expression is initially detected in the fetal testis at 11.5 dpc (Fig. 2). Later, in the adult testis, *Dhh* is found specifically in Sertoli cells. *Dhh* is also detected in other tissues of the developing mouse embryo, including Schwann cells, the endothelium of the vasculature, and the endocardium *(99)*. The highly restricted expression pattern of *Dhh* in the somatic cells of the male gonad suggests that *Dhh* may be an important regulator of the male phenotype.

Dhh mutant mice have been generated by gene targeting in ES cells *(100)*. Male *Dhh* mutants were found to be viable yet sterile because of a disruption to spermatogenesis. Interestingly, the expression of *Patched* (*Ptc*), which encodes the receptor for hedgehogs, is normally detected in the Leydig cells. In the *Dhh* mutants, *Ptc* was not detected. These findings suggest that DHH produced by Sertoli cells interacts with Leydig cells to support spermatogenesis. Therefore, the defects in spermatogensis may be caused by alterations in androgens that are required for male germ-cell development. This would have to be an androgen deficiency that only affects spermatogenesis because the *Dhh* mutant males (at least on the genetic backgrounds analyzed) were normally virilized. These results also indicate that *Dhh* is not involved in sex determination, but rather sex differentiation and spermatogenesis. The initiation of *Dhh* expression in the somatic cells of the male fetal gonad occurs soon after the initiation of *Sry* expression, suggesting that SRY regulates *Dhh*, either directly or indirectly. Therefore, it will be interesting to determine the *cis*- and *trans*-acting factors required to direct Sertoli-cell-specific transcription of *Dhh*.

TRANSGENIC MARSUPIALS: A NEW FRONTIER

Sexual development in marsupials is fundamentally similar to that described for eutherians mammals such as the mouse and human. However, there are some significant differences in the morphology of the reproductive structures and in the timing of events. It is our goal to exploit these differences in order to gain a more detailed insight into mammalian sexual development from an alternative perspective.

In marsupials, as in eutherians, the Y chromosome is testis-determining *(101,102)*, presumably under the direction of the marsupial homolog of *SRY (12)*. Testicular differentiation precedes ovarian differentiation—as is the case in eutherians—and the development of the internal reproductive structures is similarly under the influence of MIS and steroids. However, in marsupials, these events occur after birth. In addition, there are some sexually dimorphic structures in the marsupial that appear to be under direct genetic—rather than secondary gonadal—hormonal control. In at least four species of marsupials, including the South American grey opossum (*Monodelphis domestica*), North American Virginia opossum (*Didelphis virginiana*), and the Australian brushtail possum (*Trichosuras vulpecula*) and tammar wallaby (*Macropus eugenii*), clearly defined scrotal and mammary anlagen are visible several days before gonadal sex differentiation *(103–106)*. Many species of marsupials develop a pouch in the female. In these species, pouch development similarly appears to be independent of gonadal sex.

Descriptions of spontaneously occurring intersexes provide further evidence for the direct genetic control of these sexual dimorphisms in marsupials. Several studies *(102,107–109)* together describe five intersexes from five different marsupial species with either an XXY, XXY/XX/XY, or aberrant XY karyotype. Each individual possessed functionally normal testes, as evidenced by the presence of a normal male reproductive tract, and yet lacked a scrotum, while possessing a pouch and mammary glands. Taken together, these observations provide compelling evidence that the sexually dimorphic development of the scrotum, pouch, and mammary glands is under direct genetic control, rather than hormonally mediated, as is the case in eutherians. In short, development of the scrotum occurs when only one X chromosome is present (XY, XO), while in the presence of two X chromosomes (XX, XXY) a pouch and mammary glands form *(107,110,111)*.

Using the South American grey opossum, it is possible to rigorously test this hypothesis by injecting *M. domestica* eggs with an *M. domestica SRY* transgene. XX transgenic animals would be expected to have testes, but the key issue will be whether or not they develop a scrotum. If a scrotum does form, this would suggest that scrotal development can be induced by the presence of a testis, thus refuting the current theory that scrotal development in marsupials is independent of gonadal sex.

Research interest in marsupials is not only restricted to the field of sexual differentiation. Marsupials are also unique in comparison to eutherian mammals in their preimplantation development, placentation, the development of their reproductive organs, and X chromosome imprinting (reviewed in *112–114*). By establishing transgenic techniques in marsupials, we will generate an exceptional resource for comparative studies of the genetic control of mammalian pre- and postimplantation development.

CONCLUSION

The progression from indifferent gonad to testis or ovary is dependent upon a suite of cellular migration, differentiation, and endocrine events. The challenge in understanding how the testis and ovary come into being is in piecing together and integrating what we know at the morphological and molecular levels. From a morphological perspective, it is becoming clearer that differences between XX and XY indifferent gonads in the proliferation of coelomic epithelial cells, the migration of cells into the gonad from the mesonephros, and the cross-communication between germ cells and somatic cells, are all important factors in the development of the testis and ovary. As we learn more about

Table 1

Gene Functions and References to Studies of Transgenic Mice and Natural Mutants

Gene	*Function*	*Transgenic studies*
Sry	Sertoli cell fate	*(4,14)*
Wt1	Gonadogenesis/Regulation of *Mis* expression	*(24,27)*
M33	Gonadogenesis	*(28)*
Sox9	Sertoli-cell fate/Regulation of *Mis* expression	*(42)*
Mis	Müllerian-duct regression	*(54,73–75,84,85)*
	Leydig-cell differentiation	
	Steroidogenesis	
MIS Type II Receptor	MIS signaling	*(83,84)*
AR (Tfm)	Androgen signaling	*(86,89,90–92)*
Wnt4	Müllerian-duct formation	*(95)*
	Suppression of Leydig-cell differentiation in ovary	
Wnt7a	Müllerian-duct regression and differentiation	*(96)*
Dhh	Spermatogenesis	*(100)*
Sf1	Gonadogenesis	*(44–47)*
	Sertoli-cell and Leydig-cell endocrine function	
Dax1	Leydig-cell endocrine function	
	Granulosa-cell fate and endocrine function	*(60,61)*

the discrete morphological changes that comprise the grand process of sexual differentiation, it becomes an easier prospect to isolate and place into order candidate genes and the functions of their protein products.

At the molecular level, SRY is undoubtedly testis-determining, but we still don't know whether it acts as a dominant testis-inducer or as an inhibitor of ovarian development, and so, by default, is an indirect testis-determinant. SF1 and WT1 appear to each play multiple roles in both the early stages of gonadogenesis, and later in the regulation of testicular and ovarian endocrine function (Table 1). SOX9 appears to be essential not only in determining Sertoli-cell fate, but also in regulating their production of MIS. Importantly, *SOX9* expression in the embryo is not restricted to the testis and so the testis-specific expression of MIS must also require the collaboration of other factors. In addition to its long-known role in inducing the regression of the Müllerian ducts, MIS has more recently been credited with additional functions in regulating the proliferation of Leydig cells and their production of testosterone. The function of MIS in the postnatal ovary is speculative at best. Indeed, our knowledge of the molecular events required for granulosa-cell differentiation is sorely lacking. DAX1 may be important to granulosa cell fate by acting as an antagonist to SRY, while postulated roles for SOX3 remain to be supported by loss- or gain-of-function analyses in humans and mice. Thus, the picture of the molecular aspects of sexual differentiation that begins to emerge is one of an exceedingly complex meshing of factors—many with multiple functions—rather than a straightforward linear hierarchy of interactions.

The difficulty in piecing together a model of molecular events to explain the morphological changes observed in the differentiating gonad is that we know relatively few of the genes which must be involved in this process. As new candidate genes are identified and put to the test by gain- and loss-of-function studies in mice, we come closer to being able to integrate the molecular with the morphological. Comparative studies between different eutherian species and marsupials will allow us to test the generality of models extrapolated from the mouse.

ACKNOWLEDGMENTS

We are grateful to Soazik Jamin and Alex Arango for their thoughtful suggestions during the preparation of this manuscript. Work from Richard R. Behringer's laboratory discussed herein was supported by a grant from the NIH (HD30284).

REFERENCES

1. Burgoyne PS, Buehr M, Koopman P, Rossant J, McLaren A. Cell-autonomous action of the testis-determining gene:Sertoli cells are exclusively XY in XX´XY chimaeric mouse testes. Development 1988;102:443–450.
2. Zamboni L, Upadhyay S. Germ cell differentiation in mouse adrenal glands. J Exp Zool 1983;228:173–193.
3. Palmer S, Burgoyne PS. In situ analysis of fetal, prepuberal and adult XX´XY chimaeric mouse testes:Sertoli cells are predominantly, but not exclusively, XY. Development 1991;112:265–268.
4. Lovell-Badge R, Robertson E. XY female mice resulting from a heritable mutation in the primary testis-determining gene, Tdy. Development 1990;109:635–646.
5. Page DC, Mosher R, Simpson EM, Fisher E, Mardon G, Pollack J, et al. The sex-determining region of the human Y chromosome encodes a finger protein. Cell 1987;51:1091–1094.
6. Sinclair AH, Berta P, Palmer MS, Hawkins JR, Griffiths BL, Smith MJ, et al. A gene from the human sex-determining region encodes a protein with homology to a conserved DNA-binding motif. Nature 1990;346:240–244.
7. Gubbay J, Collignon J, Koopman P, Capel B, Economou A, Münsterberg A, et al. A gene mapping to the sex determining region of the mouse Y chromosome is a member of a novel family of embryonically expressed genes. Nature 1990;346:245–250.
8. Koopman P, Münsterberg A, Capel B, Vivian N, Lovell-Badge R. Expression of a candidate sex-determining gene during mouse testis differentiation. Nature 1990;248:450–452.
9. Hacker A, Capel B, Goodfellow P, Lovell-Badge R. Expression of Sry, the mouse sex determining gene. Development 1995;121:1603–1614.
10. Just W, Rau W, Vogel W, Akhverdian M, Fredga K, Graves JA, et al. Absence of Sry in species of the vole Ellobius. Nat Genet 1995;11:117,118.
11. Soullier S, Hanni C, Catzeflis F, Berta P, Laudet V. Male sex determination in the spiny rat Tokudaia osimensis (Rodentia: Muridae) is not Sry dependent. Mamm Genome 1998;9:590–592.
12. Foster JW, Brennan FE, Hampikian GK, Goodfellow PK, Sinclair AH, Lovell-Badge R, et al. Evolution of sex determination and the Y chromosome: SRY-related sequences in marsupials. Nature 1992;359:531–533.
13. Harley VR, Jackson DI, Hextall PJ, Hawkins JR, Berkovitz GD, Sockanathan S, et al. DNA binding activity of recombinant SRY from normal males and XY females. Science 1992;255:453–456.
14. Koopman P, Gubbay J, Vivian N, Goodfellow P, Lovell-Badge R. Male development of chromosomally female mice transgenic for Sry. Nature 1991;351:117–121.
15. Capel B. The role of Sry in cellular events underlying mammalian sex determination. Curr Top Dev Biol 1996;32:1–37.
16. Jamieson RV, Zhou SX, Wheatley SC, Koopman P, Tam PP. Seroli cell differentiation and Y-chromosome activity: a developmental study of X-linked transgene activity in sex-reversed X/XSxra mouse embryos. Dev Biol 1998;199:235–244.
17. Eicher EM, Washburn LL, Whitney JB, Morrow KE. Mus poschiavinus Y chromosome in the C57BL/6J murine genome causes sex reversal. Science 1982;217:535–537.
18. Nagamine CM, Taketo T, Koo GC. Studies on the genetics of tda-1 XY sex reversal in the mouse. Differentiation 1987;33:223–231.
19. Eicher EM, Washburn LL, Schork NJ, Lee BK, Shown EP, Xu X, et al. Sex-determining genes on mouse autosomes identified by linkage analysis of C57BL/6- YPOS sex reversal. Nat Genet 1996;14:206–209.

20. Albrecht KH, Eicher EM. DNA sequence analysis of *Sry* alleles (subgenus *Mus*) implicates misregulation as the cause of C57BL/6J-Y(POS) sex reversal and defines the SRY functional unit. Genetics 1997;147:1267–1277.
21. Schedl A, Hastie N. Multiple roles for the Wilms' tumour suppressor gene, WT1 in genitourinary development. Mol Cell Endocrinol 1998;140:65–69.
22. Pelletier J, Schalling M, Buckler AJ, Rogers A, Haber DA, Housman D. Expression of the Wilms' tumor gene *WT1* in the murine urogenital system. Genes Dev 1991;5:1345–1356.
23. Armstrong JF, Pritchard-Jones K, Bickmore WA, Hastie ND, Bard JB. The expression of the Wilms' tumour gene, *WT1*, in the developing mammalian embryo. Mech Dev 1992;40:85–97.
24. Kreidberg JA, Sariol H, Loring JM, Maeda M, Pelletier J, Housman D, et al. WT-1 is required for early kidney development. Cell 1993;74:679–691.
25. Nachtigal MW, Hirokawa Y, Enyeart-VanHouten DL, Flanagan JN, Hammer GD, Ingraham HA. Wilms' tumor 1 and Dax-1 modulate the orphan nuclear receptor SF-1 in sex-specific gene expression. Cell 1998;93:445–454.
26. Little M, Wells C. A clinical overview of *WT1* gene mutations. Hum Mutat 1997;9:209–225.
27. Patek CE, Little MH, Fleming S, Miles C, Charlieu J-P, Clarke AR, et al. A zinc finger truncation of murine WT1 results in the characteristic urogenital abnormalities of Denys-Drash syndrome. Proc Natl Acad Sci USA 1999;96:2931–2936.
28. Katoh-Fukui Y, Tsuchiya R, Shiroishi T, Nakahara Y, Hashimoto N, Noguchi K. et al. Male-to-female sex reversal in *M33* mutant mice. Nature 1998;393:688–692.
29. Karl J, Capel B. Sertoli cells of the mouse testis originate from the coelomic epithelium. Dev Biol 1998;203:323–333.
30. Foster JW, Dominguez-Steglich MA, Guioli S, Kowk G, Weller PA, Stevanovic M, et al. Campomelic dysplasia and autosomal sex reversal caused by mutations in an *SRY*-related gene. Nature 1994;372:525–530.
31. Wagner T, Wirth J, Meyer J, Zabel B, Held M, Zimmer J, Pasantes J, et al. Autosomal sex reversal and campomelic dysplasia are caused by mutations in and around the *SRY*-related gene *SOX9*. Cell 1994;79:1111–1120.
32. Tommerup N, Schempp W, Meinecke P, Pederson S, Bolund L, Brandt CCG, et al. Assignment of an autosomal sex reversal locus (SRA1) and campomelic dysplasia (CMPD1) to 17q24.3-q25.1. Nat Genet 1993;4:170–173.
33. Mansour S, Hall CM, Pembrey ME, Young ID. A clinical and genetic study of campomelic dysplasia. J Med Genet 1995;32:415–420.
34. Houston CS, Opitz JM, Spranger JW, Macpherson RI, Reed MH, Gilbert EF, et al. The campomelic syndrome: review, report of 17 cases, and follow-up on the currently 17-year-old boy first reported by Maroteaux et al. in 1971. Am J Med Genet 1983;15:3–28.
35. Südbeck P, Schmitz ML, Baeuerle PA, Scherer G. Sex reversal by loss of the C-terminal transactivation domain of human SOX9. Nat Genet 1996;13:230–232.
36. Kent J, Wheatley SC, Andrews JE, Sinclair AH, Koopman P. A male-specific role for *Sox9* in vertebrate sex determination. Development 1996;122:2813–2822.
37. Morais da Silva S, Hacker A, Harley V, Goodfellow P, Swain A, Lovell-Badge R. *Sox9* expression during gonadal development implies a conserved role for the gene in testis differentiation in mammals and birds. Nat Genet 1996;14:62–68.
38. Oreal E, Pieau C, Mattei M-G, Josso N, Picard J-Y, Carre-Eusebe D, Magre S. Early expression of *AMH* in chicken embryonic gonads precedes testicular *SOX9* expression. Dev Dyn 1998;212:522–532.
39. Spotila LD, Spotila JR, Hall SE. Sequence and expression analysis of *WT1* and *Sox9* in the red-eared slider turtle, *Trachemys scripta*. J Exp Zool 1998;281:417–427.
40. Western PS, Harry JL, Marshall Graves JA, Sinclair AH. Temperature dependent sex determination in the American alligator:*AMH* precedes *SOX9* expression. Dev Dyn 1999;216:411–419.
41. Bi W, Deng JM, Zhang Z, Behringer RR, de Crombrugghe B. *Sox9* is required for cartilage formation. Nat Genet 1999;22:85–89.
42. Bishop CE, Whitworth DJ, Qin Y, Agoulnik A, Harrison W, Agoulnik I, Harrison W, Behringer RR, et al. A transgenic insertion upstream of Sox9 is associated with dominant XX sex reversal in the mouse. Nat Genet 2000;26:490–494.
43. Parker KL, Schimmer BP. The roles of the nuclear receptor steroidogenic factor 1 in endocrine differentiation and development. Trends Endocrinol. Metab. 1996;7:203–207.
44. Ingraham HH, Lala DS, Ikeda Y, Luo X, Shen WH, Nachtigal MW, et al. The nuclear receptor steroidogenic factor 1 acts at multiple levels of the reproductive axis. Genes Dev 1994;8:2302–2312.

45. Luo X, Ikeda Y, Parker KL. A cell-specific nuclear receptor is essential for adrenal and gonadal development and sexual differentiation. Cell 1994;77:481–490.
46. Sadovsky Y, Crawford PA, Woodson KG, Polish JA, Clements MA, Tourtellotte LM, et al. Mice deficient in the orphan receptor steroidogenic factor 1 lack adrenal glands and gonads but express P450 side-chain-cleavage enzyme in the placenta and have normal embryonic serum levels of corticosteroids. Proc Natl Acad Sci USA 1995;92:10,939–10,943.
47. Shinoda K, Lei H, Yoshii H, Nomura M, Nagano M, Shiba H, et al. Developmental defects of the ventromedial hypothalamic nucleus and pituitary gonadotroph in the Ftz-F1 disrupted mice. Dev Dyn 1995;204:22–29.
48. Ikeda YW, Shen HA, Ingraham HA, Parker KL. Developmental expression of mouse steroidogenic factor 1 an essential regulator of the steroid hydroxylases. Mol Endocrinol 1994;7:852–860.
49. Muscatelli F, Strom TM, Walker AP, Zanaria E, Recan D, Meindl A, et al. Mutations in the *DAX-1* gene give rise to both X-linked adrenal hypoplasia congenita and hypogonadotropic hypogonadism. Nature 1994;372:672–676.
50. Ikeda Y, Swain A, Weber TJ, Hentges KE, Zanaria E, Lalli E, et al. Steroidogenic Factor 1 and *Dax-1* co-localize in multiple cell lineages: potential links in endocrine development. Mol Endocrinol 1996;10:1261–1272.
51. Swain A, Zanaria E, Hacker A, Lovell-Badge R, Camerino G. Mouse *Dax-1* expression is consistent with a role in sex determination as well as in adrenal and hypothalamus function. Nat Genet 1996;12:404–409.
52. Pilon N, Behdjani R, Daneaul Lussier JG, Silversides DW. Porcine steroidogenic factor-1 gene (pSF-1) expression and analysis of embryonic pig gonads during sexual differentiation. Endocrinology 1998;139:3803–3812.
53. Shen WH, Moore CC, Ikeda Y, Parker KL, Ingraham HA. Nuclear receptor steroidogenic factor 1 regulates the Müllerian inhibiting substance gene: a link to the sex determination cascade. Cell 1994;77:651–661.
54. Giuili G, Shen WH, Ingraham HA. The nuclear receptor SF-1 mediates sexually dimorphic expression of Müllerian Inhibiting Substance, in vivo. Development 1997;124:1799–1807.
55. Ito M, Yu RN, Jameson JL. DAX-1 inhibits SF-1 mediated transactivation via a carboxy-termminal domain that is deleted in adrenal hypoplasia congenita. Mol Cell Biol 1997;17:1476–1483.
56. Zazopoulos E, Lalli E, Stocco DM, Sassone-Corsi P. DNA binding and transcriptional repression by DAX-1 blocks steroidogenesis. Nature 1997;390:311–315.
57. Crawford PA, Dorn C, Sadovsky Y, Milbrandt J. Nuclear receptor DAX-1 recruits nuclear receptor corepressor N-CoR to Steroidogenic Factor 1. Mol Cell Biol 1998;18:2949–2956.
58. Bardoni B, Zanaria E, Guioli S, Floridia G, Worley KC, Tonini G, et al. A dosage sensitive locus at chromosome Xp21 is involved in male to female sex reversal. Nat Genet 1994;7:497–501.
59. Zanaria E, Bardoni B, Dabovic B, Calvari V, Fraccaro M, Zuffardi O, et al. An unusual member of the nuclear hormone receptor superfamily responsible for X-linked adrenal hypoplasia congenita. Nature 1994;372:635–641.
60. Swain A, Narvaez V, Burgoyne P, Camerino G, Lovell-Badge R. *Dax1* antagonises *Sry* action in mammalian sex determination. Nature 1998;391:761–767.
61. Yu RN, Ito M, Saunders TL, Camper SA, Jameson JL. Role of *Ahch* in gonadal development and gametogenesis. Nat Genet 1998;20:353–357.
62. Stevanovic M, Lovell-Badge R, Collignon L, Goodfellow PN. *SOX3* is an X-linked gene related to *SRY*. Hum Mol Genet 1993;3:2013–2018.
63. Collignon J, Sockanathan S, Hacker A, Cohen-Tannoudji M, Norris D, Rastan S, et al. A comparison of the properties of *Sox-3* with *Sry* and two related genes, *Sox-1*and *Sox-2*. Development 1996;122:509–520.
64. Foster JW, Graves JAM. An *SRY*-related sequence on the marsupial X chromosome—implications for the evolution of the mammalian testis determining gene. Proc Natl Acad Sci USA 1994;91:1927–1931.
65. Graves JAM. Interactions between *SRY* and *SOX* genes in mammalian sex determination. BioEssays 1998;20:264–269.
66. Uwanogho D, Rex M, Cartwright EJ, Pearl G, Healy C, Scotting PJ, Sharpe PT. Embryonic expression of the chicken Sox2, Sox3 and Sox11 genes suggests an interactive role in neuronal development. Mech Dev 1995;49:23–36.

67. Josso N, Cate RL, Picard JY, Vigier B, di Clemente N, Wilson C, et al. Anti-Müllerian hormone: the Jost factor. Recent Prog Horm Res 1993;48:1–59.
68. Tran D, Muesy-Dessolle N, Josso N. Anti-Müllerian hormone is a functional marker of foetal Sertoli cells. Nature 1977;269:411,412.
69. Münsterberg A, Lovell-Badge R. Expression of the mouse anti-Müllerian hormone gene suggests a role in both male and female sexual differentiation. Development 1991;113:613–624.
70. Tran D, Muesy-Dessolle N, Josso N. Waning of anti-Müllerian activity: an early sign of Sertoli cell maturation in the developing pig. Biol.Reprod 1981;24:923–931.
71. Baker ML, Hutson JM. Serum levels of Müllerian inhibiting substance in boys throughout puberty and in the first two years of life. J Clin Endocrinol Metab 1993;76:245–247.
72. Rey R, Lordereau-Richard I, Carel JC, Barbet P, Cate RL, Roger M, et al. Anti-Müllerian hormone and testosterone serum levels are inversely related during normal and precocious pubertal development. J Clin Endocrinol Metab 1993;77:1220–1226.
73. Behringer RR, Cate RL, Froelick GJ, Palmiter RD, Brinster RL. Abnormal sexual development in transgenic mice chronically expressing Müllerian inhibiting substance. Nature 1990;345:167–170.
74. Lyet L, Louis F, Forest MG, Josso N, Behringer RR, Vigier B. Ontogeny of reproductive abnormalities induced by deregulation of anti-Müllerian hormone expression in transgenic mice. Biol Reprod 1995;52:444–454.
75. Racine C, Rey R, Forest MG, Louis F, Ferre A, Huhtaniemi I, et al. Receptors for anti-Müllerian hormone on Leydig cells are responsible for its effects on steroidogenesis and cell differentiation. Proc Natl Acad Sci USA 1998;95:594–599.
76. Behringer RR, Finegold MJ, Cate RL. Müllerian-inhibiting substance function during mammalian sexual development. Cell 1994;79:415–425.
77. Massague J. TGF-beta signal transduction. Annu Rev Biochem 1998;67:753–791.
78. Mishina Y, Tizard R, Deng JM, Pathak BG, Copeland NG, Jenkins NA, et al. Sequence, genomic organization, and chromosomal location of the mouse Müllerian-inhibiting substance type II receptor gene. Biochem Biophys Res Commun 1997;237:741–746.
79. Baarends WM, van Helmond MJL, Post M, van der Schoot PJCM, Hoogerbrugge JW, de Winter JP, et al. A novel member of a transmembrane serine/threonine kinase receptor family is specifically expressed in the gonads and in mesenchymal cells adjacent to the Müllerian duct. Development 1994;120:189–197.
80. Teixeira J, He WW, Shah PC, Morikawa N, Lee MM, Catlin EA, et al. Developmental expression of a candidate Müllerian inhibiting substance type II receptor. Endocrinology 1996;137:160–165.
81. di Clemente N, Wilson C, Faure E, Bouissin L, Carmillo P, Tizard R, et al. Cloning, expression, and alternative splicing of the receptor for anti- Müllerian hormone. Mol Endocrinol 1994;8:1006–1020.
82. Imbeaud S, Carre-Eusebe D, Rey R, Belville C, Josso N, Picard J-Y. Molecular genetics of the persistent Müllerian duct syndrome: a study of 19 families. Hum Mol Genet 1994;3:125–131.
83. Mishina Y, Rey R, Finegold MJ, Matzuk MM, Josso N, Cate RL, et al. Genetic analysis of the Müllerian-inhibiting substance signal transduction pathway in mammalian sexual differentiation. Genes Dev 1996;10:2577–2587.
84. Mishina Y, Whitworth DJ, Racine C, Behringer RR. High specificity of Müllerian-inhibiting substance signaling in vivo. Endocrinology 1999;140:2084–2088.
85. Arango NA, Lovell-Badge R, Behringer RR. Targeted mutagenesis of the endogenous mouse Müllerian inhibiting substance gene promoter: in vivo definition of genetic pathways of vertebrate sexual development. Cell 1999;99:409–419.
86. Lyon MF, Hawkes SG. X-linked gene for testicular feminization. Nature 1970;227:1217–1219.
87. Drews U. Direct and mediated effects of testosterone:analysis of sex reversed mosaic mice heterozygous for testicular feminization. Cytogenet Cell Genet 1998;80:68–74.
88. Couse JF, Korach KS. Exploring the role of sex steroids through studies or receptor deficient mice. J Mol Med 1998;76:497–511.
89. Charest NJ, Zhou Z-X, Lubahn DB, Olsen KL, Wilson EM, French FS. A frameshift mutation destabilizes androgen receptor messenger RNA in the *Tfm* mouse. Mol Endocrinol 1991;5:573–581.
90. Gaspar M-L, Meo T, Bourgarel P, Guenet J-L, Tosi M. A single base deletion in the *Tfm* androgen receptor gene creates a short-lived messenger RNA that directs internal translation initiation. Proc Natl Acad Sci USA 1991;88:8606–8610.
91. He WW, Kumar MV,Tindall DJ. A frame-shift mutation in the androgen receptor gene causes complete androgen insensitivity in the testicular-feminized mouse. Nucleic Acids Res 1991;19:2373–2378.
92. Wilson JD. Syndromes of androgen resistance. Biol Reprod 1992;46:168–173.

93. Murphy L, O'Shaughnessy PJ. Testicular steroidogenesis in the testicular feminized (*Tfm*) mouse: loss of 17α-hydroxylase activity. J Endocrinol 1991;131:443–449.
94. Cadigan KM, Nusse R. Wnt signaling: a common theme in animal development. Genes Dev 1997;11:3286–3305.
95. Vainio S, Heikkila M, Kispert A, Chin N, McMahon AP. Female development in mammals is regulated by *Wnt-4* signalling. Nature 1999;397:405–409.
96. Parr BA, McMahon AP. Sexually dimorphic development of the mammalian reproductive tract requires *Wnt-7a*. Nature 1998;395:707–710.
97. Miller C, Sassoon DA. *Wnt-7a*maintains appropriate uterine patterning during the development of the mouse female reproductive tract. Development 1998;125:3201–3211.
98. Hammerschmidt M, Brook A, McMahon, AP. The world according to hedgehog. Trends Genet 1997;13:14–21.
99. Bitgood MJ, McMahon AP. *Hedgehog* and *Bmp* genes are coexpressed at many diverse sites of cell-cell interaction in the mouse embryo. Dev Biol 1995;172:126–138.
100. Bitgood MJ, Shen L, McMahon AP. Sertoli cell signaling by Desert hedgehog regulates the male germline. Curr Biol 1996;6:298–304.
101. Hayman DL, Martin PG. Sex chromosome mosaicism in the marsupial genera *Isoodon* and *Parameles*. Genetics 1965;52:1201–1206.
102. Sharman GB, Robinson ES, Walton SM, Berger PJ. Sex chromosomes and reproductive anatomy of some intersexual marsupials. J Reprod Fertil 1970;21:57–68.
103. O W-S, Short RV, Renfree MB, Shaw G. Primary genetic control of somatic sexual differentiation in a mammal. Nature 1988;331:716–717.
104. Renfree MB, Robinson ES, Short RV, VandeBerg JL. Mammary glands in male marsupials: 1. Primordia in neonatal opossums *Didelphis virginians*and *Monodelphis domestica*. Development 1990;110:385–390.
105. Robinson ES, Renfree MB, Short RV, VandeBerg JL. Mammary glands in male marsupials: 2. Development of teat primordia in *Didelphis virginians*and *Monodelphis domestica*. Reprod Fertil Dev 1991;3:295–301.
106. Ullmann SL. Differentiation of the gonads and initiation of mammary gland and scrotum development in the brushtail possum *Trichosurus vulpecula* (Marsupialia). Anat Embryol 1993;187:475–484.
107. Sharman GB, Hughes RK, Cooper DW. The chromosomal basis of sex differentiation in marsupials. Aust J Zool 1990;37:451–456.
108. Cooper DW, Edwards C, James E, Sharman GB, VandeBerg JL, Graves JAM. Studies on metatherian sex chromosomes. VI. A third state of an X-linked gene: partial activity for the paternally derived Pgk-A allele in cultured fibroblasts of *Macropus giganteus* and *M. parryi*. Aust J Biol Sci 1977;30:431–443.
109. Cooper DW, Johnston PG, VandeBerg JL, Robinson ES. X-chromosome inactivation in marsupials. Aust J Zool 1990;37:411–417.
110. Renfree MB, Short RV. Sex determination in marsupials: evidence for a marsupial-eutherian dichotomy. Phil Trans R Soc Lond B 1988;322:41–53.
111. Shaw G, Renfree MB, Short RV. Primary genetic control of sexual differentiation in marsupials. Aus J Zool 1990;37:443–450.
112. Tyndale-Biscoe CH, Renfree MB. Reproductive Physiology of Marsupials. Cambridge University Press, Cambridge, UK, 1987.
113. VandeBerg JL, Robinson ES, Samollow PB, Johnston P. X-linked gene expression and X-chromosome inactivation: marsupials, mouse and man compared. In: Isozymes: Current Topics in Biological and Medical Research. Vol 15, Academic Press, San Diego, CA, 1987, pp. 225–253.
114. Renfree MB. Ontogeny, genetic control, and phylogeny of female reproduction in monotreme and therian mammals. In: Szalay FS, ed. Mammal Phylogeny: Mesozoic Differentiation, Multituberculates, Monotremes, Early Therians, and Marsupials. Springer-Verlag, Berlin, 1993, pp. 4–20.
115. McElreavey K, Vilain E, Herskowitz I, Fellous M. A regulatory cascade hypothesis for mammalian sex determination: SRY represses a negative regulator of male development. Proc Natl Acad Sci USA 1993;90:3368–3372.

3 The In Vivo Function of Müllerian-Inhibiting Substance During Mammalian Sexual Development

Yuji Mishina, PhD

Contents

INTRODUCTION

In mammals, both XX and XY individuals develop one pair of undifferentiated gonads and two pairs of genital ducts, the Müllerian ducts and the Wolffian ducts, associated with the mesonephroi during development (Fig. 1). The undifferentiated gonads are bipotent, and will give rise to either testes or ovaries depending on the sex-chromosome genotype. The Müllerian ducts have the potential to give rise to female reproductive organs, including the uterus, oviducts, and upper portion of the vagina. The Wolffian ducts are the primordia of male reproductive organs, which are the vas deferens, epididymis, and seminal vesicles. Therefore, regardless of sex-chromosome genotype, each individual has the potential to develop male and female reproductive systems. The presence of the Y chromosome determines that an individual becomes a male because of a gene termed *Sry*. There must be a mechanism to select only one duct system, depending on the presence or absence of Sry. Two gonadal hormones—Müllerian-inhibiting substance (MIS)—and testosterone, play essential roles during this selection process (Fig. 1) *(1–3)*. This chapter, examines the results of in vivo approaches using transgenic and targeted mutant mice to study MIS function.

From: *Contemporary Endocrinology: Transgenics in Endocrinology*
Edited by: M. Matzuk, C. W. Brown, and T. R. Kumar © Humana Press Inc., Totowa, NJ

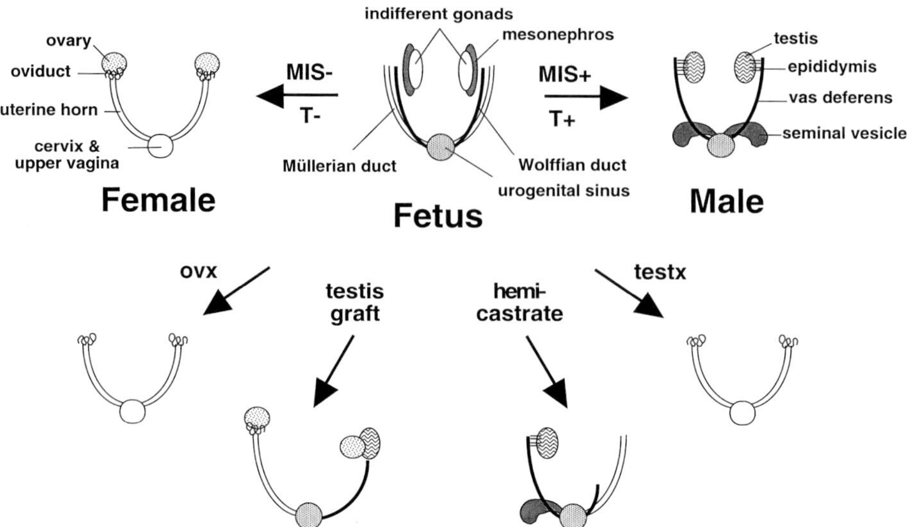

Fig. 1. Schematic representation of mammalian sexual development and its surgical alteration. The Müllerian ducts give rise to the uterus, oviducts, and upper portion of the vagina *(white)*. The Wolffian ducts give rise to the epididymis, vas deferens, and seminal vesicles *(black)*. MIS produced by the Sertoli cells of the fetal testes causes the regression of the Müllerian ducts and testosterone produced by Leydig cells induces the differentiation of the Wolffian duct system. The absence of both of these hormones during female fetal development permits the development of the Müllerian duct system while the Wolffian ducts passively regress. When ovaries or testes of a fetus are removed (ovx, testx), female development occurs. Unilateral testicular graft to a female fetus causes local regression of the Müllerian duct and persistence of the Wolffian duct (testis graft). Hemicastration of a male fetus causes a local persistence of the Müllerian duct (hemicastrate).

MÜLLERIAN-INHIBITING SUBSTANCE

The Jost Factor: A Short History of MIS

In the middle of the twentieth century, Alfred Jost performed pioneering experiments that suggested that the presence of two gonadal hormones is required for the selection of either the Müllerian or Wolffian duct system *(4,5)*. He surgically removed the gonads of fetal rabbits during sexual differentiation. Removal of both the ovaries from female fetuses and the testes from male fetuses resulted in female development (Fig. 1). In both cases, the Müllerian ducts persisted and differentiated, while the Wolffian duct system regressed. Unilateral removal of the fetal testes resulted in regression of the Wolffian ducts and persistence of the Müllerian ducts only on the removed side. A testicular graft to a female fetus caused local regression of the Müllerian duct, as well as a local differentiation of the Wolffian ducts *(3)*.

These observations led to the following hypothesis: 1) the fetal testes secrete two types of hormones—one that stimulates the differentiation of the Wolffian ducts, and the other causing regression of the Müllerian ducts, 2) the second hormone has a limited distance of action. When a crystal of synthetic androgen is introduced into the abdominal cavity of a castrated male fetus, the development of male structures is stimulated, but the Müllerian ducts are not inhibited. This result suggests that the first hormone is androgen

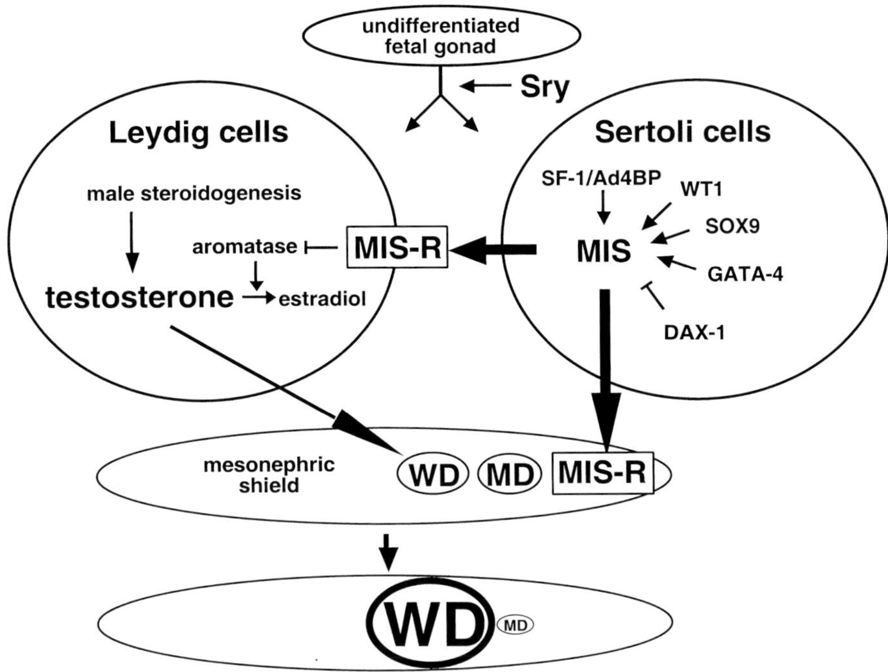

Fig. 2. Summary of the mechanism of sexual dimorphic development. The expression of *Sry* initiates testicular differentiation to allow male steroidogenesis in Leydig cells and SF-1/AdBP4 expression in Sertoli cells. SF-1/Ad4BP activates MIS gene with WT1, SOX9 and GATA-4. DAX-1 antagonizes this process. MIS binds an MIS receptor located in the surrounding mesenchyme of the Müllerian ducts to cause regression of the Müllerian ducts. Testosterone generated by Leydig cells induces the differentiation of the Wolffian ducts toward male specific reproductive tissues. MIS also binds the MIS receptor in Leydig cells to reportedly repress biosynthesis of aromatase that converts testosterone to estradiol. WD, the Wolffian duct; MD, the Müllerian duct; MIS-R, MIS receptor.

and the second hormone is separate from the masculinizing activities of androgens. The second hormone is currently termed Müllerian-inhibiting substance (MIS) or anti-Müllerian hormone (AMH) *(2)*.

The expression of Sry initiates testicular differentiation into two types of cells, Leydig cells and Sertoli cells (Fig. 2). In Sertoli cells, transcription factors that include steroidogenic factor-1 (SF-1), alternatively known as adrenal 4-binding protein (Ad4BP), initiate MIS expression *(6–8)*. Recent studies indicate that transcription factors such as Wilms' tumor 1 (WT1), DAX-1, SOX9, and GATA-4 are also involved in this process *(9–11,93–94)*. Testosterone is produced in Leydig cells. MIS may be involved in this process via inhibition of biosynthesis of aromatase that converts testosterone to estradiol *(12–15)*. MIS actively induces the regression of the Müllerian ducts, thereby preventing the development of female reproductive organs. Subsequently, testosterone induces the differentiation of the Wolffian ducts. During female development, ovaries do not express MIS, creating a permissive environment for the differentiation of the Müllerian ducts. In addition, the lack of testosterone leads to the passive regression of the Wolffian ducts. Thus, MIS and testosterone mediate a switch between the differentiation of the male and female extragonadal reproductive organs (Figs. 1 and 2).

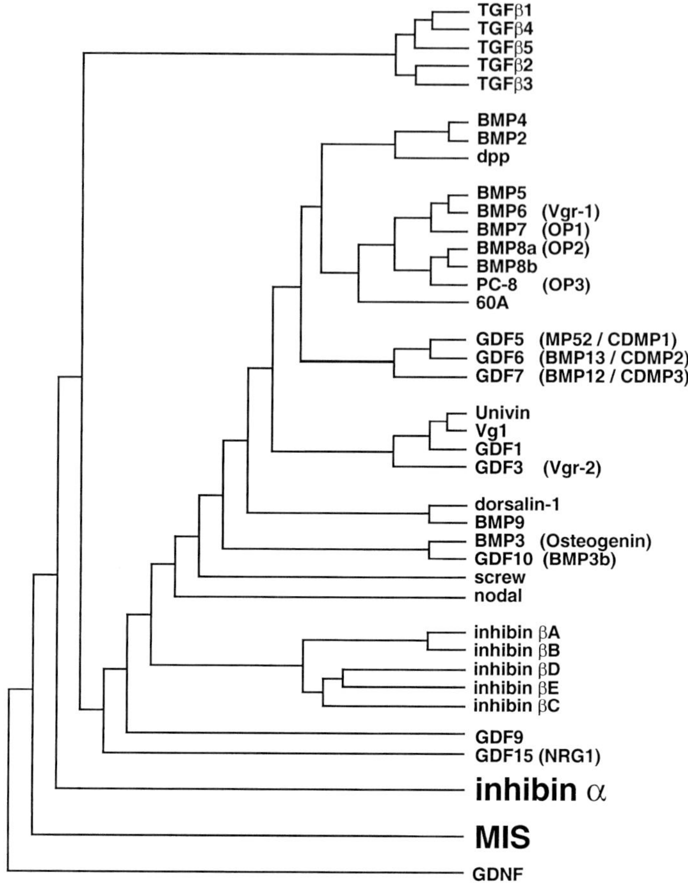

Fig. 3. Dendrogram of the TGF-β superfamily. The mature peptide sequence MIS is highly divergent from other members of the TGF-β superfamily.

MIS is a Member of the TGF-β Superfamily

The MIS cDNA was cloned by two independent groups by different approaches *(16,17)*. One approach was to sequence tryptic peptides of bovine MIS and to make degenerate oligonucleotide probes to screen a newborn bovine testis cDNA library *(16)*. The other approach was to screen a newborn bovine testis cDNA expression library, using antisera raised against bovine MIS *(17)*. Since then, the MIS gene has been isolated in humans, the cow, rat, and mouse *(16–19)*.

DNA sequencing of MIS revealed that MIS was a member of the transforming growth factor-β (TGF-β) gene superfamily *(16)* (Fig. 3). The members of this large gene family include activins, inhibins, bone morphogenetic proteins (BMPs), and growth-differentiation factors (GDFs) *(20)*. MIS is encoded by five exons encompassing 2.8 kb. MIS is a 140-kDa dimeric glycoprotein composed of two identical 70-kDa subunits *(21)*. Like other members of the TGF-β family, the 140-kDa MIS homodimer requires proteolytic processing to generate an N-terminal domain 110-kDa homodimer and a C-terminal domain 25-kDa homodimer *(22,23)*. PC5—one of the kex2/subtilisin-like enzymes—

may play a critical role for MIS processing; PC5 can cleave the MIS precursor in vitro and is also expressed in developing testes, suggesting an important physiological function *(24)*. Although only the C-terminal domain of TGF-β is required for its activity *(25)*, both the separated N- and C-terminal domains of MIS are essential to restore its regression for biological function *(26)*.

During embryogenesis, MIS expression is detected only in the male. Expression is restricted to Sertoli cells, and the highest levels of MIS are detected during the period of Müllerian-duct regression *(19,23)*. After birth, MIS is detected in Sertoli cells and granulosa cells in the ovary *(19,27–29)*. MIS protein is detected most abundantly in granulosa cells that contact the oocyte and line the antrum, suggesting that an oocyte-secreted growth factor may regulate its activity in the ovary. However, the levels of MIS in the ovary after birth are 0.1% of the levels produced by the fetal testes *(30)*. The expression pattern observed in male and female gonads suggests that, in addition to its Müllerian-duct inhibitory activity, MIS may regulate gonadal function and gametogenesis, topics to be discussed later in this chapter.

Molecular Cloning of the Type II MIS Receptor

In the case of TGF-β family ligands, two types of membrane bound serine/threonine kinases form heterodimers upon ligand binding to transduce a signal (type I and type II receptors, with mol wt of ~55 kd and 75–85 kd, respectively) *(20)*. Type II receptors are cloned by their binding ability for activin *(31)* or TGF-β *(32)*. Type I receptors have been cloned by sequence similarity of the kinase domain with type II receptors *(33–35)*. Type I receptors have a characteristic serine/glycine box at the juxtamembrane region that provides phosphorylation sites for type II receptor kinases and type I receptors are believed to phosphorylate downstream proteins for their signal transduction *(36)*. The specificity of ligand binding is primarily determined by type II receptors, although the specificity of signaling is primarily determined by type I receptors *(37)*.

Recently, a type II receptor gene for MIS was isolated from a rat Sertoli cell cDNA library as a testosterone-induced gene *(38)*, and from a rabbit fetal ovary cDNA library using a degenerated oligonucleotide probe from the conserved kinase domain *(39)*. Specific binding for MIS has been demonstrated by a transfection study, with a dissociation constant of 2.48–2.55 n*M* *(39,40)*. By *in situ* hybridization, the MIS receptor gene was shown to be expressed in the mesenchymal cells adjacent to the Müllerian ducts during male and female embryogenesis *(38,39,41)* (Fig. 2). This is an finding was intriguing, because the receptor is not expressed in the duct epithelium itself. This suggests that MIS most likely alters the surrounding mesenchyme to elicit Müllerian-duct regression. The MIS type II receptor is also expressed in Sertoli cells and granulosa cells in fetal and adult testes and ovaries, respectively, and in the gravid uterus *(38,39,42–44)*. Recently, the expression of the receptor has been confirmed in Leydig cells suggesting that MIS signaling may directly regulate Leydig-cell function *(14,15)*. Like the expression pattern of the MIS ligand, expression of the MIS receptor in the gonads suggests multiple functions of MIS.

In the case of other TGF-β family ligands, both type I and type II receptors are essential for their signal transduction. It is likely that MIS requires a type I receptor in addition to the type II receptor for signaling. A type I receptor termed Alk2 (alternatively known as activin type IA receptor (ActRIA), R1, Tsk7L, or SKR-1) is one candidate for the type I receptor for MIS *(34,45–47)*, although direct evidence is needed. Alk2 is less widely expressed then other type II receptors, including those in the mesenchyme surrounding the Müllerian ducts *(45)*. It has been reported that a point mutation in the serine/threonine kinase domain makes the kinase activity ligand-independent (constitutive active form, ca) *(48)*. When HEK-293 (human embryonic kidney) cells are transfected with ca-Alk2, the aromatase-promoter activity is repressed *(49)*. This suggests that ca-Alk2 can transduce a signal similar to the MIS signal, because MIS has the ability to repress aromatase biosynthesis *(12,13)*. Direct evidence using transgenic animals or biochemical studies is awaited.

Mechanisms of Duct Regression

The question of how MIS causes regression of the Müllerian ducts is the source of great debate. Despite the long history of MIS, little is known about its molecular mechanism of regression. Two distinctive mechanisms are proposed: 1) conversion of ductal epithelial cells to mesenchymal cells, and 2) apoptosis of ductal epithelial cells. The morphology of regression has been studied in the mouse *(50)*, the rat *(51)*, the chick *(52)*, and the alligator *(53)*. These studies indicate that the processes are remarkably well-conserved. The first event of regression is loss of the basement membrane underlying the epithelial and mesenchymal cells to allow direct contact of two populations *(54)*. Surrounding mesenchymal cells that lose the basement membrane condense to form an epithelial cuff. The epithelium subsequently shrinks by loss of cells.

Phagocytic activity is observed during the duct regression in male mice, suggesting that cell death is a major mechanism of regression. However, similar activity is also observed in the developing duct in females *(50)*. This has allowed researchers to propose a different mechanism—reentry of the ductal epithelium to the mesenchymal cell population. Ultrastructural studies have shown that there are viable epithelial cells even in highly regressing Müllerian ducts *(54)*. Interestingly, some of the cells in the surrounding mesenchyme actively metabolize hyaluronate and glycosaminoglycans, suggesting that the origin of these cells is epithelial and they have been migrating away from the duct *(55)*. Indeed, migration of ductal epithelial cells has been demonstrated in alligator embryos, using the fluorescent cell marker DiI *(53)*.

Ultrastructural studies also suggest that the Müllerian duct undergoes apoptosis. During regression, an increased number of lysosomes and infiltration of macrophages into the epithelium are observed in the rat and the rabbit *(51,56)*. Recently, it was demonstrated using the TUNEL assay that the Müllerian ducts in rat embryos undergo apoptosis *(41)*. The male genital ridge and the female genital ridge exposed to MIS show TUNEL-positive cells in ductal epithelium. It is known that caspase plays a major role in apoptosis. Interestingly, Boc-D-FMK, the caspase inhibitor, prevents MIS-induced apoptosis *(41)*. The duct epithelium alone in organ culture, or recombined tissues consisting of the duct epithelium and the mesenchyme from the embryonic kidney, do not undergo apoptosis *(41)*. The mesenchyme from the embryonic kidney does not express MIS type II receptor, and this finding proves that MIS acts specifically through the MIS type II receptor to alter the surrounding mesenchyme and thereby induces apoptosis indirectly.

These two mechanisms are not exclusive of each other. There may be several different mechanisms occurring simultaneously, or major mechanisms for regression may differ from species to species. Interestingly, regression occurs cranially to caudally, but incompletely. Transgenic and mutated mice will provide valuable insight into the precise molecular mechanisms of regression, and will provide the opportunity to identify downstream targets of MIS signaling.

IN VIVO FUNCTIONAL ANALYSIS OF MIS SIGNALING

Transgenic Mice that Chronically Express MIS: Gain-of-Function Study of the Transgenic Mouse

To investigate the in vivo functions of MIS, a gain-of-function study was initially performed in transgenic mice. The human MIS gene was expressed widely in both males and females, using the mouse metallothionein (MT) promoter *(57)*. Transgenic lines that possessed circulating levels of human MIS in plasma, ranging on average from 40 to 4400 ng/mL, were established.

Female transgenic mice lacked a uterus and oviducts, thus demonstrating the function of MIS as a Müllerian inhibitor in vivo (Table 1) *(57)*. The Müllerian ducts in transgenic females began to regress at embryonic d 13 (E 13), the same stage as in normal males *(58)*. In addition to lacking Müllerian duct-derived tissues, transgenic females showed abnormalities in their ovaries. Ovaries were present in newborn transgenic females, but germ cells were subsequently lost, and the somatic components of the ovary reorganized into structures reminiscent of the seminiferous tubules of the male gonad *(57)*. Reduction of the volume of ovaries in the transgenic females was detectable at E 14. The number of germ cells in the ovaries of transgenic mice was decreased from E 16, a time when oogonia in wild-type mice cease to proliferate, and the progression of the meiotic prophase was delayed *(58)*. It is not clear whether MIS acts directly on female germ cells. However, the results imply that an excess amount of MIS alters the function of granulosa cells, causing depletion of the germ cells *(59)*. The subsequent degeneration of the ovaries may be a nonspecific response to the loss of germ cells *(60)*.

Most of the male MT-MIS transgenic mice were phenotypically normal and fertile. However, abnormal phenotypes were observed in a portion of the males (5/21) from the highest-expressing lines (Table 1). Externally, these transgenic males were feminized, exhibiting mammary-gland development. Internally, Wolffian-duct differentiation was arrested, and the testes were undescended *(57)*. Serum testosterone levels of these transgenic mice were reduced, suggesting that a high level of MIS affected Leydig cell function *(58)*. Further, more expression of steroidogenic proteins, such as P450 cholesterol side chain-cleavage enzyme and P450 hydroxylase/C17-20 lyase, are down-regulated in transgenic male mice *(14,15)*. The number of Leydig cells is also decreased *(59)*. Because the type II receptor for MIS is expressed in Leydig-cells, it is likely that MIS signaling directly regulates Leydig cell function.

Based on these observations, one of the most striking conclusions is that MIS can act in vivo as the Müllerian inhibitor. In addition, MIS can alter the function of Leydig cells to influence Wolffian-duct development. MIS may also play roles in gonadal differentiation and germ-cell development. Finally, the observation that altered levels of MIS resulted in abnormal testicular descent implies that regulated levels of MIS are required for proper descent of the testes *(61)*.

MIS-Deficient Mice: Loss-of-Function Study

From the gain-of-function study described in the previous section, we have learned what MIS can do during male and female sexual differentiation. It is also important to

Table 1

Summary of Gain- and Loss-of-Function Studies of MIS

	Male	*Female*
MIS-overexpressing transgenic mice	Externally female Impaired development of the Wolffian duct Undescended testes Fewer germ cells	No uteri No oviducts Small ovaries
MIS-deficient mice	Presence of uteri Majority infertile Hyperplasia of Leydig cells Testicular atrophy	Normal
MIS-receptor deficient mice	Presence of uteri Majority infertile Hyperplasia of Leydig cells Testicular atrophy	Normal

know the purpose of MIS during normal development. Gene targeting in mouse embryonic stem (ES) cells makes it possible to generate mice carrying mutations in specific genes. Therefore, to understand the required functions of MIS during sexual differentiation and germ-cell development, the first two exons of MIS were deleted in ES cells and MIS-deficient mice were generated *(62)*.

As summarized in Table 1, all of the female homozygous mutants possessed a uterus with oviducts and ovaries that were morphologically normal. In addition, all of the females were fertile. Therefore, although MIS is expressed in a regulated manner in the ovary of wild-type mice after birth, there is apparently no requirement for MIS expression for normal ovarian function. One possible explanation is that related molecules proteins expressed in granulosa cells may provide redundant or compensatory functions in the absence of MIS. Candidates for such related molecules include activins and inhibins.

Morphological abnormalities of the reproductive tract were found only in male homozygous mutants. Testes were morphologically normal and completely descended in these males *(62)*. The Wolffian-duct systems were fully differentiated, and a uterus also developed (Table 1). While no coiled oviducts were found in these animals, oviductal tissue was detected histologically at the distal regions of the uterine horns. Since these MIS-deficient males have testes and both Wolffian and Müllerian duct-derived tissues, they are considered to be male pseudohermaphrodites (Fig. 4).

Approximately 90% of the MIS-deficient males were infertile. These males were able to plug females, but sperm were rarely detected in the uteri of the recipient females. Normal numbers of motile sperm were detected in the vas deferens and epididymis of the mutant males. These were shown to be capable of fertilizing oocytes in vitro. Thus, we concluded that although MIS is not required for male germ-cell development, the presence of the Müllerian and Wolffian duct-derived systems structurally interfered with the transfer of the sperm through the Wolffian duct-derived structures and into the reproductive tract of females.

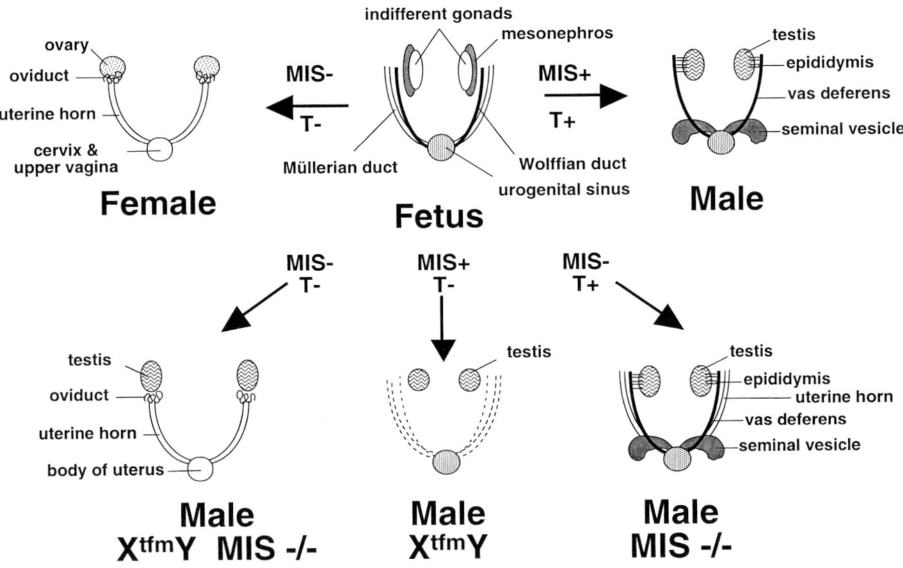

Fig. 4. Schematic representation of male pseudohermaphroditism in XY individuals that lack MIS. Male mice that only lack MIS ("MIS –/–" or "MIS–") differentiate both the Müllerian and Wolffian duct systems and are male in appearance. The presence of both types of reproductive organs severely hinders fertility. The physical association of the resulting oviductal tissue with the Wolffian duct derivatives blocks oviduct coiling. Male mice that carry the *tfm* mutation ($X^{tfm}Y$) are insensitive to androgens ("T–") leading to a passive regression of the Wolffian duct in addition to the active regression of the Müllerian duct, since MIS continues to be synthesized. Male mice that lack MIS and are insensitive to androgens differentiate the Müllerian duct system, and the insensitivity to androgens prevents differentiation of the Wolffian duct system. While these mice have testes, they are female in appearance. Since the Wolffian duct system has been eliminated in these mice the oviductal tissue assumes its normal coiled morphology, despite the presence of a testis.

Histological examination of the testes of the MIS-deficient mice revealed Leydig-cell hyperplasia and a testicular tumor of Leydig-cell origin *(62)*. However, no histological abnormalities or tumors were detected in MIS-deficient females. The development of tumors in MIS-deficient male mice was intriguing, because a targeted deletion of the related a-inhibin gene in mice leads to the development of testicular and ovarian tumors *(63)*. Thus, like inhibin, MIS also appears to function as a gonadal tumor-suppressor, but in a sexual dimorphic manner.

Together, these loss-of-function studies demonstrate that MIS is the Müllerian inhibitor, and that regression of the Müllerian duct system during fetal male development is important for male fertility. In addition, MIS is not required for male or female gametogenesis. Furthermore, the Leydig-cell hyperplasia and development of tumors suggests that MIS functions in the male gonad to influence Leydig-cell proliferation. The effect on Leydig cells is particularly interesting, because Leydig-cell numbers are decreased and function of the cells is altered in a subset of transgenic males from the highest-expressing lines *(57,59)*.

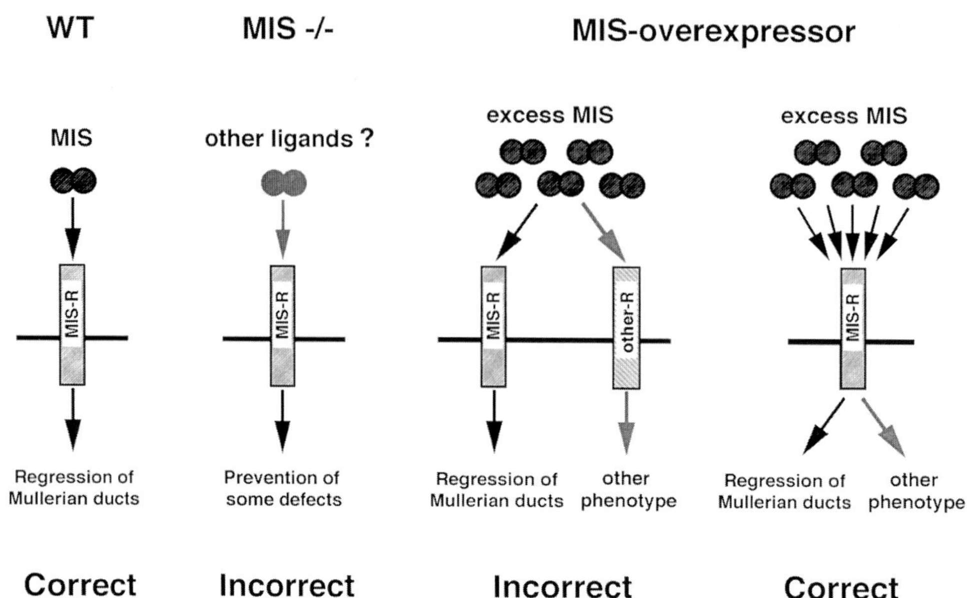

Fig. 5. Possible models to explain the contrasting results of gain-of-function and loss-of-function experiments of MIS. During normal embryogenesis, MIS interacts with the MIS receptor to regress the Müllerian ducts in male fetuses *(left)*. In the MIS ligand mutant mice, absence of MIS may be partly rescued by interaction of other TGF-β ligands with the MIS receptor (MIS-R) *(center left)*. However, this is not likely because the MIS receptor mutant mice are the phenocopy of the MIS ligand mutant mice (*see* Table 1). In the MIS overexpressing mice, excess amount of MIS may interact with other TGF-β receptors to transduce abnormal signal *(center right)*. Alternatively, excess amount of MIS may interact specifically with the MIS receptor to transduce an abnormal signal *(right)*. The generation of MIS overexpressor mice lacking the MIS receptor distinguishes the last two possibilities and shows that MIS interacts specifically with the MIS receptor (*see* text for details).

MIS Type II Receptor-Deficient Mice

Both the MIS gain-of-function and loss-of-function studies demonstrate that MIS acts as a Müllerian inhibitor in vivo. However, there are some contrasting results, particularly regarding gonadogenesis and germ-cell development, as summarized in Table 1. Two explanations are possible (Fig. 5). One is that in the MT-MIS transgenic mice, an excess amount of MIS interacts with receptors for other TGF-β family members to transduce an abnormal signal. The other is that in the MIS-deficient mice, other TGF-β family members interact with the MIS receptor to rescue some of the phenotype. To distinguish between these possibilities, we generated mice mutant for the MIS type II receptor, which lacked both the MIS ligand and receptor *(65)*. If the latter possibility is the case, MIS receptor-mutant mice would show a more severe phenotype than the MIS ligand-deficient mice.

The MIS type II receptor is encoded by 11 exons in a gene located on chromosome 15 in the mouse *(64)* and chromosome 12 in the human *(75)*. The first six exons which encode the signal peptide, the entire extracellullar region, and part of the kinase domain were deleted in ES cells to generate the MIS type II receptor-deficient mice *(65)*. Males and females homozygous for the MIS-receptor mutations were recovered from heterozygous intercrosses at the predicted Mendelian frequencies. As summarized in Table 1, the MIS-receptor-deficient mice were a phenocopy of the MIS-ligand mutant mice. All of the female homozygous mutant mice were normal and fertile. All of the male homozygous mutant mice were male pseudohermaphrodites—i.e., they had a uterus and oviducts superimposed upon the Wolffian-duct-derived tissues *(65)*. Like MIS-ligand mutant mice, the majority of the MIS type II receptor-deficient males were infertile, probably a result of the physical blockage of sperm transfer to females by persistence of the Müllerian-duct-derived tissues *(65)*.

In addition to the Müllerian-duct phenotype, the MIS type II-receptor-mutant mice showed hyperplasia in Leydig cells, and some of the males developed Leydig-cell tumors similar to the MIS-ligand mutant males *(65)*. MIS ligand/MIS type II-receptor-compound-homozygous mutant mice showed the same phenotype as either one of the single homozygous mutant male mice *(65)*. These findings strongly demonstrate that MIS is the only ligand for the MIS type II receptor, and that the MIS type II receptor is the only receptor for MIS. These findings also indicate that the rescue of a portion of the phenotype by other TGF-β superfamily ligands through the MIS receptor system does not occur (Fig. 5).

MIS Overexpression Mice in the MIS Receptor Null Background

Because the MIS type II-receptor-deficient mice phenocopy the MIS ligand-deficient mice, it is likely that the ovarian phenotype observed in female MT-MIS transgenic mice is caused by abnormal signals resulting from the excess amount of MIS. However, two possibilities remain, as shown in Fig. 5. One is that excess MIS interacts with other TGF-β-family receptors to transduce abnormal signals, and the other is that the MIS type II receptor itself transduces an abnormal signal.

To address this question, we generated female mice carrying the MT-MIS transgene that were also homozygous for a targeted deletion of the MIS type II receptor *(66)*. These females possessed normal female reproductive tracts and ovaries, and were fertile. In the absence of the receptor, despite the high serum level of MIS, these females are normal and fertile *(66)*. These results indicate that the abnormalities in the reproductive tract and ovaries of the MT-MIS transgenic females are caused by an excess of MIS signaling through the MIS type II receptor (Fig 5). They also indicate that other TGF-β family-receptor-signaling systems are not activated by high levels of MIS. Thus, the MIS type II receptor is the only receptor for MIS, even in the presence of pharmacologic levels of MIS. This is the only case where receptor-deficient mice are the phenocopy of the ligand-deficient mice among TGF-β superfamily members that have been studied. For example, BMP2-deficient mice die at E 8.5 with defects of the heart and extraembryonic tissues *(67)* and BMP4-deficient mice die at E 9.5 with defects of the posterior portion of body, eye and germ cells *(68–70)*. However, mice deficient for a type I receptor for BMP2 and BMP4 (BMP type IA receptor, *Alk3*) die at E 7.0 without mesoderm formation *(71)*. Thus, the specific ligand-receptor interaction of MIS is very unique among the TGF-β superfamily.

INTERACTION OF MIS SIGNALING AND OTHER SIGNALING PATHWAYS

*MIS/*Tfm *Double-Mutant Mice*

One of the advantages of a genetic approach is the ability to generate compound mutant mice to uncover redundant functions and to explore genetic interactions. After formation of the testis, male sexual differentiation is controlled by two hormones, testosterone and MIS (Fig. 2) (*see* Chapter 2). Therefore, we generated mutant mice that lack both MIS and androgens by interbreeding mice with the classical mutation known as testicular feminization (*Tfm*) and MIS ligand-deficient mice. *Tfm* is an X-chromosome-linked mutation of the androgen receptor *(72,73)*. Thus, $X^{tfm}Y$ males are insensitive to androgen and become feminized, lack Wolffian-duct differentiation, and have small and incompletely descended testes in which spermatogenesis is blocked at meiotic prophase. $X^{tfm}Y$ males do, however, produce MIS, as evidenced by the regression of the Müllerian-duct system (Fig. 4).

MIS-deficient $X^{tfm}Y$ males were generated by interbreeding the *Tfm* and MIS mutants *(62)*. These animals were overtly feminized with improperly descended testes and a vaginal opening. Also, Wolffian-duct differentiation was eliminated, and a uterus was developed. Interestingly, coiled oviducts were present with an infundibulum, whereas no coiled oviducts were found in the MIS-deficient male pseudohermaphrodites *(62)* (Fig. 4). These results suggested that elimination of the Wolffian ducts during female development may be required for normal morphogenesis of the oviducts. As discussed in the previous section, regression of the Müllerian ducts in XY individuals by MIS is essential for transfer of sperm through the male reproductive tract, and regression of the Wolffian ducts in XX individuals lacking testosterone is essential for normal morphogenesis of oviducts. Taken together, these studies indicate that MIS and testosterone play critical roles in the generation of the functional reproductive tracts by removal of one of the two anlagen.

MIS/Inhibin Double-Mutant Mice

Inhibins and activins are members of the TGF-β family and are functional antagonists (Fig. 3). Inhibins are heterodimers of inhibin a and activin βA or βB, whereas activins are homodimers or heterodimers of activin βA and βB. Both male and female mice lacking inhibin α-, fail to produce inhibins, are viable, and develop gonadal tumors of sex cord-stromal cell origin (Fig. 6) *(63)*. These findings suggest that inhibin is a negative regulator of gonadal stromal-cell proliferation. Like MIS, inhibin expression in the gonads is most abundant in Sertoli and granulosa cells. Prior to tumor formation in the inhibin mice, germ cell development was normal, indicating that inhibin was also not required for male or female germ-cell development. Since MIS and inhibins are related hormones produced by the same gonadal-cell types, and mutant mice for each develop gonadal tumors, it seems reasonable that there may be cross-communication between MIS and inhibin signals in the gonads. Therefore, we generated double-mutant mice lacking both MIS and inhibin α (inhα) *(65,74)*.

Both male and female MIS/inhα double-mutant mice were generated at the expected Mendelian ratios. The most prominent phenotype observed in males exhibited a fluid-filled uterus and tumor-filled testes that had retracted from their scrotal position into the abdomen (Fig. 6). Fluid secretion into the lumen of the uterus was explained by high levels of serum estrogen produced by the tumors. Injections of these MIS/inhα double mutant males with an estrogen antagonist eliminated the fluid-filled uterine phenotype *(74)*. Tumors in MIS/inhα double-mutant mice were larger, but less hemorrhagic than those in inha mutant mice (Fig. 6). Histologically, the MIS/inhα double-mutant testicular tumors were Sertoli-/granulosa-cell tumors and Leydig-cell neoplasia *(74)*. In addition, the tumors in double mutants developed at a significantly faster rate than the tumors

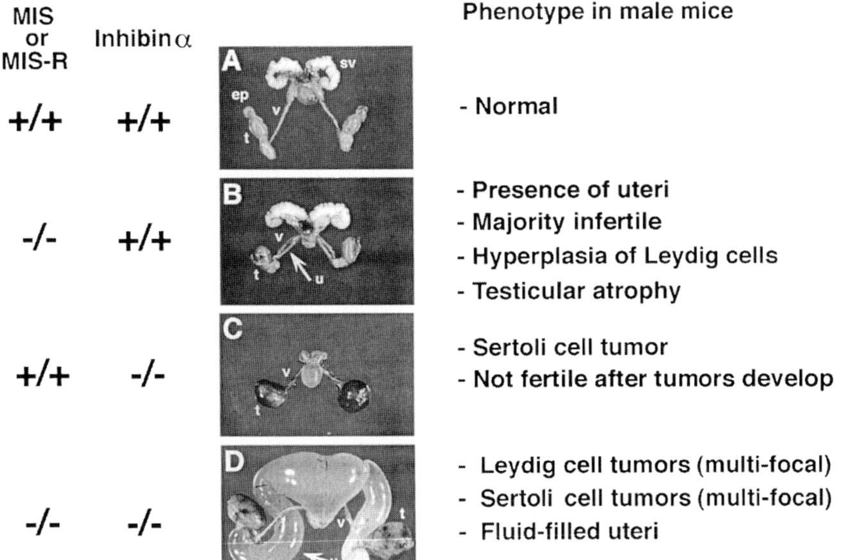

Fig. 6. Genetic interaction of MIS and inhibin α. Male sexual tracts of different genotype are shown. Testes from the MIS or MIS receptor mutant (–/–) are macroscopically normal, but hyperplasia of Leydig cells is found. The inhibin α-deficient testes are enlarged and hemorrhagic because of an invasive granulosa/Sertoli tumor. The enlarged testes from MIS/inhα and MIS receptor/inhα double mutant males are less hemorrhagic. The testicular tumors in the MIS/inhα double mutant males demonstrate multifocal Leydig-cell neoplasia and multifocal granulosa/Sertoli cell tumors. Because of the excess amount of estrogen produced by the tumor, the uterus is dilated. ep, epidydimis; sv, seminal vesicle; t, testis; u; uterus; v, vas deferens.

that arose in mice mutant for inhibin alone. Leydig-cell neoplasia was observed as early as 7 d after birth *(74)*. These observations suggested that MIS and inhibin signaling influence each other in a synergistic manner to regulate testicular cell growth.

To understand the level of cross-communication, we generated MIS-receptor/inhα double-mutant mice. The phenotype of the double-mutant males is identical to that of the MIS ligand/inhα double-mutant mice *(65)*. The testicular tumors in the MIS receptor/inhα double-mutant males and the MIS ligand/inhα double-mutant males are identical; this provides further evidence that MIS is the only ligand for the MIS receptor (Fig. 7). In addition, these results suggest that the synergy between the MIS and inhibin signaling pathways occurs downstream of each receptor, not between ligands and receptors (Fig. 7). Taken together, these observations suggest that MIS signaling can influence the development of Sertoli-/granulosa-cell tumors initiated by the absence of inhibins, and that inhibin signaling can influence the development of Leydig-cell neoplasia initiated by the absence of MIS signaling. We have not detected a synergistic phenotype in the double-homozygous mutant females, suggesting that the synergistic interaction of MIS and inhibin signaling is male-specific.

CONCLUSIONS AND PERSPECTIVES

Human males who lack MIS develop as internal pseudohermaphrodites with uterine and oviductal tissues, a condition known as persistent Müllerian-duct syndrome (PMDS)

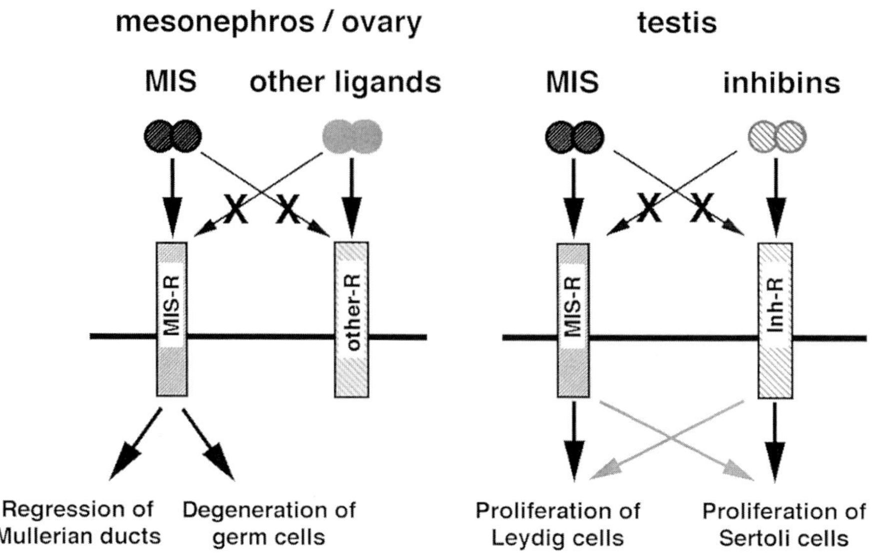

Fig. 7. High specificity of MIS signaling. MIS acts as a Müllerian inhibitor. MIS is the only ligand for the MIS type II receptor and the type II receptor is the only receptor for MIS *(left panel)*. In the MIS-overexpressing mice, an excess amount of MIS transduces abnormal signal through the MIS receptor to cause an abnormalities such as germ cell loss. MIS signaling and inhibin signaling cooperate together to regulate proliferation of Leydig cells and Sertoli cells in a negative manner *(right panel)*. This interaction most likely occurs downstream of their receptors *(gray arrow)*. The inhibin signal may play important roles in the transition from Leydig-cell hyperplasia to neoplasia.

(76). In humans, the molecular basis of the PMDS syndrome is heterogeneous *(77)*. In "MIS-negative" cases, MIS serum levels are low, and mutations of the MIS gene have been detected *(78–80)*. In contrast, "MIS positive" patients have normal serum levels of MIS, and a mutation of the MIS type II-receptor has been detected in the majority of these cases *(40,77)*. Like the mouse models discussed in this chapter, patients with MIS or MIS type II receptor gene mutations share the same clinical phenotype. Thus, the mouse and human studies suggest that the MIS signaling pathway is simple and is conserved in mammals.

However, little is known about the signaling pathway of MIS. As discussed earlier, Alk2 is the strongest candidate for an MIS type I receptor based on the expression pattern and ability to repress biosynthesis of aromatase *(45,49)*. Mutant mice that lack *Alk2* show embryonic lethality much earlier than the stage of Müllerian-duct regression *(81,82)*. To overcome this problem, we will perform a conditional mutagenesis of Alk2 to generate absence of Alk2 in the surrounding mesenchyme of the Müllerian duct. Smads are fairly well characterized gene family that act as intracellular signaling proteins for TGF-β family members. Pathway-specific Smads are directly phosphorylated by type I receptors to allow translocation of these proteins along with Smad4 (a common Smad) into nuclei to alter gene expression *(83)*. Among them, smad5 is expressed in the mesonephric shield, suggesting that it may be in the MIS signal transduction pathway *(84)*. Since Smad5 mutant embryos die at E 9.5 *(85,86)*, much earlier than sexual differentiation, conditional mutagenesis of smad5 will address its role in MIS signaling.

Humans with PMDS who are MIS positive and have apparently no alterations in their MIS type II receptors may have mutations in genes encoding downstream effectors of MIS signaling. Indeed, whereas MIS maps to chromosome 19 *(87)* and the MIS type II-receptor maps to chromosome 12 *(40,75)*, X-chromosome-linked PMDS has been reported *(88,89)*. The MIS ligand and MIS type II-receptor mutant mice will serve as valuable genetic resources to isolate the downstream targets of this differentiation pathway of male sexual development.

Reproductive tracts are one of the targets of environmental toxicants. It is known that the clinical use of diethylstilbestrol (DES) by pregnant women causes the presistence of Müllerian-duct remnants in their sons *(90)*. Organ culture studies indicate that the inhibitory effect of DES is mainly caused by a decrease in responsiveness of the target tissues to MIS *(91)*. Interestingly, the expression level of MIS and MIS receptor is increased *(92)*. Understanding of the MIS signaling mechanism should help us to solve this puzzle.

ACKNOWLEDGMENTS

I thank Dr. Richard R. Behringer for providing me an opportunity to work on MIS and the MIS receptor. I also thank Drs. E. Mitch Eddy, Retha R. Newbold and Trisha M. Castranio for helpful comments of the manuscript; and Yoshiko and Kanade H. Mishina for encouragement. While in the Behringer laboratory, this work was supported, in part, by grants from the National Institutes of Health (HD30284), National Cancer Institute CA16672, and the Sid W. Richardson Foundation.

REFERENCES

1. Cate RL, Wilson CA. Müllerian-inhibiting substance. In: Gwatkin, RBL, ed. Genes in Mammalian Reproduction. Wiley-Liss, New York, 1993, pp. 185–205.
2. Josso N, Cate RL, Picard J-Y, Vigier B, di Clemente N, Wilson C, et al. Anti-Müllerian hormone: the Jost factor. Rec Prog Horm Res 1993;48:1–59.
3. Mishina Y, Behringer, RR. The in vivo function of Müllerian-inhibinting substance during mammalian sexual development. In: Wassarman, PM, ed. Advances in Developmental Biology, vol. 4. JAI Press, Greenwich, CT, 1994, pp. 1–25.
4. Jost A. Recherches sur la differenciation sexuelle de l'embryon de lapin. Arch Anat Micro Morph Exp 1947;36:271–315.
5. Jost A. Problems of fetal endocrinology: the gonadal and hypophyseal hormones. Rec Prog Horm Res 1953;8:379–418.
6. Hatano O, Takayama K, Imai T, Waterman MR, Takakusu A, Omura T, et al. Sex-dependent expression of a transcription factor, Ad4BP, resulting steroidogenic P-450 genes in the gonads during prenatal and postnatal rat development, Development 1994;120:2787–2797.
7. Shen WH, Moore CCD, Ikeda Y, Parker KL, Ingraham HA. Nuclear receptor steroidgenic factor 1 regulates the Müllerian inhibiting substance gene: a link to the sex determination cascade. Cell 1994;77:1–20.
8. Giuili G, Shen WH, Ingraham HA. The nuclear receptor SF-1 mediates sexually dimorphic expression of Müllerian Inhibiting Substance, in vivo. Development 1997;124:1799–1807.
9. Nachtigal MW, Hirokawa Y, Enyeart-VanHouten DL, Flanagan JN, Hammer GD, Ingraham, HA. (1998) Wilms' tumor 1 and Dax-1 modulate the orphan nuclear receptor SF-1 in sex-specific gene expression. Cell 1998;93:445–454.
10. Kent J, Whitely SC, Andrew JE, Sinclair AH, Koopman P. A male-specific role for SOX9 in vertebrate sex determination. Development 1996;122:2813–2822.
11. Parker KL, Schedl A, Schimmer, BP. Gene interactions in gonadal development. Annu Rev Physiol 1999;61:417–433.
12. Vigier B, Forest MG, Eychenne B, Bézard J, Garrigou O, Robel P, et al. Anti-Müllerian hormone produces endocrine sex-reversal of fetal ovaries. Proc Natl Acad Sci USA 1989;8:3684–3688.

13. di Clemente N, Ghaffari S, Pepinsky RB, Pieau C, Josso N, Cate RL, et al. A quantitative and interspecific test for biological activity of anti-Müllerian hormone: the fetal ovary aromatase assay. Development 1992;114:721–727.
14. Racine C, Rey R, Forest MG, Louis F, Ferre A, Huhtaniemi I, Josso N, di Clemente, N. Receptors for anti-müllerian hormone on Leydig cells are responsible for its effects on steroidogenesis and cell differentiation. Proc Natl Acad Sci USA 1998;95:594–599.
15. Rouiller-Fabre V, Carmona S, Merhi RA, Cate RL, Habert R, Vigier B. Effect of anti-Müllerian hormone on Sertoli and Leydig cell functions in fetal and immature rats. Endocrinology 1998; 139:1213–1220.
16. Cate RL, Mattaliano RJ, Hession C, Tizard R, Farber NM, Cheung A, et al. Isolation of the bovine and human genes for Müllerian inhibiting substance and expression of the human gene in animal cells. Cell 1986;45:685–698.
17. Picard JY, Benarous R, Guerrier D, Josso N, Kahn, A. Cloning and expression of cDNA for anti-Müllerian hormone. Proc Natl Acad Sci USA 1986;83:5464–5468.
18. Haqq CM, Lee, MM, Tizard R, Wysk M, DeMarinis J, Donahoe PK, et al. Isolation of the rat gene for Müllerian inhibiting substance. Genomics 1992;12:665–669.
19. Münsterberg A, Lovell-Badge R. Expression of the mouse anti-Müllerian hormone gene suggests a role in both male and female sexual differentiation. Development 1991;113:613–624.
20. Hogan BLM. Bone morphogenetic proteins: multifunctional regulators of vertebrate development. Genes Dev 1996;10:1580–1594.
21. Budzik GP, Powell SM, Kamagata S, Donahoe PK. Müllerian inhibiting substance fractionation by dye affinity chromatography. Cell 1983;34:307–314.
22. Pepinsky RB, Sinclair LK, Chow EP, Mattaliano RJ, Manganaro TF, Donahoe PK, Cate RL. Proteolytic processing of Müllerian inhibiting substance produces a transforming growth factor-β-like fragment. J Biol Chem 1988;263:18,961–18,964.
23. Cate RL, Donahoe PK, MacLaughlin DT. Müllerian inhibiting substance. In: Sporn MB, Roberts AB, eds. Peptide Growth Factors and Their Receptors II. Springer-Verlag, Berlin, 1990, pp. 179–210.
24. Nachtigal MW, Ingraham HA. Bioactivation of Müllerian inhibiting substance during gonadal development by a kex2/subtilisin-like endoprotease. Proc Natl Acad Sci USA 1996;93:7711–7716.
25. Pircher R, Jullien P, Lawrence DA. b-transforming growth factor is stored in human blood platelets as a latent high molecular weight complex. Biochem Biophys Res Comm 1986;136:30–37.
26. Wilson CA, di Clemente N, Ehrenfels C, Pepinsky RB, Josso N, Vigier B, et al. Müllerian inhibiting substance requires its N-terminal domain for maintenance of biological activity, a novel finding within the transforming growth factor-b superfamily. Mol Endocrinol 1993;7:247–257.
27. Takahashi M, Koide SS, Donahoe PK. Müllerian inhibiting substance as oocyte meiosis inhibitor. Mol Cell Endorinol 1986;47:225–234.
28. Bézard J, Vigier B, Tran D, Mauléon P, Josso N. Immunocytochemical study of anti-Müllerian hormone in sheep ovarian follicles during fetal and post-natal development. J Reprod Fertil 1987;80:509–516.
29. Ueno S, Takahashi M, Manganaro TF, Ragin RC, Donahoe PK. Cellular localization of Müllerian inhibiting substance in the developing rat ovary. Endocrinology 1989;125:1060–1066.
30. Josso N. Anti-Müllerian hormone: new perspectives for a sexist molecule. Endocr Rev 1986;7:421–433.
31. Mathews L, Vale WW. Expression cloning of an activin receptor, a predicted transmembrane serine kinase. Cell 1991;65:973–982.
32. Lin HY, Wang X, Ng-Eaton E, Weinberg RA, Lodish HF. Expression cloning of the TGF-b type II receptor, a functional transmembrane serine/threonine kinase. Cell 1992;68:775–785.
33. Attisano L, Carcamo J, Ventura F, Weis FMB, Massagué J, Wrana JL. Identification of human activin and TGF-b type I receptors that form heterodimeric kinase complexes with type II receptors. Cell 1993;75:671–680.
34. Ebner R, Chen RH, Shum L, Zioncheck TF, Lee AR, Derynck R. Cloning of a type I TGF-b receptor and its effect on TGF-b binding to the type II receptor. Science 1993;260:1344–1348.
35. Franzen P, ten Dijke P, Ichijo H, Yamashita H, Schulz P, Heldin CH, et al. Cloning of a TGF beta type I receptor that forms a heteromeric complex with the TGF beta type II receptor. Cell 1993;75:681–692.
36. Wrana JL, Attisano L, Wieser R, Ventura F, Massagué J. Mechanism of activation of the TGF-β receptor. Nature 1994;370:341–347.
37. Cárcamo J, Weis FM, Ventura F, Wieser R, Wrana JL, Attisano L, et al. Type I receptors specify growth-inhibitory and transcriptional responses to transforming growth factor beta and activin. Mol Cell Biol 1994;14:3810–3821.

38. Baarends WM, van Helmond MJL, Post M, van der Schoot JCM, Hoogerbrugghe JW, de Winter JP, et al. A novel member of the transmembrane serine/threonine kinase receptor family is specifically expressed in the gonads and in mesenchymal cells adjacent to the Müllerian duct. Development 1994;120:189–197.
39. di Clemente N, Wilson C, Faure E, Boussin L, Carmillo P, Tizard R, et al. Cloning, expression and alternative splicing of the receptor for anti-Müllerian hormone. Mol Endocrinol 1994;8:1006–1020.
40. Imbeaud S, Faure E, Lamarre I, Mattéi MG, di Clemente N, Tizzad R, et al. Insensitivity to anti-Müllerian hormone due to a mutation in the human anti-Müllerian hormone receptor. Nature Genet 1995;11:382–388.
41. Roberts LM, Hirokawa Y, Nachtigal MW, Ingraham HA. Paracrine-mediated apoptosis in reproductive tract development. Dev Biol 1999;208:110–122.
42. Baarends WM, Uilenbroek JTJ, Kramer P, Hoogerbrugghe JW, de Winter JP, Karels B, et al. Anti-Müllerian hormone and anti-Müllerian hormone type II receptor messenger ribonucleic acid expression in rat ovaries during postnatal development, the estrous cycle, and gonadotropin-induced follicle growth. Endocinology 1995;136:4951–4962.
43. Baarends WM, Hoogerbrugghe JW, Post M, Visser JA, de Rooij DG, Parvinen M, et al. Anti-Müllerian hormone and anti-Müllerian hormone type II receptor messenger ribonucleic acid expression during postnatal testis development and in the adult testis of the rat. Endocinology 1995;136:5614–5622.
44. Teixeira J, He WW, Shah PC, Morikawa N, Lee MM, Catlin EA, et al. Developmental expression of a candidate Müllerian inhibiting substance type II receptor. Endocrinology 1996;137:160–165.
45. He WW, Gustafson ML, Hirobe S, Donahoe PK. Developmental expression of four novel serine/threonine kinase receptors homologous to the activin/transforming growth factor-beta type II receptor family. Dev Dyn 1993;196:133–142.
46. Matsuzaki K, Xu J, Wang F, McKeehan WL, Krummen L, Kan M. A widely expressed transmembrane serine/threonine kinase that does not bind activin, inhibin, transforming growth factor β, or bone morphogenic factor. J Biol Chem 1993;268:12,718–12,723.
47. ten Dijke P, Ichijo H, Franzen P, Schulz P, Saras J, Toyoshima H, et al. Activin receptor-like kinases: a novel subclass of cell-surface receptors with predicted serine/threonine kinase activity. Oncogene 1993;8:2879–2887.
48. Wieser R, Wrana JL, Massague J. GS domain mutations that constitutively activate T beta R-I, the downstream signaling component in the TGF-beta receptor complex. EMBO J 1995;15:2199–2208.
49. Visser J., eds. Anti-Müllerian Hormone: Molecular Mechanism of Action. Eburon, 1998.
50. Dyche WJ. A comparative study of the differentiation and involution of the Müllerian duct and Wolffian duct in the male and female fetal mouse. J Morphol 1979;162:175–209.
51. Price JM, Donahoe PK, Ito Y, Hendren WH 3d. Programmed cell death in the Müllerian duct induced by Müllerian inhibiting substance. Am J Anat 1977;149:353–375.
52. Forsberg J, Olivecrona H. Degeneration processes during the development of the Müllerian ducts in alligator and chicken embryos. Z Anat Entwiklungsgesch 1963;124:83–96.
53. Austin HB. DiI analysis of cell migration during Mullerian duct regression. Dev Biol 1995;169:29–36.
54. Trelstad RL, Hayashi A, Hayashi K, Donahoe PK. The epithelial-mesenchymal interface of the male rate Müllerian duct: loss of basement membrane integrity and ductal regression. Dev Biol 1982;92:27–40.
55. Hayashi A, Donahoe PK, Budzik GP, Trelstad RL. Periductal and matrix glycosaminoglycans in rat Müllerian duct development and regression. Dev Biol 1982;92:16–26.
56. Djehiche B, Segalen J, Chambon Y. Ultrastructure of Müllerian and Wolffian ducts of fetal rabbit in vivo and in organ culture. Tissue Cell 1994;26:323–332.
57. Behringer RR, Cate RL, Froelick GJ, Palmiter RD, Brinster RL. Abnormal sexual development in transgenic mice chronically expressing Müllerian inhibiting substance. Nature 1990;345:167–170.
58. Lyet L, Louis F, Forest MG, Josso N, Behringer RR, Vigier B. Ontogeny of reproductive abnormalities induced by deregulation of anti-Müllerian hormone expression in transgenic mice. Biol Reprod 1995;52:444–454.
59. Josso N, Racine C, di Clemente N, Rey R, Xavier F. The role of anti-Mullerian hormone in gonadal development. Mol Cell Endocrinol 1998;145:3–7.
60. Whitworth DJ. XX germ cells: The difference between an ovary and a testis. Trends Endocrinol Metab 1998;9:2–6.
61. Hutson JM, Donahoe PK. The hormonal control of testicular descent. Endocr Rev 1986;7:270–283.
62. Behringer RR, Finegold MJ, Cate RL. Müllerian-inhibiting substance function during mammalian sexual development. Cell 1994;79:415–425.

63. Matzuk MM, Finegold MJ, Su JG, Hsueh AJ, Bradley A. Alpha-inhibin is a tumour-suppressor gene with gonadal specificity in mice. Nature 1992;360:313–319.
64. Mishina Y, Tizard R, Deng JM, Pathak BG, Copeland NG, Jenkins NA, et al. Sequence, genomic organization, and chromosomal location of the mouse Mullerian-inhibiting substance type II receptor gene. Biochem Biophys Res Commun 1997;237:741–746.
65. Mishina Y, Rey R, Finegold MJ, Matzuk MM, Josso N, Cate RL, Behringer RR. Genetic analysis of the Müllerian-inhibiting substance signal transduction pathway in mammalian sexual differentiation. Genes Dev 1996;10:2577–2587.
66. Mishina Y, Whitworth DJ, Racine C, Behringer RR. High specificity of Mullerian-inhibiting substance signaling in vivo. Endocrinology 1999;140:2084-2088.
67. Zhang H, Bradley A. Mice deficient for BMP2 are nonviable and have defects in amnion/chorion and cardiac development. Development 1996;122:2977–2986.
68. Winnier G, Blessing M, Labosky PA, Hogan BLM. Bone morphogenetic protein-4 (BMP-4) is required for mesoderm formation and patterning in the mouse. Genes Dev 1995;9:2105–2116.
69. Furuta Y, Hogan BLM. BMP4 is essential for lens induction in the mouse embryo. Genes Dev 1998;12:3764–3775.
70. Lawson KA, Dunn NR, Roelen BA, Zeinstra LM, Davis AM, Wright CV, et al. Bmp4 is required for the generation of primordial germ cells in the mouse embryo. Genes Dev 1999;13: 424–436.
71. Mishina Y, Suzuki A, Ueno N, Behringer RR. Bmpr encodes a type I bone morphogenetic protein receptor that is essential for gastrulation during mouse embryogenesis. Genes Dev 1995;9:3027–3037.
72. He WW, Kumar MV, Tindall DJ. A frame-shift mutation in the androgen receptor gene causes complete testosterone insensitivity in the testicular-feminized mouse. Nucleic Acids Res 1991;19:2373–2378.
73. Charest NJ, Zhou ZX, Lubahn DB, Olsen KL, Wilson EM, French FS. A frameshift mutation destabilizes androgen receptor messenger RNA in the Tfm mouse. Mol Endocrinol 1991;5: 573–581.
74. Matzuk MM, Finegold MJ, Mishina Y, Bradley A, Behringer RR. Synergistic effects of inhibins and Müllerian-inhibiting substance on testicular tumorigenesis. Mol Endocrinol 1995;9:1337–1345.
75. Visser JA, McLuskey A, van Beers T, Weghuis DO, van Kessel AG, Grootegoed JA, Themmen AP. Structure and chromosomal localization of the human anti-mullerian hormone type II receptor gene. Biochem Biophys Res Commun 1995;215:1029–1036.
76. Guerrier D, Tran D, Vanderwinden JM, Hideux S, Van Outryve L, Legeai L, et al. The persistent Müllerian duct syndrome: a molecular approach. J Clin Endocrinol Metab 1989;68:46–52.
77. Imbeaud S, Belville C, Messika-Zeitoun L, Rey R, di Clemente N, Josso N, et al. A 27 base-pair deletion of the anti-mullerian type II receptor gene is the most common cause of the persistent mullerian duct syndrome. Hum Mol Genet 1996;5:1269–1277.
78. Knebelmann B, Boussin L, Guerrier D, Legeai L, Kahn A, Josso N, Picard JY. Anti-Müllerian hormone Bruxelles: a non-sense mutation in the last exon of the anti-Müllerian hormone gene associated with the persistent Müllerian duct syndrome in three brothers. Proc Natl Acad Sci USA 1991;88:3767–3771.
79. Carré-Eusebe, D, Imbeaud S, Harbison M, New MI, Josso N, Picard JY. Variants of the anti-Müllerian hormone gene in a compound heterozygotes with the persistent Müllerian duct syndrome and his family. Hum Genet 1992;90:389–394.
80. Imbeaud S, Carré-Eusebe D, Rey R, Belville C, Josso N, Picard JY. Molecular genetics of the persistent Müllerian duct syndrome: a study of 19 families. Hum Mol Genet 1994;3:125–131.
81. Gu Z, Reynolds EM, Song J, Lei H, Feijen A, Yu L, et al. The type I serine/threonine kinase receptor ActRIA (ALK2) is required for gastrulation of the mouse embryo. Development 1999;126:2551–2561.
82. Mishina Y, Crombie R, Bradley A, Behringer, RR. Multiple roles for Activin-Like Kinase-2 signaling during mouse embryogenesis. Dev Biol 1999;213:314–326.
83. Heldin CH, Miyazono K, ten Dijke P. TGF-beta signaling from cell membrane to nucleus through SMAD proteins. Nature 1997;390:465–471.
84. Meersseman G, Verschueren K, Nelles L, Blumenstock C, Kraft H, Wuytens G, et al. The C-terminal domain of Mad-like signal transducers is sufficient for biological activity in the Xenopus embryo and transcriptional activation. Mech Dev 1997;61:127–140.
85. Yang X, Castilla LH, Xu X, Li C, Gotay J, Weinstein M, et al. Angiogenesis defects and mesenchymal apoptosis in mice lacking SMAD5. Development 1999;126:1571–1580.
86. Chang H, Huylebroeck D, Verschueren K, Guo Q, Matzuk MM, Zwijsen A. Smad5 knockout mice die at mid-gestation due to multiple embryonic and extraembryonic defects. Development 1999;126:1631–1642.
87. Cohen-Haguenauer, O, Picard, JY, Mattei, MG, Serero, S, Nguyen VC, de Tand MF, et al. Mapping of the gene for anti-Müllerian hormone to the short arm of human chromosome 19. Cytogenet Cell Genet 1987;44:2–4.

88. Sloan WR, Walsh, PC. Familial persistent Müllerian duct syndrome. J Urol 1976;115:459–461.
89. Naguib KK, Teebi AS, Al-Awadi SA, El-Khalifa MY, Mahfouz ES. Familial uterine hernia syndrome: report of an Arab family with four affected males. Am J Hum Genet 1989;33:180,181.
90. Newbold RR. Influence of estrogenic agents on mammalian male reproductive tract development. In: Korach KS, ed. Reproductive and developmental toxicology, Marcel Dekker, New York, NY, 1998, pp. 531–551.
91. Newbold RR, Suzuki Y, McLachlan JA. Müllerian duct maintenance in heterotypic organ culture after in vivo exposure to diethylstilbestrol. Endocrinology 1984;115:1863–1868.
92. Visser JA, McLuskey A, Verhoef-Post M, Kramer P, Grootegoed JA, Themmen AP. Effect of prenatal exposure to diethylstilbestrol on Mullerian duct development in fetal male mice. Endocrinology 1998;139:4244–4251.
93. Shimamura R, Fraizer GC, Trapman J, Lau YC, Saunders GF. The Wilms' tumor gene WT1 can regulate genes involved in sex determination and differentiation: SRY, Müllerian-inhibiting substance, and the androgen receptor. Clin Cancer Res 1997;12:2571–2580.
94. Tremblay, JJ, Viger, RS. Transcription factor GATA-4 enhances Müllerian inhibiting substance gene transcription through a direct interaction with the nuclear receptor SF-1. Mol Endcrinol 1999;13:1388–1401.

4 Control of Ovarian Function

Julia A. Elvin, MD, PhD
and Martin M. Matzuk, MD, PhD

Contents

INTRODUCTION

Female fertility depends on prenatal development of the fetal gonad into an ovary and the complex interactions between the intraovarian and extraovarian factors that regulate the postnatal process of folliculogenesis. At birth, the ovary has a finite oocyte population. Folliculogenesis initiates when some oocytes within primordial follicles begin to grow in response to undiscovered intragonadal factors; other oocytes will remain quiescent until later in life, resulting in a prolonged period of fertility. Follicular development is controlled locally by paracrine factors and at a distance by endocrine hormones, such as the pituitary hormones follicle-stimulating hormone (FSH) and luteinizing hormone (LH). Intragonadal factors initiate growth of the follicle and coordinate development of the oocyte, granulosa cells, and thecal cells *(1)*. Extragonadal factors, particularly hormones from the pituitary, synchronize granulosa cell and theca cell function later in folliculogenesis to initiate puberty and integrate the reproductive system with overall female physiology *(2)*.

To understand the regulation of ovarian function in humans, it is crucial to have physiological models that mimic events occurring during human ovarian prenatal and postnatal development. Transgenic mouse technology has created many such models to study ovarian function. Transgenic mouse technology allows for the specific and reproducible alteration of gene expression, and the subsequent observation of its effect on the

From: *Contemporary Endocrinology: Transgenics in Endocrinology*
Edited by: M. Matzuk, C. W. Brown, and T. R. Kumar © Humana Press Inc., Totowa, NJ

development of a specific tissue or the entire organism *(3)* (*see* Chapter 1). The use of the transgenic approach to study gonadogenesis and the regulation of folliculogenesis in mice is expanding our understanding of ovarian development and physiology, and is helping to reveal possible causes of human infertility. This chapter summarizes several key events in female gonadogenesis and folliculogenesis in mice that have been elucidated by spontaneous or induced mutations yielding infertile or subfertile mouse models. The mouse models presented in this chapter are divided into two basic categories on the basis of whether the mutation affects prenatal ovarian formation (Tables 1–3) or postnatal ovarian function (Tables 4–9). Models with prenatal ovarian defects are subdivided into those in which the defect affects gonad formation (Table 1), germ-cell proliferation or migration (Table 2), or germ-cell survival (Table 3). Models with postnatal ovarian effects are subdivided on the basis of the stage of folliculogenesis affected by the mutation as follows: primordial follicle development and preantral follicle growth (Table 4), antrum formation and later stages of follicle growth resulting from extraovarian (Table 5) and intraovarian defects (Table 6), or ovulation and/or corpus luteum formation caused by extraovarian (Table 7) and intraovarian defects (Table 8). Models with female infertility or subfertility caused by nonfollicular defects such as fertilization, implantation, and early embryonic development (Table 9) are not described in detail in this chapter. The transgenic mouse approach has led to insights into the intricacies of gonadal development, germ-cell survival, proliferation, and migration, and control of postnatal ovarian function at all stages of folliculogenesis, ovulation, and fertilization.

PRENATAL OVARIAN DEVELOPMENT

The female gonad develops during early fetal life when primordial germ cells (PGCs), derived from the inner-cell mass of the blastocyst and residing extraembryonically, migrate to the gonadal ridges of the mesonephros of the developing embryo *(4)*. Several genes expressed in the urogenital ridge play important roles in early gonadal development (Table 1; *see* Chapter 2). Mice lacking the Wilms' tumor associated gene, which encodes the transcription factor WT-1, show normal germ cell migration but the urogenital ridge fails to develop, leading to both kidney and gonadal agenesis *(5)*. Steroidogenic factor-1 (SF-1), an orphan member of the nuclear receptor superfamily, is expressed in the hypothalamus, pituitary, all primary steroidogenic tissues, and the urogenital ridge at embryonic day (E) 9–9.5. Absence of SF-1 also leads to complete gonadal agenesis *(6)*. Although SF-1 is known to be a key regulator of steroidogenic enzyme transcription, the defects in SF-1-deficient gonadal development do not result from the lack of steroid production, because steroid-deficient rabbit models with normal SF-1 function show normal gonad formation *(7)*. Additionally, steroidogenic acute regulatory (StAR) protein knockout mice, which have severely impaired prenatal steroidogenesis caused by loss of cholesterol transport from the cytoplasm into mitochondria, also show normal prenatal ovarian development *(8,9)*. Instead, SF-1 may regulate genes important to gonadal development directly, and functional SF-1 sites are present in several other genes that play a major role in reproduction, such as the oxytocin, prolactin receptor, and Müllerian-inhibiting substance (MIS) genes *(10–12)*; (for further discussion on SF-1 and StAR, *see* Chapter 8).

Chromosomal sex determines gonadal sex in the mouse between E 10 and E 12.5. Specifically, precursor genital-ridge somatic cells carrying the sex-determining region Y (*Sry*) gene on the Y chromosome direct the development of the gonad into a testis,

Table 1

Defects in Gonad Formation

Transgenic/mutant mouse	*Major reproductive findings*	*Refs*
Dax1 transgenic	Male to female sex reversal in XY mice with a weak *Sry* allele	*(15)*
MT-Müllerian-inhibiting substance transgenic[a]	Females lack Müllerian-duct derivatives; germ cells degenerate; feminization of high expressor males	*(100)*
SF-1 (*Ftz-F1*) knockout	Failure of gonads to develop leading to complete agenesis; female internal and external genitalia	*(6,123)*
Sry transgenic	Presence of *Sry* results in formation of testis in XX embryo; no spermatogenesis	*(13)*
WT-1 knockout	Failure of gonadal development; normal germ-cell migration	*(5)*

[a]MT, metallothionein promoter.

Table 2

Defects in Germ-Cell Proliferation/Migration

Transgenic/mutant mouse	*Major reproductive findings*	*Refs*
Atrichosis (*at*) mutant	Spontaneous mutation causes marked decrease in primordial germ cell number	*(24)*
Connexin 43 knockout	Decreased germ cells from E11.5 onwards; in vitro defects in folliculogenesis after primary follicle stage	*(124)*
Germ-cell-deficient (*gcd*) transgenic	Transgene insertion causes drastic decrease in primordial germ-cell number	*(25)*
Kit ligand–*steel mutants* (deficiency)	Ovaries lack germ cells because of defects in migration and proliferation	*(19)*
TIAR knockout	PGCs present at E 11.5 but disappear by E 13.5; required for PGC proliferation	*(22)*
White spotting (*W*) mutant (c-kit deficiency)	Ovaries lack germ cells because of defects in migration and proliferation	*(18,20)*
Zfx knockout	Normal germ-cell migration; defects in mitotic proliferation	*(23)*

while a lack of expression of *Sry* allows the default pathway of ovary formation *(4)*. XX transgenic mice carrying a 14-kb DNA fragment containing the mouse *Sry* gene develop testes and male secondary sex characteristics, but lack spermatozoa and thus are infertile *(13)*. This block in spermatogenesis in the transgenic *Sry*-positive XX male mice confirms that other genes on the Y chromosome are important for spermatogenesis.

Dax1 (*Ahch*), an unusual member of the nuclear hormone receptor superfamily, has also been implicated in sex determination and gonadal differentiation. *Dax1* is initially expressed at the same time as *Sry* in both XX and XY embryos, and persists throughout ovarian development, but is downregulated during testis development. Loss of *Dax1* does not affect ovarian development or female fertility, but does cause progressive

Table 3

Defects in Germ Cell Survival

Transgenic/mutant mouse	*Major reproductive findings*	*Refs*
Ataxia telangiectasia (*Atm*)-knockout mice	Male and female infertility; apoptotic germ cell death around birth	*(35,36)*
Caspase-2-knockout	Increased number of oocytes at postnatal d 4 due to deceased apoptosis; oocytes show decreased sensitivity to doxorubicin	*(125)*
Dazla-knockout	Ovaries lack germ cells because of prenatal degeneration of oocytes after proliferation	*(126)*
Dmc1-knockout	Infertility in males and females; block in spermatogenesis in males; loss of germ cells beginning in utero with complete absence of oocytes by adulthood	*(30,31)*
Fanconi anemia complementation group C (*fac*)-knockout	Majority of females infertile; few oocytes present at birth; increased chromosome breakage and instability	*(33,34)*
Msh5-knockout	Infertility; ovaries devoid of oocytes; ovarian cysts in adults	*(28,29)*
Wnt-4-knockout	Infertility; few oocytes present at birth; Leydig cells found in ovary; Müllerian duct fails to form	*(26)*

Table 4

Mouse Models of Preantral Follicle Development

Transgenic/mutant mouse	*Major reproductive findings*	*Refs*
A. Defects in Primordial Follicles		
bax-knockout	Extended reproductive lifespan; increased primordial follicles at 2 yr of age	*(45)*
bcl-2-knockout	Fertile; reduced number of primordial follicles	*(46)*
FIGα-knockout	Infertile; primordial follicles fail to form after birth; oocytes subsequently die	*(40)*
B. Defects in Preantral Follicle Growth		
Growth differentiation factor-9 (*Gdf9*)-knockout	Infertility; defect in folliculogenesis at one-layer follicle stage	*(60)*
Kit Ligand – *Steel*^{panda} mutant	Infertility; reduced germ-cell number and defect in folliculogenesis at one-layer follicle stage	*(51,53)*
Kit Ligand – *Steel*^{t} mutant	Infertility; defect in folliculogenesis at one-layer follicle stage	*(52,127)*

testicular degeneration *(14)*. Overexpression of *Dax 1* also does not affect ovary development, but causes male-to-female sex reversal in XY mice with a weak *Sry* allele *(15)*. Thus, Dax1 does not function as an ovary determinant, as was initially hypothesized, but instead appears to antagonize testis development, and is responsible for the dosage-sensitive sex-reversal X-chromosome syndrome in humans.

Table 5

Defects in Antral Follicle Growth:Extraovarian Defects

Transgenic/mutant mouse	*Major reproductive findings*	*Refs*
Activin-receptor type IIA knockout	Infertility in females; delayed fertility in males; small gonads	*(90)*
Bovine glycoprotein hormone α-subunit promoter-DT[a]	Infertile; hypogonadal	*(128)*
Common glycoprotein hormone α-subunit knockout	Infertile; hypogonadal, hypothyroid	*(129)*
Copper/zinc superoxide dismutase (*Sod1*) knockout	Subfertility in females; decreased serum gonadotropin concentrations and embryonic death	*(130,131)*
Follicle stimulating hormone β-subunit knockout	Female infertility; folliculogenesis block before antral-follicle stage; males fertile, but decreased testis size	*(47)*
Human GnRH promoter-SV40 T-antigen*b*	Infertility caused by an arrest in GnRH neuron migration; block prior to antral-follicle formation	*(66)*
Hypogonadal (*hpg*) mouse (GnRH deletion)[a]	Infertility; small gonads; block prior to antral-follicle formation	*(48,65)*
Neuronal helix-loop-helix 2 (*Nhlh2*) knockout	Infertile, hypogonadal, and obese; females fertile if reared with males	*(132)*
Obese *(ob/ob)* mouse (leptin deficiency)	Infertility; perturbation of the hypothalamic-pituitary axis; block prior to antral-follicle formation	*(133)*

[a]DT, diphtheria-toxin A chain; *b*GnRH, gonadotropin-releasing hormone.

Table 6

Defects in Antral Follicle Formation:Intraovarian Defects

Transgenic/mutant mouse	*Major reproductive findings*	*Refs*
α-inhibin knockout	Infertility in females; secondary infertility in males; granulosa-/Sertoli-cell tumors; cachexia-like syndrome	*(83,84,134)*
Cyclin D2 knockout	Female infertility secondary to granulosa-cell defect; males fertile but decreased testis size	*(77,79)*
Follicle stimulating hormone receptor knockout	Female infertility; folliculogenesis blocked before antrum formation	*(71)*
Insulin-like growth factor 1 (*Igf1*) knockout	Hypogonadal and infertile; folliculogenesis block before antral-follicle stage	*(74,75)*
Insulin-receptor substrate 2 knockout	Rarely fertile; similar to Igf1 knockout	*(76)*
P450 aromatase knockout	Infertility; antral-follicle formation, but no CLs	*(93)*

For an ovary or testis to be truly functional later in life, primordial germ cells must migrate into the developing genital ridge. Mitotic proliferation of primordial germ cells occurs during this migration, and upon reaching the gonad, the germ cells form important

Table 7

Extra Ovarian Regulation of Ovulation and/or Corpus Luteum Formation

Transgenic/mutant mouse	*Major reproductive findings*	*Refs*
Bovine glycoprotein hormone; α promoter-βLHb-CTP[a]	Infertility, polycystic ovaries, granulosa-cell tumors	*(89,135)*
Estrogen receptor α knockout	Infertility; hemorrhagic ovarian cysts	*(95,97, 136,137)*
Growth hormone receptor knockout	Delayed puberty and age of first conception; prolonged pregnancy	*(75)*
Prolactin (*Prl)* knockout	Infertility; irregular and prolonged estrus cycles	*(138)*
Transcription-factor NGFI-A knockout	Infertility; luteinizing-hormone suppression causing no corpora lutea	*(103)*

[a]bLHβ-CTP, bovine luteinizing hormone β subunit—human chorionic gonadotropin carboxyl-terminal peptide fusion.

associations with somatic cells that will last for the duration of folliculogenesis. Unlike spermatogenesis in males, in which spermatogonia (stem cells) constantly divide and produce gametes, the ovary has a finite supply of oocytes. Therefore, the size and the rate of depletion of this pool of oocytes determine the duration of female fertility. Several mutations have been shown to disrupt either early germ-cell migration, proliferation, or both (Table 2). The tyrosine kinase receptor, c-kit, encoded by the *W* locus, is expressed on the surface of germ cells *(16)* and its ligand, stem-cell factor (SCF) (kit ligand), encoded by the *Sl* locus, is expressed by cells along the germ-cell migratory pathway *(17)*. Gonads of the white-spotting (*W*) and Steel (*Sl*) mutant mice contain few, if any, germ cells *(18–20)*, because of defects in the migration and death of germ cells. Thus, the interaction of c-kit with its ligand is required for prenatal migration, proliferation, and survival of the primordial germ cells *(21)* and also for postnatal folliculogenesis.

Similar to c-kit, TIAR, an RNA recognition motif/ribonucleoprotein-type RNA-binding protein, is highly expressed in primordial germ cells, and is essential for regulating PGC proliferation and survival *(22)*. TIAR belongs to a family of proteins that function in splicing, transport, translation, and stability of mRNA, and play multiple, key developmental functions in *Drosophila melanogaster*, such as sex determination. At E 11.5, a severely reduced number of PGCs are observed in the genital ridge of male and female mice lacking TIAR, and by E 13.5 the PGCs have completely disappeared. Postnatally, these mutant female mice essentially have streak ovaries. Because TIAR-deficient embryonic stem (ES) cells do not proliferate in vitro in the absence of leukemia-inhibitory factor, TIAR may play a role in regulating cell proliferation and survival under suboptimal conditions. Perhaps in vivo, suboptimal conditions are experienced by PGCs during migration to the gonad, providing an explanation for why proliferation and survival during this period is TIAR-dependent.

The *Zfx* gene encodes a putative zinc-finger transcription factor located on the X-chromosome *(22a)*. Female Zfx-deficient mice are subfertile and have a dramatically shortened reproductive lifespan, reminiscent of the human syndrome known as premature ovarian failure *(23)*. The number of oocytes in the perinatal ovary was <25% of controls. Although germ cell migration did not appear to be disrupted, further investigation revealed <50% the normal number of primordial germ cells in the gonad at E 11.5.

Table 8

Intra Ovarian Regulation of Ovulation and/or Corpus Luteum Formation

Transgenic/mutant mouse	*Major reproductive findings*	*Refs*
C/EBPβ (CCAAT/enhancer-binding protein β) knockout	Infertility; reduced ovulation and block in "mature" corpora lutea formation	*(113)*
Connexin 37 knockout	Infertility; defect in folliculogenesis at the Graafian follicle stage; oocytes are meiotically incompetent	*(61,120)*
Cyclin-dependent kinase 4 knockout	Infertile; defects in hypothalamic-pituitary-gonadal axis leading to prolonged cycle length and failure to form corpora lutea	*(139)*
Cyclooxygenase 2 (prostaglandin endoperoxide synthase-2) knockout	Largely infertile; absence of corpora lutea due to apparent ovulation defect	*(106)*
Estrogen receptor α and β double knockout	Infertile; no CLs, few healthy follicles, follicular nests, degenerating oocytes, few granulosa cells and Sertoli-like cells	*(99)*
Estrogen receptor β knockout	Subfertile; fewer litters and fewer pups/litter; decreased ovulation efficiency	*(98)*
$p27^{Kip1}$ knockout	Female infertility; corpora lutea defects; males fertile but increased testis size	*(117,118)*
Progesterone receptor knockout	Infertility; defects in all reproductive tissues, no ovulation but corpora lutea formation	*(116)*
Prolactin receptor knockout	Infertility; reduced ovulation and fertilization; preimplantation development blocked	*(140)*
Prostaglandin F receptor knockout	Inability to undergo parturition secondary to failure of corpora luteum to undergo luteolysis, preventing induction of oxytocin receptor in uterus	*(141)*
Stat5a/Stat5b double knockouts	Infertility due to failure to form corpora lutea	*(142)*
Steroidogenic acute regulatory (*StAR*) protein knockout	Congenital adrenal hyperplasia; males and females have female external genitalia; by puberty have lipid deposits, luteinized stromal cells and no corpora lutea	*(8,9)*
Transcription factor NGFI-A-LacZ knockin	Infertility; luteinizing hormone deficiency, LH receptor suppression, causing no ovulation and no corpora lutea	*(143)*

Based on the mutant mouse phenotype, *Zfx*-deficiency has been proposed as a model for premature ovarian failure, and its location on the X chromosome suggests that it is also involved in some reproductive defects seen in Turner's syndrome (*45*, XO). Similarly, a spontaneous mutation at the atrichosis (*at*) *(24)* locus and mutation of the germ-cell deficient (*gcd*) locus *(25)* by transgene insertion also causes infertility because of significantly reduced primordial germ-cell populations in the developing gonad. However, the identity and functions of the genes at the *at* and *gcd* loci are unknown.

Survival of germ cells, once they reach the genital ridge, is also a key determinant of the postnatal oocyte pool size. Abnormalities in survival have been noted in several transgenic models (Table 3). At 14.5 d postcoitum (dpc), there are similar numbers of

Table 9

Nonfollicular Fertility Defects (Selected)

Transgenic/mutant mouse	*Major reproductive findings*	*Refs*
Basigin knockout	Infertility in both sexes; males show block in spermatogenesis; females show reduced fertilization and implantation defects	*(144,145)*
c-mos knockout	Decreased fertility in females only resulting from parthogenetic activation; ovarian cysts and teratomas	*(121,122)*
EP2 Prostaglandin E2 receptor knockout	Subfertile because of decreased fertilization of eggs; defects in cumulus expansion	*(108–110)*
Heatshock transcription-factor 1 (*Hsf1*) knockout	Infertile; pre- and postimplantation defects	*(146)*
Interleukin 11 knockout	Female infertility caused by decreased implantation reaction and failed decidualization	*(147)*
Leukemia inhibitory factor (LIF) knockout	Infertility; embryo implantation does not occur	*(148)*
Mlh1 knockout	Male and female infertility; defective meiosis at pachytene stage (males) and failure to complete meiosis II (females) and genome instability	*(32,74)*
Osteopetrotic (colony-stimulating factor-1 mutant) mice	Male and female mice subfertile; reduced testosterone (males); implantation and lactation defects (females)	*(149)*
SR-BI (scavenger-receptor, class B, type I) knockout	Female infertility caused by defects in oocyte maturation and arrested early embryonic development	*(150,151)*
Steroid 5α-reductase type I knockout	Reduced litter size; parturition defects (i.e., fetal death caused by excess estrogens)	*(152,153)*
Wnt7a knockout	Female infertility caused by abnormalities in oviduct and uterus development	*(27)*
ZP protein 1 (*Zp1*) knockout	Thinner zona matrix; decreased fertility caused by fertilization defect	*(41,44)*
ZP protein 2 (*Zp2*) knockout	Infertile; no ZP; fragile oocytes do not survive well in oviduct; no fertilization	*(41)*
ZP protein 3 (*Zp3*) knockout	Infertile; no ZP, fragile oocytes do not survive in oviduct, no fertilization	*(42,43)*

germ cells in the *Wnt-4*-deficient and control female gonad, indicating that proliferation and migration had occurred normally *(26)*. However, at birth, the *Wnt-4*-deficient ovary contains only 10% the normal number of oocytes, indicating that massive oocyte death had occurred. In females, Wnt-4 deficiency also causes absence of the Müllerian duct, development of the Wolffian duct, and the presence of testosterone producing Leydig cells within the ovary. Oocyte death in these mice is not a result of increased production of ectopic androgens by the Leydig cells (since chronic administration of testosterone does not impair oocyte development), but likely reflects a direct role of Wnt-4 in maintaining the female germline. In contrast, Wnt-7a, which is expressed in the Müllerian-duct epithelium at E 12.5 to E 14.5, has no effect on ovarian development or function, but instead only regulates Müllerian-duct differentiation in both males and females *(27)*. Wnt-7a-deficient females have normal folliculogenesis and ovulation, but are infertile because of abnormalities in the oviduct and uterus (Müllerian-duct derivatives).

By postnatal d 2 (P2), oocytes arrest in prophase of the first meiotic division, and do not complete meiosis I until ovulation. Several proteins—particularly mismatch-repair proteins—are required for meiosis. For example, mutations in the genes *Msh5*, *Dmc1*, and *Mlh1* result in selective loss of germ cells, and often secondary effects on the somatic cells. Msh5 is a homolog of the *Escherichia coli* protein MutS, responsible for recognizing DNA replication errors and binding to the mismatched bases. Msh5 is expressed in the ovary from E 16 to P1, the time in females when germ-cell meiosis begins. *Msh5*-deficient females are infertile, and adult ovaries are devoid of follicles and are cystic *(28,29)*.

Migration, proliferation, and embryonic survival of germ cells are unaffected up to E 18, and *Msh5* knockout female embryos have normal numbers of germ cells *(28)*. However, the germ-cell population diminishes quickly and by postnatal d 25, only 1–3 oocytes are observed per ovary. Interestingly, these oocytes are in antral follicles, indicating that they have the capability to participate in folliculogenesis if still viable. Further studies have indicated that the oocytes were blocked at the zygotene stage of oogenesis, when problems with chromosomal pairing and synapsis lead to apoptotic oocyte death. Similarly, deficiency of the germline-specific RecA homolog Dmc1 results in infertility. This defect causes meiotic prophase arrest, resulting from failure of chromosome pairing and synapsis. It also leads to oocyte loss beginning in the fetus, and eventually results in a rudimentary ovary in adulthood *(30,31)*. This finding suggests that Dmc1 acts slightly later than Msh5, or that there is partial redundancy for Dmc1 function. In contrast to the Msh5-deficient phenotype, Mlh1-deficient mice are infertile, yet have completely normal ovary development, folliculogenesis, and ovulatory capacity *(32)*. Mlh1 is a MutL homolog, which in *E. coli* interacts with MutS to activate the endonuclease MutH in mismatch repair. Superovulation of *Mlh1*-deficient females yields normal numbers of oocytes. However, in vitro fertilization of these oocytes has demonstrated that they were never able to complete meiosis II after fertilization, as indicated by the complete absence of a second polar body and an inability to progress beyond the one-cell embryo stage. Thus, although the Msh5, DMC1, and Mlh1 proteins function in mismatch repair, they play unique and separate functions within the mammalian ovary. Similar to *Zfx* and the aforementioned genes, these mismatch-repair genes should be examined closely for mutations in humans, which lead to premature ovarian failure and/or cancer.

Two human diseases with impaired double-strand-break repair, Fanconi's anemia (FA) and ataxia telangiectasia (AT), have many symptoms that include subfertility or infertility. Fanconi's anemia is characterized by increased sensitivity to DNA crosslinking agents, and is caused by five distinct complementation groups. The mouse gene defective in complementation group C (*Fac*) has been cloned and disrupted in mice *(33,34)*. 73% of female *Fac –/–* mice are infertile, while the remainder had a few litters of small size (1–2 pups/litter). This phenotype is similar to female FA patients, who typically have irregular menstruation with menopause occurring at approx 30 yr of age. In humans, ataxia telangiectasia is caused by mutations in the nuclear ataxia telangiectasia mutated (ATM) protein—a member of a protein family involved in cell cycle regulation, monitoring of telomere length, meiotic recombination, and DNA repair—and is associated with extreme sensitivity to ionizing radiation. ATM protein is present at high levels in the cytoplasm of oocytes in developing follicles, and infertility is a common feature of AT in humans. Similarly, deficiency of ATM in mice phenocopies the human syndrome, with infertility caused by a complete loss of oocytes and absence of follicles by P11 *(35,36)*. At E 12.5, normal numbers of germ cells were observed in *Atm –/–* genital ridges, indicating that germ-cell migration and proliferation do not require ATM protein. However, by E 16.5, many oocytes in the ATM-deficient ovaries undergo apoptosis, leading to perinatal ovarian degeneration. Similarly, in mutant males, spermatogenesis is blocked at the zygotene stage of meiosis I, and seminiferous tubules are populated with spermatogonia and spermatocytes undergoing apoptosis. In males homozygous for a second mutation in p53 or p21, spermatogenesis can proceed to the pachytene stage, and the level of apoptosis is reduced by up to 70% *(35,36)*. Thus, it is proposed that ATM participates in monitoring and regulating meiotic progression.

The mouse models described here clearly demonstrate that prenatal ovarian development determines postnatal reproductive capacity. Factors expressed by the somatic cells along the PGC migration route and in the developing ovary, as well as the germ cells themselves, are critical for ovarian development and oocyte survival during the fetal and perinatal period (Tables 2 and 3).

POSTNATAL OVARIAN FUNCTION: FOLLICULOGENESIS

The follicle is the basic functional unit of the ovary, consisting of an oocyte surrounded by granulosa and theca cells. During normal folliculogenesis, there is coordination of oocyte growth and maturation, and granulosa and theca cell proliferation and development, within each follicular unit *(37)*. Distinct morphological and molecular changes occur in each of these components, and reflect evolving functional capabilities crucial for the continued development of the follicle and eventually, the successful completion of this developmental program. Staggered recruitment of follicles into the growing pool and consistent growth rates ensure that mature follicles are produced with every cycle. Follicular responsiveness to circulating pituitary hormones, and resultant production and secretion of steroid and peptide hormones, coordinate the release of the mature oocyte with alterations in female physiology conducive to mating, fertilization, and support of an embryo. The various mouse models discussed in the following sections are summarized in Tables 4–8.

INITIATION OF FOLLICULOGENESIS AND PREANTRAL FOLLICLE GROWTH

By postnatal d 2, meiotically arrested oocytes associate with somatic cells to form primordial follicles, the first stage of folliculogenesis *(4)*. Although the signals that trigger formation of primordial follicles and the eventual recruitment of a dormant primordial follicle into the growing pool are still unknown, the genes critical for these processes are being identified, and their functions are being analyzed (Table 4A). The subsequent period of preantral follicle growth, consisting of granulosa cell proliferation and oocyte growth, is relatively slow, and only a few mitotic figures are observed in granulosa cells at this stage *(38)*. Initiation of follicle growth in the mouse is not restricted to sexual maturity, but in fact begins within the first wk postnatally. Since there is no evidence for a reserve pool of larger follicles, it appears that once a follicle enters the growing pool, it is normally committed to a program of growth and differentiation, culminating in either apoptotic death of the granulosa cells (atresia) or ovulation of the mature oocyte *(39)*.

Several proteins are known to be expressed by mouse primordial follicles and follicles during the initiation period. The oocyte of the newly recruited follicle begins to secrete its unique extracellular glycoprotein matrix, called the zona pellucida (ZP). ZP formation at this early stage of folliculogenesis suggests that it may be important for oocyte-granulosa cell coupling, or may play a role in continued follicle development *(37)*. One of the genes involved in regulating the expression of all the three mouse ZP genes is *Fig*α (Factor In the Germline α). Figα is a basic helix-loop-helix transcription factor that is first expressed in oocytes prior to primordial germ cell formation, and is also expressed in oocytes of primary and later follicles *(40)*. Although *Fig*α (with a heterodimeric partner) binds to an E-Box in the ZP-1, ZP-2, and ZP-3 genes to regulate transcription of these genes, *Fig*α must also function prior to this point, because mice without *Fig*α have a normal number of oocytes at birth, but subsequently fail to form primordial follicles and demonstrate dramatic apoptosis of these oocytes over the next few days *(41)*. Thus, *Fig*α not only regulates expression of the ZP genes, but may also induce the expression of an oocyte gene involved in "recruitment" of pregranulosa cells to surround the oocytes and form primordial follicles.

To study the function of the ZP, all three of the ZP genes have been mutated in mice. Interestingly, mice deficient in any of the three major components of the ZP (ZP1, ZP2, or ZP3) show no defects at any stage of folliculogenesis *(41–44)*. ZP1-knockout mice are subfertile, with litter sizes 50% of wild-type controls, whereas ZP2- and ZP3-knockout mice are absolutely infertile. The reasons for these differences in fertility are secondary to the differences in the formation of the ZP in these knockouts. ZP2-knockout and ZP3-knockout mice fail to form a ZP; this absence of a zona leads to failure of fertilization and progression to the two-cell stage. In contrast, ZP1 knockout mice have a thin ZP, and the integrity of this ZP is compromised; these structural defects have resulted in a 80% reduction in the number of two-cell embryos recovered after superovulation. Thus, ZP absence (in the case of ZP2- and ZP3-knockouts) or alteration (in the case of ZP1 knockout), decreases ovulation and fertilization in vivo, confirming an important predicted role of the ZP in female reproduction.

Apoptotic cell death occurs in oocytes and granulosa cells of both primordial follicles and growing follicles. Members of the Bcl2-related protein family play either positive or negative roles in regulating apoptosis *(45,46)* (*see* Chapter 6). Bcl2 and $Bclx_L$ protect against apoptosis, while Bax, which can heterodimerize with Bcl2 and $Bclx_L$, counters their protective effect and promotes cell death when overexpressed. Bax is expressed in granulosa cells and oocytes, and plays a critical role in regulating ovarian-cell death; 6-wk old Bax-deficient mice have three times more primordial follicles than controls and one-half the number of atretic primordial follicles. This difference in the rate of follicular-pool depletion results in the presence of growing, functional follicles at 640 d of age. Despite the presence of growing follicles, no corpora lutea or pregnancies have been seen in these very old mice. Ovulation could be induced by injection of exogenous gonadotropins, indicating that reproductive senescence is caused by a combination of follicular depletion and pituitary dysfunction. Thus, Bax inactivation produces a surplus of nonatretic follicles, extends the function of the ovary into advanced chronological age, and may provide additional insight into the molecular basis of oocyte depletion associated with menopause in humans.

Growth of preantral follicles, corresponding to type 3b to type 5b follicles in the mouse, is gonadotropin-independent, and is regulated primarily by intraovarian and intrafollicular mechanisms. In the FSHβ knockout mouse *(47)* or in the *hypogonadal* (*hpg*) mouse, in which a naturally occurring mutation in the gonadotropin-releasing-hormone-gene markedly reduces the synthesis of both FSH and LH from the pituitary *(48)*, preantral follicle growth proceeds normally, confirming the gonadotropin-independence of this stage of development. However, several mouse models have clearly demonstrated that signaling from the granulosa cells to the oocyte, as well as from the oocyte to the granulosa cells, is necessary for preantral follicle growth and does not require extragonadal input (Table 4B). For example, intrafollicular signaling of kit ligand from granulosa cells to c-kit on the oocyte is critical for preantral follicle development. All mouse oocytes express c-kit, whereas granulosa cells of one-layered growing follicles, preantral follicles, and the outer (mural) layers of preovulatory follicles express kit ligand *(49,50)*. Two hypomorphic kit ligand alleles at the Sl locus, *Slt* and *Slpanda*, permit prenatal ovarian development (in contrast to the other alleles mentioned earlier) and initiation of follicular growth, but result in follicular arrest before the two-layer follicle stage *(51–53)*. In *Slpanda* homozygous mutant ovaries, kit ligand expression is virtually absent *(53)*, because of a large paracentric inversion located 115 kb 5' of the kit ligand coding sequences *(54)*. In addition, blocking antibodies to the c-kit receptor administered to mice during the first 2 wk after birth inhibit ovarian follicular development beyond the one-layer primary-follicle stage *(55)*, further supporting the essential role of kit ligand/c-kit signaling at this stage of follicular development.

Growth-differentiation factor-9 (GDF-9 or *Gdf9*), a novel, oocyte-expressed member of the transforming growth factor β (TGF-β) superfamily of secreted growth factors, is also necessary for early preantral follicle growth. *Gdf9* is first expressed by oocytes of type 3a follicles, and its expression persists in the oocyte through ovulation *(56–59)*. We have generated a GDF-9-deficient mouse model by deleting exon 2, which encodes the entire GDF-9 mature region *(60)*. Heterozygotes of both sexes and males homozygous for the deletion are fertile, but homozygous mutant females are completely infertile. Although follicular recruitment and initiation of growth is grossly normal, no follicles with two or more symmetric or concentric layers of granulosa cells are evident, indicating that folliculogenesis is blocked at the type 3b (primary) follicle stage in these mice *(60)*. This defect in follicular development is not rescued by treatment of the mice with exogenous gonadotropins. In addition, kit ligand and α inhibin mRNA are elevated in these one-layer primary follicles *(58)*. The increased signaling of kit ligand through its receptor, c-kit, in the oocyte, is a probable cause of the increased oocyte size in the *Gdf9* knockout ovary *(61)*, further supporting the importance of modulation of kit ligand/c-kit signaling in the postnatal ovary. This finding suggests that GDF-9 is a direct negative regulator of this signaling pathway.

At the early stages of oogenesis in the *Gdf9* knockout ovary, the oocytes appear to be fairly normal. However, the absence of GDF-9 signaling, and potentially the resultant increase in kit ligand signaling, eventually leads to defects in oocyte meiotic competence and abnormal germinal vesicle breakdown, and spontaneous parthenogenetic activation of the oocytes, in addition to the increased rate of growth of the oocyte. At the electron-microscopic level, the GDF-9-deficient oocytes have several unusual features, including Golgi complexes composed of single lamellae instead of stacks and a decreased number of cortical granules. Additionally, cell-cell contacts between the oocyte and granulosa cells are unusual in that oocyte microvilli are clustered next to abnormal, large processes from surrounding somatic cells. These follicle cells subsequently invade the perivitelline space, and are associated with a loss of oocyte viability *(60,61)*. The combined oocyte and granulosa cell abnormalities lead to eventual death of the oocyte, resulting in a follicular nest with granulosa cells surrounding a collapsed ZP remnant. The cells of the majority of these follicular nests are steroidogenic; these cells appear vacuolated because of the large number of lipid droplets, have an increased number of mitochondria *(60)*, and express P450 side-chain cleavage, P450 aromatase, LH receptor, and α inhibin mRNA *(58)*. Thus, although these nests of cells often resemble small corpora lutea, these follicles express both luteal (i.e., p450 side-chain cleavage and LH receptor) and nonluteal (i.e., p450 aromatase and α inhibin) markers, suggesting that the early loss of the oocyte (and possibly the absence of GDF-9) alters the differentiation program of these granulosa cells.

In the periovulatory follicle, the oocyte secretes important growth factors, which stimulate the synthesis of hyaluronic acid necessary for cumulus expansion and repress the synthesis of LH receptor and urokinase plasminogen activator (uPA) *(59)*. Because of the early block in the growth of the *Gdf9* knockout ovary, and based on the continued expression of GDF-9 beyond ovulation, we have studied the biological actions of GDF-9 in the periovulatory period, using pregnant mare serum gonadotropin (PMSG)-induced mouse granulosa cells cultured with recombinant mouse and human GDF-9 protein. Recombinant mouse GDF-9 induces hyaluronan synthase 2 (*Has2*), cyclooxygenase 2 (*Cox2*), and steroidogenic acute regulator protein (*StAR*) mRNA synthesis, and suppresses urokinase plasminogen activator and luteinizing hormone receptor (LHR) mRNA synthesis *(57)*. In addition, GDF-9 stimulates in vitro cumulus expansion of oocytecto–mized cumulus cell-oocyte complexes (i.e., complexes in which the oocyte has been microsurgically removed) *(57)*. Thus, GDF-9 is essential for granulosa-cell growth and function at early stages, and is also the oocyte-secreted factor responsible for modulating the expression of a number of cumulus cell genes critical during the periovulatory period.

The theca layer forms when the follicle achieves two layers of granulosa cells, and provides a source of aromatizable androgen to the adjacent granulosa cells, which is crucial for follicular estrogen production by enzymatic conversion *(62)*. Theca cells differentiate from mesenchymal or stromal precursors adjacent to developing follicles. Theca-interstitial cell culture experiments in vitro show that rat preantral follicles with 2–5 layers of granulosa cells (but not one-layer or antral follicles) secrete a factor in the absence of gonadotropins that induces theca layer differentiation, including expression of cytochrome P450 17α-hydroxylase-C17-20 lyase *(63)*. In GDF-9-deficient mice, a theca layer fails to form, despite the presence of increased FSH and LH *(58,60)*. However, an identifiable theca layer is formed around the multilayer preantral follicles in the FSH-deficient ovary model *(47)*. Taken together, these data support the presence of a paracrine, inductive signal secreted from preantral follicles with two or more layers of granulosa cells, which is necessary for theca layer development. GDF-9 is possibly an important direct or indirect regulator of these theca cell "recruitment/differentiation" factors.

Primordial and small preantral follicles (type 3a and 3b) represent a minute fraction of total ovarian cells, making it difficult to isolate genes and proteins involved in the earliest stages of follicular development. In addition, these stages have generally not proven amenable to extraovarian manipulation. However, development of these mouse models, in which folliculogenesis is arrested at an early stage, have defined some key factors involved in initiation and early preantral follicle growth.

ANTRAL FOLLICLE DEVELOPMENT: SENSITIVITY TO EXTRA OVARIAN AND INTRA OVARIAN REGULATION

During follicular development in mice, follicular antrum formation represents a transition from primarily intrafollicular regulation to a combination of intraovarian and extraovarian regulation. Follicles enter a rapid period of growth, and synthesize peptide and steroid hormones that impact on the reproductive axis. Multiple positive and negative feedback loops between the hypothalamus, pituitary, and ovaries coordinate follicle maturation with sexual behavior and physiological preparation for pregnancy. The hypothalamus produces and releases gonadotropin-releasing hormone (GnRH) in a pulsatile manner directly into the pituitary blood supply via the pituitary portal vessels *(64)*. GnRH pulse frequency is modulated by the endocrine status of the animal; estrogen increases pulse frequency and increases the sensitivity of the anterior pituitary to GnRH. In response to GnRH, the anterior pituitary releases the heterodimeric glycoprotein hormones, FSH and LH.

Defects in the hypothalamic-pituitary-gonadal axis have a dramatic effect on antral-follicle development, and there are multiple mouse models that exhibit infertility or subfertility because of defects at the levels of the hypothalamus and pituitary (Table 5). Loss of pituitary stimulation by hypothalamic GnRH because of a gene deletion, as in the hypogonadal (*hpg*) mouse *(65)*, or resulting from migration arrest of GnRH neurons and tumor formation, as in the GnRH-SV40 T-antigen transgenic mouse (66), prevents release of follicle-stimulating hormone (FSH) and luteinizing hormone (LH). In the GnRH mouse models, follicles progress normally to the multilayer preantral-follicle stage, but are unable to form significant antra, similar to the FSHβ-deficient mouse model *(47)*. Additionally, the ovaries are very small, with little interstitial tissue, which may be caused by the lack of LH or a direct effect of GnRH on the ovary. Mutations in the GnRH receptor *(67)* also cause recessively inherited hypogonadism, primary amenorrhea, and infertility in humans.

FSH, in conjunction with locally produced estradiol, functions primarily to promote follicular growth *(68)*, causing the mitotic index of the late preantral to early antral granulosa cells to peak *(37,69)*, and to induce the granulosa cell gene expression necessary for follicle maturation. Mutations leading to FSH deficiency, misregulation, or insensitivity disrupt antral-follicle development. In the FSHβ-deficient mouse, folliculogenesis progresses normally to the multilayer preantral follicle stage, but further granulosa cell proliferation and antrum formation is blocked, and these follicles subsequently undergo atresia *(47)*. To further study evolutionarily conserved functions of human and mouse FSH, two approaches have been taken to genetically rescue these mouse FSHβ mutants *(70)*. In the first approach, a human FSHβ transgene, which is expressed exclusively in pituitary gonadotropes, was introduced into the mouse FSHβ-deficient line leading to the production of an interspecies hybrid FSH (i.e., mouse α:human FSHβ hormone). FSHβ-deficient mice carrying the human FSHβ transgene resumed normal folliculogenesis, were fertile, and delivered normal sized litters. Thus, human FSHβ can combine with the mouse glycoprotein hormone α to form a functional (hybrid) heterodimer. In the second approach, two metallothionein-I promoter-driven transgenes were introduced into the mouse FSHβ-deficient line, leading to the ectopic expression of both human α and human FSH, predominantly in the liver. Ectopic production of human FSH rescued the fertility of only 30% of the females. Similarly, human

FSH can bind to mouse FSH receptors to restore fertility, but pituitary control of FSH synthesis and secretion are necessary for normal reproductive function in females.

Other mutations at the level of the ovary also lead to infertility becuase of defects in antrum formation (Table 6). For example, impaired responsiveness to FSH leads to ovarian failure similar to that observed in the FSH ligand knockout. Targeted disruption of the mouse FSH receptor (FSHR) *(71)* also leads to female infertility and a block in folliculogenesis before antral follicle formation, similar to that observed in the mouse FSHβ-deficient mouse. Based on the uterine morphology (i.e., uteri +/+ > +/– > –/–), it appears that there is also an effect of the FSHR mutation on heterozygotes, suggesting that although fertile, follicular estrogen production may be reduced because of decreased FSH binding. Additionally, homozygous FSHR deficiency leads to 15-fold elevated serum FSH levels. Significantly, these findings phenocopy humans with mutations in the FSH receptor *(72)*, which causes recessively inherited hypergonadotropic hypogonadism, primary amenorrhea, and infertility. Recently, a woman with secondary amenorrhea and very high serum FSH concentrations was found to be a compound heterozygote for two different mutations in the FSHR *(73)*. Ultrasonography revealed normal-sized ovaries and antral follicles up to 5 mm in diameter, but further follicular development was blocked. However, further analysis demonstrated that these FSHR mutations resulted in only partial functional impairment (i.e., a hypomorphic state), and suggested that a limited FSH effect is sufficient to promote follicular development to the small antral stage, but further development requires significant FSH stimulation (*see* activin-receptor-type II knockout model).

Insulin-like growth factor I (Igf1) knockout mice *(74,75)* have an almost identical ovarian phenotype to the FSH knockout mice (for more details, *see* Chapter 17). Igf1 has been shown to enhance proliferation of many cell types, and Igf1 and Igf1 receptor are expressed in follicles that appear healthy, suggesting that they may be markers for follicular selection. Igf1 expression, however, is not dependent upon FSH stimulation, because Igf1 continues to be expressed in FSH-deficient follicles *(75)*. Instead, granulosa cell Igf1 expression is initiated by intrafollicular signaling, possibly from the oocyte. The importance of Igf1 signaling in ovarian function is further supported by the reproductive defects in the insulin receptor substrate-2 (IRS-2)-deficient mouse *(76)*. IRS proteins function as downstream signaling proteins to mediate the cellular actions of insulin and Igf1. IRS-2-deficient females rarely become pregnant, lack corpora lutea in their ovaries, and do not respond to superovulation. Both FSH and Igf1 augment the expression of FSHR, as evidenced by a 50% decrease in FSHR concentration in the Igf1-deficient ovaries or after elimination of gonadotropins by hypophysectomy. Decreased FSHR expression, leading to relative FSH insensitivity in the Igf1-deficient follicles, appears to be at least part of the mechanism for follicular arrest in the Igf1-deficient model.

Although FSH has long been recognized to induce proliferation of granulosa cells, only recently has one of the molecular mechanisms of its mitogenic action been suggested. Granulosa-cell proliferation appears to be mediated by FSH and estradiol induction of cyclin D2 *(77)*. The D-type cyclins are known to positively regulate entry into the cell cycle by binding cyclin-dependent kinases (CDK) 4 and 6 and allowing phosphorylation of the complex by CDK-activating kinase. This activated cyclin-CDK complex then phosphorylates a number of cellular substrates, eventually activating DNA synthesis and the transition from G1 to S phase *(78)*. While cyclin D1 and D3 localize to theca and interstitial cells, cyclin D2 is expressed specifically in granulosa cells within the ovary *(77)*. Studies of rat granulosa cells cultured in vitro showed that forskolin induced cyclin D2 comparably to FSH, indicating that FSH signal transduction through the cAMP/protein kinase A pathway activated cyclin D2 expression *(79)*. Conversely, cyclin D2 expression is dramatically reduced in nonproliferating granulosa cells. For example, cyclin D2 expression in preovulatory follicle is downregulated by human chronic gonadortropin (hCG) treatment and subsequent luteinization, but continues in smaller growing follicles that do not express receptors *(77)*. Increased expression and gene amplification of cyclin D2 has been detected in a variety of human granulosa cell tumors *(79)*, emphasizing its potential clinical importance in regulating granulosa cell proliferation.

Cyclin D2-deficient mice provide definitive evidence that cyclin D2 is functionally significant for granulosa cell proliferation *(79)*. While the number of oocytes and follicles in cyclin D2-deficient ovaries was normal, there was an obvious reduction in the number of granulosa cells surrounding each oocyte. Mutant ovaries showed minimal response to FSH administration in contrast to the rapid FSH-induced proliferation of the granulosa cell layer in wild-type ovaries. This was particularly apparent in cyclin D2-deficient antral follicles that rarely had more than four layers of granulosa cells, compared with controls containing up to ten layers. The gonadotropin signal transduction cascade was shown to be intact, as cyclin D2-deficient follicles produce estradiol in response to FSH, and expressed the periovulatory and luteal cell markers, cyclooxygenase 2, cytochrome P450, cholesterol side-chain cleavage, and progesterone receptor in response to LH *(80)*. However, cyclin D2-deficient mice fail to ovulate in response to LH, and instead, corpora lutea are formed with oocytes trapped inside. The cause of the ovulation defect is still unclear, but this phenotype emphasizes the importance of the coordination of growth and differentiation for successful completion of folliculogenesis.

FSH stimulates expression of the α and β subunits of inhibin in the pituitary and ovary, leading to the production of the peptide hormones activin and inhibin, which can function in autocrine and paracrine signaling or as endocrine factors *(81)*. Activins and inhibins are dimeric members of the TGF-β superfamily, in which the ratio of α subunits to β subunits produced by the cell determines whether the hormonal output is FSH-stimulating (activin) or FSH-suppressing (inhibin) *(82)*. The α-inhibin and the activin-receptor type IIA-deficient (ActRIIA) mice emphasize the importance of gonad-produced peptide hormone feedback and the intrapituitary effects of these peptides on FSH regulation. The α-inhibin knockout mice have increased serum FSH, confirming the known role of inhibin to decrease pituitary FSH release. Furthermore, few fertilizable oocytes could be recovered from the oviducts of PMSG/hCG-primed immature inhibin-deficient females *(83)*. This finding indicates that inhibin plays an important intra ovarian function in folliculogenesis. Consistent with this intra ovarian function, inhibin also has a novel antiproliferative and tumor-suppressive role in the gonads. 100% of male and female inhibin-deficient mice develop early onset, rapidly growing granulosa/Sertoli cell tumors, which cause death secondary to a cancer cachexia-like syndrome mediated by activins secreted from the tumors *(84,85)*. The predisposition of inhibin-deficient mice to develop gonadal tumors identifies inhibin as a secreted tumor suppressor.

The involvement of gonadotropins in promoting gonadal tumorigenesis has been the subject of considerable debate. Elevated postmenopausal levels of serum FSH have been associated with some forms of human ovarian epithelial cancer in elderly *(86)*, but no direct in vivo causal relationship between FSH and ovarian cancer development has been demonstrated until recently. To investigate the role of FSH as a component of the cascade of events leading to development of gonadal tumors in inhibin-deficient mice, double-homozygous mutant mice deficient in inhibin and FSH were created *(87)*. In contrast to mice lacking inhibin alone, in which 95% develop highly hemorrhagic ovarian tumors, cachexia, and death by 17 wk of age, 70% of double-mutant females live beyond 17 wk. Although 100% of the female double-mutants still develop ovarian tumors (in contrast to the double-mutant males where 70% are still alive at 1 yr), the ovarian tumors are slow-growing and less hemorrhagic. Additionally, reduced or delayed tumor-associated cachexia-like symptoms in the double-mutants were associated with low levels of serum estradiol (decreased 87% compared to inhibin single-mutant) and activin A (<0.078 vs 157.5 ng/mL). In contrast to inhibin/FSH double mutants, mice deficient in inhibin and GnRH, leading to suppressed FSH and LH, survive more than 1 yr, do not develop cancer cachexia-like symptoms, show only premalignant lesions in the ovary *(88)*. Consistent with an important role of LH in ovarian tumorigenesis, transgenic mice overexpressing either a bovine LHα subunit or a bovine LHβ analog in the pituitary also develop granulosa cell tumors of the ovary *(89)*. Thus, these results clearly demonstrate that gonadotropins are significant determinants of ovarian tumor phenotype and progression.

In contrast to the inhibin-deficient mice, mice lacking one of the activin receptors, activin receptor type IIA (ActRIIA), have dramatically suppressed serum and pituitary levels of FSH. These results indicate that ActRIIA is the major pituitary receptor through which activins affect FSH synthesis and secretion *(90)*. ActRIIA-deficient ovaries display a block in folliculogenesis at a slightly later developmental stage than that seen in FSH-deficient mice (i.e., the small antral follicle stage). This suggests that the block in folliculogenesis is a result only to the decreased FSH concentrations (that is, the phenotype is similar to an FSH hypomorphic allele) and not to the lack of paracrine signaling through ActRIIA in the ovary. These findings are similar to the findings in human female, with low-level activity of the FSHR in the granulosa cells *(73)*. Consistent with the primary role of ActRIIA in the pituitary, ActRIIA-deficient ovaries transplanted into ovariectomized immunocompatible wild-type hosts (females with normal pituitaries and therefore normal serum FSH levels) resumed normal folliculogenesis, including formation of ovulatory follicles and corpora lutea in the host mice (M. M. Matzuk, unpublished data). Thus, lack of signaling through ActRIIA in the pituitary suppresses serum FSH levels, resulting in impaired folliculogenesis in the ActRIIA-deficient mice.

To understand further the roles of activins (and other members of the TGF-β superfamily members) in the ovary, we have generated mice that overexpress follistatin, an activin-binding protein, using the metallothionein promoter *(91)*. Female mice from two of the transgenic lines with the highest expression of the follistatin transgene often had blocks in folliculogenesis at the preantral and antral follicle stage, resulting in infertility in the most severely affected mice. Overexpression of follistatin may block follicular development by binding and inactivating activin and possibly other TGF-β family members, since both TGF-β and activin are capable of stimulating follicle growth in vitro *(92)*, and GDF-9 is required for early stages of follicular growth.

In the rat, intrafollicular estrogen signaling enhances the granulosa cell response to FSH, augmenting granulosa cell proliferation and expression of numerous FSH-regulated genes, including inhibin α and β subunits, LH receptor, and cytochrome P450 aromatase *(2)*. Cytochrome P450 aromatase catalyzes the conversion of theca-produced androgens to estrogens demonstrating the importance of coordinated development of follicular components. The interplay between FSH and aromatase and estrogen signaling sets up a positive-feedback loop for FSH within the follicle, which is important for later stages of follicle development. Three different mouse models demonstrate the importance and complexity of estrogen action within the ovary. In the aromatase knockout mouse (ArKO), conversion of androgens to estrogen is blocked *(93,94)*. ArKO female mice have increased FSH and LH and ~10-fold elevated levels of testosterone compared to controls, leading to development of male body habitus and excessive internal fat deposition. Although ovaries develop in ArKO mice and folliculogenesis proceeds with granulosa cell proliferation and evidence of limited antrum formation, corpora lutea are absent, and the mice are infertile. By 21–23 wk of age, hemorrhagic cystic follicles develop similar to the ERα knockout model, and by 1 yr, there is a dramatic reduction in the number of secondary and antral follicles. Thus, estrogen synthesis is required for formation of large antral follicles and ovulation.

Estrogens exert their effects through interaction with estrogen receptors located in the nuclei of target cells. Two different estrogen receptors, ERα and ERβ, have been cloned and are expressed in ovarian cells. ERα has a broad expression pattern, whereas ERβ is expressed at high levels only in the ovary, prostate, epididymis, lung, and hypothalamus. Within the ovary, ERα protein is expressed in theca and interstitial cells *(95)*, while ERβ is specifically expressed in granulosa cells of small, growing, and preovulatory follicles *(95,96)*. Knockout mice for either ERα or ERβ have been created, and these mice have dramatically different phenotypes from each other and from the ArKO mouse. The ERα-knockout mouse (αERKO) develops large preovulatory follicles, but fails to subsequently ovulate and form corpora lutea. Some arrested follicles undergo atresia, while others develop into large, hemorrhagic cysts *(97)*. This ovarian phenotype is caused by failure of steroid feedback at the level of the hypothalamus and brain, leading to overstimulation of the LH receptor on theca and granulosa cells by the high LH levels in the serum *(95)*. ERβ is still detected in αERKO ovaries, and cannot completely compensate for loss of ERα, indicating that ERβ and ERα have distinct functions. On the other hand, female ERβ-knockout mice are subfertile, demonstrating reduced litter number and pups per litter because of reduced ovulation efficiency *(98)*. Pharmacologic superovulation resulted in only 20% of the oocytes produced in controls. These differences in phenotypes in the βERKO and ArKO mice vs the αERKO mice may be ascribed to ERα-mediated estrogen regulation of gonadotropin synthesis and secretion in the hypothalamus and/or pituitary.

Double-mutant mice lacking both ERα and ERβ (termed αβERKO mice) have also been described *(99)*. Prepubertal αβERKO mice exhibit prominent ovarian growth and development of follicles to the antral follicle stage. These findings are reminiscent of hypergonadotropic precocious puberty, and are probably the result of elevated gonadotropins in these prepubertal females. Adult αβERKO female mice are completely infertile, follicles never develop to the preovulatory stage, and no corpora lutea are seen. One surprise in these mice is the presence of "Sertoli" tubule-like structures that are the result of oocyte loss and the dedifferentiation of the granulosa cells into "Sertoli" cells. Consistent with this phenotype, Müllerian-inhibiting substance (MIS) mRNA is increased, and Sox9, a marker for Sertoli cells, is also increased. Similar findings have been seen in female mice overexpressing MIS *(100)* after oocyte loss, and also in α-inhibin knockout female mice that develop mixed granulosa/Sertoli cell tumors *(83,84,88)*. These findings suggest an important role for estrogens, inhibins, and possibly oocyte-secreted factors in the maintenance of the normal granulosa cells phenotype.

OVULATION AND CORPORA LUTEA FORMATION

The preovulatory follicle responds to the LH surge by releasing the oocyte and undergoing a series of functional and morphological changes in a process known as luteinization. The mural granulosa cells of the preovulatory follicle express LH receptors at high levels enabling them to sense the LH surge. LH, like FSH, stimulates adenylyl cyclase to produce cAMP and activate PKA, and may also increase inositol triphosphate and activate protein kinase C *(101)*. Activation of this additional second messenger system may explain why the follicular response to the LH surge is so radically different from its response to FSH. P450 aromatase expression is abolished, cell division is halted, and genes responsible for breakdown of the follicular wall and basement membrane begin to be expressed *(102)*. Disruption of pituitary LH synthesis, LH-receptor (LH-R) binding, or its downstream signaling are likely to block ovulation, luteinization, or both. Multiple mouse models have defects in ovulation and/or corpus luteum formation because of extra-ovarian (Table 7) or intra-ovarian (Table 8) defects.

Regulation of the LHβ subunit in the pituitary occurs during transcription, polyadenylation, and glycosylation of the protein. The LHβ promoter contains binding sites for SF-1, ERα, CREB, and the zinc-finger transcription factor, NGFI-A *(103)*. Whereas either SF-1 or NGFI-A can activate LH expression at a relatively low level, together they have been shown to synergistically activate high-level LHβ expression. Female mice carrying a targeted disruption of NGFI-A are infertile due to a block in ovulation *(103)*. In this model, serum LH concentration demonstrated a sexually dimorphic response to NGFI-A disruption: LH was decreased in males and undetectable in females. Administration of exogenous LH to the NGFI-A-deficient females resulted in normal ovulation and corpus luteum formation, indicating that the critical defect causing ovarian failure was LH deficiency. In contrast, a second model in which a lacZ marker gene was inserted into the NGFI-A exon 1 to disrupt gene function showed a deficiency of LH synthesis in both males and females *(104)*. LacZ staining was observed in corpora lutea, granulosa cells of the mature antral follicles, and oocytes of mice heterozygous for the NGFI-A-LacZ knockin. Interestingly, the anovulatory phenotype in the homozygous NGFI-A-LacZ knockin females could not be rescued by pharmacological replacement of LH, potentially because of the significantly reduced levels of LHR in the granulosa cells of preovulatory follicles. These latter findings suggest that NGFI-A regulates expression of LH receptors in granulosa cells, as well as synthesis of LH in the anterior pituitary. It is unclear why these two different mutations, which are presumably both null mutations, result in somewhat different phenotypes.

Ovulation is frequently compared to an inflammatory response. Follicular hyperemia and edema occur within a few hours of the gonadotropin surge, and are probably mediated by vasoactive agents such as histamine, kinins, and prostaglandins *(102)*. In response to the gonadotropin surge and inflammatory mediators, serine proteases and metalloproteinases, such as plasminogen activator and collagenases, also increase in ovulatory follicles, suggesting a biochemical mechanism for follicular rupture. Indomethacin, a potent nonsteroidal anti-inflammatory agent, can block ovulation potentially through inhibition of prostaglandin synthesis. Cyclooxygenase 2 (*Ptgs2* herein called COX-2) is one of two genes that catalyzes the formation of prostaglandins from arachadonic acid, and has been shown to be rapidly, but transiently, induced in granulosa cells of the preovulatory follicle after the LH surge *(2)*. Normally, COX-2 mRNA concentrations peak 4 h after administration of hCG, and return to almost undetectable concentrations by 6–8 h after treatment *(105)*. In response to hCG administration, the granulosa cells closest to the oocyte, the cumulus cells, express the highest level of COX-2 mRNA *(58)*. Recently, we have shown that the oocyte-derived growth factor, GDF-9, induces COX-2 gene expression, and also induces cumulus expansion and expression of hyaluronan synthase 2 mRNA *(57)*. Thus, both extraovarian stimulation by gonadotropins and intrafollicular signaling operate to control periovulatory gene expression.

The critical role of COX-2-mediated prostaglandin synthesis in ovulation and postovulatory events is supported by the phenotype of COX-2-deficient mice and the effects of pharmacologic inhibition of COX-2 function in vivo and in vitro. In COX-2-deficient ovaries, folliculogenesis progresses normally to the preovulatory stage (106), but the number of oocytes ovulated in a normal cycle or in response to exogenous gonadotropins was reduced to 20–40% of controls. Additionally, only 1% of ovulated oocytes from COX-2-deficient mice were successfully fertilized *(107)*. A similar, although less dramatic, fertilization defect is also seen for knockouts of the EP2 prostaglandin E2 receptor *(108,109)*. The defects in fertilization in these EP2-receptor knockout mice appears to be secondary to defective cumulus expansion *(110)*. Thus, similar to GDF-9 (57), prostaglandin E2—acting through the EP2 receptor—stimulates cumulus expansion in vivo. This suggests that at least a portion of the fertility defects in the COX-2 knockout is secondary to a decrease in prostaglandin E2 signaling through this receptor. Consistent with the COX-2 knockout phenotype and the EP2 knockout phenotype, COX-2 protein is detected in the cumulus granulosa cells that are attached to the ovulated oocyte and pharmacologic inhibition of COX-2 in wild-type cumulus-oocyte complexes reduces their rate of fertilization *(111)* and implantation *(107)*. Thus, local production of prostaglandins by cumulus cells establishes an ideal microenvironment around the oocyte essential for efficient ovulation, fertilization, and implantation.

The COX-2 promoter has a binding site for the LH surge-induced transcription factor, CCAAT/enhancer-binding protein β (C/EBPβ), suggesting that C/EBPβ is one factor that may control COX-2 transcription *(112)*. Furthermore, C/EBPβ is highly expressed in granulosa cells of late antral follicles by 7 h after hCG injection, suggesting its importance in late follicular development. C/EBPβ-deficient mice demonstrate a significant decrease in ovulation efficiency and an absolute block in "mature" corpus luteum formation *(113)*. In C/EBPβ-deficient ovaries, COX-2 is still induced by the LH surge, eliminating the possibility that the phenotypic similarities between the C/EBPβ knockout and COX-2 knockout are caused simply by an absence of COX-2. However, both COX-2 and cytochrome P450 aromatase mRNA persisted at least 7 h after the LH surge suggesting that C/EBPβ instead mediates the transcriptional attenuation of COX-2 and aromatase. Wild-type, immunocompatible females retaining one wild-type ovary and one transplanted C/EBPβ-deficient ovary mated to wild-type males produce heterozygous pups, although at a much lower frequency than wild-type pups, confirming that successful ovulation and fertilization of mutant oocytes can occur. Corpora lutea are never seen in the C/EBPβ-deficient ovaries even after confirmed ovulations have occurred, suggesting that the pregnancy is supported entirely by corpora lutea of the wild-type contralateral ovary. The absolute requirement for C/EBPβ in luteal maturation may reflect a role in transcriptional attenuation of preovulatory genes or activation of other unknown genes. It will be of interest to determine whether the phenotype of the C/EBPβ knockout mice is secondary to persistent expression of key preovulatory genes, and also to understand the role of the oocyte in the regulation of genes involved in ovulation.

Coincident with inactivation of estrogen biosynthesis through P450 aromatase loss, the LH surge activates progesterone biosynthesis by stimulating cytochrome P450 cholesterol side-chain cleavage mRNA expression in granulosa cells. Progesterone is known to play an essential role in preparing the uterus for implantation of the embryo and has physiological functions in the mammary gland, brain, and ovary. These effects are mediated through binding to the progesterone receptor (PR), a member of the nuclear-receptor superfamily of transcription factors. PR mRNA has been shown to be induced by ovulatory concentrations of LH in granulosa cells in culture and in vivo *(114,115)*. The generation of the PR-knockout (PRKO) mouse has confirmed that progesterone plays an essential physiological role in ovulation *(116)* (*see* Chapter 9). PRKO females are infertile, although the ovaries exhibit normal folliculogenesis through the preovulatory stage and demonstrate corpora lutea. However, these mice fail to ovulate, even with pharmacological treatment with PMSG and hCG. Histological examination of the ovaries reveals many unruptured follicles containing oocytes surrounded by cumulus cells that have undergone expansion. Progesterone is required for postfertilization events, and these studies demonstrate that PR directly regulates the synthesis of one or more enzymes involved in proteolysis, leading to follicular rupture.

Luteinization is the terminally differentiated state of granulosa cells, and is accompanied by cell cycle arrest. $p27^{KIP1}$ is a cell cycle regulatory protein that controls cell cycle progression by binding to and inactivating cyclin-CDK complexes in response to extracellular, anti-mitogenic signals. $p27^{KIP1}$ is widely expressed in nonproliferating cells, including the cells of the corpus luteum, but is not detectable in nonluteinized granulosa cells. In addition, $p27^{KIP1}$ has been shown to be induced in granulosa cells by LH *(77)*. $p27^{KIP1}$ knockout mice are infertile, supporting a role for $p27^{KIP1}$ in regulating follicular function *(117,118)*. $p27^{KIP1}$-deficient females have prolonged estrous cycles, infrequent ovulation, and decreased copulation. Histologic examination of the ovaries reveals intact follicular development, but a marked absence of corpora lutea. However, exogenous administration of PMSG and hCG stimulates ovulation *(117,118)* and the subsequent formation of corpora lutea capable of increasing serum progesterone *(117)*. Further characterization of the reproductive abnormalities associated with $p27^{KIP1}$ deficiency demonstrates that embryos fail to implant at E 4.5, but can be rescued by administering E2 and P4 to $p27^{KIP1}$-deficient mothers *(119)*. Additionally, unilateral, transplantation of a $p27^{KIP1}$-deficient ovary into a wild-type host results in offspring derived from oocytes of both the $p27^{KIP1}$-deficient and the remaining wild-type ovary. In contrast, bilateral transplantation of a $p27^{KIP1}$-deficient ovaries into a wild-type host disrupts the estrous cycle and compromiss fertility. Conversely, transplantation of a wild-type ovary into a $p27^{KIP1}$-deficient host does not restore estrous cyclicity *(119)*. Taken together, these results suggest that perturbed ovarian steroid production, possibly caused by the failure of the luteal cell to withdraw from the cell cycle, in combination with extra ovarian—likely pituitary—defects cause infertility in the $p27^{KIP1}$-deficient mice.

During follicular growth, the oocyte grows and matures, first acquiring competence to undergo germinal vesicle (oocyte nucleus) breakdown (GVBD) and then competence to complete meiosis I. Granulosa cell-oocyte gap junctions, composed of channels of connexins, permit diffusion of ions, metabolites and, potentially, other signaling molecules. Connexin 37 forms the oocyte-granulosa cell gap junctions, and connexin 37-deficient ovaries demonstrate a lack of junctional communication between the oocyte and granulosa cells, but not among granulosa cells. Oocyte growth is reduced in these mutant ovaries compared with controls, and >90% of the oocytes are incompetent to resume meiosis *(120)*. In addition, there are defects in the later stages of follicular development, and a failure to ovulate in these connexin 37 knockout mice. These studies demonstrate that gap junction-mediated communication is important for the later stages of both oocyte and follicle development.

The LH surge stimulates the mature oocyte to undergo GVBD and progress through meiosis I. The ovulated oocyte enters meiosis II, where it arrests in metaphase of meiosis II until fertilization with subsequent release of the second polar body. pp39^{mos}, the protein product of *c-mos*, plays an important role in this process, as *c-mos*-deficient oocytes fail to maintain meiotic arrest after maturation *(121,122)*. c-mos deficiency leads to decreased fertility resulting from parthenogenic activation of ovulated oocytes, which renders them incapable of fertilization. The small numbers of offspring that do arise from *c-mos*-deficient mothers are presumed to be derived from fertilization of eggs shortly after maturation and before parthenogenic activation occurs.

CONCLUSION

Transgenic mouse models have been generated to study each stage of ovarian development and function. The transgenic mouse approach is providing in vivo evidence at all stages of folliculogenesis to support previous in vitro studies of intra-ovarian and extra-ovarian regulators, as well as defining novel mediators of ovarian function. A number of female mice with abnormalities other than follicular defects have also been generated to study processes such as the role of the zona pellucida in ovarian physiology, the functions of specific proteins in uterine development, and so forth (Table 9). Mouse models to study prenatal ovarian development and the earliest stages of folliculogenesis are particularly important, as these periods of development have been less accessible to other methods of investigation. One challenge in interpreting the phenotype of any model is to determine the direct effects of a mutation vs secondary effects caused by compensatory mechanisms. This is particularly difficult because of the multiple, interacting positive and negative feedback loops within the hypothalamic-pituitary-gonadal axis. Intercrosses to generate mice with multiple mutations have helped in some cases to clarify primary vs indirect effects of mutations. One drawback to conventional knockout mouse technology is that only the first essential function of the gene product can be examined in ovarian development or folliculogenesis, while expression may occur at multiple stages. For example, although knockout of GDF-9 results in a block in folliculogenesis at the primary follicle stage *(60)*, we have used recombinant GDF-9 to demonstrate that GDF-9 regulates multiple key periovulatory events required for normal female reproduction *(57)*. Generation of stage-specific, tissue-specific, and inducible knockout models may be useful in further elucidating of functions of already known and soon-to-be-discovered genes which play key roles in ovarian regulation.

ACKNOWLEDGMENTS

The authors thank Ms. Shirley Baker for her expert assistance in manuscript formatting. Studies in our laboratory on ovarian development and ovarian cancer have been supported by Genetics Institute and National Institutes of Health grants HD33438, CA60651, HD32067, and the Specialized Cooperative Centers Program in Reproduction Research (HD07495). Dr. Julia A. Elvin is a student in the Medical Scientist Training Program supported by NIH Training Grants GM07330 and GM08307 and Baylor Research Advocates for Student Scientists (BRASS).

REFERENCES

1. Adashi EY. Intraovarian peptides. Stimulators and inhibitors of follicular growth and differentiation. Endocrinol Metab Clin N Am 1992;21:1–17.
2. Richards JS. Hormonal control of gene expression in the ovary. Endocr Rev 1994;15:725–751.
3. Capecchi MR. Targeted gene replacement. Sci Am 21994;70:52–59.
4. Byskov AG, Hoyer PE. Embryology of mammalian gonads. In: Knobil E, Neill J, eds., The Physiology of Reproduction, Raven Press, New York, NY, 1994, pp. 487–540.
5. Kreidberg JA, Sariola H, Loring JM, Maeda M, Pelletier J, Housman D, Jaenisch R. WT-1 is required for early kidney development. Cell 1993;74:679–691.
6. Luo X, Ikeda Y, Parker KL. A cell-specific nuclear receptor is essential for adrenal and gonadal development and sexual differentiation. Cell 1994;77:481–490.
7. Pang S, Yang X, Wang M, Tissot R, Nino M, Manaligod J, et al. Inherited congenital adrenal hyperplasia in the rabbit: absent cholesterol side-chain cleavage enzyme cytochrome P450 gene expression. Endocrinology 1992;131:181–186.
8. Caron KM, Soo S-C, Wetsel WC, Stocco DM, Clark BJ, Parker KL. Targeted disruption of the mouse gene encoding steroidogenic acute regulatory protein provides insights into congenital lipoid adrenal hyperplasia. Proc Natl Acad Sci USA 1997;94:11,540–11,545.
9. Caron KM, Soo SC, Parker KL. Targeted disruption of StAR provides novel insights into congenital adrenal hyperplasia. Endocr Res 1998;24:827–834.
10. Wehrenberg U, Goedecke Sv, Ivell R, Walther N. The orphan receptor SF-1 binds to the COUP-like element in the promoter of the actively transcribed oxytocin gene. J Neuroendocrinol 1994;6:1–4.
11. Giuili G, Shen WH, Ingraham HA. The nuclear receptor SF-1 mediates sexually dimorphic expression of Mullerian Inhibiting Substance, in vivo. Development 1997;124:1799–1807.
12. Hu Z, Zhuang L, Guan X, Meng J, Dufau ML. Steroidogenic factor-1 is an essential transcriptional activator for gonad-specific expression of promoter I of the rat prolactin receptor gene. J Biol Chem 1997;272:14,263–14,271.
13 Koopman P, Gubbay J, Vivian N, Goodfellow P, Lovell-Badge R. Male development of chromosomally female mice transgenic for Sry. Nature 1991;351:117–121.
14. Yu RN, Ito M, Saunders TL, Camper SA, Jameson JL. Role of Ahch in gonadal development and gametogenesis.
15. Swain A, Narvaez V, Burgoyne P, Camerino G, Lovell-Badge R. Dax1 antagonizes Sry action in mammalian sex determination. Nature 1998;391:761–767.
16. Manova K, Bachvarova RF. Expression of c-kit encoded at the W locus of mice in developing embryonic germ cells and presumptive melanoblasts. Dev Biol 1991;146:312–324.
17. Matsui Y, Zsebo KM, Hogan BLM. Embryonic expression of a haematopoietic growth factor encoded by the Sl locus and the ligand for c-kit. Nature 1990;347:667–669.
18. Coulombre JL, Russell ES. Analysis of the pleiotropism at the W-locus in the mouse: the effects of W and WV substitution upon postnatal development of germ cells. J Exp Zool 1954;126:277–296.
19. Bennett D. Developmental analysis of a mutation with pleiotropic effects in the mouse. J Morphol 1956;98:199–233.
20. Mintz B, Russell ES. Gene-induced embryological modifications of primordial germ cells in the mouse. J Exp Zool 1957;134:207–237.
21. Besmer P, Manova K, Duttlinger R, Huang EJ, Packer A, Gyssler C, et al. The kit-ligand (steel-factor) and its receptor c-kit/W: pleiotropic roles in gametogenesis and melanogenesis. Development (Suppl) 1993:125–137.

22. Beck ARP, Miller IJ, Anderson P, Streuli M. RNA-binding protein TIAR is essential for primordial germ cell development. Proc Natl Acad Sci USA 1998;95:2331–2336.
22a. Schneider-Gaddicke A, Beer-Romano P, Rown LG, Nussbaum R, Pge DC. ZFX has a gene structure similar to ZFY, the putative human sex determinant, and escapes X inactivation. Cell 1989;57:1247–1258.
23. Luoh S-W, Bain PA, Polakiewicz RD, Goodheart ML, Gardner H, Jaenisch R, et al. Zfx mutation results in small animal size and reduced germ cell number in male and female mice. Development 1997;124:2275–2284.
24. Handel MA, Eppig JJ. Sertoli cell differentiation in the testes of mice genetically deficient in germ cells. Biol Reprod 1979;20:1031–1038.
25. Pellas TC, Ramachandran B, Duncan M, Pan SS, Marone M, Chada K. Germ-cell deficient (gcd), an insertional mutation manifested as infertility in transgenic mice. Dev Biol 1991;88:8787–8791.
26. Vainio S, Heikkila M, Kispert A, Chin N, McMahon AP. Female development in mammals is regulated by Wnt-4 signalling. Nature 1999;397:405–409.
27. Parr BA, McMahon AP. Sexually dimorphic development of the mammalian reproductive tract requires Wnt-7a. Nature 1998;395:707–710.
28. Edelmann W, Cohen PE, Kneitz B, Winand N, Lia M, Heyer J, et al. Mammalian MutS homologue 5 is required for chromosome pairing in meiosis. Nat Genet 1999;21:123–127.
29. deVries SS, Baart EB, Dekker M, Siezen A, Rooij DGd, Boer Pd, et al. Mouse MutS-like protein Msh5 is required for proper chromosome synapsis in male and female meiosis. Genes Dev 1999;13:523–531.
30. Pittman DL, Cobb J, Schimenti KJ, Wilson LA, Cooper DM, Brignull E, et al. Meiotic prophase arrest with failure of chromosome synapsis in mice deficient for Dmc1, a germline-specific RecA homolog. Mol Cell 1998;1: 697–705.
31. Yoshida K, Kondoh G, Matsuda Y, Habu T, Nishimune Y, Morita T. The mouse RecA-like gene Dmc1 is required for homologous chromosome synapsis during meiosis. Mol Cell 1998;1:707–718.
32. Edelmann W, Cohen PE, Kane M, Lau K, Morrow B, Bennett S, et al. Meitoic pachytene arrest in MLH1-deficient mice. Cell 1996;85:1125–1134.
33. Chen M, Tomkins DJ, Auerbach W, McKerlie C, Youssoufian H, Liu L, et al. Inactivation of Fac in mice produces inducible chromosomal instability and reduced fertility reminiscent of Fanconi anaemia. Nat Genet 1996;12:448–451.
34. Whitney MA, Royle G, Low MJ, Kelly MA, Axthelm MK, Reifsteck C, et al. Germ cell defects and hematopoietic hypersensitivity to g-interferon in mice with a targeted disruption of the fanconi anemia C gene. Blood 1996;88:49–58.
35. Barlow C, Hirotsune S, Paylor R, Liyanage M, Eckhaus M, Collins F, et al. Atm-deficient mice: a paradigm of ataxia telangiectasia. Cell 1996;86:159–171.
36. Barlow C, Liyanage M, Moens PB, Tarsounas M, Nagashima K, Brown K, et al. Atm deficiency results in severe meiotic disruption as early as leptonema of prophase I. Development 1998;125:4007–4017.
37. Hirshfield AN. Development of follicles in the mammalian ovary. Intl Rev Cytol 1991;124:43–101.
38. Hirshfield AN. Granulosa cell proliferation in very small follicles of cycling rats studied by long-term continuous tritiated-thymidine infusion. Biol Reprod 1989;41:309–316.
39. Peters H, Byskov AG, Himelstein-Braw R, Faber M. Follicular growth: the basic event in the mouse and human ovary. J Reprod Fertility 1975;45:559–566.
40. Liang L-F, Soyal S, Dean J. FIGα, a germ cell specific transcription factor involved in the coordinate expression of the zona pellucida genes. Development 1997;124:4939–4947.
41. Rankin T, Soyal S, Dean j. The mouse zona pellucida: folliculogenesis, fertility, and pre-implantation development. Mol Cell Endocrinol 2000;163:21–25.
42. Liu C, Litscher ES, Mortillo S, Sakai Y, Kinloch RA, Stewart, CL, et al. Targeted disruption of the mZP3 gene results in production of eggs lacking a zona pellucida and infertility in male mice. Proc Natl Acad Sci USA 1996;93:5431–5436.
43. Rankin T, Familari M, Lee E, Ginsberg A, Dwyer N, Blanchette-Mackie J, et al. Mice homozygous for an insertional mutation in the Zp3 gene lack a zona pellucida and are infertile. Development 1996;122:2903–2910.
44. Rankin T, Talbot P, Lee E, Dean J. Abnormal zonae pellucidae in mice lacking ZP1 result in early embryonic loss. Development 1999;126:3847–3855.
45. Perez GI, Robles R, Knudson CM, Flaws JA, Korsmeyer SJ, Tilly JL. Prolongation of ovarian lifespan into advanced chronological age by Bax-deficiency. Nat Genet 1999;21:200–203.

46. Ratts VS, Flaws JA, Kolp R, Sorenson CM, Tilly JL. Ablation of bcl-2 gene expression decreases the numbers of oocytes and primordial follicles established in the post-natal female mouse gonad. Endocrinology 1995;136:3665–3668.
47. Kumar TR, Wang Y, Lu N, Matzuk MM. Follicle stimulating hormone is required for ovarian follicle maturation but not male fertility. Nat Genet 1997;15:201–204.
48. Cattanach BM, Iddon CA, Charlton HM, Chiappa SA, Fink G. Gonadotrphin-releasing hormone deficiency in a mutant mouse with hypogonadism. Nature 1977;269:338–340.
49. Motro B, Bernstein A. Dynamic changes in ovarian c-kit and steel expression during the estrous reproductive cycle. Dev Dyn 1993;197:69–79.
50. Manova K, Huang EJ, Angeles M, DeLeon V, Sanchez S, Pronovost SM, et al. The expression of the c-kit ligand in gonads of mice supports a role for the c-kit receptor in oocyte growth and in proliferation of spermatogonia. Dev Biol 1993;157:85–99.
51. Beechey CV, Loutit JF, Searle AG. Panda, a new steel allele. Mouse News Lett 1986;74:92.
52 Kuroda H, Terada N, Nakayama H, Matsumoto K, Kitamura Y. Infertility due to growth arrest of ovarian follicles in the Sl/Slt mice. Dev Biol 1988;126:71–79.
53. Huang EJ, Manova K, Packer AI, Sanchez S, Bachvarova RF, Besmer P. The murine steel panda mutation affects kit ligand expression and growth of early ovarian follicles. Dev Biol 1993;157:100–109.
54. Bedell MA, Brannan CI, Evans EP, Copeland NG, Jenkins NA. DNA rearrangements located over 100 kb 5' of the Steel (Sl)-coding region in Steel-panda and Steel-contrasted mice deregulate Sl expression and cause female sterility by disrupting ovarian follicle development. Genes Dev 1995;9:455–470.
55. Yoshida H, Takakura N, Kataoka H, Kunisada T, Okamura H, Nishikawa S-I. Stepwise requirement of c-kit tyrosine kinase in mouse ovarian follicle development. Dev Biol 1997;184:122–137.
56. McGrath SA, Esquela AF, Lee S-J. Oocyte-specific expression of growth/differentiation factor-9. Mol Endocrinol 1995;9:131–136.
57. Elvin JA, Clark AT, Wang P, Wolfman NM, Matzuk MM. Paracrine actions of growth differentiation factor-9 in the mammalian ovary. Mol Endocrinol 1999;13:1035–1048.
58. Elvin JA, Yan C, Wang P, Nishimori K, Matzuk MM. Molecular characterization of the follicle defects in the growth differentiation factor-9-deficient ovary. Mol Endocrinol 1999;13:1018–1034.
59. Elvin JA, Yan C. Matzuk MM. Oocyte-expressed TGF-β superfamily members in female fertility. Mol Cell Endocrinol 2000;159:1–5.
60. Dong J, Albertini DF, Nishimori K, Kumar TR, Lu N, Matzuk MM. Growth differentiation factor-9 is required during early ovarian folliculogenesis. Nature 1996;383:531–535.
61. Carabatsos MJ, Elvin JA, Matzuk MM, Albertini DF. Characterization of oocyte and follicle development in growth differentiation factor-9-deficient mice. Dev Biol 1998;203:373–384.
62. Magoffin DA. Regulation of differentiated functions in ovarian theca cells. Sem Reprod Endocrinol 1991;9:321–331.
63. Magarelli PC, Zachow RJ, Magoffin DA. Developmental and hormonal regulation of rat theca-cell differentiation factor secretion in ovarian follicles. Biol Reprod 1996;55:416–420.
64. Everett JW. Pituitary and Hypothalamus: perspectives and overview. In: Knobil E, Neill J, eds., The Physiology of Reproduction, Raven Press, New York, NY, 1994, pp. 1509–1526.
65. Mason AJ, Hayflick JS, Zoeller RT, III, Young WS, Phillips HS, Nikolics K, et al. A deletion truncating the gonadotropin-releasing hormone gene is responsible for hypogonadism in the hpg mouse. Science 1986;234:1366–1371.
66. Radovick S, Wray S, Lee E, Nicols DK, Nakayama Y, Weintraub BD, et al. Migratory arrest of gonadotropin-releasing hormone neurons in transgenic mice. Proc Natl Acad Sci USA 1991;88:3402–3406.
67. Layman LC, Cohen DP, Jin M, Xie J, Li Z, Reindollar RH, et al. Mutations in gonadotropin-releasing hormone receptor gene cause hypogonadotropic hypogonadism. Nat Genet 1998;18:14,15.
68. Rao MC, Midgley AR, Richards JS. Hormonal regulation of ovarian cellular proliferation. Cell 1978;14:71–78.
69. Pedersen,T. Follicle Growth in the Mouse Ovary. In: Bigger JD, Schuetz AW, eds., Oogenesis, University Park Press, Baltimore, MD, 1972, pp. 361–376.
70. Kumar TR, Low MJ, Matzuk MM. Genetic rescue of follicle-stimulating hormone β-deficient mice. Endocrinology 1998;139:3289–3295.
71. Dierich A, Sairam MR, Monaco L, Fimia GM, Gansmuller A, LeMeur M, et al. Impairing follicle-stimulating hormone (FSH) signaling in vivo: Targeted disruption of the FSH receptor leads to aberrant gametogenesis and hormonal imbalance. Proc Natl Acad Sci USA 1998;95:13,612–13,617.

72. Aittomaki K, Lucena JLD, Pakarinen P, Sistonen P, Tapanainen J, Gromoll J, et al. Mutation in the follicle stimulating hormone receptor gene causes hereditary hypergonadotropic ovarian failure. Cell 1995;82:959–968.
73. Beau I, Touraine P, Meduri G, Gougeon A, Desroches A, Matuchansky C, et al. A novel phenotype related to partial loss of function mutations of the follicle stimulating hormone receptor. J Clin Invest 1998;102:1352–1359.
74. Baker J, Hardy MP, Zhou J, Bondy C, Lupu F, Bellvé AR, et al. Effects of an Igf1 gene null mutation on mouse reproduction. Mol Endocrinol 1996;10:903–918.
75. Zhou J, Kumar TR, Matzuk MM, Bondy C. Insulin-like growth factor I regulates gonadotropin responsiveness in the murine ovary. Mol Endocrinol 1997;11:1924–1933.
76. Burks DJ, deMora JF, Schubert M, Winters DJ, et al. IRS-2 pathways integrate female reproduction and energy homeostasis. Nature 2000;407:377–382.
77. Robker RL, Richards JS. Hormone-induced proliferation and differentiation of granulosa cells: a coordinated balance of the cell cycle regulators cyclin D2 and $p27^{KIP1}$. Mol Endocrinol 1998;12:924–940.
78. Elledge SJ. Cell cycle checkpoints: preventing an identity crisis. Science 1996;274:1664–1672.
79. Sicinski P, Donaher JL, Gene Y, Parker SB, Gardner H, Park MY, et al. Cyclin D2 is an FSH-responsive gene involved in gonadal cell proliferation and oncogenesis. Nature 1996;384:470–474.
80. Robker RL, Richards JS. Hormonal control of the cell cycle in ovarian cells: proliferation versus differentiation. Biol Reprod 1998;59:476–482.
81. Vale W, Bilezikjian LM, Rivier C. Reproductive and other roles of inhibins and activins. In: Knobil E, Neill J, eds., The Physiology of Reproduction, Raven Press, New York, NY, 1994, pp. 1861–1878.
82. Meunier H, Cajander SB, Roberts VJ, Rivier C, Sawchenko PE, Hsueh AJW, et al. Rapid changes in the expression of inhibin a-, bA-, and bB-subunits in ovarian cell types during the rat estrous cycle. Mol Endocrinol 1998;2:1352–1363.
83. Matzuk MM, Kumar TR, Shou W, Coerver KA, Lau AL, Behringer RR, et al. Transgenic models to study the roles of inhibins and activins in reproduction, oncogenesis, and development. Recent Prog Hormone Res 1996;51:123–157.
84. Matzuk MM, Finegold MJ, Su J-GJ, Hsueh AJW, Bradley A. a-Inhibin is a tumor-suppressor gene with gonadal specificity in mice. Nature 1992;360:313–319.
85. Coerver KA. Activin function in Cachexia-like syndrome and gonadal tumor development in inhibin-deficient mice. In: Molecular and Human Genetics, Puglisher, Houston, TX, 1996, p. 161.
86. Gershenson DM. Ovarian germ cell and stromal tumors, In: Greer BE and Berek JS, eds., Gynecologic Oncology: Treatment Rationale and Technique. Elevier Publishing, New York, NY, 1991; pp. 167–184.
87. Kumar TR, Palapattu G, Wang P, Woodruff TK, Boime I, Byrne MC., et al. Transgenic models to study gonadotropin function: the role of follicle-stimulating hormone in gonadal growth and tumorigenesis. Mol Endocrinol 1999;13:851–865.
88. Kumar TR, Wang Y, Matzuk MM. Gonadotropins are essential modifier factors for gonadal tumor development in inhibin-deficient mice. Endocrinology 1996;137:4210–4216.
89. Risma KA, Clay CM, Nett TM, Wagner T, Yun J, Nilson JH. Targeted overexpression of luteinizing hormone in transgenic mice leads to infertility, polycystic ovaries, and ovarian tumors. Proc Natl Acad Sci USA 1995;92:1322–1326.
90. Matzuk MM, Kumar TR, Bradley A. Different phenotypes for mice deficient in either activins or activin receptor type II. Nature 1995;374:356–360.
91. Guo Q, Kumar TR, Woodruff T, Hadsell LA, DeMayo FJ, Matzuk MM. Overexpression of mouse follistatin causes reproductive defects in transgenic mice. Mol Endocrinol 1998;12:96–106.
92. Liu X, Andoh K, Abe Y, Kobayashi J, Yamada K, Mizunuma H, Ibuki Y. A comparative study on transforming growth factor-beta and activin A for preantral follicles from adult, immature, and diethylstilbestrol-primed immature mice. Endocrinology 1999;140:2480–2485.
93. Fisher CR, Graves KH, Parlow AF, Simpson ER. Characterization of mice deficient in aromatase (ArKO) because of targeted disruption of the cyp19 gene. Proc Natl Acad Sci USA 1998;95:6965–6970.
94. Britt KL, Drummond AE, Cox VA, Dyson M, Wreford NG, Jones MEE, et al. An age-related ovarian phenotype in mice with targeted disruption of the *Cyp19* (aromatase) gene. Endocrinology 2000;141:2514–2623.
95. Schomberg DW, Couse JF, Mukherjee A, Lubahn DB, Sar M, Mayo KE, Korach KS. Targeted disruption of the estrogen receptor-a gene in female mice: characterization of ovarian responses and phenotype in the adult. Endocrinology 1999;140:2733–2744.

96. Byers M, Kuiper GG, Gustafsson JA, Park-Sarge OK. Estrogen receptor-beta mRNA expression in rat ovary: down-regulation by gonadotropins. Molecular Endocrinology 1997;11:172–182.
97. Lubahn DB, Moyer JS, Golding TS, Couse JF, Korach KS, Smithies O. Alteration of reproductive function but not prenatal sexual development after insertional disruption of the mouse estrogen receptor gene. Proc Natl Acad Sci USA 1993;90:11,162–11,166.
98. Krege JH, Hodgin JB, Couse JF, Enmark E, Warner M, Mahler JF, et al. Generation and reproductive phenotypes of mice lacking estrogen receptor b. Proc Natl Acad Sci USA 1998;95:15,677–15,682.
99. Couse JF, Hewitt SC, Bunch DO, Sar M, Walker VR, Davis BJ, et al. Postnatal sex reversal of the ovaries in mice lacking estrogen receptors a and b. Science 1999;286:2328–2331.
100. Behringer RR, Cate RL, Froelick GJ, Palmiter RD, Brinster RL. Abnormal sexual development in transgenic mice chronically expressing Müllerian-inhibiting substance. Nature 1990;345:167–170.
101. Richards JS, Fitzpatrick SL, Clemens JW, Morris JK, Alliston T, Sirois J. Ovarian cell differentiation: A cascade of multiple hormones, cellular signals, and regulated genes. Recent Prog Hormone Res 1995;50:223–254.
102. Espey LL, Lipner H. Ovulation. In: Knobil E, Neill J, eds., The Physiology of Reproduction, Raven Press, New York, NY, 1994, pp. 725–780.
103. Lee SL, Sadovsky Y, Swirnoff AH, Polish JA, Goda P, Gavrilina G, et al. Luteinizing hormone deficiency and female infertility in mice lacking the transcription factor NGFI-A (Egr-1). Science 1996;273:1219–1221.
104. Topilko P, Schneider-Maunory S, Levi G, Trembleau A, Gourdji D, et al. Multiple pituitary and ovarian defects in *Krox-24 (NGFI-A, EGR-1)*-targeted mice. Mol Endocrinol 1997;12:107–122.
105. Sirois J, Simmons DL, Richards JS. Hormonal regulation of messenger ribonucleic acid encoding a novel isoform of prostaglandin endoperoxide H synthase in rat preovulatory follicles. J Biol Chem 1992;267:11,586–11,592.
106. Dinchuk JE, Car BD, Focht RJ, Johnston JJ, Jaffee BD, Covington MB, et al. Renal abnormalities and an altered inflammatory response in mice lacking cyclooxygenase II. Nature 1995;378:406–409.
107. Lim H, Paria BC, Das SK, Dinchuk JE, Langenbach R, Trzaskos JM, Dey SK. Multiple female reproductive failures in cyclooxygenase 2-deficient mice. Cell 1997;91:197–208.
108. Kennedy CRJ, Zhang Y, Brandon S, Guan Y, Coffee K, Funk CD, et al. Salt-sensitive hypertension and reduced fertility in mice lacking the prostaglandin EP_2 receptor. Nat Med 1999;5:217–220.
109. Tilley SL, Audoly LP, Hicks EH, Kim H-S, Flannery PJ, Coffman TM., et al. Reproductive failure and reduced blood pressure in mice lacking the EP2 prostaglandin E2 receptor. J Clin Invest 1999;103:1539–1545.
110. Hizaki H, Segi E, Sugimoto Y, Hirose M, Saji T, Ushikubi F, et al. Abortive expansion of the cumulus and impaired fertility in mice lacking the prostaglandin E receptor subtype EP(2). Proc Natl Acad Sci USA 1999;96:10,501–10,506.
111. Viggiano JM, Herrero MB, Cebral E, Boquet MG, Gimeno MFd. Prostaglandin synthesis by cumulus-oocyte complexes: effects on in vitro fertilization in mice. Prostaglandins, Leukotrienes and Essential Fatty Acids 1995;53:261–265.
112. Sirois J, Richards JS. Transcriptional regulation of the rat prostaglandin endoperoxide synthase 2 gene in granulosa cells: evidence for the role of a cis-acting C/EBPb promoter element. J Biol Chem 1993;268:21,931–21,938.
113. Sterneck E, Tessarollo L, Johnson PF. An essential role for C/EBPb in female reproduction. Genes Dev 1997;11:2153–2162.
114. Natraj U, Richards JS. Hormonal regulation, localization, and functional activity of the progesterone receptor in granulosa cells of rat preovulatory follicles. Endocrinology 1993;133:761–769.
115. Park-Sarge OK, Mayo KE. Regulation of the progesterone receptor gene by gonadotropins and cyclic adenosine 3',5'-monophosphate in rat granulosa cells. Endocrinology 1994;134:709–718.
116. Lydon JP, DeMayo FJ, Funk CR, Mani SK, Hughes AR, Montgomery CA, et al. Mice lacking progesterone receptor exhibit pleiotropic reproductive abnormalities. Genes Dev 1995;9:2266–2278.
117. Fero ML, Rivkin M, Tasch M, Porter P, Carow CE, Firpo E, et al. A syndrome of multiorgan hyperplasia with features of gigantism, tumorigenesis, and female sterility in p27*Kip1*-deficient mice. Cell 1996;85:733–744.
118. Kiyokawa H, Kineman RD, Manova-Todorova KO, Soares VC, Hoffman ES, Ono M, et al. Enhanced growth of mice lacking the cyclin-dependent kinase inhibitor function of p27*Kip1*. Cell 1996;85:721–732.
119. Tong W, Kiyokawa H, Soos TJ, Park MS, Soares VC, Manova K, et al. The absence of p27Kip1, an inhibitor of G1 cyclin-dependent kinases, uncouples differentiation and growth arrest during the granulosa -> luteal transition. Cell Growth Differ 1998;9:787–794.

120. Simon AM, Goodenough DA, Li E, Paul DL. Female infertility in mice lacking connexin 37. Nature 1997;385:525–529.
121. Colledge WH, Carlton MB, Udy GB,Evans MJ. Disruption of c-*mos* causes parthenogenetic development of unfertilized mouse eggs. Nature 1994;370:65–68.
122. Hashimoto N, Watanabe N, Furuta Y, Tamemoto H, Sagata N, Yokoyama M, Oet al. Parthenogenetic activation of oocytes in c-*mos*-deficient mice. Nature 1994;370:68–71.
123. Ikeda Y, Luo X, Abbud R, Nilson JH, Parker KL. The nuclear receptor steroidogenic factor-1 is essential for the formation of the ventromedial hypothalamic nucleus. Mol Endocrinol 1995;9:478–486.
124. Juneja SC, Barr KJ, Enders GC, Kidder GM. Defects in the germ line and gonads of mice lacking connexin43. Biol Reprod 1999;60:1263–1270.
125. Bergeron L, Perez GI, Macdonald G, Shi L, Sun Y, Jurisicova A, et al. Defects in regulation of apoptosis in caspase-2-deficient mice. Genes Dev 1998;12:1304–1314.
126. Ruggiu M, Speed R, Taggart M, McKay SJ, Kilanowski F, Saunders P, et al. The mouse *Dazla* gene encodes a cytoplasmic protein essential for gametogenesis. Nature 1997;389:73–76.
127. Kohrogi T, Yokoyama M, Taguchi T, Kitamura Y, Tutikawa K. Effect of the *Slt* mutant allele on the production of tissue mast cells in mice. J Hered 1983;74:375–377.
128. Kendall SK, Saunders TL, Jin L, Lloyd RV, Glode LM, Nett TM, et al. Targeted ablation of pituitary gonadotropes in transgenic mice. Mol Endocrinol 1991;5:2025–2036.
129. Kendall SK, Samuelson LC, Saunders TL, Wood RI, Camper SA. Targeted disruption of the pituitary glycoprotein hormone a-subunit produces hypogonadal and hypothyroid mice. Genes Dev 1995;9:2007–2019.
130. Matzuk MM, Dionne L, Guo Q. Kumar TR, Lebovitz RM. Ovarian function in superoxide dismutase 1 and 2 knockout mice. Endocrinology 1998;139:4008–4011.
131. Ho Y-S, Gargano M, Cao J, Bronson RT, Heimler I, Hutz RJ. Reduced fertility in female mice lacking copper-zinc superoxide dismutase. J Biol Chem 1998;273:7765–7769.
132. Good DJ, Porter FD, Mahon KA, Parlow AF, Westphal H, et al. Hypogonadism and obesity in mice with a targeted deletion of the *Nhlh2* gene. Nat Genet 1997;15:397–401.
133. Chehab FF, Lim ME, Lu R. Correction of the sterility defect in homozygous obese female mice by treatment with the human recombinant leptin. Nature Genetics 1996;12:318–320.
134. Matzuk MM, Finegold MJ, Mather JP, Krummen L, Lu H, Bradley A. Development of cancer cachexia-like syndrome and adrenal tumors in inhibin-deficient mice. Proc Natl Acad Sci USA 1994;91:8817–8821.
135. Risma KA, Hirshfield AN, Nilson JH. Elevated luteinizing hormone in prepubertal transgenic mice causes hyperandrogenemia, precocious puberty, and substantial ovarian pathology. Endocrinology 1997;138:3540–3547.
136. Korach KS. Insights from the study of animals lacking functional estrogen receptor. Science 1994;266:1524–1527.
137. Couse JF, Curtis SW, Washburn TF, Lindzey J, Golding TS, et al. Analysis of transcription and estrogen insensitivity in the female mouse after targeted disruption of the estrogen receptor gene. Mol Endocrinol 1995;9:1441–1454.
138. Horseman ND, Zhao W, Montecino-Rodriguez E, Tanaka M, Nakashima K, Engle SJ, et al. Defective mammopoiesis, but normal hematopoiesis, in mice with a targeted disruption of the prolactin gene. EMBO J 1997;16:6926–6935.
139. Rane SG, Dubus P, Mettus RV, Galbreath EJ, Boden G, Reddy EP, et al. Loss of Cdk4 expression causes insulin-deficient diabetes and Cdk4 activation results in b-islet cell hyperplasia. Nat Genet 1999;22:44–52.
140. Ormandy CJ, Camus A, Barra J, Damotte D, Lucas B, Buteau H, et al. Null mutation of the prolactin receptor gene produces multiple reproductive defects in the mouse. Genes and Development 1997;11:167–178.
141. Sugimoto Y, Yamasaki A, Segi E, Tsuboi K, Aze Y, Nishimura T, et al. Failure of parturition in mice lacking the prostaglandin F receptor. Science 1997;277:681–683.
142. Teglund S, McKay C, Schuetz E, Deursen JMV, Stravopodis D, Wang D, et al Stat5a and Stat5b proteins have essential and nonessential, or redundant, roles in cytokine responses. Cell 1998;93:841–850.
143. Topilko P, Schneider-Maunoury S, Levi G, Trembleau A, Gourdji D, Driancourt MA, et al. Multiple pituitary and ovarian defects in Krox-24 (NGFI-A, Egr-1)-targeted mice. Mol Endocrinol 1998;12:107–122.

144. Kuno N, Kadomatsu K, Fan Q-W, Hagihara MN, Senda T, Mizutani S, et al. Female sterility in mice lacking the *basigin* gene, which encodes a transmembrane glycoprotein belonging to the immunoglobulin superfamily. Fed Eur Biochem Sci 1998;425:191–194.
145. Igakura T, Kadomatsu K, Kaname T, Muramatsu H, Fan QW, Miyauchi T, et al. A null mutation in basigin, an immunoglobulin superfamily member, indicates its important roles in peri-implantation development and spermatogenesis. Dev Biol 1998;194:152–165.
146. Xiao X, Zuo X, Davis AA, McMillan DR, Curry BB, Richardson JA, Benjamin IJ. HSF1 is required for extra-embryonic development, postnatal growth and protection during inflammatory responses in mice. EMBO J 1999;18:5943–5952.
147. Robb L, Li R, Hartley L, Nandurkar HH, Koentgen F, Begley CG. Infertility in female mice lacking the receptor for interleukin II is due to a defective uterine response to implantation. Nat Med 1998;4:303–308.
148. Stewart CL, Kaspar P, Brunet LJ, Bhatt H, Gadi I, Kontgen F, Abbondanzo SJ. Blastocyst implantation depends on maternal expression of leukemia inhibitory-function. Nature 1992;359:6–79.
149. Cohen PE, Zhu L, Pollard JW. Absence of colony stimulating factor-1 in osteopetrotic ($csfm^{op}$/ $csfm^{op}$) mice disrupts estrous cycles and ovulation. Biol Reprod 1997;56:110–118.
150. Rigotti A, Trigatti BL, Penman M, Rayburn H, Herz J, Krieger M. A targeted mutation in the murine gene encoding the high density lipoprotein (HDL) receptor scavenger receptor class B type I reveals its key role in HDL metabolism. Proc Natl Acad Sci USA 1997;94:12,610–12,615.
151. Trigatti B, Rayburn H, Vinals M, Braun A, Miettinen H, Penman M, et al. Influence of the high density lipoprotein receptor SR-BI on reproductive and cardiovascular pathophysiology. Proc Natl Acad Sci USA 1999;96:9322–9327.
152. Mahendroo MS, Cala KM, Russell DW. 5a-reduced androgens play a key role in murine parturition. Molecular Endocrinology 1996;10:380–392.
153. Mahendroo MS, Cala KM, Landrum CP, Russell DW. Fetal death in mice lacking 5α-reductase type I caused by estrogen excess. Molecular Endocrinology 1997;11:1–11.

5 Mouse Models to Study the Pituitary-Testis Interplay Leading to Regulated Gene Expression

Emiliana Borrelli, PHD, *T. Rajendra Kumar*, PHD, *and Paolo Sassone-Corsi*, PHD

CONTENTS

INTRODUCTION

The past decade has seen the astounding development of transgenic animal technology, which has become the most powerful tool for the study of gene function and dysfunction in vivo. All fields of biology, including endocrinology, oncogenesis, neuroscience, and embryogenesis, have greatly advanced because of the ease in generating genetically modified animals in an increasing number of research laboratories worldwide. The capacity to explore the function of one specific gene in the living animal has particularly enriched our view of complex physiological systems, such as the neuroendocrine axis. In various cases, mutant mice have been developed to verify the presumptive function of previously studied molecules. In others, the generated mutation has revealed unexpected actions of the targeted gene. This chapter focuses on some mutations affecting the male reproductive axis, as these reveal the high complexity of the system and the interplay between the regulation of gene expression and pituitary signaling. The aim of this chapter is not to provide an exhaustive list of all mice presenting defects in male gametogenesis, but to present a selected number of representative examples.

From: *Contemporary Endocrinology: Transgenics in Endocrinology*
Edited by: M. Matzuk, C. W. Brown, and T. R. Kumar © Humana Press Inc., Totowa, NJ

A large number of transgenic mice present alterations in the differentiation of germ cells. In many cases, the phenotype observed is the result of random insertion of the transgene in the genome, with the consequent inactivation of one or more genes affecting gametogenesis function. Since 10% of all transgenic insertion mutations are associated with male infertility *(1)*, this suggests that a significant proportion of cases of human male infertility may have a genetic origin, and that a large number of genes must be implicated in the process of germ-cell differentiation. The advent of targeted mutagenesis techniques has allowed a precise analysis of the physiological role of gene dysfunction. The use of homologous recombination is now a popular approach. For example, from 1989 to 1995, over 327 genes have been "knocked-out" and the number is growing exponentially every year *(2)*. Tables 1 and 2 summarize some of the knockout mouse models with reproductive defects either only in males or in both males and females.

Gametogenesis is a complex and highly regulated process, during which stem cells undergo multiple steps of differentiation (*see* Chapter 7). This program is under tight control from multiple hormones, many of which originate from the hypothalamo-pituitary axis *(3)*. In the male, spermatogenesis occurs in the seminiferous epithelium as a finely tuned, cyclic process that can be divided into three phases: spermatogonial multiplication, meiosis, and spermiogenesis *(4,5)*. Histological, physiological, and biochemical studies have provided a wealth of information on the mechanisms controlling spermatogenesis, but many components of this differentiation cascade still remain obscure.

Several lines of evidence indicate that highly specialized transcriptional mechanisms ensure stringent stage-specific gene expression in the germ cells. Specific checkpoints correspond to the activation of transcription factors; these regulate gene promoters with a restricted pattern of activity, in a germ-cell-specific fashion *(6)*. There is also evidence that general transcription factors may be differentially regulated in germ cells. For example, TBP (TATA-binding protein) accumulates in early haploid germ cells at much higher levels than in any other somatic-cell type. It has been calculated that adult spleen and liver cells contain 0.7 and 2.3 molecules of TBP mRNA per haploid genome-equivalent, respectively, while adult testis contain 80–200 molecules of TBP transcript per haploid genome-equivalent *(7)*. In addition to TBP, TFIIB and RNA polymerase II were also found to be overexpressed in testis. These remarkable features are consistent with the potent transcriptional activity that occurs in a coordinated manner during the differentiation of germ cells.

PITUITARY HORMONES

Spermatogenesis is under the hormonal control of the hypothalamo-pituitary axis *(3)*. Gonadotropin-releasing hormone (GnRH) is released from the hypothalamus into the hypothalamo-pituitary vein, and stimulates the release of follicle-stimulating hormone (FSH) and luteinizing hormone (LH) from the gonadotroph cells of the anterior pituitary. FSH and LH bind to receptors located respectively on the somatic Sertoli and Leydig cells of the testis. Pituitary gonadotropins (FSH and LH) and thyroid-stimulating hormone thyrotropin (TSH) are heterodimers composed of a common α-subunit (α-GSU) and a unique β-subunit.

Disruption of the gene for the α-subunit has resulted in mice that did not produce FSH, LH, or TSH. The homozygous mutant mice are hypogonadal, and suffer from severe hypothyroidism resulting in dwarfism *(8)*. The α-subunit-deficient male mice are infer-

Table 1

Knockout Models with Reproductive Defects Only in Males

Knockout mouse model	*Major reproductive findings*	*Refs*
Acrosin	Delayed fertility, normal binding and penetration of ZP by sperm	*(85)*
Ahch (Dax1)	Infertile; progressive degeneration of the testicular germinal epithelium	*(86)*
Angiotensin-converting enzyme	Reduced fertility caused by decreased ability of sperm to fertilize ova	*(87)*
Apaf-1	Only 5% of the mutants survive to adulthood; males infertile; spermatogonial degeneration	*(88)*
Apolipoprotein B heterozygotes	Reduced fertility; spermatozoa fertilization defects	*(89)*
Bax	Infertile; spermatogenesis block at premeiotic stage	*(61)*
Bclw	Infertile; spermatogenesis block during late spermatogenesis; eventual loss of all germ cells and Sertoli cells	*(90)*
Bone morphogenetic protein 8A	Progressive infertility; germ-cell degeneration, spermiogenesis defects, and epididymis degeneration	*(91)*
Bone morphogenetic protein 8B	Infertile; germ-cell proliferation/depletion defects	*(92)*
BRCA-1, p53 double mutant	Infertile; meiotic failure	*(93)*
Calmegin	Infertility due to impairment of sperm binding to ZP	*(94)*
Casein kinase II alpha'	Infertile; oligospermia and globozoospermia	*(95)*
c-ros tyrosine kinase receptor	Infertile; defect in volume regulatory mechanism in mature sperm, sperm flagellar angulation	*(96)*
CREM	Infertile; block at first stage of spermiogenesis	*(75,76)*
Cyclic GMP-dependent kinase 1	Reduced fertility, failure of corpora cavernosa to relax on activation of the NO/cGMP signaling	*(97)*
Cyclin A1	Infertile; block of spermatogenesis before first meiotic division, increased germ-cell apoptosis	*(65)*
Cyritestin	Infertile; failure of sperm to bind to the ZP	*(98)*
Desert hedgehog	Infertile; defects in germ cell development	*(99)*
Fertilin b	Infertile; defects in sperm-egg adhesion, fusion and ZP binding	*(100)*
Fragile X mental retardation 1 (FMR1)	Normal fertility, macroorchidism due to increased embryonic Sertoli-cell proliferation	*(101)*
GDNF	Infertile; depletion of stem-cell reserves	*(102)*
Hormone-sensitive lipase	Infertile; vacuolated epithelial cells in tubules, oligospermia	*(103)*
Hoxd-13	Defects in formation of the seminal vesicles, ventral and dorsal prostate, and bulbourethral gland	*(104)*
HR6B Ubiquitin-conjugating enzyme	Infertile; possible defect in histone poly ubiquitination and degradation during spermatogenesis	*(105)*

(continued)

Table 1 *(continued)*

Knockout mouse model	*Major reproductive findings*	*Refs*
Hsp70	Infertile; block at meiotic prophase and increased spermatocyte apoptosis	*(82)*
Inhibin/MIS double mutants	Granulosa-/Sertoli-cell tumors; Leydig-cell neoplasia; large fluid-filled uteri; complete infertility	*(32)*
INK4d	Marked testicular atrophy, increased germ-cell apoptosis, although fertile	*(106)*
JunD	Infertile; hormonal imbalance, abnormalities in sperm head and flagellum	*(107)*
Müllerian-inhibiting substance	Uteri in males causes obstruction and secondary infertility in majority of mice	*(18)*
Müllerian-inhibiting substance receptor	Partial fertility; presence of Müllerian duct causing physical blockage	*(108)*
Na(+)-K(+)-2Cl(-) cotransporter	Infertile; reduced spermatids, defects in epididymal transport of sperm	*(109)*
Osp-11 (Claudin-11)	Infertile; absence of intramembranous tight junctions between Sertoli cells	*(110)*
Ovo	Reduced fertility; hypogenitalism	*(111)*
P2X1 receptor	Reduced fertility, oligospermia, defects in contraction of the vas deferens	*(112)*
PC4	Infertile; impaired fertilizing ability of spermatozoa	*(113)*
Pi3'-kinase (cKit receptor-induced)	Infertile; decreased proliferation and enhanced apoptosis of spermatogonial cells	*(114)*
PMS2 DNA mismatch repair enzyme	Infertile; meiosis defects leading to abnormal spermatozoa	*(115)*
Protein phosphatase 1cγ	Infertile; defects in spermiogenesis	*(116)*
Retinoic acid receptor β	Male infertility secondary to germ-cell mutation defects and tubular degeneration	*(46)*
Retinoic acid receptor γ	Male sterility secondary to squamous metaplasia of the seminal vesicles and prostate	*(117)*
Retinoid receptor α	Male infertility secondary to seminiferous tubule degeneration	*(44)*
SCP3	Infertile; defects in synapsis, apoptosis during meiotic prophase	*(118)*
Sp4	Infertility caused by defects in male reproductive behavior	*(119)*
Sperm-1	Subfertile despite normal testicular morphology and sperm number	*(120)*
TLS	Infertile; defects in pairing in premeiotic spermatocytes	*(121)*
Tnp1	Reduced fertility; abnormal pattern of chromatin condensation and a severe reduction in sperm motility	*(122)*
Tyro-3 family receptors	Infertile; progressive death of differentiating germ cells, absence of mature sperm	*(123)*
Vasa	Infertile; defects in proliferation and differentiation of primordial germ cells, absence of sperm in the testes	*(124)*

Table 2

Knockout Models with Reproductive Defects in Both Sexes

Knockout mouse model	*Major reproductive findings*	*Refs*
α-inhibin	Infertility in females; secondary infertility in males; granulosa-/Sertoli-cell tumors; cachexia-like syndrome	*(27,30)*
Activin receptor type II	Infertility in females; delayed fertility in males; small gonads	*(125)*
A-myb	Male infertility; pachytene stage arrest of germ cells; nursing defects in females due to underdevelopment of mammary glands	*(126)*
Ataxia telangiectasia (Atm)	Male and female infertility; complete absence of germ cells	*(127,128)*
β1,4-Galactosyltransferase	Male and female infertility caused by abnormal glycoprotein hormone glycosylation	*(129)*
Centromere protein B	Males hypogonadal, decreased sperm number strain-dependent uterine defects in females, disrupted luminal and glandular epithelium in the uterus, reduced fertility	*(130,131)*
Cyclin D2	Female infertility secondary to a block in folliculogenesis; males fertile but with decreased testis size	*(64)*
Cyp 19	Progressive infertility, spermiogenic defects, Leydig-cell hyperplasia in males, females infertile, increased follicular atresia prior to ovulation, defects in mammary gland development	*(132,133)*
Dazla	Male and female infertility; loss of germ cells and complete absence of gamete production	*(134)*
Dmc1	Arrest of spermatogenesis at zygotene stage in males; no oocytes in the adult ovary	*(135,136)*
Emx2	Accelerated degeneration of Wolffian-duct and mesonephric tubules without the formation of the Müllerian duct	*(137)*
ERβ	No defect in male fertility, prostate hyperplasia in old males; decreased fertility in females	*(50)*
Estrogen receptor α (ERα)	Uterine/ovarian defects in females; small testes, reduced number of spermatozoa in males	*(47,49,51)*
Estrogen receptor α/β double knockout	Male phenotypes similar to ERα mice, sex-reversal at the gonad level in females	*(51)*
Follicle-stimulating hormone β subunit	Female infertility; folliculogenesis block prior to antral follicle stage; males fertile but with decreased testis size	*(9)*
Glycoprotein hormone α-subunit	Infertile; hypogonadal and hypothyroid	*(8)*
Hoxa 11	Partial homeotic transformation of vas deferens to epididymis; failure of testicular descent; absence of uterine stromal, decidual, and glandular cells in females	*(138)*

(continued)

Table 2 *(continued)*

Knockout mouse model	*Major reproductive findings*	*Refs*
Hoxa10	Variable infertility in males and females caused by cryptorchidism and preimplantation embryonic loss, respectively	*(139)*
Insulin-like growth factor (IGF-1)	Hypogonadal and infertile; pre-antral block in folliculogenesis in females	*(140)*
MLH1 DNA mismatch repair enzyme	Male and female infertility; Defective meiosis at pachytene stage (males) and failure to complete meiosis II (females)	*(141)*
Msh5	Male and female infertility; defects in zygotene stage in both sexes, characterized by impaired and aberrant chromosome synapsis, apoptotic cell death	*(142)*
Neuronal helix-loop-helix 2 (Nhlh2)	Males infertile; females fertile only in presence of males; hypothalamic defect	*(143)*
$p27^{Kip1}$ CDK inhibitory protein	Female infertility; corpus luteum defects; males fertile and increased testis size	*(144–146)*
Prolactin receptor	Female infertility caused by multiple abnormalities including irregular estrous cycles and implantation defects; males infertile or subfertile of unknown origin	*(147)*
Rho GDIalpha	Male and female infertility; impaired spermatogenesis with vacuolar degeneration of seminiferous tubules in males; postimplantation defects in females	*(148)*
Telomerase	Progressive infertility in males and females; increased apoptosis in testicular germ cells, and reduced testis size; decreased number of oocytes and uterine abnormalities	*(149)*
TIAR	Infertility; complete absence of primordial germ cells by E 13.5 leading to absence of spermatogonia and oogonia	*(150)*
Zfx	Reduced germ-cell number in both sexes resulting from defective proliferation	*(151)*

tile and exhibit prepubertal external genitalia. The testes are severely reduced in size, but the epididymis and vas deferens are present. The presence of normal prepubertal genitalia support the hypothesis that the differentiation of these structures from the Wolffian duct is testosterone-dependent. LH stimulates testosterone secretion from Leydig cells, and serum-testosterone concentrations are severely reduced in α-GSU$^{-/-}$ mice. However, these low concentrations of testosterone are still sufficient to induce sexual differentiation. Histological examination of α-GSU$^{-/-}$ testis showed that the seminiferous tubules are reduced in diameter, and that spermatogenesis is blocked at the first meiotic division. Therefore, these results indicate that gonadotropins are necessary for postnatal testicular differentiation, but that testis development proceeded normally during the fetal period. This view has been validated by other mouse models.

Mice carrying a targeted mutation in the FSHβ subunit gene have been generated *(9)*. Mutant females are infertile because of a block in folliculogenesis prior to antral-follicle formation. Importantly, and in contrast to the classical view of the FSH requirement for spermatogenesis and Sertoli-cell growth, FSHβ-deficient males are fertile, despite their small testes. The critical role played by FSH signaling is illustrated by the effect of FSH-R mutations in humans *(10–12)*. An inactivating mutation (Ala189Val) found in females with pure ovarian dysgenesis leads to a disease characterized by normal karyotype, high gonadotropins, and streaky gonads associated with primary amenorrhea. More recently, additional mutations have been described (Asp224Val and Leu601Val) that are associated to a similar pathological condition *(13)*. These mutations lie either in the extracellular domain (Ala189Val and Asp224Val) or in the third extracellular loop (Leu601Val) of the FSH-R, and are believed to modify protein folding. Importantly, males with the Ala189Val mutation display various degrees of spermatogenic failure, without azoospermia or absolute infertility *(14)*. Thus, the same inactivating mutation differentially influences reproduction in males or females.

A more recent approach, aimed at altering FSH signaling at the target tissue, has been to mutate the gene encoding the FSH receptor *(15)*. Similar to the FSHβ mutant mice, FSH-R-deficient males display small testes, partial spermatogenic failure, and reduced fertility. Thus, it appears that FSH signaling is not essential for initiating spermatogenesis, but is required to sustain adequate viability and motility of the sperm. The phenotype of mutant females is much more severe. These display thin uteri and small ovaries, and are sterile as a result of a block in folliculogenesis prior to antral-follicle formation. Drastic changes have been found in pituitary hormone levels, especially FSH, which is increased 15- to 20-fold in females and about threefold in males. This dramatic increase in FSH levels verifies the classical view of FSH signaling retroinhibition, underscoring the apparent simplicity of the system in which no alternative retroinhibitory routes seem to be activated by the lack of FSH signaling *(15)*.

Additional hormonal changes include a significant decrease in the levels of testosterone in the males. This result indicates that low testoterone levels are sufficient to sustain sex accessories, and indicates a link between FSH signaling and testosterone production. This link could involve an intracellular communication pathway that would be compromised, despite normal LH levels, in the FSH-R mutants. At the level of the pituitary gland, there is a moderate but significant enlargement in the anterior lobe, accompanied by a drastic increase of FSH-positive cells. These animals have been considered as possible models for the study of the physiological link between gonads and pituitary, and hypergonadotropic ovarian dysgenesis and infertility.

A mouse model to study aberrant LH signaling has not yet been developed. However, mice with a targeted mutation in the LH-receptor gene would constitute an invaluable tool to explore the link between pituitary-hormonal signaling and sexual differentiation. In this respect, it is important to note that homozygous missense mutations in the LH-R gene in humans are tightly associated with male pseudohermaphroditism *(16)*.

TRANSFORMING GROWTH FACTOR-β FAMILY

Müllerian-Inhibiting Substance

During mammalian embryogenesis, the Müllerian ducts have the potential to differentiate into the oviducts, uterus, and upper vagina of the female reproductive tract, while the Wolffian ducts differentiate into the vas deferens, epididymis, and seminal vesicles in the male. During male development, the Sertoli cells of the testis produce Müllerian-inhibiting substance (MIS), a protein that actively represses the differentiation of the Müllerian ducts and prevents the development of female reproductive organs *(17)*. Subsequently, testosterone produced by the Leydig cells induces the differentiation of the Wolffian ducts into male external genitalia. The MIS protein is a member of the transforming growth factor-β (TGFβ) gene superfamily which also includes the activin and inhibin genes.

The MIS gene has been mutated in the mouse by homologous recombination. The testis of $MIS^{-/-}$ males descend normally, their size is normal, and the Wolffian duct system differentiates properly *(18)*. Histological analysis of the testis shows no obvious anomalies, and there is no apparent difference in the spermatogenesis from wild-type and $MIS^{+/-}$ males. However, these males also develop Müllerian-duct-derived tissues such as a uterus, oviducts, and a vagina. About 85% of the males are infertile. Although MIS is not necessary for normal germ-cell development, the infertility of $MIS^{-/-}$ males probably results from a diversion of the sperm from its normal pathway. Finally, MIS appears to play an antitumor role in the testis, because about 25% of MIS-deficient males develop Leydig-cell hyperplasia and neoplasia.

Activin and Inhibin

Activins and inhibins are members of the TGFβ gene superfamily. Inhibins were initially isolated from mammalian follicular fluid on the basis of their ability to inhibit the release of FSH from anterior pituitary cells *(19)*. Side fractions that activated FSH release led to the isolation of activins. Activins and inhibins are protein dimers composed of a unique α-subunit and common but homologous βA or βB subunits. In adult animals, the highest levels of activin and inhibin transcripts are found in the Sertoli cells of the testis and the granulosa cells of the ovaries *(20–22)*. Interestingly, recent evidence points to the regulation of inhibin α gene expression by some isoforms of CREM *(23)*. In the testis, activin stimulates spermatogonial proliferation and androgen biosynthesis by Leydig cells, whereas inhibin has the opposite activity. Activin interacts with type I and type II cell-surface receptors, which have serine/threonine kinase activity. Two type II activin receptors––activin-receptor type IIA (ActRIIA) and activin-receptor type IIB (ActRIIB)—have been identified *(24)*. The ActRIIA isoform has been detected in specific populations of male germ cells, suggesting that this hormone can act as a gonadal paracrine and/or autocrine regulator *(20,25,26)*.

A previous study indicates that α-inhibin functions as a tumor suppressor with gonadal specificity. Male mice with a targeted mutation of the α-inhibin gene initially appeared healthy, and had normal external genitalia. However, male mutants were sterile, and analysis of the testis showed the presence of mixed or incompletely differentiated gonadal stromal tumors *(27)*. Spermatogenesis was initially active in the first 5–7 wk of life, but regression was evident with the enlargement of the tumor masses. Serum FSH concentrations were elevated two- to threefold in the homozygous mutants as compared to wild-type littermates, which confirmed the inhibitory role of inhibin on FSH release. A regulatory mechanism between activin and inhibin has been described by showing that $\alpha\text{-inhibin}^{-/-}$ mice have a 200-fold overexpression of the activin βA subunit and a threefold reduction of ActRIIA mRNA transcript in the testis *(28)*. These results suggest that inhibin is not necessary for the normal differentiation of embryonic gonads and spermatogenesis, but plays an important autocrine and/or paracrine function as a tumor suppressor in the gonads.

Activin-βA-deficient mice develop to term, but die within 24 h. These mice present multiple cranio-facial deformities, such as a lack of whiskers and lower incisors *(29)*. Mice with a mutation in the ActRcIIA gene have also been generated *(24)*, and were expected to possibly mimic the phenotype of the activin-$\beta A^{-/-}$ mice. Although some $ActRcIIA^{-/-}$ mice suffered from cranio-facial anomalies, most of them developed into adults. The ActRcIIA-deficient mice suffered from a complete suppression in FSH synthesis, and their reproductive ability was altered. Male $ActRcIIA^{-/-}$ mice reached puberty later than wild-type animals, but their stages of spermatogenesis were seemingly normal. Seminiferous-tubule diameter and volume were reduced, which could result from an overall decrease in the number of Sertoli cells. Since activin stimulates FSH synthesis in the pituitary, the small testis of the $ActRcIIA^{-/-}$ animals may result from the decreased levels of FSH, although though LH levels were normal. Mice with a double mutation of either the α-inhibin and activin βB genes or of the α-inhibin and ActRcIIA genes also develop testicular tumors *(30)*.

Disruption of the common activin/inhibin βB subunit gene produces mice deficient in activin B, activin AB, and inhibin B *(31)*. Homozygous mutants present defects in eyelid development, and $\beta B^{-/-}$ females manifest a profoundly impaired reproductive ability. However, the reproductive organs from $\beta B^{-/-}$ males appeared normal, and these mice bred normally.

MIS and Inhibin

The gonadal tumors of inhibin-deficient mice are first detected around 4 wk of age. It could be hypothesised that MIS, which is synthesized in the same gonadal cells as inhibin, may act as a tumor suppressor at this early stage. It has been demonstrated that inhibin/MIS double-mutant male mice develop testicular tumors at an even earlier age. These tumors are different from those observed in either α-$inhibin^{-/-}$ or $MIS^{-/-}$ mice *(32)*. They grow more rapidly, are less hemorrhagic, and produce less estradiol as compared to the tumors of inhibin-deficient male mice. These results suggest that inhibin and MIS synergize to function as gonadal-tumor suppressors.

NUCLEAR RECEPTORS

Retinol Receptors

Vitamin A, or retinol, is absolutely essential for spermatogenesis. Rats fed a vitamin A-deficient diet become sterile and show a drastic reduction in testis weight *(33,34)*. In animals on a vitamin A-deficient diet, only Sertoli cells and spermatogonia are apparent, while meiotic and postmeiotic germ cells degenerate.

There are two known classes of retinol receptors, retinoic-acid receptors (RARα, β, and γ) that bind 13-*cis*-retinoic acid, and all-*trans*-retinoic acid *(35,36)*, and the retinoic X receptors (RXRα, β and γ), which have a high affinity for 9-*cis*-retinoic acid *(37)*. All retinoid receptors are member of the nuclear-receptor superfamily. The RARα, RXRα, and RXRβ isoforms are widely expressed in the embryo and adult tissues *(38–41)*, while the expression of the RARγ gene is restricted to the skin. These receptors can homodimerize and heterodimerize with a wide variety of nuclear receptors, and are believed to be responsible for a wide variety of signaling pathways *(42)*.

The RARα gene encodes two isoforms. The major isoform, RARα1, is expressed ubiquitously, while the RA-inducible isoform RARα2 has a more restricted expression pattern *(43)*. Knocking out the RARα2 gene has no apparent deleterious effect since RARα2-deficient mice are healthy and fertile. However, high postnatal lethality is observed in mice homozygous for a mutation of the entire RARα gene *(44)*. Although most RARα$^{-/-}$ mutants survive at birth, the survival rate at 24 h of age is 40%, and only 12% of homozygous animals are alive after 1–2 mo. Mice surviving over 2 mo of age appear normal, but none of the males are able to sire any litters. Histological analysis of the testis has showed severe degeneration of the germinal epithelium, although some of the tubules appear normal. Vacuolization of the Sertoli cells is evident, and cytoplasmic expansion of these cells often partially fills the lumen. The epidymal duct appears normal, but contains very few spermatozoa. The degeneration observed in RXRα$^{-/-}$ testis is almost identical to that observed in animals kept on a vitamin A-deficient diet. This observation suggests that retinoic acid, and not retinol, is required for the maintenance of spermatogenesis. This hypothesis is supported by the observation that high doses of retinoic acid can restore the germ-cell degeneration induced by the vitamin A-deficient diet *(45)*.

The RXRβ gene is expressed mostly in Sertoli cells. Targeted mutation of the RXRβ gene *(46)* reveals that approx 50% of RXRβ$^{-/-}$ mutants die in utero or shortly after birth, for unknown reasons. Homozygote females are fertile, but RXRβ-deficient males are sterile. Histological analysis of the epididymis of males RXRβ$^{-/-}$ mice shows low levels of spermatozoa, most of which (95%) remains immotile. A majority of spermatozoa from RXRβ-deficient mice exhibit a coiling of the tail, and 30% have an acrosome that is indented or partially detached from the nuclear envelope. It has been suggested that the high frequency of such defects in mutant spermatozoa results from an impaired attachment of the acrosomal membrane to the nucleus. The diameter of the seminiferous tubules is normal, and the proportion and length of stages of the cycle is apparently normal. However, some of the late spermatids fail to align at the luminar side of the tubules. Moreover, remnants of spermatid heads are located inside the cytoplasm of the Sertoli cells. In young animals, lipid droplets are apparent in the cytoplasm of Sertoli cells—droplets which become more and more apparent as the animal gets older. In 6-mo-old RXRβ$^{-/-}$ animals, the lipid droplets are larger than the Sertoli-cell nuclei, and the tubule have a reduced diameter with variable degrees of cell loss. By the age of 12 mo over one-half of the tubules are replaced by tubular ghosts consisting of a thickened and convoluted basement membrane filled with lipids. These results demonstrate that disruption of RXRβ gene results in alterations of Sertoli-cell function and underscore the crucial role played by retinols in the germ-cells differentiation process.

Estrogen Receptors

Two types of estrogen receptors exist, ERα and ERβ. These receptors have significant homology, and both belong to the steroid-receptor superfamily *(47,48)*. These receptors bind 17β-estradiol, the female-sex steroid that plays a critical role in female sexual development. Targeting of the ERα has been achieved, while a full description of the anatomical and physiological features of mice mutated for ERβ is ongoing *(49,50)*. Surprisingly, inactivation of ERα gene affects male fertility. The mutant males exhibit impaired sexual behavior, including decreased intromission and ejaculation. ERα$^{-/-}$ males have low fertility with reduced testis size, and a 90% reduction in sperm number. In contrast to the severe reproductive phenotypes of ERα-deficient male mice, ERβ-deficient male mice are fertile and have no testicular defects. Older males demonstrate epithelial hyperplasia in the prostate and the bladder *(50)*. Double-mutant male mice that are deficient in both ERα and ERβ are infertile *(51)* and essentially demonstrate the ERα-deficient phenotypes, i.e., infertility and reduction in sperm number and motility.

PROLIFERATION AND APOPTOSIS

The p53 Protein

In contrast to the great body of information on the regulation of the mitotic-cell cycle, much less is known about the molecular mechanisms involved in the regulation of meiosis. Yet, the stringent control in the differentiation and proliferation timing of the germ cells indicates the presence of critical checkpoints. An analysis of cyclin-dependent kinases (cDKs) has revealed that their expression occurs at specific steps of the meiotic-cell cycle, and suggests a role in the differentiation program *(52)*. Normally, a proportion of germ cells undergo apoptosis in the seminiferous epithelium. This number increases dramatically in some pathological conditions, including idiopathic infertility caused by spermatogenic arrest in human males. Interestingly, somatic-cell-cycle regulation by the Cdk family of genes is modulated by the pleiotropic action of the tumor suppressor p53 *(53)*.

The activity of the p53 protein has been associated with both apoptosis and cell differentiation *(53–55)*. The pattern of expression of the p53 gene has been studied in mice by developing transgenic animals in which the bacterial gene encoding chloramphenicol acetyltransferase (CAT) is under the control of the p53-gene promoter *(56)*. This study showed that the testis is the organ with the highest levels of p53-gene expression in the adult. In the testis, p53-gene expression is mostly restricted to primary spermatocytes at the pachytene phase of meiosis, just before they develop into haploid spermatids. A giant cell testicular degeneration syndrome is occasionally seen in transgenic mice carrying a p53-promoter-CAT fusion, which results in decreased p53 levels in the testes *(57)*. These mice seem to recapitulate a degenerative syndrome, probably resulting from the inability of the tetraploid primary spermatocytes to complete meiotic division. The severity of this syndrome is variable from one strain to another, and can be correlated to the reduction in p53 levels. The role of p53 in the regulation of the cell cycle and in the apoptotic pathway is further emphasized by the finding of the robust overexpression of wild-type p53 protein in most testicular tumors *(58)*.

The mouse p53 gene is knocked out in ES cells and susbsequently p53-deficient mice are generated *(59)*. Mutant animals initially appear healthy and fertile. However, most p53-deficient animals die by 3–6 mo of age from multiple neoplasms. Spermatogenesis appears normal, although giant cells can be found in some tubules. Testicular section of p53-deficient mice of the 129 background exhibit a high incidence of seminomas and undifferentiated teratocarcinomas. In fact, this strain of mice seems to have a profound effect on the testicular phenotype, because $p53^{-/-}$ mice of pure 129 strain are sterile, whereas p53-deficient mice of other strains and mixed background are fertile *(57)*.

The Bax Protein

A direct connection with the germ-cell apoptotic pathway is provided by the finding that p53 acts as a transcriptional activator of the Bax gene *(60)*. Bax is a dimerization partner of Bcl-2, a protein with the potential to interfere with programmed cell death in response to a number of apoptotic stimuli. Importantly, Bax has an opposing function to Bcl-2—its ablility to accelerate apoptosis. The deletion of the Bax gene by homologous recombination results in a complete block of the spermatogenesis (*see* Chapter 6). Bax-deficient male mice are infertile, with atrophied testes and empty epididymis and vas deferens. Histological analysis of the seminiferous tubules shows spermatogenic arrest, accumulation of premeiotic spermatocytes, abnormal mitotic or meiotic figures, and multinucleated giant cells. Round spermatids are rare, and elongated spermatids are completely absent. Electron microscopy analysis has also revealed a disordered maturation scheme. Premeiotic cells have an atypical distribution of decondensed chromatin, and an irregular size and shape not typical of spermatogonia or preleptotene spermatocytes. Flow cytometry of testicular cells shows an elevated number of 2N cells, which reflects the abnormal premeiotic-cell expansion noted in the testicular section. A small proportion of 1N cells representing the round spermatids is present, but the more condensed 1N population represented by elongating spermatids and spermatozoa is completely absent. Multinucleated and pyknotic cells are also present in the testes of Bax-deficient mice, suggesting increased apoptosis in this tissue. Thus, in Bax-null mice, the premeiotic germ cells appear atypical, and instead of differentiating, enter a pathway of programmed cell death *(61)*. Deletion of the genes encoding the Bax partners Bcl-2 and Bcl-x reveals massive cell death in the lymphoid-cell lineage, but does not cause germ-cell aberrations *(62,63)*. Interestingly, Bax-null mice also show lymphoid hyperplasia *(61)*. Thus, the same molecule may act positively or negatively in the apoptotic pathway of different cell lineages.

Cyclin D2

The D-type cyclins D1, D2, and D3 are critical controllers of the G1 phase of the mammalian cell cycle. The three cyclins are expressed in overlapping, apparently redundant fashion in various proliferating tissues. Homologous recombination at the cyclin D2-gene locus has been achieved in mouse ES cells (64). Male mice carrying a cyclin-D2 mutated gene are fertile but display hypoplastic testes, suggesting a role for this cyclin in the regulation of testis growth. Interestingly, the expression of the cyclin D2 gene, and not of the D1 or D3 genes, was found to be FSH-inducible via the cAMP-dependent pathway. This finding indicates that the regulation of the various cyclin genes is under the control of various intracellular signaling pathways. The hypoplasia observed in testes correlates with the observation that some human testicular tumors display a high cyclin-D2 expression.

Cyclin A1

There are two mammalian members of cyclin A family—cyclin A1 and cyclin A2. Cyclin A2 exhibits a more widespread expression, and cyclin A2-deficient mice die embryonically. In contrast, cyclin A1 is expressed extensively in the male germ-cell lineage in mice *(65)*. Cyclin A1-deficient male mice are infertile because of a block in spermatogenesis before the first meiotic division. This is accompanied by increased germ-cell apoptosis defects in desynapsis, and a 80% reduction in cdc2 kinase activation at the end of the meiotic prophase *(65)*. The residual cyclin B1/cdc2 activity in the complete absence of cyclin A1 has been found to be insufficient for the progression of meiotic divisions. Thus, cyclin A1 represents a novel male meiotic lineage-specific class of cyclins.

TRANSCRIPTION FACTOR CREM

A Molecular Master-Switch

Cyclic AMP-dependent signaling is known to play an important role in spermatogenesis *(2)*. The receptors for LH and FSH are coupled to Gs proteins, and activate the adenylyl cyclase, further leading to an activation of the cAMP-dependent protein kinase A (PKA). Many genes that are expressed during spermatogenesis possess cAMP-responsive elements (CRE) in their promoter *(66)*. These sequences are targets for CRE-binding proteins, such as CREB or CREM, that can be phosphorylated and activated by the PKA as well as other signaling pathways *(67)*. CRE-binding proteins are relatively ubiquitous and uninducible *(68)*. However, in adult male germ cells, the activator CREM is expressed at levels that are hundreds of times higher than those in other tissues *(69)*. The CREM activator has been shown to function as a regulator of gene expression in haploid cells. CREM levels are regulated during germ-cell differentiation and by the FSH-signaling pathway *(70)*. CREM proteins are expressed in haploid spermatids, where they activate multiple genes, such as transition proteins, angiotensin-converting enzyme (ACE), calspermin, and cholesterogenic lanosterol 14α-demethylase (CYP51) *(66,71–74)*. A summary of cAMP signaling in the testis and expression patterns of key regulators of this pathway is schematically represented in Fig. 1.

The crucial role of CREM during spermatogenesis has been confirmed by its targeted mutation in the mouse *(75,76)*. In male CREM$^{+/-}$ mice, there is a 50% reduction in the number and motility of spermatazoa. There is also a twofold increase in the number of spermatozoa with an aberrant structure. In CREM$^{-/-}$ animals, the females are fertile, but the males are sterile and produce no spermatozoa. Histological analysis of seminiferous tubules reveals a complete arrest of spermatogenesis at the first stage of spermiogenesis (Fig. 2). A 10-fold increase in apoptotic germ cells is also observed, and in many cases, these apoptotic bodies acquire the shape of multinucleated giant cells *(75)*. Finally, serum concentrations of LH, FSH, and testosterone are not reduced in CREM-deficient males, indicating that the phenotype observed in these animals does not result from hormonal alterations.

ACT, A Testis-Specific Coactivator

Crucial steps in transcriptional activation by factors of the CREB/CREM class are phosphorylation at a specific serine regulatory site and the subsequent recruitment of the coactivator, CBP (CREB-binding protein) *(65,66)*. Thus, the phosphorylation event is considered to be the key event, leading to transcriptional activation in response to induction of a specific signaling route. Surprisingly, CREM is found to be unphosphorylated in male germ cells. Thus, activation by CREM must occur independently of phosphorylation, and therefore of binding of CBP. A yeast two-hybrid screen of a testis-derived cDNA library, using the CREM activation domain as bait, has revealed the presence of a novel protein, activator of CREM in testis (ACT) *(77)*. The distinctive feature of ACT is the presence of four complete LIM motifs and another half motif at the N-terminus. LIM domains comprise a conserved cysteine- and histidine-rich structure that forms two adjacent zinc fingers. This structural motif was first identified in the protein products of three genes, *Lin-11*, *Isl-1*, and *Mec-3*. The LIM domain functions as a protein-protein interaction domain. LIM domains can be present with other functional protein motifs, such as homeobox and kinase domains, but ACT belongs to the class of the LIM-only proteins (LMO), and contains no other structural motif.

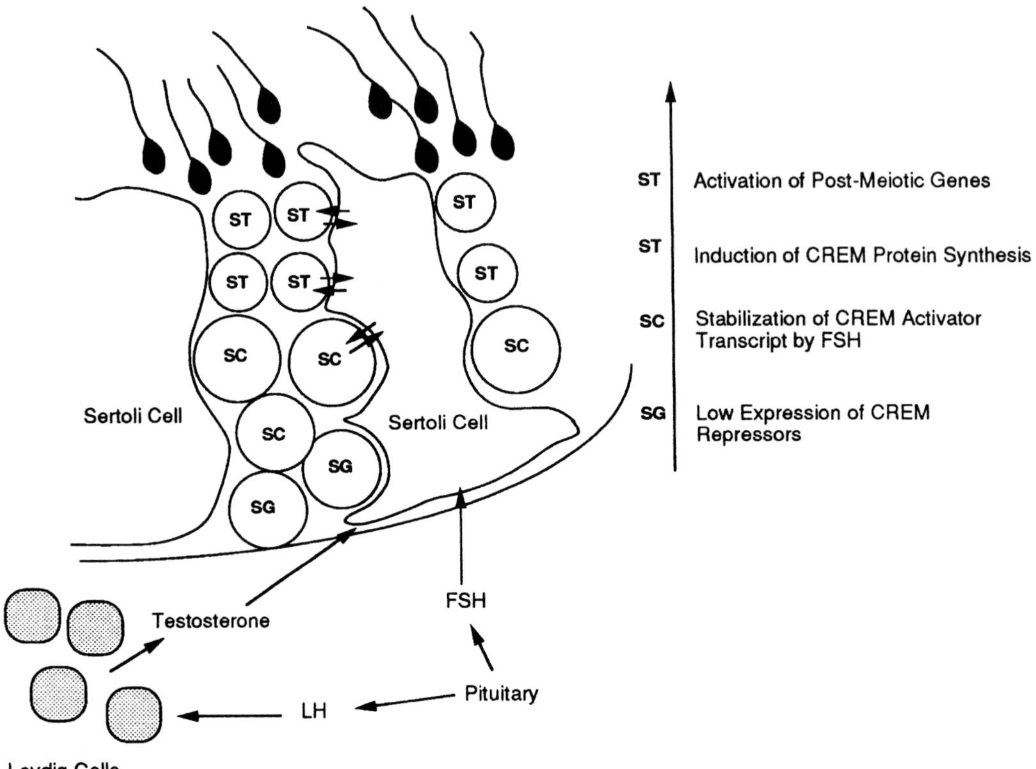

Fig. 1. Schematic representation of a section of a seminiferous tubule where the CREM expression patteren is indicated. CREM expression is regulated at multiple levels during spermatogenesis. Premeiotic germ cells spermatogonia (SG) express a low level of CREM repressor isoforms. During meiotic prophase, the pituitary follicle-stimulating hormone (FSH) is responsible for the stabilization of CREM activator transcripts in spermatocytes (SC); CREM protein, on the other hand, is detected only after meiosis in haploid spermatids (ST). Note the strict relationships between the Sertoli and germ cells *(arrows)*. In the haploid spermatids, CREM proteins activate a number of cellular genes expressed specifically during spermtid maturation.

Several lines of evidence point to the coordinated expression of CREM and ACT. ACT is abundantly and exclusively expressed in testis; ACT colocalizes with CREM in spermatids; and ACT and CREM exhibit the same expression pattern during testis development. CREM and ACT efficiently associate; the biological significance of this is that ACT has an intrinsic transactivation capacity and can convert CREM into a powerful transcriptional activator *(77)* (Fig. 3). Most importantly, co-activation through ACT can occur also in yeast, which lacks CBP and TAF130 homologs. Thus, ACT can bypass the need for CREM or CREB phosphorylation. Indeed, ACT converts an inactive CREM mutant (with the serine phosphoacceptor site mutated into alanine) into a transcriptionally active molecule, both in yeast and in mammalian cells. Thus, in male germ cells, ACT provides a novel, tissue-specific phosphorylation-independent route for transactivation by members of the CREB family *(77)*. A general model of CREM interacting with the general transcriptional machinery is depicted in Fig. 4.

Wild Type

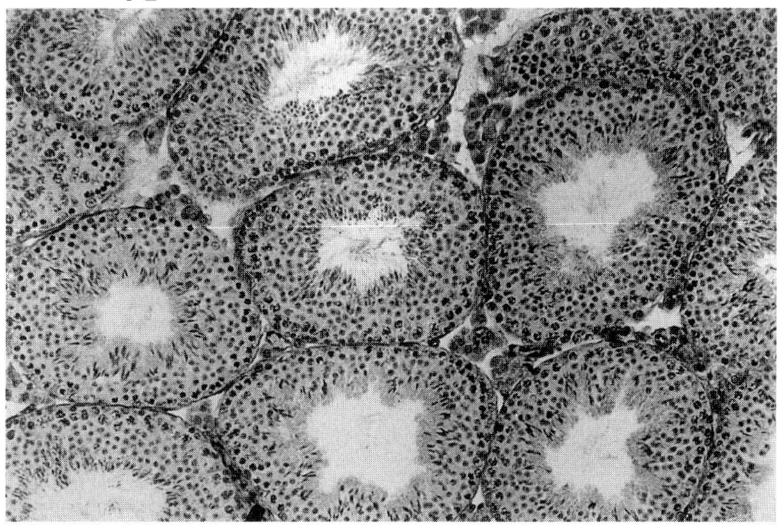

CREM -/-

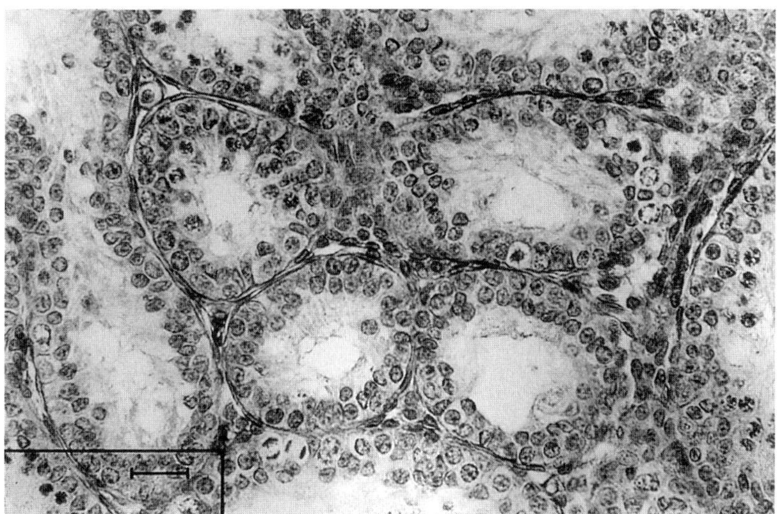

Fig. 2. CREM deficiency causes spermiogenesis arrest and make germ-cell apoptosis. Testes from a 8-wk-old homozygous mutant (–/–) and a wild-type (+/+) mouse littermate. Histological analysis of testis sections. The tubules from the CREM-deficient mice show impared spermatogenesis and some multinucleated apoptotic cells.

CHAPERONE HSP70-2

Members of the 70-kDa heat shock protein (HSP70) family are chaperones that assist in the folding, transport, and assembly of protein in the cytoplasm, mitochondria, and endoplasmic reticulum *(78)* HSP70-2 is a testis-specific gene that is expressed at high levels in pachytene spermatocytes during the meiotic phase of spermatogenesis *(79,80)*. The developmentally regulated expression of HSP70-2 during spermatogenesis implies

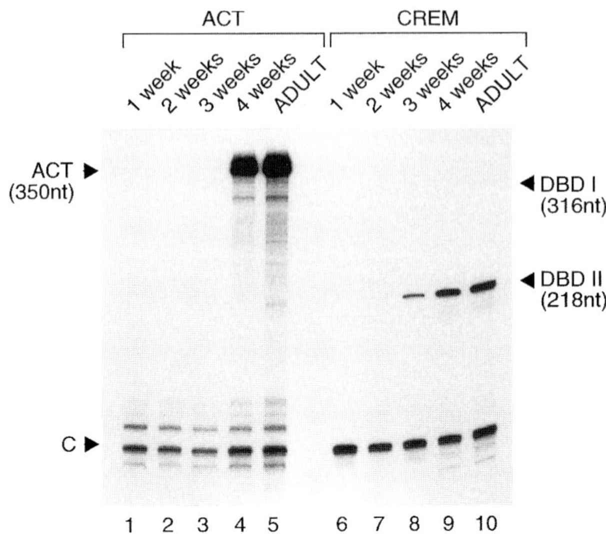

Fig. 3. ACT is exclusively in testis. Co-expression of ACT and CREM druring testis development. RNA was extracted from testes of mice at different ages and analyzed by RNase protection assay, using ACT- and CREM-specific riboprobes. C indicates a b-actin protected frgament used as an intenal control. DBD I and DBD II refer to the two alternative DNA-binding domains of CREM.

that it performs a specialized function during meiosis. This protein has been identified as a component of the synaptonemal complex in prophase nuclei of spermatogenic cells *(81)*.

HSP70-2 mutant males produce no spermatozoa and are infertile *(82)*. Spermatogonia and pachytene spermatocytes, mainly with aberrant structures, are present in HSP70-$2^{-/-}$ testis, but postmeiotic spermatids are completely absent. Pachytene spermatocytes with condensed nuclei are observed, and there is a major increase in the level of apoptotic cells in HSP70-2 mutant testis. Although typical-appearing synaptonemal complexes are observed in pachytene cells from HSP70-$2^{-/-}$ testis, synaptonemal complex development beyond the middle to late pachytene stages is not observed *(82)*. These observations suggest that HSP70-2 is not necessary for synaptonemal complex assembly, but is required during synapsis, which allows progression to the subsequent meiotic divisions.

CONCLUSIONS

The use of genetically modified mice has brought a wealth of information on the genetic control of gametogenesis, yet additional questions have arisen. Much more will be revealed by the homologous recombination approach in reproductive biology as many other animal models will be generated. Importantly, not all gene inactivations believed to influence the germ-cell differentiation program have led to the anticipated sterile phenotype. For example, the normal fertility of the acrosin-mutant mice suggests that this endoprotease is not essential for sperm penetration of the oocyte zona pellucida (ZP) or fertilization *(83)*. Important considerations include the finding that analysis of testicular function is often complicated by deleterious or lethal consequences of a specific gene inactivation. On the other hand, it is evident that many crucial elements are involved in the regulation of gametogenesis, some of which have not been considered.

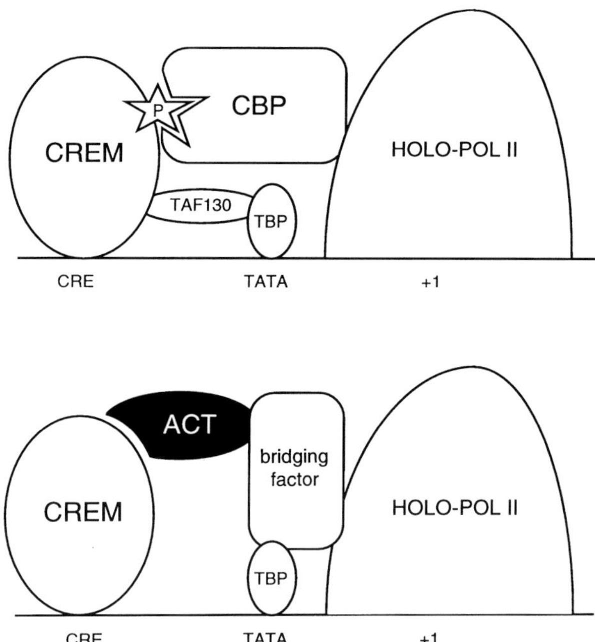

Fig. 4. CREM-mediated transcription is promoted by interaction with different co-activators. (Top) A schematic representation of the classical view by which, through interaction with CBP, activators as CREB and CREM elicit their function. A key event in this scenario is phosphorylation at Ser117 P, since it is required for binding to CBP and subsequent transcriptional activation. Interaction with TAF130 is constitutive, and occurs via the Q2 domain of CREB/CREM. (Bottom) Representation of how ACT may elicit its co-activator function via interaction with CREM. In yeast, CREB and CREM are inactive because of the lack of CBP and TAF130. ACT elicits its function and interacts with CREM, also in the absence of Ser117 phosphorylation. Thus, ACT provides an alternative activation pathway that appears to work in a signaling-independent manner. A hypothetical bridging factor, linking ACT to the basal transcription machinery, is represented.

Special attention should be given to the pathway of programmed cell death of germ cells. Very little is known, as little research has been done on the meiosis cycle as compared to the mitotic cell cycle. The increased proportion of apoptotic bodies in many mutated animals with testicular alterations indicates that apoptosis must play an important, but poorly defined, role in the spermatogenic cascade. Further studies will focus on the precise role played by well-known mitotic apoptosis-related proteins, and possibly on the discovery of novel, meiosis-specific, cell-death molecules. One interesting approach has been the screen of lines of mutant mice created using a retroviral gene-trap system for male infertility *(84)*. This approach has led to the finding that Bclw-deficient mice have testicular degeneration *(84)*.

Analysis of some of the mutant mice suggests previously unrecognized relationships. Of particular interest is the very close testicular phenotype observed in the CREM, HSP70-2, and BAX-deficient mice, suggesting that an interplay of these genes may place them on the same, or related, signaling cascades. Future work will take advantage of multiple mutations, and of conditional homologous recombination, to remarkably improve our understanding of gametogenesis.

ACKNOWLEDGMENTS

We would like to thank all the members of our laboratories for discussions. Research in the laboratories of Emiliana Borrellu and Paola Sassone-Corsi is supported by grants from CNRS, INSERM, CHUR, FRM, La Ligue and the Association pour la Recherche sur le Cancer. T. Rajendra Kumar acknowledges financial support from The Moran Foundation, Texas.

REFERENCES

1. Al Shawi R, Burke J, Bishop JO, Mullins JJ, Sharpe RM, Lathe R, et al. Transgenesis and infertility. In: Hiller SG, ed., Gonadal Development and Function . Raven Press, New York, NY, 1992; pp. 195–206.
2. Brandon EP, Idzerma RL, McKnight GS. Targeting the mouse genome: a compendium of knock-outs. Curr Biol 1995;5:625–634.
3. Veldhuis JD. The hypothalamic-pituitary-gonadal axis. In: Yen SSC, Jaffe RB, eds., Reproductive Endocrinology. Saunders, Philadelphia, PA, 1991; pp. 409–459.
4. Skinner MK. Cell-cell interactions in the testis. Endocrine Rev 1991;12:45–77.
5. Parvinen M. Regulation of the seminiferous epithelium. Endocrine Rev 1992;13:404–417.
6. Sassone-Corsi P. Transcriptional checkpoints determining the fate of male germ cells. Cell 1997;88:163–166.
7. Schmidt EE, Schibler U. High accumulation of components of the RNA polymearse II transcription machinery in rodent spermatids. Development 1995;121:2373–2383.
8. Kendall SK, Samuelson LC, Saunders TL, Wood RI, Camper SA. Targeted disruption of the pituitary glycoprotein hormone a-subunit produces hypogonadal and hypothyroid mice. Genes Dev 1995;9:2007–2019.
9. Kumar TR, Wang Y, Lu N, Matzuk M. Follicle stimulating hormone is required for ovarian follicle maturation but not male fertility. Nature Genet 1997;15:201–204.
10. Aittomäki K, Herva R, Stenman U, Juntunen K, Ylöstalo P, Hovata O, et al. Clinical features of primary ovary failure caused by a point mutation in the follicle stimulating hormone receptor gene. J Clin Endocr Metab 1996;81:3722–3726.
11. Aittomäki K, Dieguez Lucena JL, Pakarinen P, Sistonen P,Tapanainen J, Lehväslaiho H, et al. Mutation in the follicle-stimulating hormone receptor gene causes hereditary hypergonadotropic ovarian failure. Cell 1995;82:959–968.
12. Gromoll J, Simoni M, Nieschlag E. An activating mutation of the follicle-stimulating hormone receptor autonomously sustains spermatogenesis in a hypophysectomized man. J Clin Endocr Metab 1996;81:1367–1370.
13. Touraine P, Beau I, Gougeon A, Meduri G, Desroches A, Pichard C, et al. New natural inactivating mutations of the follicle-stimulating hormone receptor: correlations between receptorfunction and phenotype. Mol Endocrinol 1999;13:1844–1854.
14. Tapanainen JS, Aittomaki K, Min J, Vaskivuo T, Huhtaniemi IL. Men homozygous for an inactivating mutation of the follicle-stimulating hormone (FSH) receptor gene prosent variable suppression of spermatogenesis and fertility. Nature Gen 1997;15:205,206.
15. Dierich A, Sairam MR, Monaco L, Fimia GM, Gansmuller A, LeMeur M, et al. Impairing follicle-stimulating hormone signaling in vivo: targeted disruption of the FSH receptor leads to aberrant gametogenesis and hormonal imbalance. Proc Natl Acad Sci USA 1998;95:13,612–13,617.
16. Kremer H, Kraaij R, Toledo S, Post M, Fridman J, Hayashida C, et al. Male pseudoheramphroditism due to a homozygous missense mutation of the luteinizing hormone receptor gene. Nature Genet 1995;9:160–164.
17. Cate RL, Donahoe PK, MacLaughlin J. Müllerian-inhibiting substance. In: Sporn MB, Roberts AB, eds., Peptide Growth Factors andTheir Receptors, Vol. 2. Springer-Verlag, Berlin, Germany, 1990; pp. 179–210.
18. Behringer RR, Finegold MJ, Cate RL. Müllerian-inhibiting substance funstion during mammalian sexual development. Cell 1994;79:415–425.
19. Vale W, Hsueh A, Rivier C, Yu J. The inhibin/activin family of hormones and growth factors. In: Sporn MB, Roberts AB, eds., Peptide Growth Factors and their Receptors: Handbook of Experimental Pharmacology. Springer-Verlag, Berlin, Germany, 1990; pp. 211–248.

20. Mather JP, Woodruff TK, Krummen LA. Paracrine regulation of reproductive function by inhibin and activin. Proc Soc Exp Biol Med 1992;201:1–15.
21. Moore A, Krummen LA, Mather JP. Inhibins, activins, their binding proteins and receptors: interactions underlying paracrine activity in the testis. Mol Cell Endocrinol 1994;100:81–86.
22. Jaffe RB, Spencer SJ, Rabinovici J. Activins and inhibins: gonadal peptides during prenatal development and adult life. Ann NY Acad Sci 1993;687:1–9.
23. Mukherjee A, Urban J, Sassone-Corsi P, Mayo KE. Gonadotropins regulate inducible cyclic adenosine 3',5'-monophosphate early repressor in the rat ovary: implications for inhibin α subunit gene expression. Mol Endocrinol 1998;12:785–800.
24. Matzuk MM. Functional analysis of mammalian members of the transforming growth factor-b. Trends Endocrinol Metab 1995;6:120–127.
25. de Jong F, Grootenhuis AJ, Klaij IA, Van Beurden W. Inhibin and related proteins: localization, regulation and effects. Adv Exp Med Biol 1990;274:271–293.
26. Findlay JK. An update on the roles of inhibin, activin and follistatin as regulators of folliculogenesis. Biol Reprod 1993;48:15–23.
27. Matzuk MM, Finegold MJ, Su JJ, Hsueh AJW, Bradley A. α-inhibin is a tumor suppressor gene with gonadal specificity in mice. Nature 1992;360:313–319 .
28. Trudeau VL, Matzuk MM, Haché RJG, Renaud V. Overexpression of activin-βA subunit mRNA is associated with decreased activin type II receptor mRNA levels in the testes of α-inhibin deficient mice. Biochem Biophys Res Comm 1994;203:105–112.
29. Matzuk MM, Kumar TR, Vassali A, Bickenbach JR., Roop DR, Jaenisch R, et al. Functional analysis of activins during mammalian development. Nature 1995;374:354–356.
30. Coerver KA, Woodruff TK, Finegold MJ, Mather J, Bradley A, Matzuk MM. Activin signaling through activin receptor type II causes the cachexia-like symptoms in inhibin-deficient mice. Mol. Endocrinol. 1996;10:534–543.
31. Vassali A, Matzuk MM, Gardner HAR, Lee KF, Jaenisch R. Activin/inhibin βB subunit gene disruption leads to defects in eyelid development and female reproduction. Genes Dev 1994;8:414–427.
32. Matzuk MM, Finegold MJ, Mishina Y, Bradley A, Behringer RR. Synergistic effects of inhibins and Müllerian-inhibiting substance on testicular tumorigenesis. Mol Endocrinol 1995;9:1337–1345.
33. Huang HFS, Hembree WC. Spermatogenic response to vitamin A in vitamin A deficient rats. Biol Reprod 1979;21:891–904.
34. Morales CR, Griswold MD. Retinol-induced stage synchronization in seminiferous tubules of the rat. Endocrinology 1987;121:432–434.
35. Giguère V, Ong ES, Segui P, Evans RM. Identification of a receptor for the morphogen retinoic acid. Nature 1987;330:624–629.
36. Petkovich M, Brand NJ, Krust A, Chambon P. A human retinoic acid receptor which belongs to the family of nuclear receptors. Nature 1987;330:444–450.
37. Heyman RA, Mangelsdorf DJ, Dyck JA, Stein RB, Eichele G, Evans RM., et al. 9-cis retinoic acid is a high affinity ligand for the retinoid X receptor. Cell 1992;68:397–406.
38. Dollé P, Fraulob V, Kastner P, Chambon P. Developmental expression of murine retinois X receptor (RXR) genes. Mech Dev 1994;45:91–104.
39. Mangelsdorf DJ, Borgmeyer U, Heyman R, Zhou JY, Ong E, Oro A, et al. Characterization of three RXR genes that mediate the action of 9-cis retinoic acid. Genes Dev 1992;6:329–344.
40. Leroy P, Krust A, Zelent A, Mendelson C, Garnier JM, Kastner P, et al. Multiple isoforms of the mouse retinoic acid receptor alpha are generated by alternative splicing and differential induction by retinoic acid. EMBO J 1991;10:59–69.
41. Ruperte E, Dollé P, Chambon P, Morriss-Kay G. Retinoic acid receptors and cellular retinoid binding proteins. II. Their differential pattern of transcription during early morphogenesis in mouse embryos. Development 1991;111:45–60.
42. Mangelsdorf DJ, Thummel C, Beato M, Herrlich P, Schutz G, Umesono K, et al. The nuclear receptor superfamily: the second decade. Cell 1995;83:835–839.
43. Leroy P, Nakshatri H, Chambon P. Mouse retinoic acid receptor alpha 2 isoform is transcribed from a promoter that contains a retinoic acid response element. Proc Natl Acad Sci USA 1991;88:10,138–10,142.
44. Lufkin T, Lohnes D, Mark M, Dierich A, Gorry P, Gaub M-P, et al. High postnatal lethality and testis degeneration in retinoic acid receptor a mutant mice. Proc Natl Acad Sci USA 1993;90:7225–7229.
45. Van Pelt AM, De Rooij DG. Retinoic acid is able to reinitiate spermatogenesis in vitamin A-deficient rats and high replicate doses support the full development of spermatogenic cells. Endocrinology 1991;128:697–704.

46. Kastner P, Mark M, Leid M, Gansmuller A, Chin W, Grondona JM, et al. Abnormal spermatogenesis in RXR beta mutant mice. Genes Dev 1996;10:80–92.
47. Korach KS. Insights from the study of animals lacking functional estrogen receptor. Science 1994;266:1524–1527.
48. Kuiper GG, Enmark E, Pelto-Huikko M, Nilsson S, Gustafsson JA. Cloning of a novel estrogen receptor expressed in rat prostate and ovary. Proc Natl Acad Sci USA 1996;93:5925–5930.
49. Lubahn DB, Moyer JS, Golding TS, Couse JF, Korach KS, Smithies O. Alteration of reproductive funstion but not prenatal sexual development after insertional disruption of the mouse estrogen receptor gene. Proc Natl Acad Sci USA 1993;90:11,162–11,166.
50. Ogawa S, Chan J, Chester AE, Gustafsson JA, Korach KS, Pfaff DW. Survival of reproductive behaviors in estrogen receptor beta gene-deficient (betaERKO) male and female mice. Proc Natl Acad Sci USA 1996;96:12,887–12,892.
51. Couse JF, Hewitt SC, Bunch DO, Sar M, Walker VR, Davis BJ, et al. Postnatal sex reversal of the ovaries in mice lacking estrogen receptors alpha and beta. Science 1999;286:2328–2331.
52. Rhee K, Wolgemuth DJ. Cdk family genes are expressed not only in dividing bu also in terminally differentiated mouse germ cells, suggesting their possible function during both cell division and differentiation. Dev Dyn 1995;204:406–420.
53. Donehower LA, Bradley A. The tumor suppressor p53. Biochim Biophys Acta 1993;1155:181–205.
54. Levine A. The tumor supressor genes. Annu Rev Biochem 1993;62:623-651.
55. Ko LJ, Prives C. p53: puzzle and paradigm. Genes Dev 1996;10:1054–1072.
56. Almon ET, Goldfinger N, Kapon A, Schwartz D, Levine AJ, Rotter V. Testicular tissue-specific expression of the p53 suppressor gene. Dev Biol 1993;156:107–116.
57. Rotter V, Schwartz D, Almon E, Goldfinger N, Kapon A, Meshorer,A, et al. Mice with reduced levels of p53 protein exibit the testicular giant-cell degenerative syndrome. Proc Natl Acad Sci USA 1993;90:9075–9079.
58. Chresta CM, Hickman JA Oddball p53 in testicular tumors. Nature Med 1996;2:744–745.
59. Donehower LA, Harvey M, Slagle BL, McArthur MJ, Montgomery CA, Butel JS, et al. Mice deficient for p53 are developmentally normal but susceptible to spontaneous tumors. Nature 1992;356:215–221.
60. Miyashita T, Reed JC. Tumor suppressor p53 is a direct transcriptional activator of the human Bax gene. Cell 1995;80:293–299.
61. Knudson CM, Tung KSK, Tourtellotte WG, Brown GAJ, Korsmeyer SJ. Bax-deficient mice with lymphoid hyperplasia and male germ cell death. Science 1995;270:96–99.
62. Motoyama N, Wang F, Roth KA, Sawa H, Nakayama K, Nakayama K, et al. Massive cell death of immature hematopoietic cells and neurons in Bcl-x-deficient mice. Science 1995;267:1506–1510.
63. Nakayama K, Nakayama K, Negishi I, Kuida K, Shinkai Y, Louie MC, et al. Disappearance of the lymphoid system in Bcl-2 homozygous mutant chimeric mice. Science 1993;261:1584–1588.
64. Sicinsky P, Donaher J, Geng Y, Parker S, Gardner H, Park MY, et al. Cyclin D2 is an FSH-responsive gene involved in gonadal cell proliferation and oncogenesis. Nature 1996;384:470–474.
65. Liu D, Matzuk MM, Sung WK, Guo Q, Wang P, Wolgemuth DJ. Cyclin A1 is required for meiosis in the male mouse. Nat Genet 1998;20:377–380.
66. Delmas V, van der Hoorn F, Mellström B, Jégou B, Sassone-Corsi P. Induction of CREM activator proteins in spermatids: downstream targets and implications for haploid germ cell differentiation. Mol Endocrinol 1993;7:1502–1514.
67. De Cesare D, Fimia GM, Sassone-Corsi P. Signaling routes to CREM and CREB: plasticity in transcriptional activation. Trends Biochem 1999;24:281–285.
68. Sassone-Corsi P. Transcription factors responsive to cAMP. Annu Rev Cell Dev Biol 1995;11:355–377.
69. Foulkes NS, Mellström B, Benusiglio E, Sassone-Corsi P. Developmental switch of CREM function during spermatogenesis: from antagonist to transcriptional activator. Nature 1992;355:80–84.
70. Foulkes NS, Schlotter F, Pévet P, Sassone-Corsi P. Pituitary hormone FSH directs the CREM functional switch during spermatogenesis. Nature 1993;362:264–267.
71. Kistler MK, Sassone-Corsi P, Kistler SW. Identification of a functional cAMP-response element in the 5'-flanking region of the gene for transition protein 1 (TP1), a basic chromosomal protein of mammalian spermatids. Biol Reprod 1994;51:1322–1329.
72. Sun Z, Sassone-Corsi P, Means A. Calspermin gene transcription is regulated by two cyclic AMP response elements contained in an alternative promoter in the calmodulin kinase IV gene. Mol Cell Biol 1995;15:561–571.

73. Zhou Y, Sun Z, Means AR, Sassone-Corsi P, Bernstein KE. CREMt is a positive regulator of testis ACE transcription. Proc Natl Acad Sci USA 1996;93:12,262–12,266.
74. Rozman D, Fink M, Fimia GM, Sassone-Corsi P, Waterman MR. Cyclic Adenosine 3', 5'-monophosphate (cAMP)/cAMP-responsive element modulator (CREM)-dependent regulation of cholesterogenic lanosterol 14a-demethylmase (CYP51) in spermatids. Mol Endocrinol 1999;13:1951–1999.
75. Nantel F, Monaco L, Foulkes NS, Masquilier D, LeMeur M, Henriksén K, et al. Spermiogenesis deficiency and germ cell apoptosis in CREM-mutant mice. Nature 1996;380:159–162.
76. Blendy J, Kastner K, Weinbauer G, Nieschlag F, Schutz G. Severe impairement of spermatogenesis in mice lacking the CREM gene. Nature 1996;380:163–165.
77. Fimia GM, De Cesare D, Sassone-Corsi P. CBP-independent activation of CREM and CREB by the LIM-only protein ACT. Nature 1999;398:165–169.
78. Georgopoulos C, Welch WJ. Role of major heat shock proteins as molecular chaperones. Annu Rev Cell Biol 1993;9:601–634.
79. Allen RL, O'Brien DA, Eddy EM. A novel hsp 70-like protein (P70) is present in mouse spermatogenic cells. Mol Cell Biol 1988;8:828–832.
80. Zakeri ZF, Wolgemuth DJ, Hunt CR. Identification and sequence analysis of a new member of the mouse HSP70 gene family and characterization of its unique cellular and developmental pattern of expression in the male germ line. Mol Cell Biol 1988;8:2925–2932.
81. Allen JW, Dix DJ, Collins BW, Merrick BA, He C, Selkirk JK, et al. HSP70-2 is part of the synaptonemal complex in mouse and hamster spermatocytes. Chromosoma 1996;104:414–421.
82. Dix DJ, Allen JW, Collins BW, Mori C, Nakamura N, Poorman-Allen P, et al. Targeted gene disruption of Hsp 70-2 results in failed meiosis, germ cell apoptosis, and male infertility. Proc Natl Acad Sci USA 1996;93:3264–3268.
83. Baba T, Azuma S, Kashiwabara S, Toyoda Y. Sperm from mice carrying a targeted mutation of the acrosin gene can penetrate the oocyte zona pellucida and effect fertilization. J Biol Chem 1994;269:31,845–31,849.
84. Ross AJ, Waymire KG, Moss JE, Parlow AF, Skinner MK, Russell LD., et al. Testicular degeneration in Bclw-deficient mice. Nature Genet 1998;18:251–261.
85. Adham IM, Nayernia K, Engel W. Spermatozoa lacking acrosin protein show delayed fertilization. Mol Reprod Dev 1997;46:370–376.
86. Yu RN, Ito M, Saunders TL, Camper SA, Jameson J. Role of Ahch in gonadal development and gametogenesis. Nat Genet 1998;20:353–357.
87. Krege JH, John SW, Langenbach LL, Hodgin JB, Hagaman JR, et al. Male-female differences in fertility and blood pressure in ACE-deficient mice. Nature 1995;375:146–148.
88. Honarpour N, Du C, Richardson JA, Hammer RE, Wang X, Herz J. Adult Apaf-1-deficient mice exhibit male infertility. Dev Biol 2000;218:248–258.
89. Huang L-S, Voyiaziakis E, Chen HL, Rubin EM, Gordon JW. A novel functional role for apolipoprotein B in male infertility in heterozygous apolipoprotein B knockout mice. Proc Natl Acad Sci USA 1996;93:10,903–10,907.
90. Ross AJ, Waymire KG, Moss JE, Parlow AF, Skinner MK, Russell LD, et al. Testicular degeneration in Bclw-deficient mice. Nat Genet 1998;8:251–256.
91. Zhao G-Q, Liaw L, Hogan BLM. Bone morphogenetic protein 8A plays a role in the maintenance of spermatogenesis and the integrity of the epididymis. Development 1998;125:1103–1112.
92. Zhao G-Q, Deng K, Labosky PA, Liaw L, Hogan BLM. The gene encoding bone morphogeneetic protein 8B is required for the initiation and maintenance of spermatogenesis in the mouse. Genes Dev 1996;10:1657–1669.
93. Cressman VL, Backlund DC, Avrutskaya AV, Leadon SA, Godfrey V, Koller BH. Growth retardation, DNA repair defects, and lack of spermatogenesis in BRCA1-deficient mice. Mol Cell Biol 1999;19:7061–7075.
94. Ikawa M, Wada I, Kominami K, Watanabe D, Toshimori K, Nishimune Y, et al. The putative chaperone calmegin is required for sperm fertility. Nature 1997;387:607–611.
95. Xu X, Toselli PA, Russell LD, Seldin DC. Globozoospermia in mice lacking the casein kinase II alpha' catalytic subunit. Nat Genet 1999;23:118–121.
96. Yeung CH, Sonnenberg-Riethmacher E, Cooper TG. Infertile spermatozoa of c-ros tyrosine kinase receptor knockout mice show flagellar angulation and maturational defects in cell volume regulatory mechanisms. Biol Reprod 1999;61:1062–1069.
97. Hedlund P, Aszodi A, Pfeifer A, Aim P, Hofmann F, Ahmad M, et al. Erectile dysfunction in cyclic GMP-dependent kinase I-deficient mice. Proc Natl Acad Sci USA 2000;97:2349–2354.

98. Shamsadin R, Adham IM, Nayernia K, Heinlein UA, Oberwinkler H, Engel W. Male mice deficient for germ-cell cyritestin are infertile. Biol Reprod 1991;61:1445–1451.
99. Bitgood MJ, Shen L, McMahon AP. Sertoli cell signaling by Desert hedgehog regulates the male germline. Curr Biol 1996;6:298–304.
100. Cho C, Bunch DO, Faure JE, Goulding EH, Eddy EM, Primakoff P, et al. Fertilization defects in sperm from mice lacking fertilin beta. Science 1998;281:1857–1859.
101. Bakker CE, Verheij CE, Willemsen R, van der Helm R, Oerlemans F, Vermeij M, et al. Fmr1 knockout mice: a model to study fragile X mental retardation. Cell 1994;78:23–33.
102. Meng X, Lindahl M, Hyvonen ME, Parvinen M, Rooij DGd, Hess MW, et al. Regulation of cell fate decision of undifferentiated spermatogonia by GDNF. Science 2000;287:1489–1493.
103. Osuga J, Ishibashi S, Oka T, Yagyu H, Tozawa R, Fujimoto A, et al. Targeted disruption of hormone-sensitive lipase results in male sterility and adipocyte hypertrophy, but not in obestiy. Proc Natl Acad Sci USA 2000;97:787–792.
104. Podlasek CA, Duboule D, Bushman W. Male accessory sex organ morphogenesis is altered by loss of function in Hoxd-13. Dev Dyn 1998;208:454–465.
105. Roest HP, van Klaveren J, de Wit J, van Gurp CG, Koken MHM, et al. Inactivation of HR6B ubiquitin-conjugating DNA repair enzyme in mice causes male sterility associated with chromatin modification. Cell 1996;86:799–810.
106. Zindy F, Deursen Jv, Grosveld G, Sherr CJ, Roussel MF. INK4d-deficient mice are fertile despite testicular atrophy. Mol Cell Biol 2000;20:372–378.
107. Thepot D, Weitzman JB, Barra J, Segretain D, Stinnakre MG, Babinet C, et al. Targeted disruption of the murine junD gene results in multiple defects in male reproductive function. Development 2000;127:143–153.
108. Mishina Y, Rey R, Finegold MJ, Matzuk MM, Josso N, Cate RL, et al. Genetic analysis of the Müllerian-inhibiting substance signal transduction pathway in mammalian sexual differentiation. Genes Dev 1996;10:1–11.
109. Pace AJ, Lee E, Athirakui K, Coffman TM, O'Brien DA, Koller BH. Failure of spermatogenesis in mouse lines deficient in the Na(+)-K(+)-2Cl(-) cotransporter. J Clin Invest 2000;105:441–450.
110. Gow A, Southwood CM, Li JS, Pariali M, Riordan GP, Brodie SE, et al. CNS myelin and sertoli cell tight junction strands are absent in Osp/claudin-11 null mice. Cell 1999;99:649–659.
111. Dai X, Schonbaum C, Degenstein L, Bai W, Mahowald A, Fuchs E. The ovo gene required for cuticle formation and oogenesis in flies is involved in hair formation and spermatogenesis in mice. Genes Dev 1998;12:3452–3463.
112. Mulryan K, Gitterman DP, Lewis CJ, Vial C, Leckie BJ, Cobb AL, et al. Reduced vas deferens contraction and male infertility in mice lacking P2X1 receptors. Nature 2000;403:86–89.
113. Mbikay M, Tadros H, Ishida N, Lerner CP, Lamirande ED, Chen A, et al. Impaired fertility in mice deficient for the testicular germ cell protease PC4. Proc Natl Acad Sci USA 1997;94:6842–6846.
114. Blume-Jensen P, Jiang G, Hyman R, Lee KF, O'Gorman S, et al. Kit/stem cell factor receptor-induced activation of phosphatidylinositol 3'-kinase is essential for male fertility. Nat Genet 2000;24:157–162.
115. Baker SM, Bronner CE, Zhang L, Plug AW, Robatzek M, Warren G, et al. Male mice defective in the DNA mismatch repair gene PMS2 exhibit abnormal chromosome synapsis in meiosis. Cell 1995;82:309–319.
116. Varmuza S, Jurisicova A, Okano K, Hudson J, Boekelheide K, Shipp EB. Spermiogenesis is impaired in mice bearing a targeted mutation in the protein phosphatase 1c gamma gene. Dev Biol 1999;205:98–110.
117. Lohnes D, Kastner P, Dierich A, Mark M, LeMeur M, Chambon P. Function of retinoic acid receptor g in the mouse. Cell 1993;73:643–658.
118. Yuan L, Liu JG, Zhao J, Brundell E, Daneholt B, Hoog C. The murine SCP3 gene is required for synaptonemal complex assembly, chromosome synapsis, and male fertility. Mol Cell 2000;5:73–83.
119. Supp DM, Witte DP, Branford WW, Smith EP, Potter SS. Sp4, a member of the Sp1-family of zinc finger transcription factors, is required for normal murine growth, viability, and male fertility. Dev Biol 1996;176:284–299.
120. Pearse II, RV, Drolet DW, Kalla KA, Hooshmand F, Bermingham Jr. JR, et al. Reduced fertility in mice deficient for the POU protein sperm-1. Proc Natl Acad Sci USA 1997;94:7555–7560.
121. Kuroda M, Sok J, Webb L, Baechtold H, Urano F, Yin Y, et al. Male sterility and enhanced radiation sensitivity in TSL (-/-) mice. EMBO J 2000;19:453–462.

122. Yu YE, Zhang Y, Unni E, Shirley CR, Deng JM, Russell LD, et al. Abnormal spermatogenesis and reduced fertility in transition nuclear protein 1-deficient mice. Proc Natl Acad Sci USA 2000;97:4683–4688.
123. Lu Q, Gore M, Zhang Q, Camenisch T, Boast S, Casagranda F, et al. Tyro-3 family receptors are essential regulators of mammalian spermatogenesis. Nature 1999;398:723–728.
124. Tanaka SS, Toyooka Y, Akasu R, Katoh-Fukui Y, Nakahara Y, Suzuki R, et al. The mouse homolog of Drosophila Vasa is required for the development of male germ cells. Genes Dev 2000;14:841–853.
125. Matzuk MM, Kumar TR. Bradley A. Different phenotypes for mice deficient in either activins or activin receptor type II. Nature 1995;374:356–360.
126. Toscani A, Mettus RV, Coupland R, Simpkins H, Litvin J, Orth J, et al. Arrest of spermatogenesis and defective breast development in mice lacking A-myb. Nature 1997;386:713–717.
127. Barlow C, Hirotsune S, Paylor R, Liyanage M, Eckhaus M, Collins F, et al. Atm-deficient mice: a paradigm of ataxia telangiectasia. Cell 1996;86:159–171.
128. Xu Y, Ashley T, Brainerd EE, Bronson RT, Meyn MS, Baltimore D. Targeted disruption of ATM leads to growth retardation, chromosomal fragmentation during meiosis, immune defects, and thymic lymphoma. Genes Dev 1996;10:2411–2422.
129. Lu Q, Shur BD. Sperm from beta 1,4-galactosyltransferase-null mice are refractory to ZP3-induced acrosome reactions and penetrate the zona pellucida poorly. Development 1997;124:4121–4131.
130. Hudson DF, Fowler KJ, Earle E, Saffery R, Kalitsis P, Trowell H, Wreford NG, et al. Centromere protein B null mice are mitotically and meiotically normal but have lower body and testis weights. J Cell Biol 1998;141:309–319.
131. Fowler KJ, Hudson DF, Salamonsen LA, Edmondson SR, Earle E, Sibson MC, et al. Uterine dysfunction and genetic modifiers in centromere protein B-deficient mice. Genome Res 2000;10:30–41.
132. Fisher CR, Graves KH, Parlow AF, Simpson ER. Characterization of mice deficient in aromatase (ArKO) because of targeted disruption of the cyp19 gene. Proc Natl Acad Sci USA 1998;95:6965–6970.
133. Robertson KM, O'Donnell L, Jones ME, Meachem SJ, Boon WC, Fisher CR, et al. Impairment of spermatogenesis in mice lacking a functional aromatase (cyp 19) gene. Proc Natl Acad Sci USA 1999;96:7986–7991.
134. Rugglu M, Speed R, Taggart M, McKay SJ, Kilanowski F, Saunders P, et al. The mouse Dazla gene encodes a cytoplasmic protein essential for gametogenesis. Nature 1997;389:73–77.
135. Yoshida K, Kondoh G, Matsuda Y, Habu T, Nishimune Y, Morita T. The mouse RecA-like gene Dmc1 is required for homologous chromosome synapsis during meiosis. Mol Cell 1998;1:707–718.
136. Pittman DL, Cobb J, Schimenti KJ, Wilson LA, Cooper DM, Brignull E, et al. Meiotic prophase arrest with failure of chromosome synapsis in mice deficient for Dmc1, a germline-specific RecA homolog. Mol Cell 1998;1:697–705.
137. Miyamoto N, Yoshida M, Kuratani S, Matsuo I, Aizawa S. Defects of urogenital development in mice lacking Emx2. Development 1997;124:1653–1664.
138. Hsieh-Li HM, Witte DP, Weinstein M, Branford W, Li H, Small K, et al. Hoxa 11 structure, extensive antisense transcription, and function in male and female fertility. Development 1995;121:1373–1385.
139. Satokata I, Benson G, Maas R. Sexually dimorphic sterility phenotypes in Hoxa10-deficient mice. Nature 1995;374:460–463.
140. Baker J, Hardy MP, Zhou J, Bondy C, Lupu F, Bellvé AR, et al. Effects of an Igf1 gene null mutation on mouse reproduction. Mol Endocrinol 1996;10:903–918.
141. Edelmann W, Cohen PE, Kane M, Lau K, Morrow B, Bennett S, et al. Meiotic pachytene arrest in MLH1-deficient mice. Cell 1996;85:1125–1134.
142. deVries SS, Baart EB, Dekker M, Siezen A, Rooij DGd, Boer Pd, et al. Mouse MutS-like protein Msh5 is required for proper chromosome synapsis in male and female meiosis. Genes Dev 1999;13:523–531.
143. Good D, Porter E, Mahon K, Parlow A, Westphal H, Kirsch I. Hypogonadism and obesity in mice with a targeted deletion of the Nhlh2 gene. Nat Genet 1997;15:397–401.
144. Nakayama K, Ishida N, Shirane M, Inomata A, Inoue T, Shishido N, et al. Mice lacking p27(Kip1) display increased body size, multiple organ hyperplasia, retinal dysplasia, and pituitary tumors. Cell 1996;85:707–720.
145. Fero ML, Rivkin M, Tasch M, Porter P, Carow CE, Firpo E, et al. A syndrome of multiorgan hyperplasia with features of gigantism, tumorigenesis, and female sterility in $p27^{Kip1}$-deficient mice. Cell 1996;85:733–744.

146. Kiyokawa H, Kineman RD, Manova-Todorova KO, Soares VC, Hoffman ES, Ono M, et al. Enhanced growth of mice lacking the cyclin-dependent kinase inhibitor function of p27[Kip1]. Cell 1996;85:721–732.
147. Ormandy CJ, Camus A, Barra J, Damotte D, Lucas B, Buteau H, et al. Null mutation of the prolactin receptor gene produces multiple reproductive defects in the mouse. Genes Dev 1997;11:167–178.
148. Togawa A, Miyoshi J, Ishizaki K, Tanaka M, Takakura A, Nishioka H, et al. Progressive impairment of kidneys and reproductive organs in mice lacking Rho GDIalpha. Oncogene 1999;18:5373–5380.
149. Lee H-W, Blasco MA, Gottlieb GJ, Horner II JW, Greider CW, DePinho RA. Essential role of mouse telomerase in highly proliferative organs. Nature 1998;392:569–577.
150. Beck ARP, Miller IJ, Anderson P, Streuli M. RNA-binding protein TIAR is essential for primordial germ cell development. Proc Natl Acad Sci USA 1998;95:2331–2336.
151. Luoh S-W, Bain P, Polakiewicz R, Goodheart M, Gardner H, Jaenisch R, et al. Zfx mutation results in small animal size and reduced sperm cell number in male and female mice. Development 1997;124:2275–2284.

6 Functional Analysis of the *Bcl2* Gene Family in Transgenic Mice

Andrea J. Ross, PHD
and Grant R. MacGregor, DPHIL

CONTENTS

APOPTOSIS: AN OVERVIEW

Historical Perspective

Over 150 years have elapsed since the realization that cell death is a normal feature of animal development. This fundamental concept was introduced in 1842 by Carl Vogt, following his observation of dying notochordal and cartilaginous cells in the metamorphosing tadpole *(1)*. Since this time, there have been numerous other descriptions of naturally occurring cell deaths during the development and homeostasis of a variety of metazoan species.

Initially, the term "programmed cell death" (PCD) was used to describe the phenomenon of carefully controlled deletion of specific cells at predetermined times during development *(2)*. Walter Flemming gave the first clear description of the morphology of PCD in 1885 while studying regression of follicles in the rabbit ovary *(1)*. Using the red dye safranin, Flemming noted condensation of the chromatin along the nuclear envelope and fragmentation of the nucleus in granulosa cells, a process he termed "chromatolysis." Nearly a century later, Kerr, Wyllie, and Currie used electron microscopy to revisit the morphology of dying cells in a number of different tissues. They renamed the process "apoptosis" (from the Greek term used to describe leaves falling from trees, apo = off, ptosis = falling) *(3)*. Although the term "programmed cell death" was originally used in reference to developmentally regulated cell death, with "apoptosis" used to describe the

From: *Contemporary Endocrinology: Transgenics in Endocrinology*
Edited by: M. Matzuk, C. W. Brown, and T. R. Kumar © Humana Press Inc., Totowa, NJ

morphological changes associated with cell death, the two terms are now routinely used interchangeably.

Hallmarks

Electron microscopy of dying cells has provided a detailed picture of the cellular features of apoptosis. As described by Flemming, early hallmarks of PCD include shrinkage of the cytoplasm and compaction of nuclear chromatin. This is followed by crowding of the organelles and formation of vacuoles derived from the endoplasmic reticulum, giving the cytoplasm a "bubbling" appearance. Finally, the cell is broken down into a number of membrane-bound apoptotic bodies that are phagocytosed by neighboring cells. These morphological features can be contrasted with the cellular events associated with necrosis, which usually include swelling of the mitochondria, ER, and nucleus, with eventual rupturing of the plasma membrane. This cellular lysis often stimulates an inflammatory response that is precluded when cells are removed in an apoptotic manner. Although it is convenient to categorize all cell death into one of these two classes, the validity of such classification has recently been debated, particularly with reference to neuronal *(4)* and germ-cell death *(5)*.

Functions

Programmed cell death is required for a number of fundamental processes in metazoan development and homeostasis. PCD is associated with sculpting of body structures such as the digits on the human hand or the proamniotic cavity of the early post-implantation mouse embryo *(6,7)*. PCD also mediates removal of functionally unnecessary structures, such as the tadpole's tail during amphibian metamorphosis *(8)* or the Müllerian ductal system, the anlagen of the oviduct and uterus, during male mammalian development *(9)*. PCD regulates cell number, for example, in both the neuronal and hematopoietic lineages of vertebrates where excessive cell populations must be culled to appropriate numbers *(10,11)*. Finally, PCD fulfills an important function in surveillance systems, where potentially dangerous cells (such as auto-reactive lymphocytes, cells that have incurred excessive DNA damage, or cells that are virally infected or preneoplastic) are removed *(12)*.

Improper regulation of cell death is often associated with disease. Decreased apoptosis can result in cancer or auto-immune disorders, while increased or inappropriate apoptosis is a feature of pathologies including AIDS, Alzheimer's disease, and Parkinson's disease *(13)*. It is possible that the inappropriate PCD associated with such diseases may eventually be prevented. However, an understanding of the molecular basis underlying PCD is required before such molecular therapeutics can be developed.

APOPTOSIS: PIVOTAL MOLECULAR EVENTS

The study of apoptosis is moving at a remarkable pace. A MEDLINE search with "apoptosis" as a search term identifies over 6500 published articles for 1998 alone. Indeed, by the time this chapter is in print, it is likely that significant new breakthroughs will have been made in this field. Because of the great volume of information about the regulation of apoptosis, the following discussion will be restricted to more fundamental aspects of the process. Where appropriate, the reader will be directed to more extensive review articles.

Regulation of PCD by Extracellular Signals

How does a cell know when to die? Considerable evidence suggests that many independent cell types require continuous stimuli from the environment to suppress activation of PCD. Such cell-survival stimuli can be in several forms, which include soluble factors, cell-surface molecules, and/or components of the extracellular matrix (ECM) *(14–16)*. The role of survival factors in suppression of PCD has been demonstrated convincingly by in vitro cell-culture work. Many cells in culture require sera or specific growth factors for survival. For example, the cytokine interleukin-2 promotes survival of resting T cells, while interleukin-3 is required for myeloid cell survival. Similarly, platelet-derived growth factor and insulin-like growth factor (IGF-1) enhance survival of fibroblasts in culture, while nerve-growth factor and other neurotrophins can promote survival of specific classes of neurons *(14,17)*.

Similar conclusions have been drawn from whole-animal studies. The adrenal cortex requires pituitary-derived adrenocorticotropic hormone for normal development. Removal of this survival factor in rats results in high levels of apoptosis within the adrenal cortex *(18)*. Many cells undergo apoptotic death during the course of normal kidney development in the rat. However, injection of young rats with epidermal growth factor (EGF) prevents this cell death, establishing an antiapoptotic role for EGF in this tissue *(19)*. Analysis of mice with spontaneous and induced genetic mutations has also been informative. The c-kit proto-oncogene encodes a transmembrane tyrosine-kinase receptor that is the product of the Dominant White-Spotting (*W*) locus. Mast-cell growth factor (MGF) is a ligand for this receptor that is encoded by the Steel (*Sl*) locus. Mice with mutations in either of this receptor-ligand pair display defects in three migratory cell lineages. Survival of hematopoietic stem cells (HSC), neural crest-derived melanocytes, and primordial germ cells are all affected, resulting in macrocytic anemia, reduced pigmentation, and sterility*(20)*.

Conversely, a number of extracellular signals can promote cell death. The steroid hormone ecdysone triggers apoptosis in tissues of metamorphosing insects *(21)*. In male mammals, Müllerian-inhibiting substance (MIS), a member of the transforming growth factor-β (TGF-β) family, stimulates apoptosis and regression of the Müllerian duct during embryogenesis *(9,22)*. Finally, Fas ligand can directly activate PCD through the Fas/TNF receptor pathway.

These are but a few of the many factors known to instruct a cell to survive or die. One area of currently intensive research concerns the signal transduction machinery that links the extracellular signal to the intracellular apoptotic control system. Because of its complexity, this aspect of PCD is not discussed here, but is covered in refs. *23–25*.

The Genetics of Apoptosis: Insights from C. elegans

Genetic analysis of the nematode, *Caenorhabditis elegans* (*C. elegans*), has provided significant insight into the molecular basis for PCD. It appears that the fundamental molecular events that control apoptosis have been conserved throughout metazoan evolution.

Knowledge of the precise cellular program of its development and sophisticated genetics make *C. elegans* an ideal system for studying the regulation of PCD. During normal development of a hermaphrodite, exactly 1090 somatic cells are formed, of which 131 cells always undergo PCD *(26,27)*. Genetic studies have identified over a dozen genes required for this process of cellular elimination *(28)*. Three genes required for normal developmental cell deaths to occur are *ced-3*, *ced-4* (*ced* = cell-death abnormal) and *egl-1* (*egl* = egg-laying defective) *(29,30)*. Loss-of-function alleles of any of these individual genes results in survival of cells that would otherwise normally undergo PCD. In contrast, *ced-9* is required for negative regulation of apoptosis, and protects cells that are not predestined to die. While gain-of-function alleles of *ced-9* prevent most cell deaths, loss-of-function mutations in *ced-9* result in embryonic lethality associated with excessive cell death. Excessive cell death can be suppressed by secondary mutations in *ced-4* and *ced-3,* demonstrating that *ced-9* acts upstream of both these genes *(31)*. Further analysis has shown that *ced-4* positively regulates *ced-3 (32)*, and that *egl-1* is a positive regulator of apoptosis, acting upstream of *ced-9 (30)*. A simplified diagram of the genetic pathway of apoptosis in *C. elegans* in shown in Fig. 1.

Caspases: Intracellular Executioners

As part of the apoptotic program, a variety of cellular proteins are cleaved prior to removal of the cell. These include regulatory, structural, and housekeeping proteins such as poly(ADP)-ribose polymerase, nuclear lamins, actin, fodrin, MEK kinase 1, PI-3 kinase (Akt), Raf-1, focal-adhesion kinase, and DNA fragmentation factor (DFF) *(33)*. This proteolytic cleavage is performed by a family of gene products called *caspases*. The term caspase is derived from the specificity of these enzymes, which contain a cysteine residue at their active site and cleave following aspartate residues within the substrates.

Cloning of the *ced-3* gene, which is required for cell death in *C. elegans*, provided the first insight that caspases are involved in effecting PCD. It was discovered that *ced-3* encodes a ortholog of the previously cloned mammalian protease, interleukin-1β-converting enzyme (ICE, now known as caspase-1) *(34)*. Although CED-3 is the only known caspase involved in apoptosis in *C. elegans*, at least 14 members to date have been identified in mammals (caspase-1 through 14) *(33)*. Caspases have a wide range of substrates, not all of which are directly associated with the cell-death process. For example, some are involved in activation of proinflammatory cytokines. Caspase inhibitors block apoptosis in diverse cell types, and mice deficient for specific caspases show defects in development associated with reduced PCD, confirming the role of caspases in apoptosis both in vitro and in vivo *(35–38)*.

Because of their potent activity, caspases must be tightly regulated. Reliance upon transcription alone as a means of regulating caspase production would be capricious and impractical, especially in light of the rapidity with which some cells undergo PCD following receipt of a death-signal. Indeed, caspase activity is primarily regulated at the posttranslational level. Caspases are present in cells as inactive zymogens that are cleaved to form the active configuration, which consists of two heterodimers *(39)*. Some unprocessed zymogens have low levels of proteolytic activity, suggesting one model for the initial activation of the caspases—the "induced-proximity" model *(40)*. Procaspases associate with adaptor molecules, leading to aggregation of the zymogens and their autoprocessing caused by close proximity. Procaspases can also be activated by other caspases. This proteolytic cascade is well-suited to ensuring the rapid and efficient activation of the cell-death process. What then regulates activation of these executioners?

Death Receptors: The Fast-Track to Cell Death

Perhaps the most direct pathway for induction of caspase activation is through mammalian "death receptors." The two best characterized death receptors are Fas (also known as APO-1 or CD95), and TNFR1(p55), which are members of the tumor necrosis factor

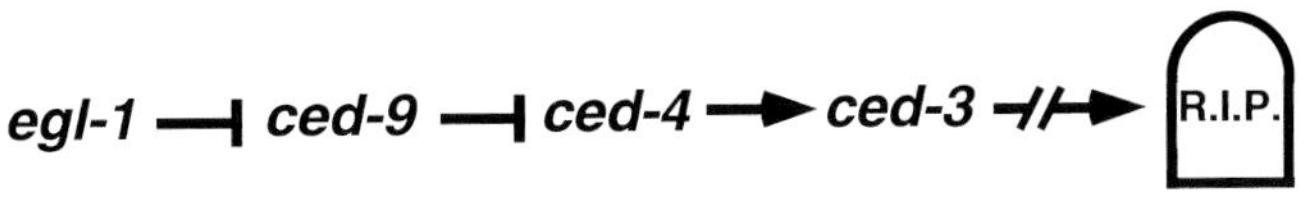

***egl-1* ⊣ *ced-9* ⊣ *ced-4* → *ced-3* -//→ R.I.P.**

Fig. 1. Summary of the genetic pathway regulating apoptosis in *C. elegans*. The slashed line between Ced-3 and RIP indicates that other gene products are required for destruction and engulfment of the dying cells.

(TNF) receptor family. Other death receptors of this family include death receptor-3 (DR-3, also known as APO-3, TRAMP, or LARD), DR-4 (TRAIL-R1, APO-2), DR-5 (TRAIL-R2, KILLER, TRICK2), the p75 nerve-growth-factor receptor, and avian CAR1 *(41,42)*. These transmembrane proteins have a number of extracellular cysteine-rich domains and share a cytoplasmic sequence called the "death-domain," which is used for recruitment of proteins involved in regulating the apoptotic pathway *(43)*. The ligands for these receptors (including FasL/CD95L, TNF, APO2L, and APO3L) are type II transmembrane proteins found on the cell surface as homotrimeric molecules, although they can also be cleaved by metalloproteases and released as soluble molecules *(44)*.

Upon ligand (FasL) binding, the monomeric Fas receptor homotrimerizes, resulting in an intracellular clustering of the death domains. FADD (Fas-associating protein with death-domain) is a cytoplasmic adaptor molecule, which also contains a death domain that can bind to these clustered Fas-receptor death domains *(45)*. FADD in turn contains another domain (called a caspase recruitment domain or CARD) that allows it to interact with the inactive form of caspase-8 (also called FLICE). CARD domains are found in some—but not all—caspases *(46)*. When procaspase-8 interacts with FADD, it is activated by low intrinsic self-cleavage and or cross-cleavage because of the proximity of other procaspase-8 molecules. The resulting mature active caspase subsequently activates downstream caspases, which are required to effect apoptosis.

Analyses of mice mutant for either Fas receptor or FasL suggest that the pathway is required for the deletion of activated mature T cells at the end of an immune response and the killing of cells (such as viral-infected or cancer cells) by cytotoxic T cells and natural killer cells *(41,47)*. TNFR1 can also activate caspase-8 in a similar pathway involving different adaptor molecules, but, to complicate matters, TNFR1 can bind alternative death-domain-containing adaptor proteins that can activate other signal transduction pathways, some of which promote cell survival *(41,42)*. Less is known about the biological functions of the other death receptors, most of which have only recently been cloned.

REGULATION OF APOPTOSIS BY THE BCL2 *GENE FAMILY*

Identification of Bcl2, a Mammalian Ortholog of ced-9

Members of the *Bcl2* gene family also control cell survival through regulation of procaspase activation. The prototypical family member *Bcl2* (B cell lymphoma gene 2) was cloned because of its involvement in t(14;18) chromosomal translocations that result in human B-cell follicular lymphoma. In such cases, the translocation breakpoint juxtaposes the *Bcl2* gene with the immunoglobulin heavy-chain transcriptional enhancer *(48,49)*. Overexpression of Bcl2 results in a novel form of cancer in which the production of excessive cells is caused by an increase in cell longevity rather than increased rate of proliferation. This mechanism was demonstrated in experiments where overexpression of BCL2 in IL-3 dependent pre-B cells prolonged survival upon cytokine withdrawal *(50,51)*.

At the time of its cloning, the peptide sequence of BCL2 provided no significant insight into the mechanism by which the protein regulates cell survival. Consequently, it was of great interest when the *ced-9* gene, which is required to prevent excessive cell death during *C. elegans* development, was shown to encode the ortholog of *BCL2*. Indeed, *BCL2* can partially compensate for loss of function of *ced-9* in *C. elegans*, *(52)*, illustrating the conservation of cell death pathways over a thousand million years of metazoan evolution.

Bcl2 *is the Founding Member of a Multi-Gene Family in Mammals*

In vertebrates, a number of proteins have been discovered that share one or more domains of homology with Bcl2. These domains have been termed BH1, BH2, BH3, and BH4 (BH = Bcl2 homology domain). As with Bcl2, some of these proteins display anti-apoptotic activity following expression in cells. However, many other related proteins actually kill cells; consequently, the Bcl2 family has been divided into two functional classes. The current death-protecting members of the family are Bcl2, $Bclx_L$ *(53)*, Bclw *(54)*, A1 *(55)*, and Mcl-1*(56)*. The death-promoting members include Bax *(57)*, Bak *(58)*, Bad *(59)*, $Bclx_S$ *(53)*, Bik *(60)*, Bid *(61)*, Bim *(62)*, Bok *(63)*, Blk *(64)*, Hrk *(65)*, Diva *(66)*, Bfl-1 *(67)*, Nix *(68)*, Nip3 *(69)*, and EGL-1 of *C. elegans (30)*. Figure 2 lists the currently known family members, and summarizes some of their structural features. All of the death-promoting family members contain a BH3 domain, and many of these members show little conservation to the rest of the gene family outside of this domain, suggesting the importance of this domain for killing activity. Several viral homologs of the anti-apoptotic members of the family also exist, including E1B 19K from adenovirus *(70)*, Bhrf1 from Epstein-Barr virus *(71)*, and Lmw5-hl from the African swine fever virus *(72)*. Expression of these viral proteins is believed to prevent host cells from undergoing apoptosis following infection with the virus.

Mechanism of **Bcl2** *Gene Function*

Several of the pro- and anti-apoptotic members of the Bcl2 family can dimerize with each other, and in many cases these interactions may be required for their respective functions. Bcl2 and Bax can both homodimerize and heterodimerize through interaction of specific BH domains. Mutations in the BH1 or BH2 domain of Bcl2 abrogate both its capacity to dimerize with Bax and prevent Bax-induced death *(73)*. Mutations in the BH3 domain of Bax block its ability to homo- and heterodimerize, and most of these mutations inhibit its pro-apoptotic activity *(74)*. Based on these findings, a "rheostat" model was proposed for Bcl2 family function, in which relative levels of Bcl2 and Bax control cell survival, and if the intracellular level of Bax:Bax homodimers exceeds that of Bax:Bcl2 heterodimers, the cell will become committed to death *(75)*.

Analysis of additional members of the Bcl2 family suggests that while the rheostat model is useful, in certain cases it may be too simplistic. For example, not all pro-apoptotic Bcl2 family members appear to function in the same manner as Bax. Many of the BH3-only proteins do not homodimerize, and can only interact with anti-apoptotic proteins *(76)*. It is possible that these BH3-only proteins promote death only by blocking the activity of Bcl2 or other death-protecting family members, whereas Bax may have additional functional roles in initiation of apoptosis.

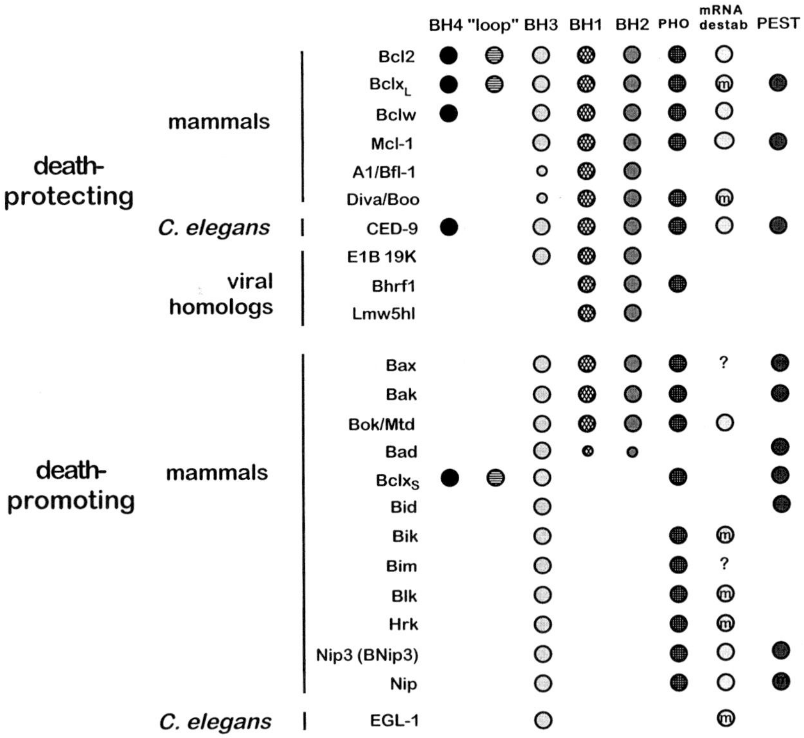

			BH4	"loop"	BH3	BH1	BH2	PHO	mRNA destab	PEST
death-protecting	mammals	Bcl2	●	●	○	●	●	●	○	
		$Bclx_L$	●	●	○	●	●	●	ⓜ	●
		Bclw	●		○	●	●	●	○	
		Mcl-1			○	●	●	●	○	●
		A1/Bfl-1			o	●	●			
		Diva/Boo			o	●	●	●	ⓜ	
	C. elegans	CED-9	●		○	●	●	●	○	●
	viral homologs	E1B 19K			○	●	●			
		Bhrf1				●	●	●		
		Lmw5hl				●	●			
death-promoting	mammals	Bax			○	●	●	●	?	●
		Bak			○	●	●	●		●
		Bok/Mtd			○	●	●	●	○	
		Bad			○	•	•			●
		$Bclx_S$	●	●	○			●		●
		Bid			○					●
		Bik			○			●	ⓜ	
		Bim			○			●	?	
		Blk			○			●	ⓜ	
		Hrk			○			●	ⓜ	
		Nip3 (BNip3)			○			●	○	●
		Nip			○			●	○	●
	C. elegans	EGL-1			○				ⓜ	

Fig. 2. The Bcl2 family of apoptosis regulators. Proteins are classified according to apoptotic activity, BH homology domains are denoted. "PHO" = hydrophobic C-terminal tail. Many members contain mRNA destabilization sequences or PEST sequences, which can effect rapid degradation of proteins, suggesting a mechanism for tightly controlling steady-state levels of family members in the cell. mRNA destabilization sequences are defined as minimal (circle containing an "m"): AUUUA, or full (solid circle): UUAUUUA(U/A)(U/A). The minimal sequence may not be adequate to mediate degradation in all circumstances *(209,210)*. PEST elements were defined using PESTfind (www.at.embnet.org/embnet/tools/bio/PESTfind/). A question mark indicates that sequence information was unavailable at the time of writing. A small circle is used if homology is minimal.

Regulation of Caspase Activation by CED-9/Bcl2 Family

Genetic studies in *C. elegans* provided the first evidence that Bcl2 is involved in regulation of caspases. Control of the apoptotic pathway in *C. elegans* has been ordered genetically, with *ced-4* acting upstream of *ced-3*, while *ced-9* negatively regulates the activity of *ced-4* (Fig. 1). Biochemical studies have demonstrated physical interactions between several of these gene products. CED-4 has been shown to interact with both CED-9 and with pro-CED-3 (the inactive form of the caspase) *(77)*, while EGL-1 can also directly associate with CED-9 *(30)*. Based on these findings, a model was proposed in which CED-4 facilitates processing of pro-CED-3 into an active caspase, subsequently effecting apoptotic death, and this activity is blocked by CED-9. This is supported by the finding that overexpression of CED-4 can kill cells, and that this effect is blocked by caspase inhibitors *(77)*. The mechanism by which CED-9 prevents activation of pro-CED-3 by CED-4 is believed to involve physical association of CED-9 with CED-4. However, it is currently unclear whether CED-9 prevents CED-4 from interacting with CED-3 directly, or perhaps holds CED-4 in a conformation in which it is unable to activate processing of CED-3. EGL-1, a BH3-only pro-apoptotic member of the Bcl2 family, appears to activate the apoptotic pathway by interacting with CED-9 to mediate release of CED-4, allowing it to initiate the caspase cascade *(30)*.

As might be anticipated, the cognate process in the mammalian system appears to be more complex. APAF-1 (Apoptosis Activating Factor-1) is a mammalian protein containing regions of homology to CED-4. APAF-1 was identified as one of three biochemically purifiable cytosolic molecules (APAF 1-3) required to activate caspase-3 in the presence of dATP in vitro *(78,79)*. APAF-2 was identified as cytochrome *c*, while APAF-3 was discovered to be caspase-9 *(80)*. Both APAF-1 and CED-4 contain a CARD domain and a nucleotide-binding oligomerization (NOD) domain. The C-terminus of APAF-1 contains a series of WD-40 domains, which are not found in CED-4. APAF-1 and procaspase-9 interact through their respective CARD domains, similar to the interaction between CED-3 and CED-4 *(80)*. It has been demonstrated that APAF-1 can bind cytochrome *c*, and in the presence of dATP forms an oligomeric complex *(81)*. These APAF-1/cytochrome *c* complexes can then bind and process procaspase-9, initiating the caspase cascade. The WD-40 repeats at the C-terminus of APAF-1 can interact with the N-terminal CED-4 homologous region of the protein, and expression of the WD-40 repeats alone can inhibit the self-association of APAF-1 and procaspase activation *(82)*. Therefore, the WD-40 repeats may function to negatively regulate procaspase-9 activation by blocking the ability of APAF-1 to oligomerize, potentially through its interaction with the CED-4 homologous region of APAF-1. Binding of cytochrome *c* and dATP hydrolysis by APAF-1 may disrupt this interaction and allow APAF-1 to undergo a conformational change and self-associate. Truncated APAF-1, lacking the WD-40 repeats, can also recruit and process procaspase-9, yet cannot release the active caspase-9 from the complex *(81)*. Thus, the WD-40 domains also appear to play a specific role in regulating release of processed, active caspases.

There is conflicting evidence regarding the role of the Bcl2 family in regulation of APAF-1 activity. $Bclx_L$ can interact with APAF-1 or CED-4 in vitro *(83,84)*, as shown by coimmunoprecipitation experiments with transfected cells, and these interactions are blocked if $Bclx_L$ is complexed with the pro-apoptotic members Bax, Bak, or Bik *(77)*. In vitro interactions between Boo and APAF-1 have also been reported *(85)*. Thus, it has been postulated that $Bclx_L$ (and Boo) may prevent cell death by inhibiting the capacity of APAF-1 to activate caspase-9, which is conceptually similar to the model for regulation of CED-4 by CED-9 in *C. elegans*. By extension, pro-apoptotic family members could then kill cells by disrupting the interactions between $Bclx_L$ and APAF-1, leaving APAF-1 free to activate caspase-9. However, independent studies failed to show any interaction between both overexpressed and endogenous APAF-1 and a number of the anti-apoptotic Bcl2 family members (Bcl2, Bclw, Mcl-1, A1, Boo, and $Bclx_L$), under conditions in which APAF-1 would interact with procaspase-9 *(86)*. Thus, in contrast to the nematode system, it is currently unclear whether the mammalian antiapoptotic Bcl2 family members actually have a role in regulating caspase activation through direct interactions with APAF-1. It is possible that inclusion of the WD-40 domains in the C-terminus of mammalian APAF-1 eliminates the requirement for negative regulation of Apaf-1 by anti-apoptotic Bcl2 family members. Other mechanisms by which the Bcl2 family may regulate caspase activation are discussed here.

The Mitochondrion as a Cellular Pandora's Box

APAF-1 requires cytochrome *c* as a cofactor for activation of caspase-9. In living cells, cytochrome *c* is usually found in the mitochondrial intermembrane space, where it functions in the electron transport chain. However, during apoptosis, cytochrome *c* is released from mitochondria into the cytosol *(87–89)*. Apoptosis-inducing factor (AIF), a flavoprotein, also resides in the mitochondrial intermembrane space in living cells. AIF can induce nuclear morphological changes associated with apoptosis in vitro, and during apoptosis it also exits the mitochondrion and translocates to the nucleus *(90)*. Release of these and perhaps other apoptogenic molecules from the mitochondrion is an important step in regulation of apoptosis, and the Bcl2 family is involved in this regulation. Addition of recombinant Bax to isolated mitochondria stimulates release of cytochrome *c* *(91)*. Overexpression of Bcl2 or $Bclx_L$ can block release of cytochrome *c* from mitochondria in cells treated with a variety of apoptotic stimuli *(87,88,92)*.

The subcellular localization of several Bcl2 family proteins suggests that they may play a direct role in the release of cytochrome *c*. Several of the family members (Fig. 2) have hydrophobic carboxy-terminal tails which can anchor the proteins in organelle membranes *(93)*. Bcl2 has been localized to the mitochondria, nucleus, and endoplasmic reticulum *(94,95)*, and $Bclx_L$ has been localized to the mitochondrial outer membrane *(96)*. Removal of the hydrophobic tail of Bcl2 changes its localization to the cytosol, and reduces its ability to block apoptosis, demonstrating that its localization is important for its function *(97)*. Although the death-promoting member Bax has a hydrophobic tail, it is localized in the cytoplasm until an apoptotic stimuli is received, when it translocates to the mitochondrial outer membrane *(98)*. This sub-cellular redistribution of Bax and the subsequent release of cytochrome *c* is associated with a conformational change in the molecule, although the mechanism which triggers this change is not yet known *(99)*. In contrast, Bcl2 is localized to mitochondria both prior to and during apoptosis *(98)*.

The manner in which Bax promotes release of cytochrome *c* and AIF from the mitochondrial intermembrane space is unclear, and may involve multiple mechanisms. One model explaining the mechanism for release of these molecules involves the pore-forming ability of several members of the Bcl2 family. Analysis of the three-dimentional structure of $Bclx_L$ revealed similarity to the pore-forming domains of the bacterial toxins, known as diptheria toxin and colicin *(100)*, and showed that Bcl2, $Bclx_L$, and Bax can all form ion channels in synthetic membranes *(101–103)*. Diptheria toxin can form pores large enough for proteins to move through lysosomal membranes, which led to the postulate that pores formed by Bcl2 family proteins might permit efflux of cytochrome *c* and other proteins *(104)*. However, no direct evidence exists that any of the Bcl2 family proteins could form a pore large enough to permit passage of cytochrome *c*.

In several cell types, apoptotic stimuli cause a dissipation of the mitochondrial inner-membrane potential ($\Delta\Psi_m$), which occurs prior to the changes seen in the nucleus and plasma membrane *(105)*. Loss of $\Delta\Psi_m$ results in uncoupling of electron transport and ATP production. The loss of $\Delta\Psi_m$ results from opening of the mitochondrial permeability transition (PT) pore, a large conductance channel in the mitochondrial membrane *(106)*. The PT pore is a multi-protein complex, and its components include the adenine nucleotide translocator (ANT), an inner-membrane protein, and voltage-dependent anion channel (VDAC), an outer-membrane protein *(107)*. Overexpression of Bax can induce mitochondrial $\Delta\Psi_m$ loss, which can be blocked by inhibitors of the PT pore, and addition of recombinant Bax or Bak to isolated mitochondria induces loss of $\Delta\Psi_m$ *(108,109)*. Bax,

Bak, and Bcl2 can physically interact with ANT in vitro, and both Bax and Bak coimmunoprecipitate with VDAC *(108,110)*. Bcl2 or Bclx can both block changes in mitochondrial membrane permeability induced by a number of stimuli, including overexpression of Bax *(111)*. These combined results indicate that the Bcl2 family can regulate activity of the PT pore, and that loss of $\Delta\Psi_m$ is a critical event in Bax-induced apoptosis.

These findings have provided another model to explain the mechanism of cytochrome *c* release from mitochondria. Upon loss of $\Delta\Psi_m$, there is an influx of water into the matrix of the mitochondrion, which leads to swelling and rupture of the smaller outer membrane. This rupture allows proteins in the intermembrane space (cytochrome *c*, AIF) to escape into the cytosol. By initiating or blocking loss of mitochondrial $\Delta\Psi_m$, the Bcl2 family proteins could regulate release of cytochrome *c*. Indeed, it has been reported that this loss of integrity of the outer membrane occurs in some apoptotic cells and results in release of cytochrome *c (111)*. Moreover, in support of this model, the mitochondrial permeability transition is required for TNFα-induced apoptosis in rat hepatocytes, and occurs prior to release of cytochrome *c (112)*. However, there are also conflicting reports that movement of cytochrome *c* from the mitochondria precedes the dissipation of $\Delta\Psi_m$ during apoptosis, and that the mitochondria remain intact *(87,88,113)*.

A recent finding may explain how Bax could stimulate release of cytochrome c without simultaneously causing a loss of $\Delta\Psi_m$. Either Bax or Bak, in combination with VDAC, can form a channel in liposome membranes, which allows passage of fluorescein labled cytochrome *c (114)*. In contrast, Bax and Bak alone cannot allow passage of cytochrome *c*. In certain situations, Bax or Bak may translocate to the mitochondria and interact with VDAC in the outer membrane to form a channel capable of releasing apoptogenic proteins from the intermembrane space. In other instances, Bax or Bak may interact with both ANT and VDAC (and potentially, other components of the PT pore) to cause a loss of mitochondrial $\Delta\Psi_m$, which subsequently leads to rupture of the mitochondrial outer membrane, release of cytochrome *c*, and apoptosis. In either case, interactions of Bcl2 family proteins with components of the PT pore would be important for execution or prevention of apoptosis. The ability of proapoptotic Bcl2 family members to induce release of cytochrome c, thereby activating a caspase cascade, and to disrupt mitochondrial function, may more thoroughly ensure a cell's commitment to death.

In summary, Bcl2 family members may initiate or prevent cell death by 1) regulating the activity of the other Bcl2 family members via dimerization, 2) controlling the release of cytochrome c and other proteins from the mitochondrial intermembrane space, and 3) affecting mitochondrial function by maintaining or disrupting the innermembrane potential. For more comprehensive information regarding mechanisms of apoptosis regulation by the Bcl2 family (*see* refs. *72,115–117*) Figure 3 shows a simplified model summarizing the mechanisms by which the Bcl2 family regulates apoptosis.

Upstream Regulation of Bcl2 Family Member Activity

Which upstream factors regulate the activity of the Bcl2 family? In many cases, it is not yet known how extracellular signals are linked to Bcl2 family proteins. However, progress has been made regarding the mechanisms by which Bid and Bad, two death-promoting Bcl2 family members, are regulated. Both of these are BH3-only proteins, lack hydrophobic tails, and are found in the cytosol.

Bad can activate apoptosis by interacting with $Bclx_L$ or Bcl2 to inhibit their death-protecting function *(59)*. In an interleukin-3 (IL-3)-dependent cell line in the presence

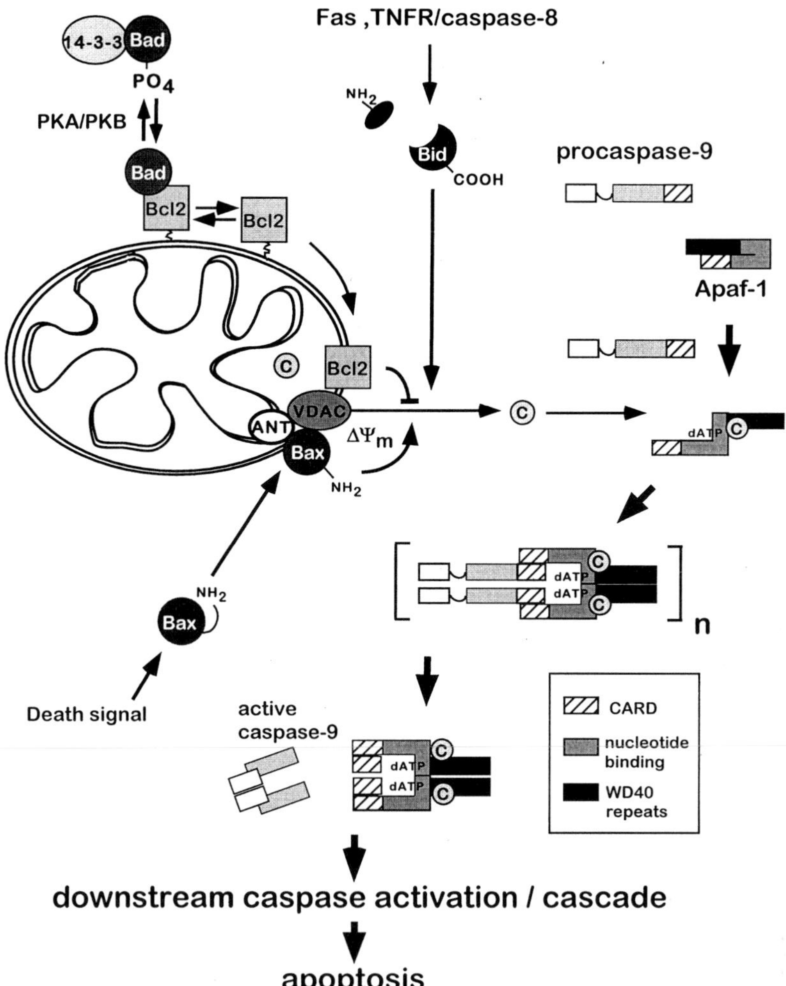

Fig. 3. Simplified model for regulation of apoptosis by the Bcl2 family. Many potential regulatory mechanisms are shown, not all of which may operate in the same cell type. Bax undergoes a conformational change, then translocates to mitochondria during apoptosis. Insertion of Bax into the mitochondrial membrane then promotes release of cytochrome c. Bid is activated by cleavage in the Fas/caspase-8 pathway, and also promotes apoptosis via stimulation of cytochrome *c* release. Bcl2, or other death-protecting members can block release of cytochrome *c* from mitochondria. Bad, when phosphorylated, is held inactive by 14-3-3 family proteins. When dephosphorylated, Bad interacts with Bcl2 or $Bclx_L$ via their BH4 domains to block their death-protecting activity. APAF-1, in the presence of cytochrome *c* and dATP (or ATP), can activate caspase-9 by facilitating cleavage of the procaspase. This may involve formation of a large multimeric complex of APAF-1 and caspase-9 molecules. For the sake of simplicity, APAF-1 is shown as a dimer in the figure. *See* text for more detailed explanation and references.

of cytokine, Bad is phosphorylated on two serine residues, 112 and 136. When phosphorylated, Bad is sequestered by 14-3-3, cytosolic proteins which prevent Bad from interacting with Bcl2 or $Bclx_L$. In the absence of survival factor, Bad is free to interact with these proteins, resulting in initiation of the apoptotic cascade *(17,118)*. Two kinases have been identified as potential regulators of Bad function. PKB/Akt can phosphorylate serine 136 in vitro and when overexpressed in cells *(17,119–121)*. PKA can phosphorylate BAD on serine 112, and inhibitors of PKA can block the phosphorylation triggered by IL-3 *(122)*. These findings suggest that there may be multiple signalling pathways that regulate the phosphorylation status of Bad and thus affect cell survival.

Bid is another pro-apoptotic, BH3-only family member, and its function is regulated by the Fas/TNFR1 signaling pathways. Bid exists in the cytosol as an inactive precursor protein. Caspase-8 activated by the FasL or TNFa pathway cleaves the N-terminus of Bid. After cleavage, the C-terminal portion of Bid (p15 Bid) translocates to the mitochondrion, where it stimulates the release of cytochrome c, thereby activating caspase-9 *(123,124)*. The release of cytochrome *c* induced by p15 Bid can be blocked by Bcl2 or $Bclx_L$ *(125)*. The cross-communication between the Fas pathway and the Bcl2 family may serve to amplify or reinforce the caspase cascade in specific cell types.

FUNCTIONAL ANALYSIS OF *BCL2* GENE FAMILY IN TRANSGENIC MICE

With the exception of studies in *C. elegans*, many of the findings discussed here result from experiments performed in mammalian-cell-culture systems. This has generated information regarding the molecular events that take place during apoptosis. However, it provides little insight about essential functions for each Bcl2 family member in vertebrates in vivo. Transgenic mouse models are ideally suited to providing such information and both gain-of-function and loss-of-function approaches have been used with success to gain insight into the in vivo function of different *Bcl2* family members.

Gain-of-Function Bcl2 *Transgenics*

A gain-of-function approach was first used to study the function of *Bcl2* in B and T lymphocytes in vivo. Overexpression of *Bcl2* extends the survival of this cell population, increasing the number of mature B cells, and eventually results in malignancy *(126–128)*. Expression of *Bcl2* in immature cortical thymocytes also rendered the cells less sensitive to a number of insults, including PMA, ionomycin, glucocorticoids, or γ-irradiation *(129)*. These experiments demonstrate the powerful utility of the transgenic system to model the effects of overexpression of *Bcl2* in human follicular lymphoma.

Bcl2 has also been expressed at high levels in a variety of tissues to determine the effect of perturbing apoptosis during normal development. Expression of *Bcl2* in neurons results in hypertrophy of the central nervouse system (CNS) associated with a reduction in cell death during early neural development *(130)*. In a similar manner, targeted expression of *Bcl2* to the lens of the eye disrupted normal fiber-cell differentiation, causing cataracts, vacuolization, disorganization, and inhibited fiber-cell enucleation *(131)*. Formation of the mouse vaginal opening at puberty requires apoptosis, and expression of *Bcl2* in the suprabasal epithelial cells and subepithelial cells of the vaginal mucosa blocks vaginal formation entirely *(132)*. Such studies demonstrate the importance of PCD during normal development, and show that *Bcl2* can function in these tissues to block apoptosis, at least when expressed at supraphysiological levels.

Ectopic expression of Bcl2 family members can also correct developmental defects associated with increased levels of apoptosis. Osteopetrotic *op/op* mice have an inactivating mutation in the macrophage colony-stimulating factor (*M-CSF*) gene. These mice have impaired development of monocytes and reduction in the macrophage and osteoclast populations. Overexpression of *Bcl2* in monocytes in *op/op* mice ameliorated loss of macrophages and mediated an almost complete reversal of the osteopetrosis *(133)*. This suggests that the M-CSF gene product is involved in survival of monocytes, potentially through a *Bcl2* -dependent pathway.

Mice with mutations in the IL-7 receptor are lymphopenic, and have severely reduced numbers of developing T cells and B cells. Overexpression of *Bcl2* rescues T-cell development and function in the IL-7R deficient mice. This is a significant finding, as severe combined immunodeficiency, an X-linked disorder in humans, may arise from defects in the IL-7 pathway *(134,135)*. These studies illustrate the potential for development of novel therapeutic regimens to treat disease resulting from an imbalance in apoptosis.

Loss-of-Function Bcl2 *Family Transgenics*

BCL2

Several groups have independently used a loss-of-function approach to identify the normal physiological roles of *Bcl2*. *Bcl2* is expressed in a wide range of tissues during mouse embryogenesis, and later in a more restricted subset of adult tissues, primarily epithelial and lymphoid stem-cell populations and neurons *(136)*. Despite this widespread embryonic expression, mice with targeted mutations in B*cl2* complete embryonic development, yet display a number of interesting postnatal phenotypes *(136,137)*.

Bcl2 null-mutant mice display severe kidney defects. At birth, the kidneys are considerably smaller than those of wild-type littermates, and these eventually become grossly enlarged and cystic *(136,137)*. These defects result from problems during kidney development. Analysis of kidneys in *Bcl2* homozygous mutant embryos using TUNEL staining during early stages of nephrogenesis revealed high levels of apoptosis of mesenchymal cells and a significant decrease in the number of nephrons. Prior to birth, the kidneys show irregular branching and convolution of the ureteric buds *(138)*. At birth, the kidneys are reduced in size and usually become cystic within the first 2 wk after birth *(136,139)*. Most *Bcl2* mutants on a congenic C57BL/6 genetic background die before 3 wk of age as a result of the kidney defect.

Bcl2 is expressed in immature cells of both B and T lineages. At birth, hematopoietic lineages are normal in the *Bcl2* mutants, but a decrease in the size of both the thymus and spleen is seen as the mice age. TUNEL staining was used to demonstrate that this is associated with apoptotic death of thymocytes and B- and T-cell populations in the spleen. Thymocytes isolated from *Bcl2* mutants were also more sensitive to both dexamethasone and γ-irradiation than thymocytes from control animals *(136,137)*. This finding indicates that *Bcl2* is important for survival of hematopoietic lineages in wild-type animals, supporting the results seen in the gain-of-function models.

The hair of *Bcl2* mutants that survive past the first few weeks turns gray. The timing of this change corresponds to the second cycle of hair-follicle growth, which produces an entire new set of hair starting at around 4 wk of age. Analysis of the hair follicles in these mice shows a reduction in both melanocytes and melanin granules indicating that *Bcl2* is important for survival of melanocytes. Significantly, epidermal melanocytes express high levels of *Bcl2 (136,137,140)*.

Bcl2 mutant mice also display phenotypes in both the intestine and nervous system. *Bcl2* is expressed in epithelial cells of the gut, and *Bcl2* mutants have abnormal villi and increased levels of apoptotic cell death in the intestinal epithelium *(137)*. Many neurons of the peripheral nervous system express *Bcl2*, and careful analysis of mutants reveals a reduction in number of both sensory and sympathetic neurons after the first postnatal week *(141)*.

These studies demonstrate multiple essential roles for *Bcl2* during mouse development and homeostasis. *Bcl2* is expressed widely during embryogenesis, yet *Bcl2*-deficient mice show relatively few developmental defects. This may be caused by redundancy between *Bcl2* and its other anti-apoptotic relatives, such as, *Bclx, Bclw,* and *A1*, many of which are also expressed during embryogenesis *(142,143)*.

Bclx

Bclx is a death-protecting member of the Bcl2 family that is most closely related to Bcl2. Two different alternatively spliced transcripts of *Bclx*—termed *BclxS* and *BclxL*—have been detected in humans *(53)*. In contrast to $Bclx_L$, $Bclx_S$—which lacks the BH1 and BH2 domains—has killing activity in vitro. However, $Bclx_S$ is expressed at extremely low levels in murine tissues, making it difficult to evaluate the role of the truncated form in regulation of apoptosis in vivo *(96)*.

Mice homozygous mutant for *Bclx* die at approx embryonic d 13.5 (E 13.5) *(144)*. By E 11.5, homozygous mutant *Bclx* embryos display extensive cell death in the differentiating neurons of the brain and spinal cord, and by E 12.5, there is massive apoptosis of neurons in the brain. *Bclx*-null embryos manifest a threefold increase in apoptotic death in hematopoietic cells of the liver as compared to control animals, and this is believed to be the primary cause of lethality at E 13.5 *(144)*. *Bclx* is normally expressed in both differentiating neurons and the hematopoietic system during embryogenesis and other death-protecting members of the *Bcl2* family appear unable to compensate for loss of *Bclx* in these tissues.

Death of mice null for *Bclx* early in embryogenesis precludes the analysis of roles of Bclx in other tissues later in embryonic or postnatal development. To circumvent this problem, chimeric mice were generated by injection of *Bclx*-null ES cells into wild-type host blastocysts. Adult chimeric mice that did not express *Bclx* in lymph nodes or spleen had reduced numbers of T and B cells, although the population distributions for both cell types was normal *(144)*. Collectively, these results demonstrate that *Bclx* is required for neural and hematopoietic development during mouse embryogenesis and hematopoietic cell survival in adult tissues.

Bax

In contrast to *Bcl2* and *Bclx, Bax* is a pro-apoptotic member of the gene family. Consequently, it may be predicted that genetic ablation of *Bax* would result in excessive cell survival. Indeed, *Bax*-null mutant mice display two major phenotypes that are both associated with cellular hyperplasia *(145)*. *Bax* mutant mice have an excess of cells in lymphoid tissues. Mutants have a 60% increase in numbers of thymocytes compared to wild-type controls, although the ratio of the various thymocyte populations is unaffected. Similarly, the spleen in *Bax* mutants is enlarged, and displays an 80% increase in numbers of B cells compared to control animals.

Bax mutant males are also sterile, and fail to produce mature sperm. Spermatogenesis is largely arrested at premeiotic stages with a major increase in the numbers of type B spermatogonia. Paradoxically, there is increased apoptosis in the male germ-cell population with only rare examples of germ cells in advanced stages of development. Possible mechanisms underlying the development of this phenotype are discussed later in this chapter.

Bclw

Bclw, another death-protecting member of the family, is also required for spermatogenesis. On a hybrid genetic background, mice mutant for *Bclw* are viable but male sterile *(146,147)*. The testes of *Bclw* mutant mice display widespread apoptosis of all stages of germ cells and a complete block in elongate haploid germ-cell development. By 6 mo of age, loss of testicular germ cells is complete. Surprisingly, Sertoli cells are also lost from the seminiferous epithelium with completion of loss at 6–8 mo of age. Subsequently, there is a dramatic reduction of the number of Leydig cells, which occurs via apoptosis and engulfment by macrophages (L. Russell et al., Biology of Reproduction 2001; in press). These findings suggest that *Bclw* is important for survival of both germ-cell and somatic cells in the testis.

Females also display reduced fertility, although this effect is subject to a genetic background effect. *Bclw* is expressed in oocytes *(148)* (A. Ross and G. MacGregor, unpublished observation), and appears to be required for survival of primordial follicles (A. Ross, J. Tilly and G. MacGregor, unpublished observations). Finally, most *Bclw*-deficient mice on a C57BL/6 genetic background can be identified because of a reduction in their body size and a facial dysmorphology. The developmental basis for these defects is currently under investigation, but is presumably associated with an increase in PCD within specific tissues. *Bclw* is widely expressed in both embryonic and adult tissues *(54)* (A. Ross and G. MacGregor, unpublished observations), yet, as with *Bcl2* deficient mice, *Bclw* mutant mice display only a limited number of overt developmental defects.

Bid

Bid is a death-promoting, BH3-only member of the Bcl2 family, which is activated upon cleavage by caspase-8 in the FasR/TNFR1 pathway *(123)*. Mice deficient for *Bid* are born at normal Mendelian frequencies, with no overt developmental defects *(149)*. This demonstrates that *Bid* is not required for normal embryonic development and survival. However, when wild-type and *Bid*-deficient mice were injected with anti-Fas antibody (which activates the Fas pathway), wild-type mice died rapidly from liver failure associated with massive hepatocellular apoptosis, while most of the *Bid*-deficient mice survived, showing little or no hepatic apoptosis. In the livers of wild-type mice injected with the antibody, downstream caspases were activated, and cytochrome *c* was released into the cytosol, but in the *Bid –/–* mice, cytochrome *c* was not released from the mitochondria and downstream caspases were not activated *(149)*. This confirms that Bid can mediate cytochrome *c* release in the FasR/TNFR1 pathway in vivo.

Double Mutants

To determine which *Bcl2* family members interact in vivo, mice were generated that have mutations in multiple *Bcl2* family genes. Mice lacking both *Bax* and *Bclx* were generated to determine whether loss of the death-promoting member *Bax* could rescue the lethality of *Bclx*-null mutants associated with excessive hematopoietic and neuronal cell loss *(150)*. Although *Bax/Bclx* double-null mutants also died during embryogenesis, a large decrease was observed in the amount of neural apoptotic death in the double mutants compared to the *Bclx* null embryos. This indicates that Bax and Bclx regulate survival of neurons in the development of the CNS. The lethality of the *Bax/Bclx* mutants is most likely caused by a failure to correct the excessive apoptosis in hematopoietic cells of the liver. A different death-promoting member of the *Bcl2* family may effect the death pathway in the hematopoietic cells of this tissue *(150)*.

Mice null-mutant for both *Bax* and *Bcl2* have also been generated *(151)*. Removal of Bax rescues the apoptosis and thymic hypoplasia in the Bcl2-deficient mice, indicating that Bax and Bcl2 both regulate cell death and survival in the thymus. In contrast, the kidney defect of the *bcl2* mutants and the testicular phenotype of *bax* mutants are not altered in the *bcl2/bax* mutants. It is possible that additional Bcl2 family members expressed in these tissues may prevent rescue of the respective mutant phenotypes.

APOPTOSIS AND *BCL2* FAMILY FUNCTION IN REPRODUCTIVE TISSUES

Functional analyses of the Bcl2 family using transgenic mice have demonstrated multiple roles for the family in regulation of apoptosis in both male and female germ cells, as well as a number of hormonally sensitive reproductive tissues. The following is an overview of known functions for Bcl2 family members in several of these organs.

Testis

Apoptosis is a normal feature of gametogenesis in both males and females. In males, there are two peak periods of germ-cell apoptosis, one during embryogenesis and the other postnatal.

PRENATAL

The primordial germ cells (PGCs) can first be identified in the mouse at E 7.5 d *post coitum* as a population of around 50 alkaline phosphatase expressing cells, which are located close to the allantoic bud *(152–154)*. Between E 8.5 and E 10.5, these cells migrate from the hindgut and into the genital ridges. The PGCs undergo several mitotic divisions during migration, until by E 13.5 there are approx 25,000 PGCs within each gonad *(153)*. PGCs in the developing ovary differentiate into oogonia, initiate meiosis to become oocytes, and by birth are arrested at prophase of the first meiotic division. Following onset of testicular differentiation the prospermatogonia continue to divide until E 14, at which point they undergo mitotic arrest *(154–156)*. Just prior to this arrest, at around E 13, large numbers of apoptotic prospermatogonia are observed, using both TUNEL and electron microscopy as criteria for identification of dying cells. By E 14, the number of apoptotic cells has dramatically decreased, and there is very little germ-cell death during the rest of the prenatal period *(157)*.

What controls survival of germ cells during embryogenesis? Several endocrine factors have been implicated in the regulation of PGC survival. *C-kit*, a tyrosine-kinase receptor encoded by the *W* locus, and a ligand, (Kit-ligand [KL]), that is encoded by the *Sl* locus are essential for migration, proliferation and survival of PGCs. This role has been demonstrated both in vivo and in vitro with mutations in either locus result in dramatic germ-cell deficiencies. KL can also promote the survival of PGCs in culture *(158)*. Other factors that may be critical for the survival of PGCs include interleukin-4, basic fibroblast growth factor, oncostatin M (OSM), leukemia-inhibitory factor (LIF) and gp130, a receptor that heterodimerizes with both OSM and LIF *(159)*.

Little is known regarding the potential role of the Bcl2 family in PGC survival. Expression of Bcl2 or Bclx can extend the survival of PGCs in culture *(160)*. However, it is currently unknown whether these members of the Bcl2 family are normally expressed in PGCs. *Bcl2*-null males are fertile, indicating that there is no requirement for Bcl2 during male germline development. However, since *Bclx*-null mutant embryos die around the time of gonadal differentiation, the status of the germline in these mutants could not be addressed. Thus, it remains possible that Bclx may have an essential role in survival of the germline during embryonic or postnatal development (160a).

Postnatal

The second peak in apoptotic death of male germ cells occurs postnatal. Shortly after birth, mitosis resumes and the first wave of spermatogenesis begins. During the first week, prospermatogonia differentiate into type A spermatogonia, the stem-cell population of the testis. This is followed by a period of spermatogonial proliferation until postnatal d 10, at which time germ cells initiate meiosis *(161)*. Many apoptotic spermatogonia and spermatocytes are seen in the testis in the second and third weeks after birth *(157,162)*. The incidence of cell death declines following the third week, when haploid cells are first found in the testis. While apoptosis is observed in male germ cells throughout adult life, the incidence of this never approaches that seen in prepubertal animals.

Considerably more is known about functions of the Bcl2 family in the postnatal and adult testis than the prenatal testis. Several Bcl2 family members are expressed in the testis, including Bclx *(163)*, Bclw *(54,147)*, Mcl-1 *(164)*, Bax *(162)*, Bak *(165)*, Bok *(166)*, Diva/Boo *(66)*, Bid (A. Ross and G. MacGregor, unpublished), Blk *(64)*, and Bod/Bim *(167)*. This makes it possible that Bcl2 family members could play many important roles in regulating cell death in the testis. Indeed, as mentioned earlier loss of function transgenic mouse studies mice have demonstrated a requirement for at least two Bcl2 family members in spermatogenesis.

Bax is expressed in spermatogonia of prepubertal and adult testes *(161)*. Analysis of prepubertal animals suggests that Bax expression in testis is maximal during the second and third weeks after birth with significantly lower levels of expression observed in the adult testis. The peak of Bax expression coincides with maximal apoptotic germ-cell deaths that occurs in the testis of wild-type animals 2–3 wk after birth. *Bax*-deficient mice have defective spermatogenesis that is associated with an accumulation of type B spermatogonia *(145)*. Testis histology of *Bax*-null mice shows that at postnatal d 9 (P9), ratios of germ cells to Sertoli cells are normal, yet by P12, there are 4–5 times as many germ cells in the knockout than in wild-type (L. D. Russell and C. M. Knudson, unpublished observations). This accumulation of spermatogonia in the *Bax* mutants coincides temporally with the peak of male germ-cell apoptosis in wild-type animals. Moreover, immunohistochemistry and TUNEL analyses indicate that many of the dying spermatogonia in testis of a 2.5-wk-old wild-type animal expressed Bax at high levels *(161)*. Collectively, these findings indicate that Bax is required for the wave of apoptotic death seen in the prepubertal testis.

Interestingly, an almost identical phenotype is seen in gain-of-function transgenic mice that overexpress either *Bcl2* or *Bclx* in germ cells in the testis *(161,168)*. In each of the latter cases, transgenic males were sterile and the seminiferous epithelium displayed a variety of defects. Some tubules had reduced numbers of spermatocytes and spermatids and increased numbers of spermatogonia. Other tubules contained either giant multinucleated cells or, in certain cases, virtually no germ cells. These phenotypes are also found in many tubules of older *Bax* mutant males. This suggests that the wave of Bax-induced apoptosis can be perturbed by ectopic and/or overexpression of either Bclx or Bcl2.

What is the function of this dramatic wave of apoptosis during prepubertal testicular development? The Sertoli cell is the somatic cell of the seminiferous epithelium that provides the immediate somatic environment in which male germ-cell development occurs *(169)*. Sertoli cells contact neighboring Sertoli cells to establish the basal and adluminal compartments, and these interactions are important for proper function of the

Sertoli cell *(170)*. Sertoli cells also form a complex network of contacts with different developmental stages of germ cells. A careful balance must be maintained between the Sertoli cells and germ cells, and this wave of apoptosis may keep germ-cell numbers at an appropriate level so as not to perturb this balance. By disrupting normal germ-cell:Sertoli-cell ratios, the accumulation of germ cells in testes of *Bax*-null mice may preclude Sertoli cells from being able to support spermatogenesis, resulting in the arrest in spermatogenesis seen in these animals, and the resulting increase in apoptotic death of germ cells. Alternatively, it is possible that *Bax*-deficient germ cells arrest in development in a cell-autonomous manner under some form of fail-safe mechanism. Finally, as Bax is expressed in mouse Sertoli cells (A. Ross and G. MacGregor, unpublished observation), it is possible that there is a fundamental developmental defect in *Bax*-deficient Sertoli cells. Discrimination between an intrinsic or extrinsic developmental defect in *Bax*-deficient germ cells, and Sertoli cells could be addressed by testicular germ-cell transplantation techniques *(171)*.

From an endocrine perspective, there is evidence that this crucial period of apoptotic death is under hormonal control. Wild-type male mice administered testosterone from 1–3 wk of age, during the normal wave of spermatogonial apoptosis, display abnormalities of the seminiferous epithelium similar to those of the *Bcl2* and *Bclx* gain-of-function and *Bax*-null transgenic mice. This suggests that testosterone can also block the early wave of apoptosis *(161)*. Conversely, androgen withdrawal in adult mice is associated with the reverse effect. Mice treated with ethane dimethanesulfonate (EDS), a Leydig-cell toxicant, showed a 24-fold increase in apoptotic germ-cell deaths *(172)*. Testicular levels of *Bax* also increased after this treatment; thus, it is possible that *Bax* expression could be regulated by androgens. In prepubertal mice, a relatively high level of apoptotic death is required to regulate the spermatogonial population, as evidenced by the phenotype of the *Bax* mutants. At this point, gonadotropin levels are considerably lower than in adults, and LH-stimulated testosterone secretion is also reduced. Lower androgen levels may facilitate increased expression of Bax, and as a result, allow the higher levels of apoptosis that are required for proper initiation of spermatogenesis.

Bclw mutant mice also show defects in spermatogenesis and are sterile, although the testicular phenotype is significantly different from that seen in the *Bax* mutants. Germ-cell development in prenatal and prepubertal testis of *Bclw* mutants appears to proceed normally through postnatal d 16 (P16). However, by P19, a sevenfold increase in degenerating spermatocytes is seen in *Bclw* mutants *(147)*. At P24, there is a significant increase in incidence of TUNEL-positive spermatocytes, and a 10-fold reduction in the number of spermatids. Subsequently, spermatids are arrested in development at step 13, and the first wave of spermatogenesis is never completed. By 2–3 mo of age, the majority of germ cells in the testis are apoptotic, and by 5–6 mo, the tubules are essentially devoid of germ cells.

In contrast to the testis in *Bax*-null animals, the Leydig-cell and Sertoli-cell populations are also affected in *Bclw* mutants. By 6 wk of age, the Sertoli population is reduced by approx 84% compared to wild-type, and by 7–8 mo the Sertoli cells are essentially absent from the tubules *(146,147)*. The Leydig-cell compartment displays hyperplasia prior to completion of loss of Sertoli cells, with an increase of as much as 50% in cell numbers compared to wild-type controls. However, the majority of Leydig cells are lost by apoptosis and phagocytosis following completion of loss of Sertoli cells, with only a relatively small number of mesenchymal cells and some phagocytes remaining in the interstitial space by 8 mo of age. At this time, the peritubular myoid cells and endothelial cells of the tubule basement are the primary cell types remaining within the testis *(146,147)*.

The primary cause of germ-cell loss in *Bclw* mutants does not appear to be caused by endocrine effects, as gonadotropin levels were initially equivalent in *Bclw* mutants and controls and, as expected, FSH levels were elevated during the period of protracted germ-cell loss *(147)*. Additionally, weights of seminal vesicles, a testosterone-sensitive tissue, were similar in young mutants and controls, suggesting that testosterone levels were not significantly affected *(146,147)*.

Bclw is expressed in both Sertoli cells and in premeiotic and meiotic male germ cells, but not in elongate germ cells as first described *(146)* (Russell et al., Biology of Reproduction 2001; in press). Thus it is likely that the arrest in elongate germ-cell development in *Bclw* mutants is not cause by cell-intrinsic effects. It is currently unknown whether Bclw is required in an autonomous manner for diploid male germ-cell survival, or whether germ-cell loss in *Bclw* mutant males is entirely an indirect consequence of *Bclw*-deficient Sertoli-cell dysfunction. This could be ascertained by transplantation of *Bclw*-deficient male germ cells into a wild-type syngenic recipient testis.

Similarly, while Sertoli cells are notoriously difficult to ablate, it is possible that the excessive loss of germ cells is indirectly responsible for Sertoli-cell attrition in *Bclw*-null males. To test this, mice deficient for both *c-kit* and *Bclw* were generated. *C-kit* deficient mice lack germ cells from birth, resulting from germ-cell-intrinsic defects. Sertoli cells are not affected, as they do not normally express *c-kit*. In testis lacking germ cells, *Bclw*-null Sertoli cells showed reduced survival compared to Sertoli cells from *Bclw* wild-type, *c-kit* deficient mice (Ross and MacGregor, submitted). This confirms that loss of *Bclw*-deficient Sertoli cells occurs independently of germ cells, and suggests that *Bclw* is required cell-intrinsically for Sertoli-cell survival.

Finally, *Bclw* is not expressed in Leydig cells *(146,147)*. Thus, it is likely that the depletion of Leydig cells in aged *Bclw* mutants is a secondary effect related to loss of Sertoli cells, which would indicate a function for Sertoli cells in support of adult Leydig-cell survival.

Bax, Bak, and Bid are all expressed in mouse Sertoli cells (Ross and MacGregor, submitted), and one or more of these proapoptotic molecules may be responsible for killing the Sertoli cells in the *Bclw* mutants. To determine whether *Bax* mediates the loss of Sertoli cells in the *Bclw*-deficient mice, *Bclw/Bax* mutant animals were generated. While spermatogenesis is still defective in the *Bclw/Bax* mutants, as it is in the *Bax* mutants alone, the loss of Sertoli cells seen in *Bclw* single mutants is suppressed (Ross and MacGregor, submitted). This demonstrates that Bclw and Bax both regulate survival of this specific cell type.

Prostate

Apoptosis occurs in a hormone-sensitive manner in the prostate. Either androgen withdrawal or castration results in increased apoptotic death in the prostate glandular epithelium *(173)*. Little is currently known about the roles of the Bcl2 family in the regulation of prostate apoptosis. While Bcl2 is not detected in normal prostate epithelial cells in humans, Bcl2 expression is detectable, and increases as prostate cancers become androgen-independent, raising the possibility that Bcl2 may have a role in neoplastic growth in this tissue *(174)*. In support of this hypothesis, mice that overexpress a *Bcl2* transgene in prostate epithelial cells have morphologically abnormal prostates, with an accumulation of cells in stromal and epithelial compartments *(175)*. Understanding Bcl2 family function in the prostate may ultimately have useful value for treatment of prostatic cancer.

Ovary

As with male germ-cell development, a significant loss of female germ cells is associated with normal ovarian development (*see* Morita and Tilly, 1999) *(176)*. Following the onset of ovarian differentiation, the oogonial population undergoes a short period of mitotic activity during which modest loss of germ cells is observed *(177)*. After exiting mitosis, oogonia progress through meiosis until they arrest at the diplotene stage of the first meiotic prophase. Many of the oocytes are subsequently surrounded by a single layer of somatic granulosa cells to form primordial follicles, and naked oocytes invariably degenerate in a process called attrition *(178)*. This is a short period of relatively high levels of germ-cell death that effectively regulates the follicular endowment of the ovary and is completed by d 4 postnatal.

The other major period of regulated cell death in the ovary occurs during follicular development. In both prepubertal and adult life, subsets of the primordial follicle endowment are recruited to develop. However, from these recruited follicles, only a small number will complete development and be ovulated. The remainder degenerate in a process called atresia, which can occur at each stage of follicular development *(179)*. Atresia results from the hormonally regulated death of granulosa cells, which are apoptotic in nature *(180,181)*.

Several endocrine factors have been identified that can regulate follicle atresia. Reduction of serum gonadotropins by hypophysectomy results in follicular atresia, which supports the role of these hormones in follicle survival in vivo *(182)*. In vitro models have also been useful in identifying factors that promote follicle survival. Both follicle-stimulating hormone (FSH) and luteinizing hormone (LH) can suppress apoptosis of isolated follicles *(183,184)*. Several growth factors produced by the follicular cells can enhance survival of isolated granulosa cells as well as survival of antral follicles including EGF, basic fibroblast growth factor (bFGF), and IGF-1 *(185,186)*. Estrogens and progesterones have also been implicated in follicular cell survival both in vivo and in vitro *(187,188)*. The basis for hormonal regulation of follicle survival is likely to be complex.

As with the testis, many members of the Bcl2 family are expressed in the ovary, including Bcl2 *(189,190)*, Bclx *(190)*, Bax *(190)*, Bad *(191)*, Bak *(165)*, Bclw *(191)*, Bok *(166)*, Bod *(167)*, and Mcl1 *(164)*. Transgenic mouse models have revealed essential roles of three of these family members, Bax, Bclw, and Bcl2, in ovarian function in vivo.

To determine whether Bcl2 could protect follicular cells from PCD, mice that overexpress *Bcl2* in granulosa cells were generated. These transgenic animals show reduced apoptosis of ovarian cells, larger litter size, and formation of teratomas in advanced age *(193)*. Conversely, analysis of ovaries from *Bcl2*-null mutants also supported a role for Bcl2 in survival of female germ cells during normal development. On a hybrid genetic background, *Bcl2* mutant female mice are fertile, and all stages of follicular development are represented. However, the total number of primordial follicles at birth was significantly reduced in the mutants compared with wild-type or heterozygous controls *(181)*. Moreover, a significant proportion of the primordial follicles did not contain an oocyte. These findings indicate that Bcl2 is important for survival of a proportion of female germ cells. It is not yet clear when the reduction in oocyte and primordial follicle number initially occurs. Bcl2 may be important for survival of oogonia during prenatal development, or it could protect oocytes during the postnatal period of extensive cell death when primordial follicles form.

Bax also functions in control of ovarian cell death. Ovaries from adult *Bax* mutant mice contain approximately three times the number of primordial follicles as control littermates. Moreover, ovaries of 20–22-mo-old *Bax*-null females that would normally have exhausted their follicle endowment have hundreds of follicles at all stages of development *(194)*. In contrast, *Bax* mutants had the same number of follicles as control animals 4 d after birth. Thus, Bax is not required for oocyte attrition during embryogenesis, but instead appears to be essential for follicular atresia postnatal. Bax probably functions by initiating death of follicular granulosa cells, in which it is expressed *(191)*. Thus, while Bcl2 function is important for early oocyte survival and establishing the primordial follicle endowment, the most important function of Bax in the ovary appears to involve rendering follicles susceptible to hormonally induced PCD.

Of clinical relevance, Bax-deficient mice were also used to study the requirement for Bax in mediating PCD of oocytes in response to the chemotherapeutic drug doxorubicin (Adriamycin). Unfertilized oocytes exposed to doxorubicin undergo apoptosis. However, oocytes from Bax-null female mice are completely resistant to this treatment *(195)*. This experimental finding could have important implications for the development of novel therapies with which to combat female infertility that results from cancer treatment.

Ovaries from 4-d-old *Bclw*-mutant females on a B6-enriched genetic background display a 40% reduction in oocyte endowment (MacGregor and J. Tilly, unpublished observations). *Bclw* is expressed in oocytes, suggesting that it is required cell-intrinsically for oocyte survival. To determine whether *Bax* mediates the reduced oocyte endowment in *Bclw* mutants, *Bclw/Bax*-deficient mice were analyzed. Removal of *Bax* restored the oocyte endowment at p4 to wild-type levels. Thus, as with the testis, both Bclw and Bax have important roles in regulation of female germ-cell survival.

Mammary Gland

Apoptosis occurs during both development of the mammary gland and its involution following withdrawal of a suckling response (weaning). During mammary-gland development, a proliferative structure called the terminal endbud is responsible for penetrating the fat pad and forming lumena and ducts from the solid tissue mass. Consistent with this developmental process, high levels of apoptosis occur in the terminal endbud, and Bclx, Bax, and Bcl2 are all expressed in this structure. To determine whether regulated expression of Bcl2 in the mammary gland was important for its development, mice were generated that expressed Bcl2 ectopically in this organ under control of the whey acidic protein (WAP) promoter. In these mice, apoptosis was reduced during mammary-gland development, and the endbud was highly disorganized. This indicates that this process requires regulation of PCD, and that this can be influenced by the Bcl2 pathway *(196)*.

PCD also occurs during mammary-gland involution, when lactational epithelial cells are removed. This hormone-sensitive process is triggered by a decrease in levels of prolactin and glucocorticoid hormone following removal of a suckling response at weaning. Many members of the Bcl2 family are expressed in mammary gland during lactation and involution, including Bcl2, Bclx, Bclw, Bax, Bak, and Bad *(197)*. Interestingly, both Bcl2 and Bclw are downregulated prior to involution. While Bclw and Bcl2 may function in mammary gland involution, Bclw is not essential for this developmental process, as *Bclw*-null mutant females do not appear to have an overt defect in mammary-gland development (A. Ross and G. MacGregor, unpublished results). It is possible that functional redundancy between Bclw and Bcl2 could mask the effects of loss of Bclw function in this tissue.

WAP-Bcl2 mice were also used to examine apoptosis during involution *(198)*. Overexpression of Bcl2 blocked apoptosis of the lactational cells following weaning. However, the alveolar collapse that is part of involution still occurred. This indicates that alveolar collapse, which is triggered by activity of matrix metalloproteases, occurs independently of apoptosis. In summary, apoptosis during mammary gland involution may involve Bcl2 family function. However, the precise role that each member plays and to what extent there is functional overlap with other family members is currently unclear.

Bax can also function as a suppressor of mammary-gland tumor formation. In an experimental mouse model of mammary tumorigenesis, expression of Bax was increased during early stages of mammary cancer development, then down regulated following transformation *(199)*. This is consistent with findings in human breast carcinomas where expression of Bax is also downregulated *(200)*. In mice mutant for one allele of *Bax*, there was a significantly increased rate in mammary cancer development and tumor growth *(199)*. These findings indicate a dose-dependent role for Bax in mediating PCD within preneoplastic mammary-gland tumors. Interestingly, the rate of tumor incidence in *Bax*-null mice was actually decreased compared with the heterozygous *Bax* mutants. However, *Bax*-null females displayed a ductal hypoplasia, which may be responsible for their reduced sensitivity to tumor formation. The ductal hypoplasia in *Bax* mutants is also of interest as it supports a role for Bax in normal development of mouse mammary glands.

Uterus

The uterine endometrium undergoes many cycles of steroid hormone-induced cell proliferation, differentiation, and death during its normal function. Estrogen depletion results in apoptotic death of murine endometrial epithelium, and estrogen, progesterone, and androgen all can suppress this epithelial cell death *(201–203)*. The manner in which these hormones regulate uterine apoptosis, and to what extent the Bcl2 family may control this process is not yet clear. Initial studies suggest a potential role for the Bcl2 family in uterine-cell deaths. *Bcl2* is expressed in human endometrium during all stages of the menstrual cycle, with its strongest expression during the proliferative phase of the cycle *(204,205)*. In a rat endometrial-cell line, apoptosis was induced by withdrawal of progesterone, or addition of antiprogestins. When these cells were treated with progesterone, levels of $Bclx_L$ in the cell increased and levels of $Bclx_S$ decreased, suggesting a possible mechanism for the protective effects of the hormone *(206)*. During uterine decidualization induced by progestin and estradiol treatment in rats, levels of Bax increased, and levels of Bcl2 decreased in stromal cells *(207)*. These studies suggest that Bcl2 family proteins could function in apoptosis during cycling of the uterine endothelium and regression of decidual zones during pregnancy, and that the levels of the family members may be under hormonal control. The death of Bcl2- or Bclx-deficient females prior to reproductive age currently precludes analysis of essential roles for Bcl2 or Bclx in uterine function in the mouse.

SUMMARY AND FUTURE PERSPECTIVES

Studies using transgenic mice have demonstrated a number of essential roles for Bcl2 family members in the regulation of apoptosis, both during embryonic development and adult tissue homeostasis. Of interest, several members of the family are involved in hormonally regulated cell deaths occurring in reproductive tissues. Both spermatogenesis and oogenesis require apoptosis of a proportion of germ cells for normal gametogenesis. Bax, a pro-apoptotic protein, and Bclw, a death-protecting protein, are both required for normal levels of apoptosis during spermatogenesis. Bclw, Bax, Bcl2, and possibly $Bclx_L$, all function in regulation of oocyte and follicular apoptosis. Many more Bcl2 family members are expressed in the ovary and testis, and it is important to determine whether these genes also play crucial roles in gametogenesis.

Loss-of-function mutations have only been generated for a handful of *Bcl2* family members. A greater understanding of the roles of the Bcl2 family in different tissues will be gained as other knockout mice are produced. Generation of mice with mutations in multiple independent *Bcl2*-family genes can be used to establish which family members act in the same pathway in specific tissues, and which members have redundant functions. Mice deficient for *Bclx* die embryonically, and other *Bcl2* relatives may have essential functions during development. Consequently, it is important to generate new conditional knockouts, such as those generated using the Cre/*lox P* system *(208)*, that are both temporally and spatially restricted, which can be used to reveal functions of these gene products in adult tissues. Future studies of the *Bcl2* gene family in mice should provide a wealth of additional insight into the mechanisms controlling apoptosis, and the physiological roles of cell death. These findings will in turn assist efforts to devise novel therapies to treat human diseases that arise from an imbalance in programmed cell death.

ACKNOWLEDGMENTS

Andrea J. Ross was supported by NIH training grant number 5 T32 GM08367. Research in Grant R. MacGregor's laboratory is supported by NIH grant number HD-36437.

REFERENCES

1. Clarke PGH, Clarke S. Historic apoptosis. Nature 1995;378:230.
2. Lockshin RA, Williams CM. Programmed cell death II: Endocrine potentiation of the breakdown of the intersegmental muscles of silkmoths. J Insect Physiol 1964;10:643.
3. Kerr JF, Wyllie AH, Currie AR. Apoptosis: a basic biological phenomenon with wide-ranging implications in tissue kinetics. Br J Cancer 1972;26:239–257.
4. Clarke PG, Posada A, Primi MP, Castagne V. Neuronal death in the central nervous system during development. Biomed Pharmacother 1998;52:356–362.
5. Russell LD. Cell loss during spermatogenesis: apoptosis or necrosis? In: Hamamah S, Mieusset R, Olivennes O, Frydman R, eds. Male Sterility and Motility Disorders: Etiological Factors and Treatment. Springer, New York, NY, 1999, pp. 203–214.
6. Saunders JW. Death in embryonic systems. Science 1966;154:604–612.
7. Coucouvanis E, Martin GR. Signals for death and survival: a two-step mechanism for cavitation in the vertebrate embryo. Cell 1995;83:279–287.
8. Kerr JF, Harmon B, Searle J. An electron-microscope study of cell deletion in the anuran tadpole tail during spontaneous metamorphosis with special reference to apoptosis of striated muscle fibers. J Cell Sci 1974;14:571–585.
9. Roberts LM, Hirokawa Y, Nachtigal MW, Ingraham HA. Paracrine-mediated apoptosis in reproductive tract development. Dev Biol 1999;208:110–122.
10. Ellis RE, Yuan JY, Horvitz HR. Mechanisms and functions of cell death. Ann Rev Cell Bio 1991;7:663–698.

11. Vaux DL, Korsmeyer SJ. Cell death in development. Cell 1999;96:245–254.
12. Stellar H. Mechanisms and genes of cellular suicide. Science 1995;267:1445–1449.
13. Thompson CB. Apoptosis in the pathogenesis and treatment of disease. Science 1995;267: 1456–1462.
14. Raff MC. Social controls on cell survival and cell death. Nature 1992;356:397–399.
15. Collins MK, Lopez Rivas A. The control of apoptosis in mammalian cells. TIBS 1993;18:307–309.
16. Bates RC, Buret A, van Helden DF, Horton MA, Burnes GF. Apoptosis induced by inhibition of cellular contact. J Cell Biol 1994;125:403–415.
17. Gajewski TF, Thompson CB. Apoptosis meets signal transduction: elimination of a BAD influence. Cell 1996;87:589–592.
18. Wyllie AH, Kerr JF, Macaskill IA, Currie AR. Adrenocortical cell deletion: the role of ACTH. J Pathol 1973;111:85–94.
19. Coles HSR, Burne JF, Raff MC. Large-scale normal cell death in the developing rat kidney and its reduction by epidermal growth factor. Development 1993;18:777–784.
20. Fleischman RA. From white spots to stem cells: the role of the Kit receptor in mammalian development. Trends Genet 1993;9:285–290.
21. Jiang C, Baehrecke EH, Thummel CS. Steroid regulated programmed cell death during Drosophila metamorphosis. Development 1997;124:4673–4683.
22. Catlin EA, MacLaughlin DT, Donahoe PK. Mullerian inhibiting substance: new perspectives and future directions. Microsc Res Tech 1993;25:121–133.
23. Anderson P. Kinase cascades regulating entry into apoptosis. Micr Mol Biol Rev 1997;61:33–46.
24. Dragovich T, Rudin CM, Thompson CB. Signal transduction pathways that regulate cell survival and cell death. Oncogene 1998;17:3207–3213.
25. Jarpe MB, Widmann C, Knall C, Schlesinger TK, Gibson S, Yujiri T, et al. Anti-apoptotic versus pro-apoptotic signal transduction: checkpoints and stop signs along the road to death. Oncogene 1998;17:1475–1482.
26. Sulston JE, Horvitz HR. Post-embryonic cell lineages of the nematode, Caenorhabditis elegans. Dev Biol 1977;56:110–156.
27. Sulston JE, Schierenberg E, White JG, Thomson JN. The embryonic cell lineage of the nematode Caenorhabditis elegans. Dev Biol 1983;100:64–119.
28. Hengartner MO. Programmed cell death in invertebrates. Curr Opin Genet Dev 1996;6:34–38.
29. Ellis HM, Horvitz HR. Genetic control of programmed cell death in the nematode C. elegans. Cell 1986;44:817–829.
30. Conradt B, Horvitz HR. The C. elegans protein EGL-1 is required for programmed cell death and interacts with the Bcl-2-like protein CED-9. Cell 1998;93:519–529.
31. Hengartner MO, Ellis RE, Horvitz HR. Caenorhabditis elegans gene ced-9 protects cells from programmed cell death. Nature 1992;356:494–499.
32. Metzstein MM, Stanfield GM, Horvitz HR. Genetics of programmed cell death in C. elegans: past, present and future. Trends Genet 1998;14:410–416.
33. Nunez G, Benedict MA, Hu Y, Inohara N. Caspases: the proteases of the apoptotic pathway. Oncogene 1998;17:3237–3245.
34. Yuan J, Shaham S, Ledoux S, Ellis HM, Horvitz HR. The C. elegans cell death gene ced-3 encodes a protein similar to mammalian interleukin-1 beta-converting enzyme. Cell 1993;75:641–652.
35. Nicholson DW, Thornberry NA. Caspases: killer proteases. TIBS 1997;22:299–306.
36. Varfolomeev EE, Schuchmann M, Luria V, Chiannilkulchai N, Beckmann JS, Mett IL, et al. Targeted disruption of the mouse Caspase 8 gene ablates cell death induction by the TNF receptors, Fas/Apo1, and DR3 and is lethal prenatally. Immunity 1998;9:267–276.
37. Kuida K, Zheng TS, Na S, Kuan C, Yang D, Karasuyama H, et al. Decreased apoptosis in the brain and premature lethality in CPP32-deficient mice. Nature 1996;384:368–372.
38. Bergeron L, Perez GI, Macdonald G, Shi L, Sun Y, Jurisicova A, et al. Defects in regulation of apoptosis in caspase-2-deficient mice. Gen Dev 1998;12:1304–1314.
39. Wilson KP, Black JA, Thomson JA, Kim EE, Griffith JP, Navia MA, et al. Structure and mechanism of interleukin-1 beta converting enzyme. Nature 1994;370:270–275.
40. Salvesen GS, Dixit VM. Caspase activation: the induced-proximity model. Proc Natl Acad Sci USA 1999;96:10,964–10,967.
41. Ashkenazi A, Dixit VM. Death receptors: signaling and modulation. Science 1998;281:1305–1308.

42. Baker SJ, Reddy EP. Modulation of life and death by the TNF receptor superfamily. Oncogene 1998;17:3261–3270.
43. Nagata S. Apoptosis by death factor. Cell 1997;88:355–365.
44. Gearing AJ, Beckett P, Christodoulou M, Churchill M, Clements J, Davidson AH, et al. Processing of tumour necrosis factor-alpha precursor by metalloproteinases. Nature 1994;370:555–557.
45. Boldin MP, Varfolomeev EE, Pancer Z, Mett IL, Camonis JH, Wallach D. A novel protein that interacts with the death domain of Fas/APO1 contains a sequence motif related to the death domain. J Biol Chem 1995;270:7795–7798.
46. Hofmann K, Bucher P, Tschopp J. The CARD domain: a new apoptotic signalling motif. TIBS 1997;22:155,156.
47. Nagata S. Mutations in the Fas antigen gene in lpr mice. Semin Immunol 1994;6:3–8.
48. Bakhshi A, Jensen JP, Goldman P, Wright JJ, McBride OW, Epstein AL, et al. Cloning the chromosomal breakpoint of t(14;18) human lymphomas: clustering around JH on chromosome 14 and near a transcriptional unit on 18. Cell 1985;41:899–906.
49. Tsujimoto Y, Gorham J, Cossman J, Jaffe E, Croce CM. The t(14;18) chromosome translocations involved in B-cell neoplasms result from mistakes in VDJ joining. Science 1985;229:1390.
50. Nunez G, London L, Hockenbery D, Alexander M, McKearn JP, Korsmeyer SJ. Deregulated Bcl-2 gene expression selectively prolongs survival of growth factor-deprived hemopoietic cell lines. J Immunol 1990;144:3602–3610.
51. Vaux DL, Cory S, Adams JM. Bcl-2 gene promotes haemopoietic cell survival and cooperates with c-myc to immortalize pre-B cells. Nature 1988;335:440–442.
52. Hengartner MO, Horvitz HR. C. elegans cell survival gene ced-9 encodes a functional homolog of the mammalian proto-oncogene bcl-2. Cell 1994;76:665–676.
53. Boise LH, Gonzalez-Garcia M, Postema CE, Ding L, Lindsten T, Turka LA, et al. bcl-x, a bcl-2-related gene that functions as a dominant regulator of apoptotic cell death. Cell 1993;74:597–608.
54. Gibson L, Holmgreen SP, Huang DC, Bernard O, Copeland NG, Jenkins NA, et al. bcl-w, a novel member of the bcl-2 family, promotes cell survival. Oncogene 1996;13:665–675.
55. Lin EY, Orlofsky A, Berger MS, Prystowsky MB. Characterization of A1, a novel hemopoietic-specific early-response gene with sequence similarity to bcl-2. J Immunol 1993;151:1979–1988.
56. Kozopas KM, Yang T, Buchan HL, Zhou P, Craig RW. MCL1, a gene expressed in programmed myeloid cell differentiation, has sequence similarity to BCL2. Proc Natl Acad Sci USA 1993;90: 3516–3520.
57. Oltvai ZN, Milliman CL, Korsmeyer SJ. Bcl-2 heterodimerizes in vivo with a conserved homolog, Bax, that accelerates programmed cell death. Cell 1993;74:609–619.
58. Chittenden T, Harrington EA, O'Connor R, Flemington C, Lutz RJ, Evan GI, et al. Induction of apoptosis by the Bcl-2 homologue Bak. Nature 1995;374:733–736.
59. Kelekar A, Chang BS, Harlan JE, Fesik SW, Thompson CB. Bad is a BH3 domain-containing protein that forms an inactivating dimer with Bcl-XL. Mol Cell Biol 1997;17:7040–7046.
60. Boyd JM, Gallo GJ, Elangovan B, Houghton AB, Malstrom S, Avery BJ, et al. Bik, a novel death-inducing protein shares a distinct sequence motif with Bcl-2 family proteins and interacts with viral and cellular survival-promoting proteins. Oncogene 1995;11:1921–1928.
61. Wang K, Yin XM, Chao DT, Milliman CL, Korsmeyer SJ. BID: a novel BH3 domain-only death agonist. Genes Dev 1996;10:2859–2869.
62. O'Connor L, Strasser A, O'Reilly LA, Hausmann G, Adams JM, Cory S, et al. Bim: a novel member of the Bcl-2 family that promotes apoptosis. EMBO J 1998;17:384–395.
63. Inohara N, Ekhterae D, Garcia I, Carrio R, Merino J, Merry A, et al. Mtd, a novel Bcl-2 family member activates apoptosis in the absence of heterodimerization with Bcl-2 and Bcl-XL. J Biol Chem 1998;273:8705–8710.
64. Hegde R, Srinivasula SM, Ahmad M, Fernandes-Alnemri T, Alnemri ES. Blk, a BH3-containing mouse protein that interacts with Bcl-2 and Bcl-xL, is a potent death agonist. J Biol Chem 1998;273:7783–7786.
65. Inohara N, Ding L, Chen S, Nunez G. harakiri, a novel regulator of cell death, encodes a protein that activates apoptosis and interacts selectively with survival-promoting proteins Bcl-2 and Bcl-X(L). EMBO J 1997;16:1686–1694.
66. Inohara N, Gourley TS, Carrio R, Muniz M, Merino J, Garcia I, et al. Diva, a Bcl-2 homologue that binds directly to Apaf-1 and induces BH3-independent cell death. J Biol Chem 1998; 273:32,479–32,486.
67. D'Sa-Eipper C, Subramanian T, Chinnadurai G. bfl-1, a bcl-2 homologue, suppresses p53-induced apoptosis and exhibits potent cooperative transforming activity. Cancer Res 1996;56:3879–3882.

68. Chen G, Cizeau J, Vande Velde C, Park JH, Bozek G, Bolton J, et al. Nix and Nip3 form a subfamily of pro-apoptotic mitochondrial proteins. J Biol Chem 1999;274:7–10.
69. Chen G, Ray R, Dubik D, Shi L, Cizeau J, Bleackley RC, et al. The E1B 19K/Bcl-2-binding protein Nip3 is a dimeric mitochondrial protein that activates apoptosis. J Exp Med 1997;186:1975–1983.
70. Subramanian T, Boyd JM, Chinnadurai G. Functional substitution identifies a cell survival promoting domain common to adenovirus E1B 19 kDa and Bcl-2 proteins. Oncogene 1995;11:2403–2409.
71. Subramanian T, Tarodi B, Chinnadurai G. Functional similarity between adenovirus E1B 19-kDa protein and proteins encoded by Bcl-2 proto-oncogene and Epstein-Barr virus BHRF1 gene. Curr Top Microbiol Immunol 1995;199:153–161.
72. Minn AJ, Swain RE, Ma A, Thompson CB. Recent progress on the regulation of apoptosis by Bcl-2 family members. Adv Immunol 1998;70:245–279.
73. Yin XM, Oltvai ZN, Korsmeyer SJ. BH1 and BH2 domains of Bcl-2 are required for inhibition of apoptosis and heterodimerization with Bax. Nature 1994;369:321–323.
74. Wang K, Gross A, Waksman G, Korsmeyer SJ. Mutagenesis of the BH3 domain of BAX identifies residues critical for dimerization and killing. Mol Cell Biol 1998;18:6083–6089.
75. Korsmeyer SJ, Shutter JR, Veis DJ, Merry DE, Oltvai ZN. Bcl-2/Bax: a rheostat that regulates an anti-oxidant pathway and cell death. Semin Cancer Biol 1993;4:327–332.
76. Kelekar A, Thompson CB. Bcl-2-family proteins: the role of the BH3 domain in apoptosis. Trends Cell Biol 1998;8:324–330.
77. Chinnaiyan AM, O'Rourke K, Lane BR, Dixit VM. Interaction of CED-4 with CED-3 and CED-9: a molecular framework for cell death. Science 1997;275:1122–1126.
78. Liu X, Kim CN, Yang J, Jemmerson R, Wang X. Induction of apoptotic program in cell-free extracts: requirement for dATP and cytochrome c. Cell 1996;86:147–157.
79. Zou H, Henzel WJ, Liu X, Lutschg A, Wang X. Apaf-1, a human protein homologous to C. elegans CED-4, participates in cytochrome c-dependent activation of caspase-3. Cell 1997;90:405–413.
80. Li P, Nijhawan D, Budihardjo I, Srinivasula SM, Ahmad M, Alnemri ES, et al. Cytochrome c and dATP-dependent formation of Apaf-1/caspase-9 complex initiates an apoptotic protease cascade. Cell 1997;91:479–489.
81. Saleh A, Srinivasula SM, Acharya S, Fishel R, Alnemri ES. Cytochrome c and dATP-mediated oligermization of Apaf-1 is a prerequisite for procaspase-9 activation. J Biol Chem 1999;274: 17,941–17,945.
82. Hu Y, Ding L, Spencer DM, Nunez G. WD-40 repeat region regulates Apaf-1 self-association and procaspase-9 activation. J Biol Chem 1998;273:33,489–33,494.
83. Hu Y, Benedict MA, Wu D, Inohara N, Nunez G. Bcl-XL interacts with Apaf-1 and inhibits Apaf-1-dependent caspase-9 activation. Proc Natl Acad Sci USA 1998;95:4386–4391.
84. Pan G, O'Rourke K, Dixit VM. Caspase-9, Bcl-XL, and Apaf-1 form a ternary complex. J Biol Chem 1998;273:5841–5845.
85. Song Q, Kuang Y, Dixit VM, Vincenz C. Boo, a novel negative regulator of cell death, interacts with APAF-1. EMBO J 1999;18:167–178.
86. Moriishi K, Huang DCS, Cory S, Adams JM. Bcl-2 family members do not inhibit apoptosis by binding the caspase activator Apaf-1. Proc Natl Acad Sci USA 1999;96:9683–9688.
87. Kluck RM, Bossy-Wetzel E, Green DR, Newmeyer DD. The release of cytochrome c from mitochondria: a primary site for Bcl-2 regulation of apoptosis. Science 1997;275:1132–1136.
88. Yang J, Liu X, Bhalla K, Kim CN, Ibrado AM, Cai J, et al. Prevention of apoptosis by Bcl-2: release of cytochrome c from mitochondria blocked. Science 1997;275:1129–1132.
89. Reed JC. Cytochrome c: can't live with it—can't live without it. Cell 1997;91:559–562.
90. Susin SA, Lorenzo HK, Zamzami N, Marzo I, Snow BE, Brothers GM, et al. Molecular characterization of mitochondrial apoptosis-inducing factor. Nature 1999;397:441–446.
91. Jurgensmeier JM, Xie Z, Deveraux Q, Ellerby L, Bredesen D, Reed JC. Bax directly induces release of cytochrome c from isolated mitochondria. Proc Natl Acad Sci USA 1998;95:4997–5002.
92. Kharbanda S, Pandey P, Schofield L, Israels S, Roncinske R, Yoshida K, et al. Role for Bcl-xL as an inhibitor of cytosolic cytochrome C accumulation in DNA damage-induced apoptosis. Proc Natl Acad Sci USA 1997;94:6939–6942.

93. Lithgow T, van Driel R, Bertram JF, Strasser A. The protein product of the oncogene bcl-2 is a component of the nuclear envelope, the endoplasmic reticulum, and the outer mitochondrial membrane. Cell Growth Differ 1994;**5**:411–417.
94. Nguyen M, Millar DG, Yong VW, Korsmeyer SJ, Shore GC. Targeting of Bcl-2 to the mitochondrial outer membrane by a COOH-terminal signal anchor sequence. J Biol Chem 1993;**268**:25,265–25,268.
95. Krajewski S, Tanaka S, Takayama S, Schibler MJ, Fenton W, Reed JC. Investigation of the subcellular distribution of the bcl-2 oncoprotein: residence in the nuclear envelope, endoplasmic reticulum and outer mitochondrial membranes. Cancer Res 1993;**53**:4701–4714.
96. Gonzalez-Garcia M, Perez-Ballestero R, Ding L, Duan L, Boise LH, Thompson CB, et al. bcl-XL is the major bcl-x mRNA form expressed during murine development and its product localizes to mitochondria. Development 1994;**120**:3033–3042.
97. Hockenberry DM, Nunez G, Milliman C, Schreiber RD, Korsmeyer SJ. Bcl-2 is an inner mitochondrial membrane protein that blocks programmed cell death. Nature 1990;**348**:334–336.
98. Wolter KG, Hsu YT, Smith CL, Nechushtan A, Xi XG, Youle RJ. Movement of Bax from the cytosol to mitochondria during apoptosis. J Cell Biol 1997;**139**:1281–1292.
99. Nechushtan A, Smith CL, Hsu Y, Youle RJ. Conformation of the Bax C-terminus regulates subcellular location and cell death. EMBO J 1999;**18**:2330–2341.
100. Muchmore SW, Sattler M, Liang H, Meadows RP, Harlan JE, Yoon HS, et al. X-ray and NMR structure of human Bcl-xL, an inhibitor of programmed cell death. Nature 1996;**381**:335–341.
101. Antonsson B, Conti F, Ciavatta A, Montessuit S, Lewis S, Martinou I, et al. Inhibition of Bax channel-forming activity by Bcl-2. Science 1997;**277**:370–372.
102. Minn AJ, Velez P, Schendel SL, Liang H, Muchmore SW, Fesik SW, et al. Bcl-x(L) forms an ion channel in synthetic lipid membranes. Nature 1997;**385**:353–357.
103. Schlesinger PH, Gross A, Yin XM, Yamamoto K, Saito M, Waksman G, et al. Comparison of the ion channel characteristics of proapoptotic BAX and antiapoptotic BCL-2. Proc Natl Acad Sci USA 1997;**94**:11,357–11,362.
104. Reed JC. Bcl-2 family proteins. Oncogene 1998;**17**:3225–3236.
105. Petit PX, Susin S, Zamzami N, Mignotte B, Kroemer G. Mitochondria and programmed cell death: back to the future. FEBS Lett 1996;**396**:7–13.
106. Xiang J, Chao DT, Korsmeyer SJ. BAX-induced cell death may not require interleukin 1 beta-converting enzyme-like proteases. Proc Natl Acad Sci USA 1996;**93**:14,559–14,563.
107. Zoratti M, Szabo I. The mitochondrial permeability transition. Biochim Biophys Acta 1995;**1241**: 139–176.
108. Narita M, Shimizu S, Ito T, Chittenden T, Lutz RJ, Matsude H, et al. Bax interacts with the permeability transition pore to induce permeability transition and cytochrome c release in isolated mitochondria. Proc Natl Acad Sci USA 1998;**95**:14,681–14,686.
109. Pastorino JG, Chen S, Tafani M, Snyder JW, Farber JL. The overexpression of Bax produces cell death upon induction of the mitochondrial permeability transition. J Biol Chem 1998;**273**:7770–7775.
110. Marzo I, Brenner C, Zamzami N, Jurgensmeier JM, Susin SA, Vieira HL, Prevost MC, Xie Z, Matsuyama S, Reed JC, Kroemer G. Bax and adenine nucleotide translocator cooperate in the mitochondrial control of apoptosis. Science 1998;**281**:2027–2031.
111. Vander Heiden MG, Chandel NS, Williamson EK, Schumacker PT, Thompson CB. Bcl-xL regulates the membrane potential and volume homeostasis of mitochondria. Cell 1997;**91**:627–637.
112. Bradham CA, Qian T, Streetz K, Trautwein C, Brenner D, Lemasters JL. The mitochondrial permeability transition is required for Tumor Necrosis Factor alpha-mediated apoptosis and cytochrome c release. Molec Cell Biol 1998;**18**:6353–6364.
113. Bossy-Wetzel E, Newmeyer DD, Green DR. Mitochondrial cytochrome c release in apoptosis occurs upstream of DEVD-specific caspase activation and independently of mitochondrial transmembrane depolarization. EMBO J 1998;**17**:37–49.
114. Shimizu S, Narita M, Tsujimoto Y. Bcl-2 family proteins regulate the release of apoptogenic cytochrome c by the mitochondrial channel VDAC. Nature 1999;**399**:483–487.
115. Newton K, Strasser A. The Bcl-2 family and cell death regulation. Curr Opin Genet Dev 1998;**8**:68–75.
116. Lincz LF. Deciphering the apoptotic pathway: all roads lead to death. Immunol Cell Biol 1998;**76**:1–19.
117. Adams JM, Cory S. The Bcl-2 protein family: arbiters of cell survival. Science 1998;**281**: 1322–1325.

118. Zha J, Harada H, Yang E, Jockel J, Korsmeyer SJ. Serine phosphorylation of death agonist BAD in response to survival factor results in binding to 14-3-3 not BCL-X(L). Cell 1996;87:619–628.
119. Datta SR, Dudek H, Tao X, Masters S, Fu H, Gotoh Y, Greenberg ME. Akt phosphorylation of BAD couples survival signals to the cell-intrinsic death machinery. Cell 1997;91:231–241.
120. del Peso L, Gonzalez-Garcia M, Page C, Herrera R, Nunez G. Interleukin-3-induced phosphorylation of BAD through the protein kinase Akt. Science 1997;278:687–699.
121. Franke TF, Cantley LC. Apoptosis. A Bad kinase makes good. Nature 1997;390:116,117.
122. Harada H, Becknell BMW, Mann M, Huang LJ, Taylor SS, Scott JD, et al. Phosphorylation and inactivation of BAD by mitochondria-anchored protein kinase A. Mol Cell 1999;3:413–422.
123. Li H, Zhu H, Xu CJ, Yuan J. Cleavage of BID by caspase 8 mediates the mitochondrial damage in the Fas pathway of apoptosis. Cell 1998;94:491–501.
124. Luo X, Budihardjo I, Zou H, Slaughter C, Wang X. Bid, a Bcl2 interacting protein, mediates cytochrome c release from mitochondria in response to activation of cell surface death receptors. Cell 1998;94:481–490.
125. Gross A, Yin X, Wang K, Wei MC, Jockel J, Milliman C, et al. Caspase cleaved Bid targets mitochondria and is required for cytochrome c release, while Bcl-xL prevents this release but not Tumor Necrosis Factor-R1/Fas death. J Biol Chem 1999;274:1156–1163.
126. Strasser A, Whittingham S, Vaux DL, Bath ML, Adams JM, Cory S, et al. Enforced BCL2 expression in B-lymphoid cells prolongs antibody responses and elicits autoimmune disease. Proc Natl Acad Sci USA 1991;88:8661–8665.
127. McDonnell TJ, Korsmeyer SJ. Progression from lymphoid hyperplasia to high-grade malignant lymphoma in mice transgenic for the t(14; 18). Nature 1991;349:254–256.
128. McDonnell TJ, Deane N, Platt FM, Nunez G, Jaeger U, McKearn JP, Korsmeyer SJ. Bcl-2-immunoglobulin transgenic mice demonstrate extended B cell survival and follicular lymphoproliferation. Cell 1989;57:79–88.
129. Strasser A, Harris AW, Cory S. bcl-2 transgene inhibits T cell death and perturbs thymic self-censorship. Cell 1991;67:889–899.
130. Martinou JC, Dubois-Dauphin M, Staple JK, Rodriguez I, Frankowski H, Missotten M, Albertini P, Talabot D, Catsicas S, Pietra C, et al. Overexpression of BCL-2 in transgenic mice protects neurons from naturally occurring cell death and experimental ischemia. Neuron 1994;13:1017–1030.
131. Fromm L, Overbeek PA. Inhibition of cell death by lens-specific overexpression of Bcl-2 in transgenic mice. Dev Genet 1997;20:276–287.
132. Rodriguez I, Araki K, Khatib K, Martinou JC, Vassalli P. Mouse vaginal opening is an apoptosis-dependent process which can be prevented by the overexpression of Bcl2. Dev Biol 1997;184: 115–121.
133. Lagasse E, Weissman IL. Enforced expression of Bcl-2 in monocytes rescues macrophages and partially reverses osteopetrosis in op/op mice. Cell 1997;89:1021–1031.
134. Akashi K, Kondo M, von Freeden-Jeffry U, Murray R, Weissman IL. Bcl-2 rescues T lymphopoiesis in interleukin-7 receptor-deficient mice. Cell 1997;89:1033–1041.
135. Maraskovsky E, O'Reilly LA, Teepe M, Corcoran LM, Peschon JJ, Strasser A. Bcl-2 can rescue T lymphocyte development in interleukin-7 receptor-deficient mice but not in mutant rag-1-/- mice. Cell 1997;89:1011–1019.
136. Veis DJ, Sorenson CM, Shutter JR, Korsmeyer SJ. Bcl-2-deficient mice demonstrate fulminant lymphoid apoptosis, polycystic kidneys, and hypopigmented hair. Cell 1993;75:229–240.
137. Kamada S, Shimono A, Shinto Y, Tsujimura T, Takahashi T, Noda T, et al. Bcl-2 deficiency in mice leads to pleiotropic abnormalities: accelerated lymphoid cell death in thymus and spleen, polycystic kidney, hair hypopigmentation, and distorted small intestine. Cancer Res 1995;55:354–359.
138. Nagata M, Nakauchi H, Nakayama K, Loh D, Watanabe T. Apoptosis during an early stage of nephrogenesis induces renal hypoplasia in Bcl-2-deficient mice. Am J Path 1996;148:1601–1611.
139. Sorenson CM, Padanilam BJ, Hammerman MR. Abnormal postpartum renal development and cystogenesis in the Bcl-2 (-/-) mouse. Am J Phys 1996;271:F184–F193.
140. Yamamura K, Kamada S, Ito S, Nakagawa K, Ichihashi M, Tsujimoto Y. Accelerated disappearance of melanocytes in Bcl-2-deficient mice. Cancer Res 1996;56:3546–3550.
141. Michaelidis TM, Sendtner M, Cooper JD, Airaksinen MS, Holtmann B, Meyer M, et al. Inactivation of Bcl-2 results in progressive degeneration of motoneurons, sympathetic and sensory neurons during early postnatal development. Neuron 1996;17:75–89.

142. Gonzalez-Garcia M, Garcia I, Ding L, O'Shea S, Boise LH, Thompson CB, et al. Bcl-x is expressed in embryonic and postnatal neural tissues and functions to prevent neuronal cell death. Proc Natl Acad Sci USA 1995;92:4304–4308.
143. Carrio R, Lopez-Hoyos M, Jimeno J, Benedict MA, Merino R, Benito A, et al. A1 demonstrates restricted tissue distribution during embryonic development and functions to protect against cell death. Am J Path 1996;149:2133–2142.
144. Motoyama N, Wang F, Roth KA, Sawa H, Nakayama K, Negishi I, et al. Massive cell death of immature hematopoietic cells and neurons in Bcl-x-deficient mice. Science 1995;267:1506–1510.
145. Knudson CM, Tung KS, Tourtellotte WG, Brown GA, Korsmeyer SJ. Bax-deficient mice with lymphoid hyperplasia and male germ cell death. Science 1995;270:96–99.
146. Print CG, Loveland KL, Gibson L, Meehan T, Stylianou A, Wreford N, et al. Apoptosis regulator bcl-w is essential for spermatogenesis but appears otherwise redundant. Proc Natl Acad Sci USA 1998;95:12,424–12,431.
147. Ross AJ, Waymire KG, Moss JE, Parlow AF, Skinner MK, Russell LD, et al. Testicular degeneration in Bclw-deficient mice. Nature Genet 1998;18:251–256.
148. Jurisicova A, Lathan KE, Casper RF, Varmuza SL. Expression and regulation of genes associated with cell death during murine preimplantation embryo development. Mol Reprod Dev 1998;51:243–253.
149. Yin X, Wang K, Gross A, Zhao Y, Zinkel S, Klocke B, Roth KA, Korsmeyer SJ. Bid-deficient mice are resistant to Fas-induced hepatocellular apoptosis. Nature 1999;400:886–891.
150. Shindler KS, Latham CB, Roth KA. Bax deficiency prevents the increased cell death of immature neurons in bcl-x-deficient mice. J Neurosci 1997;17:3112–3119.
151. Knudson CM, Korsmeyer SJ. Bcl-2 and Bax function independently to regulate cell death. Nat Genet 1997;16:358–363.
152. Ginsburg M, Snow MHL, McLaren A. Primordial germ cells in the mouse embryo during gastrulation. Development 1990;110:521–528.
153. MacGregor GR, Zambrowicz BP, Soriano P. Tissue non-specific alkaline phosphatase is expressed in both embryonic and extra-embryonic lineages during mouse embryogenesis but is not required for migration of primordial germ cells. Development 1995;121:1487–1496.
154. Lawson K, Hage W. Clonal analysis of the origin of primordial germ cells in the mouse, in germline development. In: Ciba Foundation Symposium, Wiley Press, Chichester, 1994, pp. 68–91.
155. Tam PP, Snow MH. The in vitro culture of primitive-streak-stage mouse embryos. J Embryol Exp Morphol 1980;59:131–143.
156. Hogan B, Beddington R, Costantini F, Lacy E. Manipulating the Mouse Embryo, A Laboratory Manual, Cold Spring Harbor Laboratory Press, Cold Spring Harbor, NY, 1994, pp. 497.
157. Wang RA, Nakane PK, Koji T. Autonomous cell death of mouse male germ cells during fetal and postnatal period. Biol Reprod 1998;58:1250–1256.
158. Pesce M, Farrace MG, Piacentini M, Dolci S, De Felici M. Stem cell factor and leukemia inhibitory factor promote primordial germ cell survival by suppressing programmed cell death (apoptosis). Development 1993;118:1089–1094.
159. Matsui Y. Regulation of germ cell death in mammalian gonads. Apmis 1998;106:142–147; discussion 1487,1488.
160. Watanabe M, Shirayoshi Y, Koshimizu U, Hashimoto S, Yonehara S, Eguchi Y, et al. Gene transfection of mouse primordial germ cells *in vitro* and analysis of their survival and growth control. Exp Cell Res 1997;230:76–83.
160a. Rucker EB, Dierisseau P, Wagner K-U, Garret L, Wynshaw-Boris A, Flaws J, et al. Analysis of mice with a hypomorphic mutant allele of Bclx indicates that Bclx and Bax function to regulate survival of gonocytes following their arrival in the developing gonad. Mol Endo 2000;14:1038–1046.
161. Rugh R. The Mouse, Its Reproduction and Development, Oxford University Press, Oxford, UK, 1968, pp. 430.
162. Rodriguez I, Ody C, Araki K, Garcia I, Vassalli P. An early and massive wave of germinal cell apoptosis is required for the development of functional spermatogenesis. EMBO J 1997;16:2262–2270.
163. Krajewski S, Krajewska M, Shabaik A, Wang HG, Irie S, Fong L, et al. Immunohistochemical analysis of in vivo patterns of Bcl-X expression. Cancer Res 1994;54:5501–5507.
164. Krajewski S, Bodrug S, Krajewska M, Shabaik A, Gascoyne R, Berean K, et al. Immunohistochemical analysis of Mcl-1 protein in human tissues. Differential regulation of Mcl-1 and Bcl-2 protein

production suggests a unique role for Mcl-1 in control of programmed cell death in vivo. Am J Path 1995;146:1309–1319.
165. Kiefer MC, Brauer MJ, Powers VC, Wu JJ, Umansky SR, Tomei LD, et al. Modulation of apoptosis by the widely distributed Bcl-2 homologue Bak. Nature 1995;374:736–739.
166. Hsu SY, Kaipia A, McGee E, Lomeli M, Hsueh AJ. Bok is a pro-apoptotic Bcl-2 protein with restricted expression in reproductive tissues and heterodimerizes with selective anti-apoptotic Bcl-2 family members. Proc Natl Acad Sci USA 1997;94:12,401–12,406.
167. Hsu SY, Lin P, Hsueh AJ. BOD (Bcl-2-related ovarian death gene) is an ovarian BH3 domain-containing proapoptotic Bcl-2 protein capable of dimerization with diverse antiapoptotic Bcl-2 members. Mol Endocrinol 1998;12:1432–1440.
168. Furuchi T, Masuko K, Nishimune Y, Obinata M, Matsui Y. Inhibition of testicular germ cell apoptosis and differentiation in mice misexpressing Bcl-2 in spermatogonia. Development 1996;122:1703–1709.
169. Russell LD, Griswold MD. The Sertoli Cell, Cache River Press, Clearwater FL, 1993, pp. 801.
170. Knobil E. The Physiology of Reproduction, Raven Press, 1994. .
171. Ogawa T, Arechaga JM, Avarbock MR, Brinster RL. Transplantation of testis germinal cells into mouse seminiferous tubules. Int J Dev Biol 1997;41:111–122.
172. Woolveridge I, de Boer-Brouwer M, Taylor MF, Teerds KJ, Wu FCW, Morris ID. Apoptosis in the rat spermatogenic epithelium following androgen withdrawal: changes in apoptosis-related genes. Biol Reprod 1999;60:461–470.
173. Sinowatz F, Amselgruber W, Plendl J, Kolle S, Neumuller C, Boos G. Effects of hormones on the prostate in adult and aging men and animals. Microsc Res Tech 1995;30:282–292.
174. Kiess W, Gallaher B. Hormonal control of programmed cell death/apoptosis. Eur J Endocrinol 1998;138:482–491.
175. Zhang X, Chen MW, Ng A, Ng PY, Lee C, Rubin M, et al. Abnormal prostate development in C3(1)-bcl-2 transgenic mice. Prostate 1997;32:16–26.
176. Morita Y, Tilly JL. Oocyte apoptosis: like sand through an hourglass. Dev Biol 1999;213:1–17.
177. Beaumont HM, Mandl AM. A quantitative and cytological study of oogonia and oocytes in the foetal and neonatal rat. Proc R Soc Lond B Biol Sci 1961;155:557–579.
178. Ohno S, Smith JB. Role of fetal follicular cells in meiosis of mammalian oocytes. Cytogenetics 1964;3:324–333.
179. Martimbeau S, Tilly JL. Physiological cell death in endocrine-dependent tissues: an ovarian perspective. Clin Endocrinol 1997;46:241–254.
180. Hughes FM, Gorospe WC. Biochemical identification of apoptosis (programmed cell death) in granulosa cells: evidence for a potential mechanism underlying follicular atresia. Endocrinology 1991;129:2415–2422.
181. Ratts VS, Flaws JA, Kolp R, Sorenson CM, Tilly JL. Ablation of bcl-2 gene expression decreases the numbers of oocytes and primordial follicles established in the post-natal female mouse gonad. Endocrinology 1995;136:3665–3668.
182. Braw RH, Bar-Ami S, Tsafriri A. Effect of hypophysectomy on atresia of rat preovulatory follicles. Biol Reprod 1981;25:989–996.
183. Chun SY, Billig H, Tilly JL, Furuta I, Tsafriri A, Hsueh AJ. Gonadotropin suppression of apoptosis in cultured preovulatory follicles: mediatory role of endogenous insulin-like growth factor I. Endocrinology 1994;135:1845–1853.
184. Chun SY, Eisenhauer KM, Minami S, Billig H, Perlas E, Hsueh AJ. Hormonal regulation of apoptosis in early antral follicles: follicle-stimulating hormone as a major survival factor. Endocrinology 1996;137:1447–1456.
185. Davoren JB, Kasson BG, Li CH, Hsueh AJ. Specific insulin-like growth factor (IGF) I- and II-binding sites on rat granulosa cells: relation to IGF action. Endocrinology 1986;119:2155–2162.
186. Tilly JL, Billig H, Kowalski KI, Hsueh AJ. Epidermal growth factor and basic fibroblast growth factor suppress the spontaneous onset of apoptosis in cultured rat ovarian granulosa cells and follicles by a tyrosine kinase-dependent mechanism. Mol Endocrinol 1992;6:1942–1950.
187. Peluso JJ, Pappalardo A. Progesterone and cell-cell adhesion interact to regulate rat granulosa cell apoptosis. Biochem Cell Biol 1994;72:547–551.
188. Billig H, Furuta I, Hsueh AJ. Estrogens inhibit and androgens enhance ovarian granulosa cell apoptosis. Endocrinology 1993;133:2204–2212.

189. Rodger FE, Fraser HM, Duncan WC, Illingworth PJ. Immunolocalization of bcl-2 in the human corpus luteum. Hum Reprod 1995;10:1566–1570.
190. Kugu K, Ratts VS, Piquette GN, Tilly KI, Tao X, Martimbeau S, et al. Analysis of apoptosis and expression of bcl-2 gene family members in the human and baboon ovary. Cell Death Differ 1998;5:67–76.
191. Kaipia A, Hsu SY, Hsueh AJ. Expression and function of a proapoptotic Bcl-2 family member Bcl-XL/Bcl-2-associated death promoter (BAD) in rat ovary. Endocrinology 1997;138:5497–5504.
192. Jurisicova A, Rogers I, Fasciani A, Casper RF, Varmuza S. Effect of maternal age and conditions of fertilization on programmed cell death during murine preimplantation embryo development. Mol Hum Reprod 1998;4:139–145.
193. Hsu SY, Lai RJ, Finegold M, Hsueh AJ. Targeted overexpression of Bcl-2 in ovaries of transgenic mice leads to decreased follicle apoptosis, enhanced folliculogenesis, and increased germ cell tumorigenesis. Endocrinology 1996;137:4837–4843.
194. Perez GI, Robles R, Knudson CM, Flaws JA, Korsmeyer SJ, Tilly JL. Prolongation of ovarian lifespan into advanced chronological age by Bax-deficiency. Nature Genet 1999;21:200–203.
195. Perez GI, Knudson CM, Leykin L, Korsmeyer SJ, Tilly JL. Apoptosis-associated signaling pathways are required for chemotherapy-mediated female germ cell destruction. Nat Med 1997;3:1228–1232.
196. Humphreys RC, Krajewska M, Krnacik S, Jaeger R, Weiher H, Krajewski S, et al. Apoptosis in the terminal endbud of the murine mammary gland: a mechanism of ductal morphogenesis. Development 1996;122:4013–4022.
197. Metcalfe AD, Gilmore A, Klinowska T, Oliver J, Valentijn AJ, Brown R, et al. Developmental regulation of the Bcl-2 family protein expression in the involuting mammary gland. J Cell Sci 1999;112:1771–1783.
198. Jager R, Herzer U, Schenkel J, Weiher H. Overexpression of Bcl-2 inhibits alveolar cell apoptosis during involution and accelerates c-myc-induced tumorigenesis of the mammary gland in transgenic mice. Oncogene 1997;15:1787–1795.
199. Shibata M, Liu M, Knudson MC, Shibata E, Yoshidome K, Bandey T, et al. Haploid loss of *bax* leads to accelerated mammary tumor development in C3(1)/SV40-TAg transgenic mice: reduction in protective apoptotic response at the preneoplastic stage. EMBO J 1999;18:2692–2701.
200. Krajewski S, Blomqvist C, Franssila K, Krajewska M, Wasenius VM, Niskanen E, Nordling S, Reed JC. Reduced expression of proapoptotic gene BAX is associated with poor response rates to combination chemotherapy and shorter survival in women with metastatic breast adenocarcinoma. Cancer Res 1995;55:4471–4478.
201. Jo T, Terada N, Saji F, Tanizawa O. Inhibitory effects of estrogen, progesterone, androgen and glucocorticoid on death of neonatal mouse uterine epithelial cells induced to proliferate by estrogen. J Steroid Biochem Mol Biol 1993;46:25–32.
202. Terada N, Yamamoto R, Takada T, Miyake T, Terakawa N, Wakimoto H, et al. Inhibitory effect of progesterone on cell death of mouse uterine epithelium. J Steroid Biochem 1989;33:1091–1096.
203. Terada N, Yamamoto R, Takada T, Taniguchi H, Terakawa N, Li W, et al. Inhibitory effect of androgen on cell death of mouse uterine epithelium. J Steroid Biochem 1990;36:305–310.
204. Koh EA, Illingworth PJ, Duncan WC, Critchley HO. Immunolocalization of bcl-2 protein in human endometrium in the menstrual cycle and simulated early pregnancy. Hum Reprod 1995;10:1557–1562.
205. Gompel A, Sabourin JC, Martin A, Yaneva H, Audouin J, Decroix Y, et al. Bcl-2 expression in normal endometrium during the menstrual cycle. Am J Path 1994;144:1195–1202.
206. Pecci A, Scholz A, Pelster D, Beato M. Progestins prevent apoptosis in a rat endometrial cell line and increase the ratio of bcl-XL to bcl-XS. J Biol Chem 1997;272:11,791–11,798.
207. Akcali KC, Khan SA, Moulton BC. Effect of decidualization on the expression of bax and bcl-2 in the rat uterine endometrium. Endocrinology 1996;137:3123–3131.
208. Vasioukhin V, Degenstein L, Wise B, Fuchs E. The magical touch: Genome targeting in epidermal stem cells induced by tamoxifen application to mouse skin. Proc Natl Acad Sci USA 1999;96:8551–8556.
209. Zubiaga AM, Belasco JG, Greenberg ME. The nonamer UUAUUUAUU is the key AU-rich sequence motif that mediates mRNA degradation. Molec Cell Biol 1995;15:2219–2230.
210. Lagnado CA, Brown CY, Goodall GJ. AUUUA is not sufficient to promote poly(A) shortening and degradation of an mRNA: the functional sequence within AU-rich elements may be UUAUUUA(U/A)(U/A). Mol Cell Biol 1994;14:7984–7995.

7 The Role of the C-Kit/Kit Ligand Axis in Mammalian Gametogenesis

Peter J. Donovan, PHD
and Maria P. de Miguel, PHD

CONTENTS

INTRODUCTION: THE DEVELOPMENT OF THE GERMLINE

The survival of the germline is vital to the survival of all animal species. Failure of germ cells to survive or to differentiate properly in the animal can result in reduced fertility, or in some cases, complete sterility *(1)*. In addition, defects in germline development can predispose individuals to development of cancer. For example, loss of germ cells from the ovary can be associated with premature ovarian failure, but can also dispose affected individuals to the development of ovarian cancer *(2)*. Similarly in the male, loss of germ cells from the developing testis can be associated with the development of testicular teratocarcinoma *(3)*. Testicular cancer is the most common malignancy in young men with a peak incidence from 18–35 yr of age *(4,5)*. This contrasts with the incidence of ovarian tumors, which show a higher incidence after 50 yr of age, as is the case with most other solid tumors *(6)*. Thus, even in otherwise healthy individuals, survival of the germline is an important feature of adult homeostasis.

From: *Contemporary Endocrinology: Transgenics in Endocrinology*
Edited by: M. Matzuk, C. W. Brown, and T. R. Kumar © Humana Press Inc., Totowa, NJ

Over time, the molecular mechanisms regulating germline development, differentiation, and maintenance are being elucidated. There are several reasons for this. First, the advances in our knowledge of animal genomes have allowed a number of sterile mouse mutations to be molecularly cloned. Second, genomic DNA and cDNA sequencing projects have identified many new growth factors that would have been difficult to identify by other means. Some of these growth factors have been found to be active on germline cells. Third, the revolution in mouse genetics, brought about by the ability to carry out targeted mutations in embryonic stem (ES) cells, has both deliberately and inadvertently generated a wealth of information about germline development in mammals. Understanding how survival, growth, and differentiation of germ cells is regulated will likely lead to the development of improvements in the diagnosis and treatment of human diseases including infertility and cancer. This chapter focuses on the c-kit receptor tyrosine kinase and its ligand (kit ligand or KL) and their interrelated role in the development, differentiation, and homeostasis of the mammalian germline.

Early in development, the germ cells are set aside from the somatic-cell lineages. The molecular mechanisms regulating this process and the reasons for it are poorly understood *(7)*. In most species studied to date, the germ cells arise some distance from their eventual home, the gonad. In order to reach the gonad, they must undergo a period of migration *(8)*. During the period of migration to the gonad and for a few days afterwards, the germ cells will proliferate to establish the population of cells that will give rise to the gametes. In the fully-formed mouse embryonic gonad, approx 35,000 germ cells arise from a population of about 100 cells that began migration several days earlier (*see* ref. *1*, for review). The differentiation of germ cells in the gonad depends on their sex and on the sex of the embryo. Male germ cells, now called gonocytes, enter a mitotic arrest. They resume mitosis after birth to form the testicular stem cell, the spermatogonium. The timing of when male germ cells resume mitosis depends on the species. In mice, spermatogonia resume meiosis on the day of birth, and in rats and hamsters they do so a few days after birth, while in humans, proliferation begins around three years after birth. In the prepubertal period, spermatogonia give rise to daughter cells that enter meiosis. At later points, these haploid cells undergo spermiogenesis, and at puberty they give rise to mature sperm *(9–11)*. Although there are differences in the timing of events between different species, many of the fundamental events of spermatogenesis are highly conserved in mammals. In contrast to the situation in the male, female germ cells enter meiosis in the embryo and then immediately arrest in meiotic prophase. After birth, germ cells (now called oocytes) in the mouse will resume meiosis in waves, complete meiosis I, and then arrest at metaphase of the second meiotic division (MII). At puberty, females begin to release ovulated oocytes that will complete the second meiotic division after fertilization, and these oocytes are continually released up until menopause (*see* ref. *1* for review).

During the period between entering meiosis and birth, many female germ cells are lost in a process called atresia (*see* refs. *12* and *13* for reviews). Most likely, this represents a form of programmed cell death. The reasons for this atresia and the molecular mechanisms regulating the process are poorly understood. It may represent a mechanism of quality control in which unhealthy germ cells are deleted. Alternatively, it could represent the end point of a counting mechanism in which germ-cell numbers are regulated. In males, programmed cell death is also an important part of the spermatogenic process and is regulated by p53-dependent and p53-independent mechanisms (*see* ref. *14* for review). Interestingly, blocking programmed-cell death in the testis results in aberrant spermatogenesis, and eventually infertility *(15)*. Most likely, increased numbers of differentiating spermatogonia inhibit the proliferation of undifferentiated spermatogonia resulting in sterility *(16)*. Regardless of the reasons for, and mechanisms of, apoptosis, these data suggest that programmed cell death is a requisite part of mammalian gametogenesis (*see* refs. *14* and *17* for reviews).

MUTATIONS AFFECTING FERTILITY AND GAMETOGENESIS

The molecular mechanisms regulating mammalian gametogenesis are gradually being determined. In part, this results from the wealth of information being obtained from studies of gametogenesis in model organisms including flies, worms, and yeast. Although there are clear differences between these organisms and mammals, some of the developmental mechanisms regulating germline development and meiosis are evolutionarily conserved. Furthermore, the cloning of mouse mutations that affect fertility has allowed some of the genes involved in mammalian gametogenesis and meiosis to be determined. A recent survey of the mouse mutants held at the Jackson Laboratories revealed that approx 78 mutants affecting some aspect of gametogenesis exist. In addition, a large number of knockout mice have defects in gametogenesis, thereby identifying genes involved in germline development. Two of the best-characterized sterile mouse mutants are those of the *Dominant White Spotting* (*W*) and *Steel* (*Sl*) loci.

MOLECULAR CHARACTERIZATION OF THE *STEEL/DOMINANT WHITE SPOTTING*

Mutations at the *Dominant White Spotting* (*W*) and *Steel* (*Sl*) loci are among the oldest mutations of the mouse fancy. They are characterized in part by the semidominant effect on coat color. Many, if not all, of the *W* and *Sl* mutants have a prominent head spot as well as a generalized lightening of the coat, particularly on the belly. The number of mutations identified at these loci is large, and constitutes two of the largest allelic series of mouse mutants thus far studied. Thus, a significant amount of information has been accumulated on the structure-function relationships of the proteins encoded at the *W* and *Sl* loci.

W encodes the c-kit transmembrane receptor tyrosine kinase *(18,19)* related to the colony-stimulating factor-1 (CSF-1) family of growth-factor receptors and *Sl* encodes its ligand, variously termed mast-cell growth factor (MGF), stem-cell factor (SCF), kit-ligand (KL), and steel factor (SF) *(20–23)*. For the purposes of this chapter, this factor will be known as kit-ligand or KL. KL is a transmembrane growth factor that exists in two major forms (KL-1 and KL-2), that are derived by alternate splicing *(20,24)*. KL-1 has a proteolytic cleavage site in the ectodomain (encoded in exon 6), which can be cleaved to produce a soluble factor *(25)*. KL-2 lacks the major proteolytic cleavage site encoded in exon 6. However, KL-2 is in fact proteolytically cleaved at another site (believed to be encoded in exon 7), but in a much less efficient manner *(25,26)*. Therefore, KL-2 is believed to remain largely as a membrane-bound or cell-associated factor. The cleavage of these two forms is believed to occur by distinct mechanisms, perhaps involving different metalloproteinases *(25,26)* The different KL mRNA species are expressed in distinct patterns in the adult animal, and presumably during embryonic development. Although both membrane-bound and soluble forms of KL have biological activity, there are discrete differences in the ability of cells to respond to the membrane-bound vs soluble forms. Clearly, it is possible to regulate (temporally and spatially) both the expression of different forms of KL and the enzyme responsible for their proteolytic cleavage. By these mechanisms, it should be possible to exquisitely regulate KL biological activity during development and in the adult animal. Binding of KL to the c-kit receptor causes receptor autophosphorylation, and the activation of a signaling transduction cascade required for the development of several cell lineages in the embryo and adult animal (*see* refs. *27* and *28* for reviews).

Consistent with this model of KL-c-kit interaction, mutations at the *W* and *Sl* loci have many overlapping features in addition to the effect on coat color described here. Many mutations at the *W* and *Sl* loci cause lethality when homozygous because of the role of c-kit and KL in embryonic hematopoiesis and erythropoiesis. The severe anaplastic anemia seen in these mutants is first detected at around Embryonic (E) d 15, at which time hematopoiesis is occuring in the fetal liver. Another major feature of these mutants is their drastic effect on the development of the germline. A large number of *W* and *Sl* mutants, if viable, are completely sterile, demonstrating an important role for this signaling pathway in germ-cell development (*see* ref. *29* for review). The remainder of this review will expand on the roles of the c-kit receptor and its ligand, KL, in germ-cell development, growth, and homeostasis.

A large body of evidence has accumulated showing that the c-kit receptor is expressed within the cells affected by the *W* mutation, while the KL is expressed in surrounding cells (*see* refs. *30–32*). Thus, the c-kit receptor is expressed in melanoblasts, in hemopoietic stem cells (HSC), and in germ cells. KL is expressed by epidermal cells, bone-marrow stromal cells, and, depending on the sex of the animal, Sertoli cells, or granulosa cells. These data are entirely consistent with earlier data from grafting and transplantation studies, which showed that the *W* mutation acted within the affected lineages, while the *Sl* mutation was extrinsic to them. The c-kit receptor is also expressed in other cell types, but the description of these cell types and the effect of *W* and *Sl* mutations on these cells is beyond the scope of this chapter.

As described here, a large number of mutations have been identified at the *W* and *Sl* loci, and have variable effects on development. The nature of the *Sl* mutations fall into four general classes, including lethal mutations that die around E 15.5. These mutations involve deletion of the entire coding region of KL. Lethality is most likely caused by the severe effect on fetal hematopoiesis. Since the fetal liver is a hematopoietic organ, mutants lacking KL are defective in fetal hematopoiesis that is detectable by the pale color of the fetal liver in the mutants. The original *Sl*-null mutation is representative of this class of mutation. Second, lethal mutations which die prior to E 15.5, usually around the time of implantation. This class of mutation involves deletion of the KL gene as well as other genes required for peri-implantation development. Such mutants include the Sl^{18H} and Sl^{12H} mutations. Third, there are viable mutations in which a defective form of KL is produced. This class of mutation includes the Sl^{dickie} (Sl^{d}) and Sl^{17H} mutations. Fourth, there are viable mutations in which the coding region of KL is intact, but in which the regulatory domains of the KL genomic locus are altered. This class of mutation includes the Sl^{panda} (Sl^{pan}) and $Sl^{contrasted}$ (Sl^{con}) mutations. The large number and variety of mutations at this locus have greatly contributed to the wealth of knowledge about the role of this factor in germline development and growth.

KIT LIGAND IN GERMLINE DEVELOPMENT

The infertility seen in many *W* and *Sl* mutants signifies the important role played by this signaling pathway in germline development. A growing body of evidence suggests that c-kit receptor and its ligand function at different times during the development of the germline. Moreover, the function of this signaling pathway appears to vary depending on the cell type and stage of development in which the receptor or ligand is expressed. The first stage at which these mutants affect germ-cell development is around E 9.5 in the mouse embryo, a time in which the primordial germ cells (PGC) are migrating towards the embryonic gonad. In both *W* and *Sl* mutants, PGC numbers are first altered at this time of development *(33–36)*. This suggests that prior to this time, PGC development is independent of the c-kit/KL-signaling pathway. In the fully colonized gonad of a normal embryo, there are estimated to be approx 35,000 germ cells. In *W* or *Sl* mutants, this number can be reduced by over 90%. In this situation, the numbers of germ cells that survive until birth is very small, and usually incompatible with fertility.

What is the function of KL during the period of gonad colonization? In principle, KL can regulate germ-cell survival, proliferation, migration, or differentiation, or perhaps a combination of these functions. Data from in vitro studies strongly suggests that activation of the c-kit receptor is required for PGC survival *(37–39)*. PGCs isolated from embryos will adhere to a variety of feeder layers, but will only survive on feeder layers of cells that express KL. Moreover, long-term survival of PGCs seems to depend on the expression of full-length, membrane-bound KL by feeder cells *(37)*. Confirmation of this idea comes from analysis of PGC numbers in Sl^d mice carrying a genomic deletion of sequences encoding the cytoplasmic tail and transmembrane region of KL. Because of this genomic deletion, Sl^d mice only express a soluble form of KL *(40,41)*. Since Sl^d animals are sterile, this suggests that membrane-bound KL is required for PGC survival in vivo also *(36)*. But analysis of PGC migration in Sl/Sl^d mice demonstrates that many PGCs do not reach the embryonic gonad *(36)*. These data suggest that KL may also be involved in regulating PGC migration in the developing embryo. Consistent with this idea, defective PGC migration has been observed in W^e/W^e mutant mice *(35)*. The requirement of PGC survival for membrane-bound forms of KL provides an excellent mechanism for regulating PGC migration and survival in the developing embryo. PGC migration can be controlled spatially and temporally by regulating the expression of KL by embryonic somatic cells. If the PGCs diverge away from the pathway expressing KL, they are expected to die through programmed cell death.

During the period between colonization of the gonad and birth, the c-kit/KL signal does not appear to play a significant role, since there appears to be no effect on germ-cell numbers in mutants during this period. For example, in Sl^{17H}/Sl^{17H} embryos, the numbers of PGCs are severely reduced *(42)*. But some of the germ cells survive until birth, suggesting that the germ cells that reach the embryonic gonad do not die in the intervening period *(42)*. One interesting question is whether female germ-cell atresia in the embryo is affected in *W* or *Sl* mutants, since germ-cell number is already severely reduced in these animals. Atresia may represent a simple mechanism for reducing germ-cell numbers prior to birth, or it could represent a mechanism for deleting germ cells that are inferior. Careful examination of germ-cell number in *W* or *Sl* mutant mice during this period may provide an important insight into this question.

In summary, KL seems to be required for PGC survival in the mouse embryo from E 9.5 to E 13.5. During this period, it likely acts in concert with other factors to stimulate

PGC proliferation. It may also play a role in directing migrating germ cells toward the gonad anlagen. This provides a mechanism for ensuring that only the PGCs that are in the correct position will survive and proliferate. The c-kit signal induced by KL binding also appears to have multiple functions later in germline development.

ROLE OF KIT LIGAND IN MALE GAMETOGENESIS

Once PGCs have reached the genital ridge in the male embryo, they become surrounded by somatic cells of the embryonic gonad anlagen. These somatic cells, which will eventually be called Sertoli cells, differentiate to form the seminiferous cords. The PGCs also differentiate to form the gonocyte that is morphologically distinct from the PGCs. In mice and rats, the gonocytes will proliferate for a few days, but then arrest at G_0/G_1 phase of the cell-division cycle *(11)*. In the postnatal period in the testis, the gonocytes will resume mitosis to form the testicular stem cell or spermatogonium. In rats, the timing of resumption of mitosis is strain-dependent. The gonocytes begin proliferating, and simultaneously migrate toward the basement membrane on which Sertoli cells are already established. In vitro studies suggest that expression of c-kit by gonocytes is required for the migration of gonocytes towards the basement membrane *(43)*. Ultimately, the gonocytes come to lie between Sertoli cells. Therefore, the growth and proliferation of germ cells in the testis is strictly controlled by the surrounding Sertoli cells. The division of the gonocyte likely gives rise to two cells, the true stem cell of the testis (the first undifferentiated A Spermatogonium) and the first differentiated type A spermatogonium *(44)*. The onset of division by gonocytes is considered to be the beginning of spermatogenesis. The undifferentiated type A spermatogonium actively proliferates to give rise to daughters, A2-A4, and then type B spermatogonia, which will eventually enter into meiosis (forming spermatocytes) and ultimately undergo spermiogenesis. Once the first spermatocytes have formed, the Sertoli cells form tight junctions *(45)*. The presence of the tight junctions between the Sertoli cells effectively divides the germ-cell population into two compartments. Some of the germ cells lie in the basal layer below the tight junctions, and the others lie above the tight junctions in the adluminal compartment. Potentially, this allows the Sertoli cell to control the germ-cell populations in different ways. KL could play an important role in this form of regulation because it is a membrane-bound factor, and its cleavage can be regulated temporally and spatially.

A number of important studies point to a key role for the c-kit/KL signaling pathway in the regulation of spermatogenesis. Analysis of c-kit and KL expression in the testis has given valuable insights into the function of these proteins in spermatogenesis. In the adult testes, the highest levels of c-kit expression are observed in interstitial Leydig cells *(46)*. However, c-kit is also expressed in some spermatogenic cells at the basal layer of the seminiferous cords. Immunocytochemical staining with anti-c-kit antibodies demonstrates that differentiating type A spermatogonia, intermediate type B spermatogonia, and early spermatocytes all express c-kit on the cell surface *(46)*. In some studies, c-kit was not detected on later-stage spermatocytes, spermatids, or the surrounding Sertoli cells. These data are entirely consistent with the idea that the c-kit/KL signaling pathway regulates germ-cell survival and/or proliferation in the adult gonad, but is not a regulator of meiosis. However, other studies suggest that c-kit is expressed in meiotic germ cells and could play a role in regulation of meiosis.

Mutations at both the *W* and *Sl* loci affect postnatal germ-cell development and adult spermatogenesis. This effect is observed in *W* and *Sl* mutants that are viable, and therefore carry mutations (but not complete deletions) in the c-kit and KL genes. For example, Sl^{17H} mice that produce an abnormal form of KL have only a few germ cells that survive until birth *(42)*. However, these few germ cells undergo the first round of spermatogenesis, but the differentiating germ cells are eventually lost, and the mice become sterile. At 8 wk of age, testes from Sl^{17H}/Sl^{17H} mice contained Sertoli cells and a few spermatogonia *(42)*. These data suggest that the survival and division of gonocytes in the early postnatal testis is independent of the c-kit signaling pathway. Further, they suggest that the survival of the adult spermatogonial stem cell may also be independent of this pathway. Finally, these data suggest that the growth and survival of differentiated type A spermatogonia is a c-kit-dependent process. Consistent with this data, experiments have been performed in which an anti-c-kit monoclonal antibody (ACK2) was injected into adult mice. This antibody blocks signaling via the c-kit receptor. The result of this experiment was that spermatogenesis was arrested, and the testis was depleted of differentiated germ cells *(46)*. However, undifferentiated type A spermatogonia are unaffected by such treatment, suggesting that the growth and/or survival of these cells is independent of the c-kit/KL axis *(46)*. Kit binding to its ligand leads to activation of a variety of signaling molecues, including JAK2, Src, Shc, Grb2, PLCγ, Ras, PI3K, and AKT (for review, *see 46a*). In two recent studies, the PI3K binding site on the C-Kit receptor was mutated in mice by gene replacement, and PGC survival was shown to be unaffected *(46b,46c)*. However, these studies also demonstrated that activation of the PI3K signaling pathway downstream of c-Kit is of special importance in make spermatogonial stem cells *(46b)*.

A large body of evidence suggests that KL is required for the survival of germ cells, and that along with other factors, KL can stimulate proliferation. Studies analyzing other cell systems have identified some of the genes that regulate cell survival and its counterpart, cell death (apoptosis). One of the major regulators of cell death in mammals is the *p53* tumor-suppressor gene (*see* refs. *47–49* for reviews). In normal cells, p53 levels are low, but the protein is rapidly increased in response to DNA damage or other cellular stress. The p53 signaling pathway induces cell death in damaged cells. Loss of function of the *p53* gene in mice leads to tumor formation in a variety of tissues in young adult animals *(50,51)*. This is brought about, in part, by the survival of cells that would otherwise be targeted for apoptosis by the wild-type p53 protein. Spermatogenesis in mammals is normally associated with large numbers of spermatogenic cells undergoing apoptosis. In fact, inhibition of apoptosis in the testis by forced expression of the antiapoptotic bcl-2 protein disrupts normal spermatogenesis *(15)*. Previous studies have shown that programmed cell death in the testis is regulated by both p53-dependent and p53-independent mechanisms. Most likely, binding of KL to the c-kit receptor in germ cells activates pathways that promote cell survival. Conversely, failure to activate the c-kit receptor is likely to activate pathways that induce programmed cell death or apoptosis. These pathways probably include the p53-mediated pathway. Interestingly, when sterile *W* mutant mice (W^v -bearing a defective c-kit receptor) were intercrossed with mice lacking *p53*, these mice became fertile *(52)*. Thus, by disrupting p53-mediated apoptosis, cells that lack a functional c-kit signaling pathway (and would otherwise die) are now viable. This data suggests that in the male germline, a major function of the c-kit signaling pathway is to mediate germ cell survival *(52)*. In *p53–/–*; W^v/W^v double-homozygous-mutant testes, apoptosis is still apparent, demonstrating that a p53-independent cell death is intact. However, the fact that these double-mutant mice are now fertile demonstrates the important roles of activation of the c-kit receptor by its ligand in controlling p53 activity to regulate life and death in the testis.

Germ-cell survival in the testis requires the membrane-bound form of KL rather than the soluble version. This is nicely demonstrated by the observed sterility in Sl^d mice that can only produce a soluble form of KL *(36)*. The requirement for a membrane-bound factor for germ-cell survival ensures that germ cells can only proliferate when they are in association with Sertoli cells. Thus, germ-cell growth can be strictly controlled within the testis.

While the *Sld* mutation provides evidence of the role of the membrane-bound form of KL in germ cell survival, the *Sl17H* mutation provides evidence for the specific role of the cytoplasmic tail of KL in this process. In principle, the cytoplasmic tail of KL may have a number of functions, including cytoplasmic signaling, membrane-anchorage, ectodomain cleavage, and ligand dimerization *(26,42,53)*. However, recent studies suggest that the cytoplasm tail has a unique function. In Sl^{17H} mice, a point mutation at a splice-acceptor site results in an alternative reading frame of the KL protein, and leads to the production of a KL polypeptide in which the cytoplasmic tail is altered *(42)*. The protein sequence of this mutant KL is normal from the N-terminus through the first amino acid of the cytoplasmic juxtamembrane region, but then diverges. This mutation has a drastic effect on spermatogenesis, leading to loss of germ-cell loss and male infertility, and this demonstrates an important function for the KL cytoplasmic spermatogenesis *(42)*. Recent studies on the Sl^{17H} gene product suggests that the cytoplasmic tail of KL plays an important role in localization of KL within the cell *(54)*. The Sl^{17H} mutation alters the membrane presentation of KL within the cell and affects the localization of KL on the basolateral surface of polarized epithelial cells *(54)*. Polarized epithelial cells transfected with the Sl^{17H} form of KL or mutants that lack the cytoplasmic tail express or secrete KL from their apical surface. Thus, the loss of germ cells from the testes of Sl^{17H} mice is likely caused by the inability of KL to be properly localized to the basolateral surface of Sertoli cells *(54)*.

Although many studies have suggested a role for KL in germ cell survival or proliferation, a number of studies suggest that it may also play a role in regulating male meiosis. For example, at stages VII–VIII of the mouse seminiferous epithelium, the membrane-anchored KL extends from the basal to the adluminal compartment of the Sertoli cells. Moreover, the c-kit protein is localized in germ cells that are undergoing meiosis up to pachytene stage *(55)*. In an in vitro culture system in which germ cells can transit through meiosis, an anti-c-kit antibody blocked the appearance of haploid cells and of haploid-phase gene expression. These data suggest that the c-kit/KL interaction may play a role in meiotic progression. c-Kit mRNA has also been found in round spermatids *(56)* and in the acrosomal region of sperm *(57)*. Interestingly, in haploid germ cells, a c-kit mRNA splice variant gives rise to an intracellular protein that only possesses the C-terminal portion of the kinase domain, (termed truncated c-kit or tr-kit) *(58)*. Introduction of tr-kit into metaphase II-arrested mouse oocytes causes complete oocyte activation. The oocytes underwent cortical granule exocytosis, completion of metaphase II, formation of a parthenogenetic pronucleus, and progression through the cleavage stages *(59)*. In addition, sperm undergoing the acrosome reaction demonstrate association of c-kit with the acrososme plasma membrane *(60)*, and blocking c-kit signaling in mature sperm with an anti-c-kit monoclonal antibody (mAb) prevents the acrosomal reaction *(61)*. These data support the idea that tr-kit may be the putative sperm factor required for triggering activation of mouse eggs at fertilization.

Again, it is clear that the c-kit/KL axis plays multiple roles in regulating the survival, proliferation, and differentiation of germ cells in the neonatal, prepubertal, and adult testis. Although the action of the c-kit/KL signal in some cells types is well-characterized, the role played by this signaling pathway during meiosis and in sperm function is less well understood. Understanding the role of the C-Kit signal transduction pathway in these processes represents the challenges of the next decade of research.

ROLE OF KIT LIGAND IN FEMALE GAMETOGENESIS

In female embryos, once PGCs have entered the fetal ovary, they complete mitotic divisions and enter meiosis. They eventually arrest meiosis at Prophase I, and each germ cell will become invested by a layer of flattened somatic cells that are the precursors of the follicle or granulosa cells. Therefore, in the ovary, just as in the testis, germ-cell survival, growth, and differentiation is controlled by the complex interaction between the germ cells and the surrounding somatic cells *(62,63)* (*see* Chapter 4). The structure described here, which contains a germ cell surrounded by follicle cells, is called a primordial follicle. Within a few days of birth, the ovary is filled with oocytes within these primordial follicles. But as the oocyte grows in size, the follicle cells differentiate into cuboidal cells. The appearance of these changes represents the hallmark of the next stage of follicle development, the primary follicle. About 2 wk after birth, secondary follicles containing at least two layers of follicle cells first appear in the mouse ovary. The appearance of a fluid-filled cavity (antrum) within the follicle marks the change to the tertiary- or antral-follicle stage. The follicle not only provides an environment in which the germ cell can be nursed and nourished by surrounding somatic cells but also one in which the progress of the meiotic cell cycle can be regulated. During follicle development, the oocyte grows in size so that the 12–10-μm-diameter cell present in the primordial follicle becomes an 80-μm-diameter cell in a large antral follicle. During this period, the oocytes also acquire the ability to resume meiosis following meiotic arrest (termed meiotic competence acquisition), to be fertilized, and to complete preimplantation development. Thus, the ovarian follicle is a structure in which many complex interactions take place (*see* ref. *63* and Chapter 4, for review). The granulosa cells receive and respond to signals around them, some signals from other cells within the ovary and some from outside the ovary. In addition, the granulosa cells receive and respond to signals from the developing oocytes. In this way, the development of the follicle and the maturing oocyte is coordinated to ensure that the ovulated egg is both healthy and ready (at the correct stage of the cell cycle) to receive the sperm.

The interactions that occur in the ovary are clearly complex. Oocytes and surrounding granulosa cells maintain functional gap junctions between themselves. These gap junctions account for the vast majority of metabolite influx into the oocyte required for oocyte growth. The somatic cells of the follicle are also believed to play an important role in maintaining oocytes in meiotic arrest prior to the surge in luteinizing hormone (LH). The communication between the oocyte and the follicle cells is not one-way. Compelling evidence suggests that the oocyte also signals to granulosa cells to regulate their growth and many aspects of their physiology. For example, the oocyte secretes paracrine factors such as growth-differentiation-factor-9 (GDF-9) *(64)*, which regulate granulosa-cell proliferation, suppress LH-receptor mRNA production by granulosa cells, inhibit granulosa-cell production of urokinase plasminogen activator, and regulate other functions of the somatic cells in the ovary. It is apparent that the follicle should not be regarded simply as a structure in which the nurses (granulosa cells) look after the docile patient (the oocyte), but rather one in which the patient has a telephone and call button, and it does not hesitate to use them.

What are the factors that participate in the signaling between the granulosa and the oocyte cells? The female infertility seen in some *Sl* mutants that are otherwise viable suggests that, in addition to its role during embryonic and fetal development, KL also plays an important role in adult gametogenesis. *In situ* hybridization and immunocytochemical studies of c-kit and KL mRNA and protein in the ovary give a clear picture of the cell types expressing both the receptor and its ligand *(30,65–67)*. Oocytes express c-kit mRNA and protein at all stages of follicle development, and the receptor is also expressed in ovulated eggs. The c-kit receptor is also found to be expressed in interstitial cells and theca cells *(68)*. The granulosa cells surrounding the oocyte express KL in every mammal examined thus far. In mice, KL expression is low in primordial and primary follicles, and in granulosa cells surrounding the ovulated oocyte. However, soon after the period of oocyte growth (i.e., the primary follicle through the preovulatory follicle), KL mRNA levels increase and are regulated in a hormonally dependent manner *(69)*. Taken together, these data suggest that KL could play an important role in oocyte growth. Consistent with these observations, mice carrying mutations in the *Sl* gene exhibit defects in oogenesis. For example, Sl^{panda} (Sl^{pan}) mice have DNA rearrangements in the *Sl* gene-promoter region that reduces KL expression during development and in the adult mouse *(70)*. Defective KL mRNA expression in Sl^{pan} animals causes sterility by influencing the initiation and maintenance of ovarian-follicle development. The observed arrest in follicle development occurs at the primary-follicle stage of development. Similar defects, which are found in $Sl^{contrasted}$ (Sl^{con}) mice, also affect ovarian-follicle development *(70)*. These data demonstrate a requirement for KL in follicle development.

Microinjection of the ACK2 anti-c-kit antibody into female mice disrupts oogenesis, and provides important information on the role of the c-kit signaling pathways in the ovary *(66)*. These studies suggest that c-kit is involved in the onset of primordial follicle development and primordial follicle growth consistent with the defects seen in several *Sl* mutants. In mice injected with anti-c-kit antibody every 2 d for the first 2 wk after birth, oocytes were surrounded by a single layer of granulosa cells but these cells were not synthesizing DNA *(66)*. This suggests that the granulosa cells were not actively proliferating. Nevertheless, in some cases, granulosa cells differentiated to take on a cuboidal shape. However, many follicles were observed to have degenerated, and the oocytes were lost *(66)*. When the regimen of anti-c-kit antibody injection was changed so that the antibody was not injected until 2 d after birth, primary-follicle development was only slightly disturbed. The conclusion from these studies was that c-kit function is critical for primary follicle development during the first 5 d of postnatal life *(66)*. These studies also suggested an important role for the c-kit signaling pathway in antral-follicle development and function. Thus, mice injected with the ACK-2 antibody had reduced numbers of antral follicles by comparison with matched controls. Injection of the ACK-2 antibody between postnatal d 10 and 14 inhibited granulosa-cell proliferation, and antrum formation was severely perturbed *(66)*. Once antral follicles are formed, some of these large follicles will mature to the ovulatory stage. Blocking c-kit signaling via ACK-2 injection between d 10 and d 14 caused antral-follicle degeneration and inhibition of follicular fluid formation *(66)*. This result suggests that the c-kit/KL signal is also required for the maturation of antral follicles.

Previous in vitro studies suggested that the c-kit signaling pathway plays an important role in oocyte growth. For example, when follicles are cultured in the presence of soluble KL, an increase in oocyte diameter is observed *(71)*. Detailed analysis of KL mRNA levels in developing and maturing ovarian follicles suggests that the oocyte regulates both the type and level of KL mRNA expressed by surrounding follicle cells. For example, fully grown oocytes were found to reduce the level of KL mRNA in preantral granulosa cells, and also, increased the ratio of KL-1 mRNA to KL-2 mRNA *(72)*. Other oocyte-mediated changes in KL mRNA expression were appropriate to the stage of folliculogenesis *(72)*. Therefore, the developmental regulated expression of c-kit in the oocyte and KL in the surrounding follicle cells provides a nice mechanism for crosstalk between the follicle cells and the oocyte. In response to KL, the oocyte develops or matures and sends a signal back to the follicle (granulosa) cells. Follicle cells respond by modifying the amount or type of KL they produce, which in turn affects the oocyte. In this way, the follicle cells determine which stage of development the oocyte has reached and respond accordingly, and vice versa. How does the signal transduced by the c-kit receptor, which is present in the oocyte, regulate granulosa cell function? One attractive candidate for a signal produced by oocytes and which would act on granulosa cells is GDF-9 *(73,74)*. GDF-9 is produced by the oocyte, and binds to receptors present on granulosa cells. In female mice lacking GDF-9, follicles with intact oocytes contain only a single layer of granulosa cells, similar to the *Sl* mutant mice *(75)*. Although oocyte levels of c-kit mRNA appear normal in GDF-9-deficient mice, KL levels are greatly increased *(64)*. These data suggest that KL may be regulated in a paracrine fashion by GDF-9. Interestingly, the oocytes in GDF-9 knockout mice have defects in meiotic competence (abnormal germinal-vesicle breakdown and spontaneous parthenogenetic activation), and demonstrate an increased rate of growth *(75,76)*. Although the oocyte undoubtedly plays an important role in regulating KL mRNA levels in granulosa cells, other factors not derived from oocytes are also likely to play an important role in this regulation, including signals emanating from ovarian theca cells and extragonadal factors such as FSH. This is supported by the finding that stimulation of the LH receptor in theca cells, via human chorionic gonadotropin (hCG) and equine chorionic gonadotropin (eCG) treatment results in increased levels of KL mRNA in certain granulosa-cell populations *(68)*.

A growing body of evidence suggests that the c-kit signal-transduction pathway also plays a role in regulating the meiotic cell cycle in mammal oocytes in the rat, and that activation of the c-kit receptor results in a delay in the rate of oocyte meiosis *(77)*. Thus, KL signaling may be one of the factors responsible for maintaining oocytes in meiotic arrest during their growth period. When rat oocytes are injected with c-kit antisense oligonucleotides, c-kit levels decline and the oocytes show an increased ability to resume meiosis *(78)*. In contrast, oocytes cultured in the presence of soluble KL are delayed in their ability to resume meiosis *(78)*. The changes in KL-1 and KL-2 mRNA levels and ratios in developing follicles described here may provide a mechanism for regulating meiotic resumption. It is presumed that membrane-bound forms of KL, when bound to the c-kit receptor, prevent c-kit internalization. Membrane-bound forms of KL may also send a quantitatively and qualitatively distinct signal to cells than the soluble form of KL *(79)*. Thus, a switch in production from membrane-bound to soluble c-kit by granulosa cells may be one of the signals that triggers the resumption of meiosis in mammals.

The analysis of the role of the c-kit/KL axis in female germline development reinforces the idea that these two proteins play multiple roles at different stages of germ-cell development. One of the challenges ahead will be to decipher how a single receptor, hardwired into different cell types, can give very different responses.

COULD GONADAL TUMORS BE A CONSEQUENCE OF C-KIT/KL DYSREGULATION?

While compelling evidence has been provided for the role of c-kit and KL in normal germline development, little is known about the role of these proteins in germ-cell tumor formation or in the development of gonadal tumors. Loss of germ cells from the ovary is associated with the development of epithelial ovarian tumors, including tubular adenomas. This is has been suggested to result from overproduction of pituitary gonadotrophin in response to reduced germ-cell numbers, which in turn causes overstimulation of the gonad. Such a phenotype is seen in both *W* and *Sl* mutants that are viable and sterile *(42)*. One other mechanism in which somatic cells of the ovary may become transformed is through constitutive expression of both c-kit and its ligand. In some forms of ovarian cancer (serous adenocarcinoma), both c-kit and KL proteins have been detected histochemically in the same cell type *(80)*. This could result in autocrine activation of the c-kit receptor that would be expected to drive cell-cycle progression and lead to cellular transformation. In some germ-cell tumors of the ovary (dysgerminomas), c-kit is expressed in malignant cells while the surrounding connective tissue expresses KL. Thus, the c-kit signaling pathway may play an important role—not only in normal development of the ovary, but also in development of ovarian tumors.

Curiously, some mutations at the *Sl* locus also increase the incidence of testicular teratoma and teratocarcinoma in mice *(81)*. This observation is puxxling, since KL is required for PGC survival *(37)*. Loss of one allele of *Sl* should result in loss of PGCs rather than their extended proliferation. PGC numbers in the developing embryo may be strictly regulated, so that evolutionary pressure ensures that the gonad is fully populated with germ cells. In this scenario, reduced PGC survival may cause the remaining PGCs to continue to proliferate in order to restore PGC numbers. Interestingly, in mouse mutants in which germ-cell numbers are reduced, extended proliferation of germ cells is observed. Germ-cell hyperplasia may predispose the cells to transformation, as it does in other lineages *(82)*. Some evidence for this also comes from studies in which PGC numbers were reduced by treatment of embryos with mitomycin C *(83)*. Following treatment with mitomycin C, embryo size and the number of PGCs are greatly reduced. However, after a few days of development, PGC numbers recover somewhat *(83)*. These data suggest that there is some mechanism for monitoring and counting PGC numbers in the developing gonad. In humans, testicular cancer incidence correlates with conditions in which the gonads are not well-developed (for review *see* ref. *3*), such as cryptorchidism *(84)*, gonadal dysgenesis *(85)*, androgen insensitivity syndrome *(86)*, testicular atrophy *(87)*, or infertility *(88)*. It is now well-accepted that CIS (carcinoma *in situ* of the testis) is the precursor of most germ-cell testicular cancers (for review *see* ref. *89*), with the exception of spermatocytic seminoma *(90)*. CIS cells are thought to be derived from gonocytes, and in fact, they show similar morphological and immunohistochemical characteristics *(91)*. Several reports also suggest involvement of the c-kit pathway in the origin of CIS. It has been hypothesized that overexpression (or prolonged expression) of c-kit by gonocytes may lead to abnormal cell divisions and subsequently to their transformation *(92,93)*. To confirm this hypothesis, an activating mutation in the kinase domain has been identified in the c-kit gene in germ-cell tumors in a small number of cases *(94)*, which produces a c-kit protein that is constitutively active. These data highlight the potential importance of the c-kit signal transduction pathway in gonadal tumors of both germ-cell and somatic-cell origin. Further definition of the role of this signaling pathway in gonadal tumors is necessary. Interestingly, disruption of the c-kit-signal transduction pathway may be an important method for treating testicular tumors *(95)*, and possibly other tumors in which the c-kit signal-transduction pathway is disturbed.

FUTURE PROSPECTS

The importance of fertility in human populations is likely to make the study of the role of the c-kit/KL signaling pathway in gametogenesis an enduring one. The c-kit/KL signaling pathway is undoubtedly one of the key pathways involved in regulation of gametogenesis. Moreover, it plays an important role at many different stages of gametogenesis. Several areas of research are likely to play an important role in our understanding of the role that c-kit and KL play in gametogenesis. First, studies on the signaling pathway that lies downstream of c-kit receptor will probably yield important information on how this one signaling protein can effect such different responses at different stages of germ-cell development. A key question is how the c-kit receptor is hardwired within germ cells at different stages of development. Second, targeting of the c-kit gene with different mutations via homologous recombination in ES cells will certainly identify the residues within the protein responsible for defined functions within different cell types *(26)*. Combined with cell -type-specific deletions (mediated by Lox-Cre), this approach will undoubtedly refine our understanding of the biology of the c-kit response. Third, improvements in the in vitro culture conditions for germ cells at all stages of development will further clarify the role of the c-kit/KL axis in germline development. Fourth, the further development of transplantation techniques for male germ cells *(96,97)* will allow many of the roles of KL in male spermatogonial survival and proliferation to be addressed.

The development of the mammalian germline is a complex process. The developmental processes involved include migration, survival, proliferation, growth, and differentiation. Many different types of somatic cells may be involved, and these cells may also migrate, proliferate, and differentiate in order to fulfill their function in support of the germ cells. One striking feature of these processes is the central role of the c-kit receptor and its ligand, KL. Although many other factors are involved in germline development, the study of the c-kit/KL axis is likely to continue to be a central feature of research in mammalian gametogenesis.

ACKNOWLEDGMENTS

We would like to thank Mary Bedell for many helpful discussions and Marty Matzuk for inviting us to write this review and for his patience. Work in the author's laboratory was supported in part by a Cancer Center Core Grant (P30CA56036) from the National Cancer Institute.

REFERENCES

1. McLaren A. Germ Cells and Soma: A New Look at an Old Problem. Yale Univeristy Press, New Haven, CT, 1981.
2. Bondy CA, Nelson LM, Kalantaridou SN. The genetic origins of ovarian failure. J Womens Health 1998;7(10):1225–1229.
3. Skakkebaek NE, Rajpert-De Meyts E, Jorgensen N, Carlsen E, Petersen PM, Giwercman A, et al. Germ cell cancer and disorders of spermatogenesis: an environmental connection? Apmis 1998;106(1):3–11; discussion 12.
4. Schottenfeld D, Warshauer ME, Sherlock S, Zauber AG, Leder M, Payne R. The epidemiology of testicular cancer in young adults. Am J Epidemiol 1980;112(2):232–246.
5. Osterlind A. Diverging trends in incidence and mortality of testicular cancer in Denmark, 1943-1982. Brit J Cancer 1996;53(4):501–505.
6. Bjorge T, Engeland A, Hansen S, Trope CG. Trends in the incidence of ovarian cancer and borderline tumours in Norway, 1954-1993. Int J Cancer 1997;71(5):780–786.
7. Lawson KA, Hage WJ. Clonal analysis of the origin of primordial germ cells in the mouse. Ciba Foundation Symposium, 1994;182:68–84; discussion 84–91.
8. Donovan PJ, Stott D, Cairns LA., Heasman J, Wylie CC. Migratory and postmigratory mouse primordial germ cells behave differently in culture. Cell 1986;44(6):831–838.
9. Paniagua R, Nistal M. Morphological and histometric study of human spermatogonia from birth to the onset of puberty. J Anat 1984;139(Pt 3):535–552.
10. Vergouwen RP, Jacobs SG, Huiskamp R, Davids JA, de Rooij DG. Proliferative activity of gonocytes, Sertoli cells and interstitial cells during testicular development in mice. J Reprod Fertil 1991; 93(1):233–243.
11. Hilscher B, Hilscher W, Bulthoff-Ohnolz B, Kramer U, Birke A, Pelzer H, et al. Kinetics of gametogenesis. I. Comparative histological and autoradiographic studies of oocytes and transitional prospermatogonia during oogenesis and prespermatogenesis. Cell Tissue Res 1974;154(4):443–470.
12. Tilly JL. Apoptosis and ovarian function. Rev Reprod 1996;1(3):162–172.
13. Kaipia A, Hsueh AJ. Regulation of ovarian follicle atresia. Ann Rev Physiol 1997;59:349–363.
14. Braun RE. Every sperm is sacred—or is it? Nat Genet 1998;18(3):202–204.
15. Furuchi T, Masuko K, Nishimune Y, Obinata M, Matsui Y. Inhibition of testicular germ cell apoptosis and differentiation in mice misexpressing Bcl-2 in spermatogonia. Development 1996;122(6):1703–1709.
16. De Rooij DG, Van Dissel-Emiliani FM, Van Pelt AM. Regulation of spermatogonial proliferation. Ann NY Acad Sci 1989;564:140–153.
17. Matsui Y. Regulation of germ cell death in mammalian gonads. APMIS 1998;106(1):142–147; discussion 147,148.
18. Geissler EN, Ryan MA, Housman DE. The dominant-white spotting (W) locus of the mouse encodes the c-kit proto-oncogene. Cell, 1988;55(1):185–192.
19. Chabot B, Stephenson DA, Chapman VM, Besmer P, Bernstein A. The proto-oncogene c-kit encoding a transmembrane tyrosine kinase receptor maps to the mouse W locus. Nature 1988; 335(6185):88,89.
20. Anderson DM, Lyman SD, Baird A, Wignall JM, Eisenman J, Rauch C, et al. Molecular cloning of mast cell growth factor, a hematopoietin that is active in both membrane bound and soluble forms. Cell 1990;63(1):235–243.
21. Zsebo KM, Williams DA, Geissler EN, Broudy VC, Martin FH, Atkins HL, et al. Stem cell factor is encoded at the Sl locus of the mouse and is the ligand for the c-kit tyrosine kinase receptor. Cell 1990;63(1):213–224.
22. Copeland NG, Gilbert DJ, Cho BC, Donovan PJ, Jenkins NA, Cosman D, Anderson D, Lyman SD, Williams DE. Mast cell growth factor maps near the steel locus on mouse chromosome 10 and is deleted in a number of steel alleles. Cell 1990;63(1):175–183.
23. Huang E, Nocka K, Beier DR, Chu TY, Buck J, Lahm HW, et al. The hematopoietic growth factor KL is encoded by the Sl locus and is the ligand of the c-kit receptor, the gene product of the W locus. Cell 1990;63(1):225–233.
24. Huang EJ, Nocka KH, Buck J, Besmer P. Differential expression and processing of two cell associated forms of the kit-ligand: KL-1 and KL-2. Mol Biol Cell 1992;3(3):349–362.

25. Majumdar MK, Feng L, Medlock E Toksoz D, Williams DA. Identification and mutation of primary and secondary proteolytic cleavage sites in murine stem cell factor cDNA yields biologically active, cell-associated protein. J Biol Chem 1994;269(2):1237–1242.
26. Tajima Y, Moore MA, Soares V, Ono M, Kissel H, Besmer P. Consequences of exclusive expression in vivo of Kit-ligand lacking the major proteolytic cleavage site. Proc Natl Acad Sci USA 1998;95(20):11,903–11,908.
27. Lev S, Blechman JM, Givol D, Yarden Y. Steel factor and c-kit protooncogene: genetic lessons in signal transduction. Crit Rev Oncogenesis 1994;5(2–3):141–168.
28. Besmer P, Manova K, Duttlinger R, Huang EJ, Packer A, Gyssler C, et al The kit-ligand (steel factor) and its receptor c-kit/W: pleiotropic roles in gametogenesis and melanogenesis. Development (Suppl) 1993;125–137.
29. Silvers WK. The Coat Colors of Mice. Springer-Verlag, New York, NY, 1979.
30. Matsui Y, Zsebo KM, Hogan BL. Embryonic expression of a haematopoietic growth factor encoded by the Sl locus and the ligand for c-kit. Nature 1990;347(6294):667–669.
31. Nocka K, Majumder S, Chabot B, Ray P Cervone M, Bernstein A, et al. Expression of c-kit gene products in known cellular targets of W mutations in normal and W mutant mice—evidence for an impaired c-kit kinase in mutant mice. Genes Dev 1989;3(6):816–826.
32. Keshet E, Lyman SD, Williams DE, Anderson DM, Jenkins NA, Copeland NG, et al. Embryonic RNA expression patterns of the c-kit receptor and its cognate ligand suggest multiple functional roles in mouse development. EMBO J 1991;10(9):2425–2435.
33. Mintz B. Embryological development of primordial germ cells in the mouse: influence of a new mutation, Wj. J Embryol Exp Morphol 1957;5:396–406.
34. Mintz B, Russell ES. Gene-induced embryological modifications of primordial germ cells in the mouse. J Exp Morphol 1957;134:207–237.
35. Buehr M, McLaren A, Bartley A, Darling S. Proliferation and migration of primordial germ cells in We/We mouse embryos. Dev Dyn 1993;198(3):182–189.
36. McCoshen JA, McCallion DJ. A study of the primordial germ cells during their migratory phase in Steel mutant mice. Experientia 1975;31(5):589,590.
37. Dolci S, Williams DE, Ernst MK, Resnick JL, Brannan CI, Lock LF, et al. Requirement for mast cell growth factor for primordial germ cell survival in culture. Nature 1991;352(6338):809–811.
38. Matsui Y, Toksoz D, Nishikawa S, Williams D, Zsebo K, Hogan BL. Effect of Steel factor and leukaemia inhibitory factor on murine primordial germ cells in culture. Nature 1991;353(6346):750–752.
39. Godin I, Deed R, Cooke J, Zsebo K, Dexter M, Wylie CC. Effects of the steel gene product on mouse primordial germ cells in culture. Nature 1991;352(6338):807–809.
40. Brannan CI, Lyman SD, Williams DE, Eisenman J, Anderson DM, Cosman D, et al. Steel-Dickie mutation encodes a c-kit ligand lacking transmembrane and cytoplasmic domains. Proc Natl Acad Sci USA 1991;88(11):4671–4674.
41. Flanagan JG, Chan DC, Leder P. Transmembrane form of the kit ligand growth factor is determined by alternative splicing and is missing in the Sld mutant. Cell 1991;64(5):1025–1035.
42. Brannan CI, Bedell MA, Resnick JL, Eppig JJ, Handel MA, Williams DE, et al. Developmental abnormalities in Steel17H mice result from a splicing defect in the steel factor cytoplasmic tail. Genes Dev 1992;6(10):1832–1842.
43. Orth JM, Qiu J, Jester WF Jr, Pilder S. Expression of the c-kit gene is critical for migration of neonatal rat gonocytes in vitro. Biol Reprod 1997;57(3):676–683.
44. van Haaster LH., de Rooij DG. Spermatogenesis is accelerated in the immature Djungarian and Chinese hamster and rat. Biol Reprod 1993;49(6):1229–1235.
45. Vitale R, Fawcett DW, Dym M. The normal development of the blood-testis barrier and the effects of clomiphene and estrogen treatment. Anat Rec 1973;176(3):331–344.
46. Yoshinaga K, Nishikawa S, Ogawa M, Hayashi S, Kunisada T, Fujimoto T. Role of c-kit in mouse spermatogenesis: identification of spermatogonia as a specific site of c-kit expression and function. Development 1991;113(2):689–699.
46a. Blume-Jensen P, Janknecht R, Hunter T. The kit receptor promotes cell survival via activation of PI 3-kinase and subsequent Akt-mediated phosphorylation of Bad and Ser 136. Curr Biol 1998;8(13):779–782.
46b. Blume-Jensen P, Jiang G, Hyman R, Lee KF, O'Gorman S, Hunter V. Kit/stem cell factor receptor-induced activation of phosphatildylinositol 3'-kinase is essential for male fertility. Nat Gene 2000;24(2):157–162.
46c. Kissel H, Timokhina I, Hardy MP, Rothschild G, Tjima T, Soares V, Angeles M, et al. Point mutation in kit receptor tyrosine kinase reveals essential roles for kit signaling in spermatogenesis and oogenesis without affectivn other kit responses. EMBO J 1000;19(6);1312–1326.

47. Ding HF, Fisher DE. Mechanisms of p53-mediated apoptosis. Crit Rev Oncogenesis 1998;9(1):83–98.
48. Steele RJ, Thompson AM, Hall PA, Lane DP. The p53 tumour suppressor gene. Brit J Surg 1998;85(11):1460–1467.
49. Bates S, Vousden KH. Mechanisms of p53-mediated apoptosis. Cell Mol Life Sci 1999;55(1):28–37.
50. Donehower LA, Harvey M, Slagle BL, McArthur MJ, Montgomery CA Jr, Butel JS, et al. Mice deficient for p53 are developmentally normal but susceptible to spontaneous tumours. Nature 1992;356(6366):215–221.
51. Jacks T, Remington L, Williams BO, Schmitt EM, Halachmi S, Bronson RT, et al. Tumor spectrum analysis in p53-mutant mice. Curr Biol 1994;4(1):1–7.
52. Jordan SA, Speed RM, Bernex F, Jackson IJ. Deficiency of Trp53 rescues the male fertility defects of KitW-v mice but has no effect on the survival of melanocytes and mast cells. Dev Biol 1999;215:78–90.
53. Kapur R, Cooper R, Xiao X Weiss MJ, Donovan P, Williams DA. The presence of novel amino acids in the cytoplasmic domain of stem cell factor results in hematopoietic defects in Steel(17H) mice. Blood 1999;94(6):1915–1925.
54. Wehrle-Haller B, Weston JA. Altered cell-surface targeting of stem cell factor causes loss of melanocyte precursors in Steel17H mutant mice. Dev Biol 1999;210(1):71–86.
55. Vincent S, Segretain D, Nishikawa S, Nishikawa SI, Sage J, Cuzin F, et al. Stage-specific expression of the Kit receptor and its ligand (KL) during male gametogenesis in the mouse: a Kit-KL interaction critical for meiosis. Development 1998;125(22):4585–4593.
56. Rossi P, Marziali G, Albanesi C, Charlesworth A, Geremia R, Sorrentino V. A novel c-kit transcript, potentially encoding a truncated receptor, originates within a kit gene intron in mouse spermatids. Dev Biol 1992;152(1):203–207.
57. Sandlow JI, Feng HL, Sandra A. Localization and expression of the c-kit receptor protein in human and rodent testis and sperm. Urology 1997;49(3):494–500.
58. Albanesi C, Geremia R, Giorgio M, Dolci S, Sette C, Rossi P. A cell- and developmental stage-specific promoter drives the expression of a truncated c-kit protein during mouse spermatid elongation. Development 1996;122(4):1291–1302.
59. Sette C, Bevilacqua A, Bianchini A, Mangia F, Geremia R, Rossi P. Parthenogenetic activation of mouse eggs by microinjection of a truncated c-kit tyrosine kinase present in spermatozoa. Development 1997;124(11):2267–2274.
60. Sandlow JI, Feng HL, Zheng LJ, Sandra A. Migration and ultrastructural localization of the c-kit receptor protein in spermatogenic cells and spermatozoa of the mouse. J Urol 1999;161(5):1676–1680.
61. Feng H, Sandlow JI, Sandra A. The c-kit receptor and its possible signaling transduction pathway in mouse spermatozoa. Molecular Reproduction & Development, 1998;49(3):317–326.
62. Eppig JJ, Chesnel F, Hirao Y, O'Brien MJ, Pendola FL, Watanabe S, et al. Oocyte control of granulosa cell development: how and why. Human Reprod 1997;12(11 Suppl):127–132.
63. Buccione R, Schroeder AC, Eppig JJ. Interactions between somatic cells and germ cells throughout mammalian oogenesis. Biol Reprod 1990;43(4):543–547.
64. Elvin JA, Clark AT, Wang P, Wolfman NM, Matzuk MM. Paracrine actions of growth differentiation factor-9 in the mammalian ovary. Mol Endocrinol 1999;13(6):1035–1048.
65. Manova K, Nocka K, Besmer P, Bachvarova RF. Gonadal expression of c-kit encoded at the W locus of the mouse. Development 1990;110(4):1057–1069.
66. Yoshida H, Takakura N, Kataoka H, Kunisada T, Okamura H, Nishikawa SI. Stepwise requirement of c-kit tyrosine kinase in mouse ovarian follicle development. Dev Biol 1997;184(1):122–137.
67. Keshet E, Lyman SD, et al. Embryonic RNA expression patterns of the c-kits receptor and its cognate ligand suggest multiple functional roles in mouse development. EMBO J 1991;10(9):2425–2435.
68. Parrott JA, Skinner MK. Direct actions of kit-ligand on theca cell growth and differentiation during follicle development. Endocrinology 1997;138(9):3819–3827.
69. Motro B, Bernstein A. Dynamic changes in ovarian c-kit and Steel expression during the estrous reproductive cycle. Dev Dyn 1993;197(1):69–79.
70. Bedell MA, Brannan CI, Evans EP, Copeland NG, Jenkins NA, Donovan PJ. DNA rearrangements located over 100 kb 5' of the Steel (Sl)-coding region in Steel-panda and Steel-contrasted mice deregulate Sl expression and cause female sterility by disrupting ovarian follicle development. Genes Dev 1995;9(4):455–470.
71. Packer AI, Hsu YC, Besmer P, Bachvarova RF. The ligand of the c-kit receptor promotes oocyte growth. Dev Biol 1994;161(1):194–205.
72. Joyce IM, Pendola FL, Wiggelsworth K, Eppig JJ. Oocyte regulation of Kit ligand expression in mouse ovarian follicles. Dev Biol 1999;214:342–353.

73. McPherron AC, Lee SJ. GDF-3 and GDF-9: two new members of the transforming growth factor-beta superfamily containing a novel pattern of cysteines. J Biol Chem 1993;268(5):3444–3449.
74. McGrath SA, Esquela AF, Lee SJ. Oocyte-specific expression of growth/differentiation factor-9. Mol Endocrinol 1995;9(1):131–136.
75. Dong J, Albertini DF, Nishimori K, Kumar TR, Lu N, Matzuk MM. Growth differentiation factor-9 is required during early ovarian folliculogenesis. Nature, 1996;383(6600):531–535.
76. Carabatsos MJ, Elvin J, Matzuk MM, Albertini DF. Characterization of oocyte and follicle development in growth differentiation factor-9 deficient mice. Dev Biol 1998;204:373–384.
77. Ismail RS, Okawara Y, Fryer JN, Vanderhyden BC. Hormonal regulation of the ligand for c-kit in the rat ovary and its effects on spontaneous oocyte meiotic maturation. Mol Reprod Dev 1996;43(4):458–469.
78. Ismail RS, Dube M, Vanderhyden BC. Hormonally regulated expression and alternative splicing of kit ligand may regulate kit-induced inhibition of meiosis in rat oocytes. Dev Biol 1997; 184(2):333–342.
79. Miyazawa K, Williams DA, Gotoh A, Nishimaki J, Broxmeyer HE, Toyama K. Membrane-bound Steel factor induces more persistent tyrosine kinase activation and longer life span of c-kit gene-encoded protein than its soluble form. Blood 1995;85(3):641–649.
80. Inoue M, Kyo S, Fujita M, Enomoto T, Kondoh G. Coexpression of the c-kit receptor and the stem cell factor in gynecological tumors. Cancer Res 1994;54(11):3049–3053.
81. Stevens LC. Genetic influences on teratocarcinogenesis and parthenogenesis. Prog Clin Biol Res 1981;45:93–104.
82. Noguchi T, Stevens LC. Primordial germ cell proliferation in fetal testes in mouse strains with high and low incidences of congenital testicular teratomas. J Natl Cancer Inst 1982;69(4):907–913.
83. Tam PP, Snow MH. Proliferation and migration of primordial germ cells during compensatory growth in mouse embryos. J Embryol Exp Morphol 1981;64:133–147.
84. Moller H, Prener A, Skakkebaek NE. Testicular cancer, cryptorchidism, inguinal hernia, testicular atrophy, and genital malformations: case-control studies in Denmark. Cancer Causes Control 1996;7(2):264–274.
85. Muller J, Skakkebaek NE. Gonadal malignancy in individuals with sex chromosome anomalies. Birth Defects: Original Article Series, 1990;26(4):247–255.
86. Collins GM, Kim DU, Logrono R, Rickert RR, Zablow A, Breen JL. Pure seminoma arising in androgen insensitivity syndrome (testicular feminization syndrome): a case report and review of the literature. Mod Pathol 1993;6(1):89–93.
87. Giwercman A, Lenz S, Skakkebaek NE. Carcinoma in situ in atrophic testis: biopsy based on abnormal ultrasound pattern. Brit J Urol 1993;72(1):118–120.
88. Petersen PM, Skakkebaek NE., Giwercman A. Gonadal function in men with testicular cancer: biological and clinical aspects. Apmis 1998;106(1):24–34; discussion 34–36.
89. Skakkebaek NE, Berthelsen JG, Giwercman A, Muller J. Carcinoma-in-situ of the testis: possible origin from gonocytes and precursor of all types of germ cell tumours except spermatocytoma. Int J Androl 1987;10(1):19–28.
90. Muller J. Abnormal infantile germ cells and development of carcinoma-in-situ in maldeveloped testes: a stereological and densitometric study. Int J Androl 1987;10(3):543–567.
91. Meyts ER, Jorgensen N, Muller J, Skakkebaek NE. Prolonged expression of the c-kit receptor in germ cells of intersex fetal testes. J Pathol 1996;178(2):166–169.
92. Rajpert-De Meyts E, Jorgensen N, Brondum-Nielsen K, Muller J, Skakkebaek NE. Developmental arrest of germ cells in the pathogenesis of germ cell neoplasia. Apmis 1998;106(1):198–204; discussion 204–206.
93. Jorgensen N, Giwercman A, Muller J, Skakkebaek NE. Immunohistochemical markers of carcinoma in situ of the testis also expressed in normal infantile germ cells. Histopathology 1993;22(4):373–378.
94. Tian Q, Frierson HF Jr, Krystal GW, Moskaluk CA. Activating c-kit gene mutations in human germ cell tumors. Am J Pathol 1999;154(6):1643–1647.
95. Li Q, Kondoh G, Inafuku S, Nishimune Y, Hakura A. Abrogation of c-kit/Steel factor-dependent tumorigenesis by kinase defective mutants of the c-kit receptor: c-kit kinase defective mutants as candidate tools for cancer gene therapy. Cancer Res 1996;56(19):4343–4346.
96. Brinster RL, Avarbock MR. Germline transmission of donor haplotype following spermatogonial transplantation. Proc Natl Acad Sci USA 1994;91(24):11,303–11,307.
97. Brinster RL, Zimmermann JW. Spermatogenesis following male germ-cell transplantation. Proc Natl Acad Sci USA 1994;91(24):11,298–11,302.

8 Gene Knockout Approaches to Steroidogenesis

Tomonobu Hasegawa, MD, PhD,
Liping Zhao, PhD, *Kathleen M. Caron*, PhD,
Morag Young, PhD, *and Keith L. Parker*, MD, PhD

Contents

INTRODUCTION

Genetic disorders in human patients—and to a lesser degree, in laboratory animals—have provided prismatic insights into the mechanisms of steroid hormone biosynthesis. These studies are restricted to spontaneously arising mutations, and often are hampered by ethical limitations on human experimentation. Gene targeting in mouse embryonic stem (ES) cells—ultimately producing knockout mice—has expanded considerably the number of genes whose function can be evaluated in vivo. This chapter focuses on knockout mouse studies that have defined the roles of two essential components of steroidogenesis: the transcription factor steroidogenic factor 1 (SF-1) and a protein that is required for cholesterol delivery to the steroidogenic complex, the steroidogenic acute regulatory protein (StAR).

The regulated production of steroid hormones involves complex, reciprocal interactions among the hypothalamus, anterior pituitary, and primary steroidogenic tissues, and defects at multiple levels can impair steroidogenesis. Figure 1 outlines the factors that control the production of steroid hormones in the adrenal cortex. Analyses of human patients with mutations that impair various steps in the adrenal and gonadal pathways have provided key insights into the roles of many of these components *(1,2)*. These

From: *Contemporary Endocrinology: Transgenics in Endocrinology*
Edited by: M. Matzuk, C. W. Brown, and T. R. Kumar

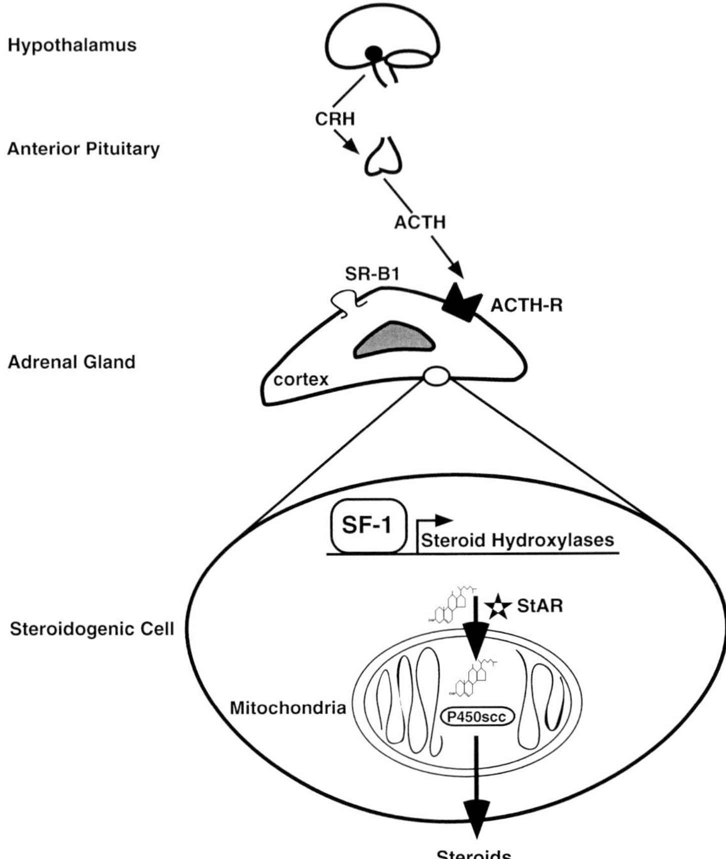

Fig. 1. Schematic overview of the multiple steps in adrenal steroidogenesis. A diagram of the different levels involved in the regulation of steroid hormone production by the adrenal cortex is shown. CRH, corticotropin-releasing hormone; ACTH, corticotropin; ACTH-R, ACTH receptor; SR-B1, high-density lipoprotein receptor; SF-1, steroidogenic factor 1; StAR, steroidogenic acute regulatory protein; P450scc, cholesterol side-chain cleavage enzyme.

human studies—coupled with cell-culture analyses in steroidogenic cells from different species—provide a framework for understanding the essential steps in steroid production. Experiments remain that cannot be performed in humans because of ethical considerations. In addition, evolving technologies for transgenesis and gene knockouts have enormously expanded our ability to examine in vivo the effects of mutating specific components of the steroidogenic complex.

StAR KNOCKOUT MICE

As shown in Fig. 1, one essential step in steroidogenesis is the translocation of cholesterol from the cytoplasm to the inner mitochondrial membrane, where the cholesterol side-chain cleavage enzyme (P450scc) catalyzes the first committed reactions in steroidogenesis. The steroidogenic acute regulatory protein (StAR), a 30-kDa mitochondrial phosphoprotein, was initially isolated because its expression within steroidogenic cells was rapidly induced by trophic hormones *(3)*. These findings suggested that StAR may contribute to steroidogenesis.

Dramatic confirmation of the essential role of StAR in these processes has come from analyses of human patients with congenital lipoid adrenal hyperplasia (lipoid CAH), an autosomal recessive disorder characterized by defects in all classes of steroid hormones (reviewed in *4*). Lipoid CAH was associated with mutations in StAR that precluded its function in cell transfection models of steroidogenesis. Certain features of the lipoid CAH phenotype were puzzling: the ratio of male to female patients (3:1) diverged from that predicted for an autosomal recessive trait. Moreover, some 46 XX patients underwent menarche with breast development at the normal age of puberty, strongly suggesting that they retained some capacity for estrogen biosynthesis *(5)*. Based in part on these sex-specific differences in gonadal pathology, Miller, Strauss, and colleagues proposed a two-hit model of the pathogenesis of StAR deficiency *(4)*. According to this model, steroidogenic cells lacking StAR initially retain some capacity for StAR-independent steroidogenesis. Over time, inadequate steroidogenesis leads to elevated levels of trophic hormones, which in turn stimulate progressive accumulation of lipids within the steroidogenic cells and ultimately cause their death. The ovaries, which are not steroidogenically active *in utero*, maintain some capacity for steroidogenesis that becomes apparent at the time of puberty.

To explore the roles of StAR in a system amenable to experimental manipulation, we used targeted gene disruption to create StAR knockout mice *(6)*. At birth, StAR knockout mice were indistinguishable from wild-type littermates, with an equal ratio of genetic males and females, but all pups had female external genitalia. A subset (~30–40%) exhibited signs of respiratory distress and died within 24 h after birth; the rest failed to grow normally and died within 2 wk after birth from adrenocortical insufficiency. Hormone assays revealed severe defects in adrenal steroids, and elevated ACTH levels consistent with a loss of negative feedback regulation at hypothalamic-pituitary levels. In contrast, gonadal hormones did not differ significantly from levels in wild-type littermates in the prepubertal state.

Histologically, the adrenal cortex of newborn StAR knockout mice contained striking lipid deposits with loss of normal cortical architecture (Fig. 2). The gonads were relatively spared, with no overt histological abnormalities, minimal lipid deposits in the steroidogenic compartment of the testis, and none in the ovary. In striking support of the two-hit model, StAR knockout mice kept alive with corticosteroid replacement therapy developed marked abnormalities of the ovaries, including a lack of corpora lutea and marked hyperplasia of lipid-engorged theca cells *(6a)*.

These StAR knockout mice may provide a useful model system for determining the mechanisms that mediate StAR's essential roles in regulated steroidogenesis. For example, immortalized cell lines derived from the steroidogenic organs of StAR knockout mice should provide an ideal system for exploring the structure-function aspects of StAR within actual steroidogenic cells. Similarly, these mice may provide a system for studying and identifying the actions of other proteins involved in cholesterol translocation within steroidogenic cells. Finally, the finding that StAR is expressed in discrete brain regions *(7)* suggests that further analyses of these knockout mice may reveal roles of StAR outside of the classical steroidogenic tissues.

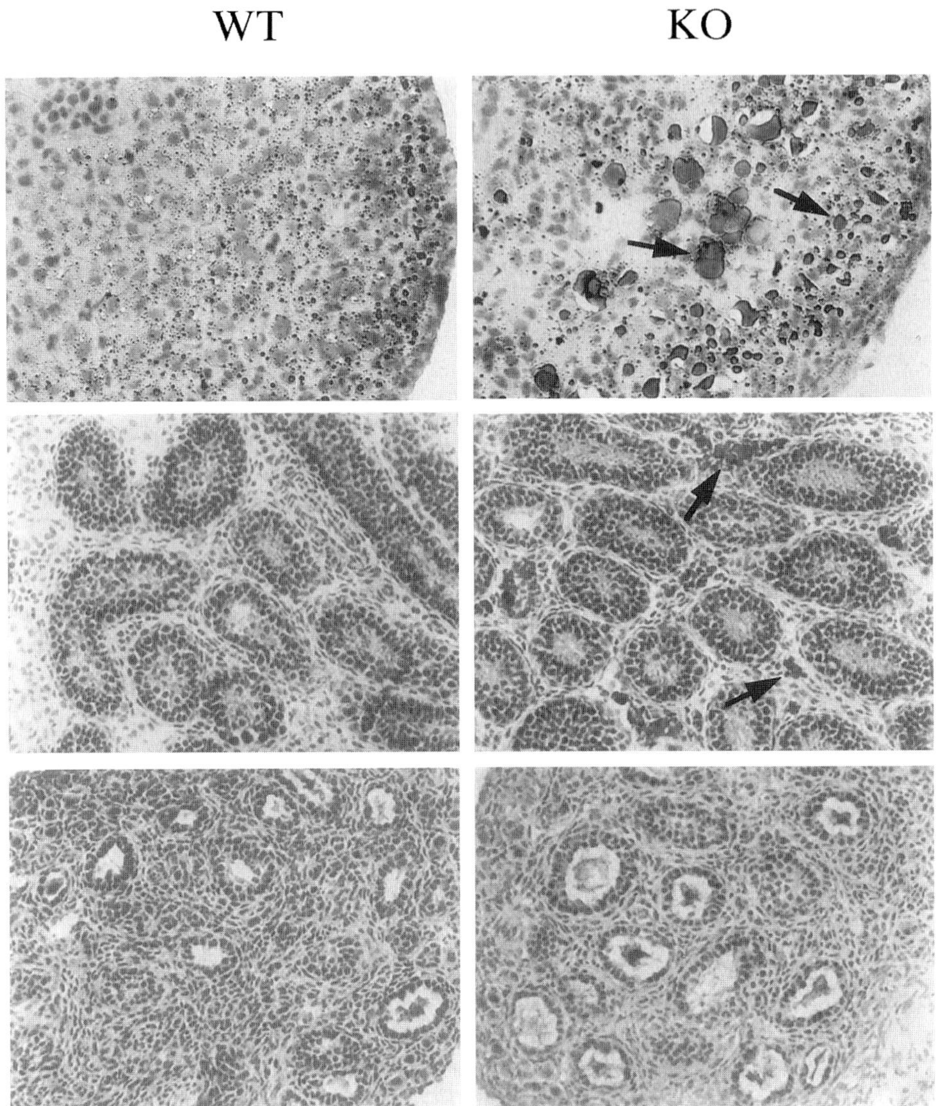

Fig. 2. Histology of the adrenal gland and gonads from newborn wild-type (WT) and StAR knockout (KO) mice. Steroidogenic organs were isolated from WT and StAR KO mice 1–6 d after birth, and sections were stained with oil red O and hematoxylin. *(Top panels)* Adrenal sections. *(Middle panels)* Testis sections. *(Bottom panels)* Ovary sections. The arrows point to areas of lipid deposits in the adrenal cortex and testis.

SF-1 KNOCKOUT MICE

SF-1 initially was identified as a transcription factor that regulated the cell-specific expression of the cytochrome P450 steroid hydroxylases that catalyze most steroidogenic conversions *(8,9)*. When a cDNA encoding SF-1 was isolated, its sequence established SF-1 as a member of the nuclear hormone-receptor family that mediates transcriptional regulation by steroid hormones, thyroid hormone, vitamin D, and retinoids. Subsequent studies have shown that SF-1 regulates adrenal and gonadal expression of many genes involved in steroidogenesis, including 3β-hydroxysteroid dehydrogenase, the ACTH receptor, StAR, and the high-density lipoprotein receptor SR-B1 (reviewed in *10*). Analyses of reporter genes driven by the Müllerian-inhibiting substance (MIS) promoter region in transfected Sertoli cells and transgenic mice suggest that SF-1 also regulates expression of the MIS gene *(11–13)*. Thus, SF-1 regulates the production of both essential hormones in male sexual differentiation: androgens and MIS.

To explore the link between SF-1 and steroidogenesis during endocrine development, the spatio-temporal profile of its expression during embryogenesis was examined *(12,14)*. SF-1 is expressed in both male and female mouse embryos from the very earliest stages of gonadogenesis, when the intermediate mesoderm condenses to form the urogenital ridge. As the testes differentiate, SF-1 expression localizes to both the interstitial region—where Leydig cells produce androgens—and the testicular cords, where Sertoli cells express MIS. In the ovaries, in contrast, SF-1 transcripts decrease just as sexual differentiation is taking place, suggesting that persistent SF-1 expression could impair female sexual differentiation. In addition to the primary steroidogenic tissues, SF-1 transcripts also are detected in the anterior pituitary and hypothalamus, suggesting that SF-1 regulates the endocrine axis at other levels.

Analyses of SF-1 knockout mice has confirmed the essential roles of SF-1 at all three levels of the hypothalamic-pituitary-steroidogenic organ axis *(15–19)*. Although the very earliest stages of gonadal development appeared intact, newborn SF-1 knockout mice exhibited adrenal and gonadal agenesis (Fig. 3). As a consequence of gonadal degeneration, before androgens and MIS are normally produced, male-to-female sex reversal of the internal and external urogenital tracts also occurred. SF-1 knockout mice also had impaired expression of a number of markers of gonadotropes—the cells in the anterior pituitary gland that regulate gonadal steroidogenesis—and they lacked the ventromedial hypothalamic nucleus (Fig. 4)—a cell group in the medial hypothalamus linked to feeding and reproductive behaviors. Although their GnRH neurons were normal in number and location, the impaired gonadotropin expression in SF-1 knockout mice raises the possibility that the VMH contributes to normal GnRH release. Finally, although the molecular basis is yet to be defined, the spleens of SF-1 knockout mice were hypoplastic and contained irregularly distributed clusters of erythrocytes, reflecting impaired erythropoiesis *(20)*.

The human SF-1 gene, located on chromosome 9q33 *(21)*, shares extensive homology with its mouse counterpart *(22,23)* and is expressed in many of the same sites *(24,25)*. A recent report described a human subject with adrenal insufficiency and 46, XY sex reversal associated with a mutation in one allele of SF-1 that precludes DNA-binding *(26)*. The mutated SF-1 protein did not exhibit dominant negative activity when coexpressed with wild-type protein, suggesting that the human disorder results from haploinsufficiency of SF-1. It remains possible, however, that the "normal" allele has a mutation that was undetected, or that a somatic mutation arose during early development. In either case, although its potential roles in the human VMH and the pituitary have not yet been established, these studies indicate that SF-1 in humans also plays essential roles in gonadal and adrenal development and function.

Another important goal for future studies is to identify the specific target genes through which SF-1 exerts its profound effects on endocrine development. Although a number of SF-1 target genes in the adrenal cortex and gonads have been identified, none of them alone can explain the loss of the adrenal glands and gonads. Moreover, no target genes

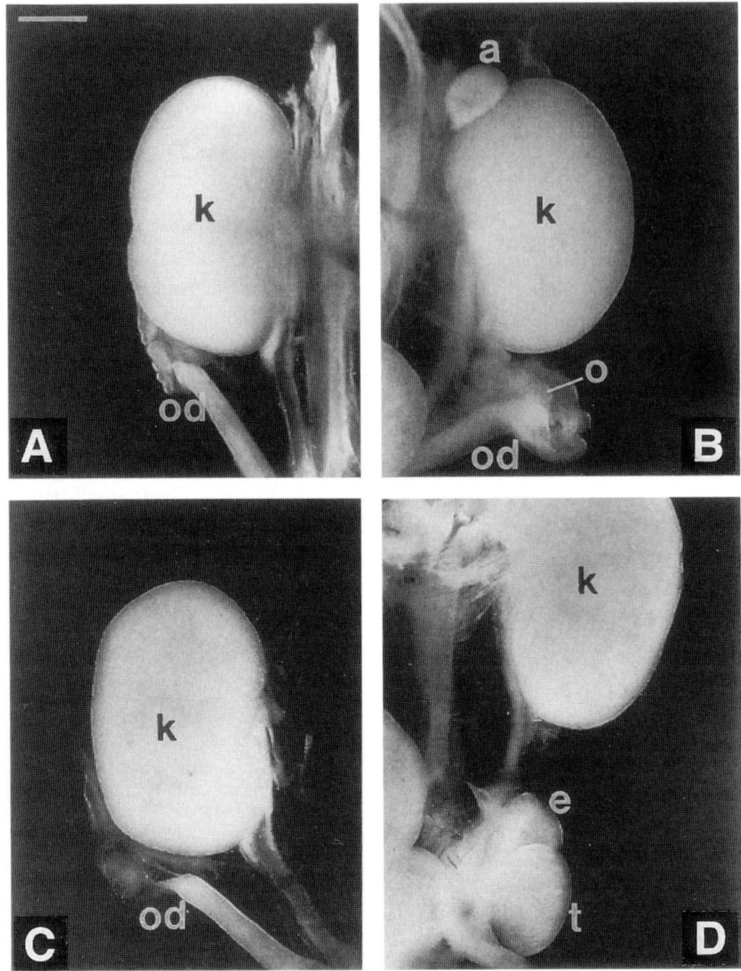

Fig. 3. Newborn SF-1 knockout mice lack adrenal glands and gonads and have female internal genitalia. SF-1 knockout mice *(left)* and wild-type littermates *(right)* were sacrificed and their genitourinary tracts were dissected. **(A)** SF-1 knockout female. **(B)** Wild-type female. **(C)** SF-1 knockout male. **(D)** Wild-type male. The scale bar = 1 mm. Reprinted with permission from ref. *(15)*. k, kidney; a, adrenal; o, ovary; t, testis; e, epididymis; od, oviduct.

of SF-1 within the VMH have been identified. The definition of additional target genes—some of which presumably impinge on the cell cycle/proliferation pathways—will undoubtedly provide important new insights into SF-1 function.

SUMMARY AND PERSPECTIVES

As outlined in this chapter, evolving technologies for creating knockout mice are providing exciting new approaches to explore the in vivo roles of various genes in the regulated production of steroid hormones. In particular, rapid advancements with tissue-specific knockouts using the bacteriophage Cre recombinase should provide major new insights into the precise roles of various genes within specific tissues. For example, VMH-specific disruption of SF-1 would provide a sophisticated system for defining the

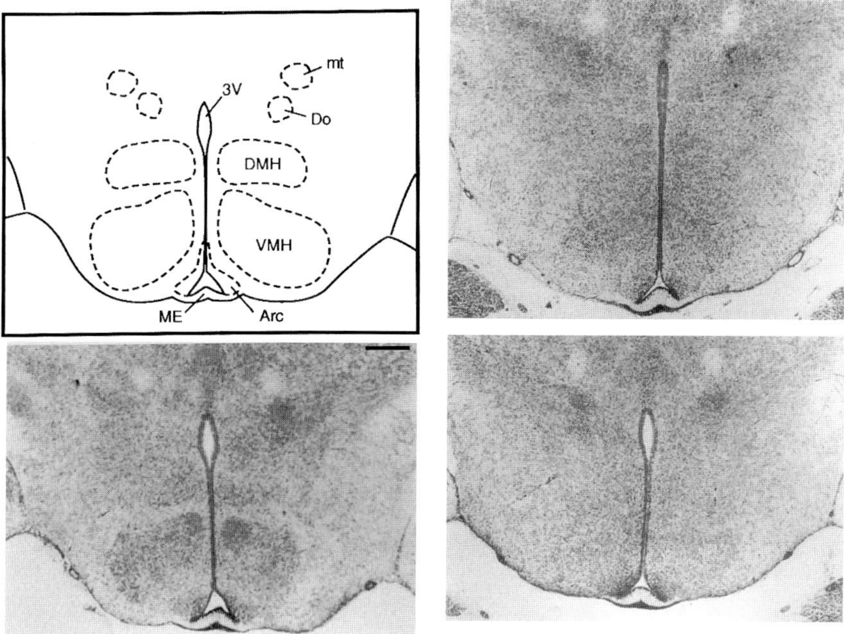

Fig. 4. SF-1 knockout mice lack the ventromedial hypothalamic nucleus. Coronal sections from wild-type *(lower left)* and SF-1 knockout male mice *(upper right)* and SF-1 knockout female mice *(lower right)* were stained with cresyl violet and photomicrographs were taken. A schematic diagram of the anatomical regions found within these sections is shown *(upper left)*. The scale bar = 200 μm. mt, mammillothalamic tract; Do, dorsal hypothalamic nucleus; 3V, third ventricle; DMH, dorsomedial hypothalamic nucleus; VMH, ventromedial hypothalamic nucleus; Arc, arcuate nucleus; ME, median eminence. Modified with permission from ref. *(19)*.

physiological functions of the VMH in the setting of normal levels of adrenal and gonadal steroids. Alternatively, transgenic rescue of the StAR knockout mice with wild-type and mutated StAR transgenes should permit a successful dissection of structure-function aspects of StAR in facilitating cholesterol translocation within steroidogenic tissues.

ACKNOWLEDGMENTS

Work in the authors' laboratory was supported by the Howard Hughes Medical Institute and by NIH grants HL48460 and DK55480.

REFERENCES

1. Donohoue PA, Parker KL, Migeon CJ. Congenital adrenal hyperplasia. In: Scriver CR, Beaudet LL, Sly WS, Valle D, eds. The Metabolic and Molecular Bases of Inherited Disease. McGraw-Hill, New York, NY, 2001; pp. 4077–4116.
2. Adashi EY, Hennebold JD. Single gene mutations resulting in reproductive dysfunction in women. N Engl J Med 1999;340:709–718.
3. Clark BJ, Wells J, King SR, Stocco DM. The purification, cloning, and expression of a novel luteinizing hormone-induced mitochondrial protein in MA-10 mouse Leydig tumor cells. Characterization of the steroidogenic acute regulatory protein (StAR). J Biol Chem 1994;269:28,314–28,322.
4. Bose HS, Sugawara T, Strauss JF 3rd, Miller WL. The pathophysiology and genetics of congenital lipoid adrenal hyperplasia. N Engl J Med 1996;335:1870–1878.

5. Bose HS, Pescovitz OH, Miller WL. Spontaneous feminization in a 46,XX female patient with congenital lipoid adrenal hyperplasia due to a homozygous frameshift mutation in the steroidogenic acute regulatory protein. J Clin Endocrinol Metab 1997;82:1511–1515.
6. Caron KM, Soo S-C, Clark BJ, Stocco DM, Wetsel W, Parker KL Targeted disruption of the mouse gene encoding the steroidogenic acute regulatory protein provides insights into congenital lipoid adrenal hyperplasia. Proc Natl Acad Sci USA 1997;94:11,540–11,545.
6a. Hasegawa T, Zhao L, Caron K, et al. Developmental roles of the steroidogenic acute regulatory protein (StAR) as revealed by StAR knockout mice. Mol Endocrinol 2000;14:1462–1471.
7. Furukawa A, Miyatake A, Ohnishi T, Ichikawa Y. Steroidogenic acute regulatory protein (StAR) transcripts constitutively expressed in the adult rat central nervous system: colocalization of StAR, cytochrome P-450SCC (CYP XIA1), and 3 beta-hydroxysteroid dehydrogenase in the rat brain. J Neurochem 1998;71:2231–2238.
8. Lala DS, Rice DA, Parker KL. Steroidogenic factor I, a key regulator of steroidogenic enzyme expression, is the mouse homolog of fushi tarazu-factor I. Mol Endocrinol 1992;6:1278–1287.
9. Morohashi K, Honda S, Inomata Y, Handa H, Omura T. A common trans-acting factor, Ad4-binding protein, to the promoters of steroidogenic P-450s. J Biol Chem 1992;267:17,913–17,919.
10. Parker KL, Schimmer BP. Steroidogenic factor 1: a key determinant of endocrine development and function. Endocr Rev 1997;18:361–377.
11. Shen W-H, Moore CCD, Ikeda Y, Parker KL, Ingraham HA. Nuclear receptor steroidogenic factor 1 regulates MIS gene expression: a link to the sex determination cascade. Cell 1994;77:651–661.
12. Hatano O, Takayama K, Imai T, Waterman MR, Takakusu A, Omura T. Morohashi K. Sex-dependent expression of a transcription factor, Ad4BP, regulating steroidogenic P-450 genes in the gonads during prenatal and postnatal rat development. Development 1994;120:2787–2797.
13. Giuili G, Shen W-H, Ingraham HA. The nuclear receptor SF-1 mediates sexually dimorphic expression of Mullerian Inhibiting Substance, in vivo. Development 1997;124:1799–1807.
14. Ikeda Y, Shen W-H, Ingraham HA, Parker KL. Developmental expression of mouse steroidogenic factor 1, an essential regulator of the steroid hydroxylases. Mol Endocrinol 1994;8:654–662.
15. Luo X, Ikeda Y, Parker KL. A cell-specific nuclear receptor is essential for adrenal and gonadal development and sexual differentiation. Cell 1994;77:481–490.
16. Sadovsky Y. Crawford PA, Woodson KG, Polish JA, Clements MA, Tourtellotte LM. et al. Mice deficient in the orphan receptor steroidogenic factor 1 lack adrenal glands and gonads but express P450 side-chain-cleavage enzyme in the placenta and have normal embryonic serum levels of corticosteroids. Proc Natl Acad Sci USA 1995;92:10,939–10,943.
17. Shinoda K, Lei H, Yoshii H, Nomura M, Nagano M, Shiba H, Sasaki H, Osawa Y, Ninomiya Y, Niwa O, Morohashi K-I. Developmental defects in the ventromedial hypothalamic nucleus and pituitary gonadotroph in the Ftz-F1 disrupted mice. Dev Dyn 1995;204:22–29.
18. Ingraham HA, Lala DS, Ikeda Y, Luo X, Shen W-H, Nachtigal MW, et al. The nuclear receptor SF-1 acts at multiple levels of the reproductive axis. Genes Dev 1994;8:2302–2312.
19. Ikeda Y, Luo X, Abbud R, Nilson JH, Parker KL. The nuclear receptor steroidogenic factor 1 is essential for the formation of the ventromedial hypothalamic nucleus. Mol Endocrinol 1995;9:478–486.
20. Morohashi K, Tsuboi-Asai H, Matsushita S, Suda, M, Nakashima M, Sasano H, Hataba Y, Li CL, Fukata J, Irie J, Watanabe T, Nagura H, Li E. Structural and functional abnormalities in the spleen of an mFtz-F1 gene-disrupted mouse. Blood 1999;93:1586–1594.
21. Taketo M, Parker KL, Howare TA, Tsukiyama R, Wong M, Niwa O, et al. Homologs of Drosophila Fushi-Tatzu Factor 1 map to mouse chromosome 2 and human chromosome 9q33. Genomics 1995;25:565–567.
22. Oba K, Yanase T, Nomura M, Morohashi K, Takayanagi R, Nawata H. Structural characterization of human Ad4BP (SF-1) gene. Biochem Biophys Res Commun 1996;226:261–267.
23. Wong M, Ramayya MS, Chrousos GP, Driggers PH, Parker KL. Cloning and sequence analysis of the human gene encoding steroidogenic factor 1. J Mol Endocrinol 1996;17:139–147.
24. Ramayya MS, Zhou J, Kino T, Segars JH, Bondy CA, Chrousos GP. Steroidogenic factor 1 messenger ribonucleic acid expression in steroidogenic and nonsteroidogenic human tissues: Northern blot and in situ hybridization studies. J Clin Endocrinol Metab 1997;82:1799–1806.
25. Hanley NA, Ball SG, Clement-Jones M, Hagan DM, Strachan T, Lindsay S, et al. Expression of steroidogenic factor 1 and Wilm's tumor 1 during early human gonadal development and sex determination. Mech Dev 1999;87:175–180.
26. Achermann JC, Ito M, Ito M, Hindmarsh PC, Jameson JL. A mutation in the gene encoding steroidogenic factor-1 causes XY sex reversal and adrenal failure in humans. Nat Genet 1999;22:125,126.

9 The Progesterone Receptor Knockout Mouse Model

New Insights into Progesterone Action In Vivo

John P. Lydon, PhD, *Selma Soyal,* PhD,
Bert W. O'Malley, MD, *and Preeti M. Ismail,* PhD

CONTENTS

INTRODUCTION

Historical Perspective

During the first decade of the twentieth centry, a series of classic experiments performed by the noted European embryologists Fraenkel, Loeb, Bouin, and Ancel unequivocally demonstrated the essential role of the corpus luteum in the establishment and maintenance of pregnancy (reviewed in *1,2*). Subsequent investigations in the 1920s revealed that organic extracts of the corpus luteum were able to elicit the distinctive histological and physiological phenotype of the endometrium (termed "progestational proliferation"), characteristic of early pregnancy in ovariectomized (OVX) rabbits (reviewed in *3*). If the animals were mated 1 d prior to ovariectomy, chronic administration of these extracts was sufficient to maintain normal development of the embryo to term. In the early 1930s, the "internal secretion" of the corpus luteum, responsible for these utero-morphic changes was identified and purified by Willard M. Allen at the University of Rochester, NY, which he names "progestrin," a substance that favors gestation *(4)*. The discovery of progestin, or progesterone, heralded a new era in

From: *Contemporary Endocrinology: Transgenics in Endocrinology*
Edited by: M. Matzuk, C. W. Brown, and T. R. Kumar © Humana Press Inc., Totowa, NJ

reproductive medicine, and it was initially envisioned that the hormone would be used to reduce or inhibit such female fertility disorders as spontaneous miscarriages in women at high risk (reviewed in *5*). Ironically, during the following decades, the use of progesterone (in derivative form) as a female contraceptive agent ("the pill") would overshadow its original promise as a fertility drug *(6,7)*.

Although progesterone's role as an indispensable steroid hormone in female fertility was firmly established during the first half of the twentieth century, defining its mechanism of action has been a preoccupation for many molecular endocrinologists for the last 30–40 yr. Influenced by the seminal investigations of Jacob and Monod in 1961 *(8)*, which set the basic tenets of gene regulation, studies on the rat uterus by Jensen and Gorski *(9–11)* and on the chick oviduct model by O'Malley et al. *(12,13)* provided the essential support for the existence of a specific intracellular binding protein ("the receptor") for estrogen and progesterone, respectively.

The identification of the progesterone receptor in the late 1960s has provided not only the molecular lynch-pin upon which our modern concepts of progesterone's mechanism of action are founded, but has had far-reaching implications for the design and application of progestin and antiprogestin clinical therapies for such target tissues as the uterus, the mammary gland, and the brain.

The Progesterone Receptor

Most of the physiological effects of progesterone are now known to be mediated by a specific intracellular transcription factor termed the progesterone receptor (PR) *(14,15)*. As for all classical steroid hormone receptors, the PR is believed to undergo an "activation step" upon ligand binding, which permits the activated ligand/receptor complex to interact effectively with specific response elements located in the promoters of target genes, the activation or repression of which manifests the progesterone-extracellular signal into an appropriate physiological response *(16,17)*. Although many steps in the mechanism of PR action have been elucidated, the specific target genes for this receptor have not yet been defined. The first cloning of PR *(18,19)* revealed it as a member of the nuclear-receptor superfamily of transcription factors, which now includes receptors for a number of potent effector molecules including steroids, retinoids, prostanoids, thyroid hormone, and vitamin D3, as well as an orphan-receptor subfamily for which ligands have not yet been assigned *(16,17,20)*. Members of this superfamily are characterized by a common structural motif that is organized into defined domains in terms of structure and function, suggesting a common underlying mechanism of action *(16,17)*; *see* Fig. 1.

An additional level of complexity concerning PR action is reflected in the existence of two naturally occurring ligand-binding forms of this receptor, termed PRA and PRB *(14,15)*; (Fig. 1). PRA and PRB are believed to arise as a result of either alternative initiation of translation from a single mRNA *(21)* or by alternate transcription from promoters within the same gene *(22,23)*. These receptor isoforms differ only in that PRB contains a short additional stretch of amino acids at the amino terminus of the receptor *(21)*. Recent in vitro studies have implicated differential transactivation functions for these receptor isoforms *(24–26)*; the net result of these differences would be to further expand the physiological responses of PR to progesterone. Although these studies have fostered much discussion concerning the individual functions of the A and B forms of PR, the evolutionary, developmental, and physiological significance of these findings is only now being elucidated. Selective ablation of PR-A function in mice

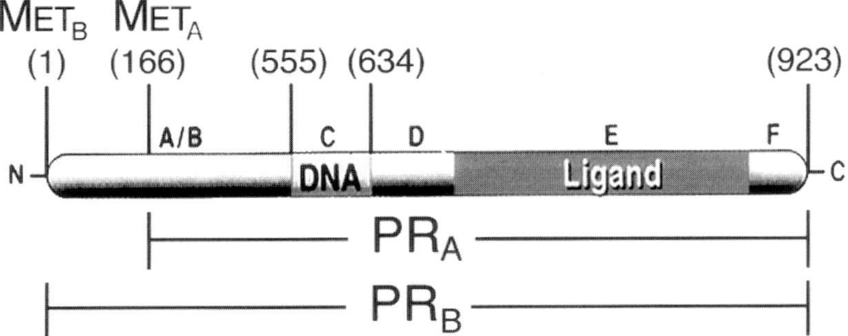

Fig. 1. The mouse progesterone-receptor (PR) contains common structural domain motifs that are shared by members of the nuclear-receptor superfamily. PR contains a long N-terminus region **(A/B)**, a short DNA-binding domain **(C)** and hinge region **(D)**, a ligand-binding domain, and a short C-terminus **(F)**. The PR is unique because it is composed of two isoforms, PR_A and PR_B. PR_B is structurally identical to PR_A, except for a 165 amino-acid extension at the N-terminus. Numbers in parentheses denote amino-acid number; MET_B and MET_A are the initiating methionines for PR_B and PR_A respectively.

recently demonstrated that the PR-B subtype regulates a subset of reproductive functions of progesterone, thereby supporing the concept that these receptor isoforms are functionally distinct in vivo *(26a)*.

In most physiological contexts, the PR is transactivated by estrogen via its cognate receptor—the estrogen receptor (ER)—implying that many of the observed physiological responses attributed to progesterone may be caused by the combined effects of progesterone and estrogen. The close temporal and spatial overlap in the functional activities of ER and PR, have made it difficult to achieve a fuller understanding of progesterone's direct involvement in many physiological systems because of the complexity of estrogen's influence. A major challenge in reproductive endocrinology has been to define and characterize those physiological responses that are specifically attributable to progesterone and/or estrogen in vivo.

The Progesterone Receptor Knockout Mouse Model

To address the physiological significance of PR function in vivo and to gain insight into progesterone's functional interrelationship with estrogen, a progesterone-receptor knockout (PRKO) mouse model was generated, in which both forms of PR (*A* and *B*) were simultaneously ablated through gene-targeting techniques *(27)*. Both male and female mice, heterozygotes and homozygotes for the PRKO mutation, developed to adulthood and at normal Mendelian frequencies. Interestingly, previous investigations have demonstrated the existence of transcripts for PR as well as for ER during the initial stages of mouse blastocyst development *(28)*, suggesting an obligate requirement for these nuclear receptors in embryogenesis. However, the first studies on the PRKO mouse revealed that embryonic-derived PR, like the ER *(29)*, is not essential for embryonic survival, and is not required for prenatal morphogenesis of the female reproductive system.

General anatomical investigations have not revealed discernible differences in organ morphology between the PRKO homozygotes and their wild-type or heterozygote siblings (PRKO homozygote = PRKO). However, as anticipated, female PRKO mice were shown to be infertile in crosses with sexually experienced wild-type males. In contrast, male PRKO mice proved to be as fertile as their wild-type and heterozygote litter mates; interestingly, the estrogen receptor-α knockout (ERKO) male has been shown to be infertile *(29)*.

As will be appreciated in the ensuing sections, apart from representing the ultimate means of dissecting estrogen vs progesterone effects in vivo, the PRKO mouse has provided a wealth of biological information concerning progesterone's role in female fertility. The expanded panoply of physiological responses attributed to progesterone as a result of these studies has underscored the essential role of PR as a central coordinator of a number of reproductive systems that collectively ensure female fertility, and ultimately, species survival. Finally, recognizing the clear phylogenetic differences between humans and mice, we believe that the PRKO mouse—a eutherian mammal like the human—holds great promise for the future as a valid experimental system with which to evaluate the controversial involvement of progesterone in human cancers such as that of the uterus and breast.

OVARIAN FOLLICULAR RUPTURE IS CRITICALLY DEPENDENT ON PR FUNCTION

Induced by the preovulatory gonadotropin surge, ovulation culminates with the rupture and subsequent luteinization of the preovulatory follicle, expulsion of the cumulus-oocyte complex, and resumption of oocyte meiosis. The biochemical and molecular events that define this process are precisely regulated throughout the hypothalamic-pituitary-ovarian axis by the individual and integrative actions of hypothalamic gonadotropin-releasing hormone (GnRH), the pituitary gonadotropins: follicle-stimulating hormone (FSH) and luteinizing hormone (LH), and the ovarian steroids: estradiol and progesterone (reviewed in *30,31*).

The PR Regulates the Generation of the LH Surge

Estradiol, secreted from the maturing follicle, was originally assumed to be the primary steroidal trigger in the manifestation of the preovulatory LH surge. However, this assumption was subsequently challenged by the observation that the estrogen-induced LH surge elicited in the OVX mouse could never attain the level or the duration normally observed at proestrus (reviewed in ref. *32*). Further experiments demonstrated that the addition of progesterone to the above hormone treatment was necessary to fully reinstate the proestrus LH surge, thereby implicating the unique importance of preovulatory progesterone in this process *(33)*. As further support for this concept, the administration of progesterone-synthesis inhibitors *(34)* or PR antagonists *(35)* to intact rats on the morning of proestrus was shown to significantly attenuate the LH surge. With the advent of the PRKO mouse *(27)*, more recent investigations have unequivocally established not only the pivotal role of PR in the generation of the LH surge, but have confirmed that this receptor mediates its regulatory effects at all levels of the hypothalamic-pituitary-ovarian axis through various mechanisms of action.

Initial investigations into the PRKO reproductive phenotype revealed that, unlike the normal mouse, the PRKO female was incapable of exhibiting the LH surge when exposed to male mouse odor *(36)*. In addition, vaginal-smear cytology revealed the absence of a normal estrous cycle in these mice. To determine whether the failure of the PRKO mouse to generate the LH surge was caused by an inherent inability of the PRKO hypothalamus

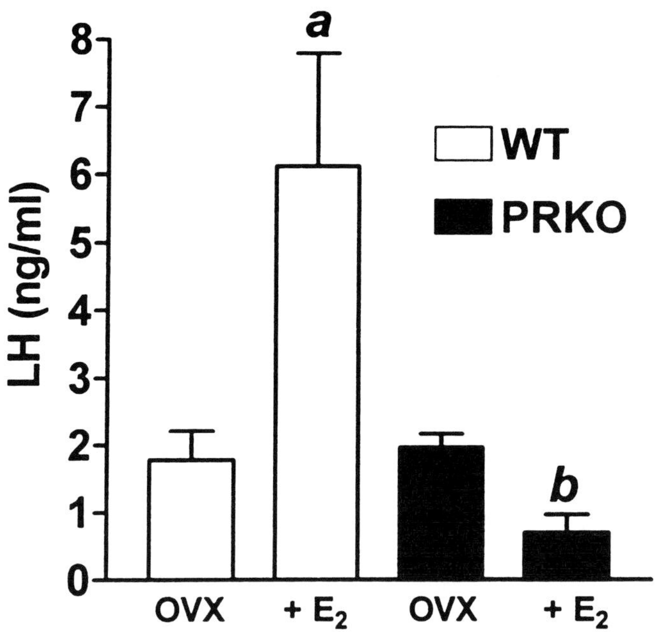

Fig. 2. Effect of estrogen priming on plasma LH levels of OVX wild-type and PRKO mice. Estrogen-primed (+ E2) OVX wild-type mice exhibited an LH surge (**a**; $p < 0.05$, $n = 8$) in comparison to unprimed OVX wild-type mice (OVX; $n = 8$). Plasma LH levels in unprimed ovariectomized PRKO mice (OVX; $n = 8$) were similar to their wild-type counterparts, whereas estrogen-priming (+ E2; $n = 8$) led to a significant (**b**; $p < 0.05$ below PRKO OVX) decrement in LH release. Reproduced with permission of Chappell et al. *(37)*.

and/or pituitary to interpret the estradiol signal, OVX PRKO and wild-type mice were evaluated for their capacity to evoke the LH surge in response to exogenous estradiol (ref. *37* and Fig. 2). While a significant estrogen-induced LH surge was elicited in the wild-type mouse, the PRKO female failed to display such a surge. Instead, LH levels in response to estrogen administration were significantly decreased in the PRKO female in comparison to no hormone treatment, suggesting that the estrogen-induced negative-feedback influence on LH secretion remained intact in the PRKO female despite the absence of estrogen-positive feedback effects. These results further suggested that one point of divergence between the negative and positive effects of estrogen on LH secretion may be mediated at the level of PR, where PR mediates estradiol-positive feedback effects and PR-independent signaling pathways are responsible for estradiol-negative-feedback actions.

Since ovariectomy resulted in the removal of assayable serum progesterone, it was logical to conclude that the LH surge observed in the estrogen-primed OVX wild-type mouse was not dependent on circulating progesterone. However, as noted previously, progesterone is required for the full proestrus LH surge to occur. Intriguingly, the PRKO mouse was incapable of exhibiting the estrogen-induced LH surge, suggesting that estrogen requires PR to elicit the LH surge, but not its ligand.

Recently, ligand-independent activation of PR by intracellular second messenger pathways has been implicated in the modulation of female rodent sexual receptivity and behavior *(38,39)*. Moreover, from recent in vitro studies, Turgeon and Waring have

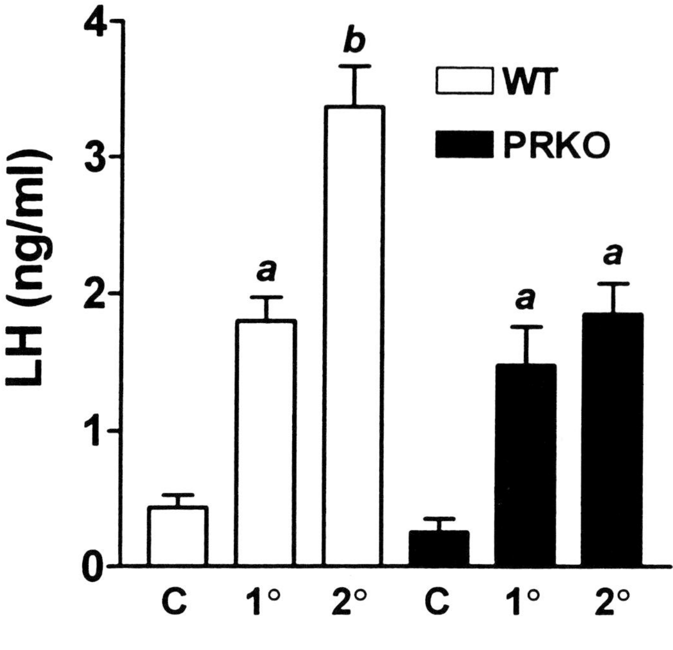

Fig. 3. Plasma LH levels in wild-type and PRKO mice given either one or two pulses of exogenous GnRH. One pulse of GnRH (1^0) given to estrogen-primed ovariectomized WT and PRKO mice elicited significant (**a**; $p < 0.001$) increases in LH above ovariectomized estrogen-primed controls in both test groups. E-primed, ovariectomized WT mice given two pulses of GnRH 60 minutes apart (2^0) exhibited a significant (**b**; $p < 0.001$) additional increase in LH, whereas no further elevation in LH was observed in estrogen-primed ovariectomized PRKO mice. Reproduced with permission of Chappell et al. *(37)*.

proposed that ligand-independent activation of PR is also essential for the GnRH self-priming mechanism known to occur in the pituitary gonadotrope *(40–43)*. The GnRH self-priming mechanism is defined as the increase in the magnitude of the pituitary gonadotrope LH response to successive GnRH stimulation, and has been shown to occur in rats *(44)*, sheep *(45)*, and humans *(46)*.

To determine whether the PRKO pituitary could mount a GnRH self-priming response, estrogen-primed OVX wild-type and PRKO mice were administered either one or two injections of GnRH *(37)*. In the case of the wild-type mouse, a significant GnRH self-priming response was elicited, as the second of two LH responses were at least twofold greater than the first (Fig. 3). Conversely, in the PRKO females, the second of the two consecutive GnRH injections failed to induce an additional increase in LH secretion. These results provided compelling in vivo evidence for an obligate requirement for ligand-independent activation of PR in the pituitary GnRH-self-priming response. Furthermore, the dependency of the GnRH-self-priming mechanism on estrogen priming underscored the physiological importance of estrogen-induced PRs in this process. Indeed, in response to increasing levels of ovarian estrogen, PR mRNA levels are elevated in the pituitary as well as in the hypothalamus, particularly in the arcuate and medial

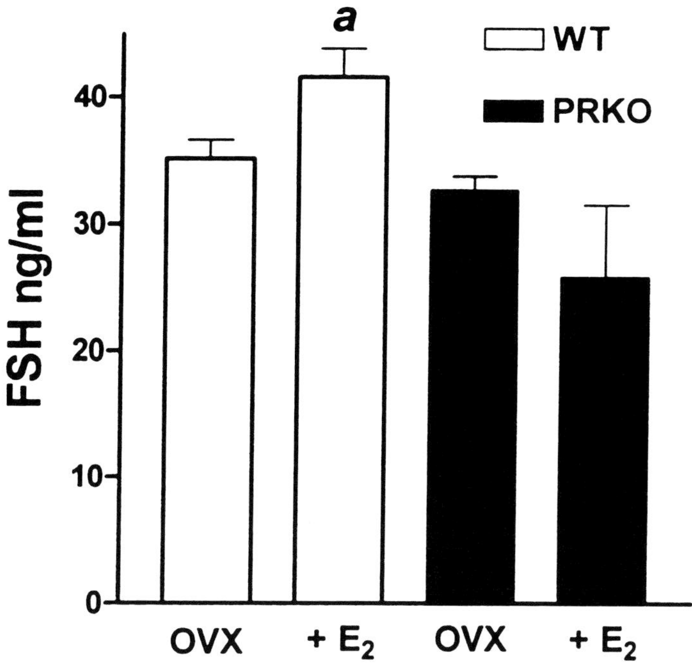

Fig. 4. Effect of estrogen-priming on plasma FSH levels of OVX wild-type and PRKO females. FSH levels were elevated in estrogen-primed OVX wild-type mice (+E_2; $p < 0.05$, $n = 8$) in comparison to unprimed OVX wild-type mice (OVX; $n = 8$). Unprimed OVX PRKO mice (OVX; $n = 8$) exhibited plasma FSH levels similar to unprimed OVX wild-type mice; additionally, no elevation was observed in estrogen-primed OVX PRKO mice (+E_2; $n = 8$) vs unprimed OVX PRKO controls. Reproduced with permission of Chappell et al. *(37)*.

preoptic nucleus *(47–51)*. Whether the defect in LH-surge secretion in the PRKO female can be explained solely by the loss of GnRH self-priming in the pituitary or the absence of hypothalamic GnRH surges, or caused by a defect in both mechanisms, awaits further investigation. Current studies have focused on determining whether the PRKO hypothalamus can exhibit an actual GnRH surge. Although PR expression has not been detected in GnRH neurons *(52)*, PR-containing neurons are known to innervate this neuronal group, suggesting that PR in combination with other potent regulators of GnRH release—i.e., neuropeptide Y (NPY) *(35,53)*, galanin *(54,55)*, catecholamines *(52,56)*, glutamate *(57)*, and gamma-aminobutyric acid *(58)*—may indirectly modulate hypothalamic GnRH secretion.

In addition to the absence of an LH surge and estrous cyclicity, PRKO-serum LH levels were found to be two- to threefold over normal basal (metestrus) values, confirming the negative-feedback control that progesterone exerts on LH secretion outside the LH-surge event *(36)*. Consistent with previous investigations, FSH levels in OVX wild-type and PRKO mice were significantly elevated, and this increase was not reduced by estrogen-priming; in accordance with studies in the rat demonstrating that estrogen alone was insufficient as a negative-feedback regulator of FSH release in the absence of ovarian-derived inhibin. Because of the absence of a full negative-feedback effect on FSH secretion, the characteristic FSH surge in the wild-type mouse was difficult to discern—result of high background levels of FSH (Fig. 4). However, a small yet significant FSH increase was observed in estrogen-primed OVX wild-type mice that was not observed in similarly treated PRKO mice, suggesting, as for the LH surge, that the estrogen-primed FSH surge requires PR (Fig. 4).

An Intraovarian Role for PR Action

In addition to PR's involvement in the manifestation of the preovulatory gonadotropin surge, an intra-ovarian role for progesterone and its receptor has been implicated in the mediation of the LH-induced ovulation event (follicular rupture). In response to the LH surge, progesterone levels increase in the proestrous follicle, and are further elevated in the corpus luteum during diestrous and pregnancy *(59)*. Antibodies to progesterone *(60)* or inhibitors of its synthesis *(61,62)* were shown to reduce the number of ovulations in the gonadotropin-induced rat model. Further studies by van der Schoot et al. *(63)* and Sanchez-Criado *(64)* have demonstrated that the PR antagonist RU486 could inhibit ovulations in the rat, implicating intra-ovarian PR in this process. Interestingly, histological examination of these ovaries revealed the presence of unruptured follicles exhibiting clear signs of luteinization. In these experiments, however, the PR antagonist also blocked the LH surge, making it difficult to dissociate any intrinsic ovarian effects from those of disrupted LH secretion. To obviate this problem, Loutradis et al. *(65)* subsequently used RU486 in combination with gonadotropin treatment, and found that ovulation was blocked by the antagonist. However, the block in ovulation did not occur if the antagonist was administered prior to, or 4 h after, gonadotropin treatment, providing a more convincing argument in favor of an intra-ovarian role for PR during the periovulatory period. As further support for this hypothesis, recent molecular approaches demonstrated that granulosa-cell PR expression was both temporally restricted to the rodent periovulatory period and dependent upon ovulatory (but not basal) levels of LH *(66,67)*.

That PR is a critical factor in the rupture of the preovulatory follicle recently has been demonstrated in vivo by the striking ovarian phenotype exhibited by the gonadotropin-primed PRKO mouse *(27)*. Despite exposure to superovulatory levels of gonadotropins, the PRKO mouse failed to ovulate, as first evidenced by the absence of oocytes in the oviduct (*see* Table 1). Closer histological examination of the PRKO ovaries revealed that although follicular development apparently progressed to the pre-ovulatory stage, follicular rupture was effectively eliminated (Fig. 5). Despite the absence of ovulation, granulosa cells within the unruptured PRKO follicle responded to the LH surge, as evidenced by cumulus expansion and subsequent expression of specific biochemical markers of the luteal-cell phenotype (i.e., the cytochrome P450 cholesterol side-chain cleavage). The presence of an apparently functional corpus luteum in the PRKO ovary may explain, in part, the equivalent levels of serum progesterone observed in the PRKO and wild-type mouse *(36)*.

The existence of oocytes in the PRKO ovary prompted the question as to whether oocyte maturation and/or function required the PR. If this was the case, PR involvement in oocyte development would probably occur by PR-initiated paracrine signals emanating from preovulatory granulosa cells, since PR expression is restricted to the granulosa cells of the rodent follicle at proestrus. To answer this question, cumulus-ooycte cell complexes (arrested in prophase 1 of meiosis), isolated from large antral follicles of PRKO and WT ovaries, were evaluated for their capacity to mature to metaphase II of meiosis and be fertilized in vitro. These experiments demonstrated that PRKO oocytes

Table 1

Oocytes and Embryos Produced Following Superovulation

Group	*Oocytes*	*1-Cell stage*	*2-Cell stage*	*N*
WT	11 ± 3	13 ± 2	1	6
PRKO	0	0	0	6

Ovarian function was assayed by determining the ovary's response to superovulatory doses of the gonadotropins PMSG and hCG *(27)*. Following hCG treatment, mice were placed overnight with sexually experienced wild-type males. Oocytes and embryos were flushed from both oviducts of each animal 24 h following hCG administration, and examined and counted using a dissecting microscope. The data are means ± standard deviation. Reproduced with permission from Lydon et al. *(27)*.

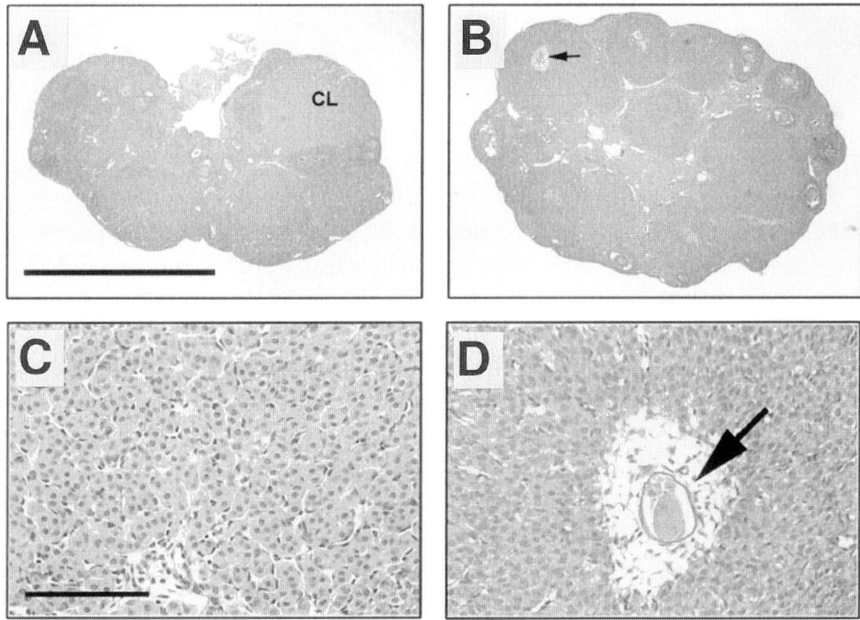

Fig. 5. Differential ovarian responses to superovulatory levels of gonadotropins in the PRKO and wild-type mouse. **(A)** A section of a typical wild-type ovary following treatment with pregnant mare serum gonadotropin (PMSG) followed by human chorionic gonadotropin (hCG) (*see* ref. *27* for more details). Note the presence of numerous corpora lutea (CL) following ovulation; scale bar: 100 µm. **(B)** A typical section through the PRKO ovary following an identical superovulatory treatment regimen. Note the unusual presence of an entrapped oocyte within a CL *(indicated by arrow)*. (**C** and **D**) Higher magnifications of the corpora lutea in **A** and **B**, respectively; scale bar: 5 µm. Reproduced with permission of Lydon et al. *(27)*.

retained the capability to be fertilized. In addition, the resultant embryos were able to progress to the 2-cell stage and blastocyst-stage at normal frequencies (Table 2). Indeed, pronuclear-stage embryos, derived from PRKO oocytes and transferred to foster mothers, were able to develop to term, demonstrating that PR function is not essential for oocyte maturation, fertilization, or early embryonic development.

Table 2

In Vitro Fertilization of Wild-Type and PRKO Oocytes[a]

Oocyte genotype	*Wild-type*	*PRKO*
Oocytes fertilized in vitro that reached the 2-cell stage		
Experiment 1	90% $n = 30$	89% $n = 300$
Experiment 2	91% $n = 43$	88% $n = 265$
Experiment 3	59% $n = 82$	92% $n = 136$
2-Cell embryos that reached the blastocyst stage		
Experiment 1	93% $n = 27$	94% $n = 267$
Experiment 2	79% $n = 39$	86% $n = 233$
Experiment 3	83% $n = 48$	90% $n = 125$

[a]Personal communication from Dr. Joanne Richards, Department of Molecular and Cellular Biology, Baylor College of Medicine.

The current challenge is to define the signaling pathways through which PR mediates the physiological effects that enable follicular rupture to occur. Evidence suggests that dissolution of the follicular wall and the discharge of the oocyte is subsequent to a prostaglandin-mediated cascade of proteolytic reactions *(68)*. Because the preovulatory increase in progesterone precedes the rise in prostaglandins, it has been suggested that progesterone may regulate prostaglandin synthesis and/or ovarian proteolytic activity as a prelude to follicular rupture *(69–71)*. Because the PRKO ovarian phenotype is restricted to a functional defect in follicular-wall rupture, this animal offers an attractive approach to evaluating the expression profiles for those ovarian-gene products previously implicated to be regulated by progesterone and proposed as having a role in follicular rupture (reviewed in *72*). Furthermore, the PRKO mouse in conjunction with differential cDNA cloning strategies, such as differential-display polymerase chain reaction (PCR), subtractive hybridization, and/or screening of gene-arrayed libraries may prove useful in identifying and isolating novel downstream PR-target gene(s) that are essential and specific for the final stages of ovulation. Unlike past attempts to develop contraceptive strategies based on directly disabling PR function *(73)*, with the inherent side effects of PR-antagonism outside the ovary, these screening approaches may provide molecular targets exclusive to the ovary, and therefore may present safer routes for new contraceptive therapeutic strategies in the future.

AN ESSENTIAL ROLE FOR PR IN SEXUAL BEHAVIOR

Female Sexual Behavior

Steroid autoradiography *(48)* and *in situ* hybridization *(74)* have demonstrated that PR expression in the female rodent brain is primarily concentrated in the pre-optic area (POA)—the hypothalamus and the amygdala. These brain areas have long been associated with a female sexual response exhibited by rodents, termed lordosis. The lordosis response is characterized by a pronounced dorsoflexive posture consisting of a concave arching of the back with the rump and head elevated, which serves to solicit and to facilitate copulation by the male. The ventral medial nucleus of the hypothalamus

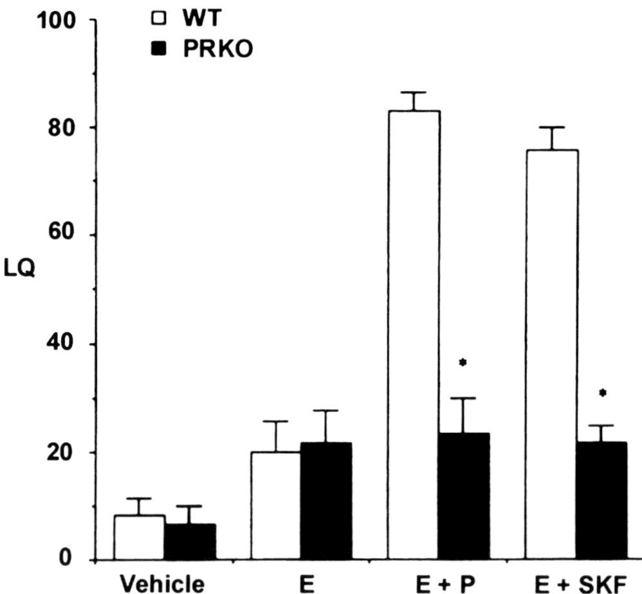

Fig. 6. Comparative responses to Estrogen (E), E+ Progesterone (P), and E+ the dopamine D1 agonist, SKF38393 (SKF) on lordosis behavior in wild-type and PRKO females in the presence of sexually experienced wild-type male mice. E, E+P, or E + SKF were directly administered into the third cerebral ventricle of wild-type and PRKO females. Control mice received vehicle (saline). Note the statistically significant differences in lordosis behavior exhibited by estrogen-primed wild-type and PRKO females in response to either P or the dopamine agonist (*, $p < 0.001$). Reproduced with permission of Mani et al. *(38)*.

(VMNH) is primarily responsible for synchronizing the induction of this behavior with the onset of proestrus, and the net result is to optimize the potential for fertilization to occur *(75)*. Both estrogen and progesterone have been shown to play pivotal roles in eliciting this behavior *(76)*. In the case of the OVX adult rat, progesterone was required under physiological estrogen-priming conditions to induce lordosis *(77)*. In these studies, PR was induced by estrogen in the VMNH, and its induction temporally coincided with the induction of lordosis. Furthermore, the inhibition of this progesterone-facilitated sexual behavior by the progesterone antagonist RU486 *(78,79)*, as well as antisense oligonucleotides to PR mRNA *(80–82)* indicated an important role for hypothalamic PRs in the manifestation of this behavior.

As a critical in vivo validation of these observations, the PRKO female was evaluated for lordosis response capability *(27,38)*. Unlike normal females, the OVX PRKO mouse, sequentially administered estrogen and progesterone, was unable to exhibit a lordosis response when placed proximal to a sexually experienced male (Fig. 6). This behavioral impairment confirmed recent findings describing the inability of progesterone to elicit a lordosis response in estrogen-primed OVX rats that were previously administered PR antisense oligonucleotides, either intracerebroventrically or into the VMNH. This PRKO study underscored the essential role played by PR in female sexual behavior, and reiterated the functional versatility of this receptor in ensuring female fertility.

Previous in vitro studies have demonstrated that PR can be transactivated in a ligand-independent manner by dopamine, a catecholamine neurotransmitter *(83)*. Using a spectrum of dopaminergic agonists, the ligand-independent activation of PR by this neurotransmitter was specifically mediated through the dopamine D1 receptor subtype. Because dopamine has been implicated in the elaboration of rodent sexual behavior *(84–86)*, recent efforts have focused on evaluating the physiological significance of ligand-independent activation of PR by dopamine in the generation of the lordosis response. Using antisense oligonucleotides to PR mRNA, administered into the third intracerebral ventricle of rats, we recently reported the elimination of dopamine-induced sexual behavior, providing strong support for the proposal that dopamine requires the presence of PR to induce lordosis *(87)*. Based on these studies, we hypothesized the existence of a cross-communication pathway between cell-membrane receptors for dopamine and intracellular PR. Furthermore, this dual mode of PR activation may represent a critical mechanism by which neurotransmitter signaling impacts steroid receptor-dependent gene expression that is essential for behavioral responses.

To unequivocally demonstrate that unoccupied PR is essential for dopamine-initiated sexual behavior in vivo, the estrogen-primed OVX PRKO female was tested for its ability to display sexual behavior in response to dopamine *(38)*. Although wild-type mice exhibited a robust sexual behavioral response as a result of dopamine administration, the PRKO female failed to elicit this behavior (Fig. 6). The inability of both progesterone and dopamine to elicit sexual behavior in the PRKO female provides definitive in vivo evidence for the proposal that PR is an important point of convergence through which steroid and neurotransmitter signaling pathways regulate sexual behavior. Although the precise molecular mechanism(s) by which dopamine modulates PR activity remain undefined, it has been suggested that phosphorylation of the receptor and/or its coactivator may be involved *(88,89)*.

The behavioral phenotype exhibited by the PRKO female provides a clear indication of an abnormality of the neuroendocrine system. Although gross histopathological defects have not been detected in the PRKO brain, we speculate that the brain lesion will be subtle and may be associated with a defect in a "PR-induced" developmental event during the establishment of neuronal networks in prenatal life *(90)*. However, we know that the pathway for lordosis itself is intact, since a vigorous response occurs following serotonin administration. Serotonin is known to operate independently of PR *(38)*. It is also quite possible that the neuroanatomical defect occurs in ongoing transient and reversible "organizational events" (synaptic plasticity), such as the fluctuations in dendritic spine density that have been reported to occur in the VMNH of the adult female in response to changes in the levels of estrogen and progesterone during the estrous cycle *(91)*.

Does the PR have a Role in Male Sexual Behavior?

Historically, progesterone has not been considered to exert significant effects in mammalian male sexual behavior. However, initial studies with birds and rodents suggested that progesterone administration may even inhibit sexual behavior in these species *(92–95)*. Indeed, these results have served as a rationale for the use of pharmacological doses of progestins in the chemical castration of sex offenders *(96,97)*. That progesterone might exhibit bona fide physiological effects in male sexual behavior was first indicated by the distinct diurnal variation in the secretion profile for progesterone in male rats *(98)* and humans *(99)*. Moreover, investigations with a number of lizard species have shown that administration of progesterone will induce typical courtship and copulatory behaviors in castrated males *(100–104)*. Using the reptilian model, investigators have demonstrated that progesterone synergizes with testosterone to induce the male sexual behavioral response *(101–103)*, in much the same way that progesterone synergizes with estrogen in the elaboration of female sexual behavior *(75)*.

Reptilian male behavioral studies have prompted a reevaluation of progesterone's involvement in male sexual behavior in the mammal. Recent investigations have demonstrated that the systemic administration of progesterone to physiological levels or the direct delivery of progesterone to the POA induced the classical male sexual behavioral response in intact and castrated rats *(105,106)*. The POA is generally considered the neuroanatomical structure primarily responsible for regulating male sexual behavioral responses. Furthermore, as observed in the lizard, the manifestation of the male sexual behavior depended upon functional synergy between testosterone and progesterone *(105)*. To date, the majority of information on progesterone's role in male sexual behavior has been drawn from the investigation of reptilian models. From these experiments, it has been hypothesized that progesterone may function by sensitizing the POA to androgens *(104)*. Indeed, the recent observation that direct administration of progesterone to the rat POA stimulates male-typical sex behavior, while the progesterone antagonist RU486 suppresses this behavioral response *(105)*, provides further support in favor of an important role for PR—particularly POA-derived PR—in male sexual behavior.

The PRKO mouse has yet to be extensively utilized to examine PR's role in male sexual behavior (as measured by mount, intromission, and ejaculation frequencies). However, initial investigations with this mouse model have indicated an important role for PR in this behavior *(107)*. Although a small but significant difference in male sexual behavior was observed between intact wild-type and PRKO males, evaluation of these behavioral measures following castration revealed a profound difference between both test groups (Fig. 7). The observed differences in the responses of the wild-type and PRKO males to castration when provided with prior sexual experience have been attributed to the theory that PR is required for neural circuits modified by "sexual experience" to bypass androgen dependence *(107)*.

Previous investigations into rat male sexual behavior have reported that those sexually experienced males manifesting a sexual behavior following castration exhibited preceding surges of dopamine in the POA, whereas castrated males that were unable to display this behavior did not *(108)*. Furthermore, local administration of dopamine agonists into the POA reinstated the copulatory behavior to previously unresponsive castrates *(109)*. As mentioned previously, we have demonstrated that dopamine can induce female sexual behavior through ligand-independent activation of PR in the VMNH *(38,81,87,110)*. By extension, we hypothesize that similar mechanisms may occur in the POA to induce male sexual behaviors. In light of our initial studies with the PRKO male, it is tempting to speculate that sexual experience may enhance dopamine release from the POA—which results in ligand-independent activation of PR—and may cause male sexual behavior to occur. The PRKO male plays an essential role in this hypothesis. Future investigations will focus on testing whether the PRKO male can respond to dopamine, and whether sexually experienced wild-type castrates display a reduction in sexual behavior when administered dopamine antagonists.

NEW PERSPECTIVES ON PROGESTERONE ACTION IN THE UTERUS

In response to embryo attachment and subsequent penetration of the uterine epithelial-cell layer and underlying basement membranes, the uterus undergoes profound

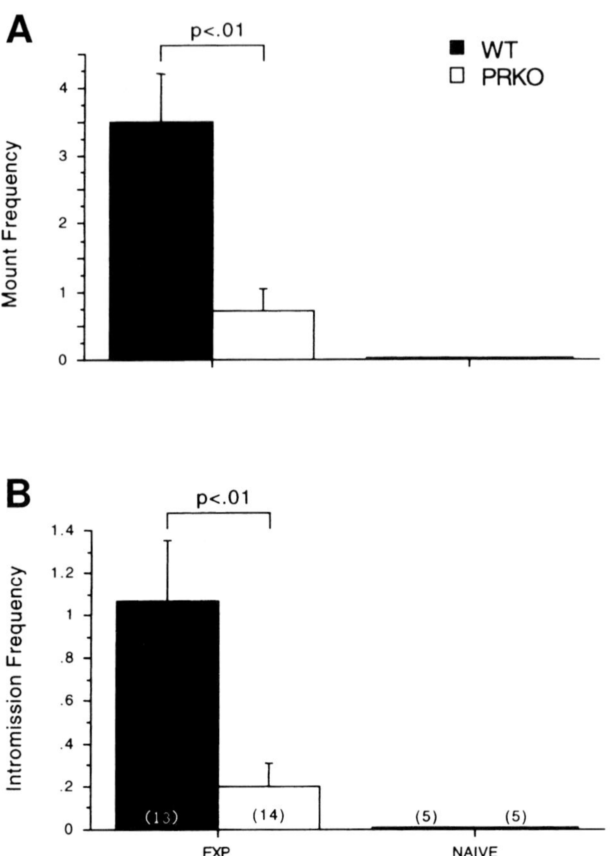

Fig. 7. The comparative effect of sexual experience on male sexual behavior in the WT and PRKO. **(A)** Mount frequencies of experienced (EXP) and naïve (NAÏVE) WT and PRKO males castrated for 3 wk. **(B)** Intromission frequencies in 3-wk castrates. Numbers in parentheses refer to sample sizes. Reproduced with permission of Phelps et al. *(107)*.

morphological, cellular, and vascular changes that are precisely regulated by the coordinate actions of progesterone and estrogen (reviewed in *111*). The remodeling of the uterus, known as the decidual response, culminates with the generation of the decidua that surrounds the developing fetus and becomes the maternal component of the placenta.

Between embryo implantation and parturition, progesterone is believed to maintain the pregnancy state by blocking the responsiveness of the myometrial smooth-muscle layer to such contractile stimuli as prostaglandins and oxytocin, in addition to inhibiting the premature softening and dilation of the cervix (cervical ripening) (reviewed in *112*). Superimposed on these effects, progesterone is also implicated in the local suppression of undesirable immunological responses that are elicited within the maternal compartment against the developing fetus *(113–115)*. Together, these distinct regulatory roles further emphasize the pleiotropic nature of progesterone in the establishment and elaboration of the maternal-fetal interface.

In defining the dynamic interplay between progesterone and estrogen, both in the development of the receptive uterus and in the subsequent induction of the decidual response, the mouse—the PRKO mouse—has proven to be an invaluable investigative tool. During the initial stages of mouse pregnancy, the distinct secretion patterns for progesterone and estrogen have provided some of the most informative clues for their selective contributions to uterine development and function *(116–120)*. For example, during the first 2 d of murine pregnancy, the proliferation of the uterine luminal epithelial-cell layer was shown to coincide with rising levels of pre-ovulatory estrogen. On the third day, uterine stromal cells were found to undergo proliferation in response to progesterone, synthesized and secreted from recently formed corpora lutea. Preimplantation (nidatory) estrogen further augmented this effect on d 4. Embryo implantation occurred later on d 4, and in response to progesterone, uterine epithelial cells switched from a proliferative to a differentiative pathway, while stromal fibroblasts embarked on a program of differentiation to become decidual cells. Collectively, these observations reveal that the spectrum of proliferative and differentiative responses elicited within the uterus is dependent on the synchronized actions of progesterone and estrogen, and that these responses are restricted to distinct cellular compartments of the uterus.

In addition to the pregnant mouse, the pseudopregnant, the delayed implantation, and the OVX steroid-treated mouse models have offered more simplified approaches to further dissecting the selective uterine effects of progesterone and estrogen *(121–123)*. In the case of the steroid-treated OVX mouse, whereas exogenous estrogen was shown to stimulate uterine epithelial-cell proliferation, the co-administration of progesterone was found to inhibit estrogen-induced epithelial proliferation in favor of stromal proliferation *(123–126)*. Similar hormone treatments applied to the PRKO mouse have recently highlighted the importance of PR in preventing estrogen-induced uterine epithelial hypertrophy and hyperplasia, stromal edema, and local proinflammatory responses (refs. *27* and *127,* and Fig. 8). Interestingly, the uterine defect in the PRKO female was shown to closely correlate with uterine aberrations that occur in rodents *(123,128)* and in humans *(129,130)* as a result of unopposed estrogen treatment, and further supports the rationale for including progestins in postmenopausal hormone-replacement therapies *(131)*.

Molecular endocrinologists have long recognized that the synergistic and antagonistic actions of estrogen and progesterone are underscored by the mutual regulation of their respective receptors *(132–134)*. In the case of the rodent uterus, *in situ* radiolabeled steroid binding and recent immunohistochemistry have localized the nuclear receptors for progesterone and estrogen to the epithelial, stromal, and myometrial compartments *(135–138)*. From these studies, an important question has arisen: In the uterus, treated with either estrogen or estrogen plus progesterone, can the expression profiles for ER and PR shed new light on how these hormones regulate uterine proliferation in a compartmentalized, specific fashion?

To address this question, immunohistochemical analysis was recently employed in conjunction with the steroid-treated OVX mouse model *(138)*. In the case of the hormonally untreated OVX mouse, uterine PR expression was predominantly localized to the luminal epithelial-cell layer. However, following estrogen administration, PR expression was significantly reduced in the luminal epithelial compartment, but increased in the stromal and myometrial compartments. While estrogen repressed PR expression in the luminal epithelium, expression in the glandular epithelium was unaffected, indicating a differential regulation of PR by estrogen in these epithelial compartments; the glandular epithelium is believed to originate from down-growths of the luminal epithelium. The inclusion of progesterone in the hormonal treatment resulted in a marked reduction in PR expression in all uterine compartments, demonstrating a general negative feedback of PR on its own expression.

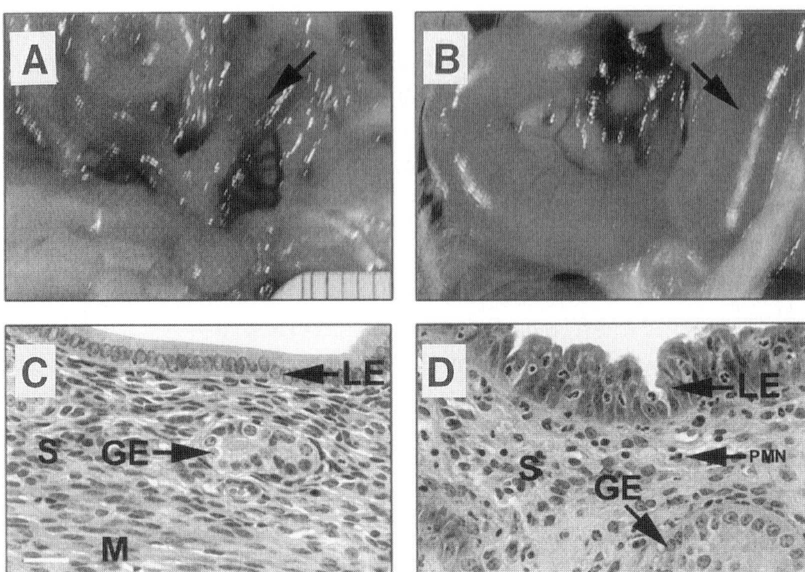

Fig. 8. Wild-type and PRKO uterine responses to E and P treatment. In situ gross anatomy of wild-type **(A)** and PRKO **(B)** uteri, following E and P treatment, is indicated by arrows. Note the marked enlarged fluid filled uterus in the PRKO (B). Histological analysis of a representative cross-section of the uterine wall of the hormonally treated wild-type mouse **(C)** shows the presence of a normal uterine architecture, luminal epithelium (LE), glandular epithelium (GE), stromal-cell layer (S) and myometrium (M); scale bar: 50 µm. Histological analysis of a typical transverse section of the uterine wall of the PRKO, treated with E and P, reveals an abnormal uterine structure. Note the hyperplastic luminal (LE) and hypertrophic glandular (GE) epithelium, loosely arranged stromal layer (S), and the presence of polymorphonuclear (PMN) leucocytes; scale bar: 50 µm. Reproduced with permission of Lydon et al. *(27)*.

The fact that progesterone inhibits estrogen-induced luminal-epithelial proliferation and that the downregulation of PR expression by estrogen in this compartment is matched by its contemporaneous upregulation in the stromal compartment has recently led to an important hypothesis that stromal PRs may exclusively regulate estrogen-induced luminal-epithelial proliferation. If this hypothesis is true, it is conceptually possible that estrogen induction of PR in the stroma sets in motion a control mechanism early on, by which estrogen self-limits its action in a compartmentalized, specific manner. Furthermore, defining the stromal PR population as the exclusive mediators of progesterone's inhibition of estrogen-induced uterine epithelial proliferation would greatly simplify future studies concerned with identifying those key effector molecules downstream of PR which are involved in this important growth-inhibitory uterine-signaling pathway.

This recently confirmed hypothesis could only have been tested by experiments utilizing the PRKO mouse in combination with uterine epithelial-stromal reciprocal transplantation approaches *(139)*. Intriguingly, parallel studies with the ERKO mouse have demonstrated that only stromal ERs are necessary for mediating estrogen's proliferative signal to the uterine luminal-epithelial compartment *(140)*. Apart from demonstrating the exclusive requirement for stromal PRs in the inhibition of estrogen-induced epithelial proliferation by progesterone, the PRKO mouse studies have raised the following provocative questions: 1) What is the functional significance of epithelial-derived PRs

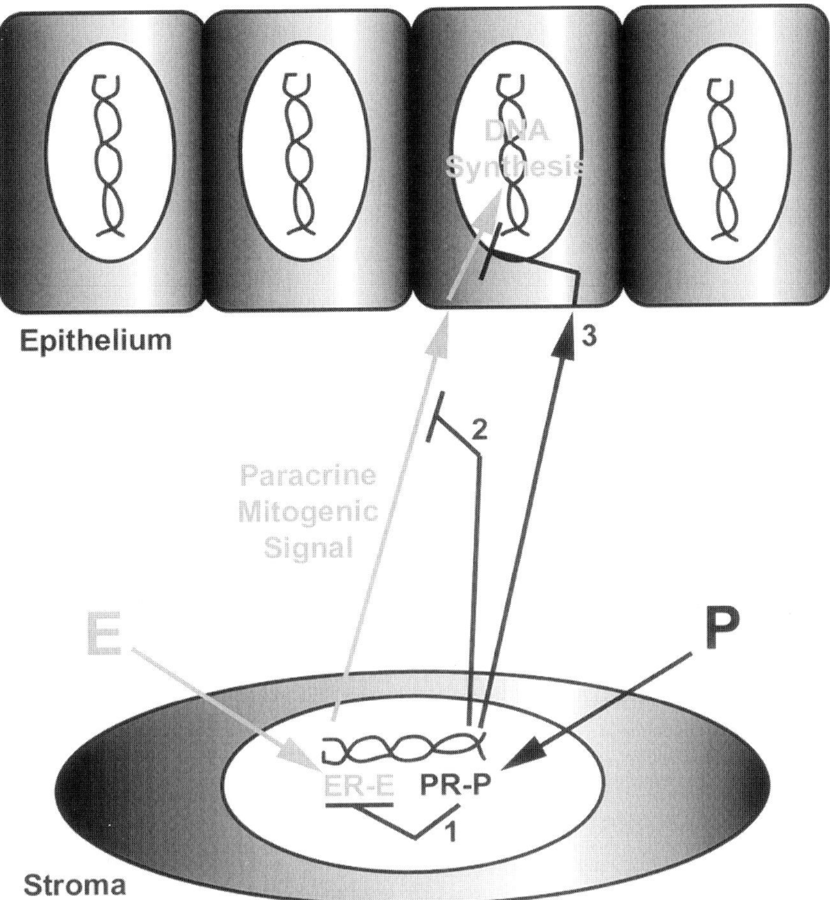

Fig. 9. Proposed mechanism of E and P action on uterine epithelial DNA synthesis. Estrogen-induced proliferative and progesterone-induced inhibitory signals are indicated by *light* and *dark arrows*, respectively. Estrogen binds to stromal ER and generates a paracrine signal which induces DNA synthesis of uterine epithelium. Progesterone binds to stromal PR, which leads to inhibition of uterine epithelial proliferation. The three possible inhibitory mechanisms are: 1) PR inhibits transcription of ER-dependent paracrine mediators; 2) progesterone-induced gene products antagonize the action of E-induced paracrine mediators through a variety of indirect mechanisms; and 3) progesterone-induced paracrine mediator is a direct inhibitor of epithelial proliferation, such as TGF-β. Reproduced with permission of Kurita et al. *(139)*.

in the uterus? 2) What are the mechanisms by which stromal PRs mediate their growth-inhibitory effects to the luminal-epithelial compartment?

The physiological relevance of epithelial-derived PR has yet to be elucidated, and may not be answered until an epithelial-specific PRKO mouse is generated. However, a number of mechanistic models have been set forth to explain how stromal PRs may inhibit estrogen-induced proliferation of the luminal-epithelial compartment *(139)*. For example, stromal PRs may directly inhibit the intracellular synthesis or secretion of an estrogen-inducible paracrine factor, activate a paracrine factor that negates the effect of estrogen-induced paracrine mediators through indirect effects, and/or induce a paracrine factor that directly inhibits luminal-epithelial proliferation (summarized in Fig. 9).

Whether one or all of these models proves to be valid awaits further experimentation; however, the steady increase in the number of potential uterine molecular targets recently reported for progesterone *(141–149)* offers renewed hope for a more comprehensive molecular explanation for progesterone's modulation of uterine proliferation and differentiation in the near future. Finally, we believe that understanding the mechanism by which the stromal progesterone-signaling pathway impacts the luminal-epithelial compartment will provide a new perspective on the molecular and cellular processes underlying the pathogenesis of such uterine disorders as endometrial hyperplasias and adenocarcinomas *(150)*.

PROGESTERONE IS ESSENTIAL FOR MAMMARY GLAND PROLIFERATION AND DIFFERENTIATION

Implications for Current Breast Cancer Investigations

Clinical and epidemiological studies have revealed a close association between breast-cancer risk and the cyclical exposure of the mammary gland to ovarian sex steroids that occurs during the premenopausal years (reviewed in *151*). This correlation is further substantiated by the fact that inhibition or reduction of such steroidal exposure, (e.g., after oophorectomy, and in late menarche and early menopause), has been demonstrated to markedly reduce breast-cancer risk *(152–155)*. The increase in breast cancer observed with advancing age (Fig. 10) is currently hypothesized to arise from ovarian sex-steroid-induced proliferation of the mammary epithelial cell, which allows for the occurrence and aggregation of genetic changes throughout the reproductive years that result in breast cancer in later life *(156)*. With a primary correlate of breast-cancer risk linked with the cyclical exposure of the mammary epithelial cell to ovarian sex steroids, breast-cancer prevention treatments based on suppressing ovarian steroidogenesis are currently being explored *(156)*.

Until recently, estrogen was assumed to be the primary ovarian steroid involved in normal proliferation of the human mammary epithelial cell, and was considered important in the advancement of this cell type to a neoplastic state (reviewed in *157*). Based in part on its established involvement in the differentiation of the endometrium, ovarian progesterone was assumed to exert anti-estrogenic, therefore antiproliferative, effects in the mammary gland, and was judged to contribute negligible effects to mammary tumorigenesis. However, these assumptions have not adequately explained why the proliferative index for the mammary epithelial compartment is highest during the progesterone-dominant luteal phase of the human menstrual cycle *(158)*, or why progestins included in combined oral contraceptives can prevent ovarian and endometrial cancer but not breast cancers *(159)*. Many of the conflicting reports concerning progesterone's role in mammary-gland development and neoplasia may be partially based on the inherent experimental difficulties in the use of the human as a model system. Despite over fifty million years of divergent evolution between the rodent and human, the developmental biology of the rodent mammary gland is remarkably similar to the human; thus, the rodent model has been established as the experimental system of choice in studying mammary-gland development and function *(160,161)*.

In the case of the murine experimental system, the PRKO mouse represents a new expanding subfamily of knockout mouse models that have recently been utilized to explore the various stages of mammary-gland development in an in vivo context

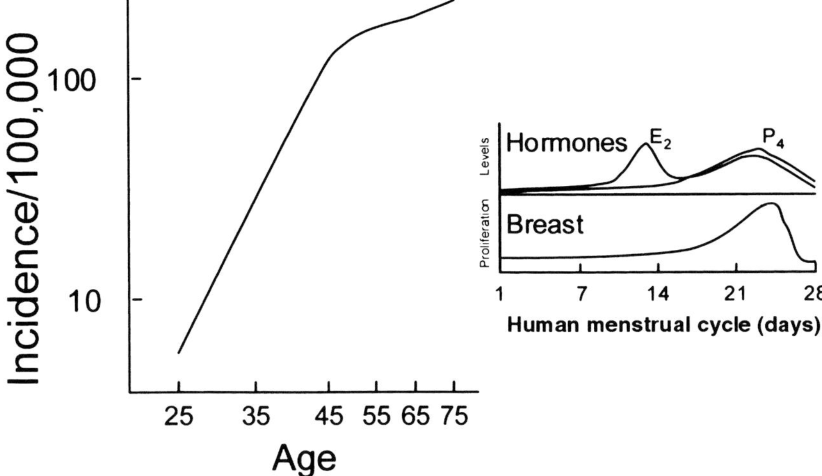

Fig. 10. Age-incidence rate for breast cancer. The log of age is plotted against the log of breast cancer incidence. The effect of menopause (~ age 50) on the rate at which breast cancer increases is clearly evident. The graph to the right displays the typical hormone profile for E and P during the menstrual cycle of a healthy premenopausal woman. Notice that the luteal phase of the cycle (d 14–28) is coincident with the progesterone peak that follows ovulation as well as the second peak of estrogen. Reproduced (in adapted form) with permission from Spicer et al., Cancer Investigation 1995;13:495–504.

(reviewed in *162*). To evaluate the importance of PR in murine mammary-gland proliferation and differentiation, comparative whole-mount analysis revealed that unlike wild-type mice, the PRKO mammary gland failed to develop the typical pregnancy-associated epithelial ductal morphogenesis that consists of extensive dichotomous branching with attendant interductal lobuloalveolar development that occurs in response to exogenous estrogen and progesterone treatment (Fig. 11 and ref. *27*). These initial gross morphological studies, in addition to recent molecular analysis (163), unequivocally demonstrate both a proliferative and a differentiative involvement for progesterone and its receptor in this tissue.

Previous binding studies using the radiolabeled progestin agonist (^{3}H)-R5020 provided evidence for the existence of progestin-binding sites both in the epithelial and stromal compartments that constitute the mammary gland of the adult virgin mouse *(164)*. Subsequent investigations revealed that these mammary epithelial- and stromal-derived progestin binding activities were distinct in their biochemistry and ontogenesis *(165)*. Like the uterus, these results suggested that PR may play an important role in reciprocal homeostatic cell-cell interactions that occur between the epithelial- and stromal-cell populations of the mammary gland that, like the uterus, are essential for normal development and function *(166–168)*. Because breakdown in communication between epithelial- and stromal-cell groups within the mammary gland could potentially lead to inappropriate epithelial-cell proliferation such as neoplasia, and because of the potential involvement of PR in this intercellular communication pathway, the PRKO mouse in combination with established mammary-gland transplantation procedures was used to evaluate the selective contributions of epithelial- and stromal-PR populations to mammary-gland ductal branching and alveologenesis.

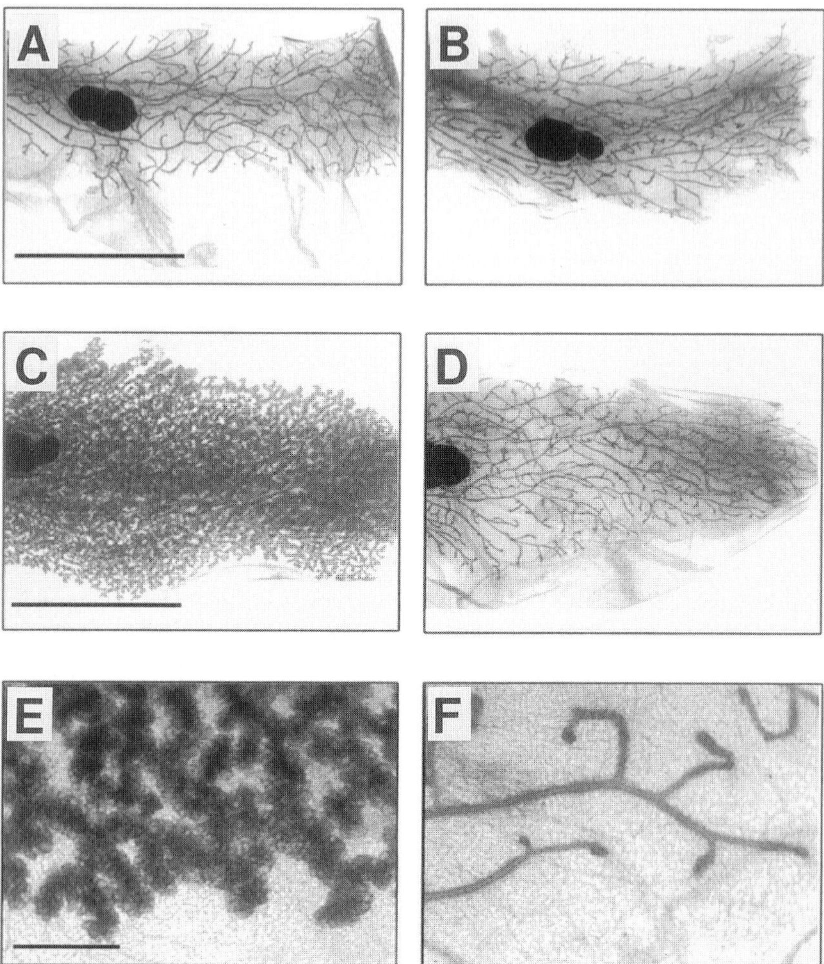

Fig. 11. Removal of PR function results in a defect in mammary gland ductal branching and alveologenesis. Adult (12-wk-old) virgin wild-type **(A)** and PRKO **(B)** mammary glands show a similar ductal morphology; scale bar in panel A is 5 mm. The PRKO mammary phenotype is clearly evident when wild-type **(C)** and PRKO **(D)** are transplanted into a host mouse that subsequently becomes pregnant; scale bar in panel C is 5 mm. Despite the presence of pregnancy hormones, the PRKO gland (D) does not exhibit the typical pregnancy-associated ductal morphology as displayed by the wild-type transplanted gland (C). (**E** and **F**) Higher magnifications of C and D, respectively; scale bar in (E) is 500 µm.

Using this technological approach, we recently demonstrated that the luminal-epithelial compartment of the murine mammary gland is the primary target for the progesterone-induced proliferative signal *(169,170)*. Importantly, the luminal epithelial cell has been considered the primary site for the initial carcinogenic insult *(161)*. Moreover, these findings have provided essential functional support for subsequent immunohistochemical studies that have localized PR expression predominantly to the luminal-epithelial cell *(171–173)*. These investigations also offer strong evidence suggesting that the PR-initiated proliferative signal(s) may impact neighboring mammary epithelial cells that lack PR through unidentified paracrine factor(s) *(170)*.

From a mechanistic perspective, it is interesting to note that while progesterone exerts its antiproliferative effects on the luminal-epithelial layer of the uterus via stromal PRs, progesterone manifests its proliferative and differentiative effects in the mammary gland through epithelial-derived PRs. These observations may eventually explain the differential actions of this steroid on these two progesterone-target tissues.

Although these recent findings represent important advances in our current understanding of progesterone's role in normal mammary-gland development, defining its controversial involvement in mammary tumorigenesis has proven to be one of the more challenging research areas in mammary-gland biology. However, a number of reports have implicated a role for progestins in the active progression of certain carcinogen-induced mammary tumors in the rat and mouse *(174–178)*. These observations have underscored the necessity to better understand progesterone's participation in mammary-gland tumor progression, in order develop a more rational basis for the current use of progestins in contraception and postmenopausal hormone replacement therapies and also to enable the design of novel diagnostic approaches and/or therapies for the future treatment and prevention of breast cancer.

Because of the uncertainties regarding the involvement of progesterone during mammary-gland tumorigenesis coupled with the widespread clinical use of progestins, the PRKO mouse model is currently being utilized to directly evaluate whether PR has an important role in mammary tumorigenesis in vivo distinct from ER's effects. Recent results clearly demonstrate the importance of progesterone-induced proliferative signals in mammary tumorigenesis. Using a chemical carcinogen-induced rodent mammary tumor model, we recently demonstrated that the PRKO mouse is significantly less susceptible to mammary tumorigenesis than normal mice, supporting a critical involvement for progesterone receptor function in this cancer *(178a)*.

Finally, we predict that the PRKO mouse model will be an essential research tool in the identification and isolation of novel molecular targets for PR in the mammary gland, with attendant implications for mammary-gland development, and mammary tumorigenesis. From a developmental perspective, branching morphogenesis is a recurrent theme in the ontogenesis of many organ systems during embryonic development. Since the expression of a number of these developmental genes has been reported in the adult mammary gland *(179)*, we anticipate that the PRKO mouse will be important in evaluating whether progesterone signaling converges with these developmental pathways to induce ductal branching and alveologenesis in the adult gland.

FUTURE PERSPECTIVES

To date, most research on progesterone action has concentrated on elucidating the mechanism by which this hormone modulates various aspects of female reproduction. However, an increasing number of reports have suggested progesterone's involvement in many diverse physiological systems that are apparently unrelated to female reproduction. Because PR's role is uncertain for many of these systems, we believe that the PRKO mouse model will be a powerful investigative tool in reaching a more comprehensive understanding of the relevant role of progesterone and its receptor in these areas.

For example, neurobiologists have previously detected PR in the adult brain of the male rat at levels equating those in the female *(180)*. A more detailed investigation has revealed that the most significant differences in PR levels between the male and female adult rodent brain are restricted to differences in estrogen-induced PR in the VMNH, a region of the female brain involved in the induction of lordosis. Moreover, the physiological significance of equivalent levels of this nuclear receptor in regions of the female and male brain, other than the VMNH, is one of the more challenging questions concerning the neurobiology of this steroid. Intriguingly, recent PR-immunohistochemistry has revealed a significantly higher level of PR in the medial pre-optic nucleus (MPN) in the male fetal and neonatal rat, as compared to the female. The MPN is considered one of the most sexually dimorphic structures in the rodent, and is involved in many sexually differentiated behaviors *(181)*. Based on these observations, it has been proposed that during pregnancy, high levels of maternally derived progesterone may exert sexual differentiative effects on the developing fetal CNS, and that PRs in the fetal MPN would mediate these effects *(181)*.

The existence of PRs in the CA1 hippocampal neurons *(182)* and the finding that progesterone in concert with estrogen may regulate synaptic plasticity in this region *(183)* are of particular interest. Because this brain region controls learning and other cognitive skills (184), functional roles for these gonadal hormones in behaviors other than lordosis (i.e., cognition, learning, and memory) cannot be discounted *(185,186)*.

In addition to evaluating the role of ovarian-derived progesterone on neuronal development and function, recent biochemical and electrophysiological investigations have underscored the physiological importance of neuronal-derived progesterone in local neuronal effects *(187)*. For example, Baulieu et al. have recently demonstrated that progesterone and its metabolic derivatives (classified as "neurosteroids") are synthesized within glial *(188)* and Schwann *(189)* cells of the central and peripheral nervous systems, respectively. Furthermore, through in vitro studies, these neurosteroids have been shown to exert changes on local neurotransmission *(188)* as well as nerve morphology *(189)* through either the classical PR-signaling pathway *(189)* or through less clearly understood rapid membrane effects (nongenomic) *(190)*. Clearly, the PRKO mouse model would be extremely useful in substantiating these observations within an in vivo context. From a clinical perspective, further studies on progesterone's role in brain development and function will be essential, as progesterone has been proposed as an alternative method to limit the extent of neuronal damage induced by delayed injury processes following cortical injury *(186,191–193)*. Progesterone's neuroprotective effects have been reported to include improved blood-brain barrier integrity, containment of free-radical-induced lipid peroxidation, and a significant reduction in cerebral edema *(186,191–194)*. Progesterone holds particular promise as a valid treatment regimen for brain injury, as it is equally effective at reducing cerebral edema in both males and females *(192)* without attendant undesirable side effects observed thus far. Whether PR mediates some or all of these progesterone-initiated neuroprotective effects is unclear. However, the utilization of the PRKO mouse model in conjunction with an ablation model of cortical injury *(195)* will be essential in addressing this question.

In addition to its proposed role in the nervous system, progesterone has been implicated to have a physiological role in cardiovascular biology. Premenopausal women have a significantly lower risk (ratio 1:10) of cardiovascular disease than men of equivalent age. However, this protection diminishes in the postmenopausal woman. At age 75, the incidence is essentially the same in both sexes, with cardiovascular disease the leading cause of mortality/morbidity in women by age 60 (reviewed in *196*). Because natural and surgical (bilateral oophorectomy) menopause accelerates the development of coronary heart disease, ovarian hormones have been implicated in protection against cardiovascular disease in premenopausal women *(197)*. Previously, estrogen replacement therapy has been shown to reduce cardiovascular mortality in premenopausal women by an average of 50% *(198–200)*. It is believed that approx 25–50% of the cardioprotection afforded by estrogen can be attributed to beneficial changes in serum lipoprotein levels, while the remainder of this protection is associated, in part, with direct effects of estrogen on the cardiovascular system *(199–201)*.

Because estrogen monotherapy has been linked to increases in the risk of endometrial cancer *(202)*, progestins have usually been included to reduce this risk *(203)*. The role of progesterone in combination with estrogen in cardioprotection is currently controversial. A number of previous investigations have proposed that progestins counteract the beneficial anitiatherogenic effects of estrogen by its negative influence on serum lipids, particularly HDL cholesterol *(204)*. However, these findings have recently been challenged *(205)*, and as mentioned previously, estrogen-induced beneficial effects on serum lipoproteins accounts for less than 50% of the total cardioprotective effects of this hormone. Recent studies have demonstrated that those postmenopausal women, using both estrogen and progesterone replacement, exhibited a significantly lower incidence of cardiovascular deaths than those using estrogen monotherapy *(206)*. These observations support a beneficial role for progesterone in cardiovascular disease.

As further support for this proposal, PRs as well as ERs have been detected in the vasculature of humans and other mammals *(207–209)*. Indeed, recent studies on the proliferation of arterial smooth-muscle cells (an important constituent of atherosclerotic plaques) suggest that progesterone exerts beneficial effects in this process *(209)*. To unequivocally define progesterone's role in atherosclerotic cardiovascular disease, the PRKO mouse will provide a powerful new approach in determining PR's role in these processes that are distinct from ER-mediated effects. We believe that the introduction of the PRKO mutation into existing mouse models for atherosclerosis *(210)*, as well as its use in combination with current carotid-artery injury paradigms *(211)*, will provide further insight into the mechanisms by which progesterone exerts its implicated atherosclerotic-protective effects.

In the case of bone homeostasis, inappropriate bone loss that occurs as a result of the onset of menopause is known to be associated with contemporaneous decreases in serum estrogen levels, as evidenced by the beneficial effects of estrogen replacement therapies in reversing this effect *(212)*. Interestingly, with the addition of progesterone to such hormonal replacement therapies, a number of biochemical and morphometric investigations have offered support for the proposal that progesterone may synergistically interact with estrogen in some aspect of bone remodeling *(213)*. The recent detection of PRs in osteoblast cells *(214)* would seem to further support this notion. However, because of the ambiguities concerning the individual roles of PR and ER in this process, we expect that the PRKO mouse will provide a meaningful investigative approach to mechanistically dissect the respective roles of PR and ER in this process.

Finally, the PRKO mouse model has provided an experimental platform for investigating the physiological functions of progesterone, as well as representing an unparalleled opportunity to examine the selective effects of estrogen and progesterone in vivo. As mentioned previously, considerable in vitro evidence has suggested that the A and B forms of the PR have distinct regulatory functions. It is conceptually plausible that the differential effects of these receptor isoforms could reflect, at the molecular level, the in vivo pleiotropic effects of progesterone action, as recently revealed by the phenotypic analysis of the PRKO mouse. While the significance of these in vitro investigations is well recognized, the in vivo functional relevance for the existence of these receptor isoforms is only now being understood. As for the seemingly intractable problem of dissecting estrogen and progesterone actions in vivo, a new generation of knockout mouse models has likewise been generated to examine the selective in vivo importance of the PRA and PRB isoforms *(26a)*.

CONCLUSIONS

Because of the PRKO mouse, many gray areas of progesterone biology have become clearer, and several new findings have been made. Although investigations on the PRKO have yielded great insights into the physiologic effects of progesterone and its receptor, these studies have underscored how little we know about those complex physiological processes that depend on progesterone action. Because of our recent successes with the PRKO mouse, we predict that new derivations of this mouse model (i.e., PRAKO, PRBKO, PR-LacZ, or GFP Knockin, conditional, tissue, and cell-lineage-specific knockouts) will collectively illuminate further the endocrinological effects of this "internal secretion" of the corpus luteum, well into the twenty-first century.

ACKNOWLEDGMENTS

For the studies described herein, the technical assistance of Gouqing Ge and Jie Li, Department of Molecular and Cellular Biology, Baylor College of Medicine, is greatly appreciated. This research was supported by NIH Grants CA-77530 (to John P. Lydon) and HD-07857 (to Bert. W. O'Malley).

REFERENCES

1. Corner GW. Oestrus, ovulation and menstruation. Physiol Rev 1923;111:457–482.
2. Frobenius W. Ludwig Fraenkel: 'spiritus rector' of the early progesterone research. Eur J Obstet Gynocol 1999;83:115–119.
3. Corner GW, Allen WM. Physiology of the corpus luteum II. Am J Physiol 1929;88:326–339.
4. Allen WM. Physiology of the corpus luteum V. Am J Physiol 1930;92:174–188.
5. Csapo A. Progesterone. Sci Am 1958;198:40–46.
6. Pincus G. Steroid labile reproductive processes in mammals. Harvey Lect. 1966-67;62:165–189.
7. Pincus G. Control of fertility in mammals by hormonal steroids. Anat Rec 1967;157:53–61.
8. Jacob F, Monod J. Genetic regulatory mechanisms in the synthesis of proteins. J Mol Biol 1961;3:318–356.
9. Jensen EV, Jacobson HI. Basic guides to the mechanism of estrogen action. Recent Prog Hormone Res 1962;18:387–414.
10. Jensen EV, Suzuki T, Kawashima T, Stumpf WE, Jungblut PW, DeSombre ER. A two-step mechanism for the integration of estradiol with rat uterus. Proc Natl Acad Sci USA 1968;59:632–638.
11. Toft D, Gorski J. A receptor molecule for estrogens: isolation from the rat uterus and preliminary characterization. Proc Natl Acad Sci USA 1966;55:1574–1581.
12. O'Malley BW, McGuire WL, Kohler PO, Korenman SG. Studies on the mechanism of steroid hormone regulation of synthesis of specific proteins. Recent Prog Hormone Res 1969;25:105–160.
13. O'Malley BW, Sherman MR, Toft DO. Progesterone "receptors" in the cytoplasm and nucleus of chick oviduct target tissue. Proc Natl Acad Sci USA 1970;67:501–508.
14. Schrader WT, O'Malley BW. Progesterone-binding components of chick oviduct. IV. Characterization of purified subunits. J Biol Chem 1972;247:51–59.
15. Horwitz KB, Alexander PS. In situ photolinked nuclear progesterone receptors of human breast cancer cells: subunit molecular weights after transformation and translocation. Endocrinology 1983;113:2195–2201.

16. Tsai M-J, O'Malley BW. Molecular mechanisms of action of steroid/thyroid receptor superfamily members. Ann Rev Biochem 1994;63:451–486.
17. Mangelsdorf DJ, Thummel C, Beato M, Herrlich G, Schutz G, Umesono K, et al. The nuclear receptor superfamily: the second decade. Cell 1995;83:835–839.
18. Conneely OM, Sullivan WP, Toft DO, Birnbaumer M, Cook RG, Maxwell BL, et al. Molecular cloning of the chicken progesterone receptor. Science 1986;233:767–770.
19. Jeltsch JM, Krozowski Z, Quirin-Stricker C, Gronemyer H, Simpson RJ, Garnier JM, et al. Cloning of the chicken progesterone receptor. Proc Natl Acad Sci USA 1986;83:5424–5428.
20. Kliewer SA, Lehmann JM, Willson TM. Orphan nuclear receptors: shifting endocrinology into reverse. Science 1999;284:757–760.
21. Conneely OM, Maxwell BL, Toft DO, Schrader WT, O'Malley BW. The A and B forms of the chicken progesterone receptor arise by alternate initiation of translation of a unique mRNA. Biochem Biophys Res Commun 1987;149:493–501.
22. Kastner P, Krust A, Turcotte B, Strupp U, Tora L, Gronemeyer H, et al. Two distinct estrogen-regulated promoters generate transcripts encoding the two functionally different human progesterone receptor forms A and B. EMBO J 1990;9:1603–1614.
23. Kraus WL, Montano MM, Katzenellenbogen BS. Identification of multiple, widely spaced estrogen-responsive regions in the rat progesterone receptor gene. Mol Endocrinol 1994;8:952–969.
24. Vegeto E, Shahbaz MM, Wen DX, Goldman ME, O'Malley BW, McDonnell DP. Human progesterone receptor A form is a cell and promoter specific repressor of human progesterone receptor B function. Mol Endocrinol 1993;7:1244–1255.
25. Tung L, Mohamed MK, Hoeffler JP, Takimoto GS, Horwitz KB. Antagonist-occupied human progesterone B-receptors activate transcription without binding to progesterone response elements and are dominantly inhibited by A-receptors. Mol Endocrinol 1993;7:1256–1265.
26. Sartorius CA, Groshong SD, Miller LA, Powell RL, Tung L, Takimoto GS, et al. New T47D breast cancer cell lines for the independent study of progesterone B- and A-receptors: only antiprogestin-occupied B-receptors are switched to transcriptional agonists by cAMP. Cancer Res 1994;54: 3868–3877.
26a. Mulac-Jericivic B, Mullinax RA, DeMayo FJ, Lydon JP, Connelly OM. Subgroup of reproductive functions of progesterone mediated by progesterone receptor-b isoform. Science 2000;289:1751–1758.
27. Lydon JP, DeMayo FJ, Funk CR, Mani SK, Hughes AR, Montgomery Jr. CA, et al. Mice lacking progesterone receptors exhibit pleiotropic reproductive abnormalities. Genes Dev 1995;9:2266–2278.
28. Hou Q, Gorski J. Estrogen receptor and progesterone receptor genes are expressed differentially in mouse embryos during preimplantation development. Proc Natl Acad Sci USA 1993;90:9460–9464.
29. Lubahn DB, Moyer JS, Golding TS, Couse JF, Korach KS, Smithies O. Alteration of reproductive function but not prenatal sexual development after insertional disruption of the mouse estrogen receptor gene. Proc Natl Acad Sci USA 1993;11:162–166.
30. Freeman ME. The neuroendocrine control of the ovarian cycle of the rat. In: Knobil E, Neill JD, editors. The Physiology of Reproduction. Raven Press, New York, NY, 1994. pp. 613–658.
31. Hotchkiss J, Knobil E. The menstrual cycle and its neuroendocrine control. In: Knobil E, Neill JD, editors. The Physiology of Reproduction. Raven Press, New York, NY, 1994. pp. 711–750.
32. Mahesh VB, Brann DW. Regulation of the preovulatory gonadotropin surge by endogenous steroids. Steroids 1998;63:616–629.
33. Krey LC, Tyrey L, Everett JW. The estrogen-induced advance in the cyclic LH surge in the rat: dependency on ovarian progesterone secretion. Endocrinology 1973;93:385–390.
34. DePaolo LV. Attenuation of preovulatory gonadotropin surges by epostane: a new inhibitor of 3 β-hydroxysteroid dehydrogenase. J Endocrinol 1988;118:59–68.
35. Bauer-Dantoin AC, Tabesh B, Norgle JR, Levine JE. RU486 administration blocks neuropeptide Y potentiation of luteinizing hormone (LH)-releasing hormone-induced LH surges in proestrous rats. Endocrinology 1993;133:2418–2423.
36. Chappell PE, Lydon JP, Conneely OM, O'Malley BW, Levine JE. Endocrine defects in mice carrying a null mutation for the progesterone receptor gene. Endocrinology 1997;138:4147–4152.
37. Chappell P, Schneider JS, Kim P, Xu M, Lydon JP, O'Malley BW, et al. Absence of LH surges and GnRH self-priming in ovariectomized (ovx), estrogen (E2)-treated, progesterone receptor knockout (PRKO) mice. Endocrinology 1999;140:3653–3658.
38. Mani SK, Allen JMC, Lydon JP, Mulac-Jericevic B, Blaustein JD, DeMayo FJ, et al. Dopamine requires the unoccupied progesterone receptor to induce sexual behavior in mice. Mol Endocrinol 1996;10:1728–1737.

39. Auger AP, Moffatt CA, Blaustein JD. Progesterone-independent activation of rat brain progestin receptors by reproductive stimuli. Endocrinology 1997;138:511–514.
40. Turgeon JL, Waring DW. Luteinizing hormone-releasing hormone-induced luteininzing hormone secretion in vitro: cyclic changes in responsiveness and self-priming. Endocrinology 1980;106:1430–1436.
41. Turgeon JL, Waring DW. Rapid augmentation by progesterone of agonist-stimulated luteinizing hormone secretion by cultured pituitary cells. Endocrinology 1990;127:773–780.
42. Turgeon JL, Waring DW. A pathway for luteinizing hormone releasing-hormone self-potentiation: cross-talk with the progesterone receptor. Endocrinology 1992;130:3275–3282.
43. Turgeon JL, Waring DW. Activation of the progesterone receptor by the gonadotropin-releasing hormone self-priming signaling pathway. Mol Endocrinol 1994;8:860–869.
44. Alyer MS, Fink G, Greig F. Changes in the sensitivity of the pituitary gland to luteinizing hormone releasing factor during the oestrous cycle in the rat. J Endocrinol 1974;60:47–54.
45. Phogat JB, Smith RF, Dobson H. Effect of ACTH on gonadotropin releasing hormone-induced luteinizing hormone secretion in vitro. Anim Reprod Sci 1997;48:53–65.
46. Urban RJ, Veldhuis JD, Dufau ML. Estrogen regulates the gonadotropin-releasing hormone-stimulated secretion of biologically active luteinizing hormone. Clin Endocrinol Metab 1991;72:660–668.
47. MacLuskey NJ, McEwen BS. Oestrogen modulates progestin receptor concentrations in some rat brain regions but not in others. Nature 1978;274:276–278.
48. MacLuskey NJ, McEwen BS. Progestin receptors in rat brain: distribution and properties of cytoplasmic progestin-binding sites. Endocrinology 1980;106:192–202.
49. Parsons B, MacLuskey NJ, Krey L, Pfaff DW, McEwen BS. The temporal relationship between estrogen-inducible progestin receptors in the female rat brain and the time course of estrogen activation of mating behavior. Endocrinology 1980;107:774–779.
50. Romano GJ, Krust A, Pfaff DW. Expression and estrogen regulation of progesterone receptor mRNA in neurons of the mediobasal hypothalamus: an *in situ* hybridization study. Mol Endocrinol 1989;3:1295–1300.
51. Hagihara K, Hirata S, Osada T, Hirai M, Kato J. Distribution of cells containing progesterone receptor mRNA in the female rat di- and telencephalon: an *in situ* hybridization study. Mol Brain Res 1992;14:239–249.
52. Fox SR, Harlan R, Shivers B, Pfaff DW. Chemical characterization of neuroendocrine targets for progesterone in the female rat brain and pituitary. Neuroendocrinology 1990;51:276–283.
53. Sahu A, Crowley WR, Kalra SP. An opioid-neuropeptide-Y transmission line to luteinizing hormone(LH)-releasing hormone neurons: a role in the induction of LH surge. Endocrinology 1990;126:876–883.
54. Brann DW, Chorich LP, Mahesh VB. Effect of progesterone on galanin mRNA levels in the hypothalamus and the pituitary: correlation with the gonadotropin surge. Neuroendocrinology 1993;58:531–538.
55. Rossmanith WG, Marks DL, Clifton DK, Steiner RA. Induction of galanin mRNA in GnRH neurons by estradiol and its facilitation by progesterone. J Neuroendocrinol 1996;8:185–191.
56. Leranth C, MacLuskey N, Shanabrough M, Naftolin F. Catecholaminergic innervation of luteinizing hormone-releasing hormone and glutamic acid decarboxylase immunopositive neurons in the rat medial preoptic area. An electron-microscopic double immunostaining and degeneration study. Neuroendocrinology 1988;48:591–602.
57. Brann DW, Mahesh VB. Endogenous excitatory amino acid involvement in preovulatory and steroid-induced surge of gonadotropins in the female rat. Endocrinology 1991;128:1541–1547.
58. Unda R, Brann DW, Mahesh VB. Progesterone suppression of glutamic acid decarboxylase (GAD67) mRNA levels in the preoptic area: correlation to the luteinizing hormone surge. Neuroendocrinology 1995;62:562–570.
59. Rothchild I. The regulation of the mammalian corpus luteum. Recent Prog Hormone Res 1981;17:183–298.
60. Mori T, Suzuki A, Nishimura T, Kambegawa A. Inhibition of ovulation in immature rats by antiprogesterone antiserum. J Endocrinol 1977;73:185–186.
61. Lipner H, Greep RO. Inhibition of steroidogenesis at various sites in the biosynthetic pathway in relation to induced ovulation. Endocrinology 1971;88:602–607.
62. Snyder BW, Beecham GD, Schane HP. Inhibition of ovulation in rats with epostane, an inhibitor of 3b-hydroxysteroid dehydrogenase. Proc Soc Exp Biol Med 1984;176:238–242.
63. Van der Schoot P, Bakker GH, Klijn JGM. Effects of the progesterone antagonist RU486 on ovarian activity in the rat. Endocrinology 1987;121:1375–1382.

64. Sanchez JE, Bellido C, Galiot F, Lopez FJ, Gaytan F. A possible mechanism of the anovulatory action of antiprogesterone RU486 in the rat. Biol Reprod 1990;42:877–886.
65. Loutradis D, Bletsa R, Aravantinos L, Kallianidis K, Michalas S, Psychoyos A. Preovulatory effects of the progesterone antagonist mifepristone (RU486) in mice. Hum Reprod 1991;6:1238–1240.
66. Park-Sarge O-K, Mayo K. Transient expression of progesterone receptor messenger RNA in ovarian granulosa cells after the preovulatory luteinizing hormone surge. Mol Endocrinol 1991;5:967–978.
67. Natraj U, Richards JS. Hormonal regulation localization and functional activity of the progesterone receptor in granulosa cells of rat preovulatory follicles. Endocrinology 1993;133:761–769.
68. Espey LL, Lipner H. Ovulation. In: Knobil E, Neill JD, editors. The Physiology of Reproduction. 2nd ed. Raven Press, New York, NY, 1994. pp. 725–780.
69. Iwamasa J, Shibata S, Tanaka N, Matsuura K, Okamura H. The relationship between ovarian progesterone and proteolytic enzyme activity during ovulation in the gonadotropin-treated immature rat. Biol Reprod 1992;46:309–313.
70. Tanaka N, Espey LL, Stacy S, Okamura H. Epostane and indomethacin actions on ovarian kallikrein and plasminogen activator activities during ovulation in the gonadotropin-primed immature rat. Biol Reprod 1992;46:665–670.
71. Espey LL. Current status of the hypothesis that mammalian ovulation is comparable to an inflammatory reaction. Biol Reprod 1994;50:233–238.
72. Richards JS, Russell DL, Robker RL, Dajee M, Alliston TN. Molecular mechanisms of ovulation and luteinization. Mol Cell Endocrinol 1998;145:47–54.
73. Spitz IM, Croxatto HB, Lahteenmaki P, Heikinheimo O, Bardin CW. Effect of mifepristone on inhibition of ovulation and induction of luteolysis. Hum Reprod 1994;9:69–76.
74. Lauber AH, Romano GJ, Pfaff DW. Sex difference in estradiol regulation of progestin receptor mRNA in rat mediobasal hypothalamus as demonstrated by *in situ* hybridization. Neuroendocrinology 1991;53:608–613.
75. Pfaff DW, Schwartz-Giblin S, McCarthy MM, Kow L. Cellular and molecular mechanisms of female reproductive behavior. In: Knobil E, Neill JD, editors. The Physiology of Reproduction. Raven Press, New York, NY, 1994. pp. 107–220.
76. Lauber AH, Romano GJ, Pfaff DW. Steroid control of higher brain functions: gene expression for estrogen and progesterone receptor mRNAs in rat brain and possible relations to sexually dimorphic functions. J Steroid Biochem Mol Biol 1991;40:53–62.
77. McEwen BS, Jones K, Pfaff DW. Hormonal control of sexual behavior in the female rat: molecular, cellular and neurochemical studies. Biol Reprod 1987;36:37–45.
78. Brown TJ, Blaustein JD. Abbreviation of the period of sexual behavior in female guinea pigs by the progesterone receptor antagonist RU38486. Brain Res 1986;373:3–113.
79. Vathy IU, Etgen AM, Barfield RJ. Actions of RU38486 on progesterone facilitation and sequential inhibition of rat estrous behavior: correlation with neural progestin receptors. Horm Behav 1989;23:43–56.
80. Pollio G, Xue P, Zanisi A, Maggi A. Antisense oligonucleotide blocks progesterone-induced lordosis behavior in ovariectomized rats. Mol Brain Res 1993;19:135–139.
81. Mani SK, Blaustein JD, Allen JMC. Inhibition of rat sexual behavior by antisense oligonucleotides to the progesterone receptor. Endocrinology 1994;135:1409–1414.
82. Ogawa S, Olazabal UE, Pfaff DW. Effects of intrahypothalamic administration of antisense DNA for progesterone receptor mRNA on reproductive behavior and progesterone immunoreactivity. J Neurosci 1994;14:1766–1774.
83. Power RF, Mani SK, Codina J, Conneely OM, O'Malley BW. Dopaminergic and ligand-independent activation of steroid hormone receptors. Science 1991;254:1636–1639.
84. Foreman MM, Moss RL. Role of hypothalamic dopaminergic receptors in the control of lordosis behavior in the female rat. Physiol Behav 1979;22:282–289.
85. Caggiula AR, Antelman SM, Chiodo LA, Lineberry CG, editors. Brain dopamine and sexual behavior: psychopharmacological and electrophysiological evidence for antagonism between active and passive components. Pergamon Press, New York, NY, 1979.
86. Pfaus JG, Damsma G, Wenkstern D, Fibiger HC. Sexual activity increases dopamine transmission in the nucleus accumbens and striatum of female rats. Brain Res 1995;693:21–30.
87. Mani SK, Allen JMC, Clark JH, Blaustein JD, O'Malley BW. Convergent pathways for steroid hormone-and neurotransmitter-induced rat sexual behavior. Science 1994;265:1246–1249.
88. O'Malley BW, Schrader WT, Mani S, Smith C, Weigel NL, Conneely OM, et al. An alternative ligand-independent pathway for activation of steroid receptors. Recent Prog Horm Res 1995;50:333–347.

89. Smith CL, Onate SA, Tsai M-J, O'Malley BW. CREB binding protein acts synergistically with steroid coactivator-1 to enhance steroid receptor-dependent transcription. Proc Natl Acad Sci USA 1996;93:8884–8888.
90. Arnold AP, Breedlove SM. Organizational, and activational effects of sex steroids on brain and behavior: a reanalysis. Horm Behav 1985;19:469–498.
91. Frankfurt M, Gould E, Woolley C, McEwen BS. Gonadal steroids modify dendritic spine density in ventromedial hypothalamic neurons: a Golgi study in the adult rat. Neuroendocrinology 1990;51:530–535.
92. Erickson CJ, Bruder RH, Komisaruk BR, Lehrman DS. Selective inhibition of androgen-induced behavior in male ring doves (Streptopelia risoria). Endocrinology 1967;81:39–44.
93. Erpino MJ. Hormonal control of courtship behavior in the pigeon (Columba livia). Anim Behav 1969;1:401–405.
94. Erpino MJ. Temporary inhibition by progesterone of sexual behavior in intact male mice. Horm Behav 1973;4:335–339.
95. Bottoni L, Lucini V, Massa R. Effect of progesterone on the sexual behavior of the male Japanse quail. Gen Comp Endocrinol 1985;57:345–351.
96. Bradford JMW. Treatment of sexual offenders with cyproterone acetate. In: Sitse JMA, editor. The pharmacology and endocrinology of sexual function. Elsevier, New York, 1988, pp. 526–536.
97. Lehne GK. Treatment of sex offenders with medroxyprogesterone acetate. In: Sitse JMA, editor. The pharmacology and endocrinology of sexual function. Elsevier, New York, 1988, pp. 516–525.
98. Kalra PS, Kalra SP. Circadian periodicities of serum androgens, progesterone, gonadotropins and luteinizing hormone-releasing hormone in male rats: The effects of hypothalamic deafferentation, castration and adrenalectomy. Endocrinology 1977;10:1821–1827.
99. Vermueulen A, Verdonck L. Radioimmunoassay of 17 b-hydroxy-5a-androstan-3-one,4-androstene-3, 17-dione, dehydroepiandrosterone, 17-hydroxyprogesterone and progesterone and its application to human male plasma. J Steroid Biochem 1976;7:1–10.
100. Lindzey J, Crews D. Hormonal control of courtship and copulatory behavior in male Cnemidophorus inornatus, a direct sexual ancestor of a unisexual, parthenogenetic lizard. Gen Comp Endocrinol 1986;64:411–418.
101. Lindzey J, Crews D. Effects of progestins on sexual behavior in castrated lizards (Cnemidophoru inornatus). J Endocrinol 1988;119:265–273.
102. Young LJ, Greenberg N, Crews D. The effects of progesterone on sexual behavior in male green anole lizards (Anolis carolinensis). Horm Behav 1991;25:477–488.
103. Lindzey J, Crews D. Interactions between progesterone and androgens in the stimulation of sex behaviors in male little striped whiptail lizards, Cnemidophorous inornatus. Gen Comp Endocrinol 1992;86:52–58.
104. Crews D, Godwin J, Hartman V, Grammer M, Prediger E, Sheppherd R. Intrahypothalamic implantation of progesterone in castrated male whiptail lizards (Cnemidophorus inornatus) elicits courtship and copulatory behavior and affects androgen receptor- and progesterone receptor-mRNA expression in the brain. J Neurosci 1996;16:7347–7352.
105. Witt D, Young L, Crews D. Progesterone modulation of androgen-dependent sexual behavior in male rats. Physiol Behav 1995;57:307–313.
106. Witt DM, Reigada LC, Wengroff BE. Intrahypothalamic progesterone regulates androgen-dependent sexual behavior in male rats. Soc Neurosci Abstr 1997;23:1357.
107. Phelps SM, Lydon JP, O'Malley BW, Crews D. Regulation of male sexual behavior by progesterone receptor, sexual experience, and androgen. Horm Behav 1998;34:294–302.
108. Hull EM, Du J, Lorrain DS, Matuszewich L. Extracellular dopamine in the medial preoptic area: implications for sexual motivation and hormonal control of copulation. J Neurosci 1995;15:7465–7471.
109. Hull EM, Du J, Lorrain DS, Matuszewich L. Testosterone, preoptic dopamine, and copulation in male rats. Brain Res Bull 1997;44:327–333.
110. Mani SK, Blaustein JD, O'Malley BW. Progesterone receptor function from a behavioral perspective. Horm Behav 1997;31:244–255.
111. Dey SK. Implantation. Lippincott-Raven, Philadelphia, PA, 1996.
112. Rothchild I. Role of Progesterone in Initiating and Maintaining Pregnancy. Raven Press, New York, NY, 1983.
113. Grossman CJ. Interactions between the gonadal steroids and the immune system. Science 1985;227:257–261.

114. Szwkeres-Bartho J, Kinsky R, Chaouat G. The effect of a progesterone-induced immunologic blocking factor on NK-mediated resorption. Am J Reprod Immun Microbiol 1990;24:105–107.
115. Morell V. Zeroing in on how hormones affect the immune response. Science 1995;269:773–775.
116. Finn CA, Martin L. Patterns of cell division in the mouse uterus during early pregnancy. J Endocrinol 1967;39:593–597.
117. Finn CA, Martin L. Hormone secretion during early pregnancy in the mouse. J Endocrinol 1969;45:57–65.
118. Finn CA, Martin L. The role of the oestrogen secreted before oestrus in the preparation of the uterus for implantation in the mouse. J Endocrinol 1970;47:431–438.
119. Finn CA, Martin L. The onset of progesterone secretion during pregnancy in the mouse. J Reprod Fert 1971;25:299,300.
120. Finn CA, Martin L. The control of implantation. J Reprod Fert 1974;39:195–206.
121. Yoshinaga K, Adams CE. Delayed implantation in the spayed, progesterone treated adult mouse. J Reprod 1966;12:593–595.
122. Psychoyos A. Endocrine control of egg implantation. In: Greep RO, Astwood EG, Geiger SR, editors. Handbook of Physiology. American Physiological Society, Washington, DC, 1973, pp. 187–215.
123. Martin L, Finn CA, Trinder G. Hypertrophy and hyperplasia in the mouse uterus after oestrogen treatment: an autoradiographic study. J Endocr 1973;56:133–144.
124. Martin L, Finn CA, Carter J. Effects of progesterone and oestradiol-17b on the luminal epithelium of the mouse uterus. J Reprod Fert 1970;21:461–469.
125. Martin L, Finn CA, Trinder G. DNA synthesis in the endometrium of progesterone-treated mice. J Endocrinol 1973;56:303–307.
126. Martin L, Finn CA. The inhibition by progesterone of uterine epithelial proliferation in the mouse. J Endocrinol 1973;57:549–554.
127. Tibbetts TA, Conneely OM, O'Malley BW. Progesterone via its receptor antagonizes the pro-inflammatory activity of estrogen in the mouse uterus. Biol Reprod 1999;60:1158–1165.
128. Quarmby VE, Korach KS. The influence of 17b-estradiol on patterns of cell division in the uterus. Endocrinology 1984;114:694–702.
129. Whitehead MI. The effects of estrogen and progesterone on the postmenopausal endometrium. Maturitas 1978;9:309–313.
130. Gelfand MM, Ferenczy A. A prospective 1-year study of estrogen and progestin in postmenopausal women: effects on the endometrium. Obstet Gynecol 1989;74:398–402.
131. Greenblatt RB, Gambrell RD, Stoddard LD. The protective role of progesterone in the prevention of endometrial cancer. Path Res Pract 1982;174:297–318.
132. Jensen EV, DeSombre ER. Mechanism of action of the female sex hormones. Annu Rev Biochem 1972;41:203–230.
133. Hsueh AJW, Peck Jr EJ, Clark JH. Progesterone antagonism of the oestrogen receptor and oestrogen-induced uterine growth. Nature 1975;254:337–339.
134. Katzenellenbogen BS. Dynamics of steroid hormone receptor action. Annu Rev Physiol 1980;42:17–36.
135. Yamashita S, Newbold RR, McLachlan JA, Korach KS. Developmental pattern of estrogen receptor expression in female mouse genital tracts. Endocrinology 1989;125:2888–2896.
136. Murakami R, Shughrue PJ, Stumpf WE, Elger W, Schulze P-E. Distribution of progestin-binding cells in estrogen-treated and untreated neonatal mouse uterus and oviduct: autoradiographic study with [125I] progestin. Histochemistry 1990;94:155–159.
137. Parczyk K, Madjno R, Michna H, Nishino Y, Schneider M. Progesterone receptor repression by estrogens in rat uterine epithelial cells. J Steroid Biochem Mol Biol 1997;63:309–316.
138. Tibbetts TA, Mendoza-Meneses M, O'Malley BW, Conneely OM. Mutual and intercompartmental regulation of estrogen receptor and progesterone receptor expression in the mouse uterus. Biol Reprod 1998;59:1143–1152.
139. Kurita T, Young P, Brody JR, Lydon JP, O'Malley BW, Cunha GR. Stromal progesterone receptors mediate the inhibitory effects of progesterone on estrogen-induced uterine epithelial cell deoxyribonucleic acid synthesis. Endocrinology 1998;139:4708–4713.
140. Cooke PS, Buchanan DL, Young P, Setiawan T, Brody J, Korach KS, et al. Stromal estrogen receptors mediate mitogenic effects of estradiol on uterine epithelium. Proc Natl Acad Sci USA 1997;94:6535–6540.
141. Das SK, Chakraborty I, Paria BC, Wang XN, Plowman G, Dey SK. Amphiregulin is an implantation-specific and progesterone-regulated gene in the mouse uterus. Mol Endocrinol 1995;9:691–705.

142. Bruner KL, Rodgers WH, Gold LI, Korc M, Hargrove JT, Matrisian LM, et al. Transforming growth factor b mediates the progesterone suppression of an epithelial metalloproteinase by adjacent stroma in the human endometrium. Proc Natl Acad Sci USA 1995;92:7362–7366.
143. Surveyor GA, Gendler SJ, Pemberton L, Das SK, Wegner CC, Dey SK, et al. Expression and steroid hormonal control of muc-1 in the mouse uterus. Endocrinology 1995;136:3639–3647.
144. Iruela-Arispe ML, Porter P, Bornstein P, Sage EH. Thrombospondin-1, an inhibitor of angiogenesis, is regulated by progesterone in the human endometrium. J Clin Invest 1996;97:403–412.
145. Rider V, Carlone DL, Foster RT. Oestrogen and progesterone control basic fibroblast growth factor mRNA in the rat uterus. J Endocrinol 1997;154:75–84.
146. Ma L, Benson GV, Hyunjung L, Dey SK, Mass RL. Abdominal B (AbdB) hoxa genes: regulation in adult uterus by estrogen and progesterone and repression in Mullerian duct by synthetic estrogen diethylstilbestrol (DES). Dev Biol 1998;197:141–154.
147. Chen GTC, Getsios S, MacCalman CD. 17b-Estradiol potentiates the stimulatory effects of progesterone on cadherin-11 expression in cultured human endometrial stromal cells. Endocrinology 1998;139:3512–3519.
148. Zhu L-J, Cullinan-Bove K, Polihronis M, Bagchi MK, Bagchi IC. Calcitonin is a progesterone-regulated marker that forecasts the receptive state of endometrium during implantation. Endocrinology 1998;139:3923–3934.
149. Kumar S, Zhu L-J, Polihronis M, Cameron ST, Baird DT, Schatz F, et al. Progesterone induces calcitonin gene expression in human endometrium within the putative window of implantation. J Clin Endocrinol Meta 1998;83:4443–4450.
150. Vellios F. Endometrial hyperplasia and carcinoma *in situ*. Gynecol Oncol 1974;2:152–161.
151. Russo IH, Russo J. Role of hormones in mammary cancer initiation and progression. J Mam Gland Biol Neoplasia 1998;3:49–61.
152. Hirayama T, Wynder EL. A study of epidemiology of cancer of the breast. II. The influence of hysterectomy. Cancer 1962;15:28–38.
153. Feinleib M. Breast cancer and artificial menopause. J Natl Cancer Inst 1968;41:315–329.
154. Trichopoulos D, MacMahon B, Cole P. Menopause and breast cancer. J Natl Cancer Inst 1972;48:605–613.
155. Henderson BE, Ross RK, Judd HL, Krailo MD, Pike MC. Do regular ovulatory cycles increase breast cancer risk? Cancer 1985;56:1206–1208.
156. Henderson BE, Ross RK, Pike MC. Hormonal chemoprevention of cancer in women. Science 1993;259:633–638.
157. Clarke CL, Sutherland RL. Progestin regulation of cellular proliferation. Endocrine Rev 1990;11:266–300.
158. Anderson TS, Ferguson JP, Raab GM. Cell turnover in the "resting" human breast : Influence of parity, contraceptive pill, age and laterality. Brit J Cancer 1982;46:376–382.
159. Pike MC, Spicer DV, editors. Contraception. Springer-Verlag, New York, NY, 1993.
160. Medina D. The mammary gland: a unique organ for the study of development and tumorigenesis. J Mammary Gland Biol Neoplasia 1996;1:5–19.
161. Cardiff RD, Wellings SR. The comparative pathology of human and mouse mammary glands. J Mammary Gland Biol Neoplasia 1999;4:105–122.
162. Hennighausen L, Robinson GW. Think globally, act locally: the making of a mouse mammary gland. Genes Dev 1998;12:449–455.
163. Said TK, Conneely OM, Medina D, O'Malley BW, Lydon JP. Progesterone, in addition to estrogen, induces cyclin D1 expression in the murine mammary epithelial cell, *in vivo*. Endocrinology 1997;138:3933–3939.
164. Haslam SZ, Shyamala G. Relative distribution of estrogen and progesterone receptors among the epithelial, adipose, and connective tissue components of the normal mammary gland. Endocrinology 1981;108:825–830.
165. Haslam SZ. The ontogeny of mouse mammary gland responsiveness to ovarian steroid hormones. Endocrinology 1989;125:2766–2772.
166. Cunha GR. Role of mesenchymal-epithelial interactions in normal and abnormal development of the mammary gland and prostate. Cancer 1994;74:1030–1044.
167. Cunha GR, Yom YK. Role of mesenchymal-epithelial interactions in mammary gland development. J Mammary Gland Biol Neoplasia 1996;1:5–19.
168. Cunha GR, Young P, Hom YK, Cooke PS, Taylor JA, Lubahn DB. Elucidation of a role for stromal steroid hormone receptors in mammary gland growth and development using tissue recombinants. J Mammary Gland Biol Neoplasia 1997;2:393–402.

169. Humphreys RC, Lydon JP, O'Malley BW, Rosen JM. Use of PRKO mice to study the role of progesterone in mammary gland development. J Mammary Gland Biol Neoplasia 1997;2:343–354.
170. Brisken C, Park S, Vass T, Lydon JP, O'Malley BW, Weinberg RA. A paracrine role for the epithelial progesterone receptor in mammary gland development. Proc Natl Acad Sci USA 1998;95:5076–5081.
171. Silberstein GB, Van Horn K, Shyamala G, Daniel CW. Progesterone receptors in the mouse mammary duct: distribution and developmental regulation. Cell Growth Differen 1996;7:945–952.
172. Shyamala G, Barcellos-Hoff MH, Toft D, Yang X. In situ localization of progesterone receptors in normal mouse mammary glands : absence of receptors in the connective and adipose stroma and a heterogeneous distribution in the epithelium. J Steroid Biochem Mol Biol 1997;63:251–259.
173. Shyamala G. Progesterone signaling and mammary gland morphogenesis. J Mammary Gland Biol Neoplasia 1999;4:89–104.
174. Jull JW. The effects of oestrogens and progesterone on the chemical induction of mammary cancer in mice of the *IF* strain. J Path Bact 1954;68:547–559.
175. Jabara AG, Harcourt AG. Effects of progesterone, ovariectomy and adrenalectomy on mammary tumors induced by 7, 12-dimethylbenz(a)anthracene in Sprague-Dawley rats. Pathology 1971;3:209–214.
176. Welsch CW. Host factors affecting the growth of carcinogen-induced rat mammary carcinomas: a review and tribute to Charles Brenton Huggins. Cancer Res 1985;45:3415–3443.
177. Robinson SP, Jordan VC. Reversal of antitumor effects of tamoxifen by progesterone in the 7, 12-dimethylbenzanthracene-induced rat mammary carcinoma model. Cancer Res 1987;47:5386–5390.
178. Russo IH, Russo J. Progestagens and mammary gland development: differentiation *versus* carcinogenesis. Acta Endocrinol 1991;125:7–12.
178a. Lydon JP, Ge G, Kittrell FS, Medina D, O'Malley BW. Murine mammary gland carcinogenesis is critically dependent on progesterone receptor function. Cancer Res 1999;59:4276–4284.
179. Daniel CW, Smith GH. The mammary gland: a model for development. J Mammary Gland Biol Neoplasia 1999;4:3–8.
180. McEwen BS. Genomic regulation of sexual behavior. J Steroid Biochem 1988;30:179–183.
181. Wagner CK, Nakayama AY, De Vries GJ. Potential role of maternal progesterone in the sexual differentiation of the brain. Endocrinology 1998;139:3658–3661.
182. Parsons B, Rainbow TC, Maclusky NJ, McEwen BS. Progestin receptor levels in rat hypothalamus and limbic nuclei. J Neurosci 1982;12:2549–2554.
183. Gould E, Woolley C, Frankfurt M, McEwen BS. Gonadal steroids regulate dendritic spine density in hippocampal pyramidal cells in adulthood. J Neurosci 1990;10:1286–1291.
184. Olton DDS. Memory functions and the hippocampus. In: Seifert W, editor. Neurobiology of the Hippocampus. Academic Press, London, UK, 1983, pp. 335–373.
185. Williams CL, Meck WH. The organizational effects of gonadal steroids on sexually dimorphic spatial ability. Psychoneuroendocrinology 1991;16:155–176.
186. Roof RL, Duvdevani R, Braswell L, Stein DG. Progesterone facilitates cognitive recovery and reduces secondary neuronal loss caused by cortical contusion injury in male rats. Exp Neurol 1994;129:64–69.
187. Akwa Y, Young J, Kaggadj K, Sancho MJ, Zucman D, Vourc'H C, et al. Neurosteroids: biosynthesis, metabolism and function of pregnenolone and dehydroepiandrosterone in the brain. J Steroid Biochem Mol Biol 1991;40:71–81.
188. LeGoascogne C, Robel P, Gouezou M, Waterman M. Neurosteroids: cytochrome p-450scc in rat brain. Science 1987;237:1212–1215.
189. Koenig HL, Schumacher M, Ferzaz B, DoThi AN, Ressouches A, Guennoun R, et al. Progesterone synthesis and myelin formation by Schwann cells. Science 1995;268:1500–1503.
190. Paul SM, Purdy RH. Neuroactive steroids. FASEB J 1992;6:2311–2322.
191. Betz AL, Coester HC. Effects of steroids on edema and sodium uptake of the brain during focal ischemia in rats. Stroke 1990;21:199–204.
192. Roof RL, Duvdevani R, Stein DG. Progesterone treatment attenuates brain edema following contusion injury in male and female rats. Rest Neurol Neurosci 1992;4:425–427.
193. Zuccarello M, Anderson D. Interaction between free radicals and excitatory amino acids in the blood-brain barrier disruption after iron injury in the rat. J Neurotrauma 1993;10:397–403.
194. Yu WH. Survival of motoneurons following axotomy is enhanced by lactation or progesterone treatment. Brain Res 1989;491:379–382.
195. Asbury ET, Fritts ME, Horton JE, Isaac WL. Progesterone facilitates the acquisition of avoidance learning and protects against subcortical neuronal death following prefrontal cortex ablation in the rat. Behav Brain Res 1998;97:99–106.

196. Bruckert E, Turpin G. Estrogens and progestins in postmenopausal women: influence on lipid parameters and cardiovascular risk. Horm Res 1995;43:100–103.
197. Godsland IF, Wynn V, Crook D, Miller NE. Sex, plasma lipoproteins, and atherosclerosis-prevailing assumptions and outstanding questions. Am Heart J 1987;114:1467–1503.
198. Stampfer MJ, Colditz GA. Estrogen replacement therapy and coronary heart disease: a quantitative assessment of the epidemiologic evidence. Prev Med 1991;20:47–63.
199. Barrett-Connor E, Bush TL. Estrogen and coronary heart disease in women. JAMA 1991;265:1861–1867.
200. Sullivan JM, Fowlkes LP. The clinical aspects of estrogen and cardiovascular system. Obstet Gynecol 1996;87(suppl.):36–43.
201. Wahl PW, Walden CE, Knapp RH, Wallace R, Rifkind B. Effect of estrogen/progesterone potency on lipid/lipoprotein cholesterol. N Engl J Med 1983;308:862–867.
202. Report. C. Estrogen replacement therapy in the menopause. JAMA 1983;249:359–361.
203. Persson I, Adami HO, Bergkvist L, Lindgreen A, Petterson B, Hoover R, et al. Risk of endometrial cancer after treatment with oestrogens alone or in conjunction with progesterone: results of a prospective study. Br Med J 1989;298:147–151.
204. Hirvonen E, Malkonen M, Manninen V. Effects of different progestogen on lipoproteins during postmenopausal therapy. N Engl J Med 1981;304:560–563.
205. Cheng W, Lau OD, Abumrad NA. Two antiatherogenic effects of progesterone on human macrophages; inhibition of cholesteryl ester synthesis and block of its enhancement by glucocorticoids. J Clin Endo Met 1999;84:265–271.
206. Grodstein F, Stampfer MJ, Manson JE, al. e. Post menopausal estrogen and progestin use and the risk of cardiovascular disease. N Engl J Med 1996;335:453–461.
207. Ingegno MD, Money SR, Thelmo W, Greene GL, Davidian M, Jaffe BM, et al. Progesterone receptors in the human heart and great vessels. Lab Invest 1988;59:353–356.
208. Knauthe R, Diel P, Hegele-Hartung C, Engelhaupt A, Fritzemeier K-H. Sexual dimorphism of steroid hormone receptor messenger ribonucleic acid expression and hormonal regulation in rat vascular tissue. Endocrinology 1996;137:3220–3227.
209. Lee W-S, Harder JA, Yoshizumi M, Lee M-E, Haber E. Progesterone inhibits arterial smooth muscle cell proliferation. Nat Med 1997;3:1005–1008.
210. Bourassa KP-A, Milos PM, Gaynor BJ, Breslow JL, Aiello RJ. Estrogen reduces atherosclerotic lesion development in apolipoprotein E-deficient mice. Proc Natl Acad Sci USA 1996;93:10,002–10,027.
211. Iafrati MD, Karas RH, Aronovitz M, Kim S, Sullivan Jr. TR, Lubahn DB, et al. Estrogen inhibits the vascular injury response in estrogen receptor a-deficient mice. Nat Med 1997;3:545–548.
212. Hutchinson T, Polansky S, Feinstein A. Post-menopausal oestrogens protect against fractures of hip and distal radius: a case-control study. Lancet 1979;2:705–709.
213. Bain SD, Jensen E, Celino DL, Bailey MC, Lantry MM, Edwards MW. High-dose gestagens modulate bone resorption and formation and enhance estrogen-induced endosteal bone formation in the ovariectomized mouse. J Bone Mineral Res 1993;8:219–229.
214. Prior JC. Progesterone as a bone-trophic hormone. Endocrine Rev 1990;11:386–398.

10 Knockout and Transgenic Mouse Models that Have Contributed to the Understanding of Normal Mammary Gland Development

Tiffany N. Seagroves, PHD
and Jeffrey M. Rosen, PHD

Contents

INTRODUCTION

The Mammary Gland as a Model System of Development

The mammary gland in mice is in many ways analogous to the *Drosophila* eye as a target organ that can be readily manipulated to dissect information about the function of genes and signaling pathways during development. The majority of mammary-gland development occurs postnatally, and therefore can be manipulated without the problems of embryonic lethality observed in many other organ systems. Classically, the mammary

From: *Contemporary Endocrinology: Transgenics in Endocrinology*
Edited by: M. Matzuk, C. W. Brown, and T. R. Kumar © Humana Press Inc., Totowa, NJ

gland has been a model system for endocrinologists. Yet, new technologies developed within the past 30 yr have transformed the mammary gland into a powerful genetic model system. Basic developmental biological processes—including ductal morphogenesis and patterning, proliferation, differentiation, and apoptosis as well as stromal-epithelial interactions—may be studied using a combination of classical biological techniques coupled with the sophisticated methods of modern mouse genetics.

There are several advantages to the use of the mammary gland as a model system to study development. First, the mammary gland is not essential for survival of individual mice. More importantly, since rodents contain multiple pairs of glands, one or multiple glands may be biopsied from the same animal over the course of development, decreasing animal-to-animal experimental variability. Second, in contrast to most organ systems, the mammary gland develops primarily after birth, eliminating the need to harvest delicate embryos at precise stages of pregnancy. Third, multiple techniques exist to assay the epithelial or stromal contribution of a particular gene to mammary development, allowing complex signaling pathways to be dissected.

Significance of Understanding the Genetic Pathways Controlling Normal Mammary Gland Development

Breast cancer is the most frequently diagnosed cancer for women in the United States, projected to affect 1 in 8 women over their lifetime *(1,2)*. In the past three decades since the initiation of the "War on Cancer," the vast majority of National Institute of Health (NIH)-funded research has been performed using established human breast-cancer cell lines, often derived from pleural effusions of breast-cancer patients. These cell lines are routinely genetically unstable, and contain a plethora of genetic alterations. In 1998, the priorities of the National Cancer Institute (NCI) were changed with regards to the funding of breast-cancer research. An emphasis on "... a more complete understanding of the normal mammary gland at each stage of development will...be a critical underpinning of continued advances in detecting, preventing and treating breast cancer...The principal scientific need in this field now is to focus studies on the early transitions from the normal to the malignant state in human and rodent model systems." Therefore, the NCI recommended: "to increase funding for projects that integrate knowledge of cell signaling with whole organ biology for the development (of) in vitro models of breast differentiation, experimental xenograft systems and transgenic/knock-out mouse models." As this chapter demonstrates, gene-deleted or "knockout" and transgenic mice have already elucidated several basic mechanisms that control normal mammary morphogenesis. Furthermore, several transgenic and knockout mouse models mimic the pathology observed in breast cancers *(3)*. The results of these and future studies will certainly impact breast cancer diagnosis, prognosis and treatment protocols.

Technologies to Investigate Mammary Gland Development

Multiple technologies exist to analyze mammary development and physiology following deletion, overexpression, or direct delivery of a gene product of interest. Many of these techniques are reviewed in ref. *4*. Often, the first clue that a gene is contributing to mammary development is the failure of knockout mice to support their litters at lactation. Several key regulators of mammary development have been serendipitously identified in mice that were created for other reasons. Genes can be targeted (and overexpressed) in the mammary epithelium under the control of the mouse mammary-tumor virus (MMTV) or milk-protein gene promoters, such as β-casein, whey acidic protein (WAP), β-lactoglobulin, or α-lactalbumin *(5)*. All of these promoters will direct transgene expression primarily during pregnancy and lactation in a temporal and hormonally regulated fashion. Occasionally, the MMTV and the WAP promoters may be expressed at sufficient levels in virgin mice to influence ductal morphogenesis *(6–8)*.

The response of the mammary gland to systemic administration of exogenous growth factors, steroid hormones, or polypeptides may be assayed by subcutaneous (sc) interscapular implantation of beeswax pellets containing the gene product of interest, or using Alzet mini-pumps. Alternatively, factors diluted into saline solutions or sesame oil may be injected subcutaneously behind the neck. To assess localized responses, smaller ethylene/vinyl acetate copolymer (Evac) pellets containing the factor of interest may be directly implanted into a mammary gland *(9)*. Implantation of blank Evac pellets within the contralateral gland of the same animal allows direct comparison of efficacy of treatment, and eliminates concerns of animal-to-animal variability.

Other technologies rely on the intrinsic ability of the mammary gland to regenerate. The mammary epithelial cells (MEC) are capable of completing multiple cycles of reproductive development, including proliferation and differentiation during pregnancy, secretion of milk at lactation, and massive apoptosis during involution. During successive pregnancies, these processes will be repeated. The dramatic increase in the number of MEC that occur during each pregnancy suggests that stem-cell populations exist in the mammary gland. Based on the results of serial dilution experiments, it is hypothesized that approx 1 in every 2,500 MEC is a stem cell *(10–12)*.

Transplantation techniques developed by DeOme et al. rely on the ability of MEC to repopulate a host fat pad that has been surgically "cleared" of its endogenous epithelium *(13)*. A schematic of the transplantation protocol is presented in Fig. 1. Following removal of the host's endogenous epithelium, small portions (1 × 1mm) of donor mammary tissue, averaging approx 4600 MEC *(10)*, are inserted into a pocket created within the host fat pad. After a period of outgrowth, typically 6–8 wk posttransplantation, the ductal network has reorganized, growing away from the site of the transplant to the edges of the fat pad (Fig. 1). Serial transplantation is usually possible for up to five generations from tissue that has regenerated from the original transplanted graft *(14)*. In 1988, Sheffield et al. described successful transplantation of human breast epithelium into cleared fat pads of athymic nude mice, providing another application of transplantation to breast-cancer research *(15)*.

This transplantation technology has provided multiple assays to analyze phenotypes observed in knockout and transgenic mouse models *(16)*. First, defects in development may be assigned to either the epithelial or stromal compartments of the mammary gland through reciprocal transplantation. For example, mammary epithelium isolated from knockout mice may be transplanted into the cleared fat pads of wild-type hosts and conversely, wild-type epithelium may be transplanted into the cleared fat pads of knockout hosts *(17,18)*. If the knockout tissue fails to develop normally in the wild-type fat pad, the defect is MEC autonomous. Conversely, if the wild-type epithelium does not develop within the knockout stroma, the defect resides in the stromal compartment. To control for defects attributed to transplantation, such as disruption of the connection to the nipple, development of the transplanted outgrowths may be compared to development of an endogenous thoracic gland present in the host. Although mature females are used most often as the donors for transplantation, the mammary anlage may also be rescued from embryos and transplanted into 3-wk old hosts as early as embryonic (E) d 12.0. This

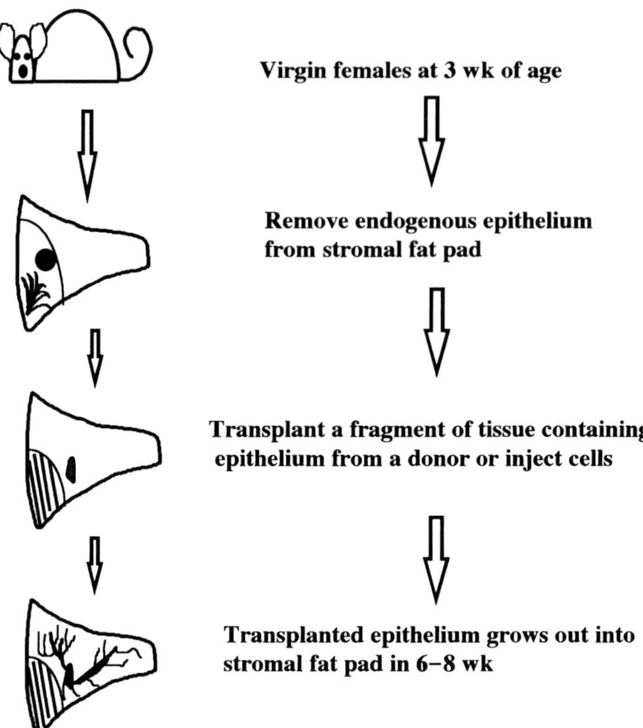

Fig. 1. Transplantation of cells or epithelium into the cleared inguinal (#4) fat pads of 3-wk-old hosts. By 3 wk of age, the ducts have grown just to the lymph node (*dark circle*, one-third down fat pad from the nipple). To "clear" the fat pad, the endogenous epithelium is cut away *(solid line)* from the remaining fat pad just beyond the lymph node. The nipple and area of skin where epithelium was removed is then cauterized *(dashed lines)*. A small piece of epithelium is placed into a small pocket of the remaining fat pad *(gray oval)*. Alternatively, MEC may be injected into the cleared fat pad. Normal epithelium will grow out to the edges of the "cleared" fat pad within 6–8 wk posttransplantation.

circumvents problems with embryonic and neonatal lethality often encountered with knockout mice. Second, transplantation of donor epithelium isolated from knockout mice into wild-type host fat pads may separate systemic from mammary-intrinsic developmental defects. Third, cleared mammary fat pads may be injected with purified preparations of MEC that have been genetically manipulated in vitro.

Reciprocal epithelial-stromal interactions may also be investigated by transplantation under the kidney capsule. This technique, developed by the Cunha laboratory *(19)*, has provided clues to several factors controlling embryonic and postnatal mammary gland development. Since athymic nude mice are used as the hosts of the recombined tissues, the genetic background of the epithelium and stroma does not have to be similar. Using these transplantation methods, complex biological questions can be addressed.

Organ and tissue-culture techniques have been developed to investigate mammary-gland development in vitro within a controlled physical environment (reviewed in *20,21*). The culture of explanted mammary tissue facilitated investigation of the specific contributions of the hormones and growth factors required for mammary-gland development and differentiation *(22)*. More recently, protocols for isolating pure preparations of MEC by collagenase/trypsin digestion of whole mammary glands of virgin or pregnant mice have been developed (reviewed in *21*). Several methods exist to experimentally manipulate primary MEC. Primary MEC transduced with retroviruses in vitro may be re-introduced into the cleared fat pads of hosts to study development in vivo. This technique is a faster, cheaper alternative to producing transgenic mice, but lack of a phenotype may result from a failure to effectively transduce the stem-cell population of MEC. In addition, primary MEC may be cultured for up to 10 d before entering crisis, facilitating investigation of cell-signaling pathways *(21)*. Although primary MEC must be prepared for each experiment, it is likely that cell-signaling events in freshly isolated primary MEC are more physiologically relevant to normal mammary development than those observed in established cell lines.

Since traditional knockout mouse models fail to address questions regarding gene deletion at particular stages of development, new technologies have been created to delete genes in a tissue-specific and/or temporal manner. The most prevalent method is based on the Cre-*lox* system described in P1 bacteriophage, which involves Cre recombinase-mediated deletion of DNA fragments that are flanked by *lox*-P sites ("*floxed*") *(23)*. *Floxed* DNA is then introduced by homologous recombination into embryonic stem (ES) cells. Gene-targeted mice heterozygous for the *floxed* allele are bred to homozygosity. Upon addition of Cre, both copies of a gene will be deleted. Once a gene has been successfully *floxed* and introduced in mice, there are two methods to induce Cre expression. First, mice that carry *floxed* alleles may be crossed to transgenic mice that express Cre via tissue-specific promoters. For example, transgenic lines of mice that express Cre recombinase via the MMTV, WAP, and β-lactoglobulin promoters have been created *(24,25)*. Alternatively, primary MEC may be prepared from *floxed* mice, transduced with Cre via adenovirus infection and transplanted into the cleared fat pad of host mice *(25a)*. Utilization of the Cre/*lox* system will further refine the ability of the mammary gland to reveal mechanisms of development.

General Morphology of Embryonic Mouse Mammary Gland Development

Mammary rudiments appear as epidermal thickenings on either side of the trunk, referred to as the "milk streak," by E 12.0 *(26)*. Dense mammary mesenchyme, composed of 2–3 layers of tightly packed fibroblasts, envelops the developing mammary bud, appearing at E 13.0. The functions of this mesenchyme are to support growth of the epithelial bud and to regulate sexual dimorphism of the gland in response to testosterone. Posterior to the dense mammary mesenchyme lies the future fat pad, which is composed of preadipocytes *(26)*. By E 14.0, the dense mammary mesenchyme expresses relatively high levels of androgen receptor (AR) *(26)*. Androgens secreted by the testes in the male induce active regression of the mammary epithelial bud *(27)*. In females the anlage continues to grow very slowly until E 16, when proliferation increases, and the cord of the mammary epithelium begins to grow into the surrounding fat pad, opening to the nipple *(26)*. By birth, a primary duct connected to the nipple containing 12–15 small branches of ductal epithelium is present. An excellent mini-review of embryonic mammary development can be found on the world wide web (www) at http://mammary.nih.gov/atlas/wholemounts/normal/slides/main.html. From birth to approx 3 wk of age, the ductal epithelium grows slowly in proportion to the increase in size of the entire animal *(28)*. By 3 wk of age, the ductal tree has extended out to the lymph node in the inguinal (#4) gland that marks the first one-third of the mammary fat pad. If cleared

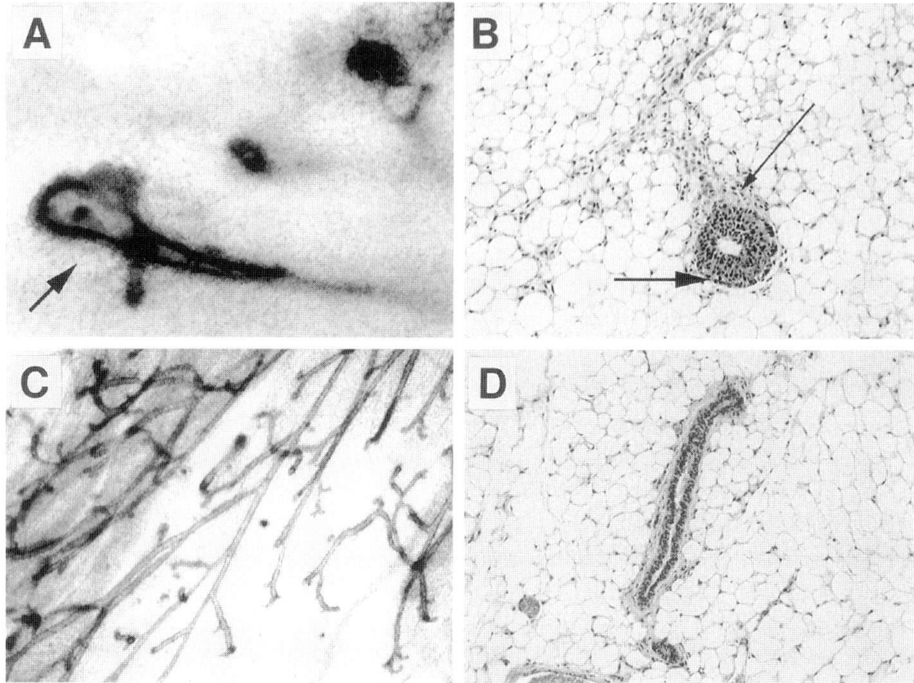

Fig. 2. Histology of ductal morphology in nulliparous female mice. Whole mounts **(A,C)** or H&E-stained sections **(B,D)** of inguinal (#4) mammary glands prepared from 6-wk-old (juvenile; A,B) or 12-wk-old (mature; C,D) pure C57/Bl6 mice were digitally captured at 10× (A,B) or 4× (C,D) microscopic magnification. The arrows indicate either the epithelial cells *(large arrows)* or the dense layer of extracellular matrix surrounding the epithelial cells *(small arrows)*.

of endogenous epithelium, the remaining two-thirds of the mammary stroma may be used as the site of transplanted tissue or cells, as previously described.

General Morphology of Postnatal Mouse Mammary Gland Development

The majority of postnatal development begins at 3 wk of age, at the onset of puberty. Postnatal development consists of four tightly regulated stages: ductal outgrowth into the stromal fat pad from 3–9 wk of age, lobuloalveolar proliferation and differentiation during pregnancy, synthesis and secretion of milk at lactation, and involution of the secretory epithelium following weaning. Each stage depends on a critical balance between proliferation, differentiation, and apoptosis. With the advent of knockout and transgenic mice models, the specific contributions of hormones, growth factors, and cell-signaling pathways began to be mapped to these stages of mammary development *(29)*.

Rapid development of the rudimentary ductal epithelium is initiated in response to increasing levels of circulating hormones synthesized by the pituitary gland and the ovary, such as estrogen, progesterone, and growth hormone (GH) *(30)*. Other growth factors, such as epidermal growth factor (EGF) and insulin-like growth factor (IGF)-I, have also been demonstrated to stimulate ductal morphogenesis *(31,32)*. Ductal morphogenesis progresses through a balance between proliferation and apoptosis with multi-layered club-shaped structures known as terminal end buds (TEB) (Fig. 2A–B). The TEB is composed of two predominant epithelial-cell types. The outermost layer of cells, known as cap cells, are in close contact with the basement membrane at the distal portion of the end bud. Proliferation occurs primarily in these distal-cell layers *(33)*. Cap cells lack differentiated features and intracellular junctions, and are not polarized *(30)*. Programmed cell death occurs within the more proximal innermost cell layers of the TEB known as body cells *(33)*, resulting in the formation of a hollow duct composed of a single layer of luminal-epithelial cells. As the ducts approach the edges of the fat pad, by approx 8–9 wk of age, the TEBs disappear, signaling the end of ductal morphogenesis (Fig. 2C). The virgin gland remains relatively quiescent until the onset of pregnancy or the administration of exogenous hormones such as estrogen and progesterone. The rapidly proliferating cap cells of the TEB are hypothesized to be targets of carcinogenic agents *(34)*, since pregnancy or the administration of hormones that accompany pregnancy are protective against cancer in both rodent and human (female) models *(1,35,36)*.

The MEC are surrounded by an outer layer of myoepithelial cells that will contract to express milk through the nipple during lactation. The function of the myoepithelial cells in the nulliparous animal is unknown *(30)*. Epithelial cells may be distinguished from myoepithelial cells in two ways. First, myoepithelial cells, which line the single layer of luminal-epithelial cells, are thin and spindle-shaped in contrast to the columnar appearance of epithelial cells. Second, they can be distinguished by expression of specific cytokeratin intermediate filaments. Ductal luminal or secretory epithelial cells express cytokeratins 8 and 18 (K8 & K18), whereas cytokeratin 14 (K14) is expressed only in myoepithelial cells *(37)*.

Pregnancy induces proliferation of the secretory units of the mammary gland—the alveoli—which originate from putative ductal progenitor cells and proliferate to eventually fill the entire stromal fat pad. The gestation period of most mouse strains is between 19–21 d and can be divided into three major stages of development: "early" (0–10 d), "mid-" (10–15 d), or "late" pregnancy (15 d until parturition). The highest rates of DNA synthesis in the ductal epithelium are observed at d 3 of pregnancy, before decreasing and being observed primarily in the developing alveoli *(38)*. Proliferation of alveoli per total number of MEC is maximal during the early pregnancy, from d 6–10 *(38)*. By d 6 of pregnancy, fine secondary/tertiary branches with clusters of alveoli are apparent (Fig. 3A,B). By 10 d of pregnancy, the alveoli have begun to appear uniformly along the ductal network (Fig. 3C,D).

Concurrent with proliferation, alveoli begin to functionally differentiate at mid-pregnancy, as assayed by the synthesis milk-protein genes, such as β-casein and WAP *(39)*. The accumulation of proteinaceous and lipid secretory products is evident in the lumen of the alveoli of hematoxylin and eosin (H&E)-stained sections prepared from 15-d pregnant female (Fig. 3E,F). At this stage of pregnancy, alveoli have begun to fill in the "spaces" between the alveoli-lined ducts. By d 18 of pregnancy, the MEC population of the mammary gland accounts for approx 90% of all cells *(40)*; the entire fat pad has become filled with alveoli.

During lactation, the secretory epithelium maximally produces and secretes milk. Accumulation of milk distends the alveoli, flattening the epithelial cells into tightly packed rings of cells with large lumens (Fig. 3G). At involution, following the weaning of pups, extensive tissue remodeling and apoptosis of the secretory epithelium occurs, until the gland contains the simple ductal network seen in the virgin. The collapsed nature and disorganization of the MEC typical of an involuting gland 4 d following forced removal of a litter is shown in Fig. 3H. These stages of mammary development

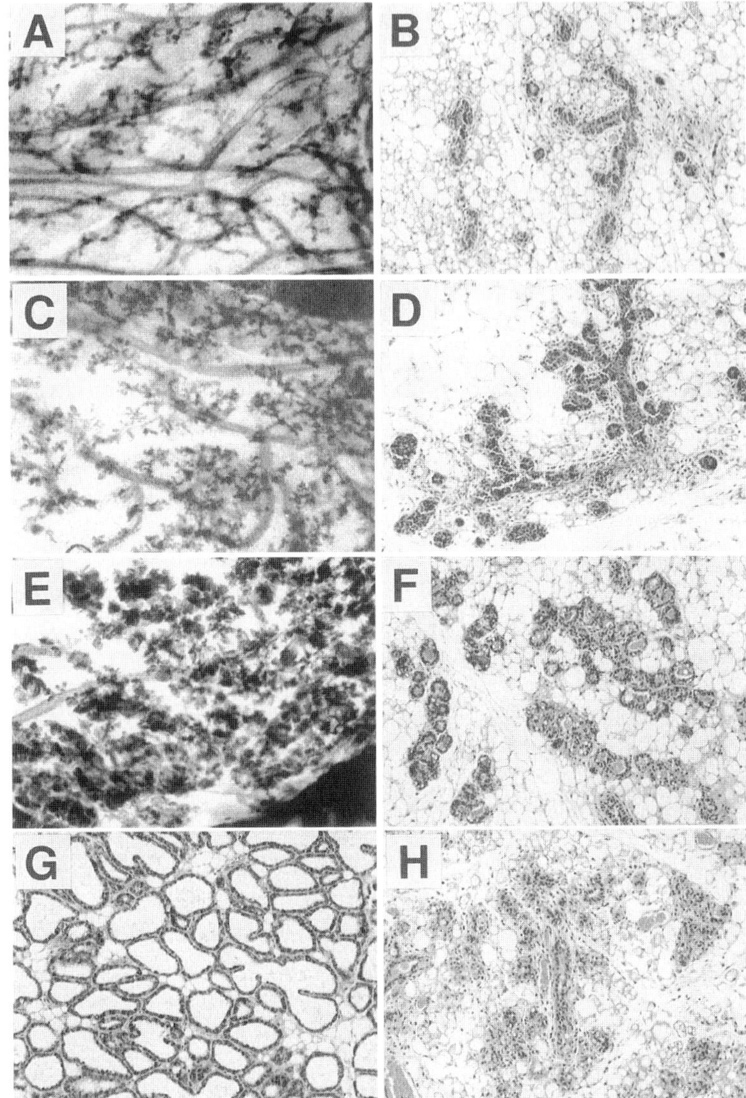

Fig. 3. Histology of the inguinal mammary gland during pregnancy, lactation and involution. Whole mounts **(A,C,E)** and/or H&E-stained sections **(B,D,F–H)** were prepared from 6-d pregnant (A,B), 10-d pregnant (C,D), 15-d pregnant (E,F), 1-d lactating (G) or 4-d forced involuted (H) pure C57BL/6 mice. Note the increase in secretory products in the lumen of alveoli at 15 d of pregnancy (F), the distended lumen during lactation (G) and the collapsed, disorganized structure of the alveoli during involution (H). All images were digitally captured at 10× (A,C,E) or 20× (B,D,F–H) magnification.

may be reviewed in more detail online on the "Biology of the Mammary Gland" home page, sponsored by the NIH at http://mammary.nih.gov.

The mammary gland is capable of repeating the process of pregnancy, lactation, and regression to a more virgin-like state multiple times. Involution of the gland naturally proceeds as neonates begin to eat solid food *(41)*, or can either be induced by forced removal of the entire litter or by physically sealing individual nipples, resulting in accumulation of milk or "milk stasis" in the sealed gland *(42)*. In the natural weaning model, milk stasis does not occur, presumably because of a systemic feedback mechanism *(42)*. By either method, the process of regression occurs in two stages: a proteinase-independent initiation of programmed cell death of epithelial cells, and a tissue remodeling resulting from increased expression of extracellular matrix degrading proteases *(42–44)*. Apoptosis is induced within 24 h of forced involution or teat sealing and is observed as early as d 16 of natural weaning *(42)*. By d 22 of natural weaning, the DNA content is one-half that observed at d 16 *(42)*.

As the epithelial cells die, they are sloughed into the lumen, and are phagocytosed by macrophages that increase at d 3 and are maximal at d 10 *(43)*. In forced involution models, by d 4, extensive tissue remodeling is obvious in histological preparations, and the alveoli have collapsed and appear disorganized (Fig. 3H). By d 8 of involution, very few clusters of alveoli remain. In most mice strains, relatively little epithelium is present by d 10 of involution *(43)*. Mice that are pregnant upon natural weaning of their first litter transiently exhibit increased levels of apoptosis, but do not undergo the second phase of involution—tissue remodeling *(42,44)*. Lactation can be restored in the forced involution model if pups are returned to the mother within 2 d, but involution proceeds when pups are removed for 3 d *(44)*. The mechanism of this commitment to involution is not well-defined.

Control of Embryonic Development

In comparison to postnatal mammary gland development, very few knockout or transgenic mice have been reported to have defects in embryonic development. Some of the genes implicated in embryonic mammary development are presented in Fig. 4. Presumably, disruption of genes that mediate epithelial-mesenchymal interactions would result in embryonic defects. Two such mouse models have been reported, *p63* –/– *(45,46)* and lymphoid-enhancer factor (*Lef*)-1 –/– *(47)* mice. *p63*, a homolog of the tumor-suppressor gene *p53* , is expressed in ectodermally-derived tissues of the embryo *(45)*. Mice lacking *p63* die shortly after birth, most likely a result of dehydration caused by poorly differentiated skin. Newborn mice do not have limbs (or have truncated limbs), and have no hair follicles or teeth. No mammary buds form in *p63* –/– mice. Interestingly, the *p63* –/– mice fail to induce expression of another gene that has been implicated in embryonic development, *Lef-1 (45)*.

LEF-1 is a sequence-specific DNA-binding protein that belongs to the T cell factor (TCF) subset of the high-mobility group (HMG) family of proteins *(48)*. *Lef-1* is expressed in several developing tissues in the embryo, but is restricted to lymphoid tissues in adult animals *(48)*. Similar to *p63* –/– mice, mice lacking *Lef-1* do not form tissues that require ectoderm-mesenchymal interactions such as teeth and whiskers *(48)*. In contrast to *p63* –/– mice, *Lef-1* –/– mice do form mammary buds; however, the buds are arrested and do not progress to ductal development *(48)*. LEF-1 mediates Wnt signaling pathways and may be activated by bone morphogenetic protein (BMP)-4 *(49)*. BMP-4 is expressed in the fetal mesenchyme surrounding the mammary bud, indicating that BMP-4 expressed in the mammary stroma may induce expression of Lef-1 in the mammary epithelium *(50)*. Intriguingly, LEF-1 and other TCF family members may directly induce cyclin D1 expression *(51)*. Therefore, mammary development may be arrested in *Lef-1* –/– mice because the mammary bud fails to proliferate to form the primary duct.

During embryogenesis, LEF-1 expression partially overlaps with expression of *Alx-4*, a protein identified in a screen for retinoblastoma (Rb) interacting proteins *(52)*.

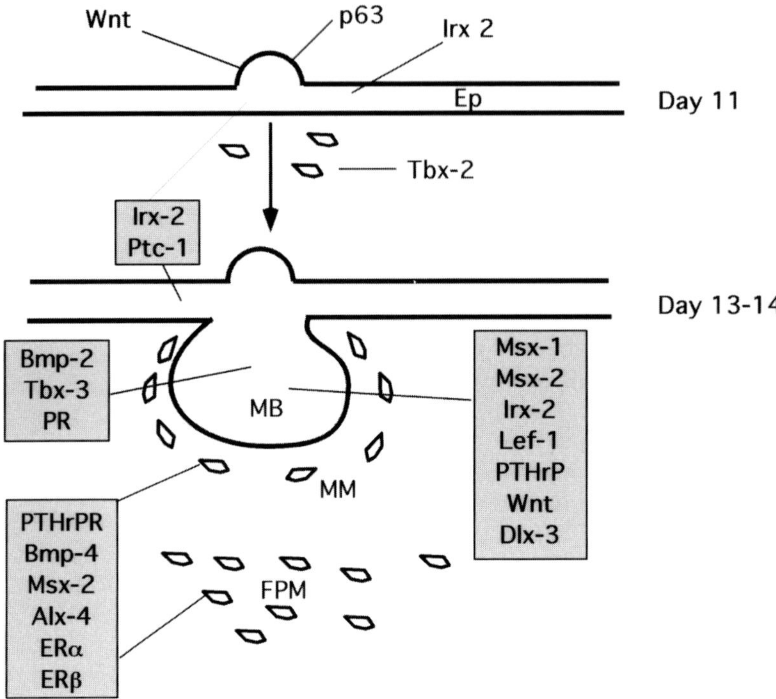

Fig. 4. Genes that are implicated in regulation of embryonic mammary gland development. Ep, epidermis; MB, mammary bud; MM, mammary mesenchyme; FPM, fat-pad mesenchyme. This figure was adapted and reprinted with permission from CW Daniel and Plenum Publishing *(63)*.

Alx-4 , a paired like (*PL*) homeodomain protein, is expressed in the mesenchyme directly surrounding the ductal epithelium and TEBs in 5-wk-old virgin mice *(53)*. Deletion of LEF-1 may also perturb *Alx-4* expression, resulting in deregulated control of proliferation. Mice lacking *Alx-4* exhibit normal mammary-gland development, possibly caused by compensation by other *PL* family members *(53)*.

In humans, inherited dominant mutations of the TBX-3, a member of the T-box (*tbx*) family of transcription factors, results in ulnar-mammary syndrome, in which the breasts are severely underdeveloped or lacking *(54)*. The expression of some *Tbx* family members has been localized to the mammary anlage during mouse embryonic development *(55)*.

Two genes related to the muscle segment homeobox (*msh*) gene, *Msx-1* and *Msx-2*, have also been implicated in epithelial-mesenchymal interactions *(56)*. *Msx-1* is expressed in the epithelial bud, whereas *Msx-2* is expressed in the epithelium and mesenchyme of the embryonic mammary gland *(50)*. Deletion of *Msx-1* does not alter mammary-gland development, as the newborn mammary gland is unaltered *(50)*. However, Msx-2 expression may be sufficient to compensate.

Parathyroid-hormone related protein (PTHrP) and its receptor, first implicated in hypercalcemia that develops in many types of cancer *(57)*, are also required for development of the embryonic mammary gland *(58)*. PTHrP mRNA is expressed in the mammary epithelial bud beginning on E 12 and is expressed in developing ducts during ductal morphogenesis *(59)*. The PTHrP receptor is expressed in the dense mammary mesenchyme from E 12 and in the fat-pad precursor cells from E 18, adjacent to developing ducts *(59)*. During bone development, PTHrP appears to be downstream of the BMPs *(60)*, as are several other targets of mammary embryogenesis. Mice lacking PTHrP develop mammary buds, but the mammary anlage fails to undergo branching morphogenesis. The epithelial cells degenerate by birth, resulting in a fat pad devoid of epithelium. Ductal morphogenesis has been partially rescued by crossing K14-PTHrP transgenic mice with the null mice. However, the ducts penetrated very slowly compared to normal litter mates, and were devoid of any fine side branching *(61)*.

The mechanisms of positioning and patterning of the multiple pairs of mammary glands in rodents is unknown. During embryonic development, homologs of *Drosophila* homeotic genes may participate in the simultaneous establishment of the pairs of mammary buds along the milk streak. Particular members of the Iroquois (*Irq*) and distal-less (*Dlx*) homeodomain-containing gene families are expressed in the developing mammary gland *(62,63)*. The Iroquois genes (*Irx1-3*) have been implicated in the development of the mouse nervous system, ear, heart, and limbs *(64)*. *Irx-2* is expressed in the mammary epithelium of the developing mouse mammary gland, but its role in development is unknown. However, recent experiments have determined that the human homolog, IRX-2 continues to be expressed during postnatal development within the epithelium of the terminal ductal lobular unit (TDLU) of the mature human breast *(65)*. Several members of the *Dlx* gene family are expressed in the developing mouse *(66,67)*. Unexpectedly, *Dlx-3* expression was localized to the developing mammary glands of transgenic mice expressing *lacZ* under the control of the *Xenopus Dlx-3* promoter. Mice lacking *Dlx-3* die at E 9.5 to E 10 of development, preventing analysis of mammary development in this model *(68)*.

Ductal Morphogenesis

To date, a majority of transgenic and gene-deleted mice that effect ductal morphogenesis exhibit retarded or delayed morphogenesis compared to their wild-type litter mates. Several of the genes implicated are growth factors or their corresponding receptors. This is not surprising, because the ductal epithelium must proliferate in order to penetrate the fat pad.

Steroid Hormones

Classic hormonal ablation experiments have demonstrated that the steroid hormones are required for ductal morphogenesis. Ovariectomy results in regression of the ductal epithelium, but outgrowth can be restored in the TEBs following administration of estrogen *(69)*. Administration of estrogen and progesterone induces DNA synthesis in both TEBs and the ductal epithelium *(70)*. Proliferation of the ductal epithelium is preceded by one round of proliferation in the mammary stroma, leading to a hypothesis as early as 1984 proposing that the mammary fat pad is the site of initial estrogen action *(71)*.

Estrogenic compounds can activate either of two estrogen receptors (ER), the classical ERα and the more recently discovered ERβ. ERβ shares 95% and 55% percent homology with the DNA-binding and ligand-binding domains of ERα, respectively *(72)*. Both ERα and ERβ mRNAs are expressed in the dense mammary mesenchyme surrounding the mammary bud beginning at d E 12.5 *(73)*. To address the contribution of ERα to reproductive development, ERα –/– (αERKO) mice have been created *(74)*. The αERKO mice are viable, but do not ovulate or form copora lutea, significantly decreasing the amount of circulating progesterone *(75)*. Confirming hormonal ablation studies, deletion of ERα (αERKO) impairs ductal morphogenesis *(74)*. At 4 mo of age, the ductal network present in αERKO females is severely compromised in comparison to wild-type littermates, and no TEBs are present.

Chimeric reconstitutions of neonatal wild-type and αERKO mammary epithelium and stroma grafted under the kidney capsule of ovary-intact mice have demonstrated that extensive ductal outgrowth occurs only in the presence of wild-type stroma *(76)*. These results indicate that stromal ERα regulates expression of factors that act in a paracrine fashion to stimulate proliferation of the ductal epithelium. Addition of exogenous steroid hormones or prolactin partially restores ductal development and induces lobuloalveolar development in αERKO mice *(75)*. Upon treatment for 21 d with estrogen and progesterone, ductal morphogenesis and alveolar development are partially rescued in ovariectomized (OVX) αERKO mice, suggesting that ERβ may be able to compensate for lack of ERα.

Partial rescue of the αERKO phenotype is also evident in pituitary-isografted-intact αERKO mice *(75)*. Pituitary isografts have induced high levels of circulating prolactin and progesterone in ovary-intact wild-type and αERKO mice *(75)*. Pituitary isografting of OVX mice, however, did not rescue ductal outgrowth, presumably because prolactin partially functions to stimulate ovarian production of progesterone *(77)*, which must synergize with estrogen to induce ductal morphogenesis.

In contrast to the distinct defects observed in the nonhormone-stimulated αERKO mouse, deletion of ERβ does not significantly impair ductal outgrowth, lobuloalveolar development, or lactation, implicating ERα as the primary mediator of estrogenic action in the mammary gland *(75)*. Further experiments will be required to determine a specific role of ERα vs ERβ in the mammary gland. It will be particularly important to perform the hormonal "rescue" experiments for αERKO/βERKO double knockout mice to determine whether estrogen is truly required for ductal morphogeneis and/or lobuloalveolar development. Although progesterone acts synergistically with estrogen to induce ductal outgrowth, deletion of the progesterone receptor (PR) does not impair ductal outgrowth *(78)*.

Local Growth Factors that Induce Ductal Morphogenesis

Epidermal growth factor (EGF) also stimulates DNA synthesis in OVX nulliparous mice *(79)*. Implantation of pellets containing EGF into the mammary glands of OVX mice induces localized expression of the of EGF receptor (EGFR) in the stroma immediately surrounding the ducts and TEBs, and in the TEB cap cells *(31)*. Similar to the phenotype of the αERKO mouse, EGFR –/– epithelial-stromal reconstitution experiments have indicated that the presence of EGFR in the stroma is required to induce ductal outgrowth *(80)*. Recent studies have demonstrated that neutralizing antibodies to EGF prevents proliferation of TEBs following ovariectomy and that similar results are obtained when the mammary glands containing antiestrogens are treated with EGF *(81)*. These results have led to the hypothesis that EGF mediates its effects on TEB proliferation partially through the ER *(81)*. In addition, estrogen treatment induces expression of EGFR in the mammary epithelium of sexually mature mice *(82)*. Since EGF- and estrogen-stimulated cell-signaling pathways appear to be closely intertwined, it is not surprising that deletion of either receptor results in retarded ductal morphogenesis.

In addition to EGF, other members of the EGF family of growth factors control ductal morphogenesis. For example, overexpression of either amphiregulin or heregulin (*neu* differentiation factor, NDF) stimulates aberrant proliferation of the ductal epithelium *(83,84)*. Similar to EGF, slow-release pellets containing AR induce ductal outgrowth in OVX mice *(83)*, and MEC manipulated to overexpresss amphiregulin form lobular hyperplasias in a transplant model *(83)*. Mammary glands of sexually mature transgenic mice that overexpress NDF under control of MMTV promoter aberrantly contain TEBs at sexual maturity, which normally disappear when ductal outgrowth is complete. Multiply bred MMTV-NDF mice form adenocarcinomas at approx 12 mo of age *(84)*. Similarly, 100% of all whey acidic protein transforming growth factor α (WAP-TGFα) transgenic mice develop mammary adenomas, and nulliparous females exhibit precocious development *(85)*.

In contrast, mice carrying single deletions of TGFα *(86)* or EGF *(87)* exhibit no gross defects in mammary development. Since mice carrying single deletions of either amphiregulin, EGF, or TGFα are fertile, it is possible to assay to mammary development of double or triple knockouts. In contrast to EGF and TGFα, deletion of amphiregulin impaired ductal outgrowth and development and approx 20% of the fat pad contained epithelium at 12 wk of age, with TEBs still present *(87)*. However, within TEBs, there were no differences in the levels of DNA synthesis and apoptosis compared to normal litter mates *(87)*; therefore, the defect in ductal elongation cannot be explained by defects in the TEB cell population. Deletion of all three ligands resulted in further impaired ductal development and a failure to complete ductal morphogenesis and fully differentiate during pregnancy—conditions that led to poor survival rates for pups *(87)*. In contrast, EGF/TGFα double-knockout mice completed lobuloalveolar development, and were capable of supporting their litters. Therefore, amphiregulin is an important regulator of ductal outgrowth, although it acts via an unknown mechanism. In single EGF or TGFα knockout mice, the overlapping expression of other EGF family members must be sufficient to compensate for their roles in development *(88)*.

Local Growth Factors that Inhibit Ductal Morphogenesis

Members of the transforming growth factor (TGF)-β superfamily are also implicated in ductal morphogenesis. Ductal elongation is locally impaired in developing mammary glands implanted with Evac pellets that contain either of the three TGF-β ligands (TGF-β1-3) *(9,89)*. In contrast, the formation of alveoli during pregnancy is not compromised *(9)*. Therefore, the TGF-β ligands function as negative regulators of ductal morphogenesis. Transgenic mice expressing TGFβ-1 under control of the MMTV promoter exhibit impaired ductal outgrowth and significantly fewer lateral branches than wild-type litter mates, but alveolar development is unaffected *(90)*, consistent with the results from the Evac pellet studies. In contrast, WAP-TGF-β1 transgenic mice failed to support their litters as a result of limited alveolar development during pregnancy *(91)*. The timing of transgene expression exerted by the MMTV vs WAP promoter may contribute to this apparent discrepancy. The expression of a dominant-negative form of the TGF-β receptor type II (DNIIR) under control of MMTV promoter results in premature development of lateral branching and alveoli capable of producing β-casein in the nulliparous mammary gland *(92)*. The overexpression of DNIIR in the mammary stroma also induces increased lateral branching *(93)*. In contrast, another member of the TGF-β superfamily positively induces ductal development. Reciprocal transplantation experiments have confirmed that the presence of inhibinβB is required in the stroma to support normal ductal morphogenesis and lobuloalveolar development *(17)*.

Other Factors

Premature lateral branching and functional differentiation of the epithelium is also observed in the mammary glands of transgenic mice that overexpress the matrix metalloproteinases (MMPs) stromelysin-1 or matrilysin-1, implicating that an intact basement membrane contributes to suppression of lobuloalveolar development in virgin mice *(7,94)*. Therefore, controlled ductal elongation and lateral branching are significantly influenced by the local concentrations of growth factors and their corresponding receptors and extracellular matrix molecules.

In contrast to these models describing retarded or accelerated growth of the ductal epithelium, deletion of the transcription factor CCAAT/enhancer-binding protein(C/EBP)β dramatically increases intraductal spacing, characterized by a severe bloating of the ducts via an unknown mechanism *(18,95)*. The C/EBPβ –/– mice also exhibit decreased lateral branching; however, there are no apparent defects in the rate of ductal outgrowth compared to wild-type litter mates. Decreased levels of lateral branching in virgin mammary glands have also been reported for several other gene-deleted mice, including the progesterone receptor (PR), prolactin, and prolactin receptor (PrlR) mouse models *(96–98)*. Since each of these molecules contributes to lateral branching and lobuloalveolar development during pregnancy, these observations are consistent. However, the principal defect resulting from deletion of each of these genes is a failure of alveoli to develop during pregnancy.

Lobuloalveolar Development, Proliferation, and Differentiation in Response to Pregnancy

Steroid Receptors and Prolactin

For several years, estrogen, progesterone, and prolactin have been implicated as mediators of alveolar development in the mammary gland *(99)*. The minimal hormones required to induce lobuloalveolar development in OVX mice are estrogen and progesterone. Addition of all three hormones synergistically enhances lobuloalveolar development in this model *(100)*. As discussed, exogenous administration of progesterone and prolactin appear to partially rescue lobuloalveolar development in αERKO mice *(75)*. In contrast, deletion of either PrlR or PR completely inhibits alveolar development and ductal lateral branching, but does not severely compromise the process of ductal outgrowth and bifurcation *(96–98)*.

Mature nulliparous females lacking prolactin do not exhibit lateral branching as extensive as their wild-type litter mates *(101)*. The persistence of TEB in animals as old as 5 mo of age is also unusual *(97)*. In addition, the females are sterile *(101)*. However, following mating progesteron-implanted PRL–/– mice exhibited extensive lobulo–alveolar development by d 18 of pregnancy *(101a)*.

It is likely that prolactin is absolutely essential for alveolar development since prolactin signals through the prolactin receptor and transplanted epithelium lacking the prolactin receptor does not develop into lobuloalveolar structures in response to a full-term pregnancy *(98)*. In fact, even prolactin-receptor +/– mice do not develop enough alveoli to support their litters of pups at first lactation *(102)*, implying that a threshold of prolactin-receptor expression is required to induce alveolar morphogenesis. At lactation, the alveoli of prolactin-receptor +/– mice are collapsed, and do not synthesize lipids and proteinaceous material associated with milk production *(98)*. During a second pregnancy, most prolactin-receptor +/– mice are capable of nursing their litters *(102)*. This type of compensation is typical of many knockout mouse models that are fertile (like the prolactin-receptor +/– mouse), and can be bred multiple times. Transplantation of prolactin-receptor –/– epithelium into the cleared fat pads of wild-type hosts has confirmed that the prolactin receptor is required for lateral branching and development of alveoli—absolutely no alveoli formed by d 1 of lactation *(98)*.

In the mammary gland, prolactin-mediated signaling induces signal transducers and activators of transcription (STAT)5a and STAT5b tyrosine phosphorylation and translocation to the nucleus through activation of the JAK (Janus kinase)-STAT pathway *(103)*. Unlike the prolactin-receptor –/– mouse, mice lacking STAT5a are fertile, and exhibit a moderate impairment of alveolar development that prevents successful lactation during the first pregnancy *(103)*. However, deletion of STAT5a does not significantly affect expression of β-casein, and only slightly reduces WAP expression, a likely result of compensation by STAT5b *(103)*. In contrast, fertility is compromised in mice lacking STAT5b *–/– (104)*, but the mice which do become pregnant do not display any gross defects in mammary gland morphology *(29)*. Analysis of development of mammary epithelium isolated from mice lacking both STAT5a and STAT5b, rescued by transplantation into normal hosts, has indicated that deletion of both STAT5 genes results in a phenotype very similar , but not identical, to the prolactin-receptor –/– mammary gland (K. Miyoshi and L. Hennighausen, personal communication).

Similar to the prolactin receptor, deletion of PR (PRKO) totally inhibits alveolar development *(96)*. In contrast to ERα –/– mice, the defect in PR-mediated development has been localized to the mammary epithelium *(105)*. Elegant PR +/+ and PRKO-*lacZ*-tagged MEC reconstitution experiments have demonstrated that PR acts via a paracrine mechanism to induce alveolar proliferation *(105)*. Alveolar development can be rescued if PRKO MEC mixed with PR +/+ MEC are reconstituted in close proximity within the cleared fat pads of syngeneic hosts (105). Recombination of PR +/+ stroma and PR –/– epithelium indicates that the stroma does not play a critical role in alveolar morphogenesis, further emphasizing the importance of epithelial-epithelial paracrine interactions, rather than epithelial-stromal interactions, in PR action *(105)*. These results support previous studies which suggested by immunohistochemistry that PR was expressed exclusively in the mammary epithelium *(106)*. These studies do not rule out the possibility that progesterone may play some role in the mammary stroma *(107)*.

Cell-Cycle Regulators and Transcription Factors

Cyclin D1 expression is controlled by extracellular mitogens, increasing and decreasing in response to the presence of growth factors *(108)*. In the mammary gland, synergistic induction of cyclin D1 expression is observed in mice treated with estrogen and progesterone *(109)*. The deletion of PR eliminates this synergism, resulting in decreased cyclin D1 expression, providing one mechanism of impaired development in the PRKO mouse *(109)*. Mice lacking cyclin D1 are incapable of nursing their pups, and have relatively few alveoli compared to wild-type litter mates *(110,111)*. When analyzed by transplantation in pregnant hosts, the defect in the proliferation of alveoli was localized to the epithelium *(112)*. The mammary phenotype is completely rescued if cyclin E is expressed under control of the endogenous cyclin D1 promoter, implying that cyclin E is directly downstream of cyclin D1 *(113)*.

The transcription factor C/EBPβ is also required for development of alveoli in response to estrogen and progesterone *(18,95)*. Reminiscent of the PRKO and prolactin-receptor

–/– mice, mice lacking C/EBPβ are infertile *(114)* and C/EBPβ, like PR, acts in an epithelial-cell autonomous manner *(18,95)*. Coupled with the marked inhibition of lobuloalveolar development, a transient decrease in proliferation of C/EBPβ –/– epithelium transplanted into the cleared fat pads of C/EBPβ +/+ mice at d 6 and 16 of pregnancy was previously reported *(18)*. Based on these observations, the expression and localization of PR and its relationship to proliferation was determined in C/EBPβ –/– mice. Unexpectedly, PR mRNA and protein were upregulated threefold compared to wild-type litter mates *(114a)*. In addition, the cellular distribution of PR was altered from the normal punctate pattern in wild-type mice to a more uniform pattern in mature C/EBPβ –/– mice. The aberrant localization and expression of PR correlated with a dramatic 10-fold decrease in alveolar proliferation in response to treatment with exogenous estrogen and progesterone. Preliminary analysis has also indicated that like PR, expression and cellular localization of prolactin-receptor may be increased/disrupted in C/EBPβ –/– mice, implicating C/EBPβ as an important mediator of alveolar progenitor-cell fate decisions *(114a)*.

Control of Lactation

In most transgenic and knockout models analyzed to date, the failure of a mother to successfully nurse her pups has been directly related to the extent of alveolar development and differentiation of the mammary epithelium observed during pregnancy. However, there are at least five genes which, when deleted, result in impaired secretion of milk from otherwise completely developed alveoli: colony-stimulating factor-1 (CSF-1), oxytocin (OT), relaxin, the receptor-like tyrosine phosphatase LAR, and the winged helix/forkhead transcription factor *Mf3*. At lactation, oxytocin stimulates contraction of the myoepithelial cells in order to express milk through the nipple. The deletion of oxytocin impairs ejection of milk from alveoli, resulting in premature, milk stasis-induced apoptosis *(115)*. The phenotype of mice lacking Mf3 is very similar to that observed for OT, and injections of OT will restore secretion of milk *(116)*.

The LAR receptor-like tyrosine phosphatase is an integral membrane protein that contains an intracellular phosphatase domain. The extracellular ligands that activate LAR are unknown. However, it is known that LAR may associate with focal adhesion complexes through its interactions with LAR-interacting protein (LIP)1 *(117)*. During pregnancy, LAR mRNA is expressed at higher levels than other protein tyrosine phosphatases (PTP), such as PTPs, and expression of LAR increases dramatically d 9 to d 16 of pregnancy *(118)*. The number and size of alveoli present in LAR –/– are only slightly reduced compared to LAR +/+ mice, but the alveoli that do form in LAR –/– mice fail to exhibit secretory activity, although the alveolar lumens are not collapsed *(118)*. The ability of LAR to be sequestered to focal adhesion complexes suggests that disruption of LAR may interfere with extracellular to nuclear-matrix signals proposed to be important in the regulation of milk-protein gene expression *(119)*. Deletion of relaxin, implicated in nipple development and myoepithelial-cell contraction *(120)*, results in a phenotype very similar to lack of LAR, except that relaxin –/– mice exhibit dilated ducts and decreased nipple size *(121)*.

CSF-1 is a growth factor circulated in the serum that is produced by the uterus at high levels during pregnancy *(122)*. In contrast to most knockout mouse models, in which decreased development is associated with a failure to lactate, mice lacking CSF-1 (*op/op*) display significantly increased numbers of alveoli at lactation compared to wild-type litter mates, but fail to lactate *(123)*. The CSF-1 –/– mice are capable of producing normal levels of milk-protein genes, but the alveoli remain collapsed and the products are not secreted; therefore, the mothers are not able to support their litters *(123)*. Exogenous administration of CSF-1 to mice only partially rescues the lactation defect, most likely caused by the inability to induce circulating levels as high as those produced by the uterus during a normal pregnancy *(123)*.

Regression and Remodeling During Involution

The Role of Glucocorticoids

Serum levels of glucocorticoids remain high throughout lactation, but decrease from 34 ng/L to 14 ng/L by d 3 of involution *(124)*. It has been established for several years that administration of exogenous hydrocortisone to lactating rats inhibits regression of the mammary epithelium *(125)*. The tissue-remodeling phase, but not the apoptosis phase, of involution can be delayed for up to 10 d by injecting mice daily with glucocorticoids *(43)*. These observations are consistent with induction of several tissue-remodeling proteinases following glucocorticoid withdrawal, including stromelysin-1 *(124)*, urokinase-like plasminogen activator (uPA), and gelatinase A *(43)*. Mammary glands of transgenic mice overexpressing stromelysin-1 undergo premature apoptosis during pregnancy *(7)*, suggesting that the glucocorticoid-mediated suppression of apoptosis during pregnancy and lactation can be overcome by overexpression of factors that mediate the tissue-remodeling phase of involution.

Cell Death Inducers and Inhibitors

During the first phase of involution, expression of several factors associated with cell death are induced, including p53, bax, and bcl-x-$_{short}$ *(42,44)*. The deletion of p53 in Balb/C inbred mice results in delayed involution and decreased rates of apoptosis *(126)*, but these effects may be strain-specific, because lack of p53 does not significantly impair involution or apoptosis in C57/Bl6 × 129 females *(127)*. Individual members of the bcl-2 family may either enhance or inhibit apoptosis (reviewed in ref. *128*). The ratios of pro- to anti-apoptotic factors has been demonstrated to be critical to modulate apoptosis *(129)*. In particular, bcl-2 is anti-apoptotic, whereas bax and bcl-x-$_{short}$ promote cell death *(128)*. The mammary glands of mice lacking bcl-2 develop normally *(130)*, but, as expected, overexpression of bcl-2 under control of the WAP promoter delays apoptosis during involution *(131)*. In addition, a decrease in the percentage of apoptotic cells in the TEB has been reported in juvenile WAP-bcl-2 transgenic mice *(33)*. The effects of Cre-mediated deletion of bcl-x in the mammary gland resulted in accelerated apoptosis during involution *(131a)*.

Transcription Factors

The Cre-*lox*-mediated deletion of STAT3 in the mammary gland also results in delayed involution *(132a)*, consistent with previous observations of enhanced STAT3 phosphorylation in sealed, but not in unsealed mammary glands *(44)*. The factors downstream of STAT3 are unknown but IGFBP-5 expression is markedly reduced during involution in the mammary gland of the STAT3 conditional knockout mice *(132a)*. However, expression of the transcription factor C/EBPδ, which is dramatically upregulated within 24 h of involution *(132)*, is reported to be directly regulated by STAT3 following stimulation by interleukin-6 (IL-6) *(133)*. Mice lacking C/EBPδ have been created, and are viable and fertile *(134)*, but analysis of development of the mammary gland has not been reported.

Growth Factors

Overexpression of several growth factors delays or impairs involution, including TGF-α and IGF-I *(85,135–137)*. These observations are not surprising, because IGF-I acts as a cell-survival factor in mammary epithelial cells *(138,139)*. In fact, multiply bred WAP-IGF-I transgenic mice eventually develop mammary tumors, presumably because fewer cells enter apoptosis, allowing accumulation of spontaneous mutations acquired during multiple rounds of proliferation during pregnancy *(139a)*. Tumors have also been observed in transgenic mice overexpressing IGF-II after multiple pregnancies *(140)*.

The IGF binding proteins (IGFBPs) may either enhance or attenuate the mitogenic effects of IGF-I or IGF-II *(141)*. Two IGFBPs—IGFBP-3 and IGFBP-5—have been implicated in regulation of involution. The level of IGFBP-5 secreted in milk increases 50-fold during forced involution, and increases 5– to 10-fold specifically in sealed teats compared to open teats *(142)*. WAP-driven overexpression of IGFBP-3 delays involution, presumably because of the augmentation of IGF-I biological activity, although the delay is not as severe as observed in WAP-IGF-I mice *(136)*.

These observations may correlate with the expression of IGFBP-5 and IGFBP-3 during normal virgin development. In the TEB of the juvenile mammary gland, relatively high levels of both IGFBP-5 and IGFBP-3 mRNA are expressed *(143)*. The expression of IGFBP-5 mRNA is uniform in all cell layers, whereas expression of IGFBP-3 mRNA appears to be restricted to the outer cell layers that may be coincident with the proliferative cap-cell layer *(143)*. IGFBP-5 may associate with the extracellular matrix *(144)*, presumably to regulate IGF-I biological activity over a short range. Secreted IGFBP-5 in milk may have activities distinct from the IGFBP-5 bound to the extracellular matrix, since bound IGFBP-5 is reported to enhance IGF-I mitogenic activity *(144)*, whereas expression of IGFBP-5 in milk is correlated with induction of apoptosis during involution *(142)*. Alternatively, the effects of IGFBP-5 may be tissue-specific. It is possible that the IGFBP-5 present in the TEB inhibits IGF-I action, as during involution, in order to determine which cell layers proliferate. If the restricted expression pattern of IGFBP-3 mRNA in the TEB does mediate proliferation of the cap-cell layers, the observation of delayed involution in WAP-IGFBP-3 mice would be consistent with this model.

In general, deletion of individual genes does not significantly alter involution. In contrast, overexpression of growth factor and anti-apoptotic genes is sufficient to delay—but not halt—involution. Interestingly, some of the same genes that are implicated as factors regulating virgin ductal morphogenesis in TEBs—such as IGFBP-5, IGFBP-3, and bcl-2—also appear to be important for the induction of apoptosis during involution.

Summary of Knockout and Transgenic Models

In summary, the mouse models described here have confirmed the results derived from classical endocrinology experiments demonstrating that hormones control mammary-gland development and differentiation. In addition, they have also provided clues

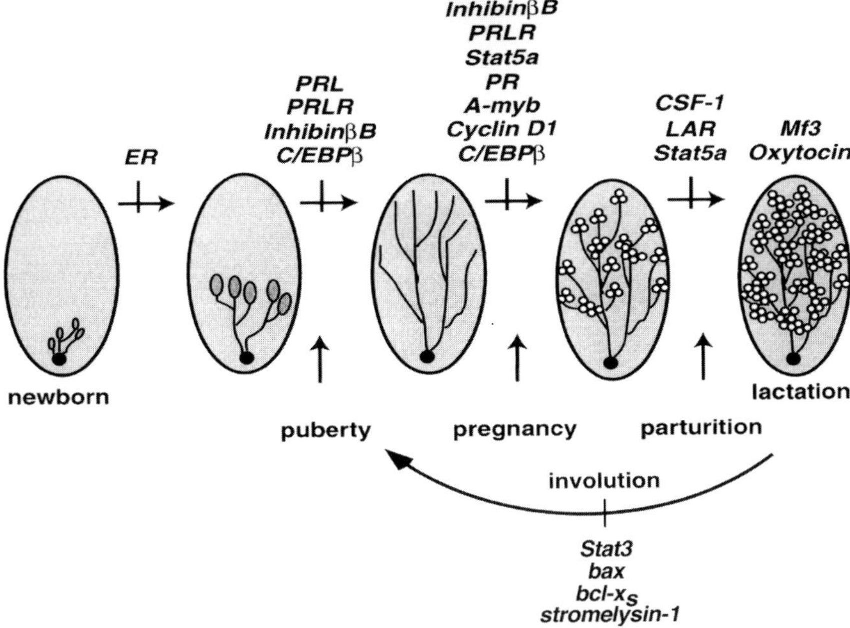

Fig. 5. A partial list of genes regulating the progression of mouse mammary-gland development, differentiation, lactation, and remodeling. The large arrow indicates regression of epithelial cells during involution to a more virgin-like state in which the mammary gland contains primarily ductal epithelium. This figure was adapted and reprinted with permission from Cold Spring Harbor Laboratory Press *(29)*.

to the some of the cell-signaling pathways that mediate mammary-gland development. A partial list of some the factors controlling postnatal mammary-gland development is presented in Fig. 5.

This is an exciting time in mammary-gland biology. We now have the techniques required to answer critical and mechanistic questions. The many technical advantages that the mammary gland offers in combination with the sophisticated genetic manipulations available in the field of mammary gland research should provide an impetus for more investigators to pursue questions of development and cell-signaling pathways in this model system.

The authors apologize to those investigators whose recent transgenic and knockout models were omitted from this review. The current address for T.N.S. is the Department of Biology, University of California, San Diego; tseagroves@ucsd.edu.

REFERENCES

1. Love S, Parker B, Ames M, Taylor C, Figlin RA. Practice guidelines for breast cancer. Cancer J Scient Am 1996;2:S7–S21.
2. Parker SL, Tong T, Bolden S, Wing PA. Cancer Statistics. CA-A Cancer J Clinic 1997;47:5–51.
3. Cardiff RD, Wellings SR. The comparative pathology of human and mouse mammary glands. J Mammary Gland Biol Neoplasia 1999;4:105–122.
4. Medina D, Daniel C, eds. Experimental models of development, function, and neoplasia. J Mammary Gland Biol Neoplasia 1996;1:1–136.

5. Hennighausen L, Westphal C, Sankaran L, Pittius CW. Regulation of expression of genes for milk proteins. Biotechnology 1991;16:65–74.
6. Mok E, Golovkina TV, Ross SR. A mouse mammary tumor virus mammary gland enhancer confers tissue-specific but not lactation-dependent expression in transgenic mice. J Virol 1992;66:7529–7532.
7. Sympson CJ, Talhouk RS, Alexander CM, et al. Targeted expression of stromelysin-1 in mammary gland provides evidence for a role of proteinases in branching morphogenesis and the requirement for an intact basement membrane for tissue-specific gene expression. J Cell Biol 1994;125:681–693.
8. Robinson GW, McKnight RA, Smith GH, Hennighausen L. Mammary epithelial cells undergo secretory differentiation in cycling virgins but require pregnancy for the establishment of terminal differentiation. Development 1995;121:2079–2090.
9. Daniel CW, Silberstein GB, Van Horn K, Strickland P, Robinson S. TGF-beta 1-induced inhibition of mouse mammary ductal growth: developmental specificity and characterization. Dev Biol 1989;135:20–30.
10. Smith GH, Medina D. A morphologically distinct candidate for an epithelial stem cell in mouse mammary gland. J Cell Sci 1988;90:173–183.
11. Kordon EC, Smith GH. An entire functional mammary gland may comprise the progeny from a single cell. Development 1998;125:1921–1930.
12. Chepko G, Smith GH. Mammary epithelial stem cells: our current understanding. J Mammary Gland Biol Neoplasia 1999;4:35–52.
13. DeOme KB, Fauklin LJ, Bern HA, Blair PB. Development of mammary tumors from hyperplastic alveolar nodules transplanted into gland-free mammary fat pads of female C3H mice. Cancer Res 1959;78:515–520.
14. Daniel CW, DeOme KB, Young JT, Blair PB, Fauklin LJ. The in vivo life span of normal and preneoplastic mouse mammary glands: a serial transplantation study. Proc Natl Acad Sci USA 1968;61:53–60.
15. Sheffield LG, Welsch CW. Transplantation of human breast epithelia to mammary-gland-free fat-pads of athymic nude mice: influence of mammotrophic hormones on growth of breast epithelia. Int J Cancer 1988;41:713–719.
16. Edwards PAW, Abram CL, Bradbury JM. Genetic manipulation of mammary epithelium by transplantation. J Mammary Gland Biol Neoplasia 1996;1:75–90.
17. Robinson GW, Hennighausen L. Inhibins and activins regulate mammary epithelial cell differentiation through mesenchymal-epithelial interactions. Development 1997;124:2701–2708.
18. Robinson GW, Johnson PF, Hennighausen L, Sterneck E. The C/EBPbeta transcription factor regulates epithelial cell proliferation and differentiation in the mammary gland. Genes Dev 1998;12:1907–1916.
19. Cunha GR, Young P, Christov K, et al. Mammary phenotypic expression is induced in epidermal cells by embryonic mammary mesenchyme. Acta Anat 1995;152:195–204.
20. Ip M, Darcy KM. Three-dimensional mammary primary culture systems. J Mammary Gland Biol Neoplasia 1996;1:91–110.
21. Pullan SE, Streuli CH. The mammary epithelial cell. In: Harris A, ed. Epithelial cell culture. Cambridge, UK, Cambridge University Press, 1997, pp. 97–121.
22. Juergens WG, Stockdale FE, Topper YJ, Elias JJ. Hormone-dependent differentiation of mammary gland in vitro. Proc Natl Acad Sci USA 1965;54:629–634.
23. Sauer B. Inducible gene targeting in mice using the Cre/*lox* system. METHODS: a companion to Methods in Enzymology 1998;14:381–392.
24. Wagner KU, Wall RJ, St-Onge L, et al. Cre-mediated gene deletion in the mammary gland. Nucleic Acids Res 1997;25:4323–4330.
25. Selbert S, Bentley DJ, Melton DW, et al. Efficient BLG-Cre mediated gene deletion in the mammary gland. Transgenic Res 1998;7:387–396.
25a. Rijnkels M, Rosen JM. Adenovirus-Cre mediated recombination in mammary epithelial progenitor cells. J Cell Sci 2001; in press.
26. Sakakura T. Mammary embryogenesis. In: Neville MC, Daniel CW, eds. The Mammary Gland. Plenum Press,New York, NY, 1987, pp. 37–63.
27. Kratochwil K, Schwartz P. Tissue interaction in androgen response of embryonic mammary rudiment of mouse: identification of target tissue for testosterone. Proc Natl Acad Sci USA 1976;73:4041–4044.
28. Knight CH, Peaker M. Development of the mammary gland. J Reprod Fertil 1982;65:521–536.

29. Hennighausen L, Robinson GW. Think globally, act locally: the making of a mouse mammary gland. Genes Dev 1998;12:449–455.
30. Daniel Cw, Silberstein GB. Postnatal development of the rodent mammary gland. In: Neville MC, Daniel CW, eds. The Mammary Gland. Plenum Press, New York, NY, 1987, pp.3–36.
31. Coleman S, Silberstein GB, Daniel CW. Ductal morphogenesis in the mouse mammary gland: evidence supporting a role for epidermal growth factor. Dev Biol 1988;127:304–315.
32. Kleinberg DL. Role of IGF-I in normal mammary development. Breast Cancer Res Treat 1998;47:201–208.
33. Humphreys RC, Krajewska M, Krnacik S, et al. Apoptosis in the terminal endbud of the murine mammary gland: a mechanism of ductal morphogenesis. Development 1996;122:4013–4022.
34. Russo J, Russo IH. The etiopathogenesis of breast cancer prevention. Cancer Lett 1995;90:81–89.
35. Guzman RC, Yang J, Rajkumar L, Thordarson G, Chen X, Nandi S. Hormonal prevention of breast cancer: mimicking the protective effect of pregnancy. Proc Natl Acad Sci USA 1999;96:2520–2525.
36. Yang J, Yoshizawa K, Nandi S, Tsubura A. Protective effects of pregnancy and lactation against N-methyl-N- nitrosourea-induced mammary carcinomas in female Lewis rats. Carcinogenesis 1999;20:623–628.
37. Taylor-Papadimitriou J, Lane EB. Keratin expression in the mammary gland. In: Neville MC, Daniel CW, eds. The Mammary Gland. Plenum Press, New York, NY, 1987, pp. 181–215.
38. Borst DW, Mahoney WB. Mouse mammary gland DNA synthesis during pregnancy. J Exp Zool 1982;221:245–250.
39. Rosen JM, Wyszomierski SL, Hadsell D. Regulation of milk protein gene expression. Annu Rev Nutr 1999;19:407–436.
40. Munford RE. Changes in mammary glands of rats and mice during pregnancy, lactation and involution. J Endocrinol 1963;28:1–15.
41. Shipman LJ, Docherty AH, Knight CH, Wilde CJ. Metabolic adaptations in mouse mammary gland during a normal lactation cycle and in extended lactation. Q J Exp Physiol 1987;72:303–311.
42. Quarrie LH, Addey CV, Wilde CJ. Programmed cell death during mammary tissue involution induced by weaning, litter removal, and milk stasis. J Cell Physiol 1996;168:559–569.
43. Lund LR, Romer J, Thomasset N, et al. Two distinct phases of apoptosis in mammary gland involution: proteinase-independent and -dependent pathways. Development 1996;122:181–193.
44. Li M, Liu X, Robinson G, et al. Mammary-derived signals activate programmed cell death during the first stage of mammary gland involution. Proc Natl Acad Sci U S A 1997;94:3425–3430.
45. Mills AA, Zheng B, Wang XJ, Vogel H, Roop DR, Bradley A. p63 is a p53 homologue required for limb and epidermal morphogenesis. Nature 1999;398:708–713.
46. Yang A, Schweitzer R, Sun D, et al. p63 is essential for regenerative proliferation in limb, craniofacial and epithelial development. Nature 1999;398:714–718.
47. van Genderen C, Okamura RM, Farinas I, et al. Development of several organs that require inductive epithelial- mesenchymal interactions is impaired in LEF-1-deficient mice. Genes Dev 1994;8:2691–2703.
48. Travis A, Amsterdam A, Belanger C, Grosschedl R. LEF-1, a gene encoding a lymphoid-specific protein with an HMG domain, regulates T-cell receptor alpha enhancer function (published erratum appears in Genes Dev 1991 Jun;5[6]:following 1113). Genes Dev 1991;5:880–894.
49. Kratochwil K, Dull M, Farinas I, Galceran J, Grosschedl R. Lef1 expression is activated by BMP-4 and regulates inductive tissue interactions in tooth and hair development. Genes Dev 1996;10:1382–1394.
50. Phippard DJ, Weber-Hall SJ, Sharpe PT, et al. Regulation of Msx-1, Msx-2, Bmp-2 and Bmp-4 during foetal and postnatal mammary gland development. Development 1996;122:2729–2737.
51. Shtutman M, Zhurinsky J, Simcha I, et al. The cyclin D1 gene is a target of the beta-catenin/LEF-1 pathway. Proc Natl Acad Sci U S A 1999;96:5522–5527.
52. Wiggan O, Taniguchi-Sidle A, Hamel PA. Interaction of the pRB-family proteins with paired-like homeodomains. Oncogene 1998;16:227–236.
53. Hudson R, Taniguchi-Sidle A, Boras K, Wiggan O, Hamel PA. Alx-4, a transcriptional activator whose expression is restricted to sites of epithelial-mesenchymal interactions. Dev Dyn 1998;213:159–169.
54. Bamshad M, Root S, Carey JC. Clinical analysis of a large kindred with Pallister ulnar-mammary syndrome. Am J Med Genet 1996;65:325–331.
55. Chapman DL, Garvey N, Hancock S, et al. Expression of the T-box family of genes, *Tbx1-Tbx5*, during early mouse development. Dev Dyn 1996;206:379–390.
56. Foerst-Potts L, Sadler TW. Disruption of Msx-1 and Msx-2 reveals roles for these genes in craniofacial, eye, and axial development. Dev Dyn 1997;209:70–84.

57. Wysolmerski JJ, Broadus AE. Hypercalcemia of malignancy: the central role of parathyroid hormone-related protein. Annu Rev Med 1994;45:189–200.
58. Dunbar ME, Young P, Zhang JP, et al. Stromal cells are critical targets in the regulation of mammary ductal morphogenesis by parathyroid hormone-related protein. Dev Biol 1998;203:75–89.
59. Dunbar ME, Wysolmerski JJ. Parathyroid hormone-related protein: a developmental regulatory molecule necessary for mammary gland development. J Mammary Gland Biol Neoplasia 1999;4:21–34.
60. Zou H, Wieser R, Massague J, Nisswander L. Distinct roles of type I bone morphogenetic protein receptors in the formation of and differentiation of cartilage. Genes & Dev 1997;11:2191–2203.
61. Wysolmerski JJ, Philbrick WM, Dunbar ME, Lanske B, Kronenberg H, Broadus AE. Rescue of the parathyroid hormone-related protein knockout mouse demonstrates that parathyroid hormone-related protein is essential for mammary gland development. Development 1998;125:1285–1294.
62. Morasso MI, Mahon KA, Sargent TD. A Xenopus distal-less gene in transgenic mice: conserved regulation in distal limb epidermis and other sites of epithelial-mesenchymal interaction. Proc Natl Acad Sci USA 1995;92:3968–3972.
63. Daniel CW, Smith GH. The mammary gland: a model for development. J Mammary Gland Biol Neoplasia 1999;4:3–8.
64. Bosse A, Zulch A, Becker MB, et al. Identification of the vertebrate Iroquois homeobox gene family with overlapping expression during early development of the nervous system. Mech Dev 1997;69:169–181.
65. Lewis MT, Ross S, Strickland PA, Snyder CJ, Daniel CW. Regulated expression patterns of IRX-2, an Iroquois-class homeobox gene, in the human breast. Cell Tissue Res 1999;296:549–554.
66. Bulfone A, Kim HJ, Puelles L, Porteus MH, Grippo JF, Rubenstein JL. The mouse Dlx-2 (Tes-1) gene is expressed in spatially restricted domains of the forebrain, face and limbs in midgestation mouse embryos (published erratum appears in Mech Dev 1993 Aug;42[3]:187). Mech Dev 1993;40:129–140.
67. Papalopulu N, Kintner C. Xenopus Distal-less related homeobox genes are expressed in the developing forebrain and are induced by planar signals. Development 1993;117:961–975.
68. Morasso MI, Grinberg A, Robinson G, Sargent TD, Mahon KA. Placental failure in mice lacking the homeobox gene Dlx3. Proc Natl Acad Sci USA 1999;96:162–167.
69. Nandi S. Endocrine control of mammary gland development and function in the C3H/Crgl mouse. J Natl Cancer Inst 1958;21:1029–1063.
70. Bresciani F. Topography of DNA synthesis in the mammary gland of the C3H mouse and its control by ovarian hormones: an autoradiographic study. Cell Tissue Kinet 1968;1:51–63.
71. Shyamala G, Ferenczy A. Mammary fat pad may be a potential site for initiation of estrogen action in normal mouse mammary gland. Endocrinology 1984;115:1078–1081.
72. Kuiper GG, Enmark E, Pelto-Huikko M, Nilsson S, Gustafsson JA. Cloning of a novel receptor expressed in rat prostate and ovary. Proc Natl Acad Sci USA 1996;93:5925–5930.
73. Lemmen JG, Broekhof JL, Kuiper GG, Gustafsson JA, van der Saag PT, van der Burg B. Expression of estrogen receptor alpha and beta during mouse embryogenesis. Mech Dev 1999;81:163–167.
74. Lubahn DB, Moyer JS, Golding TS, Couse JF, Korach KS, Smithies O. Alteration of reproductive function but not prenatal sexual development after insertional disruption of the mouse estrogen receptor gene. Proc Natl Acad Sci USA 1993;90:11,162–11,166.
75. Couse JF, Korach KS. Estrogen receptor null mice: what have we learned and where will they lead us? Endocr Reviews 1999;20:358–417.
76. Cunha GR, Young P, Hom YK, Cooke PS, Taylor JA, Lubahn DB. Elucidation of a role of stromal steroid hormone receptors in mammary gland growth and development by tissue recombination experiments. J Mammary Gland Biol Neoplasia 1997;2:393–402.
77. Galosy S, Talamantes F. Luteotropic actions of placental lactogens at midpregnancy in the mouse. Endocrinology 1995;136:3993–4003.
78. Lydon JP, DeMayo FJ, Conneely OM, O'Malley BW. Reproductive phenotpes of the progesterone receptor null mutant mouse. J Steroid Biochem Mol Biol 1996;56:67–77.
79. Haslam SZ, Counterman LJ, Nummy KA. Effects of epidermal growth factor, estrogen, and progestin on DNA synthesis in mammary cells in vivo are determined by the developmental state of the gland. J Cell Physiol 1993;155:72–78.
80. Wiesen JF, Young P, Werb Z, Cunha GR. Signaling through the stromal epidermal growth factor receptor is necessary for mammary ductal development. Development 1999;126:335–344.
81. Ankrapp DP, Bennett JM, Haslam SZ. Role of epidermal growth factor in the acquisition of ovarian steroid hormone responsiveness in the normal mouse mammary gland. J Cell Physiol 1998;174:251–260.

82. Haslam SZ, Counterman LJ, Nummy KA. EGF receptor regulation in normal mouse mammary gland. J Cell Physiol 1992;152:553–557.
83. Kenney NJ, Smith GH, Rosenberg K, Cutler ML, Dickson RB. Induction of ductal morphogenesis and lobular hyperplasia by amphiregulin in the mouse mammary gland. Cell Growth Differ 1996;7:1769–1781.
84. Krane IM, Leder P. NDF/heregulin induces persistence of terminal end buds and adenocarcinomas in the mammary glands of transgenic mice. Oncogene 1996;12:1781–1788.
85. Sandgren EP, Schroeder JA, Qui TH, Palmiter RD, Brinster RL, Lee DC. Inhibition of mammary gland involution is associated with transforming growth factor alpha but not c-myc-induced tumorigenesis in transgenic mice. Cancer Res 1995;55:3915–3927.
86. Luetteke NC, Qiu TH, Peiffer RL, Oliver P, Smithies O, Lee DC. TGF alpha deficiency results in hair follicle and eye abnormalities in targeted and waved-1 mice. Cell 1993;73:263–278.
87. Luetteke NC, Qiu TH, Fenton SE, et al. Targeted inactivation of the EGF and amphiregulin genes reveals distinct roles for EGF receptor ligands in mouse mammary gland development. Development 1999;126:2739–2750.
88. Schroeder JA, Lee DC. Dynamic expression and activation of ERBB receptors in the developing mouse mammary gland. Cell Growth Differ 1998;9:451–464.
89. Robinson SD, Silberstein GB, Roberts AB, Flanders KC, Daniel CW. Regulated expression and growth inhibitory effects of transforming growth factor-β isoforms in mouse mammary development. Development 1991;113:867–878.
90. Pierce Jr DF, Johnson MD, Matsui Y, et al. Inhibition of mammary duct development but not alveolar outgrowth during pregnancy in transgenic mice expressing active TGF-β1. Genes Dev 1993;7:2308–2317.
91. Jhappan C, Geiser AG, Kordon EC, et al. Targeting expression of a transforming growth factor beta 1 transgene to the pregnant mammary gland inhibits alveolar development and lactation. EMBO J 1993;12:1835–1845.
92. Gorska AE, Joseph H, Derynck R, Moses HL, Serrra R. Dominant-negative interference of the transforming growth factor β type II receptor in mammary gland epithelium results in alveolar hyperplasia and differentiation in virgin mice. Cell Growth Differ 1998;9:220–239.
93. Gorska JH, Sohn P, Moses HL, Serra R. Overexpression of a kinase-deficient transforming growth factor-beta type II receptor in mouse mammary stroma results in increased epithelial branching. Mol Biol Cell 1999;10:1221–1234.
94. Rudolph-Owen LA, Cannon P, Matrisian LM. Overexpression of the matrix metalloproteinase matrilysin results in premature mammary gland differentiation and male infertility. Mol Biol Cell 1998;9:421–435.
95. Seagroves TN, Krnacik S, Raught B, et al. C/EBPbeta, but not C/EBPalpha, is essential for ductal morphogenesis, lobuloalveolar proliferation, and functional differentiation in the mouse mammary gland. Genes & Dev 1998;12:1917–1928.
96. Lydon JP, DeMayo FJ, Funk CR, et al. Mice lacking progesterone receptor exhibit pleiotropic reproductive abnormalities. Genes Dev 1995;9:2266–2278.
97. Horseman ND. Prolactin and mammary gland development. J Mammary Gland Biol Neoplasia 1999;4:79–88.
98. Brisken C, Kaur S, Chavarria TE, et al. Prolactin controls mammary gland development via direct and indirect mechanisms. Dev Biol 1999;210:96–106.
99. Imagawa W, Yang J, Guzman R, Nandi S. Control of mammary gland development. In: Knobil E, Neil JD, eds. The Physiology of Reproduction, 2nd ed. Raven Press, New York, NY, 1994, pp. 1033–1063.
100. Nandi S, Bern HA. Relation between mammary-gland responses to lactogenic hormone combinations and tumor susceptibility in various strains of mice. J Natl Cancer Inst 1960;24:907–931.
101. Horseman N, Zhao W, Montecino-Rodriquez E, et al. Defective mammopoiesis, but normal hemtaopoiesis in mice with targeted disruption of the prolactin gene. EMBO J 1997;16:101–110.
101a. Vomachka AJ, Pratt SL, Lockefeer JA, Horseman ND. Prolactin gene-disruption arrests mammary gland development and retards T-antigen induced tumor growth. Oncogene 2000;19:1077–1084.
102. Ormandy CJ, Camus A, Barra J, et al. Null mutation of the prolactin receptor gene produces multiple reproductive defects in the mouse. Genes Dev 1997;11:167–178.
103. Liu X, Robinson GW, Wagner KU, Garrett L, Wynshaw-Boris A, Hennighausen L. Stat5a is mandatory for adult mammary gland development and lactogenesis. Genes Dev 1997;11:179–186.
104. Udy GB, Towers RP, Snell RG, et al. Requirement of STAT5b for sexual dimorphism of body growth rates and liver gene expression. Proc Natl Acad Sci USA 1997;94:7239–7244.

105. Brisken C, Park S, Vass T, Lydon JP, O'Malley BW, Weinberg RA. A paracrine role for the epithelial progesterone receptor in mammary gland development. Proc Natl Acad Sci USA 1998;95:5076–5081.
106. Shyamala G, Barcellos-Hoff MH, Toft D, Yang X. In situ localization of progesterone receptors in normal mouse mammary glands: absence of receptors in the connective and adipose stroma and a heterogeneous distribution in the epithelium. J Steroid Biochem Mol Biol 1997;63:251–259.
107. Humphreys RC, Lydon J, O'Malley BW, Rosen JM. Mammary gland development is mediated by both stromal and epithelial progesterone receptors. Mol Endocrinol 1997;11:801–811.
108. Sherr CJ. Mammalian G1 cyclins. Cell 1993;73:1059–1065.
109. Said TK, Conneely OM, Medina D, O'Malley BW, Lydon JP. Progesterone, in addition to estrogen, induces cyclin D1 expression in the murine mammary epithelial cell, in vivo. Endocrinology 1997;138:3933–3939.
110. Fantl V, Stamp G, Andrews A, Rosewell I, Dickson C. Mice lacking cyclin D1 are small and show defects in eye and mammary gland development. Genes & Dev 1995;9:2364–2372.
111. Sicinski P, Donaher JL, Parker SB, et al. Cyclin D1 provides a link between development and oncogenesis in the retina and breast. Cell 1995;82:621–630.
112. Fantl V, Edwards PA, Steel JH, Vonderhaar BK, Dickson C. Impaired mammary gland development in Cyl-1(-/-) mice during pregnancy and lactation is epithelial cell autonomous. Dev Biol 1999;212:1–11.
113. Geng Y, Whoriskey W, Park MY, et al. Rescue of cyclin D1 deficiency by knockin cyclin E. Cell 1999;97:767–777.
114. Sterneck E, Tessarollo L, Johnson PF. An essential role for C/EBPbeta in female reproduction. Genes Dev 1997;11:2153–2162.
114a. Seagroves TN, Lydon JP, Hovey RC, Vonderhaar BK, Rosen JM. C/EBPβ controls cell fate determination during mammary gland development. Mol Endocrinol 2000;14:359–368.
115. Wagner K-U, Young III WS, Liu X, et al. Oxytocin and milk removal are required for post-partum mammary gland development. Genes Funct 1997;1:233–244.
116. Labosky PA, Winnier GE, Jetton TL, et al. The winged helix gene, Mf3, is required for normal development of the diencephalon and midbrain, postnatal growth and the milk-ejection reflex. Development 1997;124:1263–1274.
117. Serra-Pages C, Kedersha NL, Fazikas L, Medley Q, Debant A, Streuli M. The LAR transmembrane protein tyrosine phosphatase and a coiled-coil LAR-interacting protein co-localize at focal adhesions. EMBO J 1995;14:2827–2838.
118. Schaapveld RQ, Schepens JT, Robinson GW, et al. Impaired mammary gland development and function in mice lacking LAR receptor-like tyrosine phosphatase activity. Dev Biol 1997; 188:134–146.
119. Roskelley CD, Srebrow A, Bissell MJ. A hierarchy of ECM-mediated signalling regulates tissue-specific gene expression. Curr Opin Cell Biol 1995;7:736–747.
120. Hwang JJ, Lee AB, Fields PA, Haab LM, Mojonnier LE, Sherwood OD. Monoclonal antibodies specific for rat relaxin. Passive immunization with monoclonal antibodies throughout the second half of pregnancy disrupts birth in intact rats. Endocrinology 1991;131:3034–3042.
121. Zhao L, Roche PJ, Gunnersen JM, et al. Mice without a functional relaxin gene are unable to deliver milk to their pups. Endocrinology 1999;140:445–453.
122. Pollard JW. Role of colony-stimulating factor-1 in reproduction and development. Mol Reprod Dev 1997;46:54–60; discussion 60,611.
123. Pollard JW, Hennighausen L. Colony stimulating factor 1 is required for mammary gland development during pregnancy. Proc Natl Acad Sci USA 1994;91:9312–9316.
124. Feng Z, Marti A, Jehn B, Altermatt HJ, Chicaiza G, Jaggi R. Glucocorticoid and progesterone inhibit involution and programmed cell death in the mouse mammary gland. J Cell Biol 1995;131:1095–1103.
125. Johnson RM, Meites J. Effects of cortisone acetate on milk production and mammary involution in parturient rats. Endocrinology 1958;63:290–294.
126. Jerry DJ, Kuperwasser C, Downing SR, et al. Delayed involution of the mammary epithelium in BALB/c-p53 ^{null} mice. Oncogene 1998;17:2305–2312.
127. Li M, Hu J, Heermeier K, Hennighausen L, Furth PA. Apoptosis and remodeling of mammary gland tissue during involution proceeds through p53-independent pathways. Cell Growth & Differ 1996;7:13–20.
128. White E. Life, death, and the pursuit of apoptosis. Genes Dev 1996;10:1–5.
129. Knudson CM, Korsmeyer SJ. Bcl-2 and Bax function independently to regulate cell death. Nat Genet 1997;16:358–363.

130. Veis DJ, Sorenson CM, Shutter JR, Korsmeyer SJ. Bcl-2-deficient mice demonstrate fulminant lymphoid apoptosis, polycystic kidneys, and hypopigmented hair. Cell 1993;75:229–240.
131. Jager R, Herzer U, Schenkel J, Weiher H. Overexpression of Bcl-2 inhibits alveolar cell apoptosis during involution and accelerates c-myc-induced tumorigenesis of the mammary gland in transgenic mice. Oncogene 1997;15:1787–1795.
131a. Walton KD, Wagner K-U, Ruder E, Shillingford J, Hennighausen L. Conditional deletion of the bcl-x gene from mouse mammary epithelium results in accelerated apoptosis during involution, but does not compromise cell function during lactation. Development 2001; in press.
132. Gigliotti AP, DeWille JW. Lactation status influences expression of CCAAT/enhancer binding protein isoform mRNA in the mouse mammary gland. J Cell Physiol 1998;174:232–239.
132a. Chapman RS, Lourenco PC, Tonner E, Flint DJ, Selbert S, Takeda K, et al. Suppression of epithelial apoptosis and delayed mammary gland involution in mice. Genes Dev 1999;13:2604–2616.
133. Cantwell CA, Sterneck E, Johnson PF. Interleukin-6-specific activation of the C/EBPdelta gene in hepatocytes is mediated by Stat3 and Sp1. Mol Cell Biol 1998;18:2108–2117.
134. Tanaka T, Yoshida N, Kishimoto T, Akira S. Defective adipocyte differentiation in mice lacking the C/EBPβ and/or C/EBPδ gene. EMBO J 1997;16:7432–7443.
135. LeRoith D, Neuenschwander S, Wood TL, Henninghausen L. Insulin-like growth factor-I and insulin-like growth factor binding protein-3 inhibit involution of the mammary gland following lactation: studies in transgenic mice. Prog Growth Factor Res 1995;6:433–436.
136. Neuenschwander S, Schwartz A, Wood TL, Roberts Jr CT, Henninghausen L, LeRoith D. Involution of the lactating mammary gland is inhibited by the IGF system in a transgenic mouse model. J Clin Invest 1996;97:2225–2232.
137. Hadsell DL, Greenberg NM, Fligger JM, Baumrucker CR, Rosen JM. Targeted expression of des(1-3) human insulin-like growth factor I in transgenic mice influences mammary gland development and IGF-binding protein expression. Endocrinology 1996;137:321–330.
138. Rosfjord EC, Dickson RB. Growth factors, apoptosis, and survival of mammary epithelial cells. J Mammary Gland Biol Neoplasia 1999;4:229–237.
139. Deeks S, Richards J, Nandi S. Maintenance of normal rat mammary epithelial cells by insulin and insulin-like growth factor 1. Exp Cell Res 1988;174:448–460.
139a. Hadsell DL, Murphy KL, Bonnette SG, Reece N, Laucirica R, Rosen KM. Cooperative interaction between mutant p53 and des(1–3)IGF-1 accelerates mammary tumorigenesis. Oncogene 2000;19:889–898.
140. Bates P, Fisher R, Ward A, Richardson L, Hill DJ, Graham CF. Mammary cancer in transgenic mice expressing insulin-like growth factor II (IGF-II). Br J Cancer 1995;72:1189–1193.
141. Clemmons DR. Role of insulin-like growth factor binding proteins in controlling IGF actions. Mol Cell Endocrinol 1998;140:19–24.
142. Tonner E, Barber MC, Travers MT, Logan A, Flint DJ. Hormonal control of insulin-like growth factor-binding protein-5 production in the involuting mammary gland of the rat. Endocrinology 1997;138:5101–5107.
143. Wood TL, Richert M, Stull M, Allar M. Insulin-like growth factors and insulin-like growth factor binding proteins during postnatal development of murine mammary glands. J Mammary Gland Biol Neoplasia 2000;5:31–42.
144. Jones JI, Gockerman A, Busby Jr WH, Camacho-Hubner C, Clemmons DR. Extracellular matrix contains insulin-like growth factor binding protein-5: potentiation of the effects of IGF-I. J Cell Biol 1993;121:679–687.

11 Prolactin and the Prolactin Receptor

Nelson D. Horseman, PhD

Contents

INTRODUCTION

In mammals, prolactin is both an explicit inducer of mammary-gland development and lactation, and an integrator of multiple physiological adaptations during the post-mating phase of the reproductive cycle. These general features of the physiology of mammalian prolactin appear to have evolved from the effects of prolactin on parental physiology and osmoregulation in nonmammalian vertebrates. Recent application of mouse genetic technologies to studies of prolactin have clarified several controversial concepts, and provided systems for studying basic prolactin biology and clinical questions in areas such as pituitary tumors, infertility, breast cancer, and prostate neoplasia. This chapter focuses primarily on results from mouse genetic experiments after a basic introduction to the biology and pathobiology of prolactin signaling.

PROLACTIN SECRETION, RECEPTORS, AND SIGNAL TRANSDUCTION

Pituitary lactotrophs (prolactin-secreting cells) are the last hormone-producing cell type to differentiate in the rat pituitary gland *(1)*. Mammalian lactotrophs are unique because they require tonic inhibitory control by dopamine (prolactin-inhibiting factor), which is synthesized by neuronal cell bodies in the arcuate nucleus, and secreted from termini in the hypothalamic median eminence *(2,3)* (Fig. 1). A variety of peptide-releasing factors, including vasoactive intestinal peptide, thyrotrophin-releasing hormone,

From: *Contemporary Endocrinology: Transgenics in Endocrinology*
Edited by: M. Matzuk, C. W. Brown, and T. R. Kumar © Humana Press Inc., Totowa, NJ

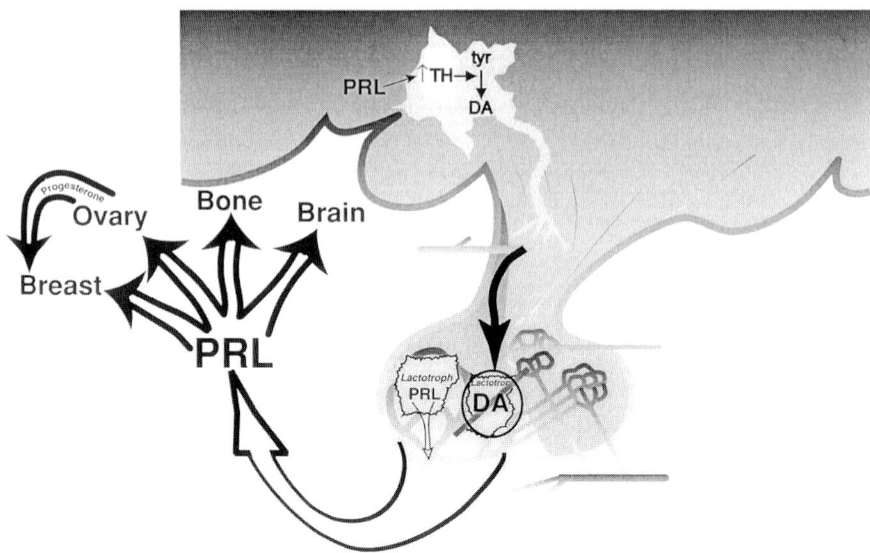

Fig. 1. Summary of the basic physiology of prolactin, which is secreted from lactotrophs in the anterior pituitary gland and circulates in the bloodstream to peripheral target tissues, such as the ovaries, mammary glands, and bone, and to the central nervous system. The dominant inhibitory regulator of prolactin secretion is dopamine, which is secreted at the median eminence. Dopamine is synthesized in tuberoinfundibular neurons with cell bodies in the arcuate nucleus of the hypothalamus. Dopamine synthesis is mediated by the conversion of tyrosine (tyr) to dopamine through the catalytic action of tyrosine hydroxylase (TH). TH levels are increased, presumably by transcriptional and posttranscriptional mechanisms, when prolactin levels are elevated.

galanin, and oxytocin stimulate prolactin release from lactotrophs *(3)*. In mammals, these factors play a subsidiary role to that of dopamine, but in nonmammals, positive releasing factors are the dominant regulators of prolactin secretion *(4)*. The differentiation of lactotrophs, and the biosynthesis of prolactin mRNA, is positively regulated by the transcription factor Pit-1 (also known as GHF-1) *(5)*. Pit-1 is a member of the homeobox protein superfamily, which includes several families of developmental regulators that function in both invertebrates and vertebrates. Pit-1 mediates not only the initial differentiation of a lineage of pituitary cells that gives rise to lactotrophs, somatotrophs, and thyrotrophs, but also the stabilization of the differentiated state of lactotrophs and somatotrophs *(6)*.

Prolactin receptors have been identified and cloned from a variety of mammalian and nonmammalian species *(7–12)*. The receptors for prolactin, growth hormone, and a variety of hematopoietic cytokines belong to a conserved superfamily of single-transmembrane-spanning proteins that share both structural and biochemical properties *(9,13,14)*. These receptors couple with and activate noncovalently associated tyrosine kinases, which phosphorylate various effector proteins. In the case of prolactin, the particular kinase that is activated by receptor binding is Janus kinase 2 (JAK2). The primary downstream effector for prolactin is a member of a family of related transcription factors, called STAT (signal transducer and activator of transcription) proteins *(13,15)*. Phosphorylation of STAT proteins converts them from an inactive, latent form to active, dimeric complexes that translocate to the nucleus and bind to

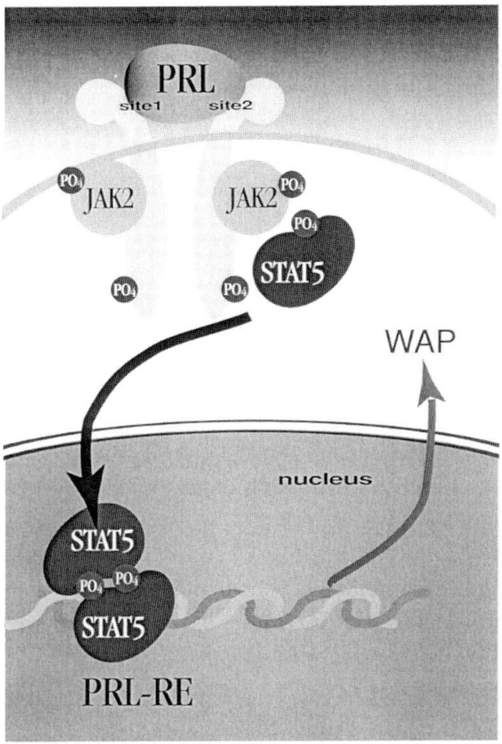

Fig. 2. Signal transduction from the prolactin receptor. This simplified diagram depicts only the main signal transduction pathway that is proven for prolactin. In this mechanism, prolactin associates with prolactin receptors through two binding sites, and thereby mediates the formation of a productive dimeric configuration. In this configuration, JAK2 is activated, leading to the tyrosine phosphorylation of residues on the prolactin receptor, JAK2 and STAT5. STAT5 dimerizes and translocates to the nucleus, where it associates with prolactin-response elements (prolactin-RE) in the promoters of regulated genes. This activation of a latent transcription factor results in increased expression of genes such as those for milk proteins like whey acidic protein (WAP).

specific DNA promoter elements. STAT5a and STAT5b are mediators of prolactin actions in vivo (*16*). This pathway is depicted in Fig. 2, along with ancillary signaling pathways that branch from JAK2 activation to couple with alternative effector systems, including MAP kinases, PI3 kinase, and protein kinase C, each of which has been suggested to be activated when prolactin binds to its receptor *(9)*.

The genes for all of the main elements of the prolactin-signaling pathway, the ligand, the receptor, the specific kinases and transcription factors have been recently knocked out in mice *(17–22)*. The initial studies of these mouse strains have provided clarification and valuable insight into the physiology and pathophysiology of prolactin, and shed light on intracellular prolactin-signaling mechanisms in the context of the entire animal.

KNOWLEDGE OF PROLACTIN FUNCTION FROM STUDIES OF KNOCKOUT AND TRANSGENIC MICE

The availability of knockout mice for each component of the prolactin signaling pathway has provided an opportunity to compare phenotypes among these animals.

These comparisons provide both concordances that support general conclusions about prolactin function, and dissimilarities that point to promising future research areas. The only component of the prolactin-signaling pathway that is an "essential" gene is JAK2 protein-tyrosine kinase. JAK2 knockout homozygous mice die *in utero* from severe defects of hematopoiesis *(22)*. Other than the JAK2 knockouts, all other mice with defects in prolactin-signaling molecules are born normally and survive to adulthood without obvious pathological consequences. In adulthood, various problems become evident in these mice.

Reproductive Biology

The obvious general observation from knockouts of genes in the prolactin pathway has been that prolactin signaling is necessary for several aspects of female reproduction in mice. The particular functional defects in the female knockouts are less obvious, and will require much more experimental work to elucidate. In contrast to the severe and multiple defects in female mice, male mice with knockouts of prolactin-signaling molecules have normal fertility, and are not severely affected in any aspect of reproductive physiology or behavior *(23)*. Although prolactin deficiency in these genetic models does not lead to any severe male reproductive problems, transgenic overexpression of prolactin (gain-of-function) causes extreme prostate hyperplasia *(24)*.

Female mice with knockouts of either the prolactin ligand or receptor are completely infertile *(17,18)*. Detailed analyses of the receptor knockouts showed that these mice reach puberty and undergo estrous cycles similar to normal controls, although the estrous cycles are less regular than normal. The prolactin-R-knockout females ovulate and mate, but do not become pregnant. One aspect of the defective reproduction in these mice is their inability to support implantation of otherwise normal embryos. The defect of implantation can be explained by the lack of sufficient progesterone caused by a failure to form functional corpora lutea. Mating failed to induce pseudopregnancy in prolactin-R-knockout mice, which is necessary for maintaining a uterine environment receptive to for implantation *(18)*. When normal embryos from wild-type females were transplanted into prolactin-R-knockout females they did not implant, although embryos from knockout females could be successfully transplanted to wild-type host females.

There seems to be an additional defect contributing to infertility in prolactin-R-knockout mice. Embryos were unable to implant in the uterine wall, and there was also a reduction in the efficiency of embryonic development in the prolactin-R-knockout mice within the first hours and days after fertilization. Embryos flushed from the oviducts of prolactin-R-knockout females progressed through cell divisions more slowly, and many fewer reached the blastocyst stage on schedule. The low efficiency of embryonic development in the prolactin-R-knockout oviductal environment suggests that prolactin signaling is important for secretion of maternal growth factors, cell-adhesion molecules, or other oviductal products that provide the optimal embryonic environment. This apparent role of prolactin signaling is completely unexplored. The fact that the inefficiency of embryonic development in the knockout females appears within the first few hours after mating argues for a role of the proestus surge of prolactin, rather than the postmating surges, in the promotion of an optimal oviductal environment *(18)*.

Prolactin-ligand knockout females are also infertile *(17)*. Although the characterization of reproductive function for the ligand-deficient females has been less extensive than that done in the prolactin-R-knockouts, the defects are qualitatively similar. That is, the ovaries in these mice failed to luteinize, embryos did not implant, and early embryonic development was inefficient in the prolactin-knockout maternal environment (unpublished observations). In the case of female reproductive function, there is a remarkable degree of concordance between prolactin-ligand and receptor-knockout mice.

Whereas the reproductive consequences of knocking out the prolactin gene or its receptor are dramatic in females, this is not the case in male mice. The fertility and reproductive behaviors of male prolactin-knockout mice were indistinguishable from normal controls *(17)*. Prolactin-R-knockout males were initially described as subfertile compared with normal controls *(18)*, but additional data collected subsequently did not support that conclusion, although the receptor knockout mice may mature somewhat more slowly than normal (P. Kelly, personal communication). Detailed analysis of male reproductive physiology in prolactin-knockout mice showed them to have normal pituitary contents of both LH and FSH, though plasma LH was about one-third lower than normal litter mates. Although LH secretion trended lower, testosterone secretion from testes of prolactin-knockout mice was normal both in vivo and in vitro. Therefore, the only detectable effect of disruption of the prolactin gene on male reproductive hormone secretion was an apparent change in the relationship of LH and testosterone secretion, suggesting a modest increase in the sensitivity of the hypothalamus to the feedback-inhibitory action of testosterone *(23)*.

Transgenic models of prolactin overexpression have received limited use, presumably because hyperprolactinemia can be produced in several other simpler ways, including injections of purified prolactin, minipump infusions, injection of dopamine D2 receptor agonists (to increase pituitary prolactin secretion), and engraftment of donor pituitary glands *(25)*. The latter approach, in which one or more pituitary glands from donor animals are grafted under the kidney capsule, has been a staple research technique because prolactin, unlike other pituitary hormones, is hypersecreted from the implanted gland once it is removed from the direct influence of hypothalamic dopaminergic inhibition. Prolactin levels at least one order of magnitude higher than normal virgin female serum concentrations, and similar to lactating levels, may be produced by pituitary grafts during chronic implantation experiments, and these levels are maintained for many months *(25)*.

Two transgenic mouse models have been used to produce high levels of prolactin bioactivity in vivo. Human GH transgenic mice display various phenotypes that are attributable to the prolactin-like bioactivity of hGH *(26–28)*. In particular, hGH transgenic mice have a high rate of spontaneous breast cancers, whereas transgenic overexpresssion of bovine GH, which is not lactogenic, does not lead to breast cancer *(27)*.

Mice that overexpress the rat prolactin gene, driven by a metallothionein promoter, have been observed to develop prostate enlargement that closely mimics human benign prostate hyperplasia *(24)*. The enlargement of the prostate gland in prolactin-transgenic mice resulted from both increased secretory material and cellular proliferation, particularly of the interstitial cells. The relative contributions of direct prolactin effects on the prostate, and indirect actions mediated by androgens in these transgenic mice, have not yet been determined.

Mammary-Gland Biology

Disruption of the gene for either prolactin or its receptor results in a complex phenotype of defective mammary-gland development *(17,18)*. In homozygous prolactin-

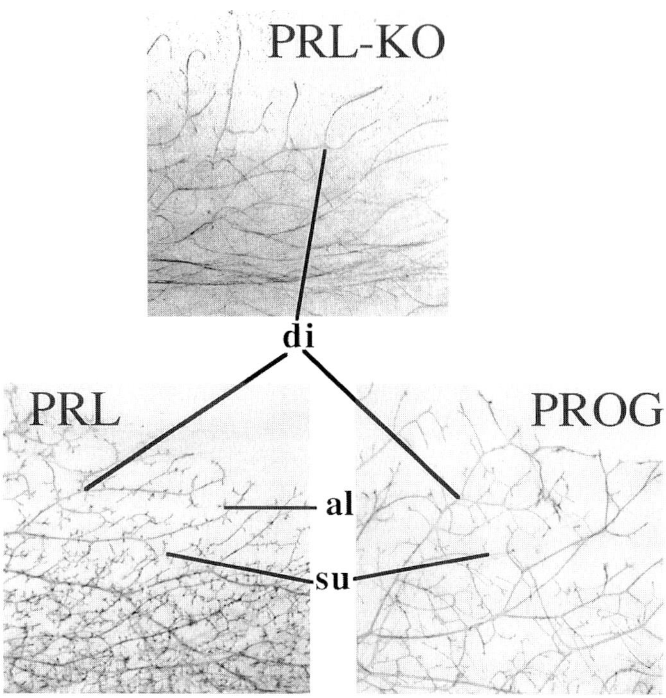

Fig. 3. Hormonal induction of mammary-gland maturation. The mammary glands of mice with a disrupted prolactin gene (prolactin-knockout) do not mature past a simple, dichotomous (di) branching duct system. The two hormones that contribute to the development of the mature mammary-gland structures are progesterone (PROG), which drives the growth of subordinate (su) branches that sprout from the main ductal tree, and prolactin, which drives the differentiation of alveolar buds along the ductal system. The frames represent whole mounts of mammary glands stained with safranin O. Each of the mice carried the null mutation of the prolactin gene, and were either untreated (prolactin-knockout), implanted with a 25-mg pellet of PROG for 18 d (PROG), or given two normal pituitaries, grafted under the kidney capsule (prolactin).

knockout mice, the mammary gland arrests in a "pubertal" state, characterized by the presence of a simple dichotomously branching epithelial-duct system, with persistent terminal end buds *(17)*. In heterozygous litter mates, or wild-type controls the mammary gland matures to a complex epitheial network, including not only a dichotomous primary-ductal system, but also closely spaced subordinate branches and alveolar buds associated with the ductal system (Fig. 3). The arrested phenotype of prolactin-knockout mouse mammary glands is fully rescued by grafting a normal pituitary under the kidney capsule to expose the animal to sufficient prolactin to drive mammary-gland maturation. To determine the relative contributions of primary prolactin deficiency and secondary progesterone deficiency to the phenotype in the prolactin-knockout mammary gland, progesterone alone (by implanted pellet) was supplied to prolactin-knockout mice. In this case, the subordinate branches along the primary ductal system were rescued, but not the alveolar budding that is seen in normal mice. Mammary glands of prolactin-R-knockout mice were transplanted into normal hosts, which were then mated. In this experiment the prolactin-R-deficient glands underwent subordinate branching, but not alveolar differentiation, indicating that prolactin acts both directly and indirectly on the mammary glands *(29)*. These results reinforce those of Lyons *(30)*, who used hypophysectomy and hormone replacement to show that prolactin was necessary for alveolar growth and differentiation, whereas progesterone was needed for the finely branched architecture of the mature adult mammary gland. Disruption of the progesterone-receptor (*PR*) gene resulted in stunted subordinate branching *(31)*, consistent with progesterone acting as a growth factor for these epithelial structures. Whereas a single functional copy of the prolactin gene was sufficient for normal mammary-gland development *(17)*, two copies of the prolactin-R gene were necessary. In prolactin-R hemizygous (+/–) mice mammary gland development was delayed so that adequate lactation occurred only after either multiple pregnancies, or if the first pregnancy was postponed until the females were 20 wk old *(18)*.

STAT5 Knockout Models

Disruption of genes for the primary downstream effectors of prolactin action, STAT5a, and STAT5b has provided insights into the relative roles of these transcription factors in prolactin signaling, and the convergence of signaling from other hormones on these effector proteins. Disruption of either STAT5a or STAT5b alone resulted in only partial defects in functions mediated by prolactin *(19,20)*. STAT5a-deficient mice did not lactate, although they bred normally and the mammary glands partially developed during pregnancy *(19)*. STAT5b-deficient mice were subfertile, but mammary-gland development was not affected *(20)*. STAT5b-deficient mice showed partial GH resistance, as well as the reproductive deficit *(20)*. Combined disruption of both the STAT5a and STAT5b genes resulted in an overt phenotype that closely resembled a combination of prolactin and GH deficiency. The double-STAT5 knockout mice were dwarf and IGF-1 levels were low, resembling profound GH deficiency; and the females were both infertile and displayed arrested mammary gland development, resembling prolactin deficiency *(21)*. In addition to the defects in GH and prolactin signaling, mice with combined STAT5 disruptions had partial defects in responses to IL-3, IL-5, IL-7, and GM-CSF. These effects on hematopoiesis were modest compared with the profound defects in growth and reproductive physiology caused by STAT5 deficiency. These results lead to the natural inference that STAT5 isoforms have overlapping roles in mediating prolactin and GH actions, and have remained tightly coupled with the actions of these related hormones throughout their evolution.

Pituitary-Gland Tumors

The pituitary gland of PROLACTIN–KO mice grows abnormally, and eventually undergoes adenomatous transformation *(32)*. The lifespan of prolactin-knockout mice is shortened by several months because of the mass effects of the pituitary tumors, but this is the only overt pathology that was observed in these animals. The growth of pituitary tumors in prolactin-knockout mice results from proliferation of the lactotroph lineage, which, because of the gene disruption, does not produce bioactive prolactin. The functionally disrupted prolactin gene in these mice directs the synthesis of a nonbioactive N-terminal peptide, and the cells that produce this mutant polypeptide (pseudo-lactotrophs) are the source of the hyperplasia and adenomas in prolactin-knockout pituitaries. Injections of bromocriptine (dopamine D2 receptor agonist) caused regression of the hyperplasia in prolactin-knockout pituitaries, leading to the conclusion that dysinhibition of lactotroph proliferation is the primary cause of pituitary adenomas in these mice. The phenotype of hyperplasia and pituitary tumors in prolactin-knockout mice was similar to the phenotype of mice in which the dopamine D2 receptor gene was knocked out *(33,34)*. In D2 receptor knockout mice, there is also hyperprolactinemia, because these mice have the normal prolactin gene. Pituitary tumors developed in both models, the hyperprolactinemic D2 receptor knockouts, and the prolactin-absent prolactin knockouts; therefore, bioactive prolactin is not necessary to promote the growth and/or transformation of lactotrophs, although it may, if present, act as an autocrine growth factor *(35)*.

WHY DO THE MOUSE PHENOTYPES IN THE PROLACTIN PATHWAY CONFOUND US?

Results from the knockout models have provided new experimental windows into aspects of prolactin physiology that have previously been obscure, difficult to study, or controversial. Among these, the most intriging information available thus far is in the areas of maternal behavior, bone regulation, hematopoiesis, and immune function. These areas offer some of the most interesting challenges and opportunities for future research using contemporary functional genetic approaches.

Maternal Behavior

Analyses of maternal behavior*s* in knockouts of the prolactin ligand and the prolactin receptor have yielded directly contradictory results *(17,36)*. Resolving these contradictions will be important for understanding the ways in which prolactin may influence neurobehavioral physiology. To understand the possible meaning of prolactin actions on maternal behavior, it is important to consider the differences in maternal behavior between laboratory mice and other animal systems. Laboratory rats, hamsters, and mice from wild populations display maternal behaviors such as nest-building, pup retrieval, nursing postures, or pup grooming only when they have been through a pregnancy and delivery, or had a period of training and habituation *(37–39)*. In contrast, laboratory strains of mice spontaneously demonstrate complex, stereotypic maternal behaviors when virgin females (or to a lesser extent, males) are exposed to foster pups. It is likely that the evolutionary basis of the propensity of laboratory mouse strains to display maternal behaviors is the intentional selection for docility, sociability, and high reproductive potential during domestication of laboratory mice. Previous studies, mostly in rats, have shown that prolactin injections or intracerebroventricular infusions significantly reduced the time required for inexperienced females to begin showing maternal behaviors *(39)*. The relative roles of prolactin and reproductive steroids in the mediation of hormonal effects on maternal behavior have been controversial. However, data showing that direct brain infusion of prolactin influences behavior, and that substantial amounts of prolactin exists in the brain *(40,41)*, have been used to support the concept that prolactin itself has a direct influence on maternal behaviors.

Genetic knockout mice have provided new challenges and opportunities to address prolactin actions on maternal behavior. Results in prolactin-knockout mice may support the concept that prolactin has no direct effect on maternal behavior. Virgin mice with a disruption of the prolactin gene display maternal behaviors that are indistinguishable to those of their normal siblings *(17)*. In sharp contrast, mice with a disruption of the prolactin-receptor gene are profoundly deficient in these behaviors *(36)*. prolactin-R-knockout mice failed to retrieve foster pups, did not assume a nursing posture, and did not engage in grooming behaviors (such as maternal anogenital licking) as did their normal counterparts. Given the relative consistency between the reproductive phenotypes of prolactin-knockout and prolactin-R-knockout mice, what explains the dramatic difference in maternal behaviors between these models? Two classes of hypotheses may serve to explain these observations; one may hypothesize either that receptor nonspecificity may come into play, or that there is a developmental difference between the strains of mice.

To reconcile the behavioral phenotypes of prolactin-knockout and prolactin-R-knockout mice, one might propose that some ligand other than prolactin interacts with the prolactin-R in mice where the prolactin ligand is knocked out, and this alternative ligand would stimulate maternal behaviors. Candidates for such an alternative ligand are easy to imagine, but difficult to reconcile with the available data. Rodent growth hormones do not productively interact with the prolactin-R at physiological concentrations *(42)* although primate GH does, and there is no evidence that prolactin-knockout mice hypersecrete GH at the levels that would be required for cross activation (about three orders of magnitude). The targeting construct for the prolactin ligand knockout *(17)* was designed so that an N-terminal fragment of the gene is synthesized in the prolactin-knockout mice, and this fragment may activate the prolactin-R. However, the binding mechanism for prolactin and its receptor are well understood, and binding requires discontinuous sequences located in both the N- and C-terminal halves of the molecule *(43)*, so the disrupted N-terminal fragment could not mimic normal prolactin-receptor binding. Moreover, direct evidence of the lack of bioactivity, and the prolactin-deficient phenotypic characteristics of the prolactin-knockout mice *(17)* make it clear that this hypothesis would not be true. A third candidate for alternative prolactin-R ligands are the placental lactogens *(44)*. Placental lactogens would activate the prolactin-R in prolactin-knockout mice if they were synthesized and secreted, so it may be reasonable to hypothesize, although no evidence yet exists, that placental lactogens are made in the brain and activate the prolactin-R in prolactin-knockout mice.

If an alternative ligand does not account for the difference in maternal behavior in prolactin-knockout and prolactin-R-knockout mice, it is possible that differences in the developmental environment of these strains could provide an answer. One major difference between the models of prolactin and prolactin-R gene disruption stems from the fact that mice are exposed to maternal prolactin bioactivity during gestation and early postnatal life through the amniotic fluid and milk *(44,45)*. So, if the receptor is intact as in prolactin-KO mice, exogenous ligand might affect the development of mice even if the endogenous ligand is knocked out. In contrast, if the receptor is knocked out, the fetal and neonatal mice will be unable to respond to maternally transmitted prolactin, as well as endogenously secreted prolactin. It is conceivable that this early exposure to exogenous prolactin may allow development of certain phenotypic traits. In the case of maternal behavior, prolactin-knockout mice may be predisposed to respond maternally to foster pups because their nervous system is appropriately "programmed" by exposure to maternal prolactin early in life. If this hypothesis is correct, it raises fundamental questions about how laboratory strains of mice can be programmed by perinatal prolactin exposure, whereas wild-mice and members of other species do not become maternally programmed without hormones and experience gained later in life.

Bone Regulation

Two sets of circumstantial evidence have stimulated interest in the notion that prolactin (or prolactin-like hormones) may be an important regulator of bone metabolism. First, pregnancy and lactation require massive calcium mobilization to support fetal growth and to provide for calcium secretion into milk. While vitamin D and other calciotropic hormones are clearly involved in these processes, it is likely that prolactin acts specially as an integrator of maternal physiology to promote changes in bone and calcium metabolism during pregnancy and lactation. Clinical studies have correlated prolactin levels with bone-calcium mobilization *(46)*. Second, a high level of prolactin receptor gene expression has been demonstrated in fetal skeletal elements in mice, rats, and humans *(47–50)*, suggesting that skeletal tissues may be targets for placental lactogens during development.

Prolactin-R-knockout mice show a phenotype of reduced bone formation rate and bone mineralization *(51)*, suggesting that prolactin signaling is needed to either directly stimulate bone-mineral deposition, or to sensitize bone cells to other osteogenic hormone signals. prolactin-R gene expression was detected in osteoblasts, supporting the notion that prolactin may directly stimulate the function of bone-forming cells. These data are contrary to both clinical and experimental correlations regarding prolactin and bone homeostasis, and to the intuitive expectation that a lactogenic hormone should increase bone mobilization. Thus, a loss of prolactin signaling may result in higher levels of bone mineral. In prolactin-knockout mice, we have begun to examine bone growth and mineralization, and these preliminary studies show a phenotype that is different from that of the prolactin-R-knockout mice. The prolactin ligand-deficient mice show no evidence of reduced bone formation that would be similar to the receptor-deficient mice. The pattern generally points to higher, rather than lower bone formation in the prolactin-knockout mice compared with their normal counterparts (unpublished studies).

The possible difference between the bone phenotypes of prolactin-R-knockout and prolactin-knockout mice may be attributable to strain differences, since the receptor-negative mice were of a mixed genetic background (129Sv/C57Bl/6) *(51)* and the prolactin-negative mice were congenic on a C57Bl/6J background. In addition, it is likely that the bone defects in the prolactin-receptor knockouts are partly caused by development defects. Proper skeletal development may require signaling from placental lactogens acting on fetal prolactin receptors. In contrast, the phenotype in the prolactin ligand knockouts would include only direct and indirect physiological consequences of prolactin acting on a normally developed system. Further studies of these complementary models are certain to provide important information regarding hormone actions on bone metabolism.

Hematopoiesis and Immune Function

Several observations have contributed to the idea that prolactin may be an important regulator of hematopoiesis and immune function in mammals. First, both the prolactin receptor and ligand have been identified in cells of the hematopoietic system *(52–56)*. Second, pharmacological studies in animals and clinical correlations in humans have shown that prolactin alters parameters of hematopoiesis and immune function *(57–61)*. Third, certain cells cultured from hematopoietic lineages respond to prolactin in vitro *(62)*. Fourth, the prolactin receptor and downstream signaling molecules are homologous with proteins that mediate the actions of a variety of hematopoietic cytokines *(13)*. Despite these types of evidence, it has remained difficult to reconcile the idea that prolactin is an essential immunoregulatory hormone with the fundamental requirements of its role in reproduction. prolactin is secreted at much higher levels in females than in males, and at highly variable levels at different phases of the reproduction/lactation cycle. Despite the dramatic fluctuations in prolactin secretion in these situations, the hematopoietic and immune systems undergo no dramatic changes in function corresponding to these differences in prolactin secretion. It is possible to invoke any number of possible explanations for this basic contradiction, but there is insufficient evidence to accept any of these explanations at this time.

To determine whether prolactin is required for any aspect of hematopoiesis of immune function the phenotypes of prolactin-knockout and prolactin-R-knockout mice have been analyzed in detail *(17,63)*. In prolactin-knockout mice, we have showed that the primary development of lymphoid, myeloid, and erythroid lineages was unaltered when prolactin-deficient mice were compared with their normal counterparts. Subsequently, we have examined secondary immune responses, and shown similarly that prolactin-knockout mice are normal with respect to immune functions (unpublished data). Similarly, prolactin-R-knockout mice have normal development of immune cells, and normal responses in terms of antibody synthesis, and mitogen- and antigen-induced proliferation, natural killer (NK) cell cytotoxicity, and other measures of innate and acquired immunity *(63)*.

It is conceivable that there is functional redundancy between prolactin and one or more other hormones which regulate the immune system, and that there are compensatory adjustments for the defects provoked by the absence of prolactin. Another way of looking at the actions of prolactin on the immune system is to consider that these effects may not be specific to the immune system. From this perspective, the impact of prolactin on immunity may be part of a broader role of prolactin as a modulator of stress responses *(64)*. Prolactin acting as a systemic integrator of stress responses, rather than a specific regulator of immune function, would be consistent with its role during pregnancy and lactation, which place high demands on all physiological systems.

SUMMARY AND PERSPECTIVE

Prolactin is essential to the survival of the mammal because of its multiple roles in reproduction. Although we understand a great deal about the biochemistry, neuroendocrinology, and signaling mechanisms for prolactin, certain aspects of prolactin physiology and cell biology still remain unclear. The phenotypes of transgenic and knockout mice point to several important areas of future prolactin research. The mechanisms underlying female infertility and maternal behavior remain unknown; and in the male, the possibility that prolactin plays an important role in prostate growth needs investigation. Novel experimental approaches need to be developed to study the bone as a direct and/or indirect target of prolactin and placental lactogen action during development and adulthood. In addition, reconciling the immune-system phenotypes of these mice with the published information on prolactin and immune function will require more work.

The primary target of prolactin action in the mammal is the mammary gland, where profound defects in development occur when any of the genes in the prolactin pathway are disrupted. Although prolactin has been considered to be essential for mammary-gland development since it was discovered six decades ago, we do not know any of the genes that actually mediate developmental changes induced by prolactin. Understanding the molecular and cellular biology of developmental changes brought about by prolactin in the mammary gland is likely to further not only our understanding of normal mammary-gland development, but also our understanding of aberrant mammary-gland growth in breast cancer. The use of mouse genetics to complement other experimental approaches will continue to provide new avenues to study the biology of prolactin and its related hormones.

ACKNOWLEDGMENTS

Thanks to Candice Arnold for help preparing the manuscript, and Stacy Shipman for helping with the figures. The studies were supported by grants from the National Institutes of Health and by funding from the Shriners Hospitals for Children.

REFERENCES

1. Cooke NE, Liebhaber SA. Molecular biology of the growth hormone-prolactin gene system. Vita Horm 1995;50:385–459.
2. Ben-Jonathan N, Arbogast LA, Hyde JF. Neuroendocrine regulation of prolactin release. Prog Neurobiol 1989;33:399–447.
3. Ben-Jonathan N. Regulation of prolactin secretion. In: Imura H, ed. The Pituitary Gland, Second Edition, Raven Press, New York, NY, 1994, pp. 261–283.
4. Lea RW, Talbot RT, Sharp PJ. Passive immunization against chicken vasoactive intestinal polypeptide suppresses plasma prolactin and crop sac development in incubating ring doves. Horm Behav 1991;25:283–294.
5. Ingraham HA, Chen R, Mangalam HJ, Elsholtz HP, Flynn SE, Lin CR, et al. A tissue-specific transcription factor containing a homeodomain specifies a pituitary phenotype. Cell 1988;55:519–529.
6. Simmons DM, Voss JW, Ingraham HA, Holloway JM, Broide RS, Rosenfeld MG Swanson L, W. Pituitary cell phenotypes involve cell-specific Pit-1 mRNA translation and synergistic interactions with other classes of transcription factors. Genes Dev 1990;4:695–711.
7. Boutin J-M, Jolicoeur C, Okamura H, Gagnon J, Edery M, Shirota M, et al. Cloning and expression of the rat prolactin receptor, a member of the growth hormone/prolactin receptor gene family. Cell 1988;53:69–77.
8. Kelly PA, Djiane J, Postel-Vinay M-C, Edery M. The prolactin/growth hormone receptor family. Endocrine Rev 1991;12:235–251.
9. Bole-Feysot C, Goffin V, Edery M, Binart N, Kelly PA. Prolactin (PRL) and its receptor: actions, signal transduction pathways and phenotypes observed in PRL receptor knockout mice. Endocr Rev 1998;19:225–268.
10. Chen X, Horseman ND. Cloning, expression, and mutational analysis of the pigeon prolactin receptor. Endocrinology 1994;135:269–276.
11. Tanaka M, Maeda K, Okubo T, Nakashima K. Double antenna structure of chicken prolactin receptor deduced from the cDNA sequence. Biochem Biophys Res Commun 1992;188:490–496.
12. Sandra O, Sohm F, De Luze A, Prunet P, Edery M, Kelly PA. Expression cloning of a cDNA encoding a fish prolactin receptor. Proc Natl Acad Sci USA 1995;92:6037–6041.
13. Horseman ND, Yu-Lee L-Y. Transcriptional regulation by the helix bundle peptide hormones: GH, PRL, and hematopoietic cytokines. Endocr Rev 1994;15:627–649.
14. Cosman D. The hematopoietin receptor superfamily. Cytokine 1993;5:95–106.
15. Darnell Jr JE, Kerr IM, Stark GR. Jak-Stat pathways and transcriptional activation in response to IFNs and other extracellular signaling proteins. Science 1994;264:1415–1421.
16. Liu S, Robinson GW, Gouilleux F, Groner B, Henninghausen L. Cloning and expression of Stat 5 and an additional homologue (Stat 5b) involved in prolactin signal transduction in mouse mammary tissue. Proc Natl Acad Sci USA 1995;92:8831–8835.
17. Horseman ND, Zhao W, Montecino-Rodriguez E, Tanaka M, Nakashima K, Engle SJ, et al. Defective mammopoiesis, but normal hematopoiesis, in mice with a targeted disruption of the prolactin gene. EMBO J 1997;16:6926–6935.

18. Ormandy C, J., Camus A, Barra J, Damotte D, Lucas B, Buteau H, et al. Null mutation of the prolactin receptor gene produces multiple reproductive defects in the mouse. Genes Devel 1997;11:167–178.
19. Liu X, Robinson GW, Wagner K-U, Garrett L, Wynshaw-Boris A, Hennighausen L. Stat5a is mandatory for adult mammary gland development and lactogenesis. Genes Deve 1997;11:179–186.
20. Udy GB, Towers RP, Snell RG, Wilkins RJ, Park S-H, Ram PA, Waxman DJ, Davey HW. Requirement of STAT5b for sexual dimorphism of body growth rates and liver gene expression. Proc Natl Acad Sci USA 1997;94:7239–7244.
21. Teglund S, McKay C, Schuetz E, van Deursen JM, Stravopodis D, Wang D, Brown M, Bodner S, Grosveld G, Ihle JN. Stat5a and Stat5b proteins have essential and nonessential, or redundant, roles in cytokine responses. Cell 1998;93:841–850.
22. Parganas E, Wang D, Stravopodis D, Topham D, Marine J-C, Teglund S, et al. Jak2 is essential for signaling through a variety of cytokine receptors. Cell 1998;93:385–395.
23. Steger RW, Chandrashekar V, Zhao W, Bartke A, Horseman ND. Neuroendocrine and reproductive functions in male mice with targeted disruption of the prolactin gene. Endocrinology 1998;139:3691–3695.
24. Wennbo H, Kindblom J, Isaksson OG, Tornell J. Transgenic mice overexpressing the prolactin gene development dramatic enlargement of the prostate gland. Endocrinology 1997;138:4410–4415.
25. Adler RA. Anterior piruitary-grafted rat: a valid model of chronic hyperprolactinemia. Endo Rev 1986;7:302–313.
26. Milton S, Cecim M, Li YS, Yun JS, Wagner TE, Bartke A. Transgenic female mice with high human growth hormone levels are fertile and capable of normal lactation without even having been pregnant. Endocrinology 1992;131:536–538.
27. Cecim M, Bartke A, Yun JS, Wagner TE. Expression of human, but not bovine growth hormone genes promotes development of mammary tumors in transgenic mice. Transgenics 1994;1:431–437.
28. Wennbo H, Gebre-Medhin M, Gritli-Linde A, Ohlsson C, Isaksson OG, Tornell J. Activation of the prolactin receptor but not the growth hormone receptor is important for induction of mammary tumors in transgenic mice. J Clin Invest 1997;100:2744–2751.
29. Briskin C, Kaur S, Chavarria TE, Binart N, Sutherland RL, Weinberg RA Kelly PA, Ormandy CJ. Prolactin controls mammary gland development via direct and indirect mechanisms. Dev Biol 1999;210:96–106.
30. Lyons W, Li CH, Johnson RE. Hormonal control of mammary growth and lactation. Rec Prog Horm Res 1958;14:219–254.
31. Lydon JP, MeMayo FJ, Funk CR, Mani SK, Hughes AR, Montgomery Jr CA, et al. Mice lacking progeserone receptor exhibit pleiotropic reproductive abnomalities. Genes Dev 1995;9:2266–2278.
32. Shipman SL, Scheiber MD, Horseman ND. Immunohistochemical analysis of prolactin gene expression in mice carrying a targeted mutation of the prolactin structural gene. Program of the 81st Annual Meeting of The Endocrine Society, San Diego, CA, 1999, p. 384
33. Saiardi A, Bozzi Y, Bail J-H, Borrelli E. Antiproliferative role of dopamine: loss of D_2 receptors causes hormonal dysfunction and pituitary hyperplasia. Neuron 1997;19:115–126.
34. Kelly M, Rubinstein M, Asa S, Zhang G, Saez C, Bunzow J, et al. Pituitary lactotroph hyperplasia and chronic hyperprolactinemia in dopamine D2 receptor-deficient mice. Neuron 1997;19:103–113.
35. Krown KA, Wang Y-F, Ho TWC, Kelly PA, Walker AM. Prolactin isoform 2 as an autocrine growth factor for GH_3 cells. Endocrinology 1992;131:595–602.
36. Lucas BK, Ormandy CJ, Binart N, Bridges RS, Kelly PA. Null mutation of the prolactin receptor gene produces a defect in maternal behavior. Endocrinology 1998;139:4102–4107.
37. McCarthy MM, Curran GH, Siegal HI. Evidence for the involvement of prolactin in the maternal behavior of the hamster. Physiol Behav 1994;55:181–184.
38. McCarthy MM, vom Saal FS. Influence of reproductive state on infanticide by wild female house mice (Mus musculus). Physiol Behav 1985;35:843–849.
39. Bridges RS. The role of lactogenic hormones in maternal behavior in female rats. Acta Pædiatr Suppl 1994;397:33–39.
40. Emanuele NV, Jurgens JK, Halloran MM, Tentler JJ, Lawrence AM, Kelley MR. The rat prolactin gene is expressed in brain tissue: detection of normal and alternatively spliced prolactin messenger RNA. Mol Endocrinol 1992;6:35–42.
41. DeVito WJ. Distribution of immunoreactive prolactin in the male and female brain: effects of hypophysectomy and intraventricular administration of colchicine. Neuroendocrinology 1988;47:284–289.
42. Niall HD, Hogan ML, Tregear GW, Segre GV, Hwang P, Friesen H. The chemistry of growth hormone and the lactogenic hormones. Rec Prog Horm Res 1973;29:387–404.
43. Wells JA, de Vos AM. Hematopoietic receptor complexes. Annu Rev Biochem 1996;65:609–634.

44. Soares MJ, Muller H, Orwig KE, Peters TJ, Dai G. Uteroplacental prolactin family and pregnancy. Biol Reprod 1998;58:273–284.
45. Kacsóh B, Veress Z, Tóth BE, Avery LM, Grosvenor CE. Bioactive and immunoreactive variants of prolactin in milk and serum of lactating rats and their pups. J Endocrinol 1993;138:243–257.
46. Klibanski A, Neer RM, Beitins IZ, Ridgway EC, Zervas NT, McArthur JW. Decreased bone density in hyperprolactinemic women. N Engl J Med 1980;303:1511–1514.
47. Freemark M, Nagano M, Edery M, Kelly PA. Prolactin receptor gene expression in the fetal rat. J Endocrinol 1995;144:285–292.
48. Royster M, Driscoll P, Kelly PA, Freemark M. The prolactin receptor in the fetal rat: cellular localizaton of messenger ribonucleic acid, immunoreactive protein, and ligand-binding activity and induction of expression in late gestation. Endocrinology 1995;136:3892–3900.
49. Freemark M, Driscoll P, Maaskant R, Petryk A, Kelly PA. Ontogenesis of prolactin receptrs in the human fetus in early gestation. J Clin Invest 1997;99:1107–1117.
50. Tzeng S, Linzer D. Prolactin receptor expression in the developing mouse embryo. Mol Reprod Dev 1997;48:45–52.
51. Clément-Lacroix P, Ormandy C, Lepescheux L, Ammann P, Damotte D, Goffin V, et al. Osteoblasts are a new target for prolactin: analysis of bone formation in prolactin receptor knockout mice. Endocrinology 1999;140:96–105.
52. Dardenne M, Kelly PA, Bach J-F, Saving W. Identification and functional activity of prolactin receptors in thymic epithelial cells. Proc Natl Acad Sci USA 1991;88:9700–9704.
53. O'Neal KD, Schwarz LA, Yu-Lee L-Y. Prolactin receptor gene expression in lymphoid cells. Mol Cell Endocrinol 1991;82:127–135.
54. Gagnerault MC, Touraine P, Savino W, AKP, Dardenne M. Expression of proalctin receptors in murine lymphoid cells in normal and autoimmune situations. J Immunol 1993;150:5673–5681.
55. Touraine P, do Carmo Leite de Moraes M, Dardenne M, Kelly PA. Expression of short and long forms of prolactin receptor in murine lymphoid tissues. Mol Cell Endocrinol 1994;104:183–190.
56. O'Neal K, Montgomery DW, Truong TM, Yu-Lee L-Y. Prolactin gene expression in human thymocytes. Mol Cell Endocrinol 1992;87:19–23.
57. Matera L, Cesano A, Bellone G, Oberholtaer E. Modulatory effect of prolactin on the resting and mitogen-induced activity of T, B, and NK lymphocytes. Brain Behav Immun 1992;6:409–417.
58. Murphy WJ, Durum SK, Longo DL. Differential effects of growth hormone and prolactin on murine T cell development and function. J Exp Med 1993;178:231–236.
59. Hooghe R, Delhase M, Vergani P, Malur A, Hooghe-Peters EL. Growth hormone and prolactin are paracrine growth and differentiation factors in the haemopoietic system. Immunol Today 1993;14:212–214.
60. Koojiman R, Hooghe-Peters EL, Hooghe R. Prolactin, growth hormone, and insulin-like growth factor-I in the immune system. Adv Immunol 1996;63:377–454.
61. Walker SE, Allen SH, McMurray RW. Prolactin and autoimmune disease. Trends Endocrinol Metab 1993;4:147–151.
62. Gout PW, Beer CT, Noble RL. Prolactin-stimulated growth of cell cultures established from malignant Nb rat lymphomas. Canc Res 1980;40:2433–2436.
63. Bouchard B, Ormandy C, Di Santo JP, Kelly PA. Immune system development and function in prolactin receptor-deficient mice. J Immunol 1999;163:576–582.
64. Dorshkind K, Horseman ND. The roles of prolactin, growth hormone, insuline-like growth factor-I, and thyroid hormones in lymphocyte development and function: insights from genetic models of hormone and hormone receptor deficiency. Endocrine Rev 2000;21:292–312.

12 Transgenic Models for Oxytocin and Vasopressin

Larry J. Young, PhD *and Thomas R. Insel,* MD

Contents

INTRODUCTION

Oxytocin (OT) and arginine vasopressin (AVP) were among the first peptide hormones to be isolated and sequenced *(1)*. Both OT and AVP consist of nine amino acids that form a ring structure from disulfide bonds, bridging two cysteine residues. Although the two hormones are similar in structure, differing at only two positions, their functions appear to be quite distinct. Oxytocin has been implicated in the induction of labor during parturition, milk ejection during lactation *(2)*, and the control of reproductive and maternal behaviors *(3)*. Arginine vasopressin, also known as antidiuretic hormone, plays a crucial role in maintaining osmotic homeostasis and vascular tone, and has more recently been implicated in cognition and social behaviors *(4–6)*.

OT and AVP are neuropeptides, synthesized primarily in the hypothalamus. OT is found principally in the supraoptic (SON) and paraventricular nuclei (PVN) *(7)*. AVP is found in both these nuclei, as well as in the suprachiasmatic nucleus *(7)*. Even in those cell groups synthesizing both hormones, OT and AVP are found in separate neurons *(8)*. The magnocellular OT and AVP cells of the PVN and SON project to the posterior pituitary, where OT and AVP are stored prior to release into the general circulation *(9)*. Both peptides are also synthesized in parvocellular neurons of the PVN, which project to other brain regions. In addition, oxytocin is also expressed in the uterus *(10)*, ovary *(11)*, heart *(12)* and, in some species, the testis *(13)*.

From: *Contemporary Endocrinology: Transgenics in Endocrinology*
Edited by: M. Matzuk, C. W. Brown, and T. R. Kumar © Humana Press Inc., Totowa, NJ

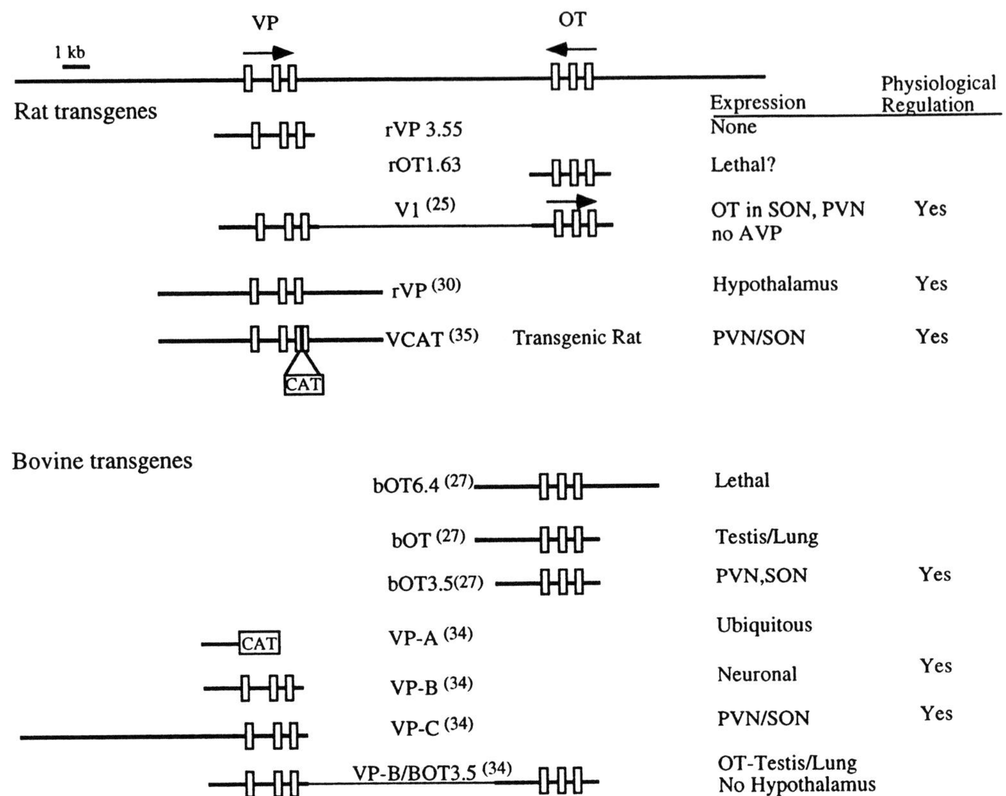

Fig. 1. Schematic of the OT/AVP gene loci and transgene constructs used to examine the regulation of OT and AVP gene expression. The reference describing each transgene is provided in parentheses. The pattern of transgene expression and physiological regulation are indicated in the right two columns. Figure modeled after Ho et al. *(26)*.

The genes for OT and AVP were first cloned by Ivell and Richter *(14,15)*. The OT and AVP genes share a similar structure, and are likely to have evolved from a common ancestral gene. mRNA derived from both of these genes are translated into preprohormones, comprised of a signal peptide, the OT or AVP sequence, and a selective neurophysin transport protein. Both OT and AVP genes are composed of three exons, with the hormone sequence encoded by exon 1, and the neurophysin encoded by exons 1 through 3. The precise role of neurophysin is unknown, but it is believed to be involved in posttranslational processing and transport. The OT and AVP genes are linked in a tail-to-tail arrangement, with intergenic regions ranging from 3 kb in mice *(16)* to 12 kb in rat *(17)* and human *(18)* (Fig. 1). In the hypothalamus, OT and AVP genes are expressed at very high levels, and are regulated by specific physiological stimuli, such as vaginocervical stimulation or plasma osmolarity. Indeed, analysis of transcript abundance in the hypothalamus has revealed that OT and AVP mRNA are some of the most abundant messenger RNAs in the hypothalamus, after subtraction of transcripts found in the cerebellum and hippocampus *(19)*.

In the hypothalamus, OT and AVP mRNA expression often seem to be regulated in parallel by similar physiological stimuli. Infusion of hypertonic saline, i.e., an osmotic challenge, results in a several-fold increase in both OT and AVP mRNA in the hypothalamus *(20)*. Both OT and AVP mRNA in the SON increase threefold during the 24 h prior to delivery, and remain elevated throughout lactation *(21)*. OT mRNA levels increase up to 100-fold in the rat uterus just prior to delivery, and drop precipitously after parturition *(10)*. During this time, the total amount of OT mRNA in the uterus is estimated to be 70-fold that of the entire hypothalamus *(22)*.

OT and AVP signals are transduced by four membrane-bound, G-protein-coupled receptors identified as OT, V1a, V1b, and V2 receptors. Each of these receptors has been cloned, sequenced, and fully characterized, as recently reviewed by Barberis and Tribollet *(23)*. The four receptors share roughly 45% homology, and are not entirely selective for either hormone. For instance, it is clear that both hormones bind with high affinity to the OT receptor. Under most conditions, AVP has a higher affinity than OT for the V1a, V1b, and V2 subtypes. OT receptors are found in the uterus, kidneys, mammary glands, and brain. OT-receptor expression in several tissues is regulated by estrogen and is elevated several-fold in the uterus at the onset of parturition *(24,25)*. The AVP-receptor subtypes are differentially expressed, and subserve different functions. The V1a receptor is found in the liver, vasculature, and in the central nervous system (CNS). The V1b receptor is found in the anterior pituitary as well as in the brain, and the V2 receptor is found in the kidney, where it regulates urine concentration.

The tissue specificity and complex physiological regulation of OT, AVP, and their receptors make these genes useful models for investigating the molecular mechanisms regulating gene expression using transgenic approaches. This chapter reviews the results of pronuclear injections of various OT and AVP transcripts, initial studies of viral-vector gene transfer, and several recent studies using OT knockout mice. We also describe early attempts to develop transgenic models overexpressing OT and AVP receptors. Although these various transgenic studies are still at an early stage, the results to date have already yielded some important lessons about OT and AVP functions.

OT AND AVP TRANSGENIC MICE: HOW IS TRANSCIPTION REGULATED?

Several groups have created transgenic mice, using constructs containing various regions of the OT and AVP genes in order to determine the locations of the regulatory elements necessary for the tissue-specific expression and the regulation of the *OT* and *AVP* genes. These experiments have demonstrated that regulatory elements surrounding the OT/VP locus are sufficient to confer cell-specific expression and physiological regulation within the magnocellular neurons of the hypothalamus. The transgene constructs used in several of these experiments are illustrated in Fig. 1, and the results of these studies are summarized here.

Young et al. created a rat minigene locus (V1) containing 1.63 kb of the *OT* gene spliced in a tail-to-head orientation with a 3.55-kb fragment of the AVP gene *(26)*. Mice transgenic for this construct were found to express the rat OT gene in PVN and SON. The rat OT mRNA was detected only in OT neurons of the mouse hypothalamus with 90% of the OT neurons expressing the transgene. Rat OT mRNA was not detected in AVP cells of the PVN or SON. Rat AVP transcripts were not detected in any tissue. The rat OT transgene was regulated by physiological stimuli appropriately, since transcripts increased by threefold during lactation. Using antibodies which recognize the rat

OT-neurophysin—not that of the mouse—it was demonstrated that the products of the rat transgene were translated and transported to posterior pituitary *(27)*.

The results from this experiment indicate that regulatory elements contained within this construct are sufficient to promote cell-specific expression and regulation of the OT gene, but not the rat *AVP* gene. It is interesting to note that similar constructs containing only the rat OT or AVP locus failed to yield expression, suggesting an interaction between elements on the AVP and OT loci.

Ho et al. used several bovine OT constructs to investigate the molecular regulation of cell-specific expression of the *OT* gene. A construct (bOT3.5) containing 0.6 kb of 5' and 1.9 kb of 3' sequence of the *bOT* gene resulted in appropriate neuron-specific expression and physiological regulation in mice *(28)*. Mice transgenic for this construct expressed the bovine OT in oxytocinergic magnocellular neurons of the PVN and SON, but expression was generally excluded from AVP neurons in these areas. Furthermore, hyperosmotic stimulation increased the bOT expression in the SON but not in the PVN, a pattern similar to that of endogenous OT. This expression pattern indicates that regulatory elements in close proximity to the bovine OT gene are sufficient to drive cell-specific expression.

Comparison of the expression pattern of bOT3.5 and other constructs provide further clues to the regulation of the *OT* gene. A second bovine construct (bOT) was identical to bOT3.5, except that it contained an additional 0.7 kb of downstream sequence, and was expressed in the lung and testis *(13,28)* rather than the hypothalmus. The authors suggest that a repressor of hypothalamic expression may be located in additional downstream sequence of the bovine *OT* gene. This repressor may normally be inhibited by additional elements located outside of the sequences found in this construct. Interestingly, a third, larger construct (bOT6.4) containing 3 kb upstream and 2.5 kb downstream sequence appeared to be lethal. Despite the transfer of over 1000 injected embryos, no pups transgenic for this construct were born. A final construct (VP-B/bOT3.5), in which bOT3.5—which itself is expressed in a cell-specific manner—was spliced to the bovine AVP locus, was used to create transgenic mice. Cell-specific expression of the bovine *OT* gene in the hypothalamus was lost with the addition of the AVP sequence, again suggesting an interaction between the OT and AVP loci.

One particularly interesting observation based on the studies using the bovine OT transgene was the expression of the bOT3.5 transgene in the Sertoli cells of testis, and the induction of bovine OT expression in the ovary during the onset of parturition *(28,29)*. The endogenous mouse *OT* gene is not expressed in the testis or the ovary at any stage of the estrus cycle, pregnancy, or lactation. However, in cattle, the *OT* gene is expressed in the sertoli cells of the testis *(13)*, and expression in the bovine ovary increases with parturition *(30)*. Results with bOT3.5 in mice suggest that species-specific patterns of expression may be determined by regulatory elements surrounding the gene.

Studies focusing on AVP transgenes have yielded results similar to those of the OT studies. Mice transgenic for a rat AVP transgene (rVP, Fig. 1) consisting of 3 kb of 5' and 3 kb of 3' flanking sequence express the transgene in the hypothalamus, although it was not determined whether the expression was localized within the PVN or SON *(31)*. In the rat, osmotic stimulation results in an increase in both the amount of endogenous AVP mRNA and in the transcript length by increasing the length of the poly-A tail by up to 150 bp *(32)*. The expression of the rat *AVP* gene in the transgenic mice described here was increased several-fold by water deprivation, but the length of the mRNA was not affected. This suggests that the regulatory elements required for hypothalamus-specific expression and regulation by osmotic stimuli are located within 3 kb of the AVP coding sequence, but mechanisms controlling poly (A) tail length are unclear. Interestingly, in the mouse lengthening of the endogenous AVP transcript as does not occur as in the rat, suggesting the regulation of AVP poly (A) tail length is inherent to the host's cellular mechanisms rather than specific sequences on the gene. Plasma and urine concentration of AVP, as determined by radioimmunoassay, was elevated in homozygous rVP transgenic mice compared to wild-type mice, suggesting that mature protein was produced, transported to the pituitary and released into the bloodstream *(33)*. Behavioral analysis of these mice indicated that elevated central release of AVP in homozygotes increased attention and alertness *(34)*, an effect consistent with the cognitive effects of AVP previously reported.

Using a series of bovine AVP constructs, Ang et al. have provided further insight into the cell-specific regulation of *AVP* gene expression in the hypothalamus *(35)*. A reporter construct (VP-A, Fig. 1) driven by 1.25 kb of the bovine VP 5' flanking sequence is ubiquitously expressed in both peripheral and brain tissues. However, expression of a similar construct (VP-B, Fig. 1), with the same 1.25 kb 5' flanking region with the reporter gene replaced by the structural *AVP* gene, was restricted to the brain with little expression detected in peripheral tissues. Expression of a third construct (VP-C) with 9 kb of the 5' flanking region and the structural AVP sequence was cell-type-specific in the hypothalamus and exhibited physiological regulation in response to osmotic challenge. It was not determined whether the bovine AVP was expressed exclusively in AVP neurons of the PVN and SON. The expression patterns of these three constructs suggest that the 1.25-kb promoter confers general expression in both the periphery and the brain and that a silencer element in the structural gene of VP-B restricts expression to the brain, while a second silencer within the 9 kb of 5' flanking sequence of VP-C further restricts expression to the PVN and SON.

Recently, transgenic *rats* have been created using a modified rat AVP minigene *(36)*. The construct, consisting of 5 kb of 5' and 3 kb of 3' sequence, contains a modified sequence derived from chloramphenicol acetyl transferase (CAT) which results in a protein tag, allowing the immunological differentiation of the endogenous and transgene-derived rVP proteins *(37)*. This transgene is expressed in AVP magnocellular neurons in PVN and SON, and is excluded from OT neurons. The level of expression of this transgene was 2.85% of the endogenous AVP expression, but increased 10- to 15-fold in response to osmotic challenge. Little or no expression was detected in parvocellular neurons. Immunocytochemical analysis of the tagged sequence revealed that the rVP transgene mRNA was translated into protein, and that this protein was transported to the posterior pituitary *(37)*.

Perhaps the most intriguing OT/VP transgenic studies are those with rats transgenic for the vasotocin/isotocin locus of the pufferfish *Fugu rubripes (38)*. Fugu has a highly compact genome—approx 400 Mb—with relatively few repetitive sequences and a modal intron length of 80 bp *(39)*. Vasotocin (VT) and isotocin (IT) are the teleost homologs of AVP and OT. VT and IT are expressed in separate magnocellular neurons of the preoptic nucleus in fish *(40)*, and their expression is regulated by osmotic stimuli *(41)*, a pattern similar to that of AVP and OT in mammals. Unlike the mammalian OT/AVP locus, VT and IT are arranged in a head-to-tail array (rather than tail-to-tail), separated by a 24.4-kb sequence containing several genes *(38)*. A 43-kb *Fugu* cosmid containing the VT/IT locus was used to create transgenic rats. Although no VT expression was detected, reminiscent of the rat and bovine OT/VP-linked transgenes discussed here, *Fugu* IT mRNA was expressed in magnocellular neurons of the SON and PVN in the transgenic rats. Double-labeled *in situ* hybridization revealed that in the magnocellular neurons of the hypothalamus, the *Fugu* IT mRNA was co-localized with the rat's endogenous OT mRNA, but was excluded from magnocellular neurons containing AVP. Furthermore, the *Fugu* IT mRNA was increased sixfold with dehydration, although the poly A tail length was unchanged. These surprising results imply that the molecular mechanisms controlling cell-specific OT gene expression and physiological regulation in the hypothalamus arose early in vertebrate evolution and have remained highly conserved across vertebrate taxa.

VIRAL VECTOR-MEDIATED GENE TRANSFER: A TRANSGENIC APPROACH TO FUNCTION

The experiments described in the previous section utilized standard embryonic pronuclear injection techniques to transfer OT and AVP genes. These studies have been useful in gaining insights into the control of OT/VP gene expression, but have contributed less in terms of understanding the physiological roles of OT and AVP. Viral vectors are becoming useful biological vehicles to transfer genes into specific tissues adult animals *(42)*. An adenoviral vector has recently been used to transfer and express the rat AVP gene in the SON of the Brattleboro rat *(43)*. The Brattleboro rat has a single-base mutation in the neurophysin-coding sequence of the AVP gene which results in a frameshift and premature stop upon translation *(44)*. Homozygous Brattleboro rats are unable to regulate urine concentration in the kidneys, and have been used as a model of hypothalamic *diabetes insipidus*. An adenoviral vector containing the rat AVP gene driven by a CMV promoter, injected into the adult rat hypothalamus, induced AVP mRNA expression for up to 6 mo after injection *(43)*. Although viral-derived AVP mRNA was detected in the SON as well as the substantia innominata after site-specific injection, protein was present only in the SON, presumably because the substantia innominata lacks the cellular machinary required for processing the AVP protein. Analysis of water intake, urine volume, and urine concentration demonstrated that the virally derived AVP was capable of partially ameliorating the symptoms of diabetes insipidus in these rats. Viral-vector technologies offer great promise in defining the timing and spatial pattern of transgene expression, as well as extending the range of species and model systems for functional studies of gene expression.

OT KNOCKOUT MICE: WHAT ARE THE CONSEQUENCES OF GENE DELETION?

There have been no reports of an AVP knockout (KO) mouse, probably because of the availability of the naturally occurring AVP deficient Brattleboro rat model. However, there have been two independently derived OT KO mice reported in detail. Nishimori et al. *(45)* created a complete OT KO by a targeted deletion of the first exon, including sequence encoding the OT peptide. *In situ* hybridization demonstrated a complete loss of OT mRNA, but preserved AVP mRNA in the hypothalamus of the KO mice (Fig. 2). Immunocytochemical analysis of the hypothalamus also found no detectable OT peptide in homozygous KO mice, and a reduction of OT peptide in heterozygous KO mice *(46)*.

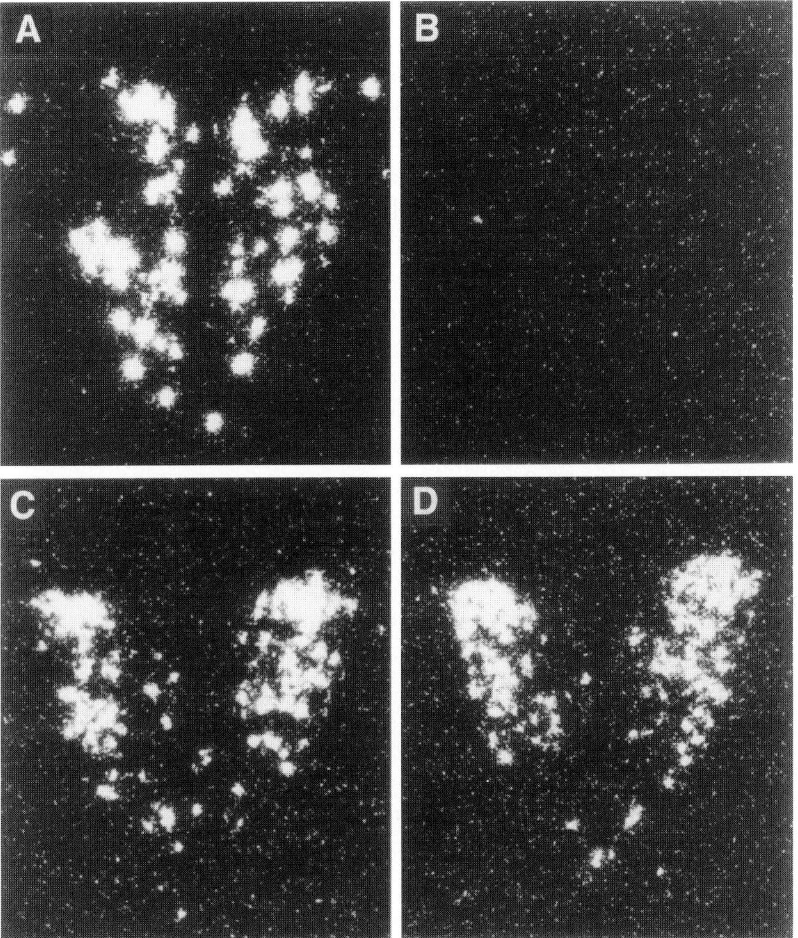

Fig. 2. Oxytocin **(A,B)** and vasopressin **(C,D)** mRNA in the paraventricular nucleus of the hypothalamus of heterozygous (A,C) and homozygous (B,D) oxytocin KO mice. No oxytocin mRNA is detected in the brain of the homozygous KO mice. AVP mRNA expression is unaffected by the oxytocin knockout allele. Reproduced with permission from Nishimori et al. *(44)*.

The distribution of oxytocin receptors in the brain was identical in animals with and without OT, suggesting that neuropeptide innervation plays a limited role in the expression of the neuropeptide receptor.

Young et al. targeted most of the first, second, and third exons of the *OT* gene, resulting in the deletion of the OT-neurophysin-coding sequence, but leaving the OT coding sequence intact *(47)*. This approach reduced OT mRNA expression by 99%, with only trace amounts of OT mRNA and no processed OT-immunoreactivity was detected in homozygous KO mice. This loss of OT mRNA expression in the neurophysin knockout mice demonstrates the importance of this region of the gene for regulating expression. These mice also exhibited a 30% decrease in AVP mRNA in the PVN and SON.

These two strains of mice, as well as a third independently created OT KO mouse, (reported in ref. *48*) exhibit the same basic phenotype: all mutant mice are fertile and female mice demonstrate normal parturition and reproductive and maternal behaviors, but fail to nurse because of the inability to eject milk. Pups from OT KO dams die within 24 h, but can be kept alive by injecting the mother with exogenous OT peptide *(41,45,47)*. Analysis of mammary tissue revealed normal milk production, but suckling-induced milk ejection was apparently absent *(45,47)*. These lines of OT KO mice have been subsequently used for investigating the role OT in behavior *(46,48–50)*, and parturition *(48)*, and have been useful in investigating the regulation of mammary-gland involution *(50)*.

Behavioral analysis of the OT KO mice has yielded surprising and interesting results. A number of studies in rats have implicated oxytocin in the facilitation of female sexual behavior and maternal behavior. In the female rat, high levels of estrogen are secreted during proestrus *(51)*, and injection of exogenous estrogen *(52)* results in increased oxytocin receptor expression in the ventromedial nucleus of the hypothalamus (VMN), a region critical for the expression of lordosis behavior *(53)*. Infusion of oxytocin into the posterior VMN facilitates *(54)* and an oxytocin receptor antagonist inhibits *(55)* the expression of lordosis behavior in female rats, suggesting that OT is necessary for the expression of female sexual behavior. Infusion of OT into the brain of virgin rats facilitates the expression of maternal behavior *(56)*, while an oxytocin receptor antagonists inhibits maternal behavior *(57)*, suggesting a role for OT in the expression of maternal behavior. Female OT KO mice are fertile and engage in apparently normal sexual behavior *(45,47)*. Quantitative analysis of maternal behavior in these mice has revealed no deficits in nurturing behavior under normal conditions with the exception of an inability to nurse *(46)*. Since AVP also binds the OT receptor, it is possible that AVP may be responsible for the expression of maternal behavior in the OT KO mice. However, when OT KO mice were infused with a selective OT antagonist, which significantly blocked the OT receptors in the brain, they continued to display normal maternal behavior (Fig. 3).

Do these observations imply that OT is not required for the expression of female sexual or maternal behavior? These contradictory results may be the result of species differences in the role of OT in the regulation of reproductive behavior in rats and mice. Rats and mice have very different distributions and regulatory patterns of OT receptors in the brain *(46)*. For example, while the OT receptor is upregulated by estrogen in the VMN of rats, a process concidered to be important in the induction of behavioral estrus, estrogen downregulates OT-receptor expression in the same region of mice *(58)*. With regard to maternal behavior, female rats exhibit a dramatic shift from indifference to pups to intensive maternal care at parturition, and this change is delayed by an oxytocin antagonist. Once rats develop maternal behavior, the maintainance of maternal behavior appears to be hormone-independent. Some strains of laboratory mice, including the C57 and 129 strains used to create KO mice, exhibit high levels of maternal care spontaneously, and therefore may not require OT to shift their behavior toward nurturing.

Despite the apparent lack of effects on reproductive and maternal behaviors, OT KO mice exhibit other behavioral phenotypes different from that of nontransgenic mice. Increases and decreases in aggressive behavior have both been reported for OT-deficient mice. Male OT-deficient mice exhibited slightly less aggressive behavior when confronted with a novel male in a neutral arena *(49)*. In another study, male OT KO mice exhibited increased aggression in a resident intruder paradigm *(59)*. However, this increased aggression was only evident in obligate KO—those produced from homozygote knockout parents—but not in KO obtained by crossing heterozygous parents. The difference in behavioral phenotype between KO males from KO and heterozygous parents suggests a potential role of prenatal exposure of OT for the normal development of social behaviors.

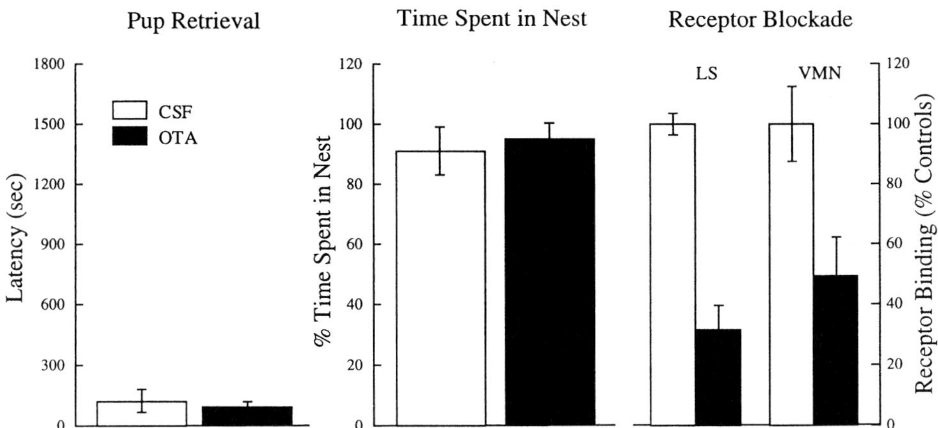

Fig. 3. Oxytocin knockout mice display normal maternal behavior, with the exception of the inability to nurse. Virgin homozygous KO mice were cannulated and infused for 5 d into the lateral ventricle, using an osmotic minipump with either artificial cerebrospinal fluid (CSF) or a specific oxytocin-receptor antagonist (OVTA, 75 ng/h). The antagonist significantly blocked oxytocin receptors throughout the brain, including the lateral septum (LS) and the ventromedial nucleus of the hypothalamus (VMN), yet maternal behavior was unaffected. This suggests that the activation of oxytocin receptors by other hormones is not responsible for the expression of maternal behavior in oxytocin KO mice.

Oxytocin and vasopressin have also been implicated in the formation of social memory *(60,61)*. OT KO mice exhibit a deficit in the formation of social memory *(59)*. Male mice typically investigate a novel mouse, i.e., a juvenile or ovariectomized (OVX) female, for a duration of time which decreases upon subsequent exposures to the same juvenile. This is interpreted as the formation of a social memory, so that the experimental male spends less time investigating a familiar mouse. OT KO mice fail to show this decrease in the amount of time investigating a novel animal upon subsequent exposures, indicating a deficit in social memory. This deficit is not caused by deficits in olfactory processing, since OT KO mice do not show difficulty in locating food in a foraging task. Spatial memory also appears intact in OT KO mice, as they show normal performance in the Morris water maze test. These results confirm a role for OT in the expression of memories specific for social stimuli.

The role of OT in the induction of labor has been further investigated using OT KO and cyclooxygenase-1 (COX-1) KO mice *(48)*. COX-1 is important for the induction of prostaglandin synthesis, which is involved in the initiation of labor in rodents. OT KO mice have normal gestation times, while COX-1 KO mice have delayed delivery (21.6 d compared to 19.6 d in wild-types). Surprisingly, double mutants for OT and COX-1 initiated labor at the same time as wildtypes, but exhibited an extended duration of labor, with pups being delivered over a 2-d period. Analysis of the corpus lutea of these mice indicated that while COX-1 KO mice had a delayed involution of the corpus luteum, OT KO and double COX-1/OT KO mice displayed normal luteolysis. These observations suggest that OT and prostaglandin may be working together in the timing of labor. OT may have luteotropic effects, which maintain the corpus luteum in late pregnancy, until prostaglandins reach a luteolytic threshold. In the COX-1 KO mouse, luteolysis is delayed because of the luteotropic effect of OT. The absence of OT in the COX-1/OT double mutants permits luteolysis in the absence of prostaglandins, thereby initiating labor. The extended labor in the COX-1/OT double mutant, which is not observed in either of the single mutants, suggests a redundancy in the roles of OT and prostaglandin in the progression of labor *(48)*.

OT- AND AVP-RECEPTOR TRANSGENES

Little is known about the molecular mechanisms regulating OT- and AVP-receptor expression. There have been few attempts to generate transgenic mice using OT- and AVP-receptor genes. We have initiated transgenic studies focusing on expression of OT receptor (OTR) and V1a receptors in the brain. The behavioral effects of OT and AVP are thought to be mediated by OT-receptor and the V1a receptor subtype, respectively. Unlike OT and AVP peptides, the receptors for OT and AVP—particularly the V1a subtype—reveal tremendous species differences in expression pattern within the brain. Even among closely related species, the OTR and V1a-receptor are expressed in different regions of brain, thus providing an opportunity to understand the evolution of gene-expression patterns. For example, in the rat, OT-receptor binding is abundant in the bed nucleus of the stria terminalis (BnST) and the VMN of the hypothalamus, and is rare in the lateral septum. In contrast, in the mouse brain the OT receptor is abundant in lateral septum, but rare in the BnST and VMN. Furthermore, estrogen increases OT-receptor expression in the VMN of the rat, and decreases expression the same region of the mouse brain. Therefore, the region-specific regulation of the receptor gene must be species-specific, in contrast to the highly conserved nature of the OT and VP gene expression.

The species differences in receptor distribution may contribute to species differences in the roles of OT and AVP in regulating behavior. Behavioral studies using species with different receptor distributions have suggested that the pattern of receptor distribution in the brain is associated with species-typical behaviors. For example, OTR and V1a receptors are involved in several social behaviors related to monogamy in voles, such as pair bonding *(62)* and paternal care *(63)*. Vole species that are monogamous, such as prairie voles and pine voles, share a common regional distribution of receptor expression that differs from that of montane voles and meadow voles, which are nonmonogamous *(64–67)*. Transgenic studies using prairie-vole OTR and V1a genes have been used to investigate the regulation of region-specific expression of these genes.

A reporter-gene construct (pvOTR-lacz) containing 5 kb of the 5' flanking region of the prairie vole OT-receptor gene placed upstream of *lac z* sequences was used to create transgenic mice. The pattern of expression of this transgene in the brain was similar to that of the prairie vole. β-gal staining was detected in the lateral septum, cortex, VMN, and amygdala of the transgenic mice *(46,68)* (Fig. 4). Low levels of ectopic expression were also found in the thalamus. The overall pattern of expression of the pvOTR-lacz was not identical to the pattern of OT receptor binding in the prairie vole. For example, β-gal was not expressed in the prelimbic cortex of the adult mouse, but was found in prepubertal animals, while the OT receptor is abundant in the prairie-vole prelimbic cortex. This expression pattern of the reporter gene driven by the prairie vole 5' flanking region demonstrates that regulatory elements found in the 5-kb region upstream of the gene are sufficient to promote expression of the OT-receptor gene in a region-specific manner. Comparison of the first 1500 bp of the 5' flanking region of the prairie vole and the rat demonstrate that the 5' flanking region of the OT receptor is less

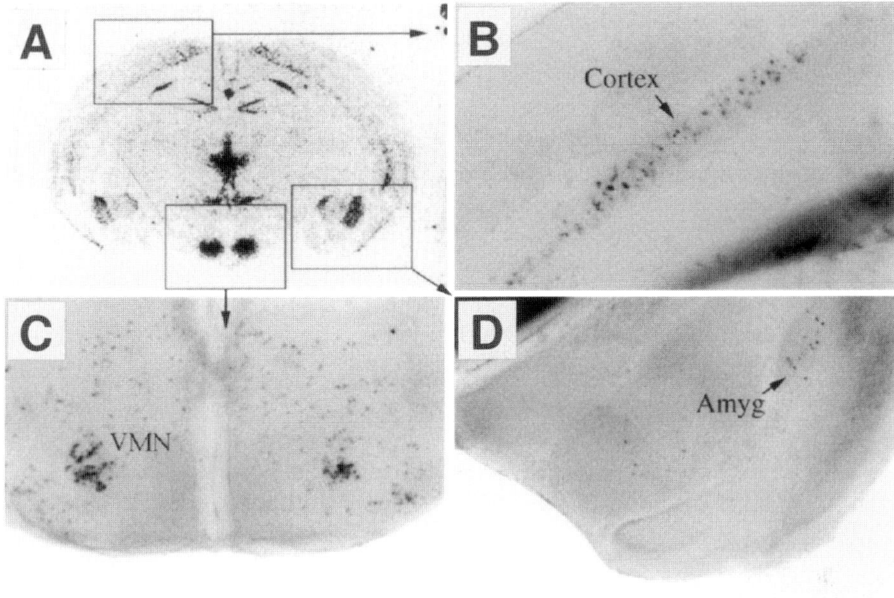

Fig. 4. Expression pattern in transgenic mice brains of a *lac Z* reporter construct spliced downstream of 5 kb of the prairie-vole oxytocin-receptor 5' flanking region **(B–D)**. The pattern of oxytocin-receptor mRNA in the prairie-vole brain is shown in **A**. Both oxytocin-receptor mRNA in the prairie vole and β-gal in the transgenic mice are found in the cortex (B), ventromedial nucleus of the hypothalamus (VMN, C) and the amygdala (D).

conserved across species than that of the *OT* gene, possibly explaining the variations in expression pattern in brain across species *(65)*.

Analysis of *V1a-receptor* gene structure and expression in voles has yielded more interesting results. The expression pattern of the *V1a-receptor* gene is strikingly different between monogamous and nonmonogamous vole species. Not surprisingly, these species exhibit quite different behavioral responses to exogenous AVP *(64)*. The coding sequence of the prairie and montane vole *V1a-receptor* genes share 99% sequence homology. However, the 5' flanking region of the prairie vole, but not the montane vole, *V1a-receptor* gene contains a 460-bp expansion of repetitive sequences located 800 bp upstream of the transcription start site *(69)*. In addition, the prairie vole gene has been duplicated, with a second copy of the gene containing a single-base deletion, resulting in a frameshift and stop codon. Thus, significant changes in gene structure between these closely related species are associated with species differences in the patterns of gene expression in the brain.

To determine whether sequences surrounding the prairie vole *V1a-receptor* gene could confer region-specific and species-specific expression patterns, we created transgenic mice using a transgene sequence containing 2.2 kb of the 5' flanking region, the coding sequence with the intron and 2.4 kb of the 3' flanking region (pvV1a) *(69)*. The transgene was expressed and translated in the mouse brain in a regional distribution strikingly similar to that of the prairie vole (Fig. 5). Receptor autoradiography using an ^{125}I labeled ligand was used to detect V1a receptor in transgenic and nontransgenic brain. In the nontransgenic mouse, the V1a receptor is detected in the diagonal band, lateral

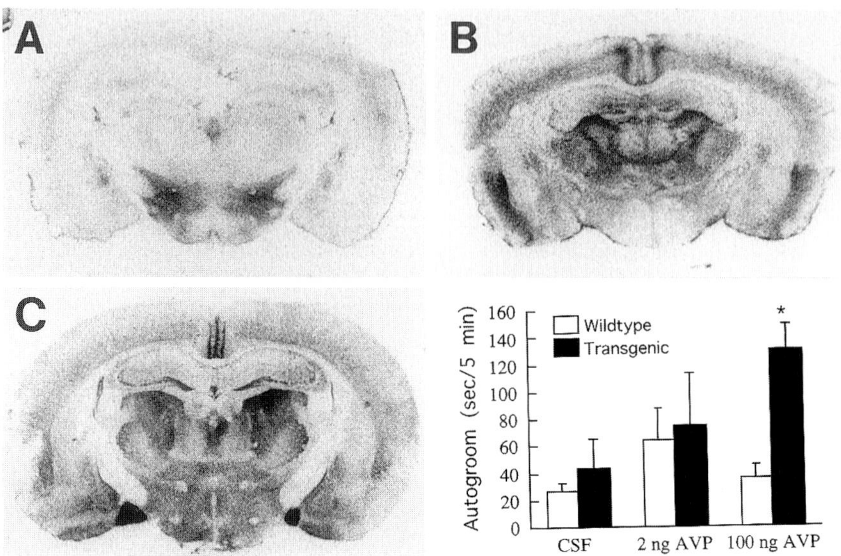

Fig. 5. Receptor autoradiograms illustrating the distribution of V1a receptors in the brains of wild-type mice **(A)**, mice transgenic for the prairie-vole *V1a-receptor* gene **(B)**, and the prairie vole **(C)**. Note the similarity in the binding pattern between the transgenic mouse and the prairie vole. Differences in behavioral response to AVP is indicated in the graph. AVP injected into the lateral ventricles of the brain results in a dose-dependent increase in autogrooming behavior in transgenic mice, but not in wild-type mice, suggesting that the transgene-encoded receptors are functional.

septum, and amygdala. In the prairie vole, and the mouse transgenic for the prairie-vole V1a receptor, V1a receptor binding is also detected in the cortex, cingulate cortex, dorsolateral and ventroposterior thalamus, and several other structures. The V1a-receptor binding is quantitatively similar in mouse and vole. Furthermore, V1a-receptor binding is not elevated in the regions of the transgenic mouse brain which normally express the receptor in mice, such as the lateral septum and diagonal band.

Since the effects of V1a-receptor activation are mediated by a G-protein second-messenger cascade, it is possible that even though receptors are expressed in novel brain regions, activation of these receptors will not produce an effect if the appropriate downstream machinery is not present. To determine whether the transgene-derived V1a receptors were functionally coupled to effector systems, mice were treated with AVP and their behavior was observed. AVP infused into the lateral ventricles of the brain produced different behavioral effects in transgenic mice and nontransgenic littermates. AVP increased autogrooming behavior in a dose-dependent manner in transgenic mice, but not in nontransgenic littermates. Furthermore, the transgenic mice displayed increased social interest after a single injection of AVP, a response similar to that of prairie voles, but not nontransgenic mice *(69)*. These results demonstrate that transgenic techniques may be useful in examining the relationship between the neuroanatomical expression pattern of neuropeptide receptors and the behavioral patterns they control. Further examination of the pvV1a transgenic mice will determine whether the behavioral roles of AVP in voles has been conferred to mice by expressing the V1a receptor in a pattern similar to that of voles.

CONCLUSION

Over the past 30 yr, studies with the oxytocin and vasopressin systems have proven to be prototypes for understanding the synthesis, storage, and secretion of neuropeptides. With the advent of powerful molecular and in vivo transgenic techniques in the past decade, mechanisms of transcription and tissue-specific expression have begun to be addressed. While this field is still young, the following conclusions can be derived from recent studies. First, the tissue-specific regulation of OT and AVP gene expression within the hypothalamus relies on a complex interaction of highly conserved promoters and repressors that surround the OT and AVP genes. Second, deletion of the OT gene in mice results in relatively few functional deficits, with a conspicuous preservation of parturition and maternal behaviors, underscoring the significance of complex interactions with complementary systems, or species differences in the physiological roles of OT. Finally, the marked species differences in OT- and AVP-receptor expression in the brain provide an interesting molecular mechanism for the evolution of species-typical behaviors that can be investigated through transgenic approaches. With these observations regarding both the peptide- and the peptide-receptor genes, it seems likely that the OT and AVP systems will continue to be informative models for understanding peptide hormone actions.

REFERENCES

1. du Vingeaud V, Ressler C, Trippett S. The sequence of amino acids in oxytocin, with a proposal for the structure of oxytocin. J Biol Chem 1953;205:949–957.
2. Gainer H, Wray W. Cellular and molecular biology of oxytocin and vasopressin. In: Knobil E, Neill JD, eds. The Physiology of Reproduction. Raven Press, New York, NY, 1994, pp. 1099–1129.
3. Insel TR, Young L, Wang Z. Central oxytocin and reproductive behaviours. Rev Reprod 1997;2:28–37.
4. de Wied D. Neuropeptides in learning and memory. Behav Brain Res 1997;83:83–90.
5. Engelmann M, Wotjak CT, Neumann I, Ludwig M, Landgraf R. Behavioral consequences of intracerebral vasopressin and oxytocin: focus on learning and memory. Neurosci Biobehav Rev 1996;20:341–358.
6. Young LJ, Wang Z, Insel TR. Neuroendocrine bases of monogamy. TINS 1998;21:71–75.
7. Swanson LW, Sawchenko PE. Hypothalamic integration: organization of the paraventricular and supraoptic nuclei. Annu Rev Neurosci 1983;6:269–324.
8. Mohr E, Bahnsen U, Kiessling C, Richter D. Expression of the vasopressin and oxytocin genes in rats occurs in mutually exclusive sets of hypothalamic neurons. FEBS Lett 1988;242:144–148.
9. Brownstein M, Russell J, Gainer H. Synthesis, transport and release of posterior pituitary hormones. Science 1980;207:373–387.
10. Lefebvre DL, Giad A, Bennet H, Lariviere R, Zingg HH. Oxytocin gene expression in the uterus. Science 1992;1553–1555.
11. Ivell R, Brackett KH, Fields MJ, Richter D. Ovulation triggers oxytocin gene expression in the bovine ovary. FEBS Lett 1985;190:263–267.
12. Jankowski M, Hajjar F, Kawas SA, Mukaddam-Daher S, Hoffman G, McCann SM, et al. Rat heart: a site of oxytocin production and action. Proc Natl Acad Sci USA 1998;95:14,558–14,563.
13. Ang HL, Ungefroren H, de Bree F, Foo NC, Carter D, Burbach JP, et al. Testicular oxytocin gene expression in seminiferous tubules of cattle and transgenic mice. Endocrinology 1991;128:2110–2117.
14. Schmale H, Heinsohn S, Richter D. Structural organization of the rat gene for the arginine vasopressin-neurophysin precursor. EMBO J 1983;2:763–767.
15. Ivell R, Richter D. Structure and comparison of the oxytocin and vasopressin genes from rat. Proc Natl Acad Sci USA 1984;81:2006–2010.
16. Hara Y, Battey J, Gainer H. Structure of mouse vasopressin and oxytocin receptor genes. Mol Brain Res 1990;8:319–324.

17. Schmitz E, Mohr E, Richter D. Rat vasopressin and oxytocin genes are linked by a long interspersed repeated DNA element (LINE): sequence and transcriptional analysis of LINE. DNA Cell Biol 1991;10:81–91.
18. Sausville E, Carney D, Battey J. The human vasopressin gene is linked to the oxytocin gene and is selectively expressed in a cultured lung cancer line. J Biol Chem 1985;260:10,236–10,241.
19. Gautvik KM, de Leeca L, Gautvik VT, Danielson PE, Tranque P, Dopazo A, et al. Overview of the most prevalent hypothalamus-specific mRNA's as identified by directional tag PCR subtraction. Proc Natl Acad Sci USA 1996;93:8733–8738.
20. Lightman SL, Young III WS. Vasopressin, oxytocin, dynorphin, enkephalin, and corticotrophin releasing factor mRNA stimulation in the rat. J Physiol 1987;394:23–39.
21. Van Tol HHM, Bolwerk ELM, Liu B, Burbach JPH. Oxytocin and vasopressin gene expression in hypothalamo-neurohypophyseal system of the rat during the estrous cycle, pregnancy, and lactation. Endocrinology 1988;122:945–951.
22. Zingg HH, Rozen F, Chu K, Larcher A, Arslan A, Richard S, et al. Oxytocin and oxytocin receptor gene expression in the uterus. Rec Prog Horm Res 1995;50:255–273.
23. Barberis C, Tribollet E. Vasopressin and oxytocin receptors in the central nervous system. Crit Rev Neurobiol 1996;10:119–154.
24. Larcher A, Neculcia J, Breton C, Arslan A, Rozen F, Russo C, et al. Oxytocin receptor gene expression in the rat uterus during pregnancy and the estrous cycle and in response to gonadal steroid treatment. Endocrinology 1995;136:5350–5356.
25. Quinones-Jenab V, Jenab S, Ogawa S, Adan RAM, Burbach JP, Pfaff DW. Effects of estrogen on oxytocin receptor messenger ribonucleic acid expression in the uterus, pituitary, and forebrain of the female rat. Neuroendocrinology 1997;65:9–17.
26. Young III WS, Reynolds K, Shepard EA, Gainer H, Castel M. Cell-specific expression of the rat oxytocin gene in transgenic mice. J Neuroendocrinol 1990;2:917–925.
27. Belenky M, Castel M, Young III WS, Gainer H, Cohen S. Ultrastructural immunolocalization of rat oxytocin-neurophysin in transgenic mice expressing the rat oxytocin gene. Brain Res 1992;583:279–286.
28. Ho MY, Carter DA, Ang HL, Murphy D. Bovine oxytocin transgenes in mice. J Biol Chem 1995;270:27,199–27,205.
29. Ho MY, Murphy D. A bovine oxytocin transgene in mice: expression in the female reproductive organs and regulation during pregnancy, parturition and lactation. Mol Cell Endocrinol 1997;136:15–21.
30. Ivell R, Rust W, Einsanier A, Hartung S, Fields M, Fuchs AR. Oxytocin and oxytocin receptor gene expression in the reproductive tract of the pregnant cow: rescue of luteal oxytocin production at term. Biol Reprod 1995;53:553–560.
31. Grant FD, Reventos J, Gordon JW, Kawabata S, Miller M, Majzoub JA. Expression of the rat arginine vasopressin gene in transgenic mice. Mol Endocrinol 1993;7:659–667.
32. Zingg HH, Lefebvre DL, Almazan G. Regulation of poly(A) tail size of vasopressin mRNA. J Biol Chem 1988;263:11,041–11,043.
33. Miller M, Kawabata S, Wiltshire-Clement M, Reventos J, Gordon JW. Increased vasopressin secretion and release in mice transgenic for the rat arginine vasopressin gene. Neuroendocrinology 1993;57:621–625.
34. Miller M, Haroutunian V, Wiltshire-Clement M. Altered alertness of vasopressin-secreting transgenic mice. Peptides 1995;16:1329–1333.
35. Ang HL, Carter DA, Murphy D. Neuron-specific expression and physiological regulation of bovine vasopressin transgenes in mice. EMBO J 1993;12:2397–2409.
36. Zeng Q, Carter DA, Murphy D. Cell specific expression of a vasopressin transgene in rats. J Neuroendocrinol 1994;6:469–477.
37. Waller S, Fairhall KM, Iain JX, Robinson CAF, Murphy D. Neurohypophyseal and fluid homeostasis in transgenic rats expressing a tagged rat vasopressin prepropeptide in hypothalamic neurons. Endocrinology 1996;137:5068–5077.
38. Venkatesh B, Si-Hoe SL, Murphy D, Brenner S. Transgenic rats reveal functional conservation of regulatory controls between *Fugu* isotocin and rat oxytocin genes. Proc Natl Acad Sci USA 1997;94:12,462–12,466.
39. Brenner S, Elgar G, Sandford R, Macrae A, Venkatesh B, Aparicio S. Characterization of the pufferfish (Fugu) genome as a compact model vertebrate genome. Nature 1993;366:265–268.

40. Goossens N, Diericks K, Vandesande F. Immunocytochemical localization of vasotocin and isotocin in the preopticohypophysial neurosecretory system of teleosts. Gen Comp Endocrinol 1977;32:371–375.
41. Hyodo S, Urano A. Changes in expression of provasotocin and proisotocin genes during adaptation to hyper- and hypo-osmotic environments in rainbow trout. J Comp Physiol B 1991;161:549–556.
42. Verma IM, Somia N. Gene therapy—promises, problems and prospects. Nature 1997;389:239–242.
43. Geddes BJ, Harding TC, Lightman SL, Uney JB. Long-term gene therapy in the CNS: reversal of hypthalamic diabetes insipidus in the Brattleboro rat by using an adenovirus expressing arginine vasopressin. Nat Med 1997;3:1402–1404.
44. Schmale H, Richter D. Single base deletion in the vaspressin gene is the cause of diabetes insipidus in Brattleboro rats. Nature 1984;308:705–709.
45. Nishimori K, Young LJ, Guo Q, Wang Z, Insel TR, Matzuk MM. Oxytocin is required for nursing but is not essential for parturition or reproductive behavior. Proc Natl Acad Sci USA 1996;93: 11,699–11,704.
46. Young LJ, Winslow JT, Wang Z, Gingrich B, Guo Q, Matzuk MM, et al. Gene targeting approaches to neuroendocrinology: oxytocin, maternal behavior, and affiliation. Horm Behav 1997;31:221–231.
47. Young III WS, Shepard E, Amico J, Hennighausen L, Wagner KU, LaMarca ME, et al. Deficiency in mouse oxytocin prevents milk ejection, but not fertility or parturition. J Neuorendocrinol 1996;8:847–853.
48. Gross GA, Imamura T, Luedke C, Vogt SK, Olson LM, Nelson DM, et al. Opposing actions of prostoglandins and oxytocin determine the onset of murine labor. Proc Natl Acad Sci USA 1998;95:11,875–11,879.
49. DeVries AC, Young III WS, Nelson RJ. Reduced aggressive behavior in mice with targeted disruption of the oxytocin gene. J Neuroendocrinol 1997;9:363–368.
50. Li M, Liu X, Robinson G, Bar-Peled U, Wagner KU, Young WS, et al. Mammary-derived signals activate programmed cell death during the first stage of mammary gland involution. Proc Natl Acad Sci USA 1997;94:3425–3430.
51. Bale TL, Dorsa DM, Johnston CA. Oxytocin receptor mRNA expression in the ventromedial hypothalamus during the estrous cycle. J Neurosci 1995;15:5058–5064.
52. Bale TL, Dorsa DM. Sex differences in and effects of estrogen on oxytocin receptor messenger ribonucleic acid expression in the ventromedial hypothalamus. Endocrinology 1995;136:27–32.
53. Pfaff DW, Schwartz-Giblin S. Cellular and molecular mechanisms of female reproductive behaviors. In: Knobil E, Neill JD, eds. The Physiology of Reproduction. Volume 2. Second Edition., Raven Press, New York, NY, 1994, pp. 107–220.
54. Schumacher M, Coirini H, Pfaff DW, McEwen BS. Behavioral effects of progesterone associated with rapid modulation of oxytocin receptors. Science 1990;250:691–694.
55. Witt DM, Insel TR. A selective oxytocin antagonist attenuates progesterone facilitation of female sexual behavior. Endocrinology 1991;128:3269–3276.
56. Pedersen CA, Prange AJ. Induction of maternal behavior in virgin rats after intracerebroventricular administration of oxytocin. Proc Natl Acad Sci USA 1979;76:6661–6665.
57. van Leengoed E, Kerker E, Swanson HH. Inhibition of postpartum maternal behaviour in the rat by injecting an oxytocin antagonist into the cerebral ventricles. J Endocrinol 1987;112:275–282.
58. Insel TR, Young L, Witt DM, Crews D. Gonadal steroids have paradoxical effects on brain oxytocin receptor. J Neuroendocrinol 1993;5:619–628.
59. Winslow JT, Young LJ, Hearn E, Gingrich B, Wang Z, Guo Q, et al. Phenotypic expression of an oxytocin peptide null mutation in mice. Adv Exp Med Biol 1998;449:241–243.
60. Popik P, Vetulani J, van Ree JM. Social memory and neurohypophyseal hormones. Eur Neuropsychopharmacol 1993;3:200,201.
61. Dantzer R, Bluthe RM, Koob GF, Le Moal M. Modulation of social memory in male rats by neurohypophyseal peptides. Psychopharmacol 1987;91:363–368.
62. Winslow J, Hastings N, Carter CS, Harbaugh C, Insel. TR. A role for central vasopressin in pair bonding in monogamous prairie voles. Nature 1993;365:545–548.
63. Wang Z, Ferris CF, De Vries GJ. Role of septal vasopressin innervation in paternal behavior in prairie voles (*Microtus ochrogaster*). Proc Natl Acad Sci USA 1994;91:400–404.
64. Young LJ, Winslow JT, Nilsen R, Insel TR. Species differences in V1a receptor gene expression in monogamous and non-monogamous voles: behavioral consequences. Behav Neurosci 1997;111:599–605.
65. Young LJ, Huot B, Nilsen R, Wang Z, Insel TR. Species differences in central oxytocin receptor gene expression: comparative analysis of promoter sequences. J Neuroendocrinol 1996;8:777–783.

66. Insel TR, Shapiro LE. Oxytocin receptor distribution reflects social organization in monogamous and polygamous voles. Proc Natl Acad Sci USA 1992;89:5981–5985.
67. Insel TR, Wang Z, Ferris CF. Patterns of brain vasopressin receptor distribution associated with social organization in microtine rodents. J Neurosci 1994;14:5381–5392.
68. Young LJ, Waymire KG, Nilsen R, Macgregor GR, Wang Z, Insel TR. The 5' flanking region of the monogamous prairie vole oxytocin receptor gene directs tissue specific expression in mice. Annu NY Acad Sci 1996;807:514–517.
69. Young LJ, Nilsen R, Waymire KG, MacGregor GR, Insel TR. Increased affiliative response to vasopressin in mice expressing the V1a receptor from a monogamous vole. Nature 1999;400:766–768.

13 Glycoprotein Hormones

Transgenic Mice as Tools to Study Regulation and Function

Ruth A. Keri, PhD *and John H. Nilson,* PhD

Contents

INTRODUCTION

Members of the glycoprotein hormone family include the gonadotropins: luteinizing hormone (LH), follicle-stimulating hormone (FSH), and chorionic gonadotropin (CG), and thyroid-stimulating hormone (TSH). These hormones are essential for the proper development and function of additional endocrine glands, and ultimately affect reproduction and metabolism. Each family member is comprised of a shared α-subunit that combines with unique β-subunits to form heterodimeric hormones. Thus, it is the β-subunit that confers biological specificity to each hormone *(1)*. All mammals synthesize and secrete the three pituitary glycoprotein hormones (LH, FSH, and TSH). LH and FSH are produced in gonadotropes, while TSH is produced by thyrotropes of the anterior pituitary. In contrast, CG is synthesized and secreted from placental syncytiotrophoblasts only in primates and equids *(2)*. LH and FSH act in concert to stimulate gonadal growth, gametogenesis, and steroidogenesis in males and females *(3,4)*. Similarly, CG acts at the level of the ovary by binding to the same receptor as LH *(5)*. This binding event is necessary for maintenance of the corpus luteum during early pregnancy in humans *(6)*. In contrast to the gonadotropins, TSH stimulates thyroid growth as well as synthesis and

From: *Contemporary Endocrinology: Transgenics in Endocrinology*
Edited by: M. Matzuk, C. W. Brown, and T. R. Kumar © Humana Press Inc., Totowa, NJ

secretion of thyroid hormone *(7)*. Expression and secretion of the pituitary glycoprotein hormones are stimulated by trophic factors from the hypothalamus, and inhibited by hormones secreted from their respective end-organs. An overview of the glycoprotein hormone axes is presented in Fig. 1.

Transgenic mouse studies over the last decade have significantly advanced our understanding of the mechanisms underlying the regulated synthesis of the glycoprotein hormones in the appropriate cell type. In addition, multiple transgenic approaches have been used to determine the key regulatory proteins involved in development of specific cell lineages, as well as cell-specific synthesis and regulation of each hormone. More recently, the focus has turned to the development of transgenic mouse models of human disease. In this chapter, we examine the progression of studies, analyzing the roles of the individual hormones in reproduction and metabolism and dissection of the regulatory cascades controlling synthesis and secretion of these hormones. New disease models are also discussed.

DISRUPTION OF GENES ENCODING GLYCOPROTEIN HORMONE SUBUNITS

Prior to the development of "knockout" technology, the question of gene function was often answered by removal of a tissue that was capable of synthesizing a particular gene product followed by replacement studies with the protein of interest. While informative, these types of approaches resulted in only crude estimates of the function of specific gene products in the entire animal. In addition, developmental studies often could not be performed with this approach, because of the obvious limitations of size of a developing organ and precision of the surgical technique. More informative are lines of mice that harbor spontaneous mutations in genes involved in various physiological processes. However, these mutants are rare, and are only informative for the limited number of genes they represent. Thus, with the advent of "knockout" technology, there has been an explosive growth in our understanding of the functional significance of hormone-encoding genes. Most importantly, there are several examples of targeted gene disruption that have discounted previous conceptions of a gene's function or necessity *(8–10)*. In this regard, α- and FSHβ-subunit genes provide particularly striking examples. This section focuses on genes encoding glycoprotein hormones and discusses how the physiological understanding of their function has been enhanced through use of mice with both natural mutations and targeted disruptions.

Targeted Disruption of the Common α-Subunit Gene

To determine the roles of the glycoprotein hormones in development as well as in adult physiology, targeted disruption of several subunit-encoding genes has been performed. Of particular interest are mice harboring a null allele for the common α-subunit of the glycoprotein hormones. Mice lacking the α-subunit have obligatory loss of all three pituitary glycoprotein hormones because of the inability to synthesize and secrete any of the intact α:β heterodimers. Although heterozygous mice are normal, α-subunit homozygous-deficient mice are infertile and dwarfed *(11)*. This is a result of the loss of both the gonadotropins and thyrotropin. Adult α-subunit-deficient mice display infantile gonads similar to those observed in prepubescent mice. In concordance with inappropriate gonadal development is the lack of gonadal androgens and estrogens. Interestingly, the development of male external genitalia occurs normally, supporting

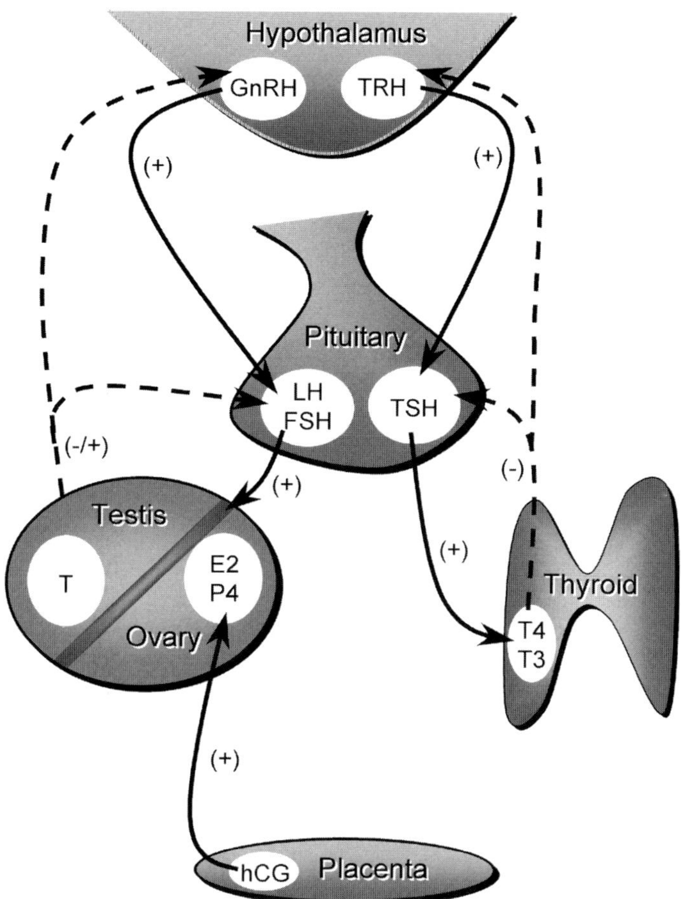

Fig. 1. Organ axes associated with the glycoprotein hormones. Pituitary expression of LH, FSH, and TSH require stimulation by the hypothalamic trophic factors GnRH (gonadotropin-releasing hormone) and TRH (thyrotropin-releasing hormone). LH and FSH then stimulate the gonads to produce the sex steroids estrogen (E2), progesterone (P4), and testosterone (T). In humans, placental production of chorionic gonadotropin (hCG) also promotes ovarian steroid synthesis and release. On the other hand, TSH stimulates thyroid production of thyroxine (T4) and triiodothyronine (T3). Thyroid hormones and sex steroids then feedback to suppress their respective axes, both at the hypothalamic and pituitary levels.

other studies that suggest that fetal production of testosterone and subsequent differentiation of secondary sex organs in mice and rats is independent of LH input *(12,13)*. This contrasts with rabbits, in which LH induction of testosterone is necessary for appropriate male development *(14,15)*.

Likewise, the absence of thyrotropin results in hypothyroid mice that have rudimentary thyroid glands and severely suppressed levels of thyroid hormone. Pituitaries from these deficient mice display thyrotrope hypertrophy and hyperplasia *(11)*; a result of the absence of negative feedback by thyroid hormone *(16)*. In addition, somatotrope and lactotrope cell numbers are reduced *(11)*. This supports the hypothesis that development of lactotropes is dependent upon free α-subunit *(17)*. However, treatment with thyroid hormone causes repopulation of the pituitary by somatotropes and lactotropes, and restores normal somatic growth *(16)*. These data implicate thyroid hormone in the development of appropriate cell proportions in the anterior pituitary gland, and conclusively show that the α-subunit is only indirectly involved (via thyroid hormone) in the appropriate development and function of lactotropes and somatotropes.

These studies reveal the impact of a combined deficiency in LH, FSH, and TSH because of the loss of the common α-subunit shared by all three hormones. Dissection of the roles of the individual hormones requires targeting of the individual β-subunits. While not performed yet for LHβ or TSHβ, the FSHβ-knockout mouse has been derived, and is discussed in the next section.

Targeted Disruption of the FSHβ Gene

Disruption of the FSHβ gene has led to both anticipated and unexpected findings regarding its role in reproductive function. Female mice lacking FSH are infertile because of an arrest of folliculogenesis at the preantral stage, and display reduced uterine size caused by the loss of significant estrogen synthesis by the ovary *(18)*. These mice phenocopy a form of primary amenorrhea in women that results from a point mutation in the FSH receptor *(19)*. Surprisingly, male mice lacking FSH are fertile, despite a reduction in testis size and sperm count *(18)*. Similar effects are observed in human males lacking appropriate FSH signaling following a mutation in its cognate receptor *(20)*. Using another approach to assess FSH function, Sassone-Corsi and colleagues specifically ablated the allele for the FSH receptor in mice. Similar to the FSHβ knockout, the female mice are infertile, while males have reduced fertility caused by smaller testis size and reduced sperm count *(21)*. These male mice also have reduced serum testosterone levels. This model mimics the human disease known as hereditary hypergonadotropic ovarian failure, which is also associated with a mutation in the human FSH-receptor gene *(19)*. Until the development of these mouse models and the identification of men harboring the inactivating mutation of the receptor, it was widely believed that FSH was required for the initiation of spermatogenesis in puberty and maintenance of normal sperm production in adults. Thus, these studies have significantly altered our view of FSH function in males, and challenge the notion that a male contraceptive targeting FSH activity would be efficacious.

Pathologies observed in the FSH-deficient mice are solely the result of an inactive FSHβ-subunit gene, because the defect can be corrected by targeted overexpression of the human FSHβ-subunit in the pituitary. Restoration of normal FSH levels was accomplished by breeding female transgenic mice harboring the 10 kb human FSHβ-subunit gene with males containing a disrupted FSHβ allele *(22)*. Progeny were intercrossed, generating a bitransgenic animal that was deficient for the endogenous murine FSHβ-subunit gene and harbored the human homolog. Female mice with the disrupted endogenous allele accompanied by the human FSHβ allele displayed restored folliculogenesis, fertility, and normal litter sizes *(22)*. In male bitransgenics, testis size and sperm count were comparable to that observed in wild-type mice *(22)*. This suggested that species-specific differences in gene regulation were not intrinsic to this gene, and that there are no species-specific differences in heterodimer interaction with its cognate receptor, because the human/mouse hybrid FSH protein was fully functional. Although pulsatile, regulated release of FSH has been implicated in the maintenance of follicular development, ectopic expression of human FSH in the liver of FSHβ-knockout mice also resulted in fertile females that delivered term pregnancies, although inefficiently *(22)*. In contrast, males with ectopic expression were completely normal. This implies that regulation of FSH release from the pituitary plays an important role in regulating female reproductive function, whereas FSH is not necessary for normal male fertility. Additional studies involving the use of transgenic mice that overexpress FSH are described in more detail in the following discussion of mouse models of human disease.

TISSUE- AND CELL-SPECIFIC EXPRESSION AND HORMONAL REGULATION: MODELS TO DISSECT PROMOTER FUNCTION

As shown previously, the use of targeted gene disruption or "knockout" technology has significantly extended our understanding of the genetic basis of glycoprotein hormone physiology beyond the level obtained by basic physiological manipulations and simple cell-culture studies. Likewise, transgenic mice have been useful in determining the molecular events that lead to gene expression. In several instances, a transgenic approach has revealed that models developed through transfection assays may not be an accurate reflection of the physiological regulation of a gene *(23–26)*. Thus, combined use of both analytical tools should result in greater information than either alone. While transfection studies facilitate more rapid and extensive molecular analysis of gene expression, transgenic technology affords a route for definitively testing models established in vitro in a physiological setting. It is with this in mind that we discuss advances made in the understanding of the molecular basis of regulation of the gonadotropin genes. We discuss inroads made into the basic mechanisms underlying expression of each of the subunits that comprise the glycoprotein hormone family, highlighting significant contributions of transgenic mice to our basic understanding of these processes. The impact of these studies is broader than simply understanding the expression of the glycoprotein hormone subunit genes. In addition to this fundamental body of knowledge, these studies have also supplied a mechanism for developing cell-specific targeting vehicles for expression paradigms in mice, allowing further delineation of the fundamental cell biology of the gonadotrope and thyrotrope.

Early studies utilizing transgenic mice in the study of glycoprotein hormone gene function were targeted at the characterization of tissue-specific expression and hormonal regulation of these genes. This involved the use of transgenes with truncated promoters linked to a variety of reporter genes such as chloramphenicol acetyl transferase (CAT), β-galactosidase (β-gal), and luciferase (luc). All five glycoprotein hormone subunit genes or promoters have been analyzed in transgenic mice to a greater or lesser extent. We will begin by discussing the α-subunit, common to all of the hormones, and then consider each of the β-subunits. The CGβ genes will be considered last, because this is the only subunit whose expression is confined to the placenta. Table 1 summarizes the transgenic analyses of all of the glycoprotein-hormone subunit promoters.

The Common α-Subunit Gene: Expression in Gonadotropes, Thyrotropes, and Trophoblasts

The α-subunit promoter was the first to be examined using a transgenic approach. Because of the absence of gonadotrope-lineage cell lines, it was essential to use transgenic mice to examine the cell-specific expression and hormonal regulation of this gene. We determined that the promoter proximal 1500 bp of the human α-subunit gene or 315 bp of its bovine counterpart were sufficient to confer pituitary-specific expression of a CAT

Table 1

Glycoprotein Hormone Promoter Usage in Transgenic Mice

Glycoprotein hormone subunit promoter	*Species*	*Length*	*Reporter*	*Tissue/cell type(s) of expression*	*Hormonal regulation*	*Refs*
Alpha	Human	5.5 Kb	SV40 T-Ag	Pituitary mature gonadotrope		*(74)*
		1.8 Kb	CAT	Pituitary gonadotrope	E2 (–) T (–) GnRH (+)	*(27–30)*
				Placenta		*(27)*
			β-gal	Pituitary gonadotrope		*(30)*
			SV40 T-Ag	Pituitary early gonadotrope		*(31)*
	Bovine	315 bp	CAT	Pituitary gonadotrope	E2 (–) T (–) GnRH (+)	*(27–30)*
			DT-A	Pituitary gonadotrope		*(36)*
			LHβ	Pituitary gonadotrope		*(179)*
	Murine	4.6 Kb	β-gal	Rathke's pouch pituitary gonadotrope thyrotrope	T3/T4	*(37)*
			LIF	Rathke's pouch pituitary		*(184)*
			DT-A tox176	Pituitary gonadotrope thyrotrope		*(38)*
		2.7 and 1.49 Kb	β-gal	Pituitary gonadotrope thyrotrope		*(37)*
		480 bp	β-gal	Pituitary gonadotrope thyrotrope		*(37)*
		–4.6/ –3.7 Kb :: –341/ +43 bp	β-gal	Pituitary gonadotrope thyrotrope		*(39)*

(continued)

reporter gene *(27)*. Soon after, it was shown that these regions also conveyed appropriate feedback regulation by the gonadal steroids 17β-estradiol *(28)*, testosterone *(29)*, and dihydrotestosterone *(29)*. In addition, these promoter regulatory regions were adequate to respond to a stimulatory gonadotropin-releasing hormone (GnRH) treatment paradigm *(30)*. The demonstration of the presence of regulatory elements that mediate the

Table 1 *(continued)*

Glycoprotein hormone subunit promoter	*Species*	*Length*	*Reporter*	*Tissue/cell type(s) of expression*	*Hormonal regulation*	*Refs*
FSHβ	Human	4 Kb	FSHβ	Pituitary gonadotrope	E2 (–) T(–) GnRH (+)	*(46–48)*
			SV40 T-Ag	Pituitary gonadotrope		*(178)*
	Bovine	2.3 Kb	HSV-TK	Pituitary gonadotrop Testis spermatocyte leydig Ovary corpus luteum Theca		*(51–53)*
LHβ	Human	1.2 Kb	SV40 T-Ag	Pituitary gonadotrope		*(59,74)*
	Bovine	776 bp	CAT	Pituitary gonadotrope	E2 (–) T (–) GnRH (+)	*(26)*
	Rat	1.8 Kb	SV40 T-Ag	Nonspecific		*(74)*
		2 Kb	Luc	Pituitary gonadotrope	E2 (–) T (–) GnRH (+)	*(58)*
	Ovine	1.9 Kb	CAT	Pituitary gonadotrope	E2 (–) P4 (–) T(–) GnRH (+)	*(56,57)*
TSHβ	Human	1.1 Kb	SV40 T-Ag	Pituitary		*(90)*
	Murine	1 Kb	TRβ Mutant	Pituitary		*(88)*
		4.6 Kb	TRβ Mutant	Pituitary		*(89)*
CGβ	Human	36 Kb Cosmid	CGβ	Placenta Pituitary Cerebral cortex Adrenal		*(92)*

effects of these hormones has spawned additional transfection studies with the gonadotrope-derived αT3-1 cell line *(31)*. Androgen-responsive elements (αBE and CRE) were identified for the human α-subunit promoter *(32)*, and elements involved in GnRH regulation (GnRHRE and PGBE) were defined for the murine gene *(33)*. Interestingly, although estradiol responsiveness has been demonstrated in the transgenic mouse, this has not been recapitulated via transfection analyses *(28,29)*. This finding suggests that estradiol regulation may involve a higher-order regulatory mechanism, possibly requiring additional tissues such as the hypothalamus. Recent work utilizing mice that hypersecrete LH supports this notion. In these mice, a loss of hypothalamic input accompanies a loss of estrogen-negative feedback, suggesting that the suppressive actions of estrogen are mediated via changes in GnRH secretion *(34)*.

Surprisingly, the 1500-bp human α promoter can only confer gonadotrope-specific expression; expression cannot be detected in thyrotropes, and the promoter is not regulated by chemical thyroidectomy using propylthiouracil and replacement therapy with thyroid hormone *(30)*. This finding indicates that the proximal promoter contains sufficient information to direct gonadotrope-specific expression, but it does not contain the necessary *cis*-acting elements required for thyrotrope expression. Further confirmation of the lack of thyrotrope-specific elements in the 315-bp bovine α promoter has been generated using this promoter linked to the diptheria toxin A (DT-A) chain. Low-level expression of DT-A is toxic to any cell in which it is produced *(35)*. Nonmosaic founder mice harboring the BαDT-A transgene were infertile, and transgene-positive F1 progeny from mosaic founders were also infertile. Development of the external genitalia and initally gonads appeared normal, but in adults, the gonads maintained a prepubertal character *(36)*. Thus, the reproductive phenotypes arose from a loss of gonadotropins that was confirmed serologically. Immunohistochemical approaches revealed that the loss of gonadotropins was caused by specific destruction of gonadotropes. Interestingly, prolactin levels were also reduced in these mice; however, this was likely a secondary effect because of the loss of estrogen synthesis in the gonads rather than a direct effect of gonadotrope loss on lactotrope development. Although gonadotropes were profoundly affected, no change was observed in thyrotrope number, and serum TSH levels were normal *(36)*. As a whole, these mice had many of the characteristics of the *hpg* (hypogonadal or GnRH-deficient) mouse, in contrast to the α-knockout described earlier, confirming that the promoter was not expressed in thyrotropes. Thus, several distinct transgenic approaches suggested that mammalian α-subunit gene promoters must contain thyrotrope-specific elements further upstream of the promoter proximal region.

To further define the minimal promoter region needed to target the pituitary primordia that give rise to mature gonadotropes and thyrotropes, mice were generated that contained a 4.6-kb fragment of the mouse α-subunit gene promoter linked to the coding sequence for β-galactosidase. This transgene conferred gonadotrope- and thyrotrope-specific expression, as well as appropriate regulation by radio-thyroidectomy and thyroid hormone replacement *(37)*. In another study, the 4.6-kb mouse α promoter was linked to an expression cassette encoding DT-A, or an attenuated form of the protein that is approx 30-fold less active. As observed with the shorter bovine α-subunit promoter, mice harboring this transgene lacked gonadotropes. However, thyrotrope ablation also occurred, further supporting the theory that this promoter is active in both cell types *(38)*. Truncation of the murine promoter to just 480 bp still resulted in gonadotrope- and thyrotrope-specific expression of a β-gal reporter gene, implying that this region contained sufficient *cis*-acting elements to direct appropriate cell-type-specific gene expression *(37)*. However, although all truncated promoters were functional, high-level expression was only observed with the full-length promoter, and truncation to 2.7 kb of the 5' flanking region resulted in a significant attenuation of activity *(37)*. This finding suggests that an important *cis*-acting element must reside within the region between –4.6 and –2.7 kb. These results also suggest that species-specific differences must exist among the α-subunit promoters in their placement and possibly composition of thyrotrope-specific elements, because the 1.5-kb region of the human α-subunit gene promoter was able only to direct gonadotrope expression *(30)*.

Subsequent studies examining the cell-specific expression of the mouse promoter revealed that it contains proximal and distal elements that must interact to yield high-level, hormonally regulated, thyrotrope expression. The distal and proximal elements were further refined and reside at positions –4.6/–3.7 kb and –341/–297 bp, respectively *(39)*. While the distal element appears to act as an enhancer of promoter activity, cell-culture studies suggest that the proximal region may be involved in promoter enhancement, as well as restricting expression of the α-subunit promoter to the pituitary *(39,40)*. The activity of the proximal region is mediated by a LIM homeoprotein-binding element referred to as the pituitary glycoprotein basal element (PGBE) *(40)*. This element was initially identified for its importance in gonadotrope-specific gene expression *(33)*, but now appears to also mediate activity of the promoter in thyrotropes *(40)*. These results indicate that the mechanisms that dictate thyrotrope or gonadotrope activation of the α-subunit promoter involve both shared components, such as PGBE, and several factors that appear to be unique to each cell type.

During the course of these initial studies, the gonadotrope-derived αT3-1 cell line was developed *(31)*. Thus, our current understanding of gonadotrope-specific expression of the α-subunit gene has stemmed from use of these cell lines, rather than additional transgenic approaches. These studies have resulted in the identification and characterization of numerous *cis*-acting elements important for the appropriate function of this promoter, such as the PGBE *(33)*, αBE-1 *(41)*, αBE-2 *(41)*, USF *(42)*, GnRHRE *(33)*, and the GSE *(43)*. It is important to underscore that the recapitulation of early data obtained from transient transfection studies in transgenic mice supports the further use of these in vitro models for assessing the cell-specific and hormonal regulation of the α-subunit gene.

Use of transgenes containing the human α-subunit promoter in mice presents a unique opportunity to assess additional species-specific differences in the cell-specific expression of this gene. While humans express the α-subunit gene in placenta as the α-subunit of CG, expression in mice is restricted to the pituitary. Previous studies using transient transfection analyses in the placentally derived BeWo *(44)*, JEG-3 *(45)*, and Jar cell lines revealed that a cAMP response element, or CRE, was essential for high-level activity of the human α-subunit promoter. In contrast, the bovine promoter, which harbors a CRE-homolog that is incapable of interacting with the cAMP-binding protein (CREB), had no detectable activity in these cells *(27)*. This finding suggests that the mechanism underlying the species-dependent expression of the α-subunit promoter is dictated by the presence or absence of a functional CRE. Alternatively, it is also possible that the placenta in species other than primates and horses lacks the necessary transcription factors to activate the α-subunit gene. To address this question directly, expression of the human and bovine α-subunit promoter-containing transgenes was assessed in the placentas of transgenic mice *(27)*. Interestingly, the human α promoter functioned in the murine placenta, while the bovine promoter did not *(27)*. These results support the idea that the presence of a CRE is essential for activating the α-subunit promoter in placenta, and that this tissue in nonexpressing species can activate a functional α-subunit gene. Since pituitary expression occurred for the bovine and human genes, these studies also revealed that pituitary-specific expression of this gene occurs independently of a consensus CRE.

Pituitary Gonadotropin Gene Expression

Follicle Stimulating Hormone β-Subunit

Although gonadotrope-derived cell lines have been developed that demonstrate high-level expression of the α-subunit gene, complementary cell lines that express the hormone-defining β-subunits have lagged significantly behind. Thus, use of transgenic mice was essential to characterize the promoters for the FSHβ and LHβ genes. FSHβ gene expression in transgenic mice was first examined using a 10-kb genomic fragment of the human gene containing 4 kb and 2 kb of the 5' and 3' flanking regions, respectively. This fragment conveyed appropriate gonadotrope-specific expression of the human FSHβ-subunit *(46)*. As a result of this increased expression, intact male mice also demonstrated elevated serum levels of FSH *(47)*. Expression in transgenic females was quite low, and alterations in pituitary content or serum levels of FSH could only be observed following ovariectomy *(47)*. Expression of the transgene was suitably regulated by the sex steroids testosterone and estradiol *(47)*. In addition, breeding experiments involving insertion of the FSHβ transgene into the *hpg* genetic background previously described, revealed that the transgene is also regulated by GnRH *(48)*. Further studies demonstrated that GnRH treatment could not reverse the suppressive effects of androgens *(48)*. This finding supports implications from previous studies in humans *(49,50)* of a direct pituitary effect of androgens in suppressing FSHβ gene expression rather than an indirect effect at the level of GnRH regulation.

In support of the fully functional role of the human FSHβ transgene, Matzuk and colleagues showed that the presence of this transgene on an FSHβ-deficient background fully restored fertility in males and females *(22)*. Restoration by the genomic FSHβ transgene was more efficient than that observed following ectopic expression of the α- and FSHβ-subunits, predominantly in the liver *(22)*. This is probably a result of maintenance of appropriate regulation of the genomic sequence in gonadotropes vs the high-level FSH expression achieved ectopically from the metallothionein promoter. From these studies, it is clear that the 10-kb genomic human FSHβ clone contains the necessary *cis*-acting elements to convey appropriate cell-specific expression to the gonadotrope, as well as appropriate hormonal regulation. Identification of specific sequences has not been accomplished; however, there is evidence that at least some of the required elements are located within the proximal promoter. This stems from several in vitro studies using ovine pituitary primary cultures, and the use of the 2.3-kb promoter of the bovine FSHβ promoter to direct expression of Herpes simplex virus thymidine kinase (HSV-TK). Mice harboring this construct express HSV-TK in pituitary, testes, and to some extent the ovary *(51,52)*, and pituitary expression is restricted to gonadotropes *(53)*. Although unexpected, expression of the transgene in gonadal tissue may be appropriate, given the detection of endogenous α and FSHβ expression in murine testis *(51)*. This assessment is still unproven, because of the reported testis-specific expression of the HSV-TK expression cassette in the absence of any promoter *(54)*. Using the nucleoside analog gancyclovir, cells that express HSV-TK can be ablated because of the blockade of DNA synthesis *(55)*. Treatment of FSHβ-HSV-TK transgenic mice with gancyclovir results in partial ablation of gonadotropin levels *(53)*. This supports the conclusion that this promoter is functional in some, but not all, gonadotropes in transgenic mice. In addition, ablation of cells following activation of the FSHβ promoter without consequence on GH-, TSH-, or prolactin-producing cells indicates that mature gonadotropes are not necessary for maintenance of other anterior-pituitary cell types. A fully penetrant FSHβ promoter, which will likely require the use of additional transgenic animals, has not yet been identified.

Luteinizing Hormone β-Subunit Gene

Similar to the FSHβ gene, mechanistic studies of LHβ gene expression within appropriate lineage cells have required the use of transgenic animals. In some respects, information gleaned from these studies has progressed beyond that for the FSHβ gene, because promoter regulatory sequences have been defined. The promoters from rat, bovine, and ovine LHβ genes have been studied in transgenic mice. All appear to mediate cell-specific expression to the gonadotrope and confer appropriate regulation by GnRH and gonadal steroids *(26,56–58)*. The sizes of the promoters examined ranges from 0.8–2 kb. The shortest corresponds to an 800-bp region from the bovine gene. Assuming minimal species-specific differences, this suggests that the minimal necessary *cis*-acting elements for appropriate expression patterns must reside within this region. As with the α and FSHβ promoter containing transgenes, it is important to underscore that all sequences analyzed to date have been incapable of preventing integration site effects on expression levels. Thus, additional sequences are probably necessary for appropriate shielding from heterochromatin in gonadotropes.

Many studies of the mechanisms underlying basal- and GnRH-stimulated expression of the LHb promoter have been performed in heterologous cell lines typically derived from kidney fibroblasts or somatotropes. While such analyses have identified several *cis*-acting elements that are theoretically involved in appropriate regulation of this promoter, confirmation of these results will require the use of the newly derived gonadotrope-lineage cell line, LβT2 *(59)*, or further use of transgenic mice. In this regard, three *cis*-acting elements that have been identified through a variety of in vitro methods have been confirmed within the bovine promoter in transgenic mice (*60* and personal observation). One such site, the gonadotrope-specific element (GSE), had been identified in the α-subunit promoter through the use of transient transfections. This site binds the orphan nuclear receptor, steroidogenic factor 1 or SF-1 *(60,61)*. The LHβ proximal promoter harbors two sequences that match an SF-1 consensus binding site. Both have been implicated in basal- and GnRH-stimulated activity of the rat promoter using transient transfection in growth-hormone-producing cells *(61–63)*. The more distal of these two sites in the bovine LHβ promoter has also been analyzed in transgenic mice. Mutation of this site substantially attenuates promoter activity and eliminates GnRH responsiveness *(60)*. Interestingly, targeted disruption of the SF-1 gene in mice reduced basal *(64)*, but not GnRH-stimulated *(65)*, activity of the endogenous mouse LHβ gene. This suggests that an additional protein other than SF-1 may regulate the LHβ promoter through the GSE. The nature of this factor has yet to be determined.

Similar to the GSE, a site for the bicoid-related, homeodomain-containing protein, Pitx1 (Ptx1, P-Otx), has been identified within the bovine promoter using cell-culture paradigms *(66,67)*. Mutation of this site in the context of the full-length LHβ promoter in transgenic mice reduces activity to nondetectable levels (CC Quirk et al., 2001; in press), confirming its importance for appropriate expression of the LHβ gene. In addition, targeted disruption of the Pitx1 gene in mice results in a diminution of gonadotropes, as well as a reduction in LHβ gene expression *(68)*. A similar protein, Otx1, has also been knocked out in mice. These mice present with transient dwarfism and hypogonadism that is fully corrected by 4 mo of age *(69)*. In the case of Otx1 gene disruption, gonadotrope cells are normal in number, but the ability of these cells to synthesize LH and FSH is significantly hampered early in life *(69)*. Results from mice deficient in Pitx1 and Otx1 suggest that these factors play important, but not essential, roles in regulating LHβ gene expression. The complete loss of activity observed after mutation of the binding site for these two factors implicates another factor that plays a more essential role in regulating LHβ gene expression, or indicates that these two factors cooperatively control this gene.

One factor that was fortuitously discovered as essential for LHβ gene expression is the early growth-response protein, or Egr-1 (Krox24, NGFI-A, zif/268). This zinc-finger protein is an immediate early transcription factor that is widely expressed and rapidly induced by stimuli promoting growth, differentiation, and apoptosis. Surprisingly, targeted ablation of this gene has resulted in a phenotype largely restricted to fertility. Two studies have been performed with conflicting data. In the first, disruption of the Egr-1 gene caused a selective reduction in LHβ gene expression in both sexes *(70)*. Surprisingly, only females suffered from infertility as a result. Using transient transfection assays in the kidney-derived CV-1 cell line, the authors also showed that Egr-1 has a direct, positive effect on the LHβ promoter acting through two consensus proximal Egr-1 sites *(70)*. This finding suggests that the loss of Egr-1 has a direct impact on transcriptional activity of the LHβ promoter. However, since Egr-1 is also expressed in the hypothalamus *(71)*, it is unknown whether this protein also plays an important role in regulating GnRH. Another independent study also revealed a loss in LHβ gene expression as well as reduced growth-hormone gene expression *(71)*. The decline in GH activity, which was caused by a decrease in somatotrope number, led to a 20% reduction in body mass. In contrast, gonadotrope number was unaffected, and the reduction in serum LH was the result of a selective loss of LHβ gene expression. In addition to the pituitary effects, these investigators also observed a reduction in ovarian LH-receptor gene expression. In this report, infertility was observed in both males and females (71). The underlying cause of the sex-specific differences observed between the two studies is unknown, but may stem from the use of different targeting strategies for disruption of the Egr-1 gene. Using transient transfection studies in a number of cell types, several investigators have shown that Egr-1 plays an important role in regulating GnRH-induced expression of the LHβ gene *(63,67,72,73)*. Mediation of the GnRH signal, however, requires synergistic interaction with either SF-1 *(63,72,73)* and/or Pitx1 *(67)*.

Finally, an additional site has been identified that binds to the CCAAT box-binding factor, NF-Y *(73a)*. Although this site is important for basal activity of the LHβ promoter in transgenic mice, it does not appear to mediate GnRH induction. Unlike other sites/factors characterized in the regulation of the LHβ promoter, this is the first example of a site that regulates only basal expression in transgenic mice.

Dissection of the LHβ promoter in transgenic mice has been productive and necessary because of the lack of appropriate gonadotrope-derived cell-culture models. Recently, a cell line known as LβT2 has been derived using targeted oncogenesis (*see* Chapter 20) *(59)*. This cell line expresses three hallmarks of LH-producing cells: α– and LHβ-subunits and GnRH receptors *(74)*. Although expression of FSH*b* subunit gene is not detected in this cell line, Low and colleagues have recently shown a dose-dependent secretion of FSH from these cells by activin A treatment. Thus, it is likely that this model will become a useful tool for further analyses of the molecular events that lead to cell-specific, hormonally regulated expression of the LHβ and FSHβ genes.

Cryptic Transgene Activity in Gonadotropes

Development of transgenic mice can occasionally lead to unexpected findings concerning the architecture of *cis*-acting elements underlying cell-specific expression of genes. This was the case for Low and colleagues, who analyzed somatostatin and growth hormone-containing transgenes under control of the metallothionein-I promoter *(75)*. Human GH 3' flanking sequences were also present in both transgenes. Surprisingly, these constructs were selectively expressed in the gonadotropes of transgenic mice. Further studies revealed that it was the GH 3' flanking sequences that contained elements that directed gonadotrope-specific expression *(75)*. Thus, this region of the GH gene is capable of dictating expression of a variety of promoters (metallothionein and cytomegalovirus) to gonadotropes. The native GH promoter must be capable of overriding these signals, because GH is not normally expressed in this cell type. The importance of these elements within the GH gene is unknown, but may reveal conserved sequences that are involved in generalized pituitary-specific expression. Removal of the native GH 5' promoter sequences may result in cellular expression that is not appropriately directed. Thus, these findings indicate that construction of novel transgenes using isolated fragments from a number of genes may lead to inappropriate expression patterns in mice. To address this possibility, expression patterns should be routinely confirmed with each transgene modification.

Thyrotropin (TSH) β-Subunit Gene Expression

Unlike the FSHβ and LHβ genes, there are numerous thyrotrope models available for study of TSHβ gene expression. These include the primary cell culture of pituitaries, thyrotropic tumors perpetuated in mice (MGH101A; TtT-97), and a cell line derived from MGH101A that is denoted αTSH. TtT-97 and primary pituitary cells are the only models that continue to express TSH. Because of the relative abundance of cells from the TtT-97 tumors, these are most commonly used. Using these models, several transcription factors have been characterized that appear to directly regulate expression of TSHβ. These include Pit-1 *(76–78)* and Pit-1T *(76,79)*, thyroid hormone receptor *(80–83)*, an AP-1-like factor *(84)*, retinoid X-receptor *(85)*, Oct-1 *(86)*, and GATA-2 *(87)*. Although the regions necessary and sufficient for activity of the TSHβ promoter have not been confirmed in transgenic mice, in at least three instances either the murine or human promoters were used to target informative transgenes to thyrotropes. Two of these involve the use of mutant thyroid hormone receptor expression cassettes *(88,89)*, and the third is a targeted oncogenesis approach *(90)*. The minimal murine region that has been used to target thyrotrope-specific expression was 1 kb in length.

To confirm the findings from tissue culture experiments, a transgenic approach was used to support and further refine the roles of Pit-1 and GATA-2 in activating this promoter, as well as expanding the thyrotrope lineage during pituitary development. Using cotransfection paradigms, multiple investigators have suggested that Pit-1 and GATA-2 act synergistically to activate the murine TSHβ promoter *(87,91)*. To assess this interaction further, Rosenfeld and colleagues produced transgenic mice that overexpress either fully functional or mutant forms of Pit-1 or GATA-2 in distinct pituitary-cell lineages *(91)*. They found that Pit-1 and GATA-2 are important for activation of the TSHβ promoter, and that these factors can act together to promote thyrotrope, and inhibit gonadotrope, development *(91)*. Mechanistically, this involves

DNA-binding dependent and independent effects of Pit-1 in thyrotropes and gonadotropes, respectively *(91)*. Thus, the complex interplay between Pit-1 and GATA-2 dictates, in part, cell specification during anterior pituitary development.

Placental Chorionic Gonadotropin β-Subunit Gene Expression

Upon observation of α-subunit promoter activity in mouse placenta, Boime and colleagues sought to determine whether the CGβ genes, which are unique to primates, could also function in this tissue. Transgenic mice were produced with a 36-kb cosmid fragment harboring the entire human CGβ gene cluster, composed of the six CGβ genes. Expression of three of the genes was observed in mouse placenta in ratios similar to that observed in the human placenta *(92)*. Onset of expression in the mouse occurred during the last one-third of the gestational period, whereas CG expression in the human placenta occurs early in gestation *(92)*. This difference may be attributed to several causes, including differences in gestation length, cell type, or specific changes in patterns of expression of other proteins required to activate the CGβ genes. Alternatively, the transgene containing the CGβ gene cluster may still lack essential elements that dictate temporal patterns of expression. Functionality of the CGβ genes in murine placenta supports the notion that the mechanism involved in their activation is evolutionarily conserved, and that the mammalian placenta is primed to express CGβ when a functional gene is present. This finding also suggests that a similar mechanism may lead to expression of other placenta-specific genes that are not unique to primates.

Surprisingly, the CGβ gene cluster is also active in the pituitary and brain. Examination of the cluster reveals that the LHβ sequence resides at its most 3' end, and evolution of the CGβ genes likely occured through multiple duplication events of the LHβ gene *(93)*. Although the LHβ sequence is removed from the transgene, *cis*-acting elements involved in its activation in the pituitary may be harbored within the cluster. Thus, transcription within the pituitary gland may depend upon residual LHβ gene *cis*-acting elements located within the transgene. Alternatively, derivation of the CGβ genes may have resulted in maintenance of some pituitary-specific elements. Expression of the transgene in the brain is also not surprising, based upon multiple reports of LHβ gene expression in the brain *(94,95)*. However, a functional significance for this expression has not yet been elucidated.

REGULATION OF GONADOTROPE AND THYROTROPE FUNCTION: LESSONS LEARNED FROM TARGETED GENE DISRUPTION

Prior to the advent of targeted recombination technology in the mouse, much of our understanding of physiological processes was obtained through indirect experiments. However, use of this approach has revealed new, and occasionally surprising, insights into the key players in regulating the function of the glycoprotein hormone-producing cells in the anterior pituitary. One of these, Egr-1, has been described in detail in the previous section. Although it was expected that removal of this factor would lead to pervasive pathological consequences, most of the effects were observed at the level of LH production and secretion. This is but one example of an outcome that was not anticipated prior to the production of the Egr-1-deficient mouse. In this section, we discuss the targeted removal of a number of other proteins via gene knockout that led to interesting results that were sometimes predictable. Several factors that regulate the

Table 2

Impact of Genetic Mutations in Mice on Expression of the Glycoprotein Hormone Subunits and/or Secretion of LH, FSH, or TSH

Gene	*Mutation type*	*Glycoprotein hormone impact*	*Pituitary impact*	*Refs*
Egr-1	Knock-out	↓ LHβ mRNA ↓ LH		*(70,71)*
GnRH	Spontaneous deletion	↓ LHβ mRNA ↓ FSHβ mRNA ↓ α mRNA ↓ LH ↓ FSH	Decreased gonadotrope number	*(98,99)*
TRH	Knock-out	↑ TSH	Decreased thyrotrope number	*(105)*
SF-1	Knock-out	↓ LHβ mRNA ↓ FSHβ mRNA ↓ α mRNA	↓ LH Content ↓ FSH Content	*(64,65)*
ERα	Knock-out	↑ LHβ mRNA ↑ FSHβ mRNA ↑ α mRNA ↑ LH	No preovulatory surge	*(128,129,131)*
Aromatase	Knock-out	↑ LH ↑ FSH		*(132)*
PR	Knock-out	↑LH	No preovulatory LH surge	*(135)*
AR	Spontaneous frame-shift	↑ α mRNA ↑ LH ↑ FSH		
T3Rα	Knock-out	↓ TSHβ mRNA ↓ TSH		*(157,158)*
T3Rβ	Knock-out	↑ TSH ↑ TSHβ mRNA ↑ α mRNA		*(156)*
T3Rα + T3Rβ	Knock-out	↑ TSH ↑ TSHβ mRNA ↑ α mRNA		*(159)*
Inhibin α	Knock-out	↑ FSH		*(166)*
Activin βB	Knock-out	↑ FSH		*(170)*
Activin type II receptor	Knock-out	↓ FSH		*(171)*

function of gonadotropes and thyrotropes are discussed. We analyze the role of hypothalamic trophic factors on cell function, and present a review of nuclear-receptor impact on these cells. Finally, we discuss the relative impact of TGFβ- and TGF-β-receptor family members on gonadotrope activity. Table 2 summarizes the impact of genetic mutants on glycoprotein hormone synthesis and/or secretion.

Trophic Factors for Gonadotropes and Thyrotropes

Gonadotropin-Releasing Hormone

Although not a transgenic model, the *hpg* mouse is particularly relevant to this topic. These mice harbor a spontaneous deletion of at least 33.5 kb *(96)* from chromosome 14 *(97)* removing the distal portion of the gene encoding a biosynthetic precursor of both gonadotropin-releasing hormone (GnRH) and GnRH-associated peptide (GAP) *(96)*. Although the mice are capable of expressing mRNA from the truncated gene, no synthesis of GnRH or GAP occurs. Following the loss of this hypothalamic trophic factor, the genes encoding all three subunits that comprise LH and FSH exhibit reduced expression *(98)*, and LH and FSH levels are also reduced in serum *(99)*. As a result, these mice have immature gonads. The pathological consequences of GnRH loss are similar to those observed in patients with Kallmann syndrome, which arises as a result of a developmental derangement of GnRH neuronal migration *(100)*. Replacement with exogenous GnRH fully restores serum gonadotropin levels in *hpg* mice, suggesting that gonadotrope differentiation occurs in the absence of GnRH and that the cells simply remain quiescent until appropriately stimulated *(101)*. Supporting this notion, quantitative electron microscopy revealed that gonadotropes are present, albeit in lower numbers and that they have less cytoplasm, rough endoplasmic reticulum, and a smaller Golgi apparatus *(102)*.

Rescue of GnRH function in *hpg* mice has been accomplished through avenues other than the pulsatile administration described here. Initially, transplantation approaches were performed involving the introduction of pre-optic tissue containing GnRH neurons from normal fetal mice into the third ventricle of the brain in *hpg* mice. This restored fertility in adult mice, indicating that GnRH neurons could develop appropriate contacts with the median eminence long after development of the brain and pituitary were complete *(103)*. Another study definitively showed that deletion of a region of the GnRH gene was responsible for the *hpg* phenotype by utilizing transgenic mice harboring a genomic clone that contained a 13.5-kb fragment of the mouse GnRH gene *(104)*. Mice that were homozygous mutant for the *hpg* mutation, but also harbored the GnRH transgene were fully fertile. This was one of the first examples of transgenic rescue of a genetic defect that brought the prospects of gene therapy into full view. The *hpg* model has also been useful in a number of studies of reproduction and cancer, as described in the following section.

Thyrotropin-Releasing Hormone

TRH is expressed in the hypothalamus as well as a number of other organs. The tripeptide was originally characterized for its ability to regulate serum levels of TSH, but it also appears to function as a neurotransmitter or neuromodulator. Although TRH is a well-known regulator of TSH synthesis and secretion, its role in regulating thyrotrope development was unknown. To assess the thyrotrope-specific effects as well as any other biological target of TRH, the gene was specifically disrupted in mice. Mice lacking the TRH gene exhibit hyperglycemia, because of a profound decrease in insulin secretion, and tertiary hypothyroidism *(105)*. Unexpectedly, although TRH levels were nondetectable in these mice, serum levels of TSH were nearly twice that of a wild-type mouse *(105)*. This suggests that negative feedback by thyroid hormones plays a more significant role than TRH in regulating TSH levels. In other words, the loss of negative regulation by thyroid hormones was sufficient to overcome any loss of TRH-mediated stimulation. Interestingly, these mice also displayed a decrease in thyrotrope number that was reversible with TRH treatment *(105)*. Thus, TRH is an important trophic factor for expansion of thyrotropes, but it is not required for their initial development.

Nuclear Receptors

Orphan Receptors

Nuclear receptors act as transcription factors to either activate or suppress gene expression *(106)*. These receptors comprise those with known ligands and those for which no ligand has been identified. The receptors without known ligands have been grouped within this superfamily of transcription factors based on nucleotide-sequence homology to known nuclear receptor genes *(106)*. While ligands continue to be identified for orphan receptors *(107–110)*, the possibility exists that some of these "receptors" act in a ligand-independent fashion, and that they are not true receptors in the pharmacological sense. Two orphan receptors have been shown to have significant impact on the expression of the glycoprotein-hormone genes.

Steroidogenic Factor-1. SF-1 (or Ad4BP) is an orphan nuclear receptor that was originally identified as a key regulator of steroidogenic enzyme genes in the adrenal cortex *(111,112)* (*see* Chapter 8). However, targeted ablation of this factor revealed developmental defects that were not restricted to the adrenal gland. In addition to the loss of adrenal glands, mice were born with gonadal agenesis *(113)*. With the loss of gonadal steroid feedback, it was anticipated that gonadotropin gene expression would be elevated in the pituitaries of these mice. However, Parker and colleagues observed a significant diminution of the α, LHβ, and FSHβ mRNAs *(64)*. Pituitary expression of the GnRH receptor gene was also significantly reduced *(64)*. This suggested that GnRH signaling may be affected in these mice. While GnRH neurons were normal in number and position, the ventromedial hypothalamus, a region believed to control GnRH release, was not formed correctly *(65)*. To directly prove that the loss of GnRH input resulted in reduced gonadotropin gene expression, null mice were treated with pulsatile doses of GnRH. Recovery of LH and FSH gene expression was observed following this treatment regimen, suggesting that the loss of SF-1 results in reduced gonadotropins because of an indirect impact on GnRH secretion *(65)*. As discussed previously, these results contrast with those from Nilson and colleagues where an SF-1-binding site within the LHβ promoter was shown to be a specific mediator of basal and GnRH-induced gene expression. This discrepancy could be a result of differences in GnRH induction paradigms used in these two studies. Alternatively, the SF-1-binding site within the LHβ promoter may actually bind a different *trans*-acting factor that mediates GnRH regulation. This factor would still be present in SF-1-deficient mice and would respond to the GnRH dosing paradigm used. Further studies are necessary to distinguish between these possibilities.

Dax-1. An inhibitor of SF-1, Dax-1 (or Ahch) is also an orphan member of the nuclear receptor superfamily. Patients with Dax-1 mutations present with adrenal hypoplasia congenita and hypogonadotropic hypogonadism *(114,115)*. This finding suggests that Dax-1 may play an important role in maintaining gonadotropin gene expression. Surprisingly, mice deficient in Dax-1 have normal serum gonadotropin levels and hence normal gonadal steroids *(116)*. Females are fully fertile; however, males are infertile as a result of progressive degeneration of the testicular germinal epithelium *(116)*. These results implicate Dax-1 as a crucial regulator of spermatogenesis in mice. The differential impact of the loss of Dax-1 between mice and humans suggests that there are significant species-specific differences in its mechanism of action. Importantly, Dax-1 does not appear necessary for gonadotropin gene expression in mice, but is important in humans. Its specific role in human gonadotropin gene expression remains to be determined, but likely involves actions at both the hypothalamus and pituitary *(117)*.

Nuclear Receptors

From a variety of whole-animal and cell-culture studies, it is clear that gonadal steroids regulate the synthesis and secretion of gonadotropins. While estrogens, androgens, and progestins can all feedback and inhibit gonadotropin gene expression, estrogens also have the unique ability to induce a surge in LH-subunit gene expression and release of LH *(118)*. The tissue sites of action, and the molecular mechanisms by which these steroids regulate gonadotropin gene expression, remain elusive. This is partly a result of the complexity of the multi-organ axis involving the pituitary (site of gonadotropin synthesis), hypothalamus (site of GnRH synthesis), and gonads (site of steroid synthesis). Mice harboring targeted gene deletions have facilitated the refinement of receptor function for all three steroids.

Estrogen Receptor. There are two known forms of estrogen receptor: ERα and ERβ. When the ERα-knockout (αERKO) mice were made, only one receptor had been identified. However, these mice retained 5% residual estrogen binding in the uterus *(119)*, suggesting that an additional receptor existed. Soon thereafter, ERβ was identified *(120–122)*. Expression of this new receptor occurs in shared and distinct cell types from ERα *(123)*. The two forms of ER show species-specific differences in pituitary expression. Although most species, including the mouse, appear to express ERα in the pituitary *(123–125)*, ERβ appears to be species-dependent, with expression occurring in rat *(125,126)* and human *(124)*, but not mouse, pituitaries *(123)*. Mice harboring a disrupted allele of ERβ have also been produced *(127)*, thus allowing direct functional comparisons between the two receptor types. While deficiency in ERα results in gross reproductive defects, including male and female infertility and attenuated mammary gland development *(128)*, the phenotypes associated with loss of ERβ are much more subtle. Male mice lacking ERβ are fully fertile, and female mice have only reduced ovulatory capacity *(127)*.

Development of ER-deficient animals allowed a direct assessment of whether ERα, ERβ, or both were required for estrogen-feedback suppression of gonadotropin gene expression. While secretion of both LH and FSH were expected to rise with loss of ER *(129)*, only serum LH levels were elevated in αERKO mice. LH was elevated approx sixfold in females and twofold in males *(128)*. Upon castration, serum LH continued to rise in αERKO males, indicating that the bulk of gonadal steroid feedback in males is androgen-directed *(130)*. The functional consequence of elevated LH in females is explored in the following section concerning mouse models of ovarian hyperstimulation. In contrast to serum concentrations, mRNAs corresponding to all three gonadotropin genes were elevated: (i.e., α, ~fourfold; LHβ, ~sevenfold; and FSHβ, ~sevenfold in female mice *[131]*), implying that FSHβ mRNA translation must be affected by the loss of ERα. LH and FSH were not elevated in the βERKO mice (Ken Korach, personal communication). This suggests that ERβ does not mediate estrogen-negative feedback on the gonadotropins. However, this interpretation must be restricted solely to the mouse, because of the potentially different mechanisms of action suggested by the differential expression of ERβ in the pituitary. Taken together, these data strongly suggest that ERα is necessary and sufficient to mediate estrogen-negative feedback on gonadotropin synthesis and secretion. Evaluating whether the mechanism of action of ERα involves a direct pituitary site of action or GnRH regulation will require further studies involving tissue-selective knockout or replacement of the receptors. If a pituitary effect is discovered, this will likely involve an indirect mechanism because of the lack of high-affinity binding sites for ER in the regions of the α *(28)* or LHβ *(26)* promoters known to mediate estradiol suppression of transcription.

While not a steroid-hormone receptor, the aromatase (*cyp19*)-deficient mouse (ArKO) is important because of its inability to synthesize estrogen. Surprisingly, many of the phenotypic consequences of the loss of aromatase differ from the loss of ERα. The ovarian phenotypes for the two models are strikingly different; ArKO mice have no gross cystic changes and lack corpora lutea *(132)*. Male ArKO mice are fertile compared to αERKO mice, which are infertile *(132)*. Serum LH and FSH were elevated sixfold to 10-fold and three- to fourfold, respectively, in female ArKO mice compared to wild-type mice *(132)*. Males, however, did not display elevated LH *(132)*. Serum FSH in ArKO males was not assessed. These data support the importance of the feedback-regulatory role of estrogen on LH levels in females, and again suggest that androgens mediate the majority of feedback suppression in males. The increased FSH levels observed in ArKO mice, but not the αERKO mice, may implicate ERβ in mediating estrogen-negative feedback on FSH synthesis and secretion.

Progesterone Receptor. Together, progesterone and estrogen play important roles in regulating gonadotropin synthesis and secretion. While the specific effects of estrogens are readily discernable, it has been much more difficult to identify progesterone-only effects on the gonadotropin genes. Thus, development of a progesterone receptor (PR)-deficient mouse model (PRKO) presents a unique opportunity to assess the impact of PR loss on gonadotropin gene activity (*see* also Chapter 9). Female PRKO mice are infertile as a result of anovulation, although there is no defect in follicular development *(133)*. Rather, the ovaries of these mice display an apparent inability to recognize ovulatory signals. They also have attenuated mating behavior, impaired mammary gland development, and uterine hyperplasia *(133,134)*. With regard to the gonadotropins, serum LH, but not FSH, was elevated approx twofold compared to wild-type mice in metestrus *(135)*. PRKO females were also unable to mount a preovulatory LH surge *(135)*, further contributing to the anovulation defect. These results suggest that progesterone plays an essential role in mediating ovarian-negative feedback on LH secretion and that its effects are vital for development of the pre-ovulatory LH surge. In support of this notion, Levine and colleagues found that PRKO females were unable to respond to LH surge-promoting doses of estradiol, and displayed an attenuated gonadotropin response to GnRH pulses when compared to wild-type mice *(136)*. These studies lead to the interesting possibility that both ER and PR are required to mediate an LH surge, and may partly reflect the complex regulation of these receptors by both estrogens and progestins *(137–139)*. Further studies addressing the complex interplay of these two receptors in regulating gonadotropin gene expression may require cell-specific removal of the individual receptors rather than disruption of the corresponding genes within the whole animal.

Androgen Receptor. The identification of a mouse harboring a spontaneous mutation of the androgen receptor has eliminated the necessity to construct an induced mutant by homologous recombination. The *tfm* (testicular feminization mutant) mouse harbors a single base deletion in the coding region of the androgen receptor, which causes a frame-shift mutation in the amino-terminal domain *(140,141)*. The mRNA encoding the mutant receptor disrupts the protein and is less stable than the wild-type message *(141)*. Since the androgen receptor resides on the X chromosome *(142,143)*, studies with this mouse have been largely limited to males. The only studies involving females require the use of heterozygous or chimeric mice. Circulating gonadotropin levels are elevated over those observed in wild-type male mice, and approach those observed following castration *(144)*. However, expression of the gonadotropin genes does not follow the same trend. While α-subunit mRNA is elevated two- to threefold, mRNA encoding the LHβ-subunit is only slightly elevated, and FSHβ mRNA is undetectable *(144)*. This supports the contention by Kaetzel and Nilson that alterations in expression of the α-subunit alone may lead to changes in circulating LH levels *(145)*. Chronic treatment with the GnRH agonist, Zoladex, fails to suppress serum LH levels, and is unable to fully suppress α and LHβ gene expression in *tfm* mice *(144)*. These results specifically implicate androgens in feedback control of LH, both at the level of secretion and in gene expression. In addition, the GnRH agonist data supports the notion that the major mechanism of action of androgen-receptor suppression is mediated directly at the pituitary, and does not involve GnRH or its receptor.

Thyroid Hormone Receptor. Similar to the gonadal steroids, thyroid hormone feeds back to suppress synthesis and secretion of TSH in cells located in the pars distalis of the pituitary. Repression of TSH by thyroid hormones (T3 or T4) is mediated both directly in the pituitary to regulate the α- *(146,147)* and TSHβ- *(80,81,148)* subunit genes as well as in the hypothalamus, involving regulation of TRH *(149,150)* gene expression. There are two T3-receptor genes: T3Rα and T3Rβ, and both encode functional receptors (T3Rα1, T3Rβ1, and T3Rβ2). A variant of the T3Rα gene (T3α2) is also expressed following alternative splicing of the mRNA, but it is incapable of binding T3, and therefore its function is not yet clear *(151)*. Unlike most of the nuclear-hormone receptors, the thyroid hormone receptor is capable of binding DNA in the absence of ligand *(152)*. Thus, this receptor can function in both a T3-dependent and independent manner. The tissue-specific expression of the α and β forms of T3R are different *(153,154)*, suggesting that these receptors may have unique functions in different tissues. To address the specific roles of these receptors, individual and combined mutagenesis approaches have been employed.

Targeted disruption of the T3Rβ gene causes a loss of auditory function *(155)*, hyperthyroidism, and goiter formation *(156)*. Although the mice have elevated levels of thyroid hormone, serum levels of TSH were also elevated approx threefold. This increase was caused by a two- to threefold increase in expression of the α- and TSHβ-subunit genes rather than an increase in thyrotrope number *(156)*. From these data, Curran and colleagues have concluded that the T3Rβ isoform must be the principal negative regulator of TSH gene expression, and that the α isoform is unable to compensate for its loss.

To directly evaluate the function of the T3Rα gene, it has also been mutated in mice. In contrast to the T3Rβ-deficient mice, those lacking T3Rα are hypothyroid, exhibit growth arrest, and die by 5 wk of age *(157)*. These mice do live long enough to measure TSHβ mRNA levels in the pituitary. Surprisingly, expression of the TSHβ gene was reduced threefold *(157)*. For unknown reasons, disruption of the T3Rα gene by another group resulted in a much milder hypothyroid phenotype, with only reduced heart rate and body temperature *(158)*. TSH levels were only slightly reduced in these mice *(158)*. These data suggest that the β form of T3R may predominate in negative regulation of TSH and that the α form may be a necessary positive regulator of TSH. This implies that the α and β forms of the receptor may cooperate to modulate circulating levels of TSH. To assess this directly, mice lacking all thyroid-hormone receptors were produced by intercrossing mice heterozygous for each disruption. The resultant homozygous mice had a more severe array of phenotypes than either disruption alone, supporting the notion that the T3R isoforms may cooperate in the entire animal. Although mice devoid of all T3R were viable, they were hypothyroid and growth-arrested, and had delayed bone maturation *(159)*. Serum TSH levels were elevated 60- to 160-fold, and expression of the α- and TSHβ-subunits was increased threefold and 26-fold, respectively *(159)*. These data conclusively show that the α and β forms of T3R cooperate to regulate appropriate expression and secretion of TSH. Interestingly, the phenotypes observed were not as severe as those that occur with severe hypothyroidism, suggesting that the loss of thyroid hormone has a greater impact than the loss of its receptors. This finding may implicate additional mechanisms for its action, or suggest that additional receptors for thyroid hormone must still be identified.

TGF-β Family Proteins and Receptors

Inhibins *(160,161)* and activins *(162,163)* were originally identified and characterized by their ability to respectively inhibit or activate, FSH release from the pituitary. Soon thereafter, it became clear that these members of the TGFβ family of ligands exhibited many effects that were unrelated to reproduction *(164)*. Inhibins and activins share a common β-subunit that either homodimerizes to form activin or heterodimerizes with an α-subunit to form inhibin. Two forms of the β-subunit have been identified: βA and βB *(164)*.

Inhibin

To firmly establish the role of inhibin in regulating reproductive function, Matzuk and colleagues used targeted recombination to disrupt the α-inhibin gene (*see* Chapter 14). These mice developed aggressive gonadal sex-cord-stromal tumors *(165)*, and died of a cachexia-like syndrome that was induced by excessive activin production consequential to the inability to synthesize inhibin *(166)*. Removal of gonadal tissue prior to tumor development prolonged the life-span of the inhibin-deficient mouse. However, these mice eventually succumbed to the same wasting syndrome that resulted from the development of adrenal tumors *(166)*. Since inhibin is a negative regulator of FSH, it was not surprising to find that FSH levels were increased two- to threefold in males and up to fourfold in females *(165)*. In light of the trophic effects of FSH, it is possible that FSH is the underlying cause of tumor formation. To address this issue, the mutant inhibin a allele was transferred onto the *hpg* background. As mentioned previously, *hpg* mice lack LH and FSH because of a mutation in the GnRH gene (99). Mice lacking both GnRH and inhibin did not form gonadal tumors, implicating LH and/or FSH as a genetic modifier of the tumor-suppressing abilities of inhibin *(167)*. Similar studies using combined mutants in the FSHβ and inhibin alleles revealed that tumor development is regulated by both LH and FSH, because tumors still formed in the absence of inhibin and FSH, although at a much slower rate *(168)*. The absence of FSH prevented excessive accumulation of activin; thus, these mice also did not develop the wasting syndrome associated with inhibin deficiency *(168)*. The requirement for activin in promoting cachexia-like symptoms in inhibin α-deficient mice was confirmed using combined knockout models deficient in the activin-receptor type IIA and inhibin α. Although gonadal tumors still developed in these mice, no wasting was observed *(168)*.

To directly assess the impact of excessive FSH on gonadal tumorigenesis, mice were constructed that either overexpressed the human FSHβ-subunit in gonadotropes or both the α- and FSHβ-subunits in liver under control of the metallothionein promoter (MT-hFSH). Both resulted in elevated serum FSH, yet no tumors were observed in these mice *(168)*. This finding suggests that elevated FSH alone was insufficient to induce the formation of gonadal tumors. The impact of elevated FSH on gonadal health is described in the section on mouse models of disease.

Activin and the Activin-Receptor Type IIA

To more fully understand the role of activins in regulating reproduction, both the activin βA- and βB-subunits have been the target of homologous recombination in mice. Unfortunately, mice deficient in activin βA died within 24 h after birth, a likely result of craniofacial defects *(169)*. In contrast, mice deficient in activin βB are viable, although several suffered from a lack of eyelid fusion at birth *(170)*. The reproductive capacity of adult female mice that lack activin βB is greatly reduced, although this is apparently not caused by defects in the ovary. Indeed, although these mice have a delay in parturition, normal litter sizes are found. However, the pups soon die because the activin-βB-deficient mothers have a defect in milk let-down *(170)*. FSH levels have been measured in male and female activin-βB-deficient animals. Both have a slight increase (~20%) in serum FSH levels *(170)*. Removal of the activin-bB-subunit causes the loss of two forms of activin (B and AB) and one form of inhibin (αβB). Since the levels of serum FSH are slightly increased, this suggests that the impact of inhibin B may predominate over activin B and activin AB in the normal mouse.

Since activin-βA- and βB-deficient mice demonstrate distinct phenotypes from each other, it is possible that the activin subtypes may cooperate in various regulatory processes. To determine whether this is true, knockout mice with no forms of activin were generated by breeding the activin-βA and -βB heterozygotes and subsequently crossing double heterozygotes. These βA/βB-deficient mice had all of the phenotypes observed in the individual knockouts but no additional defects *(169)*, indicating that no cooperativity exists between the different activins during embryogenesis. To further address this issue, mice deficient in activin-receptor type IIA were generated. Surprisingly, these mice had a phenotype that was distinct from those lacking either or both activins, suggesting that additional receptors/ligands for the activin family may exist *(171)*. Most mice deficient in the receptor survived to adulthood, and also exhibited reproductive abnormalities. Male mice lacking the activin-receptor type IIA were delayed in reaching fertility because of delayed seminiferous tubule development. Females, however, were infertile. Ovaries from these mice were smaller and had few corpora lutea. Synthesis and secretion of FSH was reduced ~threefold in males and females *(171)*, suggesting that this receptor directly mediates activin regulation of FSH.

Follistatin is an activin-binding protein that prevents activin interaction with the activin-type II receptor *(172)*. Mice deficient in follistatin die perinatally *(173)*; thus these mice have not been informative relative to the regulation of gonadotropin gene expression in the adult animal. Further assessment of its role in regulating pituitary gonadotropins postpuberty will require the use of regulated mutation of the gene using a system similar to those employing inducible cre-lox technology.

MODELING HUMAN DISEASES

Knowledge gained from the studies described here has permitted the rapid development of a host of models of human diseases. Several models that have provided novel insight into normal and pathological function are described in the following section.

Resistance to Thyroid Hormone (RTH) and Hypothyroidism

RTH patients display insensitivity to thyroid hormone regulation of serum levels of TSH. RTH usually occurs as an inherited disorder that segregates as a single dominant allele. The patient population has been subdivided into two groups: those displaying resistance at the level of the pituitary, resulting from a loss of appropriate regulation of TSH by thyroid hormone (PRTH); and those that are more generally thyroid-hormone resistant (GRTH). In most cases, the allele underlying RTH corresponds to a dominant negative mutation in the thyroid-hormone receptor *(174)*. Multiple mutations have been identified. To model PRTH, two different groups have developed transgenic mice that express mutant forms of the thyroid-hormone receptor exclusively in the pituitary. In one case, Refetoff and colleagues used the 1-kb TSHβ promoter to direct expression of the G345R mutant of TRβ1, which has been identified in humans with severe RTH *(88)*. This transgene was mostly expressed in the pituitary, although some low-level activity was also observed in the liver *(88)*. This finding suggests that the TSHβ promoter, which has not been used previously in transgenic mice, may have many of the elements necessary for spatially appropriate expression. Although high-level expression of the mutant receptor was present in the pituitary, mice harboring the transgene showed only mild PRTH, as revealed by normal TSH levels in the presence of high circulating T4 *(88)*. This is a possible consequence of the use of the TSHβ promoter, which may not be sufficiently active in transgenic mice to allow development of severe PRTH.

In a similar set of experiments, Wondisford and colleagues targeted expression of the D337T mutant form of TRβ to pituitary, using the 4.6-kb mouse α-subunit promoter *(89)*. This promoter directs high-level expression to both thyrotropes and gonadotropes *(37)*. The mutant used in this study also corresponds to a mutation found in severe RTH. Mice harboring the transgene developed profound PRTH. Although T4 levels were slightly elevated, serum levels of TSH and pituitary content of TSHβ were markedly induced *(89)*. In addition, induction of hypothyroidism resulted in an attenuated response of TSH compared to wild-type controls *(89)*. Thus, either the form of the receptor or the utilization of the α-subunit promoter resulted in a more useful model of PRTH, and demonstrated that transgenic mice can be used to accurately model this disease.

Although not a transgenic system, the *hyt/hyt* mouse that harbors a mutation in the TSH-receptor gene *(175)* has been useful as a model of hypothyroidism. In these mice, thyroid development is delayed and disorganized *in utero*. As expected, *hyt/hyt* mice have an ~100-fold elevation in TSH levels, but reduced T3 and T4 serum concentrations *(176)*. Thus, these mice represent a useful model of primary hypothyroidism.

Pituitary Tumors

Mellon and colleagues have used the transforming ability of SV40 large T antigen and the cell-specific nature of the glycoprotein hormone promoters to specifically target tumorigenesis to cells of the gonadotrope and thyrotrope cell lineages. This approach has proven highly productive because it has allowed both the dissection of cell lineage and the development of differentiated cell lines that express gonadotrope- or thyrotrope-specific markers. These cell lines have revolutionized the study of gonadotrope-specific gene expression, and may do the same for thyrotrope-specific expression. The first cell line described by this group was αT3-1, produced by using the 1.8-kb human α-subunit promoter to target expression of SV40 T antigen *(31)*. More than half of the transgenic mice harboring this transgene developed pituitary tumors. One of these mice developed a large tumor that was subsequently cultured and developed into a cell line. These cells represent an early progenitor of gonadotropes because they express the endogenous mouse α-subunit and GnRH receptor genes, but none of the β-subunit genes. They also respond to GnRH but not TRH *(31)*. When a longer region (5.5 kb) of the human α promoter was utilized, a variety of tumors developed in the resulting transgenic mice. In one mouse, a tumor developed from a primitive pituitary-cell lineage *(74)*. Although cells (αT1-1) derived from this tumor express the α-subunit, no other markers of pituitary differentiation, including GnRH receptor, are present *(74)*. This finding supports the notion that this cell line represents an even earlier progenitor cell that exists after the onset of α-subunit gene expression but prior to further differentiation *(74)*. Another mouse with this transgene developed tumors with similar characteristics to the αT3-1 cell line. Finally, one mouse developed a pituitary tumor from which a thyrotrope-lineage cell line (TαT1) was developed. These cells express the α- and TSHβ-subunits as well as the transcription factor Pit-1 *(74)* and TRβ1 and TRβ2 *(177)*. In addition, expression of the TSHβ gene is appropriately suppressed by T3 *(177)*.

To initiate LH-expressing tumors, Mellon and colleagues introduced transgenes containing either the 1.2-kb human or the 1.8-kb rat LHβ promoters directing expression of SV40 large T antigen into mice. Two-thirds of the mice that harbored the human promoter developed pituitary tumors. In contrast, mice harboring the rat promoter developed few pituitary tumors and acquired brain and pancreatic tumors instead. Cell lines (LβT2 and LβT4) were established from the pituitary tumors induced with the rat promoter, and these cells express the α- and LHβ-subunit, and GnRH-receptor genes *(59,74)*. In addition, GnRH stimulation of LβT2 cells leads to selective upregulation of endogenous LHβ gene expression *(59)*. Interestingly, LβT2 cells also secrete FSH, but only in response to activin A stimulation. Thus, these cells should provide a useful tool to study the molecular mechanisms underlying cell-specific and hormonally regulated expression of the gonadotropin β-subunit genes.

The function of the human FSHβ genomic fragment to target specific expression of transgenes was revealed by Kumar and colleagues *(178)* with the insertion of a temperature-sensitive SV40 T antigen cassette into the coding sequence of the gene. Mice harboring this construct developed slow-growing, multi-focal pituitary nodules that secrete FSH. Indeed, serum FSH levels were elevated five- to 10-fold in these mice *(178)*. Secretion of FSH from these nodules slowly tapered supporting the authors' conjecture that these mice represent a useful model of human null cell adenomas.

Another model of pituitary tumors has been developed using the 1.1-kb promoter of the human TSHβ gene to direct expression of wild-type SV40 large T antigen *(90)*. Transgenic mice representing a single line were phenotypic dwarfs. All developed pituitary tumors by 7 wk of age and died of a cachexia-like syndrome by 7–9 wk of age *(90)*. These results indicate that the human TSHβ promoter used in these studies can direct pituitary-specific expression of a reporter gene. Whether the promoter can also convey thyrotrope-specific expression remains to be determined. The tumors were devoid of

TSH, LH, GH, and Prl *(90)*, suggesting that the progenitor cells may have undergone transformation prior to cell-type commitment. However, no immunohistochemical staining was performed for ACTH, leaving open the possibility that the tumor may represent a corticotrope-derived lineage.

Ovarian Hyperstimulation Syndrome

Two models of ovarian hyperstimulation using transgenic technology have been described. Both involve the targeted overexpression of a gonadotropin. In the first model, our group used the 315-bp bovine α-subunit promoter to direct gonadotrope-specific expression of a chimeric bovine LHβ-subunit gene *(179)*. The LHβ subunit contained an in-frame fusion with the carboxyl terminal peptide (CTP) of chorionic gonadotropin. This addition resulted in a two- to threefold increase in the half-life of the heterodimeric LH in rats *(179)*. Transgenic male mice were fertile, with no elevations in LH or testosterone *(179)*. In contrast, transgenic females had circulating "LH" levels that were 10- to 15-fold higher than their nontransgenic counterparts. The sex-specific differences are the result of differential regulation of the transgene by estrogens vs androgens *(34)*. In response to increased LH, the ovarian sex steroids progesterone *(179)*, estradiol *(180)*, and testosterone *(180)* were elevated. As in women with polycystic ovarian disease, the ratio of estradiol to testosterone was disrupted because there was a sharper elevation of androgens compared to estrogens *(180)*. The chronic elevation of LH in the presence of elevated steroids was surprising, and was caused by the pathological development of insensitivity to estrogen-negative feedback and a loss of responsiveness to GnRH *(34)*. Ovaries from mice with elevated LH were severely affected. Concomitant with precocious puberty, changes in ovarian architecture were observed as early as 3 wk of age, and large fluid- and blood-filled cyst development was observed from 4–5 wk of age *(180)*. As a consequence of the excessive LH without the normal midcycle surge, female transgenics were anovulatory. Although capable of ovulation following hormonal induction, female transgenic were still infertile because of uterine receptivity defects and midgestational pregancy failure *(181)*. Surprisingly, although a significant acceleration of primordial follicle loss was observed in these mice and was suggestive of premature ovarian aging (182), the oocytes that could be induced to ovulate were normal and capable of producing live young when transferred to an appropriate maternal host (181).

Similar to this model, mice with targeted disruption of the ERα gene also have elevated LH, and display a remarkably similar ovarian phenotype *(128)*. In contrast, ERβ-deficient mice do not have elevated LH or develop cystic ovaries *(127)*. These studies support the notion that elevated LH induces polycystic changes in the ovary that are independent of ERα signaling. Assessment of the individual roles of LH and ERα will require further studies, including modulation of LH levels within the context of ERα deficiency.

The negative impact of excessive LH is further demonstrated in older transgenic mice harboring the chimeric LH transgene. Mice from the CF-1 strain that are 5 mo of age invariably develop granulosa cell tumors *(179,179a)*. These tumors are highly dependent upon the mouse strain on which the transgene is present; this suggests that the ability of the LH analog to act as tumor-promoting agent depends upon genetic modifiers that remain to be elucidated. Many of the changes in these mice mimic those seen following ovarian hyperstimulation, either through uncontrolled secretion of endogenous LH or following superovulation regimens. Thus, this model will likely prove useful for the dissection of the negative impact of excessive LH on ovarian physiology.

Another model of ovarian hyperstimulation involves the use of mice that hypersecrete human FSH from multiple tissues, including the liver. Female mice with excessive FSH from liver-specific expression of the α- and FSHβ-subunits display some of the same characteristics as those with elevated LH. The FSH overexpressing female mice are infertile and develop hemorrhagic, cystic ovaries *(183)*. In addition, all three ovarian sex steroids are profoundly elevated *(183)*. However, the ratio of androgens to estrogens is not increased, as it was in the LH overexpression model. Unfortunately, mice with elevated FSH die early in life from urinary-tract abnormalities *(183)*. Thus, the ability to assess the long-term impact of its excess on processes such as tumor formation is not possible. In this model, male mice also have elevated gonadotropin, and thus the impact of elevated FSH on male reproductive function can be evaluated. Male mice are infertile, although histological analysis of the testes reveals no abnormalities and abundant sperm production is observed. The exact cause of infertility is unknown, and awaits further investigation.

Cushing's Disease

Melmed and colleagues have recently described an unusual transgenic model for Cushing's disease *(184)*. This model involves the use of the 4.6-kb mouse α-subunit promoter to direct expression of the cytokine, leukemia-inhibitory factor (LIF). In addition to other functions, LIF regulates the hypothalamic/pituitary/adrenal axis through synergism with corticotropin-releasing factor (CRF) to activate pro-opiomelanocortin (POMC) gene expression and ACTH secretion *(185)*. LIF also appears to mediate immune-activated neuroendocrine induction of ACTH *(186)*. Studies using the GH promoter to direct expression of LIF have resulted in dwarf mice with persistent Rathke's cysts, suggesting that LIF (or a related protein) may regulate pituitary development *(187)*. Mice that express LIF under control of the α-subunit promoter display Cushingoid features, including truncal obesity, elevated corticosterone levels, thin skin, and increased intraperitoneal fat *(184)*. These are mice were also growth-retarded, and have immature gonads and attenuated thyroid function. These effects are probably the result of the pituitary impact of LIF overexpression early in the development of this gland. The transgenic pituitary is small, as a result of a profound reduction in the size of the anterior pituitary. The remaining adenohypophysis has numerous cysts lined by ciliated columnar cells that resemble nasal epithelium *(184)*. Gonadotropes, somatotropes, and lactotropes are nearly absent, whereas thyrotropes are variably affected, and corticotropes consume the majority of the remaining structure *(184)*. These results reveal that inappropriate LIF expression results in abnormal pituitary development that favors corticotrope commitment. In addition, these studies demonstrate the utility of using the promoters from the glycoprotein hormone genes to direct expression of a variety of factors to dissect protein function.

SUMMARY

In summary, transgenic technology has significantly extended our understanding of glycoprotein hormone physiology. From early studies analyzing promoter function to more recent analyses of gene knockouts and targeted expression of key regulatory factors, the use of transgenic mice has allowed the dissection of complex molecular pathways in a physiological context. Along the way, many novel and useful models of human disease have also emerged. Of course, this is not unique to the field of glycoprotein hormones. Rather, the use of transgenic mice has expanded many fields that influence our basic understanding of molecular physiology. The knowledge gained thus far will likely lead to many future discoveries using novel transgenes that will allow further manipulation of these systems.

ACKNOWLEDGMENTS

We would like to thank Christine Quirk and Kristen Lozada for thoughtful comments concerning the preparation of this manuscript. This work was supported by NIH grants DK28559 and HD34032.

REFERENCES

1. Fiddes JC, Talmadge K. Structure, expression, and evolution of the genes for the human glycoprotein hormones. Rec Prog Horm Res 1984;40:43–78.
2. Jameson JL, Hollenberg AN. Regulation of chorionic gonadotropin gene expression. Endocr Rev 1993;14:203–221.
3. Griffen JE, Wilson JD. Disorders of the testes and the male reproductive tract. In: Wilson JD, Foster DW, eds., Williams Textbook of Endocrinology, 8th ed., W.B. Saunders Company, Philadelphia, PA, 1992, pp. 799–852.
4. Carr BR. Disorders of the ovary and female reproductive tract. In: Wilson JD, Foster DW, eds., Williams Textbook of Endocrinology, 8th ed., W.B. Saunders Company, Philadelphia, PA, 1992, pp. 733–798.
5. Ascoli M, Segaloff DL. On the structure of the luteinizing hormone/chorionic gonadotropin receptor. Endocr Rev 1989;10:27–44.
6. Zeleznik AJ, Benyo DF. Control of follicular development, corpus luteum function, and the recognition of pregnancy in higher primates. In: Knobil E, Neill J, eds., The Physiology of Reproduction, 2nd ed, Raven Press, New York, NY, 1994, pp. 751–782.
7. Larson PR, Ingbar SH. The thyroid gland, In: Wilson JD, Foster DW, eds., Williams Textbook of Endocrinology, 8th ed., W.B. Saunders Company, Philadelphia, PA, 1992, pp. 357–487.
8. Tamemoto H, Kadowaki T, Tobe K, Yagi T, Sakura H, Hayakawa T, et al. Insulin resistance and growth retardation in mice lacking insulin receptor substrate-1. Nature 1994;372:182–186.
9. Nishimori K, Young LJ, Guo Q, Wang Z, Insel TR, Matzuk MM. Oxytocin is required for nursing but is not essential for parturition or reproductive behavior. Proc Natl Acad Sci USA 1996;93:11,699–11,704.
10. Garry DJ, Ordway GA, Lorenz JN, Radford NB, Chin ER, Grange RW, et al. Mice without myoglobin. Nature 1998;395:905–908.
11. Kendall SK, Samuelson LC, Saunders TL, Wood RI, Camper SA. Targeted disruption of the pituitary glycoprotein hormone alpha-subunit produces hypogonadal and hypothyroid mice. Genes Dev 1995;9:2007–2019.
12. Huhtaniemi I. Molecular aspects of the ontogeny of the pituitary-gonadal axis. Reprod Fertil Dev 1995;7:1025–1035.
13. O'Shaughnessy PJ, Baker P, Sohnius U, Haavisto AM, Charlton HM, Huhtaniemi I. Fetal development of Leydig cell activity in the mouse is independent of pituitary gonadotroph function. Endocrinol. 1998;139:1141–1146.
14. Monastirsky R, Laurence KA, Tovar E. The effects of gonadotropin immunization of prepubertal rabbits on gonadal development. Fertil Steril 1971;22:318–324.
15. Catt KJ, Dufau ML, Neaves WB, Walsh PC, Wilson JD. LH-hCG receptors and testosterone content during differentiation of the testis in the rabbit embryo. Endocrinology 1975;97:1157–1165.
16. Stahl JH, Kendall SK, Brinkmeier ML, Grecos TL, Watkins-Chow DE, Campos-Barros A, et al. Thyroid hormone is essential for pituitary somatotropes and lactotropes. Endocrinology 1999;140:1884–1892.
17. Begeot M, Hemming FJ, Dubois PM, Combarnous Y, Dubois MP, Aubert ML. Induction of pituitary lactotrope differentiation by luteinizing hormone alpha subunit. Science 1984;226:566–568.

18. Kumar TR, Wang Y, Lu N, Matzuk MM. Follicle stimulating hormone is required for ovarian follicle maturation but not male fertility. Nat Genet 1997;15:201–204.
19. Aittomäki K, Lucena JLD, Pakarinen P, Sistonen P, Tapanainen J, Gromoll J, et al. Mutation in the follicle-stimulating hormone receptor gene causes hereditary hypergonadotropic ovarian failure. Cell 1995;82:959–968.
20. Tapanainen JS, Aittomaki K, Min J, Vaskivmo T, Huhtaniemi I. Men homozygous for an inactivating mutation of the follicle-stimulating hormone (FSH) receptor gene present variable suppression of spermatogenesis and fertility. Nat. Genet. 1997;15:205,206.
21. Dierich A, Sairam MR, Monaco L, Fimia GM, Gansmuller A, LeMeur M, Sassone-Corsi P. Impairing follicle-stimulating hormone (FSH) signaling in vivo: Targeted disruption of the FSH receptor leads to aberrant gametogenesis and hormonal imbalance. Proc Natl Acad Sci USA 1998;95:13,612–13,617.
22. Kumar TR, Low MJ, Matzuk MM. Genetic rescue of follicle-stimulating hormone β-deficient mice. Endocrinology 1998;139:3289–3295.
23. Dente L, Ruther U, Tripodi M, Wagner EF, Cortese R. Expression of human alpha 1-acid glycoprotein genes in cultured cells and in transgenic mice. Genes Dev 1988;2:259–266.
24. Vassar R, Rosenberg M, Ross S, Tyner A, Fuchs E. Tissue-specific and differentiation-specific expression of a human K14 keratin gene in transgenic mice. Proc Natl Acad Sci USA 1989;86:1563–1567.
25. Zimmerman K, Legouy E, Stewart V, Depinho R, Alt FW. Differential regulation of the N-myc gene in transfected cells and transgenic mice. Mol. Cell. Biol. 1990;10:2096–2103.
26. Keri RA, Wolfe MW, Saunders TL, Anderson I, Kendall SK, Wagner T, et al. The proximal promoter of the bovine luteinizing hormone β-subunit gene confers gonadotrope-specific expression and regulation by gonadotropin-releasing hormone, testosterone, and 17β-estradiol in transgenic mice. Mol Endocrinol 1994;8:1807–1816.
27. Bokar JA, Keri RA, Farmerie TA, Fenstermaker RA, Andersen BA, Hamernik DL, et al. Expression of the glycorprotein homone α-subunit gene in the placenta requires a function cyclic AMP response element, whereas a different cis-acting element mediates pituitary-specific expression. Mol Cell Biol 1989;9:5113–5122.
28. Keri RA, Andersen BA, Kennedy GC, Hamernik DL, Clay CM, Brace AD, et al. Estradiol inhibits transcription of the human glycoprotein hormone α-subunit gene despite the absence of a high affinity binding site for estrogen receptor. Mol Endocrinol 1991;5:725–733.
29. Clay CM, Keri RA, Finicle AB, Heckert LL, Hamernik DL, Marshke KM, et al. Transcriptional repression of the glycoprotein homone α subunit gene by androgen may involve direct binding of androgen receptor to the proximal promoter. J Biol Chem 1993;268:13,556–13,564.
30. Hamernik DL, Keri RA, Clay CM, Clay JN, Sherman GB, Sawyer HR, Jr, et al. Gonadotrope- and thyrotrope-specific expression of the human and bovine glycoprotein hormone α-subunit gene is regulated by distinct cis-acting elements. Mol Endocrinol 1992;6:1745–1755.
31. Windle JJ, Weiner RI, Mellon PL. Cell lines of the pituitary gonadotrope lineage derived by targeted oncogenesis in transgenic mice. Mol Endocrinol 1990;4:597–603.
32. Heckert LL, Wilson EM, Nilson JH. Transcriptional repression of the α-subunit gene by androgen receptor occurs independently of DNA binding but requires the DNA-binding and ligand-binding domains of the receptor. Mol Endocrinol 1997;11:1497–1506.
33. Schoderbek WE, Roberson MS, Maurer RA. Two different DNA elements mediate gonadotropin releasing hormone effects on expression of the glycoprotein hormone alpha-subunit gene. J Biol Chem 1993;268:3903–3910.
34. Abbud R, Ameduri R, Rao S, Nett TM, Nilson JH. Chronic hypersecretion of luteinizing hormone in transgenic mice selectively alters responsiveness of the α-subunit gene to gonadotropin-releasing hormone and estrogens. Mol Endocrinol 1999;13:1449–1459.
35. Yamaizumi M, Mekada E, Uchida T, Okada Y. One molecule of diptheria toxin fragment A introduced into a cell can kill the cell. Cell 1978;15:245–250.
36. Kendall SK, Saunders TL, Jin L, Lloyd R, Glode LM, Nett TM, et al. Targeted ablation of pituitary gonadotropes in transgenic mice. Mol Endocrinol 1991;5:2025–2036.
37. Kendall SK, Gordon DF, Birkmeier TS, Petrey D, Sarapura VD, O'Shea KS, et al. Enhancer-mediated high level expression of mouse pituitary glycoprotein hormone α-subunit transgene in thyrotropes, gonadotropes, and developing pituitary gland. Mol Endocrinol 1994;8:1420–1433.
38. Burrows HL, Birkmeier TS, Seasholtz AF, Camper SA. Targeted ablation of cells in the pituitary primordia of transgenic mice. Mol Endocrinol 1996;10:1467–1477.

39. Brinkmeier ML, Gordon DF, Dowding JM, Saunders TL, Kendall SK, Sarapura VD, et al. Cell-specific expression of the mouse glycoprotein hormone a-subunit gene requires multiple interacting DNA elements in transgenic mice. Mol Endocrinol 1998;12:622–633.
40. Wood WM, Dowding JM, Sarapura VD, McDermott MT, Gordon DF, Ridgway EC. Functional interactions of an upstream enhancer of the mouse glycoprotein hormone α-subunit gene with proximal promoter sequences. Mol Cell Endocrinol 1998;142:141–152.
41. Heckert LL, Schultz K, Nilson JH. Different composite regulatory elements direct expression of the human α subunit gene to pituitary and placenta. J. Biol. Chem. 1995;270:26,497–26,504.
42. Jackson SM, Gutierrez-Hartman A, Hoeffler JP. Upstream stimulatory factor, a basic-helix-loop-helix-zipper protein, regulates the activity of the α-glycoprotein hormone subunit gene in pituitary cells. Mol Endocrinol 1995;9:278–291.
43. Barnhart KM, Mellon PL. The orphan nuclear receptor, steroidogenic factor-1, regulates the glycoprotein hormone alpha-subunit gene in pituitary gonadotropes. Mol Endocrinol 1994;8:878–885.
44. Silver BJ, Bokar JA, Virgin JB, Vallen EA, Milsted A, Nilson JH. Cyclic AMP regulation of the human glycoprotein hormone α-subunit gene is mediated by an 18-base-pair element. Proc Natl Acad Sci USA 1987;84:2198–2202.
45. Jameson JL, Powers AC, Gallagher GD, Habener JF. Enhancer and promoter element interactions dictate cyclic adenosine monophosphate mediated and cell-specific expression of the glycoprotein hormone α gene. Mol Endocrinol 1989;3:763–772.
46. Kumar TR, Fairchild-Huntress V, Low MJ. Gonadotrope-specific expression of the human follicle-stimulating hormone β-subunit gene in pituitaries of transgenic mice. Mol Endocrinol 1992;6:81–90.
47. Kumar TR, Low MJ. Gonadal steroid hormone regulation of human and mouse follicle stimulating hormone β-subunit gene expression. Mol Endocrinol 1993;7:898–906.
48. Kumar TR, Low MJ. Hormonal regulation of human follicle-stimulating hormone-β subunit gene expression: GnRH stimulation and GnRH-independent androgen inhibition. Neuroendocrinology 1995;61:628–637.
49. Sheckter CB, Matsumoto AM, Bremner WJ. Testosterone administration inhibits gonadotropin secretion by an effect directly on the human pituitary. J Clin Endocrinol Metab 1989;68:397–401.
50. Finkelstein JS, Whitcomb RW, O'Dea ISL, Longcope C, Schoenfeld DA, Crowley JWF. Sex steroid control of gonadotropin secretion in the human male. I. Effects of testosterone administration in normal and gonadotropin-releasing hormone-deficient men. J Clin Endocrinol Metab 1991;73:609–620.
51. Markkula M, Hämäläinen TM, Loune E, Huhtaniemi I. The follicle-stimulating hormone (FSH) β- and common α-subunits are expressed in mouse testis, as determined in wild-type mice and those transgenic for the FSH β-subunit/herpes simplex virus thymidine kinase fusion gene. Endocrinology 1995;136:4769–4775.
52. Markkula M, Kananen K, Klemi P, Huhtaniemi I. Pituitary and ovarian expression of the endogenous follicle-stimulating hormone (FSH) subunit genes and an FSHβ-subunit promoter-driven herpes simplex virus thymidine kinase gene in transgenic mice; specific partial ablation of FSH-producing cells by antiherpes treatment. J Endocrinol 1996;150:265–273.
53. Markkula M, Kananen K, Paukku T, Loune E, Pelliniemi LJ, Huhtaniemi I. Induced ablation of gonadotropins in transgenic mice expressing Herpes simplex virus thymidine kinase under the FSH β-subunit promoter. Mol Cell Endocrinol 1995;108:1–9.
54. al-Shawi R, Burke J, Wallace H, Jones C, Harrison S, Buxton D, Maley S, Chandley A, Bishop JO. The herpes simples virus type 1 thymidine kinase is expressed in the testes of transgenic mice under the control of a cryptic promoter. Mol Cell Biol 1991;11:4207–4216.
55. Wallace H, Ledent C, Vassart G, Bishop JO, al-Shawi R. Specific ablation of thyroid follicle cells in adult transgenic mice. Endocrinol. 1991;129:3217–3226.
56. Brown P, McNeilly JR, Wallace RM, McNeilly AS, Clark AJ. Characterization of the ovine LH β-subunit gene: the promoter directs gonadotrope-specific expression in transgenic mice. Mol Cell Endocrinol 1993;93:157–165.
57. McNeilly JR, Brown P, Mullins J, Clark AJ, McNeilly AS. Characterization of the ovine LH β-subunit gene: the promoter is regulated by GnRH and gonadal steroids in transgenic mice. J Endocrinol 1996;151:481–489.
58. Fallest PC, Trader GL, Darrow JM, Shupnik MA. Regulation of rat luteinizing hormone β gene expression in transgenic mice by steroids and α gonadotropin-releasing hormone antagonist. Biol Reprod 1995;53:103–109.

59. Turgeon JL, Kimura Y, Waring DW, Mellon PL. Steroid and pulsatile gonadotropin-releasing hormone (GnRH) regulation of luteinizing hormone and GnRH receptor in a novel gonadotrope cell line. Mol Endocrinol 1996;10:439–450.
60. Keri RA, Nilson JH. A steroidogenic factor-1 binding site is required for activity of the luteinizing hormone β subunit promoter in gonadotropes of transgenic mice. J Biol Chem 1996;271:10,782–10,785.
61. Halvorson LM, Kaiser U, Chin WW. Stimulation of luteinizing hormone β gene promoter activity by the orphan nuclear receptor, steroidogenic factor-1. J Biol Chem 1996;271:6645–6650.
62. Halvorson LM, Ito M, Jameson JL, Chin WW. Steroidogenic factor-1 and early growth response protein 1 act through two composite DNA binding sites to regulate luteinizing hormone β-subunit gene expression. J Biol Chem 1998;273:14,712–14,720.
63. Halvorson LM, Kaiser U, Chin WW. The protein kinase C system acts through the early growth response protein 1 to increase LHβ gene expression in synergy with steroidogenic factor-1. Mol Endocrinol 1999;13:106–116.
64. Ingraham HA, Lala DS, Ikeda Y, Luo X, Shen WH, Nachtigal MW, et al. The nuclear receptor steroidogenic factor 1 acts at multiple levels of the reproductive axis. Genes Dev 1994;8:2302–2312.
65. Ikeda Y, Luo X, Abbud R, Nilson JH, Parker KL. The nuclear receptor steroidogenic factor 1 is essential for the formation of the ventromedial hypothalamic nucleus. Mol Endocrinol 1995;9:478–486.
66. Tremblay JJ, Lanctot C, Drouin J. The pan-pituitary activator of transcription, Ptx1 (pituitary homeobox 1) acts in synergy with SF-1 and Pit1 and is an upstream regulator of the Lim-homeodomain gene Lim3/Lhx3. Mol Endocrinol 1998;12:428–441.
67. Tremblay JJ, Drouin J. Egr-1 is a downstream effector of GnRH and synergizes by direct interaction with Ptx1 and SF-1 to enhance luteinizing hormone beta gene transcription. Mol Cell Biol 1999;19:2567–2576.
68. Szeto DP, Rodriguez-Esteban C, Ryan AK, O'Connell SM, Liu F, Kioussi C, et al. Role of the bicoid-related homeodomain factor Pitx1 in specifying hindlimb morphogenesis and pituitary development. Genes Dev. 1999;13:484–494.
69. Acampora D, Mazan S, Tuorto F, Avantaggiato V, Tremblay JJ, Lazzaro D, et al. Transient dwarfism and hypogonadism in mice lacking Otx1 reveal prepubescent stage-specific control of pituitary levels of GH, FSH, and LH. Development 1998;125:1229–1239.
70. Lee SL, Sadovsky Y, Swirnoff A. H, Polish JA, Goda P, Gavrilina G, Milbrandt J. Luteinizing hormone deficiency and female infertility in mice lacking the transcription factor NGFI-A (Egr-1). Science 1996;273:1219–1221.
71. Topilko P, Schneider-Maunoury S, Levi G, Trembleau A, Gourdji D, Driancourt MA, et al. Multiple pituitary and ovarian defects in Krox-24 (NGFI-A, Egr-1)-targeted mice. Mol Endocrinol 1997;12:107–122.
72. Wolfe MW, Call GB. Early growth response protein 1 binds to the luteinizing hormone-β promoter and mediates gonadotropin-releasing hormone-stimulated gene expression. Mol Endocrinol 1999;13:752–763.
73. Dorn C, Ou Q, Svaren J, Crawford PA, Sadovsky Y. Activation of luteinizing hormone beta gene by gonadotropin-releasing hormone requires the synergy of early growth response-1 and steroidgenic factor-1. J Biol Chem 1999;274:13,870–13,876.
73a. Keri RA, Bachman DJ, Behrooz A, et al. An NF-Y binding site is important for basal, but not gonadotropin-releaseing hormone-stimulated, expression of the luteinizing hormone beta subunit gene. J Biol Chem 2000;275:13,082–13,088.
74. Alarid ET, Windle JJ, Whyte DB, Mellon PL. Immortalization of pituitary cells at discrete stages of development by directed oncogenesis in transgenic mice. Development 1996;122:3319–3329.
75. Low MJ, Goodman RH, Ebert KM. Cryptic human growth hormone gene sequences direct gonadotroph-specific expression in transgenic mice. Mol Endocrinol 1989;3:2028–2033.
76. Haugen BR, Wood WM, Gordon DF, Ridgway EC. A thyrotrope-specific variant of Pit-1 transactivates the thyrotropin beta promoter. J. Biol. Chem. 1993;268:20,818–20,824.
77. Mason ME, Friend KE, Copper J, Shupnik MA. Pit-1/GHF-1 binds to TRH-sensitive regions of the rat thyrotropin beta gene. Biochemistry 1993;32:8932–8938.
78. Gordon DF, Haugen BR, Sarapura VD, Nelson AR, Wood WM, Ridgway EC. Analysis of Pit-1 in regulating mouse TSH α promoter activity in thyrotropes. Mol Cell Endocrinol 1993;96:75–84.
79. Haugen BR, Gordon DF, Nelson AR, Wood WM, Ridgway EC. The combination of Pit-1 and Pit-1T have a synergistic stimulatory effect on the thyrotropin β-subunit promoter but not the growth hormone or prolactin promoters. Mol Endocrinol 1994;8:1574–1582.

80. Wondisford FE, Farr EA, Radovick SA, Steinfelder HJ, Moates JM, McClaskey JH, Weintraub BD. Thyroid hormone inhibition of human thyrotropin β-subunit gene expression is mediated by a cis-acting element located in the first exon. J Biol Chem 1989;264:14,601–14,604.
81. Wood WM, Kao MY, Gordon DF, Ridgway EC. Thyroid hormone regulates the mouse thyrotropin β-subunit gene promoter in transfected primary thyrotropes. J Biol Chem 1989;264:14,840–14,847.
82. Carr FE, Wong NC. Characterisitics of a negative thyroid hormone response element. J Biol Chem 1994;269:4175–4179.
83. Hollenberg AN, Monden T, Flynn TR, Boers ME, Cohen O, Wondisford FE. The human thyrotropin-releasing hormone gene is regulated by thyroid hormone through two distinct classes of negative thyroid hormone response elements. Mol Endocrinol 1995;9:540–550.
84. Kim MK, McClaskey JH, Bodenner DL, Weintraub BD. An AP-1-like factor and the pituitary-specific factor Pit-1 are both necessary to mediate hormonal induction of human thyrotropin beta gene expression. J Biol Chem 1993;268:23,366–23,375.
85. Haugen BR, Brown NS, Wood WM, Gordon DF, Ridgway EC. The thyrotrope-restricted isoform of the retinoid-X receptor-gamma1 mediates 9-cis-retinoic acid suppression of thyrotropin-β promoter activity. Mol Endocrinol 1997;11:481–489.
86. Kim MK, Lesoon-Wood LA, Weintraub BD, Chung JH. A soluble transcription factor, Oct-1, is also found in the insoluble nuclear matrix and possesses silencing activity in its alanine-rich domain. Mol Cell Biol 1996;16:4366–4377.
87. Gordon DF, Lewis SR, Haugen BR, James RA, McDermott MT, Wood WM, Ridgway EC. Pit-1 and GATA-2 interact and functionally cooperate to activate the thyrotropin β-subunit promoter. J Biol Chem 1997;272:24,339–24,347.
88. Hayashi Y, Xie J, Weiss RE, Pohlenz J, Refetoff S. Selective pituitary resistance to thyroid hormone produced by expression of a mutant thyroid hormone receptor β gene in the pituitary gland of transgenic mice. Biochem Biophys Res Commun 1998;245:204–210.
89. Abel ED, Kaulbach HC, Campos-Barros A, Ahima RS, Boers ME, Hashimoto K, Forrest D, Wondisford FE. Novel insight from transgenic mice into thyroid hormone resistance and the regulation of thyrotropin. J Clin Invest 1999;103:271–279.
90. Maki K, Miyoshi I, Kon Y, Yamashita T, Sasaki N, Aoyama S, Takahashi E, Namioka S, Hayashizaki Y, Kasai N. Targeted pituitary tumorigenesis using the human thyrotropin β-subunit chain promoter in transgenic mice. Mol Cell Endocrinol 1994;105:147–154.
91. Dasen JS, O'Connell SM, Flynn SE, Treier M, Gleiberman AS, Szeto DP, et al. Reciprocal interactions of Pit1 and GATA2 mediate signaling gradient-induced determination of pituitary cell types. Cell 1999;97:587–598.
92. Strauss BL, Pittman R, Pixley MR, Nilson JH, Boime I. Expression of the β subunit of chorionic gonadotropin in transgenic mice. J Biol Chem 1994;269:4968–4973.
93. Talmadge K, Vamvakopoulos NC, Fiddes JC. Evolution of the genes for the β subunits of human chorionic gonadotropin and luteinizing hormone. Nature 1984;307:37–40.
94. Hostetler G, Eaton A, Carnes M, Gildner J, Brownfield MS. Immunocytochemical localization of luteinizing hormone in rat central nervous system. Neuroendocrinology 1987;46:185–193.
95. Pelletier J, Counis R, de Reviers MM, Tillet Y. Localization of luteinizing hormone β-mRNA by in situ hybridization in the sheep pars tuberalis. Cell Tissue Res 1992;267:301–306.
96. Mason AJ, Hayflick JS, Zoeller T, Young WSI, Phillips HS, Nikolics K, Seeburg PH. A deletion truncating the gonadotropin-releasing hormone gene is responsible for hypogonadism in the *hpg* mouse. Science 1986;234:1366–1371.
97. Williamson P, Lang J, Boyd Y. The gonadotropin-releasing hormone (GnRH) gene maps to mouse chromosome 14 and identifies a homologous region on human chromosome 8. Somat. Cell Mol Genet 1991;17:609–615.
98. Saade G, London DR, Clayton RN. The interaction of gonadotropin-releasing hormone and estradiol on luteinizing hormone and prolactin gene expression in female hypogonadal (*hpg*) mice. Endocrinology 1989;124:1744–1753.
99. Cattanach BM, Iddon CA, Charlton HM, Chiappa SA, Fink G. Gonadotrophin-releasing hormone deficiency in a mutant mouse with hypogonadism. Nature 1977;269:338–340.
100. Schwanzel-Fukuda M, Bick D, Pfaff DW. Luteinizing hormone-releasing hormone (LHRH)-expressing cells do not migrate normally in an inherited hypogonadal (Kallman) syndrome. Mol Brain Res 1989;6:311–319.
101. Fink G, Sheward WJ, Charlton HM. Priming effect of luteinizing hormone releasing hormone in the hypogonadal mouse. J Endocrinol 1982;94:283–287.

102. McDowell IF, Morris JF, Charlton HM. Characterization of the pituitary gonadotroph cells of hypogonadal (hpg) male mice: comparison with normal mice. J Endocrinol 1982;95:321–330.
103. Gibson MJ, Krieger DT, Charlton HM, Zimmerman EA, Silverman AJ, Perlow MJ. Mating and pregnancy can occur in genetically hypogonadal mice with preoptic area brain grafts. Science 1984;225:949–951.
104. Mason AJ, Pitts SL, Nikolics K, Szonyi E, Wilcox JN, Seeburg PH, et al. The hypogonadal mouse: reproductive functions restored by gene therapy. Science 1986;234:1372–1378.
105. Yamada M, Saga Y, Shibusawa N, Hirato J, Murakami M, Iwasaki T, et al. Tertiary hypothyroidism and hyperglycemia in mice with targeted disruption of the thyrotropin-releasing hormone gene. Proc Natl Acad Sci USA 1997;94:10,862–10,867.
106. Mangelsdorf DJ, Thummel C, Beato M, Herrlich P, Schutz G, Umesono K, et al. The nuclear receptor superfamily: the second decade. Cell 1999;83:835–839.
107. Forman BM, Tzameli I, Choi HS, Chen J, Simha D, Seol W, et al. Androstane metabolites bind to and deactivate the nuclear receptor CAR-β. Nature 1999;395:612–615.
108. Wang H, Chen J, Hollister K, Sowers LC, Forman BM. Endogenous bile acids are ligands for the nuclear receptor FXR/BAR. Mol Cell 1999;3:543–553.
109. Makishima M, Okamoto AY, Repa JJ, Tu H, Learned RM, Luk A, et al. Identification of a nuclear receptor for bile acids. Science 1999;284:1362–1365.
110. Parks DJ, Blanchard SG, Bledsoe RK, Chandra G, Consler TG, Kliewer SA, et al. Bile acids: natural ligands for an orphan nuclear receptor. Science 1999;284:1365–1368.
111. Lala DS, Rice DA, Parker KL. Steroidogenic factor I, a key regulator of steroidogenic enzyme expression, is the mouse homolog of fushi tarazu-factor I. Mol Endocrinol 1992;6:1249–1258.
112. Ikeda Y, Lala DS, Luo X, Kim E, Moisan MP, Parker KL. Characterization of the mouse FTZ-F1 gene, which encodes a key regulator of steroid hydroxylase gene expression. Mol Endocrinol 1993;7:852–860.
113. Luo X, Ikeda Y, Parker KL. A cell-specific nuclear receptor is essential for adrenal and gonadal development and sexual differentiation. Cell 1994;77:481–490.
114. Muscatelli F, Strom TM, Walker AP, Zanaria E, Récan D, Meindl A, et al. Mutations in the DAX-1 gene give rise to both X-linked adrenal hypoplasia congenita and hypogonadotropic hypogonadism. Nature 1994;372:672–676.
115. Zanaria E, Muscatelli F, Bardoni B, Strom TM, Guioli S, Guo W, et al. An unusual member of the nuclear hormone receptor superfamily responsible for X-linked adrenal hypoplasia congenita. Nature 1994;372:635–641.
116. Yu RN, Ito M, Saunders TL, Camper SA, Jameson JL. Role of Ahch in gonadal development and gametogenesis. Nat Genet 1998;20:353–357.
117. Habiby RL, Boepple P, Nachtigall L, Sluss PM, Crowley Jr WF, Jameson JL. Adrenal hypoplasia congenita with hypogonadotropic hypogonadism: evidence that DAX-1 mutations lead to combined hypothalamic and pituitary defects in gonadotropin production. J Clin Invest 1996;98:1055–1062.
118. Thorner MO, Vance ML, Horvath E, Kovacs K. The anterior pituitary. In: Wilson JD, Foster DW, eds., Williams Textbook of Endocrinology, 8th ed., W.B. Saunders Company, Philadelphia, PA, 1992, pp. 221–310.
119. Lubahn DB, Moyer JS, Golding TS, Couse JF, Korach KS, Smithies O. Alteration of reproductive function but not prenatal sexual development after insertional disruption of the mouse estrogen receptor gene. Proc Natl Acad Sci USA 1993;90:11,162–11,166.
120. Mosselman S, Polman J, Dijkema R. ERβ: Identification and characterization of a novel human estrogen receptor. FEBS Lett 1996;392:49–53.
121. Kuiper GG, Enmark E, Pelto-Huikko M, Nilsson S, Gustafsson JÄ. Cloning of a novel receptor expressed in rat prostate and ovary. Proc Natl Acad Sci USA 1996;93:5925–5930.
122. Tremblay GB, Tremblay A, Copeland NG, Gilbert DJ, Jenkins NA, Labrie F, et al. Cloning, chromosomal localization, and functional analysis of the murine estrogen receptor-β. Mol Endocrinol 1997;11:353–365.
123. Couse JF, Lindzey J, Grandien K, Gustafsson JÄ, Korach KS. Tissue distribution and quantitative analysis of estrogen receptor-α (ERα) and estrogen receptor-β (ERβ) messenger ribonucleic acid in the wild -type and ERα-knockout mouse. Endocrinology 1997;138:4613–4621.
124. Shupnik MA, Pitt LK, Soh AY, Anderson A, Lopes MB, Laws ER, Jr. Selective expression of estrogen receptor-α and -β isoforms in human pituitary tumors. J Clin Endocrinol Metab 1998;83:3965–3972.

125. Mitchner NA, Garlick C, Ben-Jonathan N. Cellular distribution and gene regulation of estrogen receptors-α and -β in the rat pituitary gland. Endocrinology 1998;139:3976–983.
126. Wilson ME, Price Jr RH, Handa RJ. Estrogen receptor-β messenger ribonucleic acid expression in the pituitary gland. Endocrinology 1998;139:5151–5156.
127. Krege JH, Hodgin JB, Couse JF, Enmark E, Warner M, Mahler JF, et al. Generation and reproductive phenotypes of mice lacking estrogen receptor-β. Proc Natl Acad Sci USA 1998;95:15,677–15,682.
128. Couse JF, Korach KS. Estrogen receptor null mice: what have we learned and where will they take us? Endocr Rev 1999;20:358–417.
129. Gharib SD, Wierman ME, Shupnik MA, Chin WW. Molecular biology of the pituitary gonadotropins. Endocr Rev 1990;11:177–199.
130. Lindzey J, Wetsel WC, Couse JF, Stoker T, Cooper R, Korach KS. Effects of castration and chronic steroid treatments on hypothalamic gonadotropin-releasing hormone content and pituitary gonadotropins in male wild-type and estrogen receptor-α knockout mice. Endocrinology 1998;139:4092–4101.
131. Scully KM, Gleiberman AS, Lindzey J, Lubahn DB, Korach KS, Rosenfeld MG. Role of estrogen receptor-α in the anterior pituitary gland. Mol Endocrinol 1997;11:674–681.
132. Fisher CR, Graves KH, Parlow AF, Simpson ER. Characterization of mice deficient in aromatase (ArKO) because of targeted disruption of the cyp19 gene. Proc Natl Acad Sci USA 1998;95:6965–6970.
133. Lydon JP, DeMayo FJ, Funk CR, Mani SK, Hughes AR, Montgomery Jr CA, et al. Mice lacking progesterone receptor exhibit pleiotropic reproductive abnormalities. Genes Dev 1995;9:2266–2278.
134. Lydon JP, DeMayo FJ, Conneely OM, O'Malley BW. Reproductive phenotypes of the progesterone receptor null mutant mouse. J Steroid Biochem Mol Biol 1996;56:67–77.
135. Chappell PE, Lydon JP, Conneely OM, O'Malley BW, Levine JE. Endocrine defects in mice carrying a null mutation for the progesterone receptor gene. Endocrinol. 1997;138:4147–4152.
136. Chappell PE, Schneider JS, Kim P, Xu M, Lydon JP, O'Malley BW, et al. Absence of gonadotropin surges and gonadotropin-releasing hormone self-priming in ovariectomized (OVX), estrogen (E_2)-treated, progesterone receptor knockout (PRKO) mice. Endocrinology 1999;140:3653–3658.
137. Brown TJ, Clark AS, MacLusky NJ. Regional sex differences in progestin receptor induction in the rat hypothalamus: effects of various doses of estradiol benzoate. J Neurosci 1987;7:2529–2536.
138. Calderon JJ, Muldoon TG, Mahesh VB. Receptor-mediated interrelationships between progesterone and estradiol action on the anterior pituitary-hypothalamic axis of the ovariectomized immature rat. Endocrinology 1987;120:2428–2435.
139. Bethea CL, Brown NA, Kohama SG. Steroid regulation of estrogen and progestin receptor messenger ribonucleic acid in monkey hypothalamus and pituitary. Endocrinology 1996;137:4372–4383.
140. He WW, Kumar MV, Tindall DJ. A frame-shift mutation in the androgen receptor gene causes complete androgen insensitivity in the testicular feminized mouse. Nucleic Acids Res 1991;19:2373–2378.
141. Charest NJ, Zhou ZX, Lubahn DB, Olsen KL, Wilson EM, French FS. A frameshift mutation destabilizes androgen receptor messenger RNA in the Tfm mouse. Mol Endocrinol 1991;5:573–581.
142. Lyon MF, Hawkes SG. X-linked gene for testicular feminization in the mouse. Nature 1970;227: 1217–1219.
143. Migeon BR, Brown TR, Axelman J, Migeon CJ. Studies of the locus for androgen receptor: localization on the human X chromosome and evidence for homology with the Tfm locus in the mouse. Proc Natl Acad Sci USA 1981;78:6339–6343.
144. Scott IS, Bennett MK, Porter-Goff AE, Harrison CJ, Cox BS, Grocock CA, et al. Effects of the gonadotropin-releasing hormone agonist 'Zoladex' upon pituitary and gonadal function in hypogonadal (hpg) male mice: comparison with normal male and testicular feminized (tfm) mice. J Mol Endocrinol 1991;8:249–258.
145. Kaetzel DM, Nilson JH. Methotrexate-induced amplification of the bovine lutropin genes in Chinese hamster ovary cells. Relative concentration of the alpha and beta subunits determines the extent of heterodimer assembly. J. Biol. Chem. 1988;263:6344–6351.
146. Chatterjee VK, Lee JK, Rentoumic A, Jameson JL. Negative regulation of the thyroid-stimulating hormone a gene by thyroid hormone: receptor interaction adjacent to the TATA box. Proc Natl Acad Sci USA 1989;86:9114–9118.
147. Sarapura VD, Wood WM, Gordon DF, Ocran KW, Kao MY, Ridgway EC. Thyrotrope expression and thyroid hormone inhibition map to different regions of the mouse glycoprotein hormone α-subunit gene promoter. Endocrinology 1990;127:1352–1361.
148. Carr FE, Burnside J, Chin WW. Thyroid hormones regulate rat thyrotropin β gene promoter activity expressed in GH3 cells. Mol Endocrinol 1989;3:709–716.

149. Koller KJ, Wolff RS, Warden MK, Zoeller RT. Thyroid hormones regulate levels of thyrotropin-releasing-hormone mRNA in the paraventricular nucleus. Proc Natl Acad Sci USA 1987;84:7329–7333.
150. Guissouma H, Ghorbel MT, Seugnet I, Ouatas T, Demeneix BA. Physiological regulation of hypothalamic TRH transcription in vivo is T3 receptor isoform specific. FASEB J 1989;12:1755–1764.
151. Reginato MZJ, Lazar M. DNA-dependent and DNA-independent mechanisms regulate the differential heterodimerization of the isoforms of the thyroid hormone receptor with retinoid X receptor. J Biol Chem 1996;271:28,199–28,205.
152. Tsai MJ, O'Malley BW. Molecular mechanisms of action of steroid/thyroid receptor superfamily members. Annu Rev Biochem 1994;63:451–486.
153. Forrest D, Sjoberg M, Vennstrom B. Contrasting developmental and tissue-specific expression of alpha and beta thyroid hormone receptor genes. EMBO J 1990;9:1519–1528.
154. Strait KA, Schwartz HL, Perez-Castillo A, Oppenheimer JH. Relationship of c-erbA mRNA content to tissue triiodothyronine nuclear binding capacity and function in developing and adult rats. J Biol Chem 1990;265:10,514–10,521.
155. Forrest D, Erway LC, Ng L, Altschuler R, Curran T.Thyroid hormone receptor β is essential for development of auditory function. Nat Genet 1996;13:354–357.
156. Forrest D, Hanebuth E, Smeyne RJH, Everds N, Stewart CL, Wehner JM, Curran T. Recessive resistance to thyroid hormone in mice lacking thyroid hormone receptor β: evidence for tissue-specific modulation of receptor function. EMBO J 1996;15:3006–3015.
157. Fraichard A, Chassande O, Plateroti M, Roux JP, Trouillas J, Dehay C, et al. The T3Rα gene encoding a thyroid hormone receptor is essential for post-natal development and thyroid hormone production. EMBO J 1997;16:4412–4420.
158. Wikström L, Johansson C, Saltó C, Barlow C, Campos Barros A, et al. Abnormal heart rate and body temperature in mice lacking thyroid hormone receptor α1. EMBO J 1998;17:455–461.
159. Göthe S, Wang Z, Ng L, Kindblom JM, Campos Barros A, Ohlsson C, et al. Mice devoid of all known thyroid hormone receptors are viable but exhibit disorders of the pituitary-thyroid axis, growth, and bone maturation. Genes Dev 1999;13:1329–1341.
160. de Jong FH, Sharpe RM. Evidence for inhibin-like activity in bovine follicular fluid. Nature 1976;263:71,72.
161. Welschen R, Hermans WP, Dullart J, de Jong FH. Effects of an inhibin-like factor present in bovine and porcine follicular fluid on gonadotrophin levels in ovariectomized rats. J Reprod Fertil 1977;50:129–131.
162. Ling N, Ying SY, Ueno N, Shimasaki S, Esch F, Hotta M, Guillemin R. A homodimer of the β-subunits of inhibin A stimulates the secretion of pituitary follicle stimulating hormone. Biochem Biophys Res Commun 1996;138:1129–1137.
163. Ling N, Ying SY, Ueno N, Shimasaki S, Esch F, Hotta M, Guillemin R. Pituitary FSH is released by a heterodimer of the β-subunit from the two forms of inhibin. Nature 1986;321:779–782.
164. de Kretser DM, Robertson DM. The isolation and physiology of inhibin and related peptides. Biol Reprod 1989;40:33–47.
165. Matzuk MM, Finegold MJ, Su JG, Hsueh AJ, Bradley A. α-inhibin is a tumour-suppressor gene with gonadal specificity in mice. Nature 1992;360:313–319.
166. Matzuk MM, Finegold MJ, Mather JP, Krummen L, Lu H, Bradley A. Development of cancer cachexia-like syndrome and adrenal tumors in inhibin-deficient mice. Proc Natl Acad Sci USA 1994;91:8817–8821.
167. Kumar TR, Wang Y, Matzuk MM. Gonadotropins are essential modifier factors for gonadal tumor development in inhibin-deficient mice. Endocrinology 1996;137:4210–4216.
168. Coerver KA, Woodruff TK, Finegold MJ, Mather J, Bradley A, Matzuk MM. Activin signaling through activin receptor type II causes the cachexia-like symptoms in inhibin-deficient mice. Mol Endocrinol 1996;10:534–543.
169. Matzuk MM, Kumar TR, Vassalli A, Bickenbach JR, Roop DR, Jaenisch R, et al. Functional analysis of activins during mammalian development. Nature 1995;374:354–356.
170. Vassalli A, Matzuk MM, Gardner HAR, Lee KF, Jaenisch R. Activin/inhibin Bβ subunit gene disruption leads to defects in eyelid development and female reproduction. Genes Dev. 1994;8:414–427.
171. Matzuk MM, Kumar TR, Bradley A. Different phenotypes for mice deficient in either activins or activin receptor type II. Nature 1995;374:356–359.
172. Nakamura T, Takio K, Eto Y, Shibai H, Titani K, Sugino H. Activin-binding protein from rat ovary is follistatin. Science 1990;247:836–838.

173. Matzuk MM, Lu N, Vogel H, Sellheyer K, Roop DR, Bradley A. Multiple defects and perinatal death in mice deficient in follistatin. Nature 1995;374:360–363.
174. Refetoff S, Weiss RE, Usala SJ. The syndromes of resistance to thyroid hormone. Endocr Rev 1993;14:348–399.
175. Stein SA, Oates EL, Hall CR, Grumbles RM, Fernandez LM, Taylor NA, et al. Identification of a point mutation in the thyrotropin receptor of the hyt/hyt hypothyroid mouse. Mol Endocrinol 1994;8:129–138.
176. Stein SA, Shanklin DR, Krulich L, Roth MG, Chubb CM, Adams PM. Evaluation and characterization of the hyt/hyt hypothyroid mouse. II. Abnormalities of TSH and the thyroid gland. Neuroendocrinology 1989;49:509–519.
177. Yusta B, Alarid ET, Gordon DF, Ridgway EC, Mellon PL. The thyrotropin β-subunit gene is repressed by thyroid hormone in a novel thyrotrope cell line, mouse TαT1 cells. Endocrinology 1998;139:4476–4482.
178. Kumar TR, Graham KE, Asa SL, Low MJ. Simian virus 40 T antigen-induced gonadotroph adenomas: a model of human null cell adenomas. Endocrinology 1998;139:3342–3351.
179. Risma KA, Clay CM, Nett TM, Wagner T, Yun J, Nilson JH. Targeted overexpression of luteinizing hormone in transgenic mice leads to infertility, polycystic ovaries, and ovarian tumors. Proc Natl Acad Sci USA 1995;92:1322–1336.
179a. Keri RA, Lozada KL, Abdul-Karim FW, Nadean JH, Nilson JH. Luteinizing hormone induction of ovarian tumors: oligogenic differences between mouse strains dictates tumor disposition. Proc Natl Acad Sci USA 2000;97:383–387.
180. Risma KA, Hirshfield AN, Nilson JH. Elevated luteinizing hormone in prepubertal transgenic mice causes hyperandrogenemia, precocious puberty, and substantial ovarian pathology. Endocrinology 1997;138:3540–3547.
181. Mann RJ, Keri RA, Nilson JH. Transgenic mice with chronically elevated luteinizing hormone are infertile due to anovulation, defects in uterine receptivity, and midgestation pregnancy failure. Endocrinology 1999;140:2592–2601.
182. Flaws A, Abbud R, Mann RJ, Nilson JH, Hirshfield AN. Chronically elevated luteinizing hormone depletes primordial follicles in the mouse ovary. Biol Reprod 1997;57:1233–1237.
183. Kumar TR, Palapattu G, Wang P, Woodruff TK, Boime I, Byrne MC, Matzuk MM. Transgenic models to study gonadotropin function: the role of follicle-stimulating hormone in gonadal growth and tumorigenesis. Mol Endocrinol 1999;13:851–865.
184. Yano H, Readhead C, Nakashima M, Ren SG, Melmed S. Pituitary-directed leukemia inhibitory factor transgene causes Cushing's syndrome: Neuro-immune-endocrine modulation of pituitary development. Mol Endocrinol 1998;12:1708–1720.
185. Bousquet C, Ray DW, Melmed SA. Common pro-opiomelanocortin-binding element mediates leukemia inhibitory factor and corticotropin-releasing hormone transcriptional synergy. J Biol Chem 1997;272:10,551–10,557.
186. Wang Z, Ren SG, Melmed S. Hypothalamic, pituitary leukemia inhibitory factor gene expression in vivo: a novel endotoxin-inducible neuro-endocrine interface. Endocrinology 1996;137:2947–2953.
187. Akita S, Readhead C, Stefaneanu L, Fine J, Tampanaru-Sarmesiu A, Kovacs K, et al. Pituitary-directed leukemia inhibitory factor transgene forms Rathke's cleft cysts and impairs adult pituitary function. A model for human pituitary Rathke's cysts. J Clin Invest 1997;99:2462–2469.

14 Genetic Approaches to the Study of Pituitary Follicle-Stimulating Hormone Regulation

Daniel J. Bernard, PHD
and Teresa K. Woodruff, PHD

CONTENTS

INTRODUCTION

The pituitary glycoprotein hormone known as follicle-stimulating hormone (FSH) plays a fundamental role in folliculogenesis and spermatogenesis in mammals. Precise regulation of its synthesis and secretion is therefore vital to normal reproductive function, particularly in females. Several endocrine, paracrine, and autocrine factors within the hypothalamo-pituitary-gonadal axis have been identified as important regulators of FSH, including gonadotropin-releasing hormone (GnRH), inhibin, activin, follistatin, and the sex-steroid hormones. Over the past several years, a variety of genetic loss- and gain-of-function models have been created in mice using gene targeting and transgenic technologies. This chapter reviews the contributions of a variety of these models to our understanding of FSH synthesis, secretion, and function. Many of the models confirm results from years of physiological experimentation, but new insights into FSH regulation and action have emerged as a result of these advances in molecular genetics. Continued development of these models, along with the use of novel in vitro model systems, will significantly advance our understanding of FSH biology.

From: *Contemporary Endocrinology: Transgenics in Endocrinology*
Edited by: M. Matzuk, C. W. Brown, and T. R. Kumar © Humana Press Inc., Totowa, NJ

Normal reproductive function and sexual development are dependent upon the coordinated regulation of the pituitary gonadotropin hormones FSH and luteinizing hormone (LH). Both hormones are produced within gonadotrope cells of the anterior pituitary, and act at the level of the gonads, where they bind to cell-surface receptors to influence gametogenesis and steroidogenesis, respectively. Although they are structurally related and produced within the same cells, FSH and LH have very different functions, and are regulated in different ways.

Research over the past few decades has implicated a number of hormonal factors in the control of FSH synthesis and secretion. These include hypothalamic GnRH, the gonadal and pituitary peptides inhibin, activin, and follistatin, and the gonadal sex-steroid hormones. The general model that has emerged is that pulsatile release of GnRH stimulates production and secretion of FSH directly and/or through the stimulation of pituitary activin and follistatin, which positively and negatively regulate FSH, respectively. Once secreted, FSH acts on the gonads, where it stimulates inhibin production that then feeds back to the pituitary to negatively regulate FSH. In addition, LH promotes gonadal-steroid production, which then negatively (and sometimes positively) regulates FSH and LH via actions at the hypothalamic and pituitary levels. For detailed reviews of the roles played by each of these factors, *see* refs. *1–3*.

Here, we will focus on how modern approaches in molecular genetics have led to the development of various in vivo models that directly assess the roles of these various hormones in FSH regulation. In the following sections we review several mouse models, including the introduction of loss-of-function mutations by gene targeting in embryonic stem (ES) cells, gain-of-function mutations by gene overexpression in transgenic mice, and naturally occurring loss-of-function mutations. These genetic models span mutations in the FSH-subunit genes to signal-transduction molecules involved in conveying activin signals to the cell nucleus. The discussion of the various mutants is generally restricted to effects on FSH synthesis and secretion (*see* Table 1 for a summary), but other relevant defects are discussed where appropriate.

GONADOTROPIN-SUBUNIT DEFICIENCY

FSH, along with LH and TSH, is a member of the pituitary glycoprotein-hormone family. All of these proteins share a common α-subunit noncovalently linked to unique β-subunits, which confer biological specificity. The individual subunits do not have biological activity; therefore, biological activity is dependent upon heterodimerization *(1)*. Targeted deletion of the FSHβ-subunit in mice effectively removes circulating FSH and FSHβ mRNA from the anterior pituitary *(4)*. Interestingly, FSH-deficient males are still fertile, despite their decreased testes size and lowered epididymal sperm counts relative to wild-type controls. In contrast, FSH-deficient females are infertile, and show abnormal estrous cyclicity. Follicle maturation is arrested at the pre-antral stage, and the ovaries lack corpora lutea *(4)*.

Transgenic mice carrying a 10-kb fragment of the human FSHβ (hFSHβ) gene express the hFSHβ in a gonadotrope-specific fashion *(5)*. This suggests that the human FSHβ gene has the same regulatory elements as the endogenous mouse FSHβ gene. To determine whether the hFSHβ gene can function and be regulated similarly to the mouse FSHβ gene, these transgenic mice were intercrossed with FSHβ-deficient mice. The presence of the hFSHβ transgene completely rescues the wild-type phenotype *(6)*. Males

Table 1

FSH and FSHβ mRNA Levels in Various Genetic Mouse Models (Relative to Wild-Type Controls)

Model	*FSH Levels*	*FSHβ Levels*	*Refs*
Gonadotropin-subunit deficiency			
FSHβ-knockout	Not detectable	Not detectable	*(4)*
αGSU-knockout	Not measured	FSHβ immunoreactivity in gonadotropes (not quantitatively compared to wild-type)	*(7)*
GnRH deficiency			
hpg (GnRH-deficiency, spontaneous mutation)	Serum and pituitary levels decreased in males and females	Decreased in males and females	*(11–16)*
Inhibin, activin, follistatin deficiency or overexpression			
α-Inhibin-knockout (increased serum Activin A and B)	Serum levels increased in males and females Pituitary levels not measured	Not measured	*(31–33)*
α-Inhibin transgenic (mMT-1 Promoter)	Serum levels decreased in males and females Pituitary levels not measured	Not measured	*(40)*
Activin-βB-knockout	Serum levels increased in males, no change in females Pituitary levels not measured	Not measured	*(41)*
Activin-βA-knockout	Not viable; die shortly after birth (*see* text)		*(42)*
Activin-βA/βB-knockout	Not viable; die shortly after birth (*see* text)		*(42)*
ActRII-knockout	Serum and pituitary levels decreased in males Serum levels decreased in females	Not measured	*(45)*
ActRIIB-knockout	129/Sv inbred–die before weaning 129/Sv X C57BL/6–30% viable and fertile, but FSH levels not measured	Not measured	*(46)*
ActRI (ALK-2)-knockout	Embryonic lethal prior to gastrulation		*(47)*
ActRIB (ALK-4)-knockout	Embryonic lethal prior to gastrulation (*see* text)		*(48)*
Smad2-knockout	Embryonic lethal (*see* text)		*(49–51)*
Smad3-knockout (exon 2)	Viable and fertile, but die of colorectal cancer between 4 and 6 mo No FSH data reported		*(53)*
Smad3-knockout (exon 8)	Viable but die between 1 and 8 mo Immune function decreased No FSH data reported		*(52)*

(continued)

Table 1 *(continued)*

Model	*FSH Levels*	*FSHβ Levels*	*Refs*
Smad4-knockout	Embryonic lethal prior to gastrulation (*see* text)		*(54,55)*
Follistatin-knockout	Not viable; die shortly after birth (*see* text)		*(56)*
Follistatin transgenic (mMT-1 promoter)	Serum FSH normal or decreased in males and females depending on line. Related to degree of follistatin overexpression.	Not measured	*(57)*
Sex-steroid deficiency			
tfm (androgen receptor, null-mutant)	All animals phenotypic females Pituitary levels increased	Not measured	*(13)*
Estrogen receptor-α-knockout (ERKO)	Serum levels slightly increased in males, normal in females Pituitary levels not measured	Increased in females, normal in males	*(67–69)*
Estrogen receptor-β-knockout	Males and females fertile No FSH data reported	Not measured	*(65)*
Aromatase-knockout	Serum levels increased in females, not measured in males Pituitary levels not measured	Not measured	*(66)*
Progesterone receptor-knockout (PRKO)	Serum levels decreased in males, normal in females Pituitary levels not measured	Not measured	*(74,75)*
Other genetic models with altered FSH phenotype			
Steroidogenic factor-1 (SF-1)-knockout	Pituitary levels increased Serum levels not measured	Decreased	*(78,79)*
Cyclooxygenase-2 (COX-2)-knockout	Pituitary levels increased in females Serum levels normal in females FSH data not reported for males	Not measured	*(81,84)*
Superoxide dismutase-1 (SOD1)-knockout	Serum levels decreased in females Pituitary levels not measured in females FSH data not reported for males	Not measured	*(85)*
Growth differentiation factor-9 (GDF-9)-knockout	Serum levels increased in females Pituitary levels not measured in females FSH data not reported for males	Not measured	*(88–90)*

show increased testes mass, sperm counts, and sperm motility, while females show normal uterine and ovarian mass, restored estrous cyclicity (appearance of corpora lutea) and are fertile. Likewise, male FSHβ-deficient mice interbred with transgenic mice ectopically expressing transgenes for human α- and FSHβ-subunits have testes mass and sperm counts equivalent to controls, but females show only partial rescue of the wild-type phenotype *(6)*. Roughly 30% of these females are occasionally fertile. In addition, most have small uteri, and folliculogenesis does not proceed beyond the pre-antral stage (similar to FSH-deficient females) *(6)*. Taken together, these results indicate that: 1) the FSHβ-subunit is required for FSH synthesis; 2) normal FSH synthesis and secretion are necessary for fertility in female, but not male mice; 3) folliculogenesis, but not spermatogenesis, is dependent upon pituitary-specific FSH expression; and 4) both pituitary-specific expression and normal physiological regulation of FSH synthesis and secretion are conferred by DNA elements common to humans and mice.

Gene targeting in ES cells has also been used to disrupt the pituitary glycoprotein α subunit *(7)*. Because the α-subunit is common to FSH, LH, and TSH, the mutation compromises both reproductive and thyroid function. Male α-subunit-deficient mice have undetectably low levels of circulating testosterone, retarded seminiferous-tubule and Leydig-cell development, and an arrest of spermatogenesis at the first meiotic division. Female homozygous null-mutant mice show similar reproductive deficits. Uteri and ovaries are very small relative to controls, and folliculogenesis is blocked at the pre-antral stage (similar to FSH-deficient females). Interestingly, the pituitaries of α-subunit knockout mice up to at least 6 mo of age contain gonadotropes in normal proportions, as indicated by the presence of cells immunopositive for FSHβ and LHβ. These results demonstrate that expression of the α-subunit is unnecessary for complete gonadotrope differentiation, and confirm in vitro studies showing that FSHβ and LHβ are biologically inactive and/or are not efficiently secreted in the absence of a functional α-subunit *(8,9)*.

GONADOTROPIN-RELEASING HORMONE DEFICIENCY

GnRH is a potent regulator of gonadotropin synthesis and secretion. This decapeptide is produced in the brain, and is secreted in pulsatile fashion into the pituitary portal vasculature at the level of the median eminence. From there, the GnRH signal reaches the pituitary, where it binds to GnRH receptors on the surface of the gonadotropes. The pulsatile nature of GnRH secretion is key to its differential regulation of FSH and LH synthesis and secretion. In general, slow pulse frequencies favor FSH secretion, while more rapid pulses favor LH secretion. This provides a mechanism whereby a single releasing hormone can regulate two different hormones.

Although there is some debate regarding the existence of an FSH-specific hypophysiotropic-releasing factor (FSHRF) *(10)*, GnRH is generally regarded as the principal brain hormone regulating both FSH and LH. Perhaps the most compelling demonstration of this fact is the endocrine phenotype of the hypogonadal (*hpg*) mouse *(11)*. *hpg* is an autosomal recessive mutation in which at least 33.5 kb of the distal half of the prepro-GnRH gene is deleted. Mice homozygous for this mutation fail to produce functional GnRH protein *(11,12)*. Among the various reproductive deficits, serum and pituitary FSH levels are significantly reduced, as are FSHβ mRNA levels, relative to wild-type controls *(11–13)*. Male *hpg* mice have small undescended testes, and spermatogenesis is arrested at the diplotene stage. In *hpg* females, folliculogenesis is arrested at the pre-antral stage, and there is no corpus luteum formation. While levels of FSHβ mRNA are lower in *hpg* than in wild-type controls *(13)*, FSHβ mRNA is still detectable in the pituitary of both male and female *hpg* mice (14). These data indicate that a functional GnRH system is required for normal FSH synthesis and secretion. However, given that FSH protein and FSHβ mRNA are detectable in *hpg* mice, the FSH gene retains some basal activity in the absence of GnRH.

Transgenic mice carrying copies of the intact mouse GnRH gene have been interbred with the *hpg* mice to completely rescue reproductive function *(15)*. Serum and pituitary FSH levels are indistinguishable from control levels, and these rescued mice are fertile. In addition, treatment of *hpg* mice with exogenous GnRH significantly elevates levels of serum and pituitary FSH, as well as pituitary FSHβ mRNA levels *(14,16)*.

GnRH exerts its effects by binding to cell-surface receptors (GnRH-R) on gonadotropes. The GnRH-R is a member of the seven-transmembrane-spanning, G-protein-coupled receptor family, but is unique in lacking a C-terminal cytoplasmic domain *(17)*. Because mice and humans appear to have only one GnRH-R gene, null mutations in this gene may phenocopy the *hpg* mouse in terms of FSH and other reproductive deficits. Mutations in the human GnRH-R lead to hypogonadotropic hypogonadism *(18)*. To date, there are no published reports of naturally occurring null mutations in the murine GnRH-R gene, and null mutants have not been generated by targeted deletion in ES cells.

GnRH-R overexpressing mice would be particularly useful in the study of the differential regulation of FSH and LH. In vitro models suggest that FSHβ and LHβ transcriptional activity are differentially influenced by GnRH-R density. Thus, high GnRH-R levels favor LHβ reporter activity in the GH3 somatolactotropic cell line following GnRH agonist treatment. FSHβ reporter activity is greatest at lower GnRH-R levels *(17,19)*. If this model reflects the in vivo situation, then in GnRH-R overexpressing mice, LH levels may be elevated relative to controls, but FSH levels may not differ (or at least not to the same extent). In addition, a combination of targeted disruption of the GnRH-R and rescue with transgenic lines carrying different copy numbers of the intact GnRH-R gene could provide a powerful in vivo model system to examine the relationship between GnRH-R density and differential FSH and LH expression.

INHIBIN, ACTIVIN, AND FOLLISTATIN DEFICIENCY OR OVEREXPRESSION

While GnRH pulsatility provides one mechanism for differential FSH and LH regulation, other factors are clearly involved. A myriad of in vivo and in vitro studies clearly indicate a role for activins, inhibins, and follistatins in the selective regulation of FSH *(3)*. As their names suggest, activins stimulate, while inhibins attenuate FSH synthesis and secretion. Both activins and inhibins are members of the transforming growth factor-β (TGFβ) protein family. Activins are comprised of two closely related subunits (βA and βB) and form activin A and activin B by homodimerization (βA:βA and βB:βB) or activin AB by heterodimerization. Inhibins A and B are heterodimers (α:βA and α:βB) of a unique α-subunit and one of the two β-subunits they share with the activins. Follistatin is a single-chain peptide, structurally unrelated to the activins and inhibins. Follistatin inhibits FSH synthesis and secretion by binding activins irreversibly and blocking their actions.

Activins bind to one of two constitutively active serine/threonine kinase receptors, activin type II and type IIB receptors (ActRIIA and ActRIIB) *(20)*. Ligand binding causes recruitment and transphosphorylation of an activin type I receptor, ActRIB (also known as ALK-4). ALK-4, like the type II receptors, is a serine/threonine kinase. A second type I receptor, ALK-2 or ActRI, is expressed in pituitaries *(21)*, but recent data suggest that it does not play a role in activin signaling *(22)*. Upon phosphorylation, ALK-4 is activated and phosphorylates one of two downstream intracellular signaling molecules, Smad2 or Smad3. Phosphorylated Smad2 or Smad3 multimers then hetero-oligomerize with another member of the Smad family, Smad4, in the cytoplasm, and the complex translocates to the nucleus. Once in the nucleus, the Smad2/Smad4 or Smad3/Smad4 complex interacts with a DNA-binding partner to activate transcription by binding to *cis*-acting elements in the promoters of target genes. While it is assumed that the FSHβ gene is a downstream target of activin signaling, this has not been shown directly. However, because activin stimulates increases in FSHβ primary transcript levels within 30 min in perifused rat pituitaries, this rapid response suggests that FSHβ may be a direct target *(23)*.

How inhibins regulate FSH is less well understood. It is not known whether inhibins downregulate FSH by competing with activin for binding to activin receptors in the pituitary *(24,25)* or whether they act through their own receptor and signaling pathway. Support for the latter mechanism comes from two recent reports indicating the presence of high-affinity inhibin-binding proteins in murine ovarian tumors and in ovine pituitaries *(26,27)*. Molecular characterization of these proteins will aid substantially in the identification and delineation of the inhibin signal-transduction cascade. More detailed analyses will identify how activin and inhibin signaling interact to affect FSH synthesis and secretion.

A variety of genetic models have been created to examine the roles of the activins, inhibins, and follistatins in reproduction and development *(28–30)*. Next, we summarize how each of these models has contributed to our understanding of FSH regulation.

The inhibin α-subunit was deleted by homologous recombination in ES cells *(31)*. Elimination of the α-subunit precludes the ability to produce both inhibin A and B, while leaving production of all forms of activin intact. *A priori*, one would expect FSH levels to be elevated in these mice because of the potent inhibitory effects of inhibin on FSH. In fact, in both male and female homozygous mice, FSH levels are significantly elevated. Inhibin-deficient mice also show significant increases in both serum activin A and B *(32–34)*. Therefore, it is likely that the increases in serum FSH are attributable to decreased pituitary and gonadal inhibin and subsequent increases in serum and pituitary activin levels.

Inhibin-α also appears to be a tumor-suppressor gene, because virtually all inhibin-deficient mice develop gonadal sex-cord stromal tumors. Interestingly, elevated FSH levels may contribute to tumorigenesis in these mice. Inhibin- and GnRH-deficient mice have been created by interbreeding inhibin-α knockout mice with *hpg* mice. As reviewed above, *hpg* mice have very low levels of circulating gonadotropins. Mice homozygous for both the inhibin-α and *hpg* mutations do not develop gonadal or adrenal *(33)* tumors for at least 1 yr *(35)*. More recent data indicate that FSH modifies tumor growth rather than causing tumorigenesis *per se (36)*. That is, transgenic mice overexpressing human FSH at very high levels do not develop gonadal tumors, and inhibin-deficient/FSHβ-deficient double-mutant mice still develop ovarian and testicular tumors, although at a slower rate than mice deficient in inhibin alone.

Mice carrying a transgene with 6 kb of the mouse inhibin-α promoter driving the expression of SV40 T-antigen (TAg) develop gonadal tumors of Leydig- and granulosa-

/theca-cell origin between 5 and 8 mo of age *(37,38)*. When these mice are crossed with *hpg* mice to deplete endogenous gonadotropins, no tumors develop *(39)*. These data suggest that the gonadotropins (particularly FSH) drive tumorigenesis through activation of the inhibin-α promoter, which in turn drives TAg expression, and/or that the gonadotropins have a direct effect on the inhibin-α-TAg-positive cells. This scenario is different than the case in inhibin-deficient mice in which tumor growth occurs when inhibin is removed and gonadotropin levels are elevated.

An inhibin-α overexpression model has also been generated by fusing the rat inhibin-α precursor downstream of the mouse metallothionein-I promoter (MT-α mice) *(40)*. MT-α females have reduced serum FSH, but increased serum testosterone and LH levels. In addition, these females are fertile but produce smaller litters, in part because of a deficit in ovulation. A majority of MT-α females eventually develop unilateral or bilateral ovarian cysts. Male MT-α mice also have lower levels of FSH compared to controls, but are fertile and produce normal-sized litters. The testes decline in size as the animals age, and the seminiferous tubule volume and sperm counts are lower than in wild-type controls. No other testicular abnormalities were reported. These data indicate that, predictably, inhibin overexpression results in decreased FSH levels.

As discussed above, the inhibins and activins share β-subunits, and models of βA and βB deficiency as well as double βA/βB-knockouts have been produced. Because βB deletion results in deficiency in inhibin B and activin B and AB, it is difficult to predict *a priori* the effects on FSH synthesis and secretion. These mice are viable and are fertile, although some βB-knockout mice show defects in eyelid fusion depending on the strain background *(41)*. 129/Sv/C57BL/6 hybrid background βB-deficient males have slightly elevated serum FSH levels relative to wild-type controls. A similar trend is observed in females, but the difference is not statistically significant. These data indicate that relatively normal FSH regulation can occur in the absence of the βB-subunit. This may arise because of compensatory changes in βA production. Indeed, βB-deficient mice show a significant upregulation of βA protein in ovarian tissue through a posttranscriptional mechanism *(41)*.

Male βB-deficient mice breed normally, but females have some reproductive deficits. Ovarian morphology appears normal, but compared to heterozygotes or wild-type controls, homozygous-null females show a significant increase in gestation time and an impairment in the onset of labor. In addition, fetuses of βB-deficient females that survive birth, die shortly thereafter because of malnutrition caused by a nursing defect in the βB-deficient mothers.

Unlike the βB-null mutants, βA-deficient mice are not viable and die within the first 24 h after birth *(42)*. These animals fail to develop whiskers and show craniofacial defects, including a lack of upper incisors and lower molars. Many animals also show cleft or incomplete palate formation, and therefore fail to suckle after birth. Mice completely devoid of activins and inhibins have also been generated by interbreeding βA- and βB-null heterozygotes and then interbreeding the compound heterozygous offspring to produce compound homozygotes lacking both βA and βB *(42)*. Not surprisingly, the double mutant mice are not viable and die shortly after birth probably because of craniofacial defects similar to those observed in βA mutants. The defects appear to represent an additivity of the individual subunit mutants, including an eyelid-fusion defect, lack of whiskers, palate defects, and tooth defects. These data suggest that within the individual βA or βB mutant mice, there is little or no compensation by the preserved ligand for the deleted one. Unfortunately, because the mice die so close to parturition, the roles of activin A, activin AB, and inhibin A in FSH regulation cannot be assessed from these model systems.

As outlined above, activins act on target cells by first binding to one of two type II receptors. Both ActRIIA and ActRIIB are expressed in the pituitary gland and, therefore, both provide substrates for activin action in adulthood *(21,43)*. At least in adult rats, pituitary expression of ActRIIA greatly exceeds that of ActRIIB *(44)*; therefore, activin's the effects of activin on FSH in adult animals may be mediated principally through ActRIIA. Consistent with this hypothesis, targeted deletion of ActRIIA produces mice that are viable (although underrepresented at weaning, because of hypoplasia of the mandible in some newborns), but have reproductive defects *(45)*. Mutant males are fertile, but FSH levels in pituitary gonadotropes and in serum are significantly reduced relative to controls. ActRIIA-deficient males also show delayed fertility, small testes, and seminiferous-tubule diameter, but normal spermatogenesis. In agreement with the data presented here for the FSHβ mutants, FSH does not appear to be necessary for spermatogenesis in mice. In contrast to males, female ActRIIA-deficient mice are infertile. They show decreased levels of serum FSH, thin-walled uteri, small ovaries, higher than normal levels of follicular atresia, and a low incidence of corpora lutea. These data are consistent with the hypothesis that many of activin's actions on FSH synthesis and secretion in adult mice are mediated via ActRIIA. In addition, any compensatory changes in ActRIIB expression are insufficient to maintain normal wild-type FSH regulation.

ActRIIB-deficient mice have also been generated by targeted deletion in ES cells *(46)*. Animals homozygous for the mutation show a greater number and variety of deficits than the ActRIIA mutant mice, including cardiac and vertebral patterning defects. None of the homozygous mutant mice on an inbred 129/Sv background survive to weaning age. In contrast, 30% of the homozygous mice on a hybrid background (129/Sv/C57BL/6) are viable, and at least the males are fertile. There are no published data regarding serum FSH levels or gonadal function in these animals so it is not yet clear how ActRIIB-deficiency affects FSH regulation in adult mice.

Activin signaling is dependent upon transphosphorylation and activation of a type I receptor upon ligand binding to the type II receptor. Therefore, one would predict that mutations in the activin type I receptor may result in reproductive and other defects. The ActRIB (ALK-4) receptor has been deleted in ES cells, and homozygous mutants were generated by breeding of heterozygous mutants. ActRIB-deficient mice die during embryonic development, and the receptor appears to be required for gastrulation *(47,48)*. As a result of the lethality, the role of ALK-4 in activin regulation of FSH has not yet been confirmed. In the future, genetic models in which ActRIB is deleted selectively in gonadotropes may help to clarify this issue.

Once activated, type I receptors phosphorylate and activate the intracellular signaling proteins known as Smads. Smad2 and Smad3 have been identified as downstream phosphorylation targets of activated type I receptors in the TGFβ- and activin-signaling pathways. Homozygous Smad2-null mutant mice die during embryogenesis *(49–51)*. At least two lines of Smad3-null mutant mice have been generated by homologous recombination. Mice in which exon 8 is deleted are viable, but die between 1 and 8 mo of age because of compromised immune function *(52)*. Mice in which exon 2 of Smad3 is deleted are also viable, but develop metastatic colorectal adenocarcinomas between 4 and 6 mo of age *(53)*. The cause of the difference in phenotypes between these two lines of Smad3-null mice is not clear, but it is possible that in at least one of the lines—a truncated, but functional Smad3—may still be produced *(52)*. Once phosphorylated, Smad2 or Smad3 form complexes with a mediator Smad: Smad4. Similar to Smad2-null mutants, Smad4-deficient mice die early during embryonic development prior to gastrulation *(54,55)*.

The phenotypes of most of the Smad mutants preclude an assessment of Smad regulation of FSH synthesis in adult mice. However, Smad3-null mice lacking exon 2 are fertile *(53)*, so some assessment of FSH synthesis and secretion prior to tumor formation should be possible in these animals. The fact that female homozygotes breed successfully suggests that FSH function is not severely compromised. The generation of gonadotrope-specific Smad2- and Smad4-deficient mice and double mutants also lacking Smad3 may be helpful in assessing the roles of these signal transducers in activin-induced FSH production in adult mice.

Finally, as described above, follistatins inhibit FSH production by binding and inactivating activins. Therefore, disruption of follistatin function may be predicted to increase FSH synthesis by increasing activin availability. Follistatin null-mutants, created by homologous recombination in ES cells, display a variety of developmental defects and die shortly after birth *(56)*. Homozygous mutants have craniofacial defects, are growth-retarded, have shiny, taut skin, and show multiple skeletal and muscular defects. Most of these defects do not phenocopy those of activin-deficient mice, and therefore suggest that follistatins still have unidentified roles in development, perhaps interacting with other members of the TGFβ family. Because of their early death, effects of follistatin-deficiency on FSH function cannot be assessed in these mice.

More recently, several lines of follistatin-overexpressing mice were created by expressing the mouse follistatin gene downstream of the mouse metallothionein-I promoter (MT-FS) *(57)*. In the line with the highest and most widespread expression pattern (line 4), FSH levels are significantly decreased in both males and females relative to controls. Line 4 males also have the smallest testes and some are infertile. All females from line 4 are infertile, and have thin-walled uteri, small ovaries, and disrupted folliculogenesis. Taken alone, these data imply that overexpression of follistatin may lead to a sequestration of activin and concomitant decline in FSH, resulting in the small testes and disrupted estrous cyclicity in these mice. However, unlike other models in which FSH is decreased and males are still fertile (such as the ActRIIA- and FSHβ-null mice), over one-half of the males in MT-FS line 4 are infertile. In addition, in some other lines of mice in which FSH levels are not affected, testes mass is decreased, and fertility is compromised. These data suggest that overexpression of follistatin within the gonads (confirmed by RNA blot analysis) and local abrogation of activin action (and possibly other TGF-β family members) is the primary cause of reproductive deficits in these mice.

SEX-STEROID DEFICIENCY

Sex-steroid hormones are important regulators of hypothalamo-pituitary function, and their effects on gonadotropin synthesis and secretion have been studied extensively *(1,2)*. One of the most dramatic examples of the effects of steroid hormones on the gonadotropins is the release from negative feedback following gonadectomy. In males and females, castration and ovariectomy result in significant increases in circulating

FSH (and LH), and associated increases in pituitary gonadotropin-subunit mRNA levels. While some of the increases in FSH can be attributed to release from inhibin-negative feedback (particularly in females), hormone replacement experiments clearly show a role for sex steroids in FSH regulation.

Many actions of steroids on FSH are indirect and are mediated via effects on the GnRH system and GnRH pulsatility, but there are also direct effects of testosterone (at least) at the level of the pituitary *(58,59)*. For example, in both male and female rats, postgonadectomy elevation of circulating FSH and pituitary FSHβ mRNA levels are attenuated by testosterone treatment. However, in animals treated with a GnRH antagonist, testosterone actually increases serum FSH and FSHβ mRNA levels *(58,60)*. These data suggest that testosterone primarily inhibits FSH via regulation of hypothalamic GnRH, but can stimulate FSH directly at the pituitary level.

Loss-of-function mutations have occurred in androgen receptors of a variety of species *(61)*. In mice, testicular feminization, or *tfm*, is a loss-of-function mutation in the androgen receptor as a result of a frame-shift caused by a single base deletion (62). All of the mice, regardless of genetic sex, are born as phenotypic females, demonstrating the role of androgens in sex differentiation. Because these animals are androgen-insensitive, they provide a powerful model system in which to investigate androgen effects on FSH. Pituitary FSH levels are significantly elevated in these mice, confirming the negative-feedback effects of androgens on FSH production *(13)*.

Estrogens, like androgens, are potent regulators of FSH. They act primarily to downregulate FSH synthesis and secretion, and do so indirectly via regulation of GnRH secretion *(60,63)*. A prediction derived from these observations is that animals devoid of functional estrogen receptors or lacking an ability to synthesize estrogens should show elevated levels of FSH relative to wild-type animals. The recent development of estrogen receptor α-knockout (ERKO) *(64)*, estrogen-receptor β-knockout (βERKO) *(65)*, and P450 aromatase-knockout mice (ArKO) *(66)* provide powerful models in which to test this prediction.

In ERKO females, pituitary FSHβ mRNA levels are significantly greater than in control animals *(67)*, but serum FSH levels do not differ between the two genotypes *(68)*. Ovariectomy increases serum FSH and pituitary FSHβ mRNA levels in wild-type mice, but increases serum FSH alone in ERKO females. These effects are attenuated by estradiol replacement in wild-type, but not ERKO, animals. These data suggest that in the absence of estrogen-receptor α, FSHβ mRNA levels are elevated in females because of a lack of functional estrogen-negative feedback. Consequently, removal of endogenous estrogens by ovariectomy does not further increase FSHβ levels in ERKO mice, but serum FSH increases because of reduced negative feedback from another ovarian factor, probably inhibin.

Unlike the case in females, male ERKO and wild-type mice have equivalent basal FSHβ mRNA levels. Serum FSH levels are slightly, but significantly, higher in ERKO males than in controls. Following castration, both genotypes show significant elevation in serum FSH, but only wild-type animals display increased FSHβ mRNA levels. The postcastration elevation in serum FSH is blocked by testosterone or estradiol, but not dihydrotestosterone (DHT), treatment in wild-type controls. These data indicate that estrogens, but not androgens, negatively regulate FSH secretion in wild-type males, and that reduction in serum FSH following testosterone treatment may be attributed to aromatization to estrogen. The data in ERKO mice are consistent with this hypothesis in that

FSH levels are unchanged by DHT treatment and, similarly, levels are unaffected in mice treated with estradiol or the aromatizable testosterone.

In wild-type mice, estradiol, but not testosterone or DHT, blocks castration-induced increases in FSHβ levels *(69)*. Thus, estradiol is a potent regulator of steady-state FSHβ mRNA levels in male mice, just as it is in females. The failure of testosterone to block the increase in FSHβ mRNA levels suggests that the amount of estrogen produced via aromatization in these mice is sufficient to reduce serum FSH, but not FSHβ mRNA levels. Neither androgenic nor estrogenic steroids affect FSHβ mRNA levels in ERKO males. As a whole, the results for both males and females suggest that a gonadal factor (probably inhibin) regulates FSH secretion, but not steady-state mRNA levels in ERα-deficient mice. Estrogens downregulate both FSH synthesis and secretion in wild-type animals, but these effects are absent in animals lacking functional ERα.

Recently, a second estrogen receptor, ERβ, was cloned in mammals *(70–72)*. The function of this receptor subtype in FSH regulation is unknown. Mice deficient in ERβ have been generated, but basic endocrine characteristics have not been reported *(65)*. Nonetheless, both males and females are fertile, suggesting that FSH function is not completely compromised. However, female ERβ-deficient mice produce smaller and fewer litters than wild-type mice, and superovulation studies indicate a decreased propensity to ovulate in these mutant mice *(65)*. Because ERβ is produced within granulosa cells, where it may mediate some of estrogen's effects on FSH- and LH-receptor expression, the ovulatory defect suggests decreased responsiveness to gonadotropins rather than a defect in gonadotropin production or secretion. In fact, the data from ERKO mice indicate that ERβ may not play a role in FSH regulation. As reviewed above, ERKO mice show no changes in FSH synthesis or secretion in response to estradiol treatment, suggesting that ERα, but not ERβ, is primarily responsible for transducing estrogenic effects on FSH.

Another approach to understanding estrogen regulation of FSH is provided by the aromatase cytochrome P450-knockout mouse (ArKO), in which animals fail to synthesize estrogens *(66)*. Given the potent inhibitory effects of estrogens on FSH secretion, one would predict elevated FSH levels in ArKO animals much as is seen in gonadectomized mice. Indeed, female ArKO mice have significantly elevated FSH levels relative to wild-type litter mates, but FSH was not measured in males. These data are inconsistent with those from the ERKO mice, in which FSH levels do not differ dramatically between intact-receptor-deficient mice and controls (although there is a small increase in ERKO males). Estrogens may act in some fashion through ERβ (or another unidentified estrogen receptor) to maintain wild-type FSH levels in ERKO mice. Perhaps the best test of this hypothesis will be to examine FSH levels in compound homozygous mutants deficient in ERα and ERβ.

The effects of progesterone on FSH synthesis and secretion are less well understood. While there is some evidence that progesterone inhibits serum FSH and FSHβ mRNA levels, other studies show no effects *(1)*. Progesterone-receptor-knockout mice (PRKO) have been produced by gene targeting in ES cells *(73)*. Female PRKO mice have a variety of reproductive defects, including anovulation. Serum estrogen, progesterone, and FSH levels do not differ between PRKO and wild-type-females, but basal LH and prolactin levels are elevated in the former *(74)*. Following stimulation with male mouse bedding, PRKO females show only a modest increase in LH relative to wild-type controls, and no change at all in FSH levels. Five days postovariectomy, FSH levels are increased to a greater extent in PRKO than in wild-type females, but levels in the two genotypes are equivalent after 10 d. These data do not indicate a clear role for progesterone in FSH regulation in female mice. In contrast, male PRKO mice have significantly lower FSH levels than do their wild-type litter mates *(75)*. Because progesterone acts to decrease FSH in other systems, the mechanisms underlying decreased FSH levels in PR-deficient mice are unclear, although PRKO males show very high levels of progesterone, and the hormone may act in a PR-independent fashion to decrease FSH. Additional experiments are needed to clarify a role for progesterone in FSH synthesis and secretion in male and female mammals.

GENETIC IDENTIFICATION OF "NOVEL" FSH REGULATORS

Because many of the hormones and receptors discussed here have been implicated previously in FSH regulation, the effects of genetic mutations in their respective genes often (though not always) produce the expected alterations in FSH phenotype. It is likely, however, that all of the factors involved in FSH regulation have not yet been identified, so an examination of other genetic models may prove useful in identifying "novel" components of the FSH regulatory pathway(s). Here, we review four examples of how targeted deletion of certain genes has produced unanticipated changes in FSH regulation.

1. Steroidogenic factor-1 (SF-1) is a transcription factor implicated in the expression of steroid hydroxylases *(76,77)*. Homozygous SF-1 null mutants exhibit gonadal and adrenal agenesis, are phenotypically female (by appearance of external genitalia), and die shortly after birth because of adrenocortical insufficiency (78). However, animals treated with steroids are viable. These animals also exhibit pituitary and hypothalamic defects. In the pituitary, FSHβ mRNA and FSH protein are barely detectable (as is the case for LHβ and LH protein). The α-subunit of the gonadotropins is expressed, but at lower levels relative to controls, while expression of TSH, growth hormone, prolactin, and ACTH are normal *(79)*. The downregulation of FSH (and LH) is interesting in light of the fact that SF-1 is expressed specifically within the gonadotrope cells of the pituitary.

   How SF-1 regulates FSHβ is unclear at present. Unlike the α- and LHβ-subunit promoters, the FSHβ promoter does not contain a consensus SF-1 binding site *(77,79)*. Interestingly, in SF-1 knockouts, the ventromedial nuclei of the hypothalamus (VMNH) develop abnormally. While this defect does not impair the morphology of the GnRH neuronal system, it appears to impair GnRH secretion. Mutants treated with exogenous GnRH produce both FSH and LH in the pituitary *(80)*. Thus, the gonadotropes are functional in SF-1 knockouts, but lack appropriate stimulation to produce FSH and LH. While the data suggest that defects in the VMNH impair GnRH secretion, this alone does not account for the low levels of LH and FSH observed in these mice, because *hpg* mice, which completely lack GnRH protein, appear to have higher levels of FSH and LH than SF-1-knockouts. Thus, factors from the gonads (which are absent in SF-1-knockouts) and/or SF-1-dependent processes within the pituitary must be able to maintain a low, but basal level of FSH and LH expression in the absence of GnRH.
2. Cyclooxygenases (COX) are the rate-limiting enzymes in prostaglandin biosynthesis. One isoform, COX-2, is induced in granulosa cells in proximity to the oocyte by the preovulatory LH surge *(81–83)*. The importance of COX-2 expression in ovulation is demonstrated in COX-2-null mutants, which show normal follicular development but fail to ovulate even after superovulatory hormonal treatment *(81)*. COX-2 mutants have elevated pituitary FSH content, although there are no dramatic differences in circulating

FSH or LH levels *(84)*. The primary defect in ovulation appears to be caused by a failure of normal gonadotropin secretion to stimulate increases in PGE_2 levels in the absence of COX-2. In fact, COX-2 mutants will ovulate in response to PMSG/hCG if followed by PGE_2 or interleukin (IL)-1β treatment. Nonetheless, the mutants show increased pituitary FSH content. The mechanisms mediating this effect are unknown, but it is possible that COX-2 expression in granulosa cells may have some impact on inhibin expression and/or release (i.e., altered ovarian function is indirectly affecting FSH levels). An examination of circulating inhibins and inhibin-subunit expression in ovaries of COX-2 mutant and wild-type mice will provide a test of this idea.

3. Female mice lacking the copper/zinc superoxide dismutase gene (SOD1) are subfertile and have ovarian defects *(85)*. While mutations in this gene have been associated with amyotropic lateral sclerosis in humans *(86,87)*, the discovery of its role in normal ovarian function is novel. Female null mutants have small ovaries and fewer large antral follicles and corpora lutea compared with wild-type females. This ovarian phenotype resembles that of the FSHβ- and ActRIIA-knockout mice described above *(4,45)*. Like these two other genetic models, SOD1-deficient females have decreased FSH levels. How SOD1-deficiency affects FSH release is unknown. It may act at the level of the brain to regulate GnRH release, at the pituitary to regulate GnRH activity, at the ovary to affect inhibin or activin expression, or through some unknown mechanisms. As in many of the other models described here, males are fertile and show overtly normal testicular function.
4. Growth differentiation factor-9 (GDF-9) is another member of the TGFβ family expressed in the gonads. Specifically, it is expressed in the oocytes of primary, one-layer follicles. GDF-9-deficient females are infertile principally because folliculogenesis does not proceed beyond the primary one-layer stage *(88–90)*. These females also show a threefold increase in serum FSH relative to wild-type controls. FSH data have not been reported for males, but GDF-9-deficient males are fertile and have normal testis size. The mechanisms controlling the increased FSH levels in females have not yet been determined, but it is possible that inhibin production is attenuated in the absence of normal follicle maturation *(91,92)*.

Each of these genetic models displays altered FSH levels, and therefore provides additional means to identify FSH-regulatory mechanisms. While these mutations may have an impact on previously characterized components of the FSH system, including GnRH and inhibin, it may be the case that the affected genes form part of novel regulatory pathways. Given the staggering number of null-mutants and compound null-mutants being generated today, a variety of novel FSH-regulatory mechanisms are likely to be identified.

CONCLUSIONS AND FUTURE DIRECTIONS

In summary, more than 25 genetic mouse models exist that hypothetically should have an impact on FSH synthesis and/or secretion (*see* Table 1 for summary). These models include several null-mutants generated by gene targeting/ES cell technology, along with naturally occurring null-mutants and overexpression models in transgenic mice. The data emerging from these models both confirm and extend results from investigation of "normal" FSH physiology in mammals. In general, the role of GnRH as a positive regulator and inhibins, follistatins, and sex steroids as negative regulators of FSH have been confirmed. While decreased levels of serum FSH in ActRIIA-deficient mice are consistent with a role of activin in stimulating FSH secretion, many of the other components of the activin-signaling cascade, when "knocked out," produce developmental defects that preclude an understanding of their roles in FSH regulation in adulthood. In addition, the slight increase in serum FSH in βB-deficient mice is somewhat paradoxical, given the putative role for endogenous activin B in stimulating FSH *(93)*. Does this occur as a result of the absence of inhibin B, or because of activin A overexpression in these animals?

A clear understanding of this and other anomalies may derive from a lack of characterization of basic endocrine physiology in these mice. This is exemplified in Table 1. For many of the models, serum and/or pituitary content of FSH has been reported, but FSHβ transcription or steady-state mRNA levels have been examined in very few cases. This leaves open the very fundamental question of whether the mutations are affecting FSH synthesis, secretion, or both. While serum FSH often mirrors pituitary FSHβ mRNA levels, this is not always the case *(1)*. As a result, we cannot assume that changes in secretion reflect changes in synthesis. A more thorough characterization of the endocrine phenotypes of these animals will greatly facilitate our ability to differentiate among mechanisms that control FSH synthesis from those that control secretion.

In addition to the need for more thorough characterization of existing models is the need for additional models. For example, there are currently no naturally occurring or genetically engineered GnRH-R-deficient mice. While one may assume that such animals will phenocopy the *hpg* mouse, there is reason to believe that there may be differences. For example, GnRH antagonists have always been more effective in suppressing LH than FSH, but in the *hpg* mouse both LH and FSH secretion are severely impaired. Is this a result of incomplete blockade of GnRH receptors with existing antagonists? Can GnRH act through another receptor to affect FSH (e.g., an FSH-RF receptor)? Are low FSH levels in *hpg* mice a result of development in the absence of GnRH? A GnRH-R-deficient system may help discriminate between these and other possibilities. In addition, as more components in the system are identified (e.g., inhibin receptors and signaling molecules), additional loss- and gain-of-function models will be of great benefit.

While we have learned a great deal from these single-locus mutants, there is certainly much to be gained from compound-null mutants or from animals with a combination of loss- and gain-of-function mutations. Certainly, we have already seen excellent examples of these approaches. Gonadal tumor development in inhibin-deficient mice is blocked by suppression of gonadotropins through interbreeding with the *hpg* mouse *(35)*. Introduction of the ActRIIA mutation into the inhibin-deficient background prevents the cancer-cachexia and wasting syndrome caused by elevated activin levels in the gonadal tumors *(32)*. Transgenes have been used to rescue the defective reproductive phenotypes in both the *hpg* and FSHβ-deficient mice *(6,15)*. Similar approaches will continue to be of great value. For example, mice null for both forms of the estrogen receptor will provide more information regarding estrogen regulation of FSH. Will these animals phenocopy the ArKO animals?

Many of the models presented here highlight the need for more tissue specificity in both loss- and gain-of-function models. This is probably most obvious in the case of the activins and their associated signaling molecules. Knockouts of many of the components of this system are recessive-lethal (*see* Table 1), probably because of their critical roles in development *(3)*. Thus, the role of these molecules in FSH regulation has not yet been demonstrated clearly. For example, does activin signal through ActRIIB in the pituitary gonadotropes? Does Smad2 and/or Smad3 mediate activin's effects on the regulation of

FSH? And the list goes on... The use of tissue-specific knockouts using the Cre-lox system *(94)* and gonadotrope-specific promoters (e.g., GnRH-R, FSHβ, and LHβ) will be invaluable in this regard. Similarly, the use of gonadotrope-specific promoters to drive overexpression of molecules such as activin/inhibin βB or follistatin will contribute greatly to our understanding of the activin/follistatin autocrine/paracrine loop in FSH regulation *(95)*.

Despite the power of current (and future) genetic models, better in vitro systems must also be developed to understand FSH regulation. How do GnRH, activin, inhibin, and sex-steroid signal-transduction cascades interact to affect FSH synthesis and secretion? It is difficult to imagine how this issue can be addressed in in vivo models alone. Of course, a great impediment to our understanding of FSH and FSHβ regulation has been the lack of FSH-producing cell lines or the lack of FSHβ-reporter activity in primary pituitary cultures *(17,23,96,97)*. Perhaps novel cell lines can be developed by targeting oncogenesis to the pituitary with the FSHβ promoter, as has been accomplished with the α, LHβ, and GnRH-R promoters to produce gonadotrope cell lines (e.g., αT3-1, and LβT2) *(96,98,99)*. Alternatively, tweaking of existing cell lines may be all that is required *(96,98–100)*.

The study of FSH regulation has had a rich past, and promises to have a strong future. It is clear that GnRH, gonadal and pituitary peptides, and sex steroids all play major roles in regulating synthesis and secretion of the hormone. The goals for the future are to determine how all of these systems interact, and to establish how these interactions change as a product of various physiological and pathophysiological conditions. A combination of greater precision in in vivo loss- and gain-of-function models and development of viable in vitro systems should enable us to achieve these goals.

Since the original submission of this chapter in July, 1999, several additional mouse models showing altered FSH phenotypes have been reported. Other models have also been generated in which one would predict altered FSH regulation, but the relevant data have not yet been reported. These models include (but are not limited to) knockouts of the FSH and LH receptors, double-knockouts of the ERα and ERβ, knockout of gamma-glutamyl transpeptidase, pituitary-specific knockout of SF-1, over-expression of follistatin in inhibin α subunit knockouts, and knock-in of the inhibin βB subunit into the inhibin βA locus. Interested readeres are directed to references listed as Additional Reading.

ACKNOWLEDGMENTS

Address all correspondence to Daniel J. Bernard, PhD, Department of Neurobiology and Physiology, Northwestern University, 2153 N. Campus Drive, Evanston, IL 60208; Fax: 847-491-2224; Email: dbernard@northwestern.edu. Supported by NIH HD35708 and HD37096, D. Bernard is a Lalor Foundation Fellow.

REFERENCES

1. Gharib SD, Wierman ME, Shupnik MA, Chin WW. Molecular biology of the pituitary gonadotropins. Endocr Rev 1990;11:177–199.
2. Haisenleder DJ, Dalkin AC, Marshall JC. Regulation of gonadotropin gene expression. In: Knobil E, Neill JD, eds. The Physiology of Reproduction. Raven Press, New York, NY, 1994, pp. 1793–1813.
3. Vale W, Bilezikjian LM, Rivier C. Reproductive and other roles of inhibins and activins. In: Knobil E, Neill JD, eds. The Physiology of Reproduction. Raven Press, Ltd., New York, NY, 1994, pp. 1861–1878.
4. Kumar TR, Wang Y, Lu N, Matzuk MM. Follicle stimulating hormone is required for ovarian follicle maturation but not male fertility. Nat Genet 1997;15:201–204.

5. Kumar TR, Fairchild-Huntress V, Low MJ. Gonadotrope-specific expression of the human follicle-stimulating hormone beta-subunit gene in pituitaries of transgenic mice. Mol Endocrinol 1992;6:81–90.
6. Kumar TR, Low MJ, Matzuk MM. Genetic rescue of follicle-stimulating hormone beta-deficient mice. Endocrinology 1998;139:3289–3295.
7. Kendall SK, Samuelson LC, Saunders TL, Wood RI, Camper SA. Targeted disruption of the pituitary glycoprotein hormone alpha-subunit produces hypogonadal and hypothyroid mice. Genes Dev 1995;9:2007–2019.
8. Keene JL, Matzuk MM, Otani T, Fauser BC, Galway AB, Hsueh AJ, et al. Expression of biologically active human follitropin in Chinese hamster ovary cells. J Biol Chem 1989;264:4769–4775.
9. Matzuk MM, Spangler MM, Camel M, Suganuma N, Boime I. Mutagenesis and chimeric genes define determinants in the beta subunits of human chorionic gonadotropin and lutropin for secretion and assembly. J Cell Biol 1989;109:1429–1438.
10. Yu WH, Karanth S, Walczewska A, Sower SA, McCann SM. A hypothalamic follicle-stimulating hormone-releasing decapeptide in the rat. Proc Natl Acad Sci USA 1997;94:9499–9503.
11. Cattanach BM, Iddon CA, Charlton HM, Chiappa SA, Fink G. Gonadotrophin-releasing hormone deficiency in a mutant mouse with hypogonadism. Nature 1977;269:338–340.
12. Mason AJ, Hayflick JS, Zoeller RT, Young WSd, Phillips HS, Nikolics K, et al. A deletion truncating the gonadotropin-releasing hormone gene is responsible for hypogonadism in the hpg mouse. Science 1986;234:1366–1371.
13. Scott IS, Bennett MK, Porter-Goff AE, Harrison CJ, Cox BS, Grocock CA, et al. Effects of the gonadotrophin-releasing hormone agonist 'Zoladex' upon pituitary and gonadal function in hypogonadal (hpg) male mice:a comparison with normal male and testicular feminized (tfm) mice. J Mol Endocrinol 1992;8:249–258.
14. Kumar TR, Low MJ. Hormonal regulation of human follicle-stimulating hormone-beta subunit gene expression:GnRH stimulation and GnRH-independent androgen inhibition. Neuroendocrinology 1995;61:628–637.
15. Mason AJ, Pitts SL, Nikolics K, Szonyi E, Wilcox JN, Seeburg PH, et al. The hypogonadal mouse:reproductive functions restored by gene therapy. Science 1986;234:1372–1378.
16. Charlton HM, Halpin DM, Iddon C, Rosie R, Levy G, McDowell IF, et al. The effects of daily administration of single and multiple injections of gonadotropin-releasing hormone on pituitary and gonadal function in the hypogonadal (hpg) mouse. Endocrinology 1983;113:535–544.
17. Kaiser UB, Conn PM, Chin WW. Studies of gonadotropin-releasing hormone (GnRH) action using GnRH receptor-expressing pituitary cell lines. Endocr Rev 1997;18:46–70.
18. Layman LC, Cohen DP, Jin M, Xie J, Li Z, Reindollar RH, et al. Mutations in gonadotropin-releasing hormone receptor gene cause hypogonadotropic hypogonadism. Nat Genet 1998;18:14–15.
19. Kaiser UB, Sabbagh E, Katzenellenbogen RA, Conn PM, Chin WW. A mechanism for the differential regulation of gonadotropin subunit gene expression by gonadotropin-releasing hormone. Proc Natl Acad Sci USA 1995;92:12,280–12,284.
20. Mathews LS. Activin receptors and cellular signaling by the receptor serine kinase family. Endocr Rev 1994;15:310–325.
21. Fernandez-Vazquez G, Kaiser UB, Albarracin CT, Chin WW. Transcriptional activation of the gonadotropin-releasing hormone receptor gene by activin A. Mol Endocrinol 1996;10:356–366.
22. Willis SA, Zimmerman CM, Li LI, Mathews LS. Formation and activation by phosphorylation of activin receptor complexes. Mol Endocrinol 1996;10:367–379.
23. Weiss J, Guendner MJ, Halvorson LM, Jameson JL. Transcriptional activation of the follicle-stimulating hormone beta-subunit gene by activin. Endocrinology 1995;136:1885–1891.
24. Lebrun JJ, Vale WW. Activin and inhibin have antagonistic effects on ligand-dependent heteromerization of the type I and type II activin receptors and human erythroid differentiation. Mol Cell Biol 1997;17:1682–1691.
25. Martens JW, de Winter JP, Timmerman MA, McLuskey A, van Schaik RH, Themmen AP, de Jong FH. Inhibin interferes with activin signaling at the level of the activin receptor complex in Chinese hamster ovary cells. Endocrinology 1997;138:2928–2936.
26. Draper LB, Matzuk MM, Roberts VJ, Cox E, Weiss J, Mather JP, et al. Identification of an inhibin receptor in gonadal tumors from inhibin alpha-subunit knockout mice. J Biol Chem 1998;273:398–403.
27. Hertan R, Farnworth PG, Fitzsimmons KL, Robertson DM. Identification of high affinity binding sites for inhibin on ovine pituitary cells in culture. Endocrinology 1999;140:6–12.
28. Elvin JA, Matzuk MM. Mouse models of ovarian failure. Rev Reprod 1998;3:183–195.
29. Lau AL, Shou W, Guo Q, Matzuk MM. Transgenic approaches to study the functions of the transforming growth factor-beta superfamily members. In: Aono T, Sugino H, Vale WW, eds. Inhibin, Activin

and Follistatin:Regulatory Functions in System and Cell Biology. Springer-Verlag, New York, NY, 1997, pp. 220–243.
30. Matzuk MM, Kumar TR, Shou W, Coerver KA, Lau AL, Behringer RR, Finegold MJ. Transgenic models to study the roles of inhibins and activins in reproduction, oncogenesis, and development. Recent Prog Horm Res 1996;51:123–154.
31. Matzuk MM, Finegold MJ, Su JG, Hsueh AJ, Bradley A. Alpha-inhibin is a tumour-suppressor gene with gonadal specificity in mice. Nature 1992;360:313–319.
32. Coerver KA, Woodruff TK, Finegold MJ, Mather J, Bradley A, Matzuk MM. Activin signaling through activin receptor type II causes the cachexia-like symptoms in inhibin-deficient mice. Mol Endocrinol 1996;10:534–543.
33. Matzuk MM, Finegold MJ, Mather JP, Krummen L, Lu H, Bradley A. Development of cancer cachexia-like syndrome and adrenal tumors in inhibin-deficient mice. Proc Natl Acad Sci USA 1994;91:8817–8821.
34. Shou W, Woodruff TK, Matzuk MM. Role of androgens in testicular tumor development in inhibin-deficient mice. Endocrinology 1997;138:5000–5005.
35. Kumar TR, Wang Y, Matzuk MM. Gonadotropins are essential modifier factors for gonadal tumor development in inhibin-deficient mice. Endocrinology 1996;137:4210–4216.
36. Kumar TR, Palapattu G, Wang P, Woodruff TK, Boime I, Byrne MC, et al. Transgenic models to study gonadotropin function:The role of follicle-stimulating hormone in gonadal growth and tumorigensis. Mol Endocrinol 1999;13:851–865.
37. Kananen K, Markkula M, Rainio E, Su JG, Hsueh AJ, Huhtaniemi IT. Gonadal tumorigenesis in transgenic mice bearing the mouse inhibin alpha-subunit promoter/simian virus T-antigen fusion gene:characterization of ovarian tumors and establishment of gonadotropin-responsive granulosa cell lines. Mol Endocrinol 1995;9:616–627.
38. Kananen K, Markkula M, el-Hefnawy T, Zhang FP, Paukku T, Su JG, et al. The mouse inhibin alpha-subunit promoter directs SV40 T-antigen to Leydig cells in transgenic mice. Mol Cell Endocrinol 1996;119:135–146.
39. Kananen K, Rilianawati, Paukku T, Markkula M, Rainio EM, Huhtanemi I. Suppression of gonadotropins inhibits gonadal tumorigenesis in mice transgenic for the mouse inhibin alpha-subunit promoter/simian virus 40 T-antigen fusion gene. Endocrinology 1997;138:3521–3531.
40. McMullen ML, Cho BN, Yates JC, Mayo KE. Transgenic mice expressing the rat inhibin α subunit exhibit female subfertility and corresponding ovarian pathologies. The 1999 North American Inhibin and Activin Congress, Northwestern University, Evanston, IL, 1999, pp. 16.
41. Vassalli A, Matzuk MM, Gardner HA, Lee KF, Jaenisch R. Activin/inhibin beta B subunit gene disruption leads to defects in eyelid development and female reproduction. Genes Dev 1994;8:414–427.
42. Matzuk MM, Kumar TR, Vassalli A, Bickenbach JR, Roop DR, Jaenisch R, et al. Functional analysis of activins during mammalian development. Nature 1995;374:354–356.
43. Cameron VA, Nishimura E, Mathews LS, Lewis KA, Sawchenko PE, Vale WW. Hybridization histochemical localization of activin receptor subtypes in rat brain, pituitary, ovary, and testis. Endocrinology 1994;134:799–808.
44. Dalkin AC, Haisenleder DJ, Yasin M, Gilrain JT, Marshall JC. Pituitary activin receptor subtypes and follistatin gene expression in female rats:differential regulation by activin and follistatin. Endocrinology 1996;137:548–554.
45. Matzuk MM, Kumar TR, Bradley A. Different phenotypes for mice deficient in either activins or activin receptor type II. Nature 1995;374:356–360.
46. Oh SP, Li E. The signaling pathway mediated by the type IIB activin receptor controls axial patterning and lateral asymmetry in the mouse. Genes Dev 1997;11:1812–1826.
47. Gu Z, Nomura M, Simpson BB, Lei H, Feijen A, van den Eijnden-van Raaij J, et al. The type I activin receptor ActRIB is required for egg cylinder organization and gastrulation in the mouse. Genes Dev 1998;12:844–857.
48. Gu Z, Reynolds EM, Song J, Lei H, Feijen A, Yu L, et al. The type I serine/threonine kinase receptor ActRIA (ALK2) is required for gastrulation of the mouse embryo. Development 1999;126:2551–2561.
49. Nomura M, Li E. Smad2 role in mesoderm formation, left-right patterning and craniofacial development. Nature 1998;393:786–790.
50. Waldrip WR, Bikoff EK, Hoodless PA, Wrana JL, Robertson EJ. Smad2 signaling in extraembryonic tissues determines anterior-posterior polarity of the early mouse embryo. Cell 1998;92:797–808.

51. Weinstein M, Yang X, Li C, Xu X, Gotay J, Deng CX. Failure of egg cylinder elongation and mesoderm induction in mouse embryos lacking the tumor suppressor smad2. Proc Natl Acad Sci USA 1998;95:9378–9383.
52. Yang X, Letterio JJ, Lechleider RJ, Chen L, Hayman R, Gu H, et al. Targeted disruption of SMAD3 results in impaired mucosal immunity and diminished T cell responsiveness to TGF-beta. EMBO J 1999;18:1280–1291.
53. Zhu Y, Richardson JA, Parada LF, Graff JM. Smad3 mutant mice develop metastatic colorectal cancer. Cell 1998;94:703–714.
54. Sirard C, de la Pompa JL, Elia A, Itie A, Mirtsos C, Cheung A, et al. The tumor suppressor gene Smad4/Dpc4 is required for gastrulation and later for anterior development of the mouse embryo. Genes Dev 1998;12:107–119.
55. Yang X, Li C, Xu X, Deng C. The tumor suppressor SMAD4/DPC4 is essential for epiblast proliferation and mesoderm induction in mice. Proc Natl Acad Sci USA 1998;95:3667–3672.
56. Matzuk MM, Lu N, Vogel H, Sellheyer K, Roop DR, Bradley A. Multiple defects and perinatal death in mice deficient in follistatin. Nature 1995;374:360–363.
57. Guo Q, Kumar TR, Woodruff T, Hadsell LA, DeMayo FJ, Matzuk MM. Overexpression of mouse follistatin causes reproductive defects in transgenic mice. Mol Endocrinol 1998;12:96–106.
58. Paul SJ, Ortolano GA, Haisenleder DJ, Stewart JM, Shupnik MA, Marshall JC. Gonadotropin subunit messenger RNA concentrations after blockade of gonadotropin-releasing hormone action: testosterone selectively increases follicle-stimulating hormone beta-subunit messenger RNA by posttranscriptional mechanisms. Mol Endocrinol 1990;4:1943–1955.
59. Wierman ME, Wang C. Androgen selectively stimulates follicle-stimulating hormone-beta mRNA levels after gonadotropin-releasing hormone antagonist administration. Biol Reprod 1990;42:563–571.
60. Dalkin AC, Paul SJ, Haisenleder DJ, Ortolano GA, Yasin M, Marshall JC. Gonadal steroids effect similar regulation of gonadotrophin subunit mRNA expression in both male and female rats. J Endocrinol 1992;132:39–45.
61. French FS, Lubahn DB, Brown TR, Simental JA, Quigley CA, Yarbrough WG, et al. Molecular basis of androgen insensitivity. Recent Prog Horm Res 1990;46:1–38.
62. Charest NJ, Zhou ZX, Lubahn DB, Olsen KL, Wilson EM, French FS. A frameshift mutation destabilizes androgen receptor messenger RNA in the Tfm mouse. Mol Endocrinol 1991;5:573–581.
63. Shupnik MA, Gharib SD, Chin WW. Estrogen suppresses rat gonadotropin gene transcription in vivo. Endocrinology 1988;122:1842–1846.
64. Lubahn DB, Moyer JS, Golding TS, Couse JF, Korach KS, Smithies O. Alteration of reproductive function but not prenatal sexual development after insertional disruption of the mouse estrogen receptor gene. Proc Natl Acad Sci USA 1993;90:11,162–11,166.
65. Krege JH, Hodgin JB, Couse JF, Enmark E, Warner M, Mahler JF, et al. Generation and reproductive phenotypes of mice lacking estrogen receptor beta. Proc Natl Acad Sci USA 1998;95:15,677–15,682.
66. Fisher CR, Graves KH, Parlow AF, Simpson ER. Characterization of mice deficient in aromatase (ArKO) because of targeted disruption of the cyp19 gene. Proc Natl Acad Sci USA 1998;95:6965–6970.
67. Scully KM, Gleiberman AS, Lindzey J, Lubahn DB, Korach KS, Rosenfeld MG. Role of estrogen receptor-alpha in the anterior pituitary gland. Mol Endocrinol 1997;11:674–681.
68. Lindzey J, Couse JF, Stoker T, Wetsel WC, Cooper R, Korach KS. Steroid regulation of gonadotrope function in female wild-type (WT) and estrogen receptor-α knockout (ERKO) mice. 80th Annual Meeting of the Endocrine Society Abstracts, 1998, pp. 112.
69. Lindzey J, Wetsel WC, Couse JF, Stoker T, Cooper R, Korach KS. Effects of castration and chronic steroid treatments on hypothalamic gonadotropin-releasing hormone content and pituitary gonadotropins in male wild-type and estrogen receptor-alpha knockout mice. Endocrinology 1998;139:4092–4101.
70. Kuiper GG, Enmark E, Pelto-Huikko M, Nilsson S, Gustafsson JA. Cloning of a novel receptor expressed in rat prostate and ovary. Proc Natl Acad Sci USA 1996;93:5925–5930.
71. Mosselman S, Polman J, Dijkema R. ER beta:identification and characterization of a novel human estrogen receptor. FEBS Lett 1996;392:49–53.
72. Tremblay GB, Tremblay A, Copeland NG, Gilbert DJ, Jenkins NA, Labrie F, et al. Cloning, chromosomal localization, and functional analysis of the murine estrogen receptor beta. Mol Endocrinol 1997;11:353–365.
73. Lydon JP, DeMayo FJ, Funk CR, Mani SK, Hughes AR, Montgomery CA, Jr., et al. Mice lacking progesterone receptor exhibit pleiotropic reproductive abnormalities. Genes Dev 1995;9:2266–2278.

74. Chappell PE, Lydon JP, Conneely OM, O'Malley BW, Levine JE. Endocrine defects in mice carrying a null mutation for the progesterone receptor gene. Endocrinology 1997;138:4147–4152.
75. Schneider JS, Sleiter NC, Levine JE. Endocrine abnormalities in male mice carrying a null mutation for the progesterone receptor. 81st Annual Meeting of the Endocrine Society Abstracts, 1999, pp. 285.
76. Parker KL, Ikeda Y, Luo X. The roles of steroidogenic factor-1 in reproductive function. Steroids 1996;61:161–165.
77. Parker KL. The roles of steroidogenic factor 1 in endocrine development and function. Mol Cell Endocrinol 1998;145:15–20.
78. Luo X, Ikeda Y, Parker KL. A cell-specific nuclear receptor is essential for adrenal and gonadal development and sexual differentiation. Cell 1994;77:481–490.
79. Ingraham HA, Lala DS, Ikeda Y, Luo X, Shen WH, Nachtigal MW, et al. The nuclear receptor steroidogenic factor 1 acts at multiple levels of the reproductive axis. Genes Dev 1994;8:2302–2312.
80. Ikeda Y, Luo X, Abbud R, Nilson JH, Parker KL. The nuclear receptor steroidogenic factor 1 is essential for the formation of the ventromedial hypothalamic nucleus. Mol Endocrinol 1995;9:478–486.
81. Lim H, Paria BC, Das SK, Dinchuk JE, Langenbach R, Trzaskos JM, et al. Multiple female reproductive failures in cyclooxygenase 2-deficient mice. Cell 1997;91:197–208.
82. Sirois J, Simmons DL, Richards JS. Hormonal regulation of messenger ribonucleic acid encoding a novel isoform of prostaglandin endoperoxide H synthase in rat preovulatory follicles. Induction in vivo and in vitro. J Biol Chem 1992;267:11,586–11,592.
83. Sirois J. Induction of prostaglandin endoperoxide synthase-2 by human chorionic gonadotropin in bovine preovulatory follicles in vivo. Endocrinology 1994;135:841–848.
84. Davis BJ, Lennard DE, Lee CA, Tiano HF, Morham SG, Wetsel WC, et al. Anovulation in cyclooxygenase-2-deficient mice is restored by prostaglandin E2 and interleukin-1beta. Endocrinology 1999;140:2685–2695.
85. Matzuk MM, Dionne L, Guo Q, Kumar TR, Lebovitz RM. Ovarian function in superoxide dismutase 1 and 2 knockout mice. Endocrinology 1998;139:4008–4011.
86. Borchelt DR, Lee MK, Slunt HS, Guarnieri M, Xu ZS, Wong PC, et al. Superoxide dismutase 1 with mutations linked to familial amyotrophic lateral sclerosis possesses significant activity. Proc Natl Acad Sci USA 1994;91:8292–8296.
87. Kunst CB, Mezey E, Brownstein MJ, Patterson D. Mutations in SOD1 associated with amyotrophic lateral sclerosis cause novel protein interactions. Nat Genet 1997;15:91–94.
88. Dong J, Albertini DF, Nishimori K, Kumar TR, Lu N, Matzuk MM. Growth differentiation factor-9 is required during early ovarian folliculogenesis. Nature 1996;383:531–535.
89. Carabatsos MJ, Elvin J, Matzuk MM, Albertini DF. Characterization of oocyte and follicle development in growth differentiation factor-9-deficient mice. Dev Biol 1998;204:373–384.
90. Elvin JA, Yan C, Wang P, Nishimori K, Matzuk MM. Molecular characterization of the follicle defects in the growth differentiation factor 9-deficient ovary. Mol Endocrinol 1999;13:1018–1034.
91. Mayo KE. Inhibin and activin:Molecular aspects of regulation and function. TEM 1994;5:407–415.
92. Woodruff TK, D'Agostino J, Schwartz NB, Mayo KE. Dynamic changes in inhibin messenger RNAs in rat ovarian follicles during the reproductive cycle. Science 1988;239:1296–1299.
93. Corrigan AZ, Bilezikjian LM, Carroll RS, Bald LN, Schmelzer CH, Fendly BM, et al. Evidence for an autocrine role of activin B within rat anterior pituitary cultures. Endocrinology 1991;128:1682–1684.
94. Cohen-Tannoudji M, Babinet C. Beyond 'knock-out' mice: new perspectives for the programmed modification of the mammalian genome. Mol Hum Reprod 1998;4:929–938.
95. Besecke LM, Guendner MJ, Schneyer AL, Bauer-Dantoin AC, Jameson JL, Weiss J. Gonadotropin-releasing hormone regulates follicle-stimulating hormone- beta gene expression through an activin/follistatin autocrine or paracrine loop. Endocrinology 1996;137:3667–3673.
96. Alarid ET, Windle JJ, Whyte DB, Mellon PL. Immortalization of pituitary cells at discrete stages of development by directed oncogenesis in transgenic mice. Development 1996;122:3319–3329.
97. Turgeon JL, Kimura Y, Waring DW, Mellon PL. Steroid and pulsatile gonadotropin-releasing hormone (GnRH) regulation of luteinizing hormone and GnRH receptor in a novel gonadotrope cell line. Mol Endocrinol 1996;10:439–450.
98. Albarracin CT, Frosch MP, Chin WW. The gonadotropin-releasing hormone receptor gene promoter directs pituitary-specific oncogene expression in transgenic mice. Endocrinology 1999;140:2415–2421.
99. Mellon PL, Windle JJ, Weiner RI. Immortalization of neuroendocrine cells by targeted oncogenesis. Recent Prog Horm Res 1991;47:69–93.
100. Graham KE, Nusser KD, Low MJ. Lbeta T2 gonadotroph cells secrete follicle stimulating hormone (FSH) in response to activin A. J Endocrinol 1999;162:R1–R5.

Additional Reading

Abel MH, Wootton AN, Wilkins V, Huhtaniemi I, Knight PG, Charlton HM. The effect of a null mutation in the follicle-stimulating hormone receptor gene on mouse reproduction. Endocrinology 2000;141:1795–1803.

Brown CW, Houseon-Hawkins DE, Woodruff TK, Matzuk MM. Insertion of Inhbb into the Inhba locus rescues the Inhba-null phenotype and reveals new activin functions. Nat Genet 2000;25:453–457.

Cipriano SC, Chen L, Kumar TR, Matzuk MM. Follistatin is a modulator of gonadal tumor progression and the activin-induced wasting syndrome in inhibin-deficient mice. Endocrinology 2000;141:2319–2327.

Dupont S, Krust A, Gansmuller A, Dierich A, Chambon P, Mark M. Effect of single and compound knockouts of estrogen receptors alpha (ERalpha) and beta (ERbeta) on mouse reproductive phenotypes. Development 2000;127:4277–4291.

Kumar TR, Wiseman AL, Kala G, Matzuk MM, Lieberman MW. Reproductive defects in gamma-glutamyl transpeptidase-deficient mice. Endocrinology 2000;141:4270–4277.

Lei ZM, Mishra S, Zou W, Zu B, Foltz M, Li X, Rao CV. Targeted disruption of luteinizing hormone/human chorionic gonadotropin receptor gene. Mol Endocrinol 2001;15:184–200.

Zhang FP, Poutanen M, Wilbertz J, Huhtaniemi I. Normal prenatal but arrested postnatal sexual development of luteinizing hormone receptor knockout (LuRKO) mice. Mol Endocrinol 2001;15:172–183.

Zhao L, Bakke M, Krimkevich Y, Cushman LJ, Parlow AF, Camper SA, PArker KL. Steroidogenic factor 1 (SF1) is essential for pituitary gonadotrope function. Development 2001;128:147–154.

15 Transgenic Analysis of the Proopiomelanocortin Neuroendocrine System

Malcolm J. Low, MD, PhD, *Marcelo Rubinstein*, PhD, *and E. Cheng Chan*, PhD

CONTENTS

INTRODUCTION

Proopiomelanocortin (POMC) first came under intense scientific investigation following the demonstration that a single prohormone was a polypeptide precursor for multiple biologically active peptides following posttranslational processing. The expression of POMC in both the brain and pituitary and the potential role of adrenocorticotropin (ACTH) and opioids in the physiology of the mammalian stress response suggest that regulation of POMC gene expression and processing of the prohormone to smaller peptides are critical events in maintaining homeostasis. The endogenous opioid peptide β-endorphin, processed from the carboxyl terminus of POMC, is implicated in cognitive and emotional behaviors, analgesia, and control of GnRH neurons. POMC peptides may also have important developmental roles becuase of their early expression in both the fetal pituitary and medial basal hypothalamus *(1)*. A recent resurgence in the scientific popularity of this poly protein followed the demonstration that melanocortins, including α- and γ-melanocyte-stimulating hormone (MSH) play important functional roles in the central control of appetite and weight homeostasis, independently of regu-

From: *Contemporary Endocrinology: Transgenics in Endocrinology*
Edited by: M. Matzuk, C. W. Brown, and T. R. Kumar © Humana Press Inc., Totowa, NJ

latory effects on corticosterone production *(2–5)*. Therefore, POMC peptides may form an important link at multiple levels in coordinated physiological responses to stress, feeding, and reproduction. The physiology of this marvelous neuroendocrine system is virtually impossible to replicate in in vitro models. Even studies of POMC gene regulation have largely been limited to one cell type—pituitary corticotrophs represented by the AtT20 tumor cell line—although these cells may not be relevant to neurons. Transgenic mice, and more recently, mutant mouse strains with targeted gene deletions or modifications, have therefore been extremely informative tools for the analysis of this neuropeptide system. The experiments based on these technologies are the theme of this chapter.

POMC GENE EXPRESSION IN TRANSGENIC MICE

Tissue-Specific Expression and Developmental Regulation

The POMC gene is expressed at its highest levels in pituitary corticotrophs and melanotrophs and two circumscribed groups of neurons in the central nervous system, the arcuate nucleus of the hypothalamus, and the nucleus of the solitary tract (NTS) in the medulla *(6–8)*. Although POMC mRNA has also been detected in many other tissues, including skin, lymphocytes, lung, and testes, there is continuing controversy concerning the quantity and pattern of POMC peptides made in these extracephalic sites and their functional significance. POMC mRNA can first be detected in the mouse hypothalamus at d 10.5 of embryonic development (E 10.5) and in the ventral anterior pituitary gland at E 12.5 (9). Expression in the nascent melanotrophs of the dorsal intermediate lobe (IL) occurs 2 d later at E 14.5. The exact embryonic relationship between these three POMC-expressing cell types is an intriguing question in the area of developmental biology. Defining this relationship is one of the goals of our laboratory's work to determine the molecular basis of POMC gene expression. Sophisticated work by Le Douarin suggests that the basal hypothalamus and adjacent Rathke's pouch, which are closedly opposed without intervening mesodermal tissue, are derived from adjacent anterolateral neural-ridge cells and neuroepithelium *(10)*. The migratory paths of the two progenitor tissues diverge for a few days, and then are again inextricably linked by the dorsal invagination of Rathke's pouch and the expression of inductive signals from the diencephalon to the pouch in the form of bone morphogenetic protein 4 (BMP4) and fibroblast growth factor 8 (FGF8) *(11–13)*. Recent work from several laboratories, primarily using mutant mice with null mutations in a number of homeobox-containing transcription factors, has begun to outline the genetic program responsible for the orderly temporal and spatial appearance of various pituitary hormone phenotypes *(14,15)*. However, the pituitary corticotrophs still remain in a unique class, with no definitive characterization of the critical factors. Recent work from Melmed and colleagues suggests that overexpression of leukemia-inhibitory factor (LIF) in the developing pituitary anlagen of transgenic mice causes hyperplasia of corticotrophs in addition to the development of Rathke's cleft cysts *(16,17)*.

One important approach to the characterization of POMC gene-transcriptional mechanisms has been the expression in transfected cell lines of fusion genes consisting of POMC promoter sequences and a reporter sequence. The most commonly used cell line is one of several variants of the AtT20 tumor cells originally derived from a radiation-induced pituitary tumor *(18,19)*. These cells exhibit many characteristics of corticotrophs.

Small-cell lung cancers often ectopically express POMC, and cell lines derived from these tumors have also been used for the analysis of POMC gene expression *(20,21)*. However, it is clear that these cell lines are not adequate models for all cell types that normally express POMC, particularly neurons and probably also the melanotrophs of the pituitary intermediate lobe. Even with regard to corticotrophs, there are questions concerning the fidelity of these transformed cells to the primary cells in their natural-organ environment. For these reasons, our laboratory has concentrated on a transgenic mouse approach for the identification of key regulatory elements in the POMC gene.

The first report of transgenic mice expressing a neomycin-resistance coding sequence (*neo*) from 770 bases of the rat POMC promoter originated from the Drouin laboratory *(22)*. Although the pattern of expression in the pituitary gland and regulation by adrenalectomy and dexamethasone were all consistent with accurate cellular expression, the nature of the *neo* reporter and low levels of transcript made it difficult to directly demonstrate colocalization with an independent marker of pituitary phenotype. Our laboratory first approached this problem using a *lacZ* reporter gene encoding β-galactosidase, which allowed for precise cellular identification of transgene expression as well as quantitation by enzymatic assay with a fluorigenic substrate. These studies demonstrated that the same 770 bases of the rat POMC promoter from nucleotide –706 in the 5' flanking sequences to +64 in the 5' untranslated region of exon 1 were sufficient to selectively direct *lacZ* expression to pituitary corticotrophs and melanotrophs *(23)*. The ontogeny of transgene expression is closely approximated by the time course of endogenous POMC gene expression in the two lobes of the pituitary gland. We subsequently elaborated on this theme, using a series of truncations and selected mutations with a variety of reporter sequences to characterize the minimally required rat POMC promoter elements for pituitary-specific gene expression *(24,25)*. In all cases, accurate cell-specific colocalization was performed by double-label immunofluoresence histochemistry with appropriate antisera. The results of these studies are summarized in Fig. 1.

Truncations of the rat POMC promoter to nucleotide –323 or –234 had either a small or no discernible effect on pituitary expression, assessed by the ratio of expression-positive to total transgenic pedigrees and qualitative levels of reporter expression in the positively expressing lines. However, continued deletion of the 5' flanking sequences to –160 clearly abolished expression. TATA box sequences from the herpes simplex virus-thymidine kinase (HSV-TK) promoter were equivalent to native POMC promoter sequences in the context of upstream flanking sequences, suggesting the absence of critical pituitary-specific DNA binding sites in the region between nucleotides –34 and +64. Building on these initial transgenic results and information from DNase I protection assays and gel-shift assays, which utilized POMC oligonucleotide probes and fractionated nuclear proteins from AtT20 cells, we designed a series of more discrete mutations in the rat POMC promoter. These final transgenic studies suggested the key importance of two binding sites at nucleotide positions –262/–253 and –202/–193, denoted PP1 for putative pituitary POMC1 factor *(25)*. At least one of these sites had to be intact to support detectable expression of the transgenes. In addition, the transcription factor SP1 and its two binding sites at positions –201/–192 and –146/–136 appeared to play a supportive role in POMC transgene expression, but were unnecessary if at least one PP1 site was present. Notably, none of the transgene constructs exhibited selective expression in only the melanotrophs or corticotrophs, suggesting that both cell types share one or more essential transcription-factor binding sites. As with most transgenic-promoter

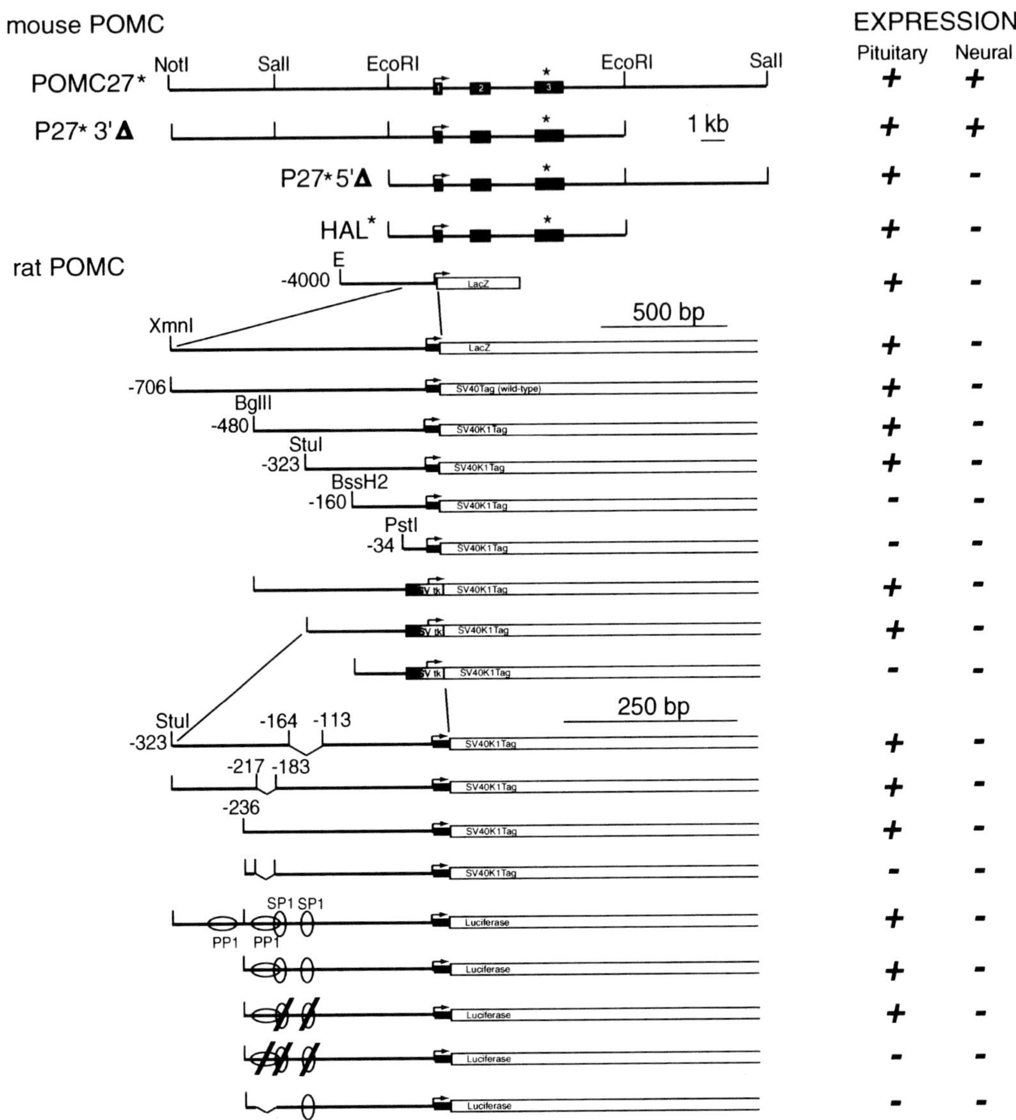

Fig. 1. Tissue-specific expression of transgenes based on mouse and rat POMC genomic regulatory elements. The four transgenes at the top of the figure are based on mouse genomic sequences encompassing the entire transcriptional unit (exons 1, 2, and 3). An oligonucleotide insertion in exon 3 designated by the * was used for detection of transgene mRNA. Pituitary expression includes cell-specific expression in both intermediate-lobe melanotrophs and AL corticotrophs. Neural expression includes cell-specific expression in both the arcuate nucleus of the hypothalamus and the nucleus of the solitary tract in the medulla. The remaining transgenes were based on rat POMC 5' flanking and promoter sequences. β-galactosidase encoded by *Lac-Z*, SV40-T antigen, and luciferase were all used as reporter sequences. Sequence deletions are indicated by gaps with nucleotide positions and site-directed mutations are indicated by slash marks over the identified transcription-factor binding sites PP1 and SP1.

analyses, we observed a large range of overall expression levels between independent pedigrees for each construct and mosaic patterns of expression within both lobes of the pituitary gland. It has been suggested that these patterns of gene expression are influenced by repressive effects of multiple-copy transgene integration in some cases *(26)*. In general, basal expression of the transgenes appeared to be greater in melanotrophs than corticotrophs, regardless of the reporter molecule used.

There are unresolved inconsistencies between the transgenic mapping of the POMC promoter and standard experiments based on transfection studies in AtT20 cells. In part, this may be a result of the more stringent requirement in transgenic mice for the definition of cell-specific expression and the importance of chromatin remodeling during in vivo development. There is strong agreement among both kinds of studies for the importance of DNA elements between nucleotides –160 and –323 in the 5' flanking domain of the POMC gene. This region also contains a site between –173/–160 that appears to mediate the transcriptional activating effects of both CRH and LIF, although it is not a binding site for CREB, AP1, or STAT proteins *(27)*. However, evidence for more distal sites binding HLH proteins and conferring a synergistic activation of POMC expression in AtT20 cells do not appear to be as important in the in vivo paradigm *(28)*.

None of the transgenes described thus far confer POMC neuronal expression in addition to expression in the pituitary cells. We conclude that additional genomic DNA sequences possibly containing a neural-specific enhancer element should exist. Modest increases, either in the form of 4 kb of rat 5' flanking sequence or a 10-kb genomic mouse POMC clone, maintained the pattern of pituitary expression and frequently produced ectopic neuronal expression *(29)*. Our most recent studies have isolated the 11-kb stretch of DNA between –13 kb and –2 kb as the location of a putative neuron-specific enhancer (Fig. 1) *(30)*. Interestingly, constructs containing these sequences have shown a 100% penetrance of transgene expression in the neurons of both the arcuate nucleus and NTS, consistent with the presence of a locus-control-type regulatory element. It remains unknown exactly how neural and pituitary cell-specific expression are related to each other. For example, are there common transcription factors and DNA-binding elements shared by both cell types, or are the transcriptional mechanisms unique apart from the basal transcriptional complex? The recent identification of a novel neuropeptide precursor gene, cocaine- and amphetamine-responsive transcript (CART), as a cotransmitter in POMC arcuate neurons also prompts the question of whether these two genes share a subset of regulatory factors *(31,32)*. Finally, it should be noted that even the largest murine POMC genomic clones tested to date appear to have inappropriately low levels of basal expression in adult corticotrophs relative to melanotrophs, or to the more robust corticotroph expression observed in development or after stimulation induced by adrenalectomy. These observations suggest that there may still be an additional corticotroph-selective enhancer more distal to the promoter that is required for a full transcriptional rate under basal conditions.

Hormonal Regulation

In addition to the identification of minimal sequences necessary for cell-specific expression, the transgenic constructs have also been useful for studying the regulation of pituitary POMC expression by hormones or physiological perturbations. Our initial studies utilizing transgenic constructs with either β-galactosidase or K1Tag reporter genes demonstrated that the same 5' flanking sequences of the rat POMC gene between –323/–34 that supported cell-specific expression were also sufficient for transcriptional activation in corticotrophs following adrenalectomy. Similarly, transgene expression

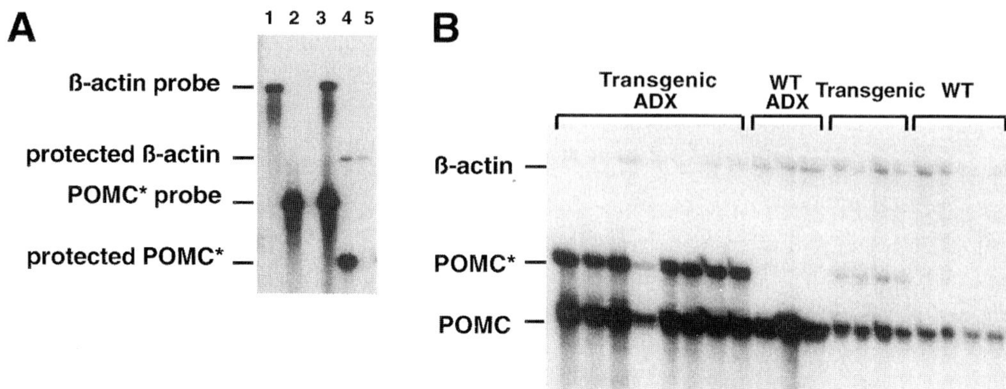

Fig. 2. RPA demonstrating the effects of adrenalectomy on POMC and POMC* mRNA levels in the anterior pituitary lobes of female wild-type and pHAL* transgenic mice. Mice were bilaterally adrenalectomized by a dorsal approach, and provided with 0.9% saline in their drinking water for 7 d prior to sacrifice. The anterior lobe of each pituitary gland was separated from the neurointermediate lobe with the aid of a dissecting microscope and total RNA was extracted from each anterior lobe by the NP-40 method. The RPA was performed using reagents from Ambion, Inc. (Austin, TX) and a mixture of [^{32}P]-labeled riboprobes for POMC/POMC* and β-actin. **(A)** The panel shows unprotected mouse β-actin probe (400 NT), protected β-actin (300 NT), unprotected POMC* probe (214 NT), and protected POMC* (199 NT). Lanes 1–3, free probes in the absence of nucleases; lane 4, both probes hybridized with transgenic pituitary RNA and nucleases; lane 5, both probes hybridized with wild-type pituitary RNA and nucleases. **(B)** The panel depicts a representative RPA experiment. Protected β-actin (300 NT), protected POMC* (199 NT), and protected POMC (185 NT).

was significantly increased in melanotrophs by administration of the dopamine D2 receptor antagonist haloperidol *(23)*. Because of the inherent difficulties in accurately comparing the relative levels of a reporter-gene product with endogenous levels of POMC mRNA, we also performed a series of regulatory experiments on transgenic mice expressing the 10-kb mouse pHAL* transgene originally described by our group in 1993 *(29)*.

pHAL* is identical to native POMC, with the addition of a random oligonucleotide sequence inserted into exon 3 to interrupt the expression of ACTH and MSH and avoid the theoretical possibility of pituitary-dependent hypercortisolism from high-level transgene expression. The position of the oligonucleotide was also designed to allow simultaneous detection and quantification of endogenous POMC mRNA and POMC* mRNA in total RNA samples, using one radiolabeled probe in an RNase protection assay (RPA) *(30)*. A representative RPA with this system is shown in Fig. 2. Wild-type pituitary glands show only the smaller protected band of 185 NT derived from POMC, while transgenic glands also show a higher mol-wt band of 199 NT derived from the protection of POMC* mRNA. Seven days after adrenalectomy, there was a marked increase in the quantity of both mRNA species in the transgenic mice, but the ratio of POMC*/POMC mRNA in the stimulated state was significantly increased compared to basal after normalization to the level of β-actin transcript (Fig. 3 and Table 1). The same result was apparent by *in situ* hybridization studies published previously *(29)*. Expression of the transgene does not compete for expression of the endogenous POMC gene (compare the levels of POMC mRNA between wild-type and transgenic mice in Fig. 3), suggesting

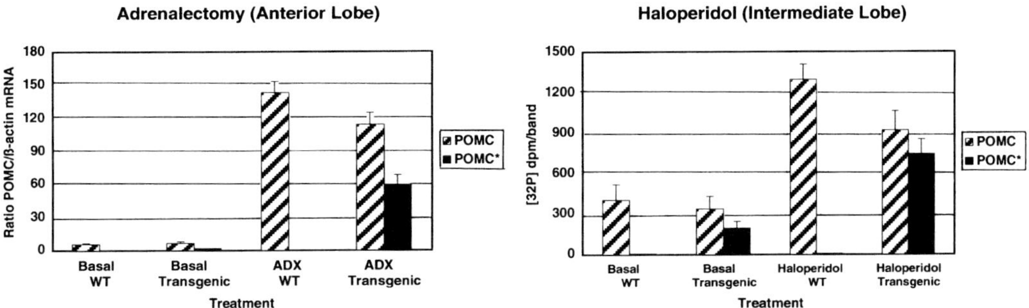

Fig. 3. Relative expression levels of POMC and POMC* mRNA in the pituitary glands of male wild-type and pHAL* transgenic mice after adrenalectomy or female mice after haloperidol treatment. Mice were bilaterally adrenalectomized by a dorsal approach and provided with 0.9% saline in their drinking water for 7 d prior to sacrifice ($n = 3$ to 8 per group). Haloperidol was administered in a dose of 10 mg/kg ip daily for 7 d prior to sacrifice ($n = 4$ to 7 per group). The neurointermediate and anterior lobes of the pituitary gland were collected separately for preparation of total RNA. The RPA for adrenalectomized mice was analyzed with a phosphoimager, and the pixel volumes for the POMC and POMC* bands were normalized as a ratio to the pixel volume for the β-actin band in each lane. The RPA for haloperidol-treated mice was analyzed by direct scintillation counting of the POMC and POMC* bands excised from the acrylamide gel after their localization by autoradiography, and the data were expressed as dpm/band. No correction was made for β-actin expression, because the levels were below the detection limit of the assay using the small amount of total RNA extracted from individual mouse neurointermediate lobes.

that there is not a limiting amount of essential transcription factors. In contrast to the results in the anterior lobe (AL), the ratio of expression of the transgene relative to endogenous POMC mRNA in melanotrophs was nearly unity in both the basal state and after stimulation by haloperidol (Fig. 3 and Table 1).

Our finding of a greater ratio of POMC*/POMC mRNA in the AL after adrenalectomy compared to sham surgery suggested that either CRH stimulation, withdrawal of direct pituitary glucocorticoid inhibition of POMC transcription, or a combination of both overcame the putative absence of a genomic element necessary for higher levels of basal expression. To further examine the relative expression levels of endogenous and POMC* mRNA in response to endocrine regulators, and to determine whether the 10-kb transgene contained all the necessary transcription-factor-binding sites for regulation by different perturbations of the HPA axis, we performed a series of experiments in the pHAL* mice. In these experiments, AL were dissected free of NIL, total RNA was extracted and subjected to RNase protection assay, and the relative expression levels of POMC and POMC* were compared and normalized to β-actin expression (Fig. 4 and Table 1). Diurnal expression levels of POMC were determined at 07:00 and 19:00 on a light-dark schedule of lights: on at 05:00 and off at 19:00. Serum corticosterone levels were fourfold higher at 19:00 compared to 07:00, while both POMC and POMC* mRNA levels were twofold higher at 07:00, with no change in the ratio between them. These data suggest that synthesis of replacement POMC peptides is higher when secretion of ACTH, and therefore corticosterone, is lower.

Osmotic stress was induced by the intraperitoneal (ip) injection of 9% NaCl. Three hours after injection, the mice showed a large increase in serum corticosterone, and

Table 1

Ratio of Transgenic POMC* mRNA to Endogenous POMC mRNA Measured by RPA in the Pituitary Anterior Lobe (AL) and Neurointermediate Lobe (NIL) of pHAL* Transgenic Mice[a]

Group and treatment	*POMC*/POMC*	*Group and treatment*	*POMC*/POMC*	*P value*[b]
Female basal 6:00 AM (AL)	0.38 ± 0.03	Female basal 8:00 PM (AL)	0.32 ± 0.02	NS[c]
Female sham (AL)	0.16 ± 0.01	Female 7 d post-ADX (AL)	0.40 ± 0.02	<0.0001
Male sham (AL)	0.21 ± 0.02	Male 7 d post-ADX (AL)	0.51 ± 0.05	0.002
Female 3 h, 0.9% NaCl i.p. (AL)	0.42 ± 0.02	Female 3 h, 9.0% NaCl i.p. (AL)	0.45 ± 0.04	NS
Female 1 h restraint stress (AL)	0.43 ± 0.02	Female 6 h restraint stress (AL)	0.47 ± 0.05	NS
Male preswim stress (AL)	0.34 ± 0.03	Male postswim stress (AL)	0.40 ± 0.03	NS
Female pre-Metyrapone (AL)	0.22 ± 0.01	Female 8 h post-Metyrapone (AL)	0.33 ± 0.04	0.002
Mean of 7 experiments (AL)	*0.31 ± 0.04*			
Female basal 6:00 AM (NIL)	0.77 ± 0.11	Female basal 8:00 PM (NIL)	0.78 ± 0.09	NS
Female basal (NIL)	0.60 ± 0.03	Female 7 d haloperidol (NIL)	0.81 ± 0.08	0.02
Female 3 h, 0.9% NaCl ip (NIL)	0.69 ± 0.05	Female 3 h, 9.0% NaCl ip (NIL)	0.74 ± 0.03	NS
Female 1 h restraint stress (NIL)	0.93 ± 0.05	Female 6 h restraint stress (NIL)	1.03 ± 0.07	NS
Male preswim stress (NIL)	1.19 ± 0.09	Male post-swimstress (NIL)	1.11 ± 0.19	NS
Mean of 5 experiments (NIL)	*0.84 ± 0.10*			*0.005*[d]

[a]Experimental procedures are described in the text and legends to Figs. 2–4. RPAs were quantified either by phosphoimager analysis, densitometry of autoradiographs, or direct scintillation counting of bands excised from the acrylamide gels.

[b]Student's t test was performed between the two POMC*/POMC ratios for each treatment pair.

[c]NS, not significant ($p > 0.1$).

[d]Student's t test was performed between the mean POMC*/POMC ratios for AL and NIL.

nearly a twofold increase in both POMC and POMC* mRNA content of the AL, with no change in the ratio between the two mRNAs compared to the control injection of isotonic 0.9% NaCl. The acute blockade of corticosterone biosynthesis by metyrapone resulted in the expected decrease in serum corticosterone levels 4–6 h after injection, and subsequently a significant increase in both POMC ($F[4,24]$, $p < 10^{-6}$) and POMC* mRNA ($F[4,20]$, $p < 10^{-5}$). Additionally, the ratio of POMC*/POMC mRNA increased significantly by 50%. Restraint stress resulted in sustained and very high levels of serum corticosterone for up to 6 h that was not accompanied by a significant change in POMC/

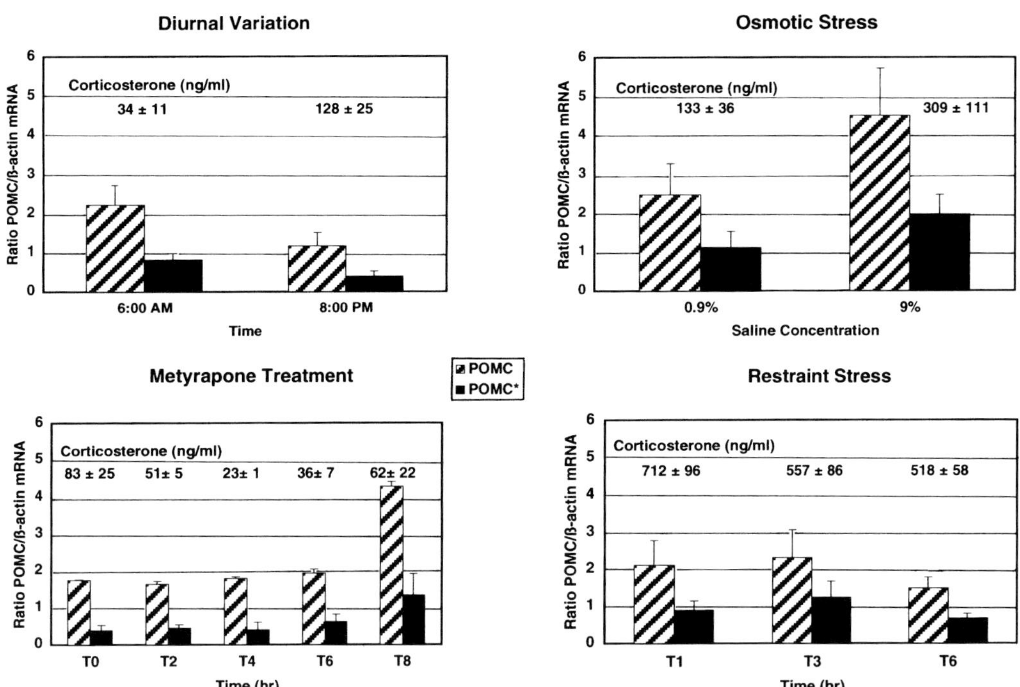

Fig. 4. Relative expression levels of POMC and POMC* mRNA in the anterior pituitary lobes of female pHAL* transgenic mice during stress-free conditions on a standard light-dark cycle or following stress. Mice were fed *ad libitum* with free access to water and housed under a light:dark cycle of 14 h:10 h with lights-on at 5:00 AM. Serum corticosterone was measured by radioimmunoassay from trunk blood collected at sacrifice; the values (mean ± SEM) are reported above each pair of columns. All experimental manipulations started between 8:00 and 9:00 AM. Osmotic stress consisted of a single ip injection of 0.9% NaCl or 9.0% NaCl (18 mL/kg) 3 h prior to sacrifice (n = 8 per dose). Metyrapone was administered in two doses of 300 mg/kg sc at times T0 and T3 to acutely inhibit the synthesis of corticosterone by the adrenal glands (n = 5 per time point). Restraint stress consisted of immobilization in vented 50-mL polypropylene conical tubes for 1, 3, or 6 h duration prior to sacrifice (n = 8 per time point).

POMC* mRNA levels or their ratio. Acute stress induced by a 5-min swim in 10°C water resulted in a transient increase in serum corticosterone levels from 16 ± 1 to 150 ±14 ng/mL after 15 min, but no change in POMC/POMC* mRNA levels (data not shown) or their ratio at time points of 15, 30, 60, 120, and 240 min after the swim.

As a whole, the results of these experiments indicate that under every condition tested there were parallel changes in the steady-state levels of POMC and POMC* mRNA, consistent with the inclusion of all necessary transcriptional *cis*-elements in the 10-kb pHAL* transgene. Depending on the physiological state, pharmacological manipulation, or type of stressor, increased POMC and POMC* expression in the AL occurred in the face of both decreased and elevated serum corticosterone. However, the ratio of POMC*/POMC mRNA increased only after adrenalectomy or metyrapone treatment. Both of these conditions are associated with decreased serum corticosterone, and therefore a reduction of the direct inhibitory effect of corticosterone on POMC gene expression at the level of the pituitary gland. The differences in response of POMC gene expression following osmotic stress or restraint stress may be explained by a synergistic activation by CRH and vasopressin in response to hyperosmolality. Presumably, the dual signaling by two stimulatory factors relieves the corticosterone inhibition, while CRH responses to restraint stress are insufficient.

MUTANT MOUSE MODELS OF MELANOTROPH AND CORTICOTROPH ADENOMAS

Transgenic Expression of SV40 T-Antigens

The targeted expression of oncogenes with tissue-specific promoters has led to new insights into the molecular basis of tumorigenesis, and has provided a source of several novel differentiated cell lines from transgenic mice *(33–36)*. Based on our characterization of the POMC gene promoter, we have expressed two forms of the simian virus SV40-Large T antigen in pituitary corticotrophs and melanotrophs. Several independent pedigrees were generated using wild-type SV40Tag, and tumors were consistently induced in the pituitary gland as predicted, and also in the thymus gland with incomplete penetrance of the phenotype *(37)*. One representative line (no. 427) has been outbred to the CD-1 genetic background for 27 consecutive generations. The incidence of thymic tumors decreased after several generations, and no such tumors have been observed in the last 10 generations. These results suggest that epistatic interactions with gene alleles from the original B6D2 genetic background were necessary for expression of the transgene in thymus. However, the pituitary-tumor phenotype has remained stable across all generations. A Kaplan-Meier survival plot demonstrated that the minimal latency to mortality from an expanding pituitary tumor is 8 wk and the median survival time is 16 wk *(38)*. Occasionally, mice will survive as long as 1 yr. This time range and the fact that transgene expression is clearly initiated long before the earliest tumors develop are consistent with the idea that stochastic secondary events likely occur in individual T-antigen expressing cells before they are transformed and lose their inhibitory growth controls. Other transgenic mice expressing the temperature-sensitive tsA58 mutant form of SV40Tag from the same POMC promoter sequences only developed pituitary tumors after 1 yr of age (our unpublished data). The longer latency was presumably a result of the expression of minimal amounts of biologically stable tsA58-Tag at the partially permissive core body temperature of the mice.

Studies demonstrating an important role of insulin-like growth factors (particularly IGF-II) on the induction and progression of pancreatic islet β-cell tumors from an insulin-Tag transgene *(39)* prompted us to determine whether those data would generalize to other endocrine cell types. Expression of IGF-I, IGF-II, the type-1 IGF receptor, and six IGF-binding proteins were examined in pituitary tumors from the POMC-Tag mice *(40)*. In contrast to the β-cell tumors, there was little expression of IGF-II in the pituitary tumors. Multiple IGF-binding proteins were expressed, but with heterogeneous patterns inconsistent with a critical role in tumorigenesis. Although not definitive, because of possible effects of circulating IGFs, these studies suggest that reactivation of IGF-II and a functional importance on tumor growth is a cell-type-specific phenomenon, and is not shared by pituitary cells.

The POMC-Tag transgene is expressed in both pituitary corticotrophs and melanotrophs, as demonstrated by *in situ* hybridization and immunohistochemistry of pituitary glands from 4-wk-old mice. However, every pituitary tumor that has developed has been of melanotroph origin based on gross anatomical findings and a variety of biochemical and molecular studies. The precise reason for the lack of corticotroph adenomas is uncertain, but may reflect either a cell-specific resistance to the transforming action of SV40 TAg or, more likely, a large discrepancy in the rate at which corticotrophs become transformed or divide after transformation so that they never effectively compete with the more rapidly dividing melanotroph tumor cells. It is also possible that an insufficient quantity of Tag is produced in corticotrophs because of the lower transcriptional efficiency of the POMC-promoter fragment in corticotrophs compared to melanotrophs. Melanotroph tumor cells are fully transformed, based on their propensity to produce secondary tumors after subcutaneous passage in nude mice and the isolation of several immortalized cell lines *(37,41,42)*. Posttranslational processing of POMC by the primary tumors was indistinguishable from normal melanotrophs, and was characterized by high proportions of acetylated and carboxyl-truncated forms of β-endorphin and acetylated α-MSH with virtually no ACTH. The tumors also expressed high levels of prohormone convertase 2 (PC2), which is characteristic of melanotrophs. However, in many mice there was a sufficient quantity of an ACTH-like peptide made and secreted from the melanotroph tumors to induce adrenal cortical hyperplasia and Cushing's disease with markedly elevated serum corticosterone. A syndrome of fatal melanotroph-dependent Cushing's disease was recently reported in mice with a null mutation in the gene encoding the neuroendocrine protein 7B2 *(43)*. 7B2 has been implicated in the activation of PC2, so it is not surprising that the 7B2-deficient mice have no functional PC2, and their melanotrophs produce ACTH instead of α-MSH and CLIP. Still unexplained, however, is the mechanism for increased ACTH release from the intermediate lobe of these mice, because a knockout of the PC2 gene itself results in high ACTH production in melanotrophs without the high constitutive secretion of the hormone *(44)*.

Two distinct cell lines have been cloned from melanotroph tumors that developed in the POMC-Tag mice *(42)*. mIL39 cells are small and bipolar, and rapidly produce secondary tumors after transplantation into nude mice. They are melanotrophs based on their expression of both POMC and the dopamine D2 receptor. Interestingly, a second-line mIL5 that was also isolated from a primary culture of a transgenic tumor fails to express either POMC or SV40Tag. The exact origin of these latter cells is therefore unknown, but they are of interest because they secrete a prolactin-regulating factor. The lack of concordance between this prolactin-regulating activity and POMC gene expression suggests that the activity is not the result of a peptide product of POMC.

In addition to the mIL39 and mIL5 cell lines, we have isolated another melanotroph-like cell line from a pituitary tumor induced by the tsA58SV40Tag *(44a)*. These cells express high levels of POMC mRNA and POMC prohormone, but they have few secretory granules and limited posttranslational processing of POMC. Preliminary studies have demonstrated the expression of several G-protein-coupled receptors, including dopamine D2, CRH R1, and $GABA_B$, that are normally expressed on melanotrophs. These receptors are functional, and therefore this cell line may be useful for analyzing the biochemical cross-communication in signaling pathways that are unique to pituitary melanotrophs.

Other strains of transgenic mice have been produced that develop POMC-expressing pituitary tumors from serendipitous expression of an oncogene. Polyoma large T-antigen expression from a polyoma early-region promoter induces large tumors that are fatal at approx 1 yr of age *(45,46)*. Although the exact cellular origin of these tumors has not been reported, they are associated with elevated plasma ACTH levels and Cushing's disease. In a similar case, SV40Tag expressed from a bovine vasopressin promoter resulted in a mouse model that mimics human multiple endocrine neoplasia syndrome, with both pancreatic β-cell and pituitary tumors *(47,48)*. A minority of the vasopressin-SV40Tag-induced tumors occured in the intermediate lobe, and were characterized by a high expression of POMC mRNA and immunoreactivity for ACTH.

Murine Melanotroph Tumors Caused by Gene Inactivation

Intermediate-lobe melanotrophs may be particularly sensitive to oncogenic mutations, leading to a loss of normal growth control. Melanotroph adenomas are common occurences in several mammalian species, including horses and dogs. This viewpoint is strengthened by the rather remarkable incidence of melanotroph tumors in several other reported mouse strains with targeted inactivation of two classes of cell-cycle control genes. The first of these is a knockout of the retinoblastoma (Rb) tumor-suppressor gene *(49–51)*. Mice that are heterozygous for the null mutation have an almost complete penetrance of the intermediate-lobe melanotroph adenoma phenotype associated with somatic loss of heterozygosity. Interestingly, Rb normally prevents entry of cells into S phase, and is one of several proteins that is known to interact with SV40Tag and to play a role in Tag's mechanism of action. Phosphorylation of Rb by cyclin-dependent kinases (CDKs) also inactivates the growth-suppressive-action of Rb, and these cdks are in turn inhibited by a group of proteins including the Cip/Kip family. Gene knockouts of $p27^{kip1}$ lead to tumorigenesis in multiple organs, but a prominent location is once again the pituitary intermediate lobe *(52–54)*. A recent study evaluating the consequences of a double mutation in both Rb and $p27^{kip1}$ suggests that the two proteins act cooperatively to suppress tumorigenesis rather than linearly in a single signaling pathway *(55)*.

We have reported a natural-strain difference in the size of the mouse intermediate lobe. 12956/SvEv mice have hyperplastic intermediate lobes that contain at least twice the cell number of inbred C57BL/6 mice *(56,57)*. This observation may be relevant to the phenotype of the mutant mice described here because they share an origin from substrain 129-derived embryonic stem (ES) cells. It is possible that epistatic effects among the targeted mutations and quantitative trait loci determining basal melanotroph proliferation rate determine the selectivity and severity of the observed intermediate-lobe tumor phenotype.

TOXIGENIC ABLATION OF MELANOTROPHS AND CORTICOTROPHS

Ablation of selected cell populations in transgenic mice using a targeted toxigenic protein has been useful in cell-lineage studies and physiological studies of tissue function within a complex biological environment *(58–61)*. The two most commonly used toxins have been diptheria toxin A chain (DTA) and herpes simplex virus type 1 thymidine kinase (HSV1-TK), although both have limitations. HSV-TK has the advantage of temporal control, because its toxicity in most mammalian cells is dependent on administration of a nucleoside-analog-substrate such as ganciclovir and subsequent incorporation of the phosphorylated product into replicating DNA strands. Recently however, DTA expression has been regulated more precisely using *Cre-loxP* technology *(62)*, and refinements of this approach may increase its utility in ablation studies of postmitotic cells including neurons.

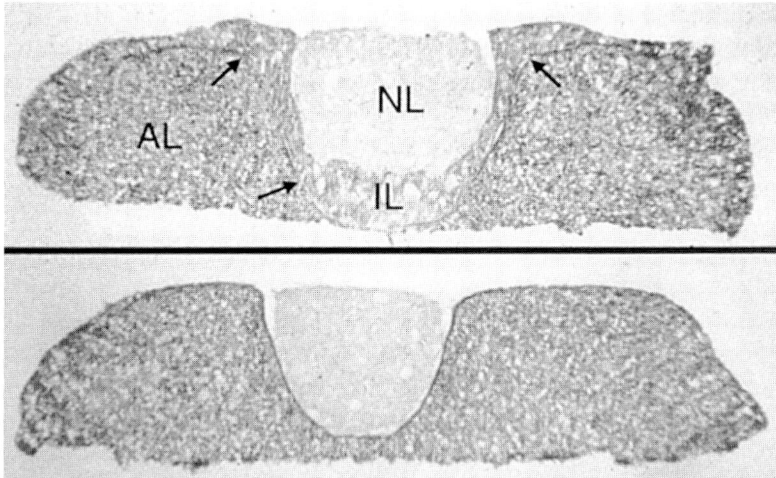

Fig. 5. Absent intermediate lobe in a POMC-TK transgenic mouse from the no. 66 pedigree without ganciclovir treatment. Frozen sections of a wild-type *(top)* and a POMC-TK transgenic mouse *(bottom)* were stained with neutral red. The transgenic pituitary gland has morphologically normal neural (NL) and anterior (AL) lobes, but no intermediate (IL) lobe indicated with arrows on the wild-type gland.

Transgenic mice that expressed a POMC-TK transgene illustrate some of the possibilities and pitfalls of the paradigm *(63)*. Of three transgenic founders generated, one produced a line with normal fertility, one with partial male infertility, and the third with extensive reproductive difficulties in both sexes that precluded any comprehensive study of phenotype because of the limited number of mice that could be generated. Line no. 401-66 with partial male infertility had high expression of TK in testis. This pattern of expression had been previously shown to be caused by cryptic transcriptional regulatory elements in the TK coding sequence *(64,65)*. HSV1-TK is also toxic to developing male germ cells in the absence of nucleoside analogs, explaining the reduced fertility *(66)*. Of interest is the fact that line 401-66 also exhibited spontaneous loss of the pituitary intermediate lobe in adult mice with variable penetrance. An example of essentially total atrophy of melanotrophs in the absence of ganciclovir treatment is shown in Fig. 5. These data demonstrate that in addition to germ cells, melanotrophs can also be susceptible to an inherent toxicity from HSV1-TK, although the mechanism is unknown.

The majority of our experiments use the 401-21 line of POMC-TK transgenic mice. The pituitary glands of these mice develop normally, and TK protein can be detected by double-immunofluoresence histochemistry in essentially all corticotrophs and melanotrophs, but in no other cell types. Treatment of adult mice with ganciclovir for 12 d has resulted in the disappearance of the entire intermediate lobe and the loss of virtually all immunoreactive corticotrophs, presumably a result of cell death. No other pituitary-hormone cell type in the anterior pituitary was noticeably affected by this treatment. Biochemically, greater than a 98% reduction of pituitary ACTH and α-MSH content was determined by radioimmunoassay. The near total ablation of corticotrophs was accompanied by dramatic reductions in serum corticosterone, and downregulation of adrenal ACTH receptor expression and p450c11β, the cytochrome enzyme specifically involved in glucocorticoid but not mineralocorticoid (aldosterone) biosynthesis.

Several important questions remain unanswered regarding POMC-TK transgenic mice. Preliminary experiments demonstrated a small but significant recovery of POMC peptide production in the anterior lobe 5 wk after the discontinuation of ganciclovir treatment *(63)*. Studies have not yet revealed whether this recovery is a result of the reappearance of differentiated corticotrophs, possibly from a stem-cell population, and if the recovery would be more substantial after longer time periods. The rate of corticotroph disappearance with ganciclovir appears to be inconsistent with a mechanism purely dependent on toxicity associated with DNA synthesis and cell division. Mitogenic activity of the pituitary gland is not as static as once believed *(67)*; however, recent estimates of normal rat pituitary-cell lifespan are on the order of 60 d with a prevelance of overt mitotic figures of 0.066% in the basal state *(68)*. This rate is substantially less than that predicted by the cell loss after ganciclovir. Unfortunately, it has not been possible to effectively use these transgenic mice for developmental studies of cell fate in the pituitary gland, because of excessive toxicity of ganciclovir to normal mouse embryos and the current unavailability of 5-FIAU used in pioneering experiments of this type *(69)*.

EXPRESSION OF ECTOPIC PROTEINS IN POMC CELLS

The identification of promoter elements sufficient for directing cell-specific gene expression has great value for the design of cell biological and physiological experiments in vivo. Targeting of SV40-Tag and HSV1-TK discussed here are two examples. POMC-promoter transgenics have also been used in several other systems. A particularly intriguing use has been the ectopic expression of proinsulin in transgenic intermediate-lobe melanotrophs *(70,71)*. Because of their complement of prohormone convertases, the proinsulin is efficiently processed to mature, biologically active insulin. Transplanted intermediate-lobe tissue from these transgenic mice into syngeneic diabetic nucleotide-binding oligomerization (NOD) mice effectively reversed the hyperglycemia and diabetic symptoms. Allografts of the islet tissue showed extensive neovascularization and enhanced viability compared to transplanted pancreatic islets. The favorable autoimmune profile of melanotroph cells compared to pancreatic β-cells suggests their further utility as a bioengineered insulin-delivery vehicle if they can be additionally altered to sense and appropriately respond to extracellular glucose concentrations.

We recently have used mouse POMC genomic elements to direct expression of enhanced green fluorescent protein (GFP) (Clontech) to pituitary corticotrophs, melanotrophs and POMC neurons in the hypothalamus and medulla *(71a)*. The potential of this technique is exemplified from similar work using GnRH-GFP transgenic mice *(72)*. Visualization of the fluorescently labeled cells in slice preparations permitted patch-clamp studies of cellular physiology in identified neurons, despite a dispersed population of rare cells. Other researchers have used GFP-labeled glial progenitors to study developmental migration patterns in explanted slices of the embryonic rodent brain *(73)*.

POMC GENE MUTATIONS

β-Endorphin-Deficient Mice

In addition to transgenic studies utilizing POMC genomic elements for cell-specific gene expression, our lab has also studied the physiological function of POMC peptides by the introduction of POMC mutations into the mouse germline. To eliminate only β-endorphin peptides and preserve the other products of the POMC prohormone, a point mutation was introduced into exon 3 of the POMC gene by homologous recombination *(74)*. The mutation caused a premature translational-stop codon at the normal amino-terminal end of β-endorphin. Mice carrying this mutation express a truncated POMC prohormone that is processed normally to ACTH and other melanocortin peptides in the pituitary and brain *(75)*. Basal activity of the hypothalamic-pituitary-adrenal axis is normal, and the β-endorphin-deficient mice have normal corticosterone responses to a variety of stressors.

β-endorphin is one of several endogenous opioid peptides produced in the brain, and interacts with both the mu and delta opioid receptors, leading us to question whether it plays any nonredundant physiological function in nociception and analgesia. The absence of β-endorphin does not affect nociceptive thresholds or the analgesic responses to morphine administered by injection over a wide dose range. However, β-endorphin-deficient mice fail to exhibit endogenous opioid-mediated analgesia in response to swim stress, and have increased levels of nonopioid (naloxone insensitive) analgesia *(75)*. Complementary studies in enkephalin-deficient mice have demonstrated that they retain endogenous opioid stress-induced analgesia *(76)*. As a whole, these data suggest that β-endorphin is a physiologically relevant mediator of antinociception.

Null Mutations in the POMC Gene

Genetic linkage studies and quantitative trait loci analyses have strongly implicated the human POMC gene locus as an important determinant of weight homeostasis in various ethnic populations, although specific alleles associated with obesity have not yet been demonstrated *(77,78)*. A small number of children with severe, early-onset obesity have been found to have mutations in the POMC gene that result in decreased gene expression *(79)*. A wealth of data indicates a critical role of POMC-derived melanocortins and the MC4-R in the anorexigenic actions of POMC, although opioid peptides also affect food consumption and food choice *(80–82)*. Interestingly, β-endorphin-deficient mice exhibit a sexually dimorphic obesity phenotype caused by increased white fat mass in males (our unpublished data). Therefore, multiple mechanisms may be involved in regulation of food intake and caloric expenditure, even from a single prohormone precursor that is expressed in a highly specific small subset of neurons.

Knockout mice with a null mutation in the POMC gene and total absence of POMC peptides exhibit a phenocopy of the human disorder with marked obesity, adrenal glucocorticoid insufficiency, and altered coat pigmentation *(83)*. Adrenal-gland development also appears to be abnormal and there is a non-Mendelian distribution of homozygote knockout mice, indicating pre- or perinatal lethality with incomplete penetrance.

SUMMARY AND CONCLUSIONS

Transgenic and targeted gene mutational studies of the POMC locus have provided many new insights to the transcriptional regulation of the gene. Targeted oncogenesis has provided valuable new melanotroph cell lines, which together with neuronal cell lines generated by the same technique should allow a definitive characterization of the molecular mechanisms underlying cell-specific expression of the POMC gene. Newer and more sophisticated molecular and genetic approaches to in vivo gene mutation will permit a detailed characterization of the physiological function of each of the biologically active peptides produced from POMC.

ACKNOWLEDGMENTS

This work was supported by grants from NIDDKD, NICHHD, and the Fogarty International Center, National Institutes of Health. We wish to thank the many members of the Low laboratory who have contributed to studies described here, including Bin Liu, Renata Hahn, Gary Hammer, Miguel Japón, Vicki Fairchild-Huntress, Marty Mortrud, Carrie Feddern, Cullen McPherson, Clark Fjeld, Kevin Nusser, Jim Smart, Suzy Appleyard, and our collaborators Richard Allen, John Pintar, and Nira Ben-Jonathan.

REFERENCES

1. Elkabes S, Loh YP, Nieburgs A, Wray S. Prenatal ontogenesis of pro-opiomelanocortin in the mouse central nervous system and pituitary gland: an in situ hybridization and immunocytochemical study. Brain Res Dev Brain Res 1989;46:85–95.
2. Fan W, Boston BA, Kesterson RA, Hruby VJ, Cone RD. Role of melanbocortinergic neurons in feeding and the agouti obesity syndrome. Nature 1997;385:165–168.
3. Huszar D, Lynch CA, Fairchild-Huntress V, Dunmore JH, Fang Q, Berkemeier LR, et al. Targeted disruption of the melanocortin-4 receptor results in obesity in mice. Cell 1997;88:131–141.
4. Elmquist JK, Elias CF, Saper CB. From lesions to leptin: hypothalamic control of food intake and body weight. Neuron 1999;22:221–232.
5. Kalra SP, Dube MG, Pu S, Xu B, Horvath TL, Kalra PS. Interacting appetite-regulating pathways in the hypothalamic regulation of body weight. Endocr Rev 1999;20:68–100.
6. Watson SJ, Akil H, Richard CWd, Barchas JD. Evidence for two separate opiate peptide neuronal systems. Nature 1978;275:226–228.
7. Palkovits M, Mezey É, Eskay RL. Pro-opiomelanocortin-derived peptides (ACTH/β-endorphin/α-MSH) in brainstem baroreceptor areas of the rat. Brain Res. 1987;436:323–328.
8. Smith,AI, Funder JW. Proopiomelanocortin processing in the pituitary, central nervous system, and peripheral tissues. Endocr Rev 1988;9:159–179.
9. Japón M, Rubinstein M, Low MJ. In situ hybridization analysis of anterior pituitary hormone gene expression during fetal mouse development. J Histochem Cytochem 1994;42:1117–1125.
10. Couly GF, Le Douarin NM. Mapping of the early neural primordium in quail-chick chimeras. I. Developmental relationships between placodes, facial ectoderm, and prosencephalon. Dev Biol 1985;110:422–439.
11. Ericson J, Norlin S, Jessell TM, Edlund T. Integrated FGF and BMP signaling controls the progression of progenitor cell differentiation and the emergence of pattern in the embryonic anterior pituitary. Development 1998;125:1005–1015.
12. Treier M, Gleiberman AS, O'Connell SM, Szeto DP, McMahon JA, McMahon AP, et al. Multistep signaling requirements for pituitary organogenesis in vivo. Genes Dev 1998;12:1691–1704.
13. Takuma N, Sheng HZ, Furuta Y, Ward JM, Sharma K, Hogan BL, et al. Formation of Rathke's pouch requires dual induction from the diencephalon. Development 1998;125:4835–4840.
14. Treier M, Rosenfeld MG. The hypothalamic-pituitary axis: co-development of two organs. Curr Opin Cell Biol 1996;8:833–843.
15. Sheng HZ, Westphal H. Early steps in pituitary organogenesis. Trends Genet 1999;15:236–240.
16. Akita S, Readhead C, Stefaneanu L, Fine J, Tampanaru-Sarmesiu A, Kovacs K, et al. Pituitary-directed leukemia inhibitory factor transgene forms Rathke's cleft cysts and impairs adult pituitary function. A model for human pituitary Rathke's cysts. J Clin Invest 1997;99:2462–2469.
17. Yano H, Readhead C, Nakashima M, Ren SG, Melmed S. Pituitary-directed leukemia inhibitory factor transgene causes Cushing's syndrome: neuro-immune-endocrine modulation of pituitary development. Mol Endocrinol 1998;12:1708–1720.
18. Furth J, Gadsden EL, Upton AC. ACTH-secreting transplantable pituitary tumors. Proc Soc Exp Biol Med 1953;84:253–254.
19. Furth J, Haran-Ghera N, Curits JJ, Buffett RF. Studies on the pathogenesis of neoplasms by ionizing radiation. I. Pituitary tumor. Cancer Res 1959;19:550–556.

20. Picon A, Bertagna X, de Keyzer Y. Analysis of proopiomelanocortin gene transcription mechanisms in bronchial tumour cells. Mol Cell Endocrinol 1999;147:93–102.
21. Picon A, Bertagna X, de Keyzer Y. Analysis of the human proopiomelanocortin gene promoter in a small cell lung carcinoma cell line reveals an unusual role for E2F transcription factors. Oncogene 1999;18:2627–2633.
22. Tremblay Y, Tretjakoff I, Peterson A, Antakly T, Zhang CX, Drouin J. Pituitary-specific expression and glucocorticoid regulation of a proopiomelanocortin fusion gene in transgenic mice. Proc Natl Acad Sci USA 1988;85:8890–8894.
23. Hammer G, Fairchild-Huntress V, Low MJ. Pituitary-specific and hormonally regulated gene expression directed by the rat proopiomelanocortin promoter in transgenic mice. Mol Endo 1990;4:1689–1697.
24. Liu B, Hammer GD, Rubinstein M, Mortrud M, Low MJ. Identification of DNA elements cooperatively activating proopiomelanocortin gene expression in the pituitary gland of transgenic mice. Mol. Cell. Biol. 1992;12:3978–3990.
25. Liu B, Mortrud M, Low MJ. DNA elements with AT-rich core sequences direct pituitary cell-specific expression of the proopiomelanocortin gene in transgenic mice. Biochemical J 1995; 312:827–832.
26. Garrick D, Fiering S, Martin DI, Whitelaw E. Repeat-induced gene silencing in mammals. Nat Genet 1998;18:56–59.
27. Bousquet C, Ray DW, Melmed S. A common pro-opiomelanocortin-binding element mediates leukemia inhibitory factor and corticotropin-releasing hormone transcriptional synergy. J Biol Chem 1997;272:10,551–10,557.
28. Therrien M, Drouin J. Cell-specific helix-loop-helix factor required for pituitary expression of the pro-opiomelanocortin gene. Mol Cell Biol 1993;13:2342–2353.
29. Rubinstein M, Mortrud M, Liu B, Low MJ. Rat and mouse proopiomelanocortin gene sequences target tissue-specific expression to the pituitary gland but not to the hypothalamus of transgenic mice. Neuroendocrinology 1993;58:373–380.
30. Young JI, Otero V, Cerdán MG, Falzone TL, Chan EC, Low MJ, et al. Authentic cell-specific and developmentally regulated expression of proopiomelanocortin genomic gragments in hypothalamic and hindbrain neurons of transgenic mice. J Neurosci 1998;18:6631–6640.
31. Elias CF, Lee C, Kelly J, Aschkenasi C, Ahima RS, Couceyro PR, et al. Leptin activates hypothalamic CART neurons projecting to the spinal cord. Neuron 1998;21:1375–1385.
32. Vrang N, Larsen PJ, Clausen JT, Kristensen P. Neurochemical characterization of hypothalamic cocaine-amphetamine- regulated transcript neurons. J Neurosci 1999;19:RC5.
33. Windle JJ, Weiner RI, Mellon PL. Cell lines of the pituitary gonadotrope lineage derived by targeted oncogenesis in transgenic mice. Mol Endocrinol 1990;4:597–603.
34. Alarid E, Windle J, Whyte D, Mellon P. Immortalization of pituitary cells at descrete stages of development by directed oncogenesis in transgenic mice. Dev 1996;122:3319–3329.
35. Adams JM, Cory S. Transgenic models of tumor development. Science 1991;254:1161–1167.
36. Efrat S, Linde S, Kofod H, Spector D, Delannoy M, Grant S, et al. β-cell lines derived from transgenic mice expressing a hybrid insulin gene-oncogene. Proc Natl Acad Sci USA 1988;85:9037–9041.
37. Low MJ, Liu B, Hammer G.D, Rubinstein M, Allen RG. Post-translational processing of proopiomelanocortin (POMC) in mouse pituitary melanotroph tumors induced by a POMC-simian virus 40 large T antigen transgene. J Biol Chem 1993;268:24,967–24,975.
38. Low MJ, Kelly MA, Graham KE, Asa SL. In: Webb S, ed., The Management of Pituitary Tumors, BioScientifica, Bristol, UK, 1998, pp. 55–69.
39. Christofori G, Naik P, Hanahan D. Deregulation of both imprinted and expressed alleles of the insulin- like growth factor 2 gene during beta-cell tumorigenesis. Nat Genet 1995;10:196–201.
40. Grewal A, Bradshaw SL, Schuller AGP, Low MJ, Pintar JE. Expression of IGF system genes during T-antigen driven pituitary tumorigenesis. Horm Metab Res 1999;31:155–160.
41. Allen DL, Low MJ, Allen RG, Ben-Jonathan N. Identification of two classes of prolactin releasing factors in intermediate lobe tumors from transgenic mice. Endocrinology 1995;136:3093–3099.
42. Hnasko R, Khurana S, Shackleford N, Steinmetz R, Low MJ, Ben-Jonathan N. Two distinct pituitary cell lines from mouse intermediate lobe tumors: a cell that produces prolactin-regulating factor and a melanotroph. Endocrinology 1997;138:5589–5596.
43. Westphal CH, Muller L, Zhou A, Zhu X, Bonner-Weir S, Schambelan M, et al. The neuroendocrine protein 7B2 is required for peptide hormone processing in vivo and provides a novel mechanism for pituitary Cushing's disease. Cell 1999;96:689–700.

44. Furuta M, Yano H, Zhou A, Rouille Y, Holst JJ, Carroll R, et al. Defective prohormone processing and altered pancreatic islet morphology in mice lacking active SPC2. Proc Natl Acad Sci USA 1997;94:6646–6651.
44a. Chromwall BM, Davis TD, Severidt MW, Wolfe SE, et al. Constitutive expression of $GABA_B$ receptors in MIL-tSA58 cells requires both $GABA_{B(1)}$ and $GABA_{B(2)}$ genes. J Neurochem 2001; in press.
45. Helseth A, Siegal GP, Haug E, Bautch VL. Transgenic mice that develop pituitary tumors. A model for Cushing's disease. Am J Pathol 1992;140:1071–1080.
46. Helseth A, Haug E, Nesland JM, Siegal GP, Fodstad O, Bautch VL. Endocrine and metabolic characteristics of polyoma large T transgenic mice that develop ACTH-producing pituitary tumors. J Neurosurg 1995;82:879–885.
47. Murphy D, Bishop A, Rindi G, Murphy MN, Stamp GW, Hanson J, et al. Mice transgenic for a vasopressin-SV40 hybrid oncogene develop tumors of the endocrine pancreas and the anterior pituitary. A possible model for human multiple endocrine neoplasia type 1. Am J Pathol 1987;129:552–566.
48. Stefaneanu L, Rindi G, Horvath E, Murphy D, Polak JM, Kovacs K. Morphology of adenohypophysial tumors in mice transgenic for vasopressin-SV40 hybrid oncogene. Endocrinology 1992;130:1789–1795.
49. Jacks T, Fazeli A, Schmitt EM, Bronson RT, Goodell MA, Weinberg RA. Effects of an Rb mutation in the mouse. Nature 1992;359:295–300.
50. Hu N, Gutsmann A, Herbert DC, Bradley A, Lee WH, Lee EY. Heterozygous Rb-1 delta 20/+mice are predisposed to tumors of the pituitary gland with a nearly complete penetrance. Oncogene 1994;9:1021–1027.
51. Nikitin A, Lee WH. Early loss of the retinoblastoma gene is associated with impaired growth inhibitory innervation during melanotroph carcinogenesis in Rb+/- mice. Genes Dev 1996;10:1870–1879.
52. Fero ML, Rivkin M, Tasch M, Porter P, Carow CE, Firpo E, et al. A syndrome of multiorgan hyperplasia with features of gigantism, tumorigenesis, and female sterility in p27(Kip1)-deficient mice. Cell 1996;85:733–744.
53. Nakayama K, Ishida N, Shirane M, Inomata A, Inoue T, Shishido N, et al. Mice lacking p27(Kip1) display increased body size, multiple organ hyperplasia, retinal dysplasia, and pituitary tumors. Cell 1996;85:707–720.
54. Kiyokawa H, Kineman RD, Manova-Todorova KO, Soares VC, Hoffman ES, Ono M, et al. Enhanced growth of mice lacking the cyclin-dependent kinase inhibitor function of p27(Kip1). Cell 1996;85:721–732.
55. Park MS, Rosai J, Nguyen HT, Capodieci P, Cordon-Cardo C, Koff A. p27 and Rb are on overlapping pathways suppressing tumorigenesis in mice. Proc Natl Acad Sci USA 1999;96:6382–6387.
56. Kelly MA, Rubinstein M, Asa SL, Zhang G, Saez C, Bunzow JR, et al. Pituitary hyperplasia and chronic hyperprolactinemia in dopamine D2 receptor-deficient mice. Neuron 1997;19:103–113.
57. Low MJ, Kelly MA, Rubinstein M, Grandy DK. Single genes and complex behaviors. Mol Psychiatry 1998;3:375–377.
58. Palmiter RD, Behringer RR, Quaife CJ, Maxwell F, Maxwell IH, et al. Cell lineage ablation in transgenic mice by cell-specific expression of a toxin gene (published erratum appears in Cell 1990 Aug 10;62(3):following 608). Cell 1987;50:435–443.
59. Evans GA. Dissecting mouse development with toxigenics. Genes Dev 1989;3:259–263.
60. Evans GA. Toxigenics. Semin Cell Biol 1991;2:71–79.
61. Heyman RA, Borrelli E, Lesley J, Anderson D, Richman DD, Baird SM, et al. Thymidine kinase obliteration: creation of transgenic mice with controlled immune deficiency. Proc Natl Acad Sci USA 1989;86:2698–2702.
62. Grieshammer U, Lewandoski M, Prevette D, Oppenheim RW, Martin GR. Muscle-specific cell ablation conditional upon Cre-mediated DNA recombination in transgenic mice leads to massive spinal and cranial motoneuron loss. Dev Biol 1998;197:234–247.
63. Allen RG, Carey C, Parker JD, Mortrud MT, Mellon SJ, Low MJ. Targeted ablation of pituitary preproopiomelanocortin (POMC) cells by Herpes Simplex Virus-1 Thymidine Kinase differentially regulates messenger RNAs encoding the ACTH receptor, and aldosterone synthase in the mouse adrenal gland. Endocrinology 1995;9:1005–1016.
64. Ellison AR, Bishop JO. Initiation of herpes simplex virus thymidine kinase polypeptides. Nucleic Acids Res 1996;24:2073–2079.
65. Ellison AR, Bishop JO. Herpesvirus thymidine kinase transgenes that do not cause male sterility are aberrantly transcribed and translated in the testis. Biochim Biophys Acta 1998;1442:28–38.

66. Braun RE, Lo D, Pinkert CA, Widera G, Flavell RA, Palmiter RD, et al. Infertility in male transgenic mice: disruption of sperm development by HSV-tk expression in postmeiotic germ cells. Biol Reprod 1990;43:684–693.
67. McNicol AM, Carbajo-Perez E. Aspects of anterior pituitary growth, with special reference to corticotrophs. Pituitary 1999;1:257–268.
68. Nolan LA, Kavanagh E, Lightman SL, Levy A. Anterior pituitary cell population control: basal cell turnover and the effects of adrenalectomy and dexamethasone treatment. J Neuroendocrinol 1998;10:207–215.
69. Borrelli E, Heyman RA, Arias C, Sawchenko PE, Evans RM. Transgenic mice with inducible dwarfism. Nature 1989;339:538–541.
70. Lipes MA, Davalli AM, Cooper EM. Genetic engineering of insulin expression in nonislet cells: implications for beta-cell replacement therapy for insulin-dependent diabetes mellitus. Acta Diabetol 1997;34:2–5.
71. Lipes MA, Cooper EM, Skelly R, Rhodes CJ, Boschetti E, Weir GC, et al. Insulin-secreting non-islet cells are resistant to autoimmune destruction. Proc Natl Acad Sci USA 1996;93:8595–8600.
71a. Cowley MA, Smart JL, Rubinstein M, Cerdán MG, Diano S, et al. Leptin activates anorexigenic POMC neurons through a neural network in the arcuate nucleus. Nature 2001; in press.
72. Spergel DJ, Kruth U, Hanley DF, Sprengel R, Seeburg PH. GABA- and glutamate-activated channels in green fluorescent protein-tagged gonadotropin-releasing hormone neurons in transgenic mice. J Neurosci 1999;19:2037–2050.
73. Kakita A, Goldman JE. Patterns and dynamics of SVZ cell migration in the postnatal forebrain: monitoring living progenitors in slice preparations. Neuron 1999;23:461–472.
74. Rubinstein M, Japon JA, Low MJ. Introduction of a point mutation into the mouse genome by homologous recomination in embryonic stem cells using a replacement type vector with a selectable marker. Nucleic Acids Res 1993;21:2613–2617.
75. Rubinstein M, Mogil JS, Japon M, Chan EC, Allen RG, Low MJ. Absence of opioid stress-induced analgesia in mice lacking β-endorphin by site-directed mutagenesis. Proc Natl Acad Sci USA 1996;93:3995–4000.
76. Konig M, Zimmer AM, Steiner H, Holmes PV, Crawley JN, Brownstein MJ, et al. Pain responses, anxiety and aggression in mice deficient in pre- proenkephalin. Nature 1996;383:535–538.
77. Comuzzie AG, Hixson JE, Almasy L, Mitchell BD, Mahaney MC, Dyer TD, et al. A major quantitative trait locus determining serum leptin levels and fat mass is located on human chromosome 2. Nat Genet 1997;15:273–276.
78. Hixson JE, Almasy L, Cole S, Birnbaum S, Mitchell BD, Mahaney MC, et al. Normal variation in leptin levels is associated with polymorphisms in the proopiomelanocortin gene, POMC. J Clin Endocrinol Metab 1999;84:3187–3191.
79. Krude H, Biebermann H, Luck W, Horn R, Brabant G, Gruters A. Severe early-onset obesity, adrenal insufficiency and red hair pigmentation caused by POMC mutations in humans. Nat Genet 1998;19:155–157.
80. Reid LD. Endogenous opioid peptides and regulation of drinking and feeding. Am J Clin Nutr 1985;42:1099–1132.
81. Hope PJ, Chapman I, Morley JE, Horowitz M, Wittert GA. Food intake and food choice: the role of the endogenous opioid peptides in the marsupial Sminthopsis crassicaudata. Brain Res 1997;764:39–45.
82. Levine AS, Grace M, Portoghese PS, Billington CJ. The effect of selective opioid antagonists on butorphanol-induced feeding. Brain Res 1994;637:242–248.
83. Yaswen L, Diehl N, Brennan MB, Hochgeschwender U. Obesity in the mouse model of pro-opiomelanocortin deficiency responds to peripheral melanocortin. Nat Med 1999;5:1066–1070.

16 Overexpression and Targeted Disruption of Genes Involved in the Control of Growth, Food Intake, and Obesity

Andrzej Bartke, PhD
and Michael Michalkiewicz, DVM, PhD

Contents

INTRODUCTION

In the early 1980s, Palmiter, Brinster, and their colleagues *(1,2)* reported dramatic effects of overexpression of a human or a rat growth hormone (GH) under control of a mouse metallothionein-I (MT) promoter in transgenic mice. Introduction of these gene constructs into the mouse genome produced striking acceleration of growth and a major increase in adult body size. Adult GH-transgenic mice are not obese, but may weigh twice as much as their normal siblings. Impressive alterations of the phenotype of these giant mice described by Palmiter et al. *(1,2)* and by other pioneers of this field *(3)* provided a demonstration of the enormous potential of transgenic technology, and undoubtedly contributed to the wide use of transgenic organisms in biological research, agriculture, and research targeted at developing gene therapies for various diseases.

From: *Contemporary Endocrinology: Transgenics in Endocrinology*
Edited by: M. Matzuk, C. W. Brown, and T. R. Kumar © Humana Press Inc., Totowa, NJ

Studies in transgenic mice overexpressing heterologous GH under control of different promoters, or oversecreting endogenous GH because of the transgenic expression of a GH-releasing hormone (GHRH) contributed a wealth of information on the consequences of lifelong GH excess, on the potential targets of GH actions, and on the mechanisms involved. More recent studies involving transgenic expression of GH antagonists (GH-Ant) and targeted disruption (knockout) of the GH receptor (GH-R) gene address the issue of the physiological role of the amounts of GH produced under normal circumstances.

This chapter, summarizes highlights of the findings in GH transgenic, GH-Ant transgenic, and GH-R-knockout mice, in the context of the current understanding of the functioning of the "somatotropic axis"—i.e., GH and insulin-like growth factor-I (IGF-I). For more information on IGF-I transgenic, IGF-knockout, and IGF-R-knockout mice, *see* Chapter 13.

The present chapter also summarizes the results of applying transgenic and targeted gene-disruption technologies to the study of other genes involved in the control of food intake, metabolism, and body composition, including neuropeptide Y (NPY), leptin, and product(s) of the genes at the Agouti locus.

PHENOTYPIC CONSEQUENCES OF OVEREXPRESSION OF GH IN TRANSGENIC MICE

Growth and IGF-I Levels/Endogenous GH Expression

Expression of heterologous human (h), bovine (b), ovine (o), or rat (r) GH under control of MT or phosphoenol pyruvate carboxykinase (PEPCK) promoter in transgenic mice involves multiple ectopic sites, including the liver and the kidney—and, for the MT promoter, also the intestine, skin, and a variety of other organs *(1–5)*. Transgene expression begins during fetal development with the MT promoter *(1–4)* and around the time of birth with the PEPCK promoter *(5)*. Transgenic pups appear normal at birth and during early postnatal development. In lines with high levels of GH expression, transgenic pups are distinguishable at weaning by their larger size, larger extremities, and somewhat different shape of the head. After weaning, transgenic mice grow faster than their normal siblings, presumably reflecting the predominant role of GH in regulating hepatic IGF-I expression and peripheral levels of IGF-I during this stage of development *(6)*. Moreover, rapid growth continues for a longer time in transgenic than normal mice *(7,8)*. The net result is an increase in adult body size, with adult body wt of these transgenic mice corresponding to between 120 and 200% of adult body wt of their normal siblings, depending on the level of GH expression. Plasma IGF-I levels in adult GH-transgenic mice are significantly elevated, and the amount of this increase corresponds to the levels of GH expression *(1,5,9)*.

Expression of endogenous (mouse) GH in the pituitary is suppressed by overexpression of heterologous GH at various ectopic sites, as expected from the well-documented negative feedback of GH and IGF-I on GH synthesis and release. This is evidenced by reduced pituitary levels of GH *(10)* and morphological evidence for the suppression of secretory activity of the somatotrophs *(11)*. The mechanisms of suppression of somatotroph function include suppression of GHRH expression and stimulation of somatostatin expression in the hypothalamus *(12,13)*. The degree of suppression of endogenous GH levels is proportional to the levels of heterologous GH in plasma *(10)*. It is important to emphasize that despite the suppression of endogenous GH secretion, the net concentrations of GH (heterologous plus endogenous) in the peripheral circulation of GH-transgenic mice is significantly—and in some lines—massively increased, thus accounting for the significant increase in plasma IGF-I levels and for the giant phenotype.

The striking impact of GH overexpression on the adult phenotype is further augmented by changes in body proportions (long legs and tail with somewhat leaner body), altered head shape with longer snout and "acromegalic" facial characteristics, loose skin, and altered pellage characteristics, most noticeable in animals with the wild-type (Agouti) coat color. Most of these changes are more pronounced in transgenic mice expressing human growth hormone (hGH) [pituitary hGH or placental hGH variant, or hGH-V] than in those overexpressing bGH. Studies of the weight of different organs in transgenic mice revealed that most organs grow "allometrically," i.e., increase in size in proportion to the increase in body wt, but some (particularly the liver) are disproportionately enlarged, while others (e.g., the brain) resemble the weights of the corresponding organs in normal mice *(7,14)*.

Different Types of GH "Transgenics"

Shanahan and his colleagues *(15)* developed a very interesting line of transgenic mice, in which oGH is expressed under control of a modified oMT promoter. In these animals, transgene expression—and thus, oGH levels—are very low. but can be increased by treating the animals with $ZnSO_4$, in keeping with the normal role of MT in heavy metal metabolism and regulation of MT expression by intake of heavy-metal ions. This allows the investigator to essentially "turn on" and "turn off" transgene expression at any time during development or adult life. Studies in these animals suggest that effects of GH overexpression in the adult resemble the effects of its overexpression throughout postnatal development *(15,16)*.

Ikeda and his colleagues *(17,18)* developed two lines of transgenic rats expressing hGH under control of mouse whey acidic protein (mWAP) promoter. In a line with high levels of serum hGH, growth was accelerated, female puberty was advanced, and ovulation was followed by pseudopregnancy, presumably representing activation of corpora lutea in response to lactogenic action of hGH. In line with low peripheral levels of hGH, growth was not affected, and female puberty was advanced, but females were sterile. Sterility appeared to be caused by suppression of endogenous PRL secretion *(17,18)*.

Other transgenic "models" for the study of GH actions include animals transfected with adenoviral vectors designed to promote transient or prolonged expression of GH in various organs *(19,20)*. This method of gene transfer produced significant levels of mouse or rat GH in genetically GH-deficient Snell dwarf (dw/dw) or little (lit/lit) mice and corrected various symptoms of GH deficiency in these animals *(19,20)*.

GH Receptors and Binding Proteins

Overexpression of GH in transgenic mice leads to a significant increase in the number of GH receptors (GH-R) in the liver *(21–25)*. The number of prolactin receptors (PRL-R) in the liver is also increased *(20,22)*. GH circulates in plasma in free form or complexed with GH-binding proteins (GHBP). In some species, including the human, GHBP is derived from enzymatic cleavage of the extracellular portion of GH-R, and thus its levels are generally related to the levels of GHR. In contrast, in mice GHBP is produced by alternative splicing of a common GH-R/GHBP pre-mRNA *(24)*, and thus the levels of

GH-R and GHBP can be independently regulated. In PEPCK-bGH transgenic mice, with high levels of bGH expression, GHBP levels in serum are markedly increased *(25)*.

Bioenergetics and Body Composition

Studies of MT-rGH transgenic mice by Rollo and his colleagues *(26)* revealed that during the rapid postweaning growth these animals consumed only as much food per g of body wt as their normal litter mates. This provides a striking demonstration of the stimulatory effects of GH on food utilization, and contributes to the evidence that GH can increase growth without increasing food intake *(27)*. In keeping with the anabolic, lipolytic, and anti-insulinemic effects of GH, GH-transgenic mice have a major increase in lean body mass *(7,14,27)*. A sustained high rate of growth and increases in the mass of muscle, bone, and viscera without a corresponding increase in food intake are associated with disproportional allocation of energy for growth, and a concomitant reduction in energy expenditures on locomotion *(26–28)*. MT-rGH-transgenic mice sleep more than normal animals, and exhibit less locomotor activity when awake *(28)*. On the basis of analysis of energy budget of MT-rGH-transgenic mice, Rollo et al. *(27)* proposed that these animals utilize an abnormally large proportion of food-derived energy for growth, and consequently experience a deficiency of energy available for maintenance and reproduction *(29,30)*. Consistent with this proposal, GH-transgenic mice have reduced reproductive potential, reduced lifespan, and various indices of reduced immune function *(31)*. In support of their hypothesis, Rollo and his colleagues *(29,30)* demonstrated that supplemental feeding of MT-rGH transgenic mice improved their reproductive performance and prolonged their lifespan. This finding is particularly impressive because the lifespan of laboratory rodents can be extended by caloric restriction rather than by increasing their caloric intake *(32)*, and indeed, supplemental feeding appeared to accelerate aging of normal mice in the same study *(29)*.

Insulin Resistance

Plasma glucose levels in transgenic mice overexpressing GH are normal, or only slightly elevated in spite of chronic elevations of both GH and corticosterone *(33,34)*. However, plasma insulin levels are markedly elevated *(33,34)*. Hyperinsulinemia and normoglycemia imply insulin resistance *(33,34)*. In PEPCK-bGH-transgenic mice, this is associated with significant reduction of the content of insulin receptors (IR) in both the particulate fraction and the solubilized membranes of the liver, corresponding to the expressed (functional) and nonexpressed (cryptic) receptors *(34)*. There were no significant changes in IR affinity, but the activity of insulin-dependent tyrosine kinase in partially purified IR preparations was markedly increased *(35)*. Subsequent studies provided evidence that both the decrease in the number of IR and the increase in their autophosphorylation activity are directly related to increased insulin levels in transgenic animals, rather than to their abnormally elevated GH. Treatment with streptozotocin or fasting for 48 h was used to suppress plasma insulin levels in transgenic mice without altering the levels of GH in their circulation. Both treatments were effective in reducing insulin, and produced an increase in IR and a decrease in their insulin-stimulated autophosphorylation *(36)*, and thus tended to normalize both the levels and the autophosphorylation of IR despite the persistence of grossly elevated GH.

Hypothalamic-Pituitary-Adrenal (H-P-A) Axis

The concentration of corticosterone—the principal adrenal glucocorticoid in many rodents, including mice—is significantly higher in GH transgenicg than in normal animals *(37)*. Elevation of plasma corticosterone levels was observed in transgenic male and female mice from lines expressing different levels of bGH, hGH, or hGH-V under basal conditions and after exposure to mild or severe stress *(37)*, and seemed unrelated to the time of the day when the samples were collected (*38*, and unpublished data). In PEPCK-bGH-transgenic mice, plasma adrenocorticotropin (ACTH) levels were significantly elevated *(38)*, while release of corticosterone from incubated adrenals and adrenal responsiveness to ACTH in vitro were unchanged *(38)*. Exposure to 15 min of immobilization stress resulted in comparable relative (i.e., % over baseline) increase in plasma corticosterone levels in GH-transgenic and normal mice, but the stress-induced corticosterone elevation persisted for a longer time in transgenic animals (Steger, unpublished). Thus, it can be concluded that overexpression of GH in transgenic mice is associated with a chronic increase in the secretion of ACTH from the pituitary, elevated plasma corticosterone levels, and reduced resilience of the H-P-A axis, i.e., a reduced ability of negative corticosterone feedback to terminate the stress response. Further studies will be necessary to relate these findings to the rates of corticosterone production and clearance in these animals and to the levels of hippocampal glucocorticoid receptors, which are responsible for the negative feedback of corticoids on corticotropin-releasing hormone (CRH)-secreting neurons and, consequently, on ACTH release.

Information on the functioning of the H-P-A axis in GH-transgenic animals is of considerable interest because chronic elevation of basal corticosterone levels and abnormally prolonged corticosterone responses to stressful stimuli may account for, or at least contribute to, many phenotypic characteristics of these animals. Thus, excessive exposure to glucocorticoids undoubtedly contributes to insulin resistance of GH-transgenic mice, and may also explain some of their reproductive deficits as well as reduced life span and various indices of premature central nervous system (CNS) aging.The well-documented immunosuppressive actions of glucocorticoids may also account for the various indications of reduced immune function in GH-transgenic mice *(30)*. However, results of recent studies in adrenalectomized bGH-transgenic mice argue against this possibility (Hall and Martinko, unpublished observations). The apparent suppression of the immune system in GH-transgenic animals deserves special emphasis because there is considerable evidence that GH is immunostimulatory rather than immunosuppressive *(39)*. Moreover, the intrinsic lactogenic activity of hGH in hGH-transgenic mice and stimulation of endogenous prolactin release in bGH-transgenic mice (*40)* could also be expected to provide additional stimulatory influence to immune function *(41)*.

Pathology and Lifespan

In addition to the expected consequences of elevated GH levels (including increased body size, splanchnomegaly, reduced fat stores, insulin resistance), GH-transgenic mice exhibit some distinct histopathological features. These include early onset and high prevalence of renal disease, glomerulosclerosis, and glomerulonephritis *(42,43)*, and major structural abnormalities of hepatocytes detectable at both the light and electron-microscopic level *(44)*. In addition, overexpression of human (hGH or hGH-V), but not bovine GH, is associated with early onset and a very high incidence of mammary tumors in females *(45)*, development of mammary tumors in some of the males (*45*, and unpublished observations), and massive age-related hypertrophy and hyperplasia of male accessory reproductive glands *(46)*. Interestingly, enlargement of seminal vesicles in aged MT-hGH males was associated with a profound reduction in the number of androgen receptors (ARs) in this tissue *(46)*.

Some of the pathological changes in GH-transgenic mice can be interpreted as symptoms of premature aging, consistent with their reduced life span *(28,42,45,47,48)*. In some instances, such as glomerulosclerosis, it is difficult to decide whether pathological alterations should be viewed as a symptom of accelerated aging or as a cause of a reduced lifespan.

It is very well-documented that transgenic mice that overexpress GH do not live as long as their normal siblings (*47,48*; review in *49*). The reduction of the lifespan is generally inversely related to the level of transgene expression, and more pronounced in animals expressing hGH than in those expressing bGH or rGH (*29,45*, and unpublished observations). Shortening of the lifespan in GH-transgenic mice can be very dramatic. In some of the lines, very few males and only an occasional female survive to the age of 1 yr, while most of their normal siblings live 2 to $2^1/_2$ yr (*45*, and unpublished data). Reduced lifespan of GH-transgenic mice is associated with various indices of premature aging, including reduced replicative potential of cells in vitro *(47)*, increased oxidative processes *(28)*, exaggerated increase in plasma corticosterone levels *(50)*, weight loss, and scoliosis with the resulting characteristic "hunched" appearance. There are also various indices of premature CNS aging in GH-transgenic mice, including increased astrogliosis, as measured by glial fibrillary acidic protein (GFAP) expression in various brain regions *(50)*, reduced turnover of norepinephrine, and dopamine in the median eminence region of the hypothalamus *(51)*, early cessation of estrous cyclicity in females, and early decline in copulatory behavior in males (*52*, Milton and Bartke, unpublished observations).

It is interesting to note that abnormally elevated plasma GH levels are also associated with reduced life expectancy in humans, namely in acromegalic patients *(53)*. However, in these individuals, reduced life span is apparently a result of the increased incidence of cardiovascular disease, diabetes, and tumors—conditions that do not seem to account for premature death in GH-transgenic mice. Although the issue of whether aging in GH-transgenic mice can be considered qualitatively normal may never be completely resolved, and cellular mechanisms responsible for their reduced life expectancy have not yet been identified, it should be pointed out that life span of these animals may be reduced because of their increased body mass. There is considerable evidence that, within a species, life expectancy is inversely related to body size (reviewed in *49*). This relationship is particularly well-documented for mice, using data derived from various selection experiments and analysis of strains and mutants differing in body size *(49)*. Rollo and colleagues recently demonstrated that reduction of the life span in MT/rGH-transgenic mice is fully consistent with the relationship of body wt to life span in different strains and populations of normal (nontransgenic) mice (Rollo, personal communication).

Comparisons of the Effects of GH Overexpression with the Effects of Overexpression of GHRH or IGF-I

Overexpression of GHRH in MT/hGHRH transgenic mice leads to chronic stimulation of the somatotrophs, progressive enlargement of the pituitary, and massive elevation of endogenous (mouse, m) GH levels in peripheral circulation *(54)*. Phenotypic characteristics of MT/hGHRH animals generally resemble those of GH-transgenic mice, and include increased body size, splanchnomegaly, reproductive deficits, alterations in hypothalamic neurotransmitters and in regulation of gonadotropin release, and reduced life span (*54,54a*; Chandrashekar and Bartke, unpublished observations). Moreover, concentrations of tachykinins in the anterior pituitary are elevated in these animals *(54b)*. The similarity in the effects of elevation of endogenous GH to the effects of overexpression of bGH in transgenic mice is consistent with the evidence that bGH binds to the murine GHR but not to the PRL-R, and thus mimics the effects of mGH *(22,55,56)*.

Studies of transgenic mice overexpressing IGF-I under control of the MT promoter *(44)* revealed some interesting differences from the phenotypes of MT-bGH and MT-GHRH Tg animals. MT-IGF-I transgenic animals did not exhibit renal glomerulosclerosis or hepatocellular hypertrophy and other abnormalities of the liver, which were characteristic of animals overexpressing bGH or mGH. In contrast, abnormalities of the skin, thickened adipose layer, and disruption of collagen bundles were evident in IGF-I transgenic but not in GH or GHRH-transgenic mice. This suggests that some of the phenotypic characteristics of GH "transgenics" may be caused by direct rather than IGF-I-mediated actions of GH, or to some physiological mechanisms that are differentially affected by GH and IGF-I, such as plasma insulin levels *(44)*.

IMPACT OF GH OVEREXPRESSION ON NEUROENDOCRINE CONTROL OF REPRODUCTION

Fertility

Reports from several laboratories that were successful in producing GH-transgenic mice in the early- and mid-1980s indicated that hemizygous transgenic males were usually fertile, transmitted transgene to approximately one-half of their progeny, and thus could be used to establish stable lines *(1–5)*. Some of of these lines are still in existence. In contrast, many of the GH-transgenic females were infertile or exhibited various reproductive deficits *(1,2,4)*. These initial observations were amply corroborated in subsequent work, and evidence for suppression of different aspects of reproductive function in transgenic mice overexpressing hGH, hGH-V, GH, oGH, or rGH was presented in numerous publications and several reviews *(57–60)*. Information on the impact of excessive levels of GH on fertility of transgenic mice can be summarized as follows:

MALES

In most of the established GH-transgenic lines, the males are fertile although their reproductive performance may be quantitatively reduced indicated by increased intervals from pairing them with females until conception, reduced proportion of females impregnated within a given period of cohabitation, and reduced proportion of males exhibiting various elements of sexual behavior (mounting the females, intromitting, and ejaculating) during a 1-h period of observation in the presence of ovariectomized (OVX) females brought into behavioral estrus by treatment with estradiol and progesterone *(58,60)*. Plasma testosterone levels in GH-transgenic males are usually normal, as is the release of testosterone from incubated testicular tissue in vitro, and testosterone responses to LH stimulation in vivo or in vitro *(58,60)*. Daily sperm production per g of testicular parenchyma in transgenic males overexpressing different GHs is normal or slightly reduced, while daily sperm production per animal is either normal or increased because of the increase in absolute weight of the testes *(32)*. Epididymal transit time is normal in GH-transgenic as compared to normal males, and epididymal sperm reserves are either increased in proportion to the increase in daily sperm production per testis or not affected *(32)*. We have not seen studies of sperm motility or other characteristics of spermatozoa in GH-transgenic males. This information is of interest because GH injections increased proportion of motile spermatozoa in genetically GH-deficient dwarf rats *(61)*. One exception to the generally adequate gonadal function and breeding performance of GH-transgenic male mice was described in a line of animals expressing hGH-V under control of the MT promoter. In this line, a significant proportion of males (approx 50%) is infertile, although they are adequately androgenized and produce functional spermatozoa *(32)*. Infertility of MT-hGH-V males is associated with and presumably caused by deficits in copulatory behavior, including reduced motivation to mate *(62)*, and quantitative alterations in some measures of copulatory behavior in those males that do mate *(62)*.

In GH-transgenic males, reproductive lifespan is shorter than in normal animals, presumably reflecting premature aging. Age-related loss of fertility in GH-transgenic mice appears to be primarily a result of deficits in male copulatory behavior *(60)*, although in mice overexpressing human GH it is also associated with marked enlargement of accessory reproductive glands *(46)*. In PEPCK-hGH males expressing very high levels of hGH, age-related decline of reproductive behavior occurs significantly earlier than in normal males from the same line, and the males become infertile before reaching 1 yr of age, although production of spermatozoa continues at a nearly normal level (Milton, Johnson, and Bartke, unpublished observations).

It is importnat to note that transgenic lines used for these studies were developed and are propagated by mating hemizygous GH-transgenic males with normal females, thus automatically selecting against male infertility. Therefore, results obtained in GH-transgenic males from established lines may underestimate the impact of lifelong excess of GH secretion on male reproductive functioning. Patients with acromegaly often experience reduced libido and impotence *(53)*.

FEMALES

Breeding performance of GH-transgenic female mice is significantly reduced in most, if not all, lines examined to date. Reproductive deficits in these animals range from failure to exhibit postpartum estrus (or to mate during postpartum estrus) with the resulting increase in interval between litters, to complete infertility of all transgenic females *(63)*. The severity of these deficits is generally related to the level of GH expression. Thus, most of the MT-bGH females from a line expressing low levels of GH are fertile, and exhibit only mild quantitative deficits in breeding performance *(63)*, while most of the PEPCK-bGH females expressing high levels of bGH are infertile *(52,63)*. We have documented that infertility of PEPCK-bGH females is caused by luteal failure that can be traced to the failure to respond to mating by increased prolactin release *(64,65)*. Pregnancy in these animals could be rescued by treatment with progesterone, prolactin, or a dopaminergic antagonist which stimulates endogenous prolactin release *(64)*. In addition to a high incidence of infertility, PEPCK-bGH females exhibit prolonged or irregular estrous cycles and a drastically shortened reproductive lifespan. A significant decline in the proportion of animals that cycle and mate can already be detected at the age of 3–4 mo *(52)* to 2–3 mo after puberty, when normal animals are attaining their peak breeding performance.

Transgenic MT-hGH females are infertile because of luteal failure *(66)*. Young females from this line cycle, ovulate and mate, fail to become pseudopregnant or pregnant, and continue mating at intervals corresponding to the length of their estrus cycle *(66,67)*. Their ova become fertilized, but are lost. Ovaries from these animals can maintain fertility when transplanted into OVX hosts *(67)*, indicating that infertility of MT-hGH transgenic mice is not caused by primary ovarian defects. Failure of activation of the corpora lutea in these animals after mating is apparently caused by suppression of prolactin release by the negative feedback action of hGH which is lactogenic in mice. In support of this interpretation, pregnancy could be rescued in these animals by treatment with prolactin-secreting ectopic pituitary transplants, leading to birth of live pups that survive to weaning and develop into normal, fertile adults *(66)*. Turnover of dopamine in the median eminence of the hypothalamus, which provides an index of activity of tuberoinfundibular dopaminergic neurons, was significantly greater in MT-hGH transgenic than in normal mice *(66)*. This suggests that hGH stimulates dopamine output by this neuronal group, and thus exerts inhibitory influence on the function of the lactotrophs and prolactin release by the pituitary.

Interestingly, PEPCK-hGH females that express much higher levels of hGH than the MT-hGH mice used for the experiments described here, can reproduce *(68)*. We interpret this finding as evidence that if levels of hGH in peripheral circulation are sufficiently high, they can directly stimulate the corpora lutea, and thus substitute for the function of endogenous prolactin and maintain pregnancy. In females from this particular line, very high levels (approx 450 ng/mL) of hGH are associated with stimulation of development and secretory activity of mammary glands, leading to lactational competence of virgin animals. Young PEPCK-hGH females that have never been pregnant exhibit maternal behavior toward foster pups, nurse them, and can raise them to weaning *(68)*. The ability of virgin PEPCK-hGH mice to lactate in response to the presence of (and suckling by) foster pups apparently represents effects of combined stimulation of their mammary glands by hGH, and by progesterone derived from functional corpora lutea that develop in these animals after every ovulation, the so-called pseudopregnancy (Milton and Bartke, unpublished observations).

Not all the effects of excessive GH levels on reproductive functioning are inhibitory. Thus, PEPCK-bGH-transgenic females attain sexual maturation significantly earlier than their normal siblings *(52)*, and the ovulation rate is increased in GH-transgenic vs normal females in each of the lines examined *(52,63)*. The effect of abnormally elevated GH levels is particularly impressive in the PEPCK-bGH line in which most of the females are infertile, but those that do become pregnant have more fetuses than normal animals from the same line *(52)*. Increased ovulation rate in transgenic mice overexpressing GH is consistent with the documented ability of IGF-I to potentiate the actions of gonadotropins on the ovary *(69)*.

Regulation of Prolactin Release

As was discussed in the preceding section of this chapter, we have traced the infertility of MT-hGH and PEPCK-bGH transgenic females to abnormalities of prolactin release. In female mice, stimuli associated with copulation induce major changes in the pattern of prolactin release, with pronounced surges that occur twice daily and are necessary for induction of active luteal phase and maintenance of luteal function during the first half of pregnancy *(70)*. These surges are apparently absent or reduced in MT-hGH females *(66)* and in most, but not all, of the PEPCK-bGH females *(63–65)*. In the search for the mechanisms of these effects, it is necessary to consider that while both hGH and bGH are somatogenic (i.e., exhibiting GH activity) in rodents, hGH is also lactogenic (i.e., exhibits activity indistinguishable from prolactin) *(55)*. Thus, hGH would be expected to suppress endogenous prolactin release by stimulating the function of tubero-infundibular dopaminergic (TIDA) neurons, and there is evidence for this effect in OVX MT-hGH females *(66)*. However, we did not detect evidence for enhanced TIDA activity in the median eminence of males from the same strain in which prolactin levels are markedly suppressed *(72)*. Moreover, we found no significant changes in TIDA activity in PEPCK-hGH transgenic mice of either sex, and no changes in plasma prolactin levels in OVX transgenic females from this line *(73)*. Preliminary data suggest that the apparent suppression of mating-induced prolactin surges in PEPCK-bGH transgenic females may be a result of alterations in serotonergic rather than dopaminergic transmission *(65)*. The ability of bGH to suppress prolactin surges in these animals was unexpected because overexpression of bGH is associated with an increase rather than suppression of the plasma prolactin level in OVX females, and in males from the same line and from other lines expressing various levels of bGH *(40,74)*.

Increase in plasma prolactin levels in transgenic mice overexpressing bGH appears to represent a previously unsuspected action of GH, and is associated with stimulation of prolactin expression in the pituitary *(75,75A)* and morphological indications of significant increases in the number and secretory activity of lactotrophs in both MT-bGH and PEPCK-bGH females *(75,75A)*. In MT-bGH females, dopamine turnover in the median eminence is reduced, suggesting a decrease in the inhibitory input of TIDA neurons *(40)*. The number of dopamine receptors in the pituitary is reduced, and the number of estrogen receptors (ERs) is elevated *(76)*. On the basis of the available data, it is difficult to determine whether these changes represent reduced responsiveness of the lactotrophs to inhibition by dopamine (thus further augmenting the effects of reduced dopamine input) and increased responsiveness of these cells to stimulation by estradiol. However, it is of considerable interest that opposite changes—i.e., increase in the number of dopamine receptors and decline in the number of ERs—were detected in transgenic mice overexpressing hGH, in which prolactin release is suppressed *(76)*.

Regulation of Gonadotropin Expression and Release

Although the occurrence of ovulatory cycles in GH-transgenic females and apparently normal spermatogenesis in GH-transgenic males suggest an adequate release of both FSH and LH in these animals, there is evidence for altered function of gonadotrophs in GH-transgenic mice. In PEPCK-bGH transgenic male mice expressing high levels of bGH, plasma FSH levels tend to be reduced, and steady-state levels of FSHβ mRNA as well as pituitary FSH content are significantly suppressed *(77)*. Steady-state levels of LHβ mRNA and LH responses to LHRH stimulation in vitro are also reduced in these animals *(77)*.

In contrast, in MT-hGH transgenic males, plasma LH levels, pituitary LH release, and expression of LHβ mRNA are significantly higher than in their normal siblings *(78,79)*. We have proposed that these effects of hGH may be related to its lactogenic activity *(50)*.

In the male mouse, hyperprolactinemia was previously shown to stimulate gonadotropin release *(80)* in contrast to the well-documented suppression of gonadotropins in hyperprolactinemic rats and humans. Elevation of LH levels in hGH-transgenic mice would thus suggest that, with respect to some of the actions hGH, signaling via the prolactin receptor can override the effects mediated via the GH receptor.

In OVX MT-bGH transgenic females, plasma levels of LH and FSH and FSH responses to LHRH in vivo were lower than the corresponding values in normal, ovariectomized females *(81)*. In transgenic females overexpressing hGH, plasma LH levels at estrus were higher, while plasma FSH levels were lower than in normal estrous animals *(82)*. However, ovulatory LH surge was significantly attenuated (Debeljuk, unpublished data).

In addition to these alterations in gonadotropin synthesis and release, transgenic mice overexpressing GH exhibit abnormalities in the feedback control of gonadotropin release. In normal animals, gonadectomy is followed by prompt and sustained increase in plasma LH and FSH levels. In contrast, in GH transgenic mice, these responses may be absent or significantly attenuated *(72,78,81–83)*. When gonadectomized MT-hGH transgenic and normal mice were injected with identical doses per g body wt of testosterone propionate (males) or estradiol (females), respectively, the suppression of plasma gonadotropin levels was smaller in transgenic than in normal animals *(78,82)*. In MT-hGH transgenic males, the ability of testosterone to suppress the postcastration rise in norepinephrine turnover in the median eminence was also reduced *(72)*. These findings imply a reduced sensitivity of the hypothalamic-pituitary system of GH transgenic mice to negative feedback of gonadal steroids.

Further studies are needed to characterize the pharmacodynamics of injected sex steroids in GH-Tg vs normal mice, and to relate the possible differences in the distribution and clearance of these hormones to their effects on gonadotropin release. There is also evidence that the effects of prolactin, NPY, and substance P on plasma gonadotropin levels can be significantly altered by overexpression of GH in transgenic mice *(84,85)*.

Role of the Central Nervous System in Mediating the Effects of GH on Reproduction

Reproductive deficits in animals that overexpress GH involve functions known to be controlled by the hypothalamus and/or other regions of the CNS. These include changes in plasma prolactin and gonadotropin levels, suppression of sexual behavior in males, and alterations in the length of the estrous cycle and in prolactin responses to copulation in females. Moreover, we have described numerous examples of altered levels and turnover rates of dopamine and norepinephrine in the median eminence and other regions of the hypothalamus in GH-transgenic as compared to normal mice *(51,58,66,72–74)*. In comparison to the hypothalami of normal mice, the hypothalami of PEPCK-bGH transgenic animals release significantly less LHRH in response to stimulation with N-methyl-D, L-aspartic acid (NMA) in vitro *(87)*.

Collectively, these findings strongly target the CNS in mediating the actions of excessive GH levels on reproduction in transgenic mice. Both GH and IGF receptors are present in various brain areas, including the hypothalamus *(88,89)*. Further studies will be needed to separate the direct and the IGF-I-mediated actions of GH on the neuronal groups involved in the control of reproductive functions and to evaluate the possible importance of other mechanisms that may mediate GH action on these neurons, including alterations in glucocorticoid and insulin levels, and in substrates for energy metabolism. Evidence for premature CNS aging in GH-transgenic mice was discussed earlier in this chapter.

TRANSGENIC AND KNOCKOUT MODELS IN THE STUDY OF GH RESISTANCE AND DEFICIENCY

In studies described in the preceding sections of this chapter, overexpression of GH or GHRH genes under control of different promoters was used to study the effects of prolonged exposure to excessive levels of GH. More recently, transgenic technology, site-directed mutagenesis and targeted gene disruption (knockout) were used to study the effects of GH resistance and GH deficiency. These studies are of particular interest because they provide information on the physiological role of the amounts of GH normally produced by the pituitary.

Transgenic Models of GH Deficiency

Behringer and colleagues *(90)* produced ablation of GH-producing cells in the mouse pituitary using a transgene composed of a rat GH promoter and a diphtheria toxin A (DTA)-chain structural gene. Absence of detectable GH in the circulation of these animals was associated with drastic reduction of body wt and significant suppression of the preprosomatostatin mRNA signal in the GH-regulating neurons of the anterior periventricular hypothalamus *(90,91)*.

In the rat, suppression of plasma GH levels and reduction of growth rate were produced by targeting an antisense RNA transgene to the GH gene *(92)*. In the same species, dominant dwarfism was produced by targeting the expression of hGH to the hypothalamic GH-releasing hormone (GHRH) neurons *(93)*. Reduced GHRH expression in the hypothalamus in these transgenic growth-retarded rats leads to reduced number of somatotrophs and reduced GHmRNA and GH content of the pituitary. Pituitary levels of ACTH and TSH are not affected, while prolactin is reduced *(93)*. Serial blood sampling reveals sexually dimorphic pattern of GH release, with regular secretory episodes only in males. Infusions of GHRH or GH releasing peptide-6 stimulate GH release and growth *(94)*.

In a recent report, Ikeda and colleagues *(95)* reported that expression of low levels of hGH in a line of mWAP-hGH transgenic rats, suppression of endogenous GH release and pulsatility of peripheral GH levels were associated with obesity and insulin resistance. Both obesity and diabetes are apparently caused by the suppression of GH release because they can be corrected by GH therapy *(95)*.

Transgenic Models of GH Resistance

In 1991, Chen et al. *(96)* reported that the GH molecule with substitution of arginine or lysine for glycine at the position 119 in the third alpha helix of bGH (or position 120 in hGH) produces GH molecules with antagonistic properties *(96,97)*. Transgenic animals overexpressing such antagonistic bGH-analog under control of the MT promoter were subsequently produced, and shown to have reduced plasma IGF-I levels, reduced body wt, and a tendency to develop obesity, i.e., exhibiting phenotypic characteristics consistent with GH resistance *(96,97)*. In these "transgenic dwarfs," the levels of GH-R in the liver and GHBP in serum are increased *(98)*, while in vitro binding of labeled wild-type bGH or hGH to serum is reduced *(98)*. Apparently, most of GHBP present in serum of MT-bGH-antagonist (Ant) mice is complexed with the antagonistic bGH analog, and thus no free GHBP is available for binding wild-type hormone. Moreover, high levels of this analog inhibit hepatic uptake of wild-type GH *(98)*. Additional mechanisms of bGH-Ant action are suggested by the evidence that this bGH antagonist apparently fails to induce the normal dimerization of GH-R *(96,98)* and occupancy of most of the GH-Rs by the antagonist interferes with formation of the complexes of one molecule of wild-type GH with two GH-Rs *(98)*. This further reduces the ability of endogenous or exogenous wild-type GH to exert their normal effects in MT-bGH-Ant animals, because dimerization of GH-Rs is required for normal GH signaling *(99)*. A striking demonstration of the effectiveness of GH antagonists to interfere with the actions of GH was provided by crossing animals that express human GH-Ant (G120R; corresponding to bovine G119R) with animals that express high levels of wild-type bGH *(100)*. Concomitant expression of hGH-Ant prevented development of glomerulosclerosis, which is characteristic of this line of bGH transgenic mice *(43,100)*.

MT-bGH-Ant mice of both sexes can reproduce, but their fertility is suppressed with deficits in litter size and postnatal survival of the pups. Turnover of dopamine in the median eminence region of the hypothalamus is reduced, but plasma prolactin levels and prolactin responses to pharmacological blockade of catecholamine synthesis are not altered (Steger and Bartke, unpublished).

Interpretation of the findings in MT-bGH-Ant transgenic mice is complicated by the possibility that antagonistic bGH analogs may exert some biological effects, including stimulation of the synthesis of GH-R and GHBP *(96,98,101)*.

Growth Hormone Receptor Knockouts

Animals with targeted disruption or knockout of the GH-R-GHBP gene—the "Laron dwarf mice"—were recently developed *(102)*. Animals homozygous for this "null mutation" (hereafter referred to as GH-R-knockout mice) exhibit characteristics of GH resistance: profound suppression of GH binding, extremely low plasma IGF-I levels, reduced postnatal (and particularly postweaning) growth, and dwarf phenotype despite significantly elevated plasma GH levels *(102)*. Although plasma insulin levels in GH-R-knockout animals are extremely low, plasma glucose levels are significantly suppressed *(102)*. Plasma corticosterone levels are normal in females and elevated in males *(102a)*.

Although most GH-R-knockout males and females can reproduce, their reproductive potential is significantly suppressed. Males exhibit reduced copulatory behavior, an increased interval between mating and conception, and increased incidence of infertility, and sire smaller litters than their normal counterparts (*102,103*, Danilovich, Wernsing, and Bartke, unpublished data). In female GH-R-knockout mice, puberty is delayed, but can be significantly advanced by injections of IGF-I *(104)*. Moreover, the estrous cycle tends to be irregular, and litter size is reduced, apparently reflecting reduced ovulation rate, and pregnancy is longer than in normal females (*102,104*). Fetal size and birthweight of pups are reduced in GH-R-knockout as compared to normal females *(104)*.

In GH-R-knockout males, plasma LH levels are normal, while prolactin levels are increased *(103)*. Acute increases in plasma LH and tetosterone levels after LHRH administration are attenuated. Moreover, testicular testosterone secretion in vitro is reduced in GH-R-knockout vs normal males, both basally and in the presence of LH in the incubation media *(103)*.

Significant delay of puberty in GH-R-knockout females and the ability of exogenous IGF-I to induce vaginal opening in these animals *(104)* are consistent with the documented involvement of the GH-IGF-I axis in the control of sexual maturation *(105,106)*. It has been proposed that IGF-I acts as a signal for the maturational activation of the hypothalamic LHRH pulse generator *(106)*. Reduced litter size in the GH-R-knockout could be expected from the well-documented ability of IGF-I to potentiate the actions of the gonadotropins on the ovary *(69)* and from the increased ovulation rate in transgenic mice overexpressing GH *(52)*. However, elevation of plasma prolactin levels in the GH-R-knockout males was unexpected and, indeed, opposite to what may have been predicted from previous findings. As mentioned earlier in this chapter, plasma prolactin levels are increased in several lines of transgenic mice that overexpress bGH.

The phenotypic consequences of GH resistance in GH-R-knockout mice are generally similar to the consequences of isolated GH deficiency in little (lit/lit) mice *(107)* and to the findings in humans with Laron dwarfism *(108)*, but are very mild in comparison to those observed in mice with null mutations of the IGF-I gene. The IGF-I-knockout mice are small at birth with poor viability, and those that survive have infantile reproductive systems totally incompatible with fertility *(109)*. Apparently, the amounts of IGF-I that can be produced in the absence of GH signaling are sufficient for fetal and postnatal survival, for qualitatively nearly normal, although quantitatively reduced fetal growth, and for reproductive development.

It is tempting to propose that the significant increase in plasma prolactin levels in GH-R-knockout males *(103)* may represent a mechanism of physiological compensation for the inability to respond to GH. There is considerable evidence that prolactin and GH can exert similar effects on many targets, including stimulation of LH binding by the Leydig cells in the testis *(110,111)*. There are also precedents for the ability of animals to compensate for targeted disruption of specific genes by utilizing alternate regulatory factors or pathways, consistent with the "redundancy" known to exist in many physiological systems. In this context, it is of interest that testicular function is apparently normal in prolactin-knockout mice *(112)* and only moderately suppressed in GH-R-knockout mice, with no or relatively mild effects on fertility *(102,103)*, while genetically dwarf mice which lack both GH and prolactin are hypogonadal and almost invariably sterile *(113)* (*see* Table 1).

TRANSGENIC AND KNOCKOUT ANIMALS IN STUDIES OF FOOD INTAKE, BODY COMPOSITION, AND OBESITY

In addition to the studies of GH described earlier in this chapter, transgenic and targeted gene-disruption technologies are used to elucidate the mechanisms controlling food intake, body composition, and the pathogenesis of obesity. Within the past few years, the applications of these genetic techniques in whole animals have identified a number of new molecules and physiologic pathways involved in the regulation of body wt, and have led to important new insights into the pathophysiology of eating disorders and obesity. These include the hormone leptin, the short and long forms of the leptin receptor, uncoupling proteins, agouti and agouti-related proteins, melanocortin-receptor isoforms, melanin-concentrating hormone, orexin, mahogany, and the proteins responsible for tub and fat, two monogenic mouse models of obesity. These efforts, in addition to characterization of several spontaneous obese mutants, provided a number of novel genetic models in which a single gene or pathway has been experimentally modified to test specific hypotheses. Both the spontaneous and experimentally generated mutants, or their intercrosses, are remarkably useful models in elucidating the

Table 1

Transgenic Models with Alterations in GH Release or Action[a]

Transgenic animals	*Phenotypes*
GH-transgenic mice with MT or PEPCK promoters *(1–3,5,15)*	Expression of heterologous (human, bovine, ovine, rat) GH genes in the liver, kidney, and other organs leads to increases in plasma IGF-I, growth and adult body size, while release of endogenous GH is suppressed. Synthesis and release of GH does not depend on physiological control mechanisms (such as GHRH and SRIF) but can be modified by composition of the diet, according to the properties of the employed promoter.
GH-transgenic mice with adenoviral vectors *(19,20)*	This form of "gene therapy" was successful in achieving transient or prolonged expression of mouse or rat GH in genetically GH-deficient mice and in partially correcting their phenotypic defects.
GH-transgenic rats *(17)*	Ectopic overexpression of human GH under control of mWAP promoter in transgenic rats produced phenotypic characteristics similar to those of MT-hGH and PEPCK-hGH transgenic mice.
GHRH-transgenic mice *(54)*	Overexpression of GHRH produces stimulation and hyperplasia of somatotrophs, progressive enlargement of the anterior or pituitary, and increases in somatic growth and adult body size. This transgenic model allows study of the effects of excessive secretion of homologous (mouse) GH of eutopic (pituitary) origin.
GH-ablated mice *(90)*	Transgenic expression of DTA-chain structural gene under control of rat GH promoter led to selective destruction of somatotrophs, absence of GH in the circulation, and reduced growth.
Growth-retarded transgenic rats *(93)*	Targeted expression of hGH in the hypothalamic GHRH neurons of transgenic rats reduced GHRH expression, presumably via negative GH feedback. The animals had fewer somatotrophs, reduced GH levels and dwarf phenotype.
GH-antagonist transgenic mice *(96)*	Overexpression of antagonistic GH analogs under control of MT promoter in transgenic mice reduces growth and adult body size, apparently by interfering with the action of endogenous GH and producing a state of GH resistance.

(continued)

Table 1 *(continued)*

Transgenic animals	*Phenotypes*
GH-R/GHBP knockouts *(102)*	Targeted disruption of the GH receptor/ GH-binding protein gene in mice produces GH resistance. Plasma GH levels are elevated, while the levels of IGF-I, postweaning growth, and adult body size are drastically reduced.

[a]Information on GH transgenics in species other than mice and rats (including domestic animals and fish) is outside the scope of this chapter.

physiology of energy metabolism, nutrient partition, the pathophysiology of obesity, and in the development of novel anti-obesity drugs.

Studies using these models lead to the following conclusions: (1) Genetic and hormonal control of food intake and body composition is extremely complex and involves interaction of multiple regulatory loops; (2) There is a great deal of redundancy in this controlling system and, similarly to many other complex biological systems, this redundant assemblage appears to provide a reliable physiological mechanism to meet an organism's constant energy needs for growth, reproduction, and maintenance; and (3) Obesity, an easily detected phenotypic indicator, is generally not a one-gene/one-disease event, but rather a result of diverse underlying metabolic and physiologic dysregulations involving a number of different molecules and independent pathways.

The intricate interactions between various hormones involved in the control of food intake and obesity are well illustrated by the relationships between GH and leptin, or between novel mahogany peptide, the agouti peptide, alpha-melanocyte-stimulating hormone (a-MSH), and melanocortin receptor 4 (MC4R). Through its potent lipolytic action, GH reduces adipose tissue mass, the source of leptin. Moreover, GH can reduce leptin expression (as measured by leptin mRNA levels) in pig adipocytes *(114)*. Pathologic overproduction of GH in acromegalic patients is associated with reduced levels of leptin *(115)*. In turn, leptin can affect GH release *(116,117)*. This includes stimulation of both basal and GHRH-induced GH release *(116)*, and blocking the inhibitory effect of fasting on the release of GH *(118)*. Interestingly, both starvation and obesity are associated with the suppression of circulating GH levels *(119,120)*. Chronic caloric restriction in rats prevents age-related decline in pulsatile GH release *(121)*.

Similarly complex interactions of various molecules and receptors have been recently discovered in the hypothalamic melanocortin signaling system, which is also involved in the control of feeding. Melanocortin signaling transduced by MC4R tonically inhibits feeding *(122)*. The agouti protein that is normally expressed in the hypothalamus *(123)* acts as an MC4R antagonist *(124)*. Increased expression of the agouti protein results in increased appetite and obesity *(125)*. MC4R-gene knockout produces a similar phenotypic change *(126)*. However, for the agouti protein to antagonize the MC4R signaling, the mahogany (attractin) protein is required *(127,128)*. It has been proposed that the role of the mahogany gene product in this pathway is to present agouti antagonist to MC4R, thereby reducing signaling. Thus, increased expression of mahogany would result in increased agouti binding and, consequently, increased appetite and body wt. Alterations in melanocortin signaling also affect the NPY-leptin pathway *(129,130)*.

As shown in the following sections, the discoveries of these intricate interrelationships would not be possible without transgenic and gene-knockout animal models.

STUDIES OF THE CENTRAL SIGNALING PATHWAYS OF METABOLIC REGULATION

Neuropeptide Y (NPY)

NPY, a 36-amino-acid neuromodulator that is widely distributed throughout the peripheral and CNS and coreleased with norepinephrine, is one of the most potent stimulators of food intake, and exerts a plethora of other physiological effects *(131)*. Overexpression of rat NPY in transgenic rats produced increases in food intake, body wt, fat deposits, and reduction in plasma leptin levels in males *(132)*. Surprisingly, effects of NPY overexpression on the same parameters in transgenic females were either less pronounced or opposite. These differences could be related to a more pronounced downregulation of NPY-receptors in NPY transgenic females (Y. Dumont, M. Michalkiewicz, and R. Quirion, unpublished observations). In NPY-transgenic rats of both sexes, spontaneous locomotor activity (wheel running) was reduced, and systolic blood pressure was elevated *(132)*. However, overexpression of the mouse NPY gene in transgenic mice did not affect body wt *(133)*. These differences between the mouse and the rat in metabolic responses to NPY overexpression may indicate the importance of species genome in phenotypic expression of the transgene. In mice, overexpression of NPY in the brain increased anxiety, but no change in food intake or body weight was observed *(134)*. This effect was presumably caused by stimulation of corticotropin-releasing factor (CRF) expression, because it could be prevented by a CRF antagonist *(134)*.

Targeted disruption of the NPY gene in mice produced no obvious phenotypic alterations, and indeed, NPY-deficient mice are remarkably normal. They grow normally, and eat and refeed normally after a fast *(135)*. Furthermore, all of their endocrine functions are normal *(136)*. The responses of NPY-null mice to diet-induced obesity, chemically induced obesity (monosodium glutamate and gold thioglucose), and genetic-based obesity (lethal yellow agouti and obese uncoupling protein-diphtheria toxin transgenics) were all also normal *(137)*. However, mice deficient in NPY are more sensitive to leptin, suggesting that NPY may normally have a tonic inhibitory action on leptin or leptin-mediated satiety signals. Exclusion of NPY did attenuate obesity in leptin-deficient ob/ob mice *(138)*. Lack of a distinct metabolic phenotype in the mouse without NPY is very surprising, considering the amount of evidence that NPY plays a central role in stimulating appetite and energy deposition. This could be a result of the fact that NPY receptors crossreact with other endogenous agonists—members of the pancreatic polypeptide family, including PYY *(139)*, or to the compensation for the NPY deletion by other orexigenic molecules produced by the same cells that produce NPY, including orexin *(140,141)*.

One way to approach this complexity was to delete the purported "appetite NPY receptors." Mice lacking the NPY Y1 *(142,143)*, Y5 *(143,144)*, and Y2 *(145)* receptor have been generated. Food intake was essentially unaffected by inactivation of the Y1 receptor, indicating that the Y1 receptor is not essential for feeding. NPY Y1 receptor knockout mice have normal feeding responses to intracerebroventricular injection of NPY. Basal feeding is only slightly reduced, whereas fast-induced refeeding is markedly reduced in these mutant mice. Interestingly, mice lacking Y1 receptor have slightly higher body wt and fat deposition. This higher-energy efficiency may result from the reduced locomotor activity of this mouse *(142)*. Similarly, studies of NPY Y5-receptor-knockout mice have not provided positive results regarding the importance of Y5 receptor in feeding *(144)*. Although, response to intracerebroventricular injection of NPY or related peptides in these mice is reduced, they feed and grow normally. Surprisingly, at a later age, the mice lacking Y5 receptor develop mild late-onset obesity characterized by increased appetite and body wt, and adiposity. These authors suggest that this unexpected observation could be caused by a compensatory elevation of NPY release from cell bodies within the arcuate nucleus *(144)*. Interestingly, these gender-dependent differences in body-wt response to genetic lesion of NPY signaling in Y1- and Y5-knockout mice corroborate the observation of reduced feeding and body wts in NPY-transgenic female rats *(132)*. Similarly, inactivation of the Y2-receptor subtype in mice resulted in hyperphagia and a mild obesity, with an attenuated response to leptin administration *(145)*.

NPY-transgenic and knockout animals provide very useful models for understanding the physiological importance of NPY in the regulation of metabolism, and for revealing the complex interactions of the NPY with other signaling pathways involved in the regulation of energy metabolism. NPY-mutant animals should also be helpful in developing novel drugs—based on the NPY antagonist—for the treatment of eating disorders and obesity.

Leptin

Leptin, a product of the adipocyte, reduces food intake and increases energy expenditure. Therefore, leptin is important in the control of body wt and composition. Mice that are genetically deficient in leptin or leptin receptor are massively obese, hyperglycemic, and insulin-resistant *(146)*. Leptin-gene therapy, using a recombinant adenovirus expressing the mouse leptin cDNA, transiently corrected the abnormal phenotype in genetically obese, leptin-deficient ob/ob mice *(147)*.

Studies using intercrosses of leptin-deficient ob/ob mice and mice with targeted disruption of the NPY gene or Y5-receptor gene *(144)* suggest that NPY may be an important mediator of leptin action. In support of this possibility, leptin receptors have been localized to NPY neurons *(148)*, ob/ob mice have elevated levels of NPY *(149)*, and leptin treatment can reduce NPY mRNA in the hypothalami of these animals *(150)*. Transgenic technology using DTA was used to produce a mouse model of leptin resistance *(151)*, which may be relevant to ethiology of obesity in the human.

Agouti and Agouti-Related Protein (AGRP)

Normal (wild-type) alleles at the agouti locus in the mouse act on the hair follicle and control deposition of pigment in the emerging hair. Deposition of yellow pigment near the tips of the hair, followed by deposition of black pigment in the remainder of the hair, produces a characteristic uneven brownish coloration of the animal, referred to as "agouti" (named after a similarly colored South American rodent). Two dominant mutations at the agouti locus, yellow (or: lethal yellow, A^y) and viable yellow (A^{vy}) produce increased growth, obesity, hyperinsulinemia, and hyperglycemia in combination with altered fur color: bright yellow in A^y and various degrees of "yellowing" in A^{vy}. These striking phenotypic alterations are believed to be mediated by ectopic expression of the agouti protein, because ubiquitous expression of either agouti protein or closely related AGRP in transgenic mice produced a phenotype closely resembling the abnormalities of the yellow (A^y) mouse *(125,152)*. A similar phenotype was produced by targeted disruption of one of the melanocortin receptors, MC4R, providing evidence that antagonistic actions at this receptor mediate the effects of both agouti protein and AGRP on body wt. Moreover, MC4R-knockout mice had altered spatial distribution of NPY expression in the hypothalamus, indicating yet another level of interaction between gene products involved in the control of food intake and body wt.

MC4R-knockout mice are extremely useful in elucidating the function of novel molecules and their complex interactions in the central melanocortin signaling involved in appetite control. Using these mutant mice, it has been demonstrated that lack of MC4R does not prevent feeding responses to a number of anorectic or orexigenic factors, including, CRH or NPY, indicating that these peptides act independently or downstream of MC4R signaling *(153)*. MC4R-knockout mice were used to localize the site of action of the novel mahogany protein within the agouti pathway. The semidominant mutant of *mahogany* locus *(mg)* suppresses the A^y-induced obesity, suggesting that for agouti protein action to induce obesity, the mahogany-gene product is required *(127,128)*. Two possible mechanisms of the mahogany-protein action include interference with agouti-peptide synthesis or its interaction with MC4R. When homozygous MC4R-knockout mice were crossbred with a strain of mice homozygous for the *mg* allele, the MC4R-knockout obese phenotype was not reduced *(127)*. Thus, these results clearly indicate that the novel mahogany-gene product acts at or upstream of MC4R.

Galanin

Neuropeptide galanin is widely distributed in the brain, especially in the hypothalamus, and the gastrointestinal tract. When administered into the hypothalamus, it stimulates appetite with preferential fat intake, reduces energy expenditure, and increases fat deposition *(154)*. However, targeted disruption of the murine galanin gene did not affect appetite or growth in the galanin-knockout mice *(155)*.

Hyperphagia and Obesity in 5-HT2C-Receptor-Knockout Mouse

Serotonin modulates numerous autonomic functions. It acts through the activation of a large family of G-protein-coupled-receptor subtypes that are widely expressed throughout the brain. The complexity of this signaling system and the paucity of selective drugs have made it especially difficult to define specific roles for 5-HT-receptor subtypes. Mutant mice lacking functional 5-HT2C receptors have been generated to elucidate the physiological function of this widely expressed receptor *(156,157)*. Unexpectedly, 5-HT2C receptor-deficient mice display substantial overweight as a result of increased appetite. This obesity is characterized by leptin and insulin resistance, impaired glucose tolerance, and increased responsiveness to high-fat feeding. Thus, these mutant mice have established a role for the 5-HT2C receptor in the serotonergic control of feeding and energy expenditure. These findings also demonstrate a dissociation of serotonin and leptin signaling in the regulation of feeding, and indicate that a perturbation of brain serotonin systems can predispose to eating disorders, obesity, and type 2 diabetes *(156,157)* (*see* Table 2).

Studies of Peripheral Mechanisms of Metabolic Regulation

Because appetite is regulated by central pathways, energy expenditure and partitioning (channeling) are also controlled by a number of molecules in peripheral tissues, such as uncoupling protein in the brown adipose tissue, lipoprotein lipase activity in the white fat and the muscles, and glucose transporter GLUT4 in the fat and the muscles. Transgenic

Table 2

Transgenic Metabolic Models with Alterations in the Central Signaling Pathways

Transgenic animal	*Metabolic phenotypes*
NPY-knockout mice *(135)*	Body wt and food intake were unchanged, while sensitivity to leptin was increased
NPY-transgenic mice *(133)*	No change in body wt or appetite
NPY-transgenic rats *(132)*	Increased food consumption and reduced free-wheel running and circulating leptin
Transgenic mice with brain expression of NPY *(134)*	No change in body wt or appetite
NPY Y1-receptor -knockout mice *(142,143)*	Reduced fast-induced refeeding and locomotor activity
NPY Y2-receptor-knockout mice *(145)*	Hyperphagia, mild obesity with an attenuated response to leptin administration
NPY Y5-receptor-knockout mice *(143,144)*	Late-onset obesity, increased appetite
Agouti *(125)* and agouti-related peptide *(152)* transgenic mice	Obesity and features of type II diabetes
Melanocortin-4-receptor-knockout mice *(126)*	Hyperphagia and obesity
Galanin-knockout mice *(155)*	No change in body wt or appetite
5-HT2C-receptor-knockout mice *(156)*	Hyperphagia, obesity, and insulin resistance

models have been generated to manipulate these peripheral pathways to understand their contribution to regulation of the energy balance and development of obesity.

Brown Adipose Tissue

The mammalian brown adipose tissue participates in the regulation of the body's energy balance. The uncoupling protein (UCP) in the mitochondrial membrane of brown adipose tissue generates heat by uncoupling oxidative phosphorylation. This heat generation process is unrelated to the performance of work and provides a means for the expedition of the body energy unrelated to the performance of work. Therefore, it has been postulated that manipulation of brown adipose tissue thermogenesis could be an effective strategy against obesity.

To test this hypothesis, several lines of transgenic mice with selective ablation of the brown adipose tissue have been generated. In these mice, the regulatory elements of the brown adipose tissue-specific UCP protein gene were used to drive expression of DTA transgene resulting in specific ablation of brown adipose tissue. These mutants are characterized by reduced energy expenditure (lower body temperature but not locomotor activity), marked obesity, increased food intake, hyperglycemia, hyperlipidemia, and increased susceptibility to a high-fat diet induced obesity *(151,158–160)*. These abnormalities are associated with insulin resistance and non-insulin-dependent diabetes mellitus (NIDDM) with both receptor and postreceptor components *(161,162)*. Interestingly, in this model, increased body wt, hyperlipidemia, late hyperphagia, and glucose homeostasis are leptin-resistant, while hypothalamic NPY and the hypothalamic-pituitary-adrenal axis remain under leptin control *(151,163)*.

These results demonstrate that brown adipose tissue is a critical organ involved in the regulation of energy homeostasis in mice, and suggest that its abnormal function may lead to development of diet-induced obesity and insulin resistance. These transgenic mice have many features in common with obesity as it appears in most humans, and therefore provide a useful model that may aid studies of the pathogenesis and treatment of human obesity, NIDDM, and their complications. In addition, studies based on these models may provide valuable insights into the mechanism for leptin resistance in human obesity.

However, mice with targeted inactivation of the gene encoding UCP-gene (lacking mitochondrial UCP) are cold-sensitive, but not hyperphagic or obese *(164)*. These UCP-deficient mice had normal body wt when fed on either a normal or a high-fat diet. However, these mice consume less oxygen after treatment with a beta3-adrenergic-receptor agonist, and they are sensitive to cold, indicating that their thermoregulation is defective. The authors propose that the loss of UCP may be compensated for by UCP2, a newly discovered homolog of UCP, because this gene is ubiquitously expressed and its expression is actually induced in the brown fat of UCP-deficient mice.

Transgenic mice with the adipocyte lipid-binding protein-2 gene promoter directing expression of the mitochondrial UCP gene in white fat were generated to examine whether increased energy dissipation in white adipose tissue can prevent obesity *(165–168)*. These transgenic mice have decreased mitochondrial-membrane potential in white adipocytes, reduced subcutaneous (sc) fat, and lower body-wt-gain response to a high-fat diet. When the transgene was expressed in Avy genetically obese mice, reductions in total body wt and sc fat stores were also observed. These observations demonstrate that mitochondrial uncoupling in white fat may prevent development of obesity and prove a potential of transgenically altered mitochondria in white fat to reduce body fat and to increase energy expenditure.

Beta-Adrenergic Receptor

Beta-adrenergic receptors are expressed predominantly in adipose tissue, and treatment with beta 3-selective agonists markedly increases energy expenditure and decreases obesity in rodents. Several transgenic models have been developed to define the role of beta 3-adrenergic receptors in regulation of energy expenditure and in development of obesity, and to investigate whether beta 3-selective agonists will be effective anti-obesity agents in humans *(169,170)*. Homologous recombination was used to generate mice that lack beta 3-adrenergic receptors. Beta 3-adrenergic receptor-deficient mice *(171)* have modestly increased fat stores (females more than males), clearly indicating that beta 3-adrenergic receptors play a role in regulating energy balance. Moreover, responses to beta 3-adrenergic receptor agonist CL 316,243 are completely lost in this mutant, suggesting that the actions of CL are mediated exclusively by the beta 3-adrenergic receptor. These beta 3-adrenergic receptor-deficient mice are a particularly useful means to a better understanding of the pharmacology of beta-adrenergic receptor.

To determine whether increasing the activity of the beta 1 adrenergic receptor in adipose tissue would affect the lipolytic rate or the development of this tissue, Soloveva et al. *(172)* used the enhancer-promoter region of the adipocyte lipid-binding protein (aP2) gene to direct expression of the human beta 1 adrenergic receptor cDNA to adipose tissue in transgenic mice. Expression of the transgene was seen only in brown and white adipose tissue. Adipocytes from transgenic mice were more responsive to beta-adrenergic receptor agonists than were adipocytes from nontransgenic mice, both in terms of cAMP production and lipolytic rates. These transgenic animals are less susceptible to diet-induced obesity. They have smaller adipose tissue depots than their nontransgenic litter mates, reflecting decreased lipid accumulation in their adipocytes. In addition to increasing the lipolytic rate, overexpression of the beta 1 adrenergic receptor induced appearance of the abundant brown fat cells in sc white adipose tissue. These results demonstrate that the beta 1-adrenergic receptor is involved in both stimulation of lipolysis and proliferation of brown fat cells in the context of the whole organism.

Rodent and human beta 3-adrenergic receptors differ with respect to expression in white vs brown adipocytes. Humans express beta 3-adrenergic receptors mRNA abundantly in brown but not white adipocytes, while rodents express beta 3-adrenergic receptors mRNA abundantly in both sites. "Humanized" mice expressing human but not murine beta 3-adrenergic receptors under the control of human gene-regulatory elements have been developed *(173)* to provide insights into mechanisms controlling human beta 3-adrenergic-receptor gene expression. In this model, 74 kb of human beta 3-adrenergic receptors genomic sequences have been transgenically introduced into gene-knockout mice lacking beta 3-adrenergic receptors. Like the situation in humans, in this mouse model human beta 3-adrenergic-receptor mRNA is expressed only in brown adipose tissue, with little or no expression in white adipose tissue. These mice are particularly useful for identification of responsible *cis*-regulatory element(s) and relevant trans-acting factor(s) controlling human beta 3-adrenergic-receptor gene expression, and are likely to provide means for development of effective anti-obesity agents in humans.

Lipoprotein Lipase

Lipoprotein lipase is the rate-limiting enzyme for hydrolysis of triglyceride, and is responsible for the import of triglyceride-derived fatty acids by muscle for utilization, and adipose tissue for storage. Therefore, it has been proposed that relative ratios of this enzyme activity in muscles or adipose tissue determine body-mass composition and play a role in the development of obesity. Increased lipoprotein lipase activity in adipose tissue could cause enhanced fat accumulation and obesity. However, transgenic mice that overexpress the human lipoprotein lipase gene or lipoprotein lipase-knockout mice have normal body wt and body-mass composition *(174,175)*. In contrast, obese ob/ob mice rendered deficient in adipose-tissue lipoprotein lipase by crossing them with the lipoprotein lipase-knockout mice had markedly diminished weight and fat mass *(174,175)*.

Generation of transgenic mice that overexpress human lipoprotein lipase in skeletal muscle allowed the examination of whether high-fat feeding-induced obesity would be prevented by diverting lipoprotein-derived triglyceride fatty acids away from storage in adipose tissue to oxidation in muscle. These mice have markedly increased lipoprotein-lipase activity in skeletal muscles with lower plasma triglycerides and carcass lipid content. The targeted overexpression of lipoprotein lipase in skeletal muscle clearly prevents a high-fat diet-induced lipid accumulation. These findings point out the possibility of preventing or treating obesity in humans by increasing lipoprotein-lipase activity in muscle by gene or drug delivery *(176)*.

Glucose Transporter

The rates of glucose utilization and partitioning between muscle and adipose tissue have important consequences for whole-body fuel metabolism, and eventually for the development of obesity. The insulin-sensitive glucose transporter, GLUT4 is the most abundant facilitative glucose transporter, and plays an important role during postpran-

Table 3

Transgenic Metabolic Models with Alterations in the Peripheral Target Organs

Transgenic animal	*Metabolic phenotypes*
Transgenic ablation of brown adipose tissue *(151,158,160)*	Hyperphagia and obesity, hyperlipidemia with resistance to insulin and leptin
Uncoupling protein-gene-knockout mice *(164)*	Increased sensitivity to cold
Uncoupling protein-transgenic mice *(165)*	Reduced body wt and sc fat
Beta 3-adrenergic receptor-knockout mice *(171)*	Increased fat stores
Beta 1-adrenergic-receptor transgenic mice *(172)*	Reduced lipid accumulation and increased lipolysis
Lipoprotein-lipase transgenic mice *(174,176)*	No change in body wt, but lower plasma triglycerides and carcass lipids content
Lipoprotein-lipase knockout mice *(175)*	No change in body wt and fat mass but their crosses with ob/ob mice had diminished weight and fat mass
Glucose transporter (GLUT4)-knockout mice *(177)*	Reduced fat deposit, increased blood glucose, and diminished sensitivity to insulin action
GLUT4-transgenic mice *(178,179)*	Increased body fat and fat-cell number

dial glucose partitioning in muscle and adipose tissue. Because GLUT4 has been shown to be dysregulated in pathological states such as diabetes and obesity, it is believed that malfunction of this transporter may cause the development of obesity.

Transgenic mouse models have been developed to directly assess how modulation of the glucose transporter may affect in vivo glucose disposal and energy metabolism. Disruption of GLUT4 gene has resulted in severely reduced adipose tissue deposits and growth retardation. Mice deficient in GLUT4 have increased blood glucose levels in the fed state, and are less sensitive to insulin action, indicating possible insulin resistance. GLUT4-null mice demonstrate that functional GLUT4 protein is not required for maintaining normal blood glucose levels, while GLUT4 is absolutely essential for sustained growth and normal cellular glucose and fat metabolism *(177)*.

Transgenic mice overexpressing the insulin-responsive glucose transporter (GLUT4) under the control of the fat-specific aP2 fatty acid-binding protein promoter/enhancer have been developed to gain insight into the role of nutrient partitioning in the development of obesity *(178,179)*. Genetic alteration of the partitioning of glucose between adipose tissue and muscle has produced a new animal model of obesity. Total body lipid is increased two- to threefold, and in vivo glucose tolerance is enhanced in transgenic mice overexpressing GLUT4 in fat. In isolated epididymal, parametrial, and sc adipose cells from these transgenic mice, basal glucose transport is over 20-fold greater than in nontransgenic litter mates. Surprisingly, fat-cell size is unaltered, and fat-cell number is markedly increased. GLUT4 overexpression in adipocytes of transgenic animals also increases whole-body insulin sensitivity. This model demonstrates that altering the partitioning of glucose between adipose tissue and muscle alters a critical step in the partitioning of lipoprotein fatty acids between these tissues, and leads to development of obesity. In these mice, obesity is entirely explained by an increase in fat-cell number without a change in fat-cell size. This animal model of obesity will advance our understanding the molecular mechanisms by which GLUT4 function in adipose tissues affects nutrient partitioning between muscle and adipose tissue, and its consequences for whole-body fuel metabolism (*see* Table 3).

ACKNOWLEDGMENTS

Studies of GH-transgenic and GH-R-knockout animals at Southern Illinois University were made possible by the generosity of Drs. Thomas Wagner and John Kopchick and Ms. June Yun, who provided us with animals to start breeding colonies, and unselfishly shared with us their experience and unpublished data. These studies were supported by NIH and USDA. Studies of NPY-transgenic rats at West Virginia University were supported by NIH, AHA, and M. Puskar. We would like to apologize to those whose work pertinent to this very broad topic was not cited because of our inadvertent omission or limitations of space.

REFERENCES

1. Palmiter RD, Brinster RL, Hammer RE, Trumbauer ME, Rosenfeld MG, Brinberg NC. Evans RM. Dramatic growth of mice that develop from eggs microinjected with metallothionein-growth hormone fusion genes. Nature1982; 300:611–615.
2. Palmiter RD, Norstedt G, Gelinas RE, Hammer RE. Brinster RL. Metallothionein-human growth hormone fusion genes stimulate growth of mice. Science 1983;222:809–814.
3. Selden RF, Wagner TE, Blethen S, Yun JS, Rowe ME, Goodman HM. Expression of the human growth hormone variant in cultured fibroblasts and transgenic mice. Proc Natl Acad Sci USA 1988;85:8241–8245.
4. Hammer RE, Pursel VG, Rexroad Jr CE, Wall RJ, Bolt DJ, Palmiter RD, Brinster RL. Genetic engineering of mammalian embryos. J Anim Sci 1986;63:269–278.
5. McGrane MM, DeVente J, Yun JS, Bloom J, Park E, Wynshaw A, Wagner T, Rottman FM, Hanson RW. Tissue-specific expression and dietary regulation of a chimeric phosphoenolpyruvate carboxykinase/ bovine growth hormone gene in transgenic mice. J Biol Chem. 1988;263:11,443–11,451.
6. D'Ercole AJ, Underwood LE. Ontogeny of somatomedin during development in the mouse; serum concentrations, molecular forms, binding proteins, and tissue receptors. Develop Biol 1980;79:33–45.
7. Shea BT, Hammer RE, Brinster RL. Growth allometry of the organs in giant transgenic mice. Endocrinology 1987;121:1924–1930.
8. Mathews LS, Hammer RE, Behringer RR, Brinster RL, Palmiter RD. Transgenic mice as experimental models for elucidating the roles of growth hormone and insulin-like growth factors for body growth. In: Isaksson O, Binder C, Hall K, Hökfelt B, eds., Growth Hormone: Basic and Clinical Aspects, Amsterdam, Elsevier, 1987.
9. Yun JS, Li Y, Wight DC, Portanova R, Selden RF, Wagner TE. The human growth hormone transgene: expression in hemizygous and homozygous mice. Proc Soc Exp Biol Med 1990;194:308–313.
10. Sotelo AI, Bartke A, Turyn D. Effects of bovine growth hormone (GH) expression in transgenic mice on serum and pituitary immunoreactive mouse GH levels and pituitary GH-releasing factor binding sites. Acta Endocrinol 1993;129:446–452.
11. Stefaneanu L, Kovacs K, Horvarth NE, Losinski E, Mayerhofer A, Wagner TE, Bartke A. An immunocytochemical and ultrastructural study of adenohypophyses of mice transgenic for human growth hormone. Endocrinology 1990;126:608–615.
12. Hurley DL, Phelps CJ. Altered growth hormone releasing hormone (GHRH) mRNA expression in transgenic mice with excess or deficient endogenous GH. Molec Cell Neurosci 1993;4:237–244.
13. Hurley DL, Bartke A, Wagner TE, Carlson SW, Wee BEF, Phelps CJ.Increased hypothalamic somatostatin expression in mice transgenic for bovine or human GH. J Neuroendocrinol 1994;6:539–548.
14. Cecim M, Bartke A, Yun J, Wagner TE. Growth allometry of transgenic mice expressing the mouse metallothionein-I/bovine growth hormone gene. Transgene 1993;1:125–132.

15. Shanahan CM, Rigby NW, Murray JD, Marshall JT, Townrow CA, Nancarrow CD, Ward KA. Regulation of expression of a sheep metallothionein Ia-sheep growth hormone fusion gene in transgenic mice. Mol Cel Biol 1989;9:5473–5479.
16. Pomp D, Nancarrow CD, Ward KA, Murray JD. Growth, feed efficiency and body composition of transgenic mice expressing a sheep metallothionein 1a-sheep growth hormone fusion gene. Livestock Prod Sci 1992;31:335–350.
17. Ikeda A, Matsuyama S, Nishihara M, Tojo H, Takahashi M. Changes in endogenous growth hormone secretion and onset of puberty in transgenic rats expressing human growth hormone gene. Endocr J 1994;41:523–529.
18. Ikeda A, Matsumoto Y, Chang KT, Nakano T, Matsuyama S, Yamanouchi K, Ohta A, Nishihra M, Tojo H, Sasaki F, Takahashi M. Different female reproductive phenotypes determined by human growth hormone (hGH) levels in hGH-transgenic rats. Biol Reprod 1997;56:847–851.
19. Hahn TM, Copeland KC, Woo SLC. Phenotypic correction of dwarfism by constitutive expression of growth hormone. Endocrinology 1996;137:4988–4993.
20. Marmary Y, Parlow AF, Goldsmith CM, He X, Wellner RB, Satomura K, Kriete MF, Robey PG, Nieman LK, Baum BJ. Construction and in vivo efficacy of a replacement deficient recombinant adenovirus encoding murine growth hormone. Endocrinology 1999;140:260–265.
21. Orian JM, Snibson K, Stevenson JL, Brandon MR, Herington AC. Elevation of growth hormone (GH) and prolactin receptors in transgenic mice expressing ovine GH. Endocrinology 1991;128:1238–1246.
22. Chen WY, White ME, Wagner TE, Kopchick JJ. Functional antagonism between endogenous mouse growth hormone (GH) and a GH analog results in dwarf transgenic mice. Endocrinology 1991;129:1402–1408.
23. Turyn D, Yun JS, Wagner TE, Bartke A. Specific somatotropic and lactogenic uptake in vivo in the livers of transgenic mice expressing bovine growth hormone gene. Growth Regul 1993;3:193–197.
24. Smith WC, Kuniyoshi J, Talamantes F. Mouse serum growth hormone (GH) binding protein has GH receptor extracellular and substituted transmembrane. Mol Endocrinol 1989;3:984–990.
25. Sotelo AI, Dominici FP, Engbers C, Bartke A, Talamantes F, Turyn D. Growth hormone-binding protein (GHBP) in normal mice and in transgenic mice expressing bovine growth hormone gene. Am J Physiol 1995;268:E745–E751.
26. Kajiura LJ, Rollo CD. The ontogeny of resource allocation in giant transgenic rat growth hormone mice. Can J Zool 1996;74:492–507.
27. Kajiura LJ, Rollo CD. A mass budget for transgenic "supermice" engineered with extra rat growth hormone genes: evidence for energetic limitation. Can J Zool 1994;72:1010–1017.
28. Rollo CD, Foss J, Lachmansingh E, Singh R. Behavioural rhythmicity in transgenic growth hormone mice: trade-offs, energetics, and sleep-wake cycles. Can J Zool 1997;75:1020–1034.
29. Rollo CD, Carlson J, Sawada M. Accelerated aging of giant transgenic mice is associated with elevated free radical processes. Can J Zool 1996;74:606–620.
30. Rollo CD, Rintoul J, Kajiura LJ. Lifetime reproduction of giant transgenic mice: the energetic stress paradigm. Can J Zool 1997;78:1336–1345.
31. Thukral, B. The effects of overproduction of bovine growth hormone (bGH) on the T-cell receptor repertoire in mice. In: PhD dissertation, Department of Microbiology, Carbondale: Southern Illinois University; 1997.
32. Weindruch R, Sohal RS. Caloric intake and aging. New Engl J Med 1997;337:986–994.
33. Bartke A, Naar EM, Johnson L, May MR, Cecim M, Yun JS, Wagner TE. Effects of expression of human or bovine growth hormone genes on sperm production and male reproductive performance in four lines of transgenic mice. J Reprod Fertil 1992;95:109–118.
34. Balbis A, Dellacha JM, Calandra RES, Bartke A. Turyn D. Down regulation of masked and unmasked insulin receptors in the liver of transgenic mice expressing bovine growth hormone gene. Life Sci 1992;51:771–778.
35. Balbis A, Bartke A, Turyn D. Overexpression of bovine growth hormone in transgenic mice is associated with changes in hepatic insulin receptors and in their kinase activity. Life Sci 1996;59:1363–1371.
36. Dominici FP, Balbis A, Bartke A, Turyn D. Role of hyperinsulinemia on hepatic insulin binding and insulin receptor autophosphorylation in the presence of high growth hormone (GH) levels in transgenic mice expressing GH gene. J Endocrinol 1998;159:15–25.
37. Cecim M, Ghosh PK, Esquifino AI, Began T, Wagner TE, Yun JS, Bartke A. Elevated corticosterone levels in transgenic mice expressing human or bovine growth hormone genes. Neuroendocrinology 1991;53:313–316.

38. Cecim M, Alvarez-Sanz M, Van de Kar L, Milton S, Bartke A. Increased plasma corticosterone levels in bovine growth hormone (bGH) transgenic mice: effects of ACTH, GH and IGF-I on in vitro adrenal corticosterone production. Trans Res 1996;5:187–192.
39. Murphy WJ, Rui H, Longo DL. Minireview-Effects of growth hormone and prolactin on immune development and function. Life Sci 1995;57:1–14.
40. Bartke A, Steger RW, Parkening TA, Collins TJ, Yun JS, Wagner TE. Influence of human and bovine growth hormones on the regulation of prolactin release in transgenic mice. In: Gupta D, Wollmann HA, Ranke MB, eds., Neuroendocrinology: New Frontiers, Brain Research Promotions, Tübigen, Germany, 1990, pp. 39–48.
41. Gagnerault MC, Touraine P, Savino W, Kelly PA, Dardenne M. Expression of prolactin receptors in murine lymphoid cells in normal and autoimmune situations. J Immunol 1993;150:5673–5681.
42. Orian JM, Lee CS, Weiss LM. Brandon MR. The expression of a metallothionein-ovine growth hormone fusion gene in transgenic mice does not impair fertility but results in pathological lesions in the liver. Endocrinology 1989;124:455–463.
43. Yang CW, Striker LJ, Pesce C, Chen WY, Paten EP, Elliot S, Doi T, Kopchick JJ. Striker GE. Glomerulosclerosis and body growth are mediated by different portions of bovine growth hormone: studies in transgenic mice. Lab Invest 1993;68:62–70.
44. Quaife CJ, Mathews LS, Pinkert CA, Hammer RE, Brinster RL, Palmiter RD. Histopathology associated with elevated levels of growth hormone and insulin-like growth factor I in transgenic mice. Endocrinology 1989;124:40–48.
45. Cecim M, Bartke A, Yun JS, Wagner TE. Expression of human, but not bovine growth hormone genes promotes development of mammary tumors in transgenic mice. Transgenics 1994;1:431–437.
46. Prins GS, Cecim M, Birch L, Wagner TE, Bartke A. Growth response and androgen receptor expression in seminal vesicles from aging transgenic mice expressing human or bovine growth hormone genes. Endocrinology 1992;131:2016–2023.
47. Pendergast WR, Li Y, Jiang D, Wolf NS. Decrease in cellular replicative potential in "giant" mice transfected with the bovine growth hormone gene correlates to shortened life span. J Cell Physiol 1993;156:96–103.
48. Wolf E, Kahnt E, Ehrlein J, Hermanns W, Brem G, Wanke R. Effects of long-term elevated serum levels of growth hormone on life expectancy of mice: lessons from transgenic animal models. Mech Age Dev 1993;68:71–87.
49. Bartke A, Brown-Borg HM, Bode AM, Carlson J, Hunter WS, Bronson RT. Does growth hormone prevent or accelerate aging? In: Bartke A, Falvo R, eds., Experimental Gerontology, Vol 33 (Proc 3rd Int. Symp on Neurbiol and Neuroendocrinol of Aging), 1998, pp. 675–687.
50. Miller DB, Bartke A, O'Callaghan JP. Increased glial fibrillary acidic protein (GFAP) levels in the brains of transgenic mice expressing the bovine growth hormone (bGH) gene. In: Bartke A, Falvo R, eds., Experimental Gerontology, Vol 30 (Proc 3rd Int. Symp on Neurobiol and Neuroendocrinol of Aging) 1995, pp. 383–400.
51. Steger RW, Bartke A, Cecim M. Premature ageing in transgenic mice expressing growth hormone genes. J Reprod Fertil Suppl 1993;46:61–75.
52. Cecim M, Kerr J, Bartke A. Effects of bovine growth hormone (bGH) transgene expression or bGH treatment on reproductive functions in female mice. Biol Reprod 1995;52:1144–1148.
53. Orme SM, McNally RJQ, Cartwright RA, Belchetz PE. Mortality and cancer incidence in acromegaly: a retrospective cohort study. J Clin Endocrinol Metab 1998;83:2730–2734.
54. Mayo KE, Hammer RE, Swanson LW, Brinster RL, Rosenfeld MG, Evans RM. Dramatic pituitary hyperplasia in transgenic mice expressing a human growth hormone releasing factor gene. Mol Endocrinol 1988;2:606–612.
54a. Debeljuk L, Steger RW, Wright JC, Mattison J, Bartke A. Effects of overexpression of growth hormone-releasing hormone on the hypothalamo-pituitary-gonadal function in the mouse. Endocrine 1999;11:171–179.
54b. Debeljuk L, Wright JC, Phelps C, Bartke A. Transgenic mice overexpressing the growth-hormone-releasing hormone gene have high concentrations of tachykinins in the anterior pituitary gland. Neuroendocrinology 1999;70:107–116.
55. Posner BI, Kelly PA, Shiu RPC, Paud R, Friesen HG. Studies of insulin, growth hormone and prolactin binding: tissue distribution, species variation and characterization. Endocrinology 1974;95:521–531.
56. Aguilar RC, Fernandez HN, Dellacha JM, Calandra RS, Bartke A, Turyn D. Identification of somatogenic binding sites in liver microsomes from normal mice and transgenic mice expressing human growth hormone gene. Life Sci 1992;50:615–620.

57. Bartke A, Naar E, Cecim M, Milton S, Liu Y-Z, Chandrashekar V, Steger RW. Effects of growth hormone on reproduction in transgenic mice. Progress in Endocrinology. The Proceedings of the Ninth International Congress of Endocrinology, 1992;19:67–70.
58. Mayerhofer A, Weis J, Bartke A, Yun JS, Wagner T. Effects of transgenes for human and bovine growth hormones on age-related changes in ovarian morphology in mice. Anat Rec 1990;117:175–186.
59. Hauser SD, McGrath MF, Collier RJ, Krivi GG. Cloning and in vivo expression of bovine growth hormone receptor mRNA. Mol Cell Endocrinol 1990;72:187–200.
60. Feldman M, Ruan W, Cunningham BC, Wells JA, Kleinberg DL. Evidence that the growth hormone receptor mediates differentiation and development of the mammary gland. Endocrinology 1993;133:1602–1608.
61. Breier BH, Vickers MH, Gravance CG, Casey PJ. Growth hormone (GH) therapy markedly increases the motility of spermatozoa and the concentration of insulin-like growth factor-I in seminal vesicle fluid in the male GH-deficient dwarf rat. Endocrinology 1996;137:4061–4064.
62. Meliska CJ, Bartke A. Copulatory behavior and fertility in transgenic male mice expressing human placental growth hormone gene. J Androl 1997;18:305–311.
63. Naar EM, Bartke A, Majumdar SS, Buonomo FC, Yun JS, Wagner TE. Fertility of transgenic female mice expressing bovine growth hormone or human growth hormone variant genes. Biol Reprod 1991;45:178–187.
64. Cecim MC, Kerr J, Bartke A. Infertility in transgenic mice overexpressing the bovine growth hormone gene: Luteal failure secondary to prolactin deficiency. Biol Reprod 1995;52:1162–1166.
65. Cecim M, Fadden C, Kerr J, Steger R, Bartke A. Infertility in transgenic mice overexpressing the bovine growth hormone gene: Disruption of the neuroendocrine control of prolactin secretion during pregnancy. Biol Reprod 1995;52:1187–1192.
66. Bartke A, Steger RW, Hodges SL, Parkening TA, Collins TJ, Yun JS, Wagner TE. Infertility in transgenic female mice with human growth hormone expression: Evidence for luteal failure. J Exp Zool 1988;248:121–124.
67. Yun JS, Wagner TE. Study of human growth hormone transgenic mice: Female reproductive system. The 10th Korea Symposium on Science and Technology 1987;3-1:279–282.
68. Milton S, Cecim M, Li YS, Yun JS, Wagner TE, Bartke A. Transgenic female mice with high human growth hormone levels are fertile and capable of normal lactation without having been pregnant. Endocrinology 1992;131:536–538.
69. Adashi EY, Resnick CE, D'Ercole AJ, Svoboda ME, Van Wyk JJ. Insulin-like growth factors as intraovarian regulators of granulosa cell growth and function. Endocr Rev 1985;6:400–418.
70. Barkley MS, Bradford GE, Geschwind II. The pattern of plasma prolactin concentration during the first half of mouse gestation. Biol Reprod 1978;19:291–296.
71. Bartke A. Differential requirement for prolactin during pregnancy in the mouse. Biol Reprod 1973;9:379–383.
72. Steger RW, Bartke A, Parkening TA, Collins T, Yun JS, Wagner TE. Neuroendocrine function in transgenic male mice with human growth hormone expression. Neuroendocrinology 1990;52:106–111.
73. Steger RW, Bartke A, Yun JS, Wagner TE. Neuroendocrine function in transgenic mice with the phosphoenolpyruvate carboxykinase/human growth hormone (PEPCK/hGH) hybrid gene and very high peripheral levels of hGH. Transgene 1993;1:19–26.
74. Steger RW, Bartke A, Parkening TA, Collins T, Cerven R, Yun JS, Wagner TE. Effects of chronic exposure to bovine growth hormone (bGH) on the hypothalamic-pituitary axis in transgenic mice: Relationship to the degree of expression of the PEPCK-bGH hybrid gene. Transgenics 1994;1:245–253.
75. Stefaneanu L, Kovacs K, Bartke A, Mayerhofer A, Wagner TE. Pituitary morphology of transgenic mice expressing bovine growth hormone. Lab Invest 1993;68:954–591.
75a. Vidal S, Stefaneanu L, Thapar K, Aminyar R, Kovacs K, Bartke A. Lactotroph hyperplasia in the pituitaries of female mice expressing high levels of bovine growth hormone. Trans Res 1999;8:191–202.
76. Vidal S, Stefaneanu L, Kovacs K, Bartke A. Gene expression of estrogen and dopamine subtype 2 receptors in transgenic mice with overproduction of heterologous growth hormones. 10th International Congress of Endocrinology, Abstract #P3-217, 1996, p. 809.
77. Tang K, Bartke A, Gardiner CS, Wagner TE, Yun JS. Gonadotropin secretion, synthesis, and gene expression in two types of bovine growth hormone transgenic mice. Biol Reprod 1993;49:346–353.
78. Chandrashekar V, Bartke A, Wagner TE. Endogenous human growth hormone (GH) modulates the effect of gonadotropin-releasing hormone on pituitary function and the gonadotropin response to the negative feedback effect of testosterone in adult male transgenic mice bearing human GH gene. Endocrinology 1988;123:2717–2722.

79. Tang K, Bartke A, Gardiner CS, Wagner TE, Yun JS. Gonadotropin secretion, synthesis, and gene expression in human growth hormone transgenic mice and in Ames dwarf mice. Endocrinology 1993;132:2518–2524.
80. Klemcke HG, Bartke A. Effects of chronic hyperprolactinemia in mice on plasma gonadotropin concentrations and testicular human chorionic gonadotropin binding sites. Endocrinology 1981;108:1763–1768.
81. Chandrashekar V, Bartke A. Influence of hypothalamus and ovary on pituitary function in transgenic mice expressing the bovine growth hormone gene and in growth hormone-deficient Ames dwarf mice. Biol Reprod 1996;54:1002–1008.
82. Chandrashekar V, Bartke A, Wagner TE. Neuroendocrine function in adult female transgenic mice expressing the human growth hormone gene. Endocrinology 1992;130:1802–1808.
83. Tang K, Bartke A, Gardiner CS, Wagner TE, Yun JS. Testosterone feedback on gonadotropin secretion and gene expression in transgenic mice expressing human growth hormone gene. J Androl 1994;15:9–14.
84. Chandrashekar V, Bartke A, Wagner TE. Interactions of human growth hormone and prolactin on pituitary and Leydig cell function in adult transgenic mice expressing the human growth hormone gene. Biol Reprod 1991;44:76–82.
85. Ghosh P, Debeljuk L, Wagner TE, Bartke A. Effect of immunoneutralization of neuropeptide Y on gonadotropin and prolactin secretion in normal mice and in transgenic mice bearing bovine growth hormone gene. Endocrinology 1991;129:597–602.
86. Esquifino AI, Arce A, Debeljuk L, Bartke A. Effects of immunoneutralization of substance P on hypothalamic neurotransmitters in normal mice and in transgenic mice expressing bovine growth hormone. Proc Soc Exp Biol Med 1998;218:68–75.
87. Mattison J, Bartke A, Steger RW. Decreased gonadotropin-releasing hormone secretory response to N-methyl-D, L-asparic acid stimulation in growth hormone (GH) transgenic male mice. Society for Neuroscience 1998;submitted.
88. Marks JL, Porte Jr D, Baskin DG. Localization of type I insulin-like growth factor receptor messenger RNA in the adult rat brain by in situ hybridization. Mol Endocrinol 1991;5:1158–1168.
89. Nyberg F, Burman P. Growth hormone and its receptors in the central nervous system - Location and functional significance. Horm Res 1996;45:18–22.
90. Behringer RR, Mathews LS, Palmiter RD, Brinster RL. Dwarf mice produced by genetic ablation of growth hormone-expressing cells. Genes Devel 1988;2:453–461.
91. Hurley DL, Phelps CJ. Hypothalamic preprosomatostatin messenger ribonucleic acid expression in mice transgenic for excess or deficient endogenous growth hormone. Endocrinology 1992;130:1809–1815.
92. Matsumoto K, Kakidani H, Anzai M, Nakagata N, Takahashi A, Takahashi Y, Miyata K. Evaluation of an antisense RNA transgene for inhibiting growth hormone gene expression in transgenic rats. Dev Genet 1995;16:273–277.
93. Flavell DM, Wells T, Wells SE, Carmignac DF, Thomas GB, Robinson IC. Dominant dwarfism in transgenic rats by targeting human growth hormone (GH) expression to hypothalamic GH-releasing factor neurons. EMBO J 1996;15:3871–3879.
94. Wells T, Flavell DM, Wells SE, Carmignac DF, Robinson IC. Effects of growth hormone secretagogues in the transgenic growth-retarded (Tgr) rat. Endocrinology 1997;138:580–587.
95. Ikeda A, Chang KT, Matsumoto Y, Furuhata Y, Nishihara M, Sasaki F, Takahashi M. Obesity and insulin resistance in human growth hormone transgenic rats. Endocrinology 1998;139:3057–3063.
96. Chen WY, Wight DC, Metha BV, Wagner TE, Kopchick JJ. Glycine 119 of bovine growth hormone is critical for growth-promoting activity. Mol Endocrinol 1991;5:1845–1852.
97. Chen WY, Chen NY, Yun J, Wagner TE, Kopchick JJ. In vitro and in vivo studies of antagonists effects of human growth hormone analogs. J Biol Chem 1994;269:15,892–15,897.
98. Sotelo AI, Bartke A, Kopchick JJ, Knapp JR, Turyn D. Growth hormone (GH) receptors, binding proteins and IGF-I concentrations in the serum of transgenic mice expressing bovine GH agonist or antagonist. J Endocrinol 1998;158:53–59.
99. Ultsch M, de Vos AM. Crystals of human growth hormone-receptor complexes. Extracellular domains of the growth hormone and prolactin receptors and a hormone mutant designed to prevent dimerization. J Mol Biol 1993;231:1133–1136.
100. Chen N, Chen WY, Striker LJ, Striker GE, Kopchick JJ. Co-expression of bovine growth hormone (GH) and human GH antagonist genes in transgenic mice. Endocrinology 1997;138:851–854.
101. Harding PA, Wang X, Okada S, Chen WY, Wan W, Kopchick JJ. Growth hormone (GH) and a GH antagonist promote GH receptor dimerization and internalization. J Biol Chem 1996;271: 6708–6712.

102. Zhou Y, Xu BC, Maheshwari HG, He L, Reed M, Lozykowski M, Okada S, Wagner TE, Cataldo LA, Coschigano K, Baumann G, Kopchick JJ. A mammalian model for Laron syndrome produced by targeted disruption of the mouse growth hormone receptor/binding protein gene (The Laron mouse). Proc Natl Acad Sci USA 1997;94:13,215–13,220.
102a. Hauck A, Hunter WS, Danilovich N, Kopchick JJ, Bartke A. Reduced levels of thyroid hormones, insulin, and glucose, and lower body core temperature in the growth hormone receptor/binding protein knockout mouse. Exp Biol Med 2001; in press.
103. Chandrashekar V, Bartke A, Coschigano KT, Kopchick JJ. Pituitary and testicular function in growth hormone receptor gene knockout mice. Endocrinology 1999;140:1082–1088.
104. Danilovich N, Wernsing D, Coschigano KT, Kopchick JJ, Bartke A. Deficits in female reproductive function in GH-R-KO mice; role of IGF-I. Endocrinology 1999;140(6):2637–2640.
105. Copeland KC, Kuehl TJ, Castracane VD. Pubertal endocrinology of the baboon: elevated somatomedin-C/insulin-like growth factor I at puberty. J Clin Endocrinol Metab 1982;55:1198–1201.
106. Hiney JK, Srivastava V, Nyberg CL, Ojeda SR, Dees WL. Insulin-like growth factor I of peripheral origin acts centrally to accelerate the initiation of female puberty. Endocrinology 1996;137:3717–3728.
107. Chubb C. Sexual behavior and fertility of little mice. Biol Reprod 1987;37:564–569.
108. Rosenfeld RG, Rosenbloom AL, Guevara-Aguirre J. Growth hormone (GH) insensitivity due to primary GH receptor deficiency. Endocr Rev 1994;15:369–390.
109. Baker J. Hardy MP, Zhou J, Bondy C, Lupu F, Bellvé AR, Efstratiadis A. Effects of an IGF-I gene null mutation on mouse reproduction. Mol Endocrinol 1996;10:903–918.
110. Bex F, Bartke A, Goldman BD, Dalterio S. Prolactin, growth hormone, luteinizing hormone receptors, and seasonal changes in testicular activity in the golden hamster. Endocrinology 1978;103:2069–2080.
111. Saez JM. Leydig cells: endocrine, paracrine, and autocrine regulation. Endocr Rev 1994;15:574–626.
112. Steger RW, Chandrashekar V, Zhao W, Bartke A, Horseman N. Neuroendocrine and reproductive functions in male mice with targeted disruption of the prolactin gene. Endocrinology 1998;139: 3691–3695.
113. Bartke A. Genetic models in the study of anterior pituitary hormones. In: Genetic Variation in Hormone Systems, CRC Press, Boca Raton, FL, 1979, pp. 113–126.
114. Spurlock ME, Ranalletta MA., Cornelius SG., Frank GR, Willis GM, Ji S, Grant AL, Bidwell CA. Leptin expression in porcine adipose tissue is not increased by endotoxin but is reduced by growth hormone. J Interferon Cytokine Res 1998;18:1051–1058.
115. Miyakawa M, Tsushima T, Murakami H, Isozaki O, Demura H, Tamnaka T. Effect of growth hormone (GH) on serum concentrations of leptin: study in patients with acromegaly and GH deficiency. J Clin Endocrinol Metab 1998;83:3476–3479.
116. Tannenbaum GS, Gurd W, Lapointe M. Leptin is a potent stimulator of spontaneous pulsatile growth hormone (GH) secretion and the GH response to GH-releasing hormone. Endocrinology 1998;39:3871–3875.
117. Barb CR, Yan X, Azain MJ, Kraeling RR, Rampacek GB, Ramsay TG. Recombinant porcine leptine reduces feed intake and stimulates growth hormone secretion in swine. Domest Anim Endocrinol 1998;15:77–86.
118. Vuaguat BA, Pierroz DD, Lalaoui M, Englaro P, Pralong FP, Blum WF, Aubert ML. Evidence for a leptin-neuropeptide Y axis for the regulation of growth hormone secretion in the rat. Neuroendocrinology 1998;67:291–300.
119. Tannenbaum GS, Epelbaum J, Colle E, Brazeau P, Martin JB. Antiserum to somatostatin reverses starvation-induced inhibition of growth hormone but not insulin secretion. Endocrinology 1978;102:1909–1914.
120. Weltman A, Weltman JY, Hartman ML, Abbott RA, Rogol AD, Evans WS, Veldhuis JD. Relationship between age, percentage body fat, fitness and 24 hour growth hormone release in healthy young adults: effects of gender. J Clin Endocrinol Metab 1994;78:543–548.
121. Sonntag WE, Xu X, Ingram RL, C'Costa A. Moderate caloric restriction alters the subcellular distribution of somatostatin mRNA and increases growth hormone pulse amplitude in aged animals. Reg Growth Hor 1995;61:601–608.
122. Bray GA. Mechanisms for development of genetic, hypothalamic and dietary obesity. In: Bray GA and Ryan DH, eds., Pennington Center Nutrition Series, Vol 5, Baton Rouge, Louisiana State University Press, 1996, pp. 3–66.
123. Haskell-Luevano C, Chen P, Li C, Chang K, Smith MS. Characterization of the neuroanatomical distribution of agouti-related protein immunoreactivity in the rhesus monkey and the rat. Endocrinology 1999;140:1408–1415.

124. Lu D, Willard D, Patel IR, Kadwell S, Overton L, Kost T, Luther M, Chen W, Woychik RP, Wilkison WO. Agouti protein is an antagonist of the melanocyte-stimulating-hormone receptor. Nature (Lond) 1994;371:799–802.
125. Klebig ML, Wilkinson JE, Geisler JG, Woychik RP. Ectopic expression of the agouti gene in transgenic mice causes obesity, features of type II diabetes, and yellow fur. Proc Natl Acad Sci USA 1995;92:4728–4732.
126. Huszar D, Lynch CA, Fairchild-Huntress V, Dunmore JH, Fang Q, Berkemeier LR, et al. Targeted disruption of the MC4R results in obesity in mice. Cell, 1997;88:131–141.
127. Nagle DL, McGrail SH, Vitale J, Woolf EA, Dussault Jr BJ, DiRocco L, Holmgren L, Montagno J, Bork P, Huszar D, Fairchild-Huntress V, Ge P, Keilty J, Ebeling C, Baldini L, Gilchrist J, Burn P, Carlson GA, Moore KJ. The mahogany protein is a receptor involved in suppression of obesity. Nature 1999;398:148–152.
128. Gunn TM, Miller KA, He L, Hyman RW, Davis RW, Azarani A, Schlossman SF, Duke-Cohan JS, Barsh GS. The mouse mahogany locus encodes a transmembrane form of human attractin. Nature 1999;398:152–156.
129. Seeley RJ, Yagaloff KA, Fisher SL, Burn P, Thiele TE, Dijk G, Baskin DG, Schwartz MW. Melanocortin receptors in leptin effects. Nature 1997;390:349.
130. Kesterson RA, Huszar D, Lynch CA, Simerly RB, Cone RD. Induction of neuropeptide Y gene expression in the dorsal medial hypothalamic nucleus in two models of the agouti obesity syndrome. Mol Endocrinol 1997;11:630–637.
131. Colmers WF, Wahlestedt C (eds). The Biology of Neuropeptide Y and Related Peptides. Humana Press, Tototwa NJ, 1993.
132. Michalkiewicz M, Michalkiewicz T., Kruelen DL, McDougall S. Increased vascular responses in neuropeptide Y transgenic rats. Am J Physiol Begul, Inegrat Comp Physiol 2001; in press.
133. Thiele TE, Marsh DJ, Ste. Marie L, Bernstein IL, Palmiter RD. Ethanol consumption and resistance are inversely related to neuropeptide Y levels. Nature 1998;396(6709):366–369.
134. Inui A, Okita M, Nakajima M, Momose K, Ueno N, Teranishi A, et al. Anxiety-like behavior in transgenic mice with brain expression of Neuropeptide Y. Proc Assoc Am Physicians 1998;110:171–182.
135. Erickson JC, Clegg KE, Palmiter RD. Sensitivity to leptin and susceptibility to seizures of mice lacking neuropeptide Y. Nature 1996;381:415–418.
136. Erickson JC, Ahima RS, Hollopeter G, Flier JS, Palmiter RD. Endocrine function of neuropeptide Y knockout mice. Regul Pept 1997;70:199–202.
137. Palmiter RD, Erickson JC, Hollopeter G, Baraban SC, Schwartz MW. Life without neuropeptide Y. Rec Prog Horm Res 1998;53:163–199.
138. Erickson JC, Hollopeter G, Palmiter RD. Attenuation of the obesity syndrome of ob/ob mice by the loss of neuropeptide Y. Science 1996;274:1704–1707.
139. Michel MC, Beck-Sickinger A, Cox H, Doods HN, Herzog H, Larhammar D, et al. XVI. International Union of Pharmacology recommendations for the nomenclature of neuropeptide Y, peptide YY, and pancreatic polypeptide receptors. Pharm Rev 1998;50:143–50.
140. Wolf G. Orexins: a newly discovered family of hypothalamic regulators of food intake. Nutr Rev 1998;56:172,173.
141. Broberger C, De Lecea L, Sutcliffe JG, Hokfelt T. Hypocretin/orexin- and melanin-concentrating hormone-expressing cells form distinct populations in the rodent lateral hypothalamus: relationship to the neuropeptide Y and agouti gene-related protein systems. J Comp Neurol 1998;402:460–474.
142. Pedrazzini T, Seydoux J, Kunstner P, Augert JF, Grouzmann E, Beermann F, Brunner HR. Cardiovascular response, feeding behavior and locomotor activity in mice lacking the NPY Y1 receptor. Nat Med 1998;4:722–726.
143. Kanatani A, Mashiko S, Murai N, Augimoto N, Ito J, Fukuroda T, et al. Key role of the Y1 receptor in NPY mediated feeding regulation: comparative studies of Y1 and Y5 receptor deficient mice. 5th International NPY Meeting, Grand Cayman, 1999, p. 31.
144. Marsh DJ, Hollopeter G, Kafer KE. Palmiter RD. Role of the Y5 neuropeptide Y receptor in feeding and obesity. Nat Med 1998;4:718–721.
145. Naveilhan P, Hassani H, Canals JM, Ekstrand AJ, Larefalk A, Chhajlani V, et al. Normal feeding behavior, body weight and leptin response require the neuropeptide Y Y2 receptor. Nat Med 1999;10:1188–1193.
146. Spiegelman BM, Flier JS. Adipogenesis and obesity; rounding out the big picture. Cell 1996;87:377–389.

147. Muzzin P, Eisensmith RC, Copeland KC, Woo SL. Correction of obesity and diabetes in genetially obese mice by leptin gene therapy. Proc Natl Acad Sci USA 1996;93:14,804–14,808.
148. Håkansson M-L, Meister B. Transcription factor STAT3 in leptin target neurons of the rat hypothalamus. Neuroendocrinology 1998;68:420–427.
149. Schwartz MW, Baskin DG, Bukowski TR, Kuijper JL, Foster D, Lasser G, et al. Specificity of leptin action on elevated blood glucose levels and hypothalamic neuropeptide Y gene expression in ob/ob mice. Diabetes 1996;45:531–535.
150. Stephens TW, Basinski M, Bristow PK, Bue-Valleskey JM, Burgett SG, Craft L, et al. The role of neuropeptide Y in the antiobesity action of the obese gene product. Nature 1995;377:530–532.
151. Mantzoros CS, Frederich RC, Qu D, Lowell BB, Maratos-Flier E, Flier JS. Severe leptin resistance in brown fat-deficient uncoupling protein promoter-driven diphtheria toxin A mice despite suppression of hypothalamic neuropeptide Y and circulating corticosterone concentrations. Diabetes 1998;47:230–238.
152. Ollmann MM, Wilson BD, Yang YK, Kerns JA, Chen Y, Gantz I, Barsh GS. Antagonism of central melanocortin receptors in vitro and in vivo by agouti-related protein. Science 1997;278:135–138.
153. Marsh DJ, Hollopeter G, Huszar D, Laufer R, Yagaloff KA, Fisher SL,Burn P, Palmiter RD. Response of melanocortin-4 receptor-deficient mice to anorectic and orexigenic peptides. Nat Genet 1999;21:119–122.
154. Leibowitz SF. Specificity of hypothalamic pepetides in the control of behavioral and physiological precesses. Ann NY Acad Sci 1994;739:12–35.
155. Wynick D, Small CJ, Bloom SR, Pachnis V. Targeted disruption of the murine galanin gene. Ann NY Acad Sci 1998;863:22–47.
156. Tecott LH, Sun LM, Akana SF, Strack AM, Lowenstein DH, Dallman MF,Julius D. Eating disorder and epilepsy in mice lacking 5-HT2c serotonin receptors. Nature 1995;374:542–546.
157. Nonogaki K, Strack AM, Dallman MF, Tecott LH. Leptin-independent hyperphagia and type 2 diabetes in mice with a mutated serotonin 5-HT2C receptor gene. Nat Med 1998;4:1152–1156.
158. Lowell BB, S-Susulic V, Hamann A, Lawitts JA, Himms-Hagen J, Boyer BB, et al. Development of obesity in transgenic mice after genetic ablation of brown adipose tissue. Nature 1993;366:740–742.
159. Hamann A, Flier JS, Lowell BB. Obesity after genetic ablation of brown adipose tissue. Z Ernahrungswiss 37 Suppl 1998;1:1–7.
160. Klaus S, Munzberg H, Truloff C, Heldmaier G. Physiology of transgenic mice with brown fat ablation: obesity is due to lowered body temperature. Am J Physiol 1998;274(2 Pt 2):R287–R293.
161. Hamann A, Benecke H, Le Marchand-Brustel Y, Susulic VS, Lowell BB, Flier JS. Characterization of insulin resistance and NIDDM in transgenic mice with reduced brown fat. Diabetes 1995;44:1266–1273.
162. Hamann A, Flier JS, Lowell BB. Decreased brown fat markedly enhances susceptibility to diet-induced obesity, diabetes, and hyperlipidemia. Endocrinology 1996;137:21–29.
163. Hamann A, Busing B, Kausch C, Ertl J, Preibisch G, Greten H, Matthael S. Chronic leptin treatment does not prevent the development of obesity in transgenic mice with brown fat deficiency. Diabetologia 1997;40(7):810–815.
164. Enerback S, Jacobsson A, Simpson EM, Guerra C, Yamashita H, Harper ME. Kozak LP. Mice lacking mitochondrial uncoupling protein are cold-sensitive but not obese. Nature 1997;387(6628):90–94.
165. Kopecky J, Clarke G, Enerback S, Spiegelman B, Kozak LP. Expression of the mitochondrial uncoupling protein gene from the aP2 gene promoter prevents genetic obesity. J Clin Invest 1995;96:2914–2923.
166. Kopecky J, Rossmeisl M, Hodny Z, Syrovy I, Horakova M, Kolarova P. Reduction of dietary obesity in aP2-Ucp transgenic mice: mechanism and adipose tissue morphology. Am J Physiol 1996;270(5 Pt 1):E776–E786.
167. Stefl B, Janovska A, Hodny Z, Rossmeisl M, Horakova M, Syrovy I, et al. Brown fat is essential for cold-induced thermogenesis but not for resistance in aP2-Ucp mice. Am J Physiol 1998;274 (3 Pt 1):E527–E533.
168. Baumruk F, Flachs P, Horakova M, Floryk D, Kopecky J. Transgenic UCP1 in white adipocytes modulates mitochondrial membrane potential. FEBS Lett 1999;444(2-3):206–210.
169. Lowell BB, Flier JS. Brown adipose tissue, beta 3-adrenergic receptors, and obesity. Annu Rev Med 1997;48:307–316.
170. Lowell BB. Using gene knockout and transgenic techniques to study the physiology and pharmacology of beta3-adrenergic receptors. Endocrine J 1998;45:S9–S13.

171. Susulic VS, Frederich RC, Lawitts J, Tozzo E, Kahn BB, Harper ME, et al. Targeted disruption of the beta 3-adrenergic receptor gene. J Biol Chem 1995;270(49):29,483–29,492.
172. Soloveva V, Graves RA, Rasenick MM, Spiegelman BM, Ross SR. Transgenic mice overexpressing the beta 1-adrenergic receptor in adipose tissue are resistant to obesity. Mol Endocrinol 1997;11(1):27–38.
173. Ito M, Grujic D, Abel ED, Vidal-Puig A, Susulic VS, Lawitts J, et al. Mice expressing human but not murine beta3-adrenergic receptors under the control of human gene regulatory elements. Diabetes 1998;47(9):1464–1471.
174. Shimada M, Ishibashi S, Yamamoto K, Kawamura M, Watanabe Y, Gotoda T, et al. Overexpression of human lipoprotein lipase increases hormone-sensitive lipase activity in adipose tissue of mice. Biochem Biophys Res Commun 1995;211(3):761–766.
175. Weinstock PH, Levak-Frank S, Hudgins LC, Radner H, Friedman JM, Zechner R. et al. Lipoprotein lipase controls fatty acid entry into adipose tissue, but fat mass is preserved by endogenous synthesis in mice deficient in adipose tissue lipoprotein lipase. Proc Natl Acad Sci USA 1997;94(19):10,261–10,266.
176. Jensen DR, Schlaepfer IR, Morin CL, Pennington DS, Marcell T, Ammon SM, Gutierrez-Hartmann A. Eckel RH. Prevention of diet-induced obesity in transgenic mice overexpressing skeletal muscle lipoprotein lipase. Am J Physiol 1997;273:R683–R689.
177. Katz EB, Stenbit AE, Hatton K, DePinho R. Charron MJ. Cardiac and adipose tissue abnormalities but not diabetes in mice deficient in GLUT4. Nature 1995;377:151–155.
178. Shepherd PR, Gnudi L, Tozzo E, Yang H, Leach F. Kahn BB. Adipose cell hyperplasia and enhanced glucose disposal in transgenic mice overexpressing GLUT4 selectively in adipose tissue. J Biol Chem 1993;268:22,243–22,246.
179. Gnudi L, Shepherd PR, Kahn BB. Over-expression of GLUT4 selectively in adipose tissue in transgenic mice: implications for nutrient partitioning. Proc Nutr Soc 1996;55:191–199.

17 Insulin and Insulin-Like Growth Factors

Targeted Deletion of the Ligands and Receptors

Carolyn A. Bondy, MD *and Domenico Accili,* MD

CONTENTS

INTRODUCTION

Insulin and insulin-like growth factors 1 and 2 (Igf1 and Igf2, respectively) are genetically and functionally related peptides with overlapping anabolic and growth-promoting effects *(1,2)*. The actions of insulin are mediated by the insulin receptor, a membrane-bound tyrosine kinase which is activated by ligand binding *(3)*. The actions of both Igf1 and 2 are mediated for the most part by the homologous Igf1 receptor, which engages many of the same intracellular signaling pathways as the insulin receptor *(4,5)*. An Igf2/mannose-6-phosphate receptor, binds Igf2 and promotes its degradation in the lyzosomal pathway *(6)*. During the past decade, each of these ligands and receptors have been subjected to targeted gene deletions, providing novel insights into growth regulation and ligand-receptor interactions in murine embryonic and postnatal development. This chapter focuses on the developmental and physiological consequences of deletions of Igf1, Igf2, insulin, and the Igf1 and insulin receptors, alone and in combination.

From: *Contemporary Endocrinology: Transgenics in Endocrinology*
Edited by: M. Matzuk, C. W. Brown, and T. R. Kumar © Humana Press Inc., Totowa, NJ

IGF1

There have been two independent Igf1 targeted deletions *(7–9)*. Igf1-null mice from both these genotypes demonstrate a birthweight approx 60–65% of normal, independent of the strain in which they are bred *(7–10)*. This differential between wild-type and Igf1-null litter mates remains constant until around postnatal d 20 (P 20), and after this time the Igf1-null mice virtually stop growing *(10)*. The litter mates continue to grow, so that ultimately, Igf1 null mice are ~30% of wild-type in body weight *(7–10)*. These findings demonstrate that Igf1 has two major phases or modes of growth promotion. Growth hormone (GH)-independent Igf1 growth augmentation occurs during fetal and early postnatal development, and is responsible for about one-third of somatic growth prior to P 20. This period is believed to be GH-independent, because mice with deletions of GH or the GH receptor demonstrate little or no dwarfism prior to P 15–20. Thus, GH-induced Igf1-mediated growth does not begin until this point *(11,12)*. Further somatic growth after P 20 depends on GH-induced Igf1 production, which doubles the mouse size between P 20 and 40. The fact that there is virtually no GH-induced growth in the Igf1 –/– mice supports the view that Igf1 mediates GH's major effects on somatic growth (*see* Chapter 16, which examines GH-transgenic mice).

The majority of Igf1-null mice die perinatally *(7–10)* with mortality rates of ~95% for both deletions in various strains. The cause of the early postnatal death of Igf1-null mice is not entirely clear. It may be caused by respiratory failure, as there is profound hypoplasia of the respiratory muscles *(7–9)* and Igf1-null lungs are disproportionately small *(10)*. In support of this theory, prominent right heart hypertrophy is seen in Igf1-null mice as early as 10 d after birth (ref. *10* and Bondy et al., unpublished data), suggesting pulmonary hypertension. There are no other obvious or gross abnormalities in organ development in the Igf1-null mice.

Igf1's Role in Longitudinal Bone Growth

Igf1 has been believed to play a major role as a mediator of GH's effects in postnatal statural or long-bone growth. To determine whether this is true, we have studied the epiphysial growth plates in Igf1-null and wild-type litter mate mice during the GH-dependent postnatal growth spurt at P 20 *(13)*. Although Igf1 was believed to promote longitudinal bone growth by stimulating growth-plate chondrocyte proliferation, we found that the number of chondrocytes was equal in wild-type and Igf1-null mice, and that their proliferation rates, as determined by BRDU incorporation, were also equal *(13)*. The size of growth-plate hypertrophic chondrocytes was significantly reduced in the Igf1-null mice. This reduction in chondrocyte size was associated with reduced glucose-transporter expression (GTs 1 & 4), reduced glycogen synthase kinase 3β phosphorylation (GSK3β), and decreased glycogen stores *(13)*. In addition, ribosomal mRNA levels were dramatically reduced in the Igf1-null chondrocytes, suggesting a global reduction in protein synthesis. In fact, the reduction in Igf1-null terminal hypertrophic chondrocyte linear dimension correlates with the approx 30% reduction in linear growth rate of these mice. The normal chondrocyte proliferation in the Igf1-null growth-plate may be due to Igf2, which is synthesized by proliferative chondrocytes *(14,15)*. This observation represents a major revision of our view of Igf1's role in longitudinal bone statural growth.

Igf1 in Brain Development

The brain of the Igf1-null mouse is actually relatively spared from the generalized dwarfism and is reduced in adult size by approx 30%, while body wt is reduced by almost 70% *(16,17)*. Cell numbers and neuroanatomical structures are preserved throughout most of the brain, with the exception of the olfactory bulb *(17)* and dentate gyrus of the hippocampal formation *(16)*. For the most part, the reduction in brain size appears to be a result of reduced neuropil—neuronal fibers and processes—with cells more densely packed in the Igf1-null cerebral and cerebellar cortices *(16,17)*. Myelination is proportionate to the reduced brain size and fiber number/diameter in both the central *(17)* and peripheral *(18)* nervous systems of the Igf1-null mouse. Furthermore, there are no functional signs of myelinopathy in the Igf1-null mouse or Igf1-null human *(19)*.

There are, however, prominent biochemical/metabolic abnormalities in the Igf1-null brain during the course of postnatal development, when Igf1 expression is usually most intense *(20)*. Glucose utilization is depressed by 40–50% in many brain regions, most notably those in which Igf1 is normally highly expressed. Interestingly, the serine/threonine kinase Akt1, which is selectively phosphorylated in Igf1-expressing cells in wild-type brains, is hypo-phosphorylated in the Igf1-null brain *(20)*. Akt hypo-phosphorylation is correlated with decreased membrane localization of GT4, which may explain the reduced glucose uptake in the Igf1-null brain. GT4 mRNA levels are also significantly reduced in the Igf1-null brain *(20)*. Akt hypo-phosphorylation is also associated with decreased phosphorylation of GSK3β and decreased glycogen stores, as seen in the Igf1-null growth plate. These findings suggest that Igf1 may act through similar signaling mechanisms in diverse tissues to achieve a true insulin-like anabolic effect on cellular somatic growth.

Igf1 in Reproduction

Igf1-null mice that survive the first postnatal week appear quite healthy and active, but both sexes are infertile *(21)*. Males have testosterone levels 18% of normal, and spermatogenesis is reduced to a similar extent. Mating behavior is absent, and distal portions of the reproductive tract (i.e., the vas deferens, seminal vesicles, and prostate) are hypoplastic, consistent with the reduced testosterone levels *(21,22)*. The molecular cause of the reduced testosterone biosynthesis in the male Igf1-null mouse remains unclear, although Leydig-cell size and numbers are decreased *(21)*.

The Igf1-null females are sexually immature, with estradiol levels reduced *(21)*. Igf1-null and wild-type ovaries have the same number of follicles, but in the Igf1-null, none of the follicles develop past the early antral stage, and the mice do not ovulate, even with pharmacological gonadotropin treatment *(21)*. We had found a one-to-one correlation between follicles expressing Igf1 and the follicle-stimulating hormone receptor (FSHR) in normal murine ovaries, and have shown that FSHR and aromatase expression were significantly reduced in the Igf1-null ovary and restored by Igf1 treatment *(23)*. These findings suggest that Igf1 normally enhances FSHR expression, and may explain why the Igf1-null follicles fail to develop past the gonadotropin-independent stage, and also fail to respond to exogenous gonadotropins.

The Igf1-null uterus is infantile in proportion *(21)*, but it is unclear whether this is a result of the essentially prepubertal status of the Igf1-null mice with low circulating estradiol levels, or the lack of Igf1 *per se*. Estrogen stimulates Igf1 expression in murine and primate uteri *(24,25)*, and the local pattern of Igf1 and Igf1-receptor expression is correlated with uterine cell proliferation in response to estrogen *(25)*. To elucidate Igf1's role in estrogen-induced uterine growth, we investigated the ability of exogenous estrogen to stimulate proliferation in the uteri of Igf1-null and wild-type litter mate mice *(26)*. Unexpectedly, given Igf1's putative role as a "G1 progression factor" *(27)*, DNA synthesis was equal in Igf1-null and WT uteri 20 h after a single estradiol dose. Cumulative labeling experiments showed that S-phase influx, duration, and efflux were not significantly different in Igf1-null and wild-type uteri, indicating that Igf1-null cells progress through G1- and S-phase normally. There was, however, a delay between S- and M-phase, since the appearance of mitotic figures was profoundly retarded in Igf1-null uteri. The mitotic index was four- to sevenfold lower in Igf1-null uterus (depending on cell type) 20 h after estradiol treatment. Normally, the mitotic response peaks 20–24 h after a dose of estradiol and returns to baseline by 48 h, when extensive apoptosis occurs in response to estradiol-withdrawal. This was the pattern observed in wild-type mice. In Igf1-null mice, however, the mitotic index was only modestly increased at 20 h, but remained significantly elevated 48 h after estradiol, suggesting that some cells were still progressing from G2 to M-phase long after the estradiol stimulus. Supporting the finding of a G2-arrest or delay, mean DNA concentration was significantly increased in Igf1-null cells 20 h after estradiol treatment *(26)*.

Extracellular signaling proteins such as hormones or growth factors have not been previously implicated in regulation of the G2 phase of the cell cycle. However, yeasts demonstrate a G2 checkpoint based on cell size and nutritional status, which may be relevant to the present observations. Within 20 h after estradiol treatment, wild-type epithelial-cell size increased by more than 200%, while Igf1-null epithelial cells increased only ~50%. Interestingly, wildtype epithelial size declined between 20 and 48 h after estradiol treatment, while Igf1-null epithelial cells continued to grow at least through the 48-h time-point. Thus, retardation in mitoses was correlated with retarded cellular somatic growth in the Igf1-null uterus, suggesting that a G2 checkpoint based on cell size may operate in this mammalian system. Interestingly, the massive apoptosis seen in wild-type uteri 48 h after estradiol withdrawal was absent in Igf1-null uteri, suggesting that completion of the full estradiol-induced mitotic cycle is required for cell death following estradiol withdrawal.

Igf1 Conditional Mutant Mice

LeRoith and colleagues have recently generated mice with variable reductions in Igf1 gene dosage using the Cre-lox system *(28)*. The variability in the extent and timing of the Igf1 deletion results in some mutant mice with greater survival rates and milder growth retardation, which is predominantly postnatal compared to the unconditional knockouts. These investigators have also developed mice with a liver-specific deletion to study hepatic Igf1 function *(29,30)*. These studies have shown that despite apparent abolition of hepatic Igf1 production, and despite a profound reduction in circulating Igf1 levels, these mice grow normally. These findings may be explained by a reduction in Igf-binding protein levels, maintaining similar levels of free Igf1 levels in the circulation.

IGF1 RECEPTOR

Targeted deletion of the Igf1 receptor (*Igf1r*) results in profound dwarfism and death at birth *(8)*. The Igf1r-null mice weigh only about 45% of wild-type at birth, and demonstrate severe hypoplasia of the muscles and skin. The newborn's lungs appear histologically normal and express surfactant, but are atelectatic and appear to have never expanded with air *(8)*. Thus, death may be caused by a failure of the respiratory muscles. As with the Igf2-knockout, the Igf1r-null mice demonstrate growth retardation beginning at E11, although placental weight is not affected.

IGF2 AND IGF2 RECEPTOR

Igf2 was the first member of the insulin/Igf family to be "knocked out," now almost a decade ago *(31)*. This experiment revealed that murine Igf2 expression is subject to genomic imprinting, and that gene is expressed from the paternal allele. As a result, deletion of the paternal allele results in dwarfism, while deletion of the maternal allele has no apparent effect *(31,32)*. Igf2 p–/+ and –/– are indistinguishable, and dwarfism is apparent shortly after placentation *(31,32)*. These mice are approx 60% smaller than wild-type litter mates from E 11, and this differential remains constant throughout life. Although it is unknown how the Igf2 deletion impairs fetal growth, defective placental growth may cause this decreased growth. Igf2 is highly expressed by the trophoblast during the time of implantation, and is also abundant in the labyrinthine zone of the placenta *(33)*. The placenta is also significantly reduced in size in the Igf2-null mouse, so it is possible that Igf2 expression augments trophoblast invasion or placental proliferation. Thus, the reduced placental size may cause the dwarfism of the Igf2-null mouse. No apparent abnormalities in the Igf2-null placenta have been reported, except that glycogen stores are reduced *(34)*. There is no increased mortality associated with paternal or homozygous Igf2 deletion, and the mice are healthy and fertile.

In mammals, Igf2 also binds to a mannose-6-phosphate receptor involved in the trafficking of lyzosomal enzymes *(6)*. This Igf2 receptor is not homologous to the insulin/Igf1 receptors, and binds only Igf2. This mannose-6-phosphate/Igf2 receptor also demonstrates genomic imprinting in the mouse that is reciprocal to that of Igf2 *(6,35)*, (i.e., it is expressed from the maternal allele and silenced in the paternal allele). Targeted deletion of this receptor results in fetal overgrowth, edema, cardiac abnormalities, and prenatal death *(35,36)*. These defects are apparently caused by Igf2 excess, since double-Igf2/Igf2 receptor mutants are viable dwarfs *(37)*.

Combined Igf1/Igf2 mutants—or Igf2/Igf1r mutants—are lethal at birth, resulting in mice born at only 30% of normal weight *(8)*. These observations led to the hypothesis that Igf2 interacts with another growth-promoting receptor during embryonic development in addition to Igf1r *(8)*. Further genetic experiments have suggested that this "X receptor" is actually the insulin receptor.

INSULIN AND INSULIN RECEPTOR

There are two insulin genes in rodents. In adult mice, insulin is expressed from the insulin 2 gene. However, ablation of either gene in mice is without consequences. In contrast, when both genes are knocked out, the mice develop diabetic ketoacidosis and die within days of birth *(38)*. Inactivation of the two insulin genes impairs embryonic growth slightly, with a 10–20% decrease in body wt *(38)*.

Mice nullizygous at the insulin receptor (IR) locus have normal features at birth, and their intrauterine growth and development appear to be normal *(39,40)*. However, upon careful measurements of embryonic weights, a small reduction (~10%)

in size may be detected. Such a reduction is statistically significant when large numbers of embryos are compared *(41)*, and provides presumptive evidence for a role of IR in embryonic growth.

Lack of IR results in severe metabolic derangement. Within a few days of birth, mutant mice die of diabetic ketoacidosis. This demonstrates that the IR is required to mediate the metabolic actions of insulin in postnatal life. Whether insulin exerts other, nonmetabolic effects, and whether these effects are also mediated by IR, cannot be addressed by this model. Likewise, it is intriguing that fetal metabolism is unaffected by the lack of IR. The early postnatal death caused by inactivation of the IR gene suggests that the functional maturation of a fuel-sensing mechanism in rodents occurs in the perinatal period, rather than during the late phase of gestation, as in humans *(42)*. These differences in developmental timing may explain some of the phenotypic differences between humans and rodents with IR gene mutations. Thus, maternal metabolism can presumably prevent the onset of metabolic abnormalities in utero. The metabolic derangement caused by lack of IR is similar to that observed in humans with acute-onset insulin-dependent diabetes, with the notable exception that in humans with acute-onset (type 1) diabetes, plasma insulin levels are undetectable, whereas in mice lacking IR, plasma insulin levels are elevated two- to 10-fold.

The notion that mice can develop and be born without functional IR had been predicted—based on the description of four human patients with leprechaunism—a genetic syndrome of extreme insulin resistance, in which children were born with homozygous non-sense mutations or deletions of the IR gene *(43–46)*. Nevertheless, there are striking differences between IR-deficient mice and humans with similar mutations. In humans, fasting hypoglycemia and retarded growth are the hallmarks of severe insulin resistance *(47,48)* (Table 1). The pathophysiologic basis of these differences remains speculative. It is especially difficult to reconcile the metabolic findings as they relate to glycemic control, since fasting hypoglycemia in children with leprechaunism is essentially a paradoxical finding. A simple explanation is that newborn mice, unlike newborn humans, are constantly nursed by their mothers. Thus, there exists a constant flow of nutrients that prevent hypoglycemia, while exacerbating hyperglycemia and ketogenesis. In support of this view, it should be noted that children with leprechaunism experience transient hyperglycemia and ketosis following meals *(43,49,50)*. An alternative explanation for the absence of hypoglycemia in mice is that insulin levels are much more elevated in humans compared to mice. In fact, insulin levels in leprechaunism are known to be as much as 1,000-fold higher than in normal controls. In IR-deficient mice, insulin levels are only two- to 10-fold higher than control mice. Thus, it is possible that, in humans, high insulin levels can cause hypoglycemia by binding to Igf1r. We have shown that Igf1r can indeed mediate hypoglycemic effects in IR-deficient mice *(51)*. An interesting point to support this hypothesis is that children with Rabson-Mendenhall syndrome, a milder variant of leprechaunism that is also associated with hypoglycemia *(52)*, generally experience an improvement of hypoglycemia in their infancy, in association with a precipitous drop in prevailing insulin concentrations.

ROLE OF IR IN EMBRYONIC DEVELOPMENT

IR are expressed in the pre-implantation mouse embryo, and insulin has been shown to stimulate glucose utilization by isolated blastocysts *(53)*. Furthermore, the lack of IR

Table 1

Phenotypic Differences Between Mice Lacking Insulin Receptors and Children Affected with Leprechaunism

	Leprechaunism	*Knockout mice*
Appearance	Dysmorphic	Normal
Growth	Poor	Near-normal
Plasma glucose	Low	High
Ketoacidosis	Transient	Constant
Hepatic steatosis	Moderate	Massive
Virilization	Present	???
Adipose tissue	Hypotrophic	Hypotrophic

in humans is associated with severe growth retardation *(43–47)*. The role of fetal IR was rather unclear prior to gene-targeting studies *(41)*. One widely held view was that IR regulated growth by regulating fuel metabolism. In addition to the growth retardation observed in cases of extreme insulin resistance, macrosomia is a well-recognized complication of fetal hyperinsulinemia, a metabolic consequence of the diabetic pregnancy *(54,55)*. An alternative explanation to the role of IR in development was that they mediate growth-promoting actions, either directly or through hybrid IR/Igf1r (Fig. 1). Targeted mutagenesis indicated that the growth and development of mouse embryos are scarcely affected by the lack of IR *(39,40)*. Taken at face value, the phenotype of IR-nullizygous mice argues against a role for IR during gestation. However, independent genetic evidence has suggested that Igf1r could not account for all the growth-promoting actions of Igf2, and has led to the hypothesis that an additional Igf2 receptor exists, distinct from the Igf2/mannose-6-phosphate receptor *(8)*, which mediates some of the growth-promoting actions of Igf2 in fetal life. This hypothesis was tested in cross-breeding experiments of mice with IR and Igf1r-null alleles *(41)* (Fig. 2).

MICE WITH COMBINED MUTATIONS OF IR AND IGF1R DEFINE A ROLE FOR IR AS IGF2-ACTIVATED EMBRYONIC GROWTH PROMOTERS

Because of the complexity of the experimental approach, it is worthwhile to summarize the conclusions of these studies before targeting specific aspects of the data *(41)* (Fig. 2). Mice lacking either Igf1 or Igf2 are small (~60% of the normal size) *(8,31,37)*. Thus, both growth factors are required for embryonic development. Remarkably, combined absence of both growth factors is compatible with embryonic growth, although the latter is greatly impaired (30% of normal size). The fact that double mutants lacking Igf1 and Igf2 are more severely affected than single mutants suggests that the two ligands act independently to stimulate growth *(8)*. Insulin, on the other hand, does not appear to play a major role during mouse embryo development, as demonstrated by the fact that mice lacking insulin are only 10–20% smaller than normal litter mates *(38)*.

Which receptor(s) mediate(s) the growth-promoting actions of Igf1 and Igf2 during embryogenesis? Studies from the Efstratiadis laboratory indicate that Igf1r is the main—and possibly the only—mediator of Igf1 actions. In contrast, Igf2 acts through three

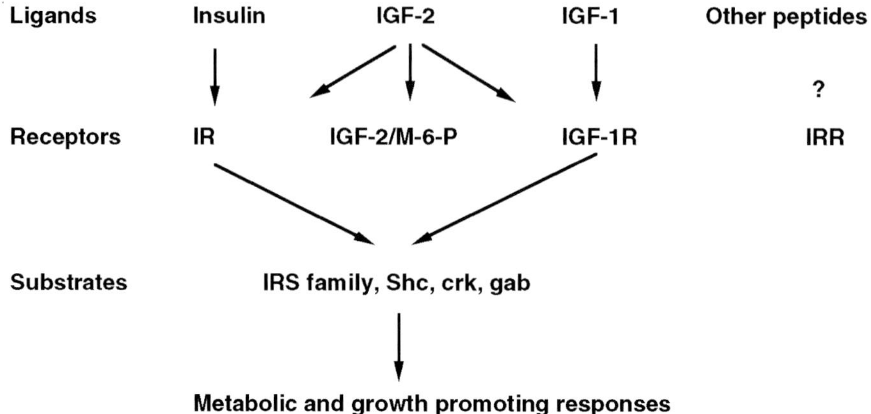

Fig. 1. Schematic of potential receptor-ligand interactions in insulin/IGF signaling pathways. Note that while insulin and IGF1 each demonstrate distinct selectivity for their cognate receptors, IGF-2 interacts with insulin and IGF-1 receptors, as well as the IGF2/M-6-P receptor.

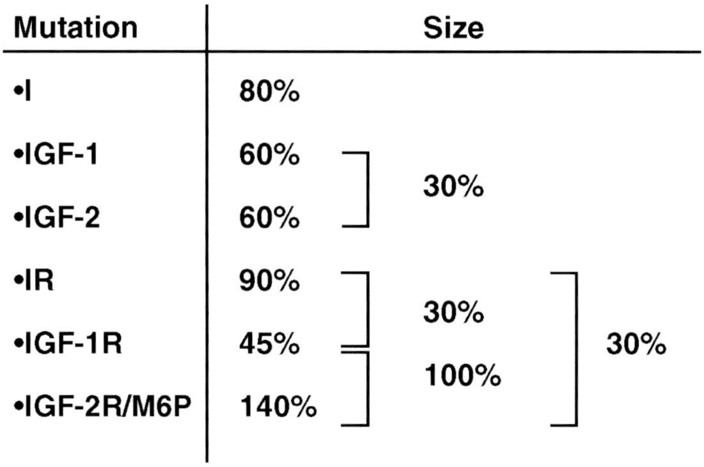

Mutation	Size		
•I	80%		
•IGF-1	60%	30%	
•IGF-2	60%		
•IR	90%	30%	
•IGF-1R	45%		30%
•IGF-2R/M6P	140%	100%	

Fig. 2. Effects of single and combined deletions of IGF system components on mouse size.

receptors: IR, Igf1r, and Igf2r. IR and Igf1r mediate the growth-promoting actions of Igf2, whereas Igf2r clear Igf2 from the circulation and thus limit its ability to act in a classic endocrine fashion *(37)* (Fig. 1). Thus, both IR and Igf1r are required to support mouse growth (Fig. 2).

Here, we present a systematic examination of the evidence supporting these conclusions. While knockouts of the insulin genes and the IR gene are associated with modest (10–20%) growth retardation and lethal metabolic abnormalities *(1–3)*, knock-outs of the Igf1 gene and its receptor are associated with severe intrauterine growth retardation (~60% of normal size in Igf1 knockouts, ~45% of normal size in Igf1r knockouts) without metabolic abnormalities *(7)* (Fig. 2). These data can be construed to suggest that Igf1r mediate growth, and IR mediate metabolic responses. Genetic crosses between IR-deficient and Igf1r-deficient mice suggest otherwise. Mice lacking both IR and Igf1r are more severely growth retarded (~30% of normal size) than mice lacking either receptor alone *(41)*. Thus, the lack of IR and Igf1r closely resembles the phenotype caused by the absence of Igf1 and Igf2, and provides genetic evidence that the two receptors mediate all the growth-promoting effects of the two ligands.

IR and Igf1r are not functionally equivalent in mediating embryonic growth and development. First, embryonic growth curves of single and double knockout mice indicate that Igf1r support embryonic growth starting in midgestation, whereas IR plays a role in late gestation *(41)*. Second, the growth-promoting actions of IR are mediated in response to Igf2, and not to insulin. This conclusion is supported by the following data: ablation of Igf1 does not alter the phenotype of Igf1r-deficient mice, suggesting that Igf1 interacts exclusively with Igf1r *(7)*. In contrast, ablation of Igf2 in Igf1r-deficient mice results in more severe growth retardation (~30% of normal), suggesting that Igf2 utilizes both Igf1r and an additional receptor to stimulate growth. The phenotype of IR/Igf1r double knockouts indicates that IR is the additional receptor used by Igf2, since the size of IR/Igf1r knockouts is identical to that of Igf2/Igf1r knockouts.

Further evidence for a role of IR as a mediator of Igf2 actions draws from experiments in which both Igf1r and Igf2r have been ablated. Lack of both Igf1r and Igf2r is associated with normal embryonic growth. The most likely explanation of this finding is that Igf1r mediates the growth-promoting actions of Igf2, whereas the Igf2r serves to clear Igf2 from the circulation. In the absence of Igf2r, Igf2 is not cleared from the circulation, its levels rise and overstimulate Igf1r, resulting in a lethal phenotype. After ablation of Igf1r and Igf2r, excess Igf2 cannot act through Igf1r, but can act through IR. The interaction with IR is not sufficiently "potent" to cause lethality, but is sufficient to rescue the phenotype *(37)*.

Analysis of skin, muscle, and bone development in mice lacking both IR and Igf1r suggests that both receptors share common cellular targets, since the development of these organs is synergistically affected by the two mutations. Independent evidence from cell-culture experiments also suggests that IR and Igf1r may stimulate Igf2-dependent growth in a concerted manner. Recently, we have been able to evaluate the relative roles of IR and Igf1r to stimulate hepatocyte growth in response to Igf2. Using hepatocytes from IR-deficient and control mice, we have shown that Igf2 stimulates cell growth through two different pathways: an IR-dependent pathway and an Igf1r-dependent pathway. Ablation of IR results in a blunted response to Igf2 that is similar to the response to Igf1. This observation is consistent with a model in which Igf2 can activate both IR and Igf1r, and the resulting signals act synergistically to promote growth *(56)*.

It is unlikely that insulin is the ligand promoting IR-mediated growth, since ablation of the insulin genes has a modest effect on embryonic growth. The growth retardation observed in insulin-deficient mice is similar to that observed in IR-deficient mice *(40)*. Indirect evidence against insulin's role in promoting fetal growth can also be derived from patients with leprechaunism caused by IR mutations. In these patients, growth is severely stunted, despite the substantial increases in circulating insulin concentrations *(45,46,57)*. Based on available evidence, it is likely that the effect of insulin on embryonic growth is exerted at the level of adipose tissue. This point is discussed in greater detail in the section on the differences between humans and mice with IR mutations.

Evidence Against a Role of Hybrid Insulin/Igf1 Receptors in Embryonic Growth

Another outstanding issue is the role of hybrid IR/Igf1r in mediating growth. The question is whether, under physiologic conditions, the growth-promoting actions of IR are mediated by holodimeric IR (composed of two α- and two β-subunits) or by heterodimeric receptors composed by an IR α/β monomer and an Igf1r-α/β monomer. Evidence that hybrid receptors do not play a significant role in growth-promoting interactions during murine embryogenesis is derived from studies of transgenic knockout mice in which kinase-inactive IR have been expressed in a null-IR background (D. Lauro and D. Accili, unpublished observations). The rationale for this experiment is that, if kinase-inactive IR engage in dimer formation with Igf1r, they should yield inactive hybrid receptors and produce a dominant negative inhibition of Igf2 action during development, resulting in dwarf mice. The results of this experiment indicate that hybrid receptors composed of an IR monomer and an Igf1r monomer are indeed present in embryos of transgenic knockout mice. Nevertheless, the growth of these mice is not impaired compared with IR-deficient mice. Thus, it is unlikely that hybrid receptors play an important role in promoting embryo growth. This may be a result of the fact that only a small fraction of receptors (~10%) is found in hybrid form.

Why are Humans—but not Mice—with IR Mutations Growth-Retarded?

As we stated earlier, humans with mutations of the IR gene are profoundly growth-retarded, while mice are hardly affected. A potential explanation of this finding is that humans and mice do not follow the same developmental timing. For example, if we compare the development of eyelids in mice and humans, it is evident that mice are born at a gestational stage corresponding to about 26 wk in human fetuses *(58)*. Likewise, if we examine body composition in newborn mice, we will find that lipid content is significantly lower in mice compared to humans (2 vs 16%, respectively) *(59)*. There are no data on the time course of growth retardation in humans with leprechaunism, but there are some data suggesting that, in patients with pancreatic agenesis, intra-uterine growth retardation is not apparent at gestational ages of 18–20 wk *(60,61)*. On the other hand, there is an abundant literature on excessive fetal growth caused by maternal hyperglycemia associated with secondary fetal hyperinsulinemia, and by primary fetal hyperinsulinemia, such as in Beckwith-Wiedemann syndrome and in persistent hyperinsulinemic hypoglycemia of infancy (PHHI). The growth patterns of fetuses of diabetic mothers suggest that the effects of fetal hyperinsulinemia on growth occur in late gestation, and that the main target tissue of these effects is adipose tissue, which is exquisitely sensitive to insulin *(54)*. One can extrapolate from these observations that mouse development is not advanced enough to show the typical phenotype of insulin-dependent growth retardation (or acceleration). A further prediction would be that formation of adipose tissue is impaired in IR-deficient mice. Indeed, a sensitive morphometric analysis of white adipose tissue development in IR-deficient mice supports this prediction. The size of dermal adipose cells is reduced by ~90%, with little effect on cell number, and no discernible effect on adipocyte differentiation *(62)*. These data are consistent with a model in which fetal IR are required for trophism, but not for differentiation of fat cells. In conclusion, IR appear to stimulate fat-cell growth in response to insulin binding, and overall embryonic growth in response to Igf2 binding.

Finally, we have been able to show that, in mice, insulin resistance resulting from targeted ablation of insulin receptors in muscle and fat using a tissue-specific knockout strategy causes only modest glucose intolerance *(57)*. These mice showed evidence of impaired insulin action in these tissues, and developed all the prodromal features of type 2 diabetes, including increased free fatty-acid concentrations, hyperinsulinemia with blunted insulin response to glucose challenge, and impaired glucose tolerance. Despite the compounded effect of peripheral insulin resistance and a mild impairment of beta-cell function, transgenic knockout mice did not become diabetic. These findings suggest that, in mice, the ability of the liver to compensate for the impairment of insulin action in muscle and fat has a protective effect against the development of diabetes. The development of tissue-specific models of insulin resistance has greatly impacted upon our understanding of the pathophysiology of type 2 diabetes, and underscores the importance of techniques of conditional manipulation of gene expression in mice.

SUMMARY

It is clear that the advent of targeted mutagenesis in embryo-derived stem cells has ushered in a new era of investigations in modern biology. Using targeted gene ablations, a number of laboratories have explored essential aspects of our understanding of the function of insulin and insulin-like growth factors. Studies in the Igf1-null mouse have provided the first convincing evidence that Igf1 plays a role in fetal growth, and have proven that Igf1 is indeed required to mediate GH's postnatal growth-promoting action. Furthermore, detailed analyses of brain and growth plate development in the Igf1-null mouse have challenged the prevailing views that Igf1 is a primary myelination factor in brain development and is responsible for chondrocyte proliferation in longitudinal bone growth. The phenotype of the Igf1-null brain suggests that Igf1 has primary effects on neuronal metabolism growth and survival, and that myelination is affected only secondarily in relation to reduced axonal number and diameter. In the Igf1-null growth plate, chondrocyte proliferation is normal, but chondrocyte hypertrophy is significantly impaired.

Interestingly, in these two very different situations, Igf1 appears to act through similar molecular mechanisms involving activation of Akt1 and downstream targets, including GLUT4 and GSK3β. The neuron and chondrocyte are both engaged in extremely rapid cellular growth, with the neuron's hypertrophy manifested in exuberant arborization, while the chondrocyte is expanding its soma and synthesizing abundant extracellular matrix. Thus in both cases, Igf1's anabolic effects may serve an essential role in support of extraordinary biosynthetic activity. Furthermore, data from the Igf1-null mouse model supports the hypothesis that Igf1 is a critical mediator of estrogen's proliferative actions in the uterus, although not as a G1 progression factor in cell-cycle control, as previously believed. In fact, entry into S-phase and DNA synthesis have been normal in every tissue examined in the Igf1-null mouse (our unpublished data). The view of Igf1 as a G1 progression factor is based on studies in immortalized cell lines subjected to artificial growth arrest, and may not reflect its role for normal cells in vivo.

Although only a single human has been described with homozygous Igf1 gene deletions *(19)*, his phenotype appears remarkably similar to that of the Igf1-null mouse. He was born small for gestational age, and demonstrated profound growth retardation during childhood. Although myelination appears normal, he is mentally retarded and hearing-impaired *(19)*. The one difference noted so far is that he appears to be virilizing relatively normally, suggesting that testicular function is not impaired to the extent shown in the Igf1-null mouse.

The lessons from the Igf2- and Igf2-receptor knockouts have been extremely enlightening in terms of revealing the epigenetic phenomena of genomic imprinting. From the

Igf2 deletion, it appears that Igf2 has an obligate role in fetal growth confined to the peri-implantation/placentation phase of development. No defects in tissue differentiation or function, or later phases of growth, are apparent in the Igf2-null mice. The consequences of the Igf2-receptor knockout have provided convincing evidence that this receptor serves primarily to metabolize and clear excess Igf2 from the developing mouse. The sophisticated genetic experiments of Efstratiadis and colleagues (reviewed in ref. *56*) have shown that Igf2 acts primarily through the Igf1 receptor, but also to a significant degree through the insulin receptor in stimulating fetal growth.

The murine insulin and insulin-receptor knockouts suggest that this ligand/receptor dyad mainly serves a postnatal metabolic role in the mouse in contrast to the situation in the human. There are still many unresolved issues, mostly because of the technical limitations of the first generation of gene-targeting experiments. For example, in the case of insulin and IGF receptors, the early demise of the nullizygous animals has prevented us from examining the role of different tissues in insulin and Igf1 actions. With the development of conditional knockout strategies, we are now able to address some of these questions.

ACKNOWLEDGMENTS

Studies in Dr. Accili's lab were partly supported through a research grant from the American Diabetes Association and a generous gift from Sigma Tau pharmaceuticals.

REFERENCES

1. Bondy CA, LeRoith D. Insulin like growth factors. In: Growth Factors and Cytokines in Health and Disease, JAI Press, Greenwich, CT, 1996, pp. 1–26.
2. Jones JI, Clemmons DR. Insulin-like growth factors and their binding proteins: biological actions. Endocr Rev 1995;16:3–34.
3. Patti ME, Kahn CR. The insulin receptor—a critical link in glucose homeostasis and insulin action. J Basic Clin Physiol Pharmacol 1998;9:89–109.
4. Sepp-Lorenzino L. Structure and function of the insulin-like growth factor I receptor. Breast Cancer Res Treat 1998;47:235–253.
5. LeRoith D. Insulin-like growth factor receptors and binding proteins. Baillieres Clin Endocrinol Metab 1996;10:49–73.
6. Braulke T. Type-2 IGF receptor: a multi-ligand binding protein. Horm Metab Res 1999;31:242–246.
7. Baker J, Liu JP, Robertson EJ, Efstratiadis A. Role of insulin-like growth factors in embryonic and postnatal growth. Cell 1993;75:73–82.
8. Liu JP, Baker J, Perkins AS, Robertson EJ, Efstratiadis A. Mice carrying null mutations of the genes encoding insulin-like growth factor I and type 1 IGF receptor. Cell 1993;75:59–72.
9. Powell-Braxton L, Hollingshead P, Warburton C, Dowd M, Pitts-Meek S, Dalton D, Gillett N, Stewart TA. Igf1 is required for normal embryonic growth in mice. Genes Dev 1993;7:2609–2617.
10. Wang J, Zhou J, Powell-Braxton L, Bondy CA. Effects of Igf1 gene deletion on postnatal growth patterns. Endocrinology 1999;140:3391–3394.
11. Donahue LR, Beamer WG. Growth hormone deficiency in 'little' mice results in aberrant body composition, reduced insulin-like growth factor-I and insulin-like growth factor-binding protein-3 (IGFBP-3). J Endocrinol 1993;136:91–104.
12. Zhou Y, Xu BC, Maheshwari HG, He L, Reed M, Lozykowski M, et al. A mammalian model for Laron syndrome produced by targeted disruption of the mouse growth hormone receptor/binding. Proc Natl Acad Sci USA 1997;94:13,215–13,220.
13. Wang J, Zhou J, Bondy CA. Igf1 stimulates longitudinal bone growth by amplifying growth plate chondrocyte hypertrophy. FASEB J. 1999;13:1985–1990.
14. Shinar DM, Endo N, Halperin D, Rodan GA, Weinreb M. Differential expression of insulin-like growth factor-I (Igf1) and IGF- II messenger ribonucleic acid in growing rat bone. Endocrinology 1993;132:1158–1167.

15. Wang E, Wang J, Chin E, Zhou J, Bondy CA. Cellular patterns of IGF system gene expression in murine chondro- and osteogenesis. Endocrinol 1995;136:2741–2752.
16. Beck KD, Powell-Braxton L, Widmer H-R, Valverde J, Hefti F. Igf1 gene disruption results in reduced brain size, CNS hypomyelination, and loss of hippocampal granule and striatal parvalbumin-containing neurons. Neuron 1995;14:717–730.
17. Cheng CM, Joncas G, Reinhardt RR, Farrer RT, Quarles R, Janssen J, et al. Insulin-like growth factor 1 and brain myelination. J Neurosci 1998;18:5673–5681.
18. Gao WQ, Shinsky N, Ingle G, Beck K, Elias KA, Powell-Braxton L. Igf1 deficient mice show reduced peripheral nerve conduction velocities and decreased axonal diameters and respond to exogenous IGF- I treatment. J Neurobiol 1999;39:142–152.
19. Woods KA, Camacho-Hubner C, Savage MO, Clark AJ. Intrauterine growth retardation and post-natal growth failure associated with deletion of the insulin-like growth factor I gene. N Engl J Med 1996;335:1363–1367.
20. Cheng C, Reinhardt RR, Lee W-H, Patel SC, Bondy CA. Igf1 regulates developing brain glucose metabolism. Proc Natl Acad Sci USA 2000;97:10,236–10,242.
21. Baker J, Hardy MP, Zhou J, Bondy CA, Lupu F, Bellve A, et al. Effects of an Igf1 null mutation on mouse reproduction. Mol Endocrinol 1996;10:903–918.
22. Ruan W, Powell-Braxton L, Kopchick JJ, Kleinberg DL. Evidence that insulin-like growth factor I and growth hormone are required for prostate gland development. Endocrinology 1999;140:1984–1989.
23. Zhou J, Kumar TR, Matzuk MM, Bondy CA. Insulin-like growth factor I regulates gonadotropin responsiveness in the murine ovary Mol Endocrinol 1997;11:1924–1997.
24. Murphy LJ, Murphy LC, Freisen HG. Estrogen induces insulin-like growth factor-I expression in the rat uterus. Mol Endocrinol 1987;1:445–450.
25. Adesanya OO, Zhou J, Bondy CA. Sex steroid regulation of igf system gene expression and proliferation in primate myometrium. J Clin Endo Metab 1996;81:1967–1974.
26. Adesanya OO, Zhou J, Powell-Braxton L, Bondy CA. IGF1 is required for G2 progression in the estradiol-induced mitotic cycle. Proc Natl Acad Sci USA 1999;96:3287–3291.
27. Pardee AB. G1 events and the regulation of cell proliferation. Science 1989;246:603–608.
28. Liu JL, Grinberg A, Westphal H, Sauer B, Accili D, Karas M, et al. Insulin-like growth factor-I affects perinatal lethality and postnatal development in a gene dosage-dependent manner: manipulation using the Cre/loxP system in transgenic mice. Mol Endocrinol 1998;12:1452–1462.
29. Sjogren K, Liu JL, Blad K, Skrtic S, Vidal O, Wallenius V, et al. Liver-derived insulin-like growth factor I (Igf1) is the principal source of Igf1 in blood but is not required for postnatal body growth in mice. Proc Natl Acad Sci USA 1999;96:7088–7092.
30. Yakar S, Liu JL, Stannard B, Butler A, Accili D, Sauer B, et al. Normal growth and development in the absence of hepatic insulin-like growth factor I. Proc Natl Acad Sci USA 1999;96:7324–7329.
31. DeChiara TM, Efstratiadis A, Robertson EJ. A growth-deficiency phenotype in heterozygous mice carrying an insulin-like growth factor II gene disrupted by targeting. Nature 1990;345:78–80.
32. DeChiara TM, Robertson EJ, Efstratiadis A. Parental imprinting of the mouse insulin-like growth factor II gene. Cell 1991;64:849–854.
33. Zhou J, Bondy CA. IgfII and its binding proteins in the placenta. Endocrinology 1992;131:1230–1240.
34. Lopez MF, Dikkes P, Zurakowski D, Villa-Komaroff L. Insulin-like growth factor II affects the appearance and glycogen content of glycogen cells in the murine placenta. Endocrinology 1996;137:2100–2108.
35. Wang ZQ, Fung MR, Barlow DP, Wagner EF. Regulation of embryonic growth and lysosomal targeting by the imprinted Igf2/Mpr gene. Nature 1994;372:464–467.
36. Lau MM, Stewart CE, Liu Z, Bhatt H, Rotwein P, Stewart CL. Loss of the imprinted Igf2/cation-independent mannose 6-phosphate receptor results in fetal overgrowth and perinatal lethality. Genes Dev 1994;8:2953–2963.
37. Ludwig T, Eggenschwiler J, Fisher P, D'Ercole AJ, Davenport ML, Efstratiadis A. Mouse mutants lacking the type 2 IGF receptor (Igf2r) are rescued from perinatal lethality in Igf2 and Igf1r null backgrounds. Dev Biol 1996;177:517–535.
38. Duvillie B, Cordonnier N, Deltour L, Dandoy-Dron F, Itier JM, Monthioux E, et al. Phenotypic alterations in insulin-deficient mutant mice. Proc Natl Acad Sci USA 1997;94:5137–5140.
39. Accili D, Drago J, Lee EJ, Johnson MD, Cool MH, Salvatore P, et al. Early neonatal death in mice homozygous for a null allele of the insulin receptor gene. Nature Genet 1996;12:106–109.

40. Joshi RL, Lamothe B, Cordonnier N, Mesbah K, Monthioux E, Jami J, et al. Targeted disruption of the insulin receptor gene in the mouse results in neonatal lethality. EMBO J 1996;15:1542–1547.
41. Louvi A, Accili D, Efstratiadis A. Growth-promoting interaction of IgfII with the insulin receptor during mouse embryonic development. Dev Biol 1997;189:33–48.
42. Girard JR, Kervan A, Soufflet E, Assan R. Factors affecting the secretion of insulin and glucagon by the rat fetus. Diabetes 1973;23:310–317.
43. Wertheimer E, Lu SP, Backeljauw PF, Davenport ML, Taylor SI. Homozygous deletion of the human insulin receptor gene results in leprechaunism. Nat Genet 1993;5:71–73.
44. Psiachou H, Mitton S, Alaghband ZJ, Hone J, Taylor SI, Sinclair L. Leprechaunism and homozygous nonsense mutation in the insulin receptor gene. Lancet 1993;1:342.
45. Krook A, Brueton L, O'Rahilly S. Homozygous nonsense mutation in the insulin receptor gene in infant with leprechaunism. Lancet 1993;342:277–278.
46. Jospe N, Kaplowitz PB, Furlanetto RW. Homozygous nonsense mutation in the insulin receptor gene of a patient with severe congenital insulin resistance: leprechaunism and the role of the insulin-like growth factor receptor. Clin Endocrinol 1996;45:229–235.
47. Accili D. Molecular defects of the insulin receptor gene. Diabetes Metab Rev 1995;11:47–62.
48. Taylor SI, Cama A, Accili D, Barbetti F, Quon MJ, Sierra M, et al. Mutations in the insulin receptor gene. Endocr Rev 1992;13:566–595.
49. Bier DM, Schedewie H, Larner J, Olefsky J, Rubenstein A, Fiser RH, et al. Glucose kinetics in leprechaunism: accelerated fasting due to insulin resistance. J Clin Endocrinol Metab 1980;51:988–994.
50. Backeljauw PF, Alves C, Eidson M, Cleveland W, Underwood LE, Davenport ML. Effect of intravenous insulin-like growth factor I in two patients with leprechaunism. Pediatr Res 1994;36:749–754.
51. Di Cola G, Cool MH, Accili D. Hypoglycemic effect of insulin-like growth factor-1 in mice lacking insulin receptors. J Clin Invest 1997;99:2538–2544.
52. Roach P, Zick Y, Formisano P, Accili D, Taylor SI, Gorden P. A novel human insulin receptor gene mutation uniquely inhibits insulin binding without impairing posttranslational processing. Diabetes 1994;43:1096–1102.
53. Schultz GA, Hogan A, Watson AJ, Smith RM, Heyner S. Insulin, insulin-like growth factors and glucose transporters: temporal patterns of gene expression in early murine and bovine embryos. Reprod Fertil Dev 1992;4:361–371.
54. Tyrala EE. The infant of the diabetic mother. Obstet Gynecol Clin N Am 1996;23:221–241.
55. Naeye RL. Infants of diabetic mothers: a quantitative, morphologic study. Pediatrics 1965;35:980–988.
56. Efstratiadis A. Genetics of mouse growth. Int J Dev Biol. 1998;42:955–976.
57. Lauro D, Kido Y, Castle AL, Zarnowski MJ, Hayashi H, Ebina Y, et al. Impaired glucose tolerance in mice with a targeted impairment of insulin action in muscle and adipose tissue. Nat Genet 1998;20:294–298.

18 Use of Transgenic Animals in Skeleton Biology

Thomas Günther, PhD, *Mary Jo Doherty,* PhD, *and Gerard Karsenty,* MD, PhD

Contents

INTRODUCTION

Three areas of skeletal biology that have benefited from the achievements in mouse genetics are skeletal patterning, cell differentiation and physiology. Skeleton patterning, or the position, type, length and shape of each individual skeletal element, is genetically controlled. Many of the genes involved have been implicated from their mutations, as identified through human genetics. However, it is only during recent mouse studies, knockout, misexpression, and overexpression that the actual function of these genes has been confirmed. In addition, a number of genes that act downstream have been identified. In terms of cell differentiation, transcription factors specific for differentiation of osteoblasts, osteoclasts, and chondrocytes (cells of the skeleton) were determined in the last decade, and it is highly probable that the identified genes represent only a small portion of the total genes involved. Finally, mouse genetics has provided clues about the function of structural proteins of the bone matrix that other biological assays could not provide.

There are two steps in bone development. First, there is patterning of each skeletal element that occurs around midgestation *(1)*. After patterning is achieved, bone develops in two different ways. In endochondral ossification, the skeleton is formed by chondrocytes and osteoblasts, which are both of mesenchymal origin *(1)*. During fetal development, chondrocytes form a blueprint of the later bone, after which they hypertrophy, calcify, and finally die. At this time, vascularization occurs, bringing in osteo-

From: *Contemporary Endocrinology: Transgenics in Endocrinology*
Edited by: M. Matzuk, C. W. Brown, and T. R. Kumar © Humana Press Inc., Totowa, NJ

blasts that replace the cartilage with bone matrix and causing the formation of the bone cavity. Osteoclast differentiation occurs next. Between the epiphyseal cartilage of the forming bone, chondrocytes become organized into the growth-plate cartilage, which together with the osteoblast provides longitudinal skeletal growth. During intramembranous ossification, some craniofacial bones and the clavicle form directly from undifferentiated mesenchymal cells *(1)*. These cells migrate into areas destined to become bone, where they form condensations that take the general shape of future skeletal elements.

Bone remodeling, or bone resorption followed by bone formation, occurs throughout life to maintain a constant bone mass *(1)*. A relative increase in bone resorption results in osteoporosis (a decrease in bone mass and high risk of fracture), whereas a relative increase in bone formation causes osteosclerosis (generalized increase in skeletal mass). This chapter discusses the following aspects of bone biology: patterning, cell differentiation, and cell function. The findings presented in this chapter originate mostly from genetic manipulations in humans and mice and studies in chick.

PATTERNING

How is the three-dimensional structure of the skeleton, position, number, and shape of the skeletal elements achieved? This process, called pattern formation, has been analyzed most in the developing limb, and we are beginning to understand how growth is coupled to the establishment of three axes: the proximal-distal (P-D; shoulder to digits), dorsal-ventral (D-V; back of hand to palm) and anterior-posterior axis (A-P; thumb to little finger). The definition of an area where the limbs are destined to form is the first step in this complex process, followed by the establishment of specified fields which, in turn ensure the linkage of growth and pattern formation. The fields are interpreted within the limb by specific gene expression, which defines the formation of the various structures *(2,3)*.

Axial patterning during embryonic development is specified by homeobox-containing transcription factors called Hox proteins *(4)*. The genes encoding these proteins are also involved in the differentiation of the lateral-plate mesoderm into forelimbs, flank, and hindlimbs *(5,6)*. Recent genetic studies suggest that the transcription factors Pitx1, Tbx4, and Tbx5 control Hox-gene expression, and that they are involved in specification of forelimb and hindlimb identity *(7–9)*. Ectopic expression of Pitx1 in the chick wing induces Tbx4, Hoxc9, 10, and 11 transcription, which results in a wing that has characteristics of a hindlimb, where the expression of all these genes is normally restricted. However, wing-specific Hoxd9 expression is repressed. Similarly, in Pitx1 mutant mice, Tbx4 expression is reduced and partial leg-to-arm transformation occurs *(10)*. The Tbx5 expression pattern is limited to the wing, and its misexpression results in wing-to-leg transformation *(8,9)*.

Limb outgrowth is believed to be initiated by fibroblast growth factor8 (Fgf8) expressed in the intermediate mesoderm *(11,12)*. Fgf8 induces Fgf10 *(13)*, which in turn causes the formation of the apical ectodermal ridge (AER). The AER is a thickened epithelial structure that forms along the most distal part of the limb bud at its A-P axis. Its main function is to mediate bud outgrowth by maintaining the mesenchyme at the limb-bud tip in an undifferentiated state called the progress zone. This ensures that the proximal bones of the limb form first, followed by the more distal structures. Thus, if the AER is removed, a truncated limb results *(14)*. Ffg2, Ffg4, and Ffg8 are expressed

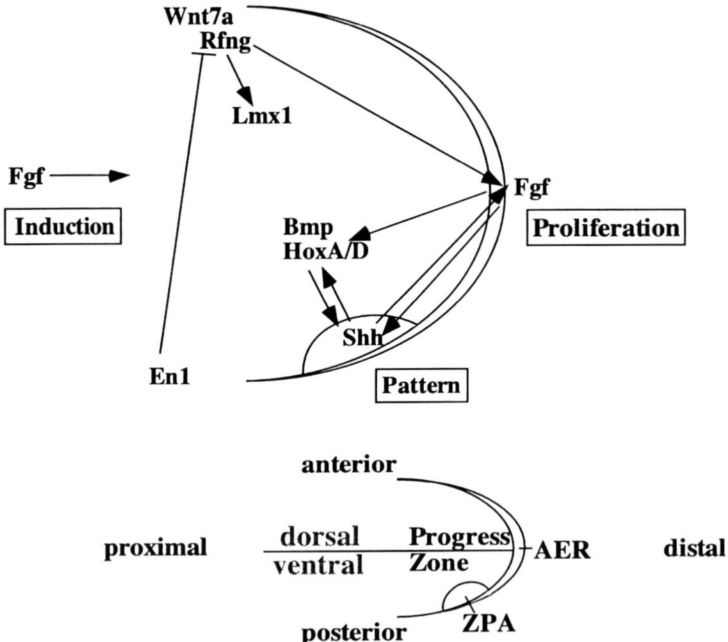

Fig. 1. A simplified model for the genetic control of limb formation. The diagram depicts a limb bud developing from left to right. The location of the genes involved in the various development stages and their interactions with each other are indicated. The lower diagram details the three different axes together with the descriptive terminology used.

in the AER, and each of them has the ability to induce limb outgrowth *(11,15,16)*. Shortly after the formation of the AER, cartilage blastemas condense, serving as a blueprint for the later bones *(17)*. To date, it is still unclear how the P-D specification is achieved. According to one model, the type of structure cells will form is specified by the length of time they spend in the progress zone *(18)*.

Molecules involved in the location of the AER are also involved in the establishment of the D-V axis (Fig. 1). The transcription factor Engrailed1 (En1), which is expressed in ventral ectoderm, restricts the expression of Wnt7a (a member of the signaling molecules encoding the Wnt gene family) and Radical fringe (Rfng) to the dorsal ectoderm *(19–21)*. The AER develops at the site of apposition between Rfng-expressing and nonexpressing cells. Misexpression of En1 and Rfng in the chick revealed that En1 represses Rfng and causes disrupted and ectopic AERs on the dorsal side, while ectopic Rfng expression induces ectopic AERs on the ventral side. Normal and ectopic AERs always correlate with viral misexpression of Rfng *(21)*. The knockout of En1 results in dorsal transformation of ventral paw structures accompanied by ventrally extended Wnt7a expression domains *(19)*. WNT7a induces the expression of the transcription factor Lmx1 in the underlying mesoderm, which in turn results in the expression of downstream genes leading to dorsalization of the limb *(22,23)*.

The A-P axis is determined by signals present in the transient cell population located at the distal posterior region of the limb bud, called the zone of polarizing activity (ZPA). Saunders and Gasseling *(24)* showed that grafting chick tissue from the posterior limb-bud mesoderm to the anterior of the host wing resulted in a mirror image duplication of the digits. It has since been demonstrated that this effect can be mimicked by the secreted protein Sonic hedgehog (SHH), which is present in the ZPA *(25–28)*. SHH acts in the mesoderm indirectly through Fgf4, produced by the AER. FGF4 induces the competence of the mesoderm to respond to SHH, which then prompts the expression of Bmp2 (a member of the TGF-β superfamily) and Hoxd, whose correct expression determines specific skeletal elements *(29–31)*. Shh expression is both AER- and Wnt7a-dependent, since absence of Wnt7a expression in the dorsal ectoderm of the limb-bud reduces the Shh-expression domain, causing a lack of posterior skeletal elements *(31)*. These studies demonstrate the intimate link that exists between all three axes through their respective signals, WNT7a, FGF4, and SHH, during limb outgrowth and patterning.

The three-dimensional structure superimposed by the AER and ZPA is interpreted, in part, by Hox genes. Hox genes are organized into four paralog clusters (Hoxa, b, c, and d), and their function has been examined extensively by deletion of Hox members by homologous recombination in the mouse *(32)*. Genes of the Hoxa and Hoxd clusters are involved in the pattern of chondrogenic condensations in the limb. For example, the homozygous triple knockout of Hoxd11, Hoxd12, and Hoxd13 resulted in small-digit primordia and disorganized cartilage pattern *(33)*. Misexpression experiments further elucidated the function of Hox and other genes. For instance, the analysis of Hoxd12 to that stage has been hampered by functional redundancy with other genes *(34,35)*. However, misexpression of Hoxd12 in the anterior part of the limb bud caused transformation of anterior to posterior digits, and showed that Hoxd12 could ectopically induce Shh *(36)*. This demonstrated that certain Hoxd genes directly amplify the posterior polarizing signal in a positive-feedback loop.

Genes of a Hox cluster that are located more 5' in the genome appear to play a dominant role when two or more of them are expressed in the same cell (posterior prevalence) *(37,38)*. This leads to a distinct pattern formation within different regions of the limb, since each gene has a different effect on proliferation and differentiation. It has also been suggested that the level of Hox-gene expression in a single cell can determine its fate *(39)*. Deletion of several Hox genes at one time results in a more severe phenotype than malformations of single knockouts. Absence of Hoxd11 or Hoxa11, for example, results in minor defects of the radius, ulna, and some of the bones of the hands and/or feet *(40–42)*. In mice missing both Hoxd11 and Hoxa11, however, the radius and ulna were almost entirely missing *(43)*.

As mentioned in the introduction, the cartilage forms as a template for bone. The size and shape of this template is determined by the number of mesenchymal precursors recruited to become chondrocytes, and their subsequent proliferation rate and deposition of extracellular matrix. The portion of condensed mesenchyme that differentiates into chondrocytes is determined by Hox genes and members of the Tgf-β superfamily *(44,45)*. For example, overexpression of Gdf5 (growth-differentiation factor), Bmp2, Bmp4, or Bmp7—all members of the TGFβ superfamily—causes an increased recruitment of mesenchymal cells resulting in the formation of longer and/or wider cartilage anlagen in chicks *(46–48)*. On the other hand, mutations in Gdf5 cause shorter distal bones in the mouse mutant brachypod *(49)*. Similarly, mutations in the human homolog of GDF-5, cartilage-derived morphogenetic protein 1, result in three allelic human conditions: brachydactyly type C, Hunter-Thompson acromesomelic chondrodysplasia, and Grebe chondrodysplasia *(50–52)*. BMPs influence mesenchymal condensations through the BMP receptor type 1A (BmpR-1a) *(53)*, and the effect of BMP on growth-plate size is in turn controlled by its endogenous antagonist noggin *(54,55)*.

CELL DIFFERENTIATION

Cartilage and bone development involves a set of interactive steps, including induction of specific precursors, promotion of cell proliferation, and differentiation. The first step is the induction of master regulatory genes in undifferentiated condensed mesenchyme that specifies an osteogenic or chondrogenic cell fate. For each cell lineage of the skeleton, a transcription factor acting as a trigger for differentiation has been identified. Sox9 has recently been demonstrated to be necessary for chondrocyte induction *(56)*. Cbfa1 is a master gene for osteoblast differentiation *(57)*, and Pu.1 is a master gene for osteoclast differentiation *(58)*. Besides these transcription factors, which act early in cell differentiation, other transcription factors and growth factors have been implicated through mouse studies to act downstream (Fig. 2).

Chondrocytes differentiate from mesenchymal condensations, where cells in the center of the condensation differentiate into proliferating chondrocytes, whereas cells in the periphery form the perichondrium. With time, chondrocytes in the center of the cartilage cease proliferating and become prehypertrophic, then hypertrophic, and finally die through apoptosis. With the invasion of blood vessels, osteoblasts arrive and deposit bone matrix, and partly control osteoclast differentiation.

Several genetic studies in humans and mice have proven that the transcription factor Sox9 is essential for chondrocyte induction. Sox9 is a transcription factor that binds directly to the promoter of collagen $\alpha 1$(II), the major cartilage-matrix protein *(59)*. Heterozygosity for Sox9 leads to campomelic dysplasia in humans, characterized by malformations of the skeleton, caused by defects in chondrocyte differentiation, and sex reversal *(60,61)*. In chimeric mice, which have both wild-type and Sox9-deficient chondrocytes, it was observed that only the wild-type cells contribute to chondrocytes *(56)*. Given the pleiotropic phenotype of the chondrocytes, it is likely that several transcription factors are involved for each subset of chondrocytes.

The length and shape of a bone is determined by the rate of endochondral ossification, and therefore is controlled both by chondrocytes and osteoblasts. A number of genes have been shown to play a role in this process, including the secreted factors Indian hedgehog (Ihh), Bmps, parathyroid hormone-related protein (PTHrP) and its receptors, and fibroblast growth factor receptor 3 (Fgfr3). PTHrP and Ihh are believed to play opposing roles in chondrocyte differentiation. PTHrP has been suggested to control the rate of differentiation of chondrocytes into hypertrophic chondrocytes, since loss of PTHrP function causes early hypertrophy and premature bone formation, whereas overexpression leads to delayed differentiation of hypertrophic chondrocytes and and an increased chondrocyte proliferation *(62–68)*. In contrast, overexpression of Ihh causes a loss of hypertrophic chondrocytes and a delay in ossification *(65)*. Histological and *in situ* studies demonstrate that cells express PTH/PTHrP receptor before differentiating into an Ihh-expressing cell type *(65)*. PTHrP, which is normally expressed in the perichondrium of the joint region of the developing cartilage, was found to be strongly upregulated after the overexpression of Ihh, suggesting that PTHrP is downstream of Ihh in regulating cartilage differentiation *(65)*.

Another growth-factor receptor FGFR3, is involved in chondrocyte differentiation. Its identification and functional characterization also arises from the field of human genetics. Indeed, FGFR3 was first isolated as the gene mutated in achondroplasia, the most frequent form of dwarfism in humans *(69,70)*. The phenotype is caused by a gain-

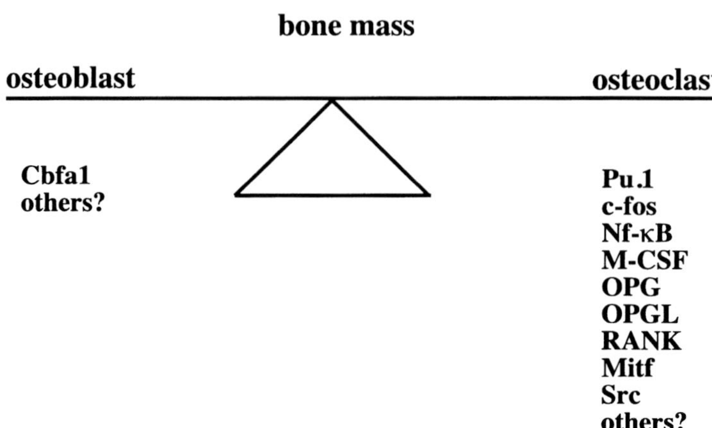

Fig. 2. Graphic representation of the factors involved in osteoblast and osteoclast differentiation and function. The balance between bone formation and resorption, also called remodeling, maintains a constant bone mass throughout life until gonadal failure. This list is not comprehensive and only the factors with known mechanisms are included. Since a number of factors are involved in osteoclast differentiation, it is also conceivable that a variety of factors are also involved in osteoblast differentiation.

of-function mutation. Consistent with this observation, Fgfr3 deletion in mice leads to increased longitudinal growth of the skeleton. These mice deficient for Fgfr3 exhibit overgrowth of long bones and vertebrae, with enlargement of the hypertrophic zone of the growth plates *(71,72)*. Thus, the role of Fgfr3 appears to be one of negative regulation by limiting chondrocyte proliferation. It has been proposed that FGFR3 inhibits growth-plate Ihh expression through the STAT1 pathway, which in turn inhibits patched (ptc) (the receptor for Ihh) and BMP expression in both growth-plate and perichondrium *(53,55,65,73–78)*.

Osteoblasts, like chondrocytes, are of mesenchymal origin. These cells are present in all bone, whether derived from endochondral or intramembranous ossification, and are responsible for secreting the bone matrix. Much of our knowledge about the transcriptional control of osteoblasts comes from human and mouse genetics. Gene-deletion studies have shown that mice deficient in Cbfa1, a member of the runt family of transcription factors, die at birth and possess an unossified skeleton *(79,80)*. Indeed, close examination of the skeletal preparations of these mice revealed a normally patterned skeleton, but for those elements formed through endochondral ossification all the skeletal elements were made of cartilage, or of mesenchymal cells for those formed through intramembranous ossification *(79,80)*.

Heterozygotes also exhibited a phenotype, with hypoplasia of the clavicle and delayed development of membranous bones. This phenotype is identical to a radiation-induced mutant called Cleidocranial dysplasia (Ccd), which is also very similar to human CCD. This, in turn, led researchers to investigate whether the Cbfa1 gene was affected in Ccd-mutant mice. Further studies demonstrated that the Cbfa1 gene is at least partially deleted in Ccd mice, and that CBFA1, like CCD, maps to chromosome 6 in humans *(80)*. It is now known that the disorder CCD is caused by CBFA1 haploinsufficiency *(80–82)*.

CBFA1 binds to and activates the promoter of most genes expressed in an osteoblast, and during development is initially expressed in every mesodermal condensation at 12.5 d

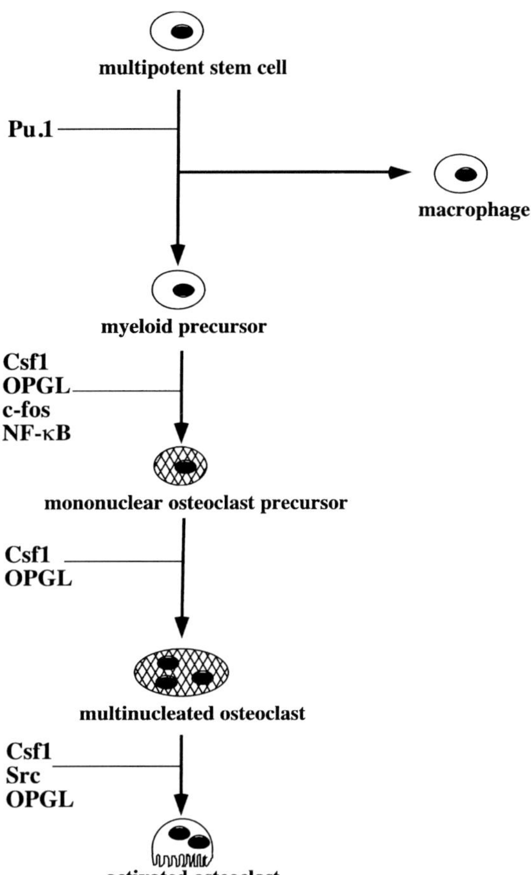

Fig. 3. Genetic control of osteoclast differentiation. The diagram shows each stage of cell differentiation, from multipotent stem cell to activated osteoclast, and the factors involved in each specific stage of differentiation.

postcoitum, marking a common progenitor for osteoblasts and chondrocytes *(83)*. Beyond embryonic development, Cbfa1 is required for osteoblast function, becuase it controls the rate of bone-extracellular-matrix deposition and therefore the rate of bone formation *(84)*.

The osteoclast, the bone-resorbing cell, is of monocyte/macrophage origin. Absence of osteoclast function leads to osteopetrosis, while increased activity causes osteoporosis. Osteoclast differentiation comprises the recruitment of new osteoclast precursors from the haematopoietic system, their development to mature osteoclasts, and finally their activation (Fig. 3). Osteoclast differentiation is generally controlled by osteoblasts.

Like chondrocytes and osteoblasts, transcription factors play important roles in osteoclast differentiation, and the most important is PU.1. PU.1 is essential for the development of myeloid and B-lymphoid cells. Since myeloid cells are osteoclast progenitors, it was hypothesized that PU.1 may be involved in osteoclast differentiation *(58)*. This was indeed shown to be the case. Macrophages expressed PU.1 mRNA, and the message was also detected in bone-marrow mononuclear cells in early coculture and at elevated levels after osteoclasts differentiated *(58)*. Mice deficient in PU.1 exhibited osteopetrosis, and were devoid of both osteoclasts and macrophages, indicating that PU.1 causes induction of osteoclast differentiation.

Several other transcription factors, such as c-Fos and NF-κB and growth factors including colony-stimulating factor 1 (CSF1, also called macrophage-colony stimulating factor 1, M-CSF1), are involved in later stages of osteoclast differentiation. Both the op/op and the c-fos null mutations arrest osteoclast development after myeloid precursors, also resulting in an osteopetrotic phenotype. The op/op mice fail to make functional CSF1, which is a defect of the osteoblast/stromal cells needed for osteoclastogenesis *(85,86)*. Analysis of the c-fos mutant bones indicated that in addition to the lack of differentiated multinucleated osteoclasts, there was an increase in the density of functional macrophages not observed in other tissues *(87)*. This implies that c-fos also affects macrophage differentiation.

A series of experiments performed by various research groups recently resulted in the identification of a number of other factors involved in differentiation of osteoclasts from hematopoietic precursors. Osteoprotegerin (OPG; also called OCIF) *(88,89)*, a secreted receptor, was initially identified by sequence homology as a novel member of the tumor-necrosis-factor receptor (TNFR) superfamily, which consists mostly of transmembrane proteins that elicit signal transduction in a variety of cells. However, OPG lacked any apparent cell-associated signals, and in an attempt to determine its function, mice were generated that overexpressed this protein *(90)*. Histological analysis of the mutant mice revealed a severe osteopetrotic phenotype with a decrease in osteoclast number. OPG-overexpressing and control animals both contained osteoclast precursors. When exogenous OPG was added to spleen cells removed from OPG-overexpressing and control animals, it was demonstrated that OPG inhibited osteoclast formation from both groups of animals. The N-terminal portion of OPG containing TNFR-like domain was found to be sufficient to inhibit osteoclastogenesis *(90)*.

From these observations, it was hypothesized that OPG may neutralize a factor that stimulates osteoclast development, thus inhibiting osteoclast maturation. Indeed, such a factor does exist. Using expression cloning, OPG ligand (OPGL; also called ODF/TRANCE/RANKL) *(91,92)* was identified as a ligand for OPG *(93)*. OPG can interact with both the soluble and membrane-bound OPGL that is located on osteoblast/stromal cells. OPGL is considered a potent osteoclast-differentiation factor, since it stimulates osteoclast formation and function *(93)*. However, in the murine osteoclast coculture model, OPGL only increased osteoclast formation in the presence of CSF1, which is provided by the stromal cells in this system *(93)*. Not surprisingly, the knockout studies of OPG and OPGL resulted in osteoporotic *(94)* and osteopetrotic *(95)* phenotypes, respectively. Interestingly, the OPG mutant mice exhibited another phenotype with marked calcification of some but not all arteries, implying that OPG plays a role in inhibiting aterial calcification *(95)*.

After the finding that the secreted receptor OPG binds to OPGL on osteoblast/stromal cells the next step was to identify an OPG receptor on cells of the osteoclast lineage. By analysis of genes expressed in a primary osteoclast precursor cell cDNA library, Hsu et al. *(96)* showed that recombinant OPGL binds specifically to a previously identified member of the TNFR family called RANK *(97)*. Like OPG overexpression, RANK overexpression causes a decrease in osteoclasts and a resultant osteopetrotic phenotype *(96)*.

Further studies showed that binding to RANK activates a signalling pathway involving interaction with the cytoplasmatic TNRF-associated factor (TRAF) proteins *(98,99)*.

These proteins in turn activate NF-κB transcription factors. NF-κB comprises a dimer of various combinations of structurally related proteins, which when activated induces the expression of a variety of genes including cytokines, adhesion molecules, and anti-apoptotic regulators *(100)*. Interestingly, when two subunits of NF-κB—p50 and p52, which are usually coexpressed—are both deleted in mice, the mutants display osteopetrosis caused by the lack of mature osteoclasts *(101,102)*.

CELL FUNCTION

The cartilagenous extracellular matrix consists of collagen types II, IX, and XI, and the link-protein aggrecan. Like osteoblasts, the chondrocyte progenitors become progressively embedded in their own matrix and differentiate into chondrocytes. Proliferation of these cells occurs for some time because of the gel-like consistency of cartilage. At the perichondrium (periphery of cartilage), mesenchymal cells continue to proliferate and differentiate, forming bone by appositional growth. Mutations in the collagens secreted in cartilage result in chondrodysplasias. For example, mutations in type II collagen cause disproportionate micromelia (Dmm) *(103)*. Mutated α1 (XI) collagen causes chondrodysplasia in cho mice *(104)* and mutated α2 (XI) collagen causes stickler syndrome *(105)*. Another chondrodysplasia, cartilage matrix deficiency (CMD), is caused by a mutation in aggrecan *(106)*, and a mutation in α1 (X) collagen is responsible for Schmidt's chondrodysplasia *(107)*.

For many proteins with a role in remodeling the extracellular matrix during endochondral development, their function has been deciphered only through mouse genetic manipulation. For instance, by targeted deletion of gelatinase B (Mmp9) Vu et al. *(108)* showed that this matrix metalloproteinase, which is a proteolytic enzyme in the extracellular matrix, controls the hypertrophic state of chondrocytes. Absence of the enzyme causes a delay in vascularization and endochondral ossification.

Matrix-Gla-protein (MGP) produced by chondrocytes and by smooth-muscle cells prevents mineralization of cartilage and blood vessels. MGP-deficient mice died a few weeks after birth from calcification of blood vessels *(109)*. Histological examination of mutant mice demonstrated that mineralization also occurred throughout the growth plate, which preceded growth-plate closure. These results suggest that calcification of the cartilage extracellular matrix may explain the closure of the growth plate at the end of puberty.

The function of the osteoblast is to form bone by producing matrix constituents, predominantly type I collagen. Mutations in the genes encoding type I collagen cause the congenital disease osteogenesis imperfecta *(110)*, characterized by a low bone mass and fragile bones. In addition to collagen type I, osteoblasts also secrete a variety of noncollagenous proteins, such as osteopontin, bone sialoprotein, decorin, osteonectin, biglycan, and osteocalcin (bone gla-protein). Absence of osteocalcin causes an increase of bone formation, indicating that this protein is a negative regulator of bone formation *(111)*. Knockout studies have also been performed on Decorin and Osteonectin without the production of a detectable skeletal phenotype *(112,113)*. However, biglycan-deficient mice (BGN) exhibit a decreased growth rate, with reduced bone mass that was not observed until 3 mo after birth *(114)*. This indicates that BGN is involved in the regulation of postnatal skeletal growth. Closer examination of mutant mice revealed that osteoblast numbers were reduced, while osteoclast numbers were normal, suggesting the reduced bone mass was caused by decreased bone formation *(114)*.

The function of the osteoclast is to resorb bone by producing a variety of different enzymes in the space created by the osteoclasts' ruffled border and the bone surface. These lysosomal enzymes are synthesized by the osteoclast and secreted across the ruffled border. In the bone-resorbing compartment, they reach a high concentration, creating an acidified environment that is reponsible for dissolving the mineral and exposing the matrix. Once exposed the enzymes degrade the matrix. Collagenase releases hydroxyapatite crystals, and the residual fibers are digested by latent collagenase or by the actions of cathepsins. Again, the use of human and mouse genetic studies have provided a vast amount of information on the function these secreted molecules play during resorption. Absence of cathepsin K, a lysosomal protease, causes the human syndrome pycnodystosis, characterized by osteosclerosis and short stature *(115,116)*. However, absence of Tartrate-resistant acid phosphatase (Acp5) results in mild osteopetrosis, and osteoclast function is only slightly affected *(117)*. Similarly, osteopontin-deficient mice (OPN) exhibit less bone resorption, but this was only detected after ovariectomy of the mice *(118–120)*. Several mutant mice have also been shown to contain osteoclasts, but interestingly, these cells lacked bone-resorbing activity. The genes Src (a cellular homolog of the avian sarcoma virus) and Mitf were responsible for the nonfunctional osteoclasts found in Src-null and microphthalmia mice, respectively *(121–125)*.

The differentiation of osteoclasts and osteoblasts is tightly connected during bone formation. This observation, and the fact that under normal circumstances bone mass is constant, led to the assumption that osteoblast and osteoclast activity is also linked during bone remodeling. This implies that a reduction in osteoblast function would automatically lead to a decreased osteoclast activity. Corall and colleagues *(126)* generated a mouse model to specifically address this question. The transgenic mouse contains an inducible construct to ablate osteoblasts. Although bone formation ceased, the level of osteoclast activity did not change, resulting in controllable osteoporosis. The results indicate that bone resorption is not controlled by bone function. Thus, the requirement for interaction between cells of the two lineages is limited to differentiation. This is further confirmed by the fact that in the absence of bone resorption, as occurs in the c-fos, OPG, and Src-knockout mice, bone formation is not slowed down. These findings and many others not included in this review, highlight the contributions of mouse genetics to the formulation of new concepts in skeleton biology.

REFERENCES

1. Karsenty G. Genetics of skeletogenesis. Dev Genet 1998;22:30–313.
2. Johnson RL, Tabin CJ. Molecular models for vertebrate limb development. Cell 1997;90:979–990.
3. Zeller R, Braun T. eds. Molecular basis of limb and muscle development. Cell Tissue Res 1999:296.
4. Gellon G, McGinnis W. Shaping animal body plans in development and evolution by modulation of Hox expression patterns. Bioessays 1998;20:116–125.
5. Cohn MJ, Pate IK, Krumlauf R, Wilkinson DG, Clarke JD, Tickle C. Hox9 genes and vertebrate limb specification. Nature 1997;387:97–101.
6. Cohn MJ, Tickle C. Developmental basis of limblessness and axial patterning in snakes. Nature 1999;399:474–479.
7. Logan M, Tabin CJ. Role of Pitx1 upstream of Tbx4 in specification of hindlimb identity. Science 1999;283:1736–1739.
8. Takeuchi JK, Koshiba-Takeuchi K, Matsumoto K, Vogel-Hopker A, Naitoh-Matsuo M, Ogura K, et al. Tbx5 and Tbx4 genes determine the wing/leg identity of limb buds. Nature 1999;398:810–814.

9. Rodriguez-Esteban C, Tsukui T, Yonei S, Magallon J, Tamura K, Izpisua-Belmonte JC. The T-box genes Tbx4 and Tbx5 regulate limb outgrowth and identity. Nature 1999;398:814–818.
10. Szeto D P, Rodriguez-Esteban C, Ryan AK, O'Connell SM, Liu F, Kioussi C, et al. Role of the Bicoid-related homeodomain factor Pitx1 in specifying hindlimb morphogenesis and pituitary development. Genes Dev 1999;13: 484–494.
11. Crossley PH, Minowada G, MacArthur CA, Martin GR. Roles for FGF8 in the induction, initiation, and maintenance of chick limb development. Cell 1996;84:127–136.
12. Vogel A, Rodriguez C, Izpisua-Belmonte JC. Involvement of Fgf-8 in initiation, outgrowth and patterning of the vertebrate limb. Development 1996;122:1737–1750.
13. Ohuchi H, Nakagawa T, Yamamoto A, Araga A, Ohata T, Ishimuru Y, et al. The mesenchymal factor, FGF10, initiates and maintains the outgrowth of the chick limb bud through interaction with FGF8, an apical ectodermal factor. Development 1997;124:2235–2244.
14. Saunders Jr JW. The proximo-distal sequence of origin of the parts of the chick wing and the role of the ectoderm. J Exp Zool 1948;108:363–403.
15. Niswander L, Tickle C, Vogel A, Booth I, Martin GR. FGF-4 replaces the apical ectodermal ridge and directs outgrowth and patterning of the limb. Cell 1993;75:579–587.
16. Fallon JF, Lopez A, Ros MA, Savage MP, Olwin BB, Simandl BK. FGF-2: apical ectodermal ridge growth signal for chick limb development. Science 1994;264:104–107.
17. Hall BK, Miyake T. The membranous skeleton: the role of cell condensations in vertebrate skeletogenesis. Anat Embryol 1992;186:107–124.
18. Summerbell D, Lewis JH, Wolpert L. Positional information in chick limb morphogenesis. Nature 1973;244:492–496.
19. Loomis CA, Harris E, Michaud J, Wurst W, Hanks M, Joyner AL. The mouse Engrailed-1 gene and ventral limb patterning. Nature 1996;382:360–363.
20. Laufer E, Dahn R, Orozco OE, Yeo CY, Pisenti J, Henrique D, et al. Expression of Radical fringe in limb-bud ectoderm regulates apical ectodermal ridge formation. Nature 1997;386:366–373.
21. Rodriguez-Esteban C, Schwabe JW, De La Pena J, Foys B, Eshelman B, Belmonte JC. Radical fringe positions the apical ectodermal ridge at the dorsoventral boundary of the vertebrate limb. Nature 1997;386:360–366.
22. Riddle RD, Ensini M, Nelson C, Tsuchida T, Jessell TM, Tabin C. Induction of the LIM homeobox gene Lmx1 by WNT7a establishes dorsoventral pattern in the limb. Cell 1995;83:631–640.
23. Vogel A, Rodriguez C, Warnken W, Izpisua-Belmonte JC. Dorsal cell fate specified by chick Lmx1 during vertebrate limb development. Nature 1995;378:716–720.
24. Saunders Jr JW, Gasseling MT. Ectoderm-mesenchymal interaction in the origins of wing symmetry. In: Fleischmajer R, Billingham RE, eds. Epithelial-Mesenchymal Interactions, Williams and Wilkins, Baltimore, MD, 1968, pp. 78–97.
25. Echelard Y, Epstein DJ, St-Jaques B, Shen L, Mohler J, McMahon JA, et al. Sonic hedgehog, a member of a family of putative signaling molecules, is implicated in the regulation of CNS polarity. Cell 1993;75:1417–1430.
26. Krauss S, Concordet JP, Ingham PW. A functionally conserved homolog of the Drosophila segment polarity gene hh is expressed in tissues with polarizing activity in zebrafish embryos. Cell 1993;75:1431–1444.
27. Riddle RD, Johnson RL, Laufer E, Tabin C. Sonic hedgehog mediates the polarizing activity of the ZPA. Cell 1993;75:1401–1416.
28. Roelink H, Augsburger A, Heemskerk J, Korzh V, Norlin S, Ruiz I Altaba A, et al. Floor plate and motor neuron induction by vhh-1, a vertebrate homolog of hedgehog expressed by the notochord. Cell 1994;76:761–775.
29. Morgan BA, Izpisua-Belmonte JC, Duboule D, Tabin CJ. Targeted misexpression of Hox-4.6 in the avian limb bud causes apparent homeotic transformations. Nature 1992;358:236–239.
30. Laufer E, Nelson CE, Johnson RL, Morgan BA, Tabin C. Sonic hedgehog and Fgf-4 act through a signaling cascade and feedback loop to integrate growth and patterning of the developing limb bud. Cell 1994;79:993–1003.
31. Yang Y, Niswander L. Interaction between the signaling molecules WNT7a and SHH during vertebrate limb development: dorsal signals regulate anteroposterior patterning. Cell 1995;80:939–947.
32. Schughart K, Kappen C, Ruddle FH. Duplication of large genomic regions during the evolution of vertebrate homeobox genes. Proc Natl Acad Sci USA 1989;86:7067–7071.
33. Zakany J, Duboule D. Synpolydactyly in mice with a targeted deficiency in the Hoxd complex. Nature 1996;384:69–71.

34. Davis AP, Capecchi MR. A mutational analysis of the 5' Hoxd genes: dissection of genetic interactions during limb development in the mouse. Development 1996;122:1175–1185.
35. Kondo T, Dolle P, Zakany J, Duboule D. Function of posterior Hoxd genes in the morphogenesis of the anal sphincter. Development 1996;122:2651–2659.
36. Knezevic V, De Santo R, Schughart K, Huffstadt U, Chiang C, Mahon KA, et al. Hoxd12 differentially affects preaxial and postaxial chondrogenic branches in the limb and regulates Sonic hedgehog in a positive feedback loop. Development 1997;124;4523–4536.
37. Lufkin T, Dierich A, LeMeur M, Mark M, Chambon P. Disruption of the Hox-1.6 homeobox gene results in defects in a region corresponding to its rostral domain of expression. Cell 1991;66:1105–1119.
38. Duboule D. Colinearity and functional hierarchy among genes of the homeotic complexes. Trends Genet 1994;10:358–364.
39. Zakany J, Fromental-Ramain C, Warot X, Duboule D. Regulation of number and size of digits by posterior Hox genes: a dose-dependent mechanism with potential evolutionary implications. Proc Natl Acad Sci USA 1997;94:13,695–13,700.
40. Davis AP, Capecchi MR. Axial homeosis and appendicular skeleton defects in mice with a targeted disruption of Hoxd11. Development 1994;120:2187–2198.
41. Favier B, Le Meur M, Chambon P, Dolle P. Axial skeleton homeosis and forelimb malformations in Hoxd11 mutant mice. Proc Natl Acad Sci USA 1995;92:310–314.
42. Small KM, Potter SS. Homeotic transformations and limb defects in Hoxa11 mutant mice. Genes Dev 1993;7:2318–2328.
43. Davis AP, Witte DP, Hsieh-Li HM, Potter SS, Capecchi MR. Absence of radius and ulna in mice lacking Hoxa11 and Hoxd11. Nature 1995;375:791–795.
44. Goff DJ, Tabin CJ. Analysis of Hoxd13 and Hoxd11 misexpression in chick limb buds reveals that Hox genes affect both bone condensations and growth. Development 1997;124:627–636.
45. Kanzler B, Kuschert SJ, Liu YH, Mallo M. Hoxa2 restricts the chondrogenic domain and inhibits bone formation during development of the branchial area. Development 1998;125:2587–2597.
46. Duprez D, Bell EJ, Richardson MK, Archer CW, Wolpert L, Brickell PM, et al. Overexpression of BMP-2 and BMP-4 alters the size and shape of developing skeletal elements in the chick limb. Mech Dev 1996;57:145–157.
47. Macias D, Ganan Y, Sampath TK, Piedra ME, Ros MA, Hurle JM. Role of BMP-2 and OP-1 (BMP-7) in programmed cell death and skeletogenesis during chick limb development. Development 1997;124:1109–1117.
48. Francis-West PH, Abdelfattah A, Chen P, Allen C, Parish J, Ladher R, et al. Mechanisms of GDF-5 action during skeletal development. Development 1999:126;1305–1315.
49. Storm EE, Huynh TV, Copeland NG, Jenkins NA, Kingsley DM, Lee SJ. Limb alterations in brachypodism mice due to mutations in a new member of the TGF beta-superfamily. Nature 1994;368:639–643.
50. Thomas JT, Lin K, Nandedkar M, Camargo M, Cervenka J, Luyten FP. A human chondrodysplasia due to a mutation in a TGF-beta superfamily member. Nat Genet 1996;12:315–317.
51. Thomas JT, Kilpatrick MW, Lin K, Erlacher L, Lembessis P, Costa T, et al. Disruption of human limb morphogenesis by a dominant negative mutation in CDMP1. Nat Genet 1997;17:58–64.
52. Polinkovsky A, Robin NH, Thomas JT, Irons M, Lynn A, Goodman FR, et al. Mutations in CDMP1 cause autosomal dominant brachydactyly type C. Nat Genet 1997;17:18,19.
53. Zou H, Wieser R, Massague J, Niswander L. Distinct roles of type I bone morphogenetic protein receptors in the formation and differentiation of cartilage. Genes Dev 1997;11:2191–2203.
54. Brunet LJ, McMahon JA, McMahon AP, Harland RM. Noggin, cartilage morphogenesis, and joint formation in the mammalian skeleton. Science 1998;280:1455–1457.
55. Pathi S, Rutenberg JB, Johnson RL, Vortkamp A. Interaction of Ihh and BMP/Noggin signaling during cartilage differentiation. Dev Biol 1999;209:239–253.
56. Bi W, Deng JM, Zhang Z, Behringer RR, de Crombrugghe B. Sox9 is required for cartilage formation. Nat Genet 1999;22:85–89.
57. Ducy P, Karsenty G. Genetic control of cell differentiation in the skeleton. Curr Opin Cell Biol 1998;10:614–619.
58. Tondravi MM, McKercher SR, Anderson K, Erdmann JM, Quiroz M, Maki R, et al. Osteopetrosis in mice lacking haematopoietic transcription factor PU.1. Nature 1997;386:81–84.
59. Bell DM, Leung KK, Wheatley SC, Ng LJ, Zhou S, Ling KW, et al. SOX9 directly regulates the type-II collagen gene. Nat Genet 1997;16:174–178.

60. Wagner T, Wirth J, Meyer J, Zabel B, Held M, Zimmer J, et al. Autosomal sex reversal and campomelic dysplasia are caused by mutations in and around the SRY-related gene SOX9. Cell 1994;79:1111–1120.
61. Foster JW, Dominguez-Steglich MA, Guioli S, Kowk G, Weller PA, Stevanovic M, et al. Campomelic dysplasia and autosomal sex reversal caused by mutations in an SRY-related gene. Nature 1994;372:525–530.
62. Karaplis AC, Luz A, Glowacki J, Bronson RT, Tybulewicz VL, Kronenberg HM, et al. Lethal skeletal dysplasia from targeted disruption of the parathyroid hormone-related peptide gene. Genes Dev 1994;8:277–289.
63. Karaplis AC, He B, Nguyen MT, Young ID, Semeraro D, Ozawa H, et al. Inactivating mutation in the human parathyroid hormone receptor type 1 gene in Blomstrand chondrodysplasia. Endocrinology 1998;139:5255–5258.
64. Schipani E, Kruse K, Jüppner, H. A constitutively active mutant PTH-PTHrP receptor in Jansen-type metaphyseal chondrodysplasia. Science 1995;268:98–100.
65. Vortkamp A, Lee K, Lanske B, Segre GV, Kronenberg HM, Tabin CJ. Regulation of rate of cartilage differentiation by Indian hedgehog and PTH-related protein. Science 1996;273:613–622.
66. Lanske B, Karaplis AC, Lee K, Luz A, Vortkamp A, Pirro A, et al. PTH/PTHrP receptor in early development and Indian hedgehog-regulated bone growth. Science 1996;273:663–666.
67. Zhang P, Jobert AS, Couvineau A, Silve C. A homozygous inactivating mutation in the parathyroid hormone/parathyroid hormone-related peptide receptor causing Blomstrand chondrodysplasia. J Clin Endocrinol Metab 1998;83:3365–3368.
68. Jobert AS, Zhang P, Couvineau A, Bonaventure J, Roume J, Le Merrer M, et al. Absence of functional receptors for parathyroid hormone and parathyroid hormone-related peptide in Blomstrand chondrodysplasia. J Clin Invest 1998;102:34–40.
69. Rousseau F, Bonaventure J, Legeai-Mallet L, Pelet A, Rozet JM, Maroteaux P, et al. Mutations in the gene encoding fibroblast growth factor receptor-3 in achondroplasia. Nature 1994;371:252–254.
70. Shiang R, Thompson LM, Zhu YZ, Church DM, Fielder TJ, Bocian M, et al. Mutations in the transmembrane domain of FGFR3 cause the most common genetic form of dwarfism, achondroplasia. Cell 1994;78:335–342.
71. Colvin JS, Bohne BA, Harding GW, McEwen DG, Ornitz DM. Skeletal overgrowth and deafness in mice lacking fibroblast growth factor receptor 3. Nat Genet 1996;12:390–397.
72. Deng C, Wynshaw-Boris A, Zhou F, Kuo A, Leder P. Fibroblast growth factor receptor 3 is a negative regulator of bone growth. Cell 1996;84:911–921.
73. Goodrich LV, Johnson RL, Milenkovic L, McMahon JA, Scott MP. Conservation of the hedgehog/patched signaling pathway from flies to mice: induction of a mouse patched gene by Hedgehog. Genes Dev 1996;10:301–312.
74. van den Heuvel M, Ingham PW. Smoothened encodes a receptor-like serpentine protein required for hedgehog signaling. Nature 1996;382:547–551.
75. Chen Y, Struhl G. Dual roles for patched in sequestering and transducing Hedgehog. Cell 1996;87:553–563.
76. Stone DM, Hynes M, Armanini M, Swanson TA, Gu Q, Johnson RL, et al. The tumour-suppressor gene patched encodes a candidate receptor for Sonic hedgehog. Nature 1996;384:129–134.
77. Naski MC, Colvin JS, Coffin JD, Ornitz DM. Repression of hedgehog signaling and BMP4 expression in growth plate cartilage by fibroblast growth factor receptor 3. Development 1998;125:4977–4988.
78. Sahni M, Ambrosetti D-C, Mansukhani A, Gertner R, Levy D, Basilico C. FGF signaling inhibits chondrocyte proliferation and regulates bone development through the STAT-1 pathway. Genes Dev 1999:13.
79. Komori T, Yagi H, Nomura S, Yamaguchi A, Sasaki K, Deguchi K, et al. Targeted disruption of Cbfa1 results in a complete lack of bone formation owing to maturational arrest of osteoblasts. Cell 1997;89:755–764.
80. Otto F, Thornell AP, Crompton T, Denzel A, Gilmour KC, Rosewell IR, Stamp GW, Beddington R S, Mundlos S, Olsen BR, Selby PB, Owen MJ. Cbfa1, a candidate gene for cleidocranial dysplasia syndrome, is essential for osteoblast differentiation and bone development. Cell 1997;89:765–771.
81. Mundlos S, Otto F, Mundlos C, Mulliken JB, Aylsworth AS, Albright S, et al. Mutations involving the transcription factor CBFA1 cause cleidocranial dysplasia. Cell 1997;89:773–779.
82. Lee B, Thirunavukkarasu K, Zhou L, Pastore L, Baldini A, Hecht J, et al. Missense mutations abolishing DNA binding of the osteoblast-specific transcription factor OSF2/CBFA1 in cleidocranial dysplasia. Nat Genet 1997;16:307–310.

83. Ducy P, Zhang R, Geoffroy Vidall AL, Karsenty G. Osf2/Cbfa1: a transcriptional activator of osteoblast differentiation. Cell 1997;89;747–754.
84. Ducy P, Starbuck M, Priemel M, Shen J, Pinero G, Geoffroy V, et al. A Cbfa1-dependent genetic pathway controls bone formation beyond embryonic development. Genes Dev 1999;13:1025–1036.
85. Yoshida H, Hayashi S, Kunisada T, Ogawa M, Nishikawa S, Okamura H, et al. The murine mutation osteopetrosis is in the coding region of the macrophage colony stimulating factor gene. Nature 1990;345:442–444.
86. Wiktor-Jedrzejczak W, Bartocci A, Ferrante AW Jr, Ahmed-Ansari A, Sell KW, Pollard JW, et al. Total absence of colony-stimulating factor 1 in the macrophage-deficient osteopetrotic (op/op) mouse. Proc Natl Acad Sci USA 1990;87:4828–4832.
87. Grigoriadis AE, Wang ZQ, Cecchini MG, Hofstette RW, Felix R, Fleisch HA, et al. c-Fos: a key regulator of osteoclast-macrophage lineage determination and bone remodeling. Science 1994;266:443–448.
88. Tsuda E, Goto M, Mochizuki S, Yano K, Kobayashi F, Morinaga T, et al. Isolation of a novel cytokine from human fibroblasts that specifically inhibits osteoclastogenesis. Biochem Biophys Res Commun 1997;234:137–142.
89. Tan KB, Harrop J, Reddy M, Young P, Terrett J, Emery J, et al. Characterization of a novel TNF-like ligand and recently described TNF ligand and TNF receptor superfamily genes and their constitutive and inducible expression in hematopoietic and non-hematopoietic cells. Gene 1997;204:35–46.
90. Simonet WS, Lacey DL, Dunstan CR, Kelley M, Chang MS, Luthy R, et al. Osteoprotegerin: a novel secreted protein involved in the regulation of bone density. Cell 1997;89:309–319.
91. Wong B., Rho J, Arron J, Robinson E, Orlinick J, Chao M, et al. TRANCE is a novel ligand of the tumor necrosis factor receptor family that activates c-Jun N-terminal kinase in T cells. J Biol Chem 1997;272:25,190–25,194.
92. Yasuda H, Shima N, Nakagawa N, Yamaguchi K, Kinosaki M, Mochizuki S, et al. Osteoclast differentiation factor is a ligand for osteoprotegerin/osteoclastogenesis-inhibitory factor and is identical to TRANCE/RANKL. Proc Natl Acad Sci USA 1998;95:3597–3602.
93. Lacey DL, Timms E, Tan HL, Kelley MJ, Dunstan CR, Burgess T, et al. Osteoprotegerin ligand is a cytokine that regulates osteoclast differentiation and activation. Cell 1998;93:165–176.
94. Bucay N, Sarosi I, Dunstan CR, Morony S, Tarpley J, Capparelli C, et al. Osteoprotegerin-deficient mice develop early onset osteoporosis and arterial calcification. Genes Dev 1999;12:1260–1268.
95. Kong YY, Yoshida H, Sarosi I, Tan HL, Timms E, Capparelli C, et al. OPGL is a key regulator of osteoclastogenesis, lymphocyte development and lymph-node organogenesis. Nature 1999;397:315–323.
96. Hsu H, Lacey DL, Dunstan CR, Solovyev I, Colombero A, Timms E, et al. Tumor necrosis factor receptor family member RANK mediates osteoclast differentiation and activation induced by osteoprotegerin ligand. Proc Natl Acad Sci USA 1999;96:3540–3545.
97. Anderson DM, Maraskovsky E, Billingsley WL, Dougall WC, Tometsko ME, Roux ER, et al. A homologue of the TNF receptor and its ligand enhance T-cell growth and dendritic-cell function. Nature 1997;390:175–179.
98. Wong BR, Josien R, Lee SY, Vologodskaia M, Steinman RM, Choi Y. The TRAF family of signal transducers mediates NF-kappaB activation by the TRANCE receptor. J Biol Chem 1998;273:28,355–28,359.
99. Lomaga MA, Yeh WC, Sarosi I, Duncan GS, Furlonger C, Ho A, Met al. TRAF6 deficiency results in osteopetrosis and defective interleukin-1, CD40, and LPS signaling. Genes Dev 1999;13:1015–1024.
100. May MJ, Ghosh S. Signal transduction through NF-kappa B. Immunol Today 1998;19:80–82.
101. Franzoso G, Carlson L, Xing L, Poljak L, Shores EW, Brown KD,et al. Requirement for NF-kappaB in osteoclast and B-cell development. Genes Dev 1997;11:3482–3496.
102. Iotsova V, Caamano J, Loy J, Yang Y, Lewin A, Bravo R. Osteopetrosis in mice lacking NF-kappaB1 and NF-kappaB2. Nat Med 1997;3:1285–1289.
103. Li Y, Olsen BR. Murine models of human genetic skeletal disorders. Matrix Biol 1997;16:49–52.
104. Li Y, Lacerda DA, Warman ML, Beier DR, Yoshioka H, Ninomiya Y, et al. A fibrillar collagen gene, Col11a1, is essential for skeletal morphogenesis. Cell 1995;80:423–430.
105. Vikkula M, Mariman EC, Lui VC, Zhidkova NI, Tiller GE, Goldring MB, et al. Autosomal dominant and recessive osteochondrodysplasias associated with the COL11A2 locus. Cell 1995;80:431–437.
106. Watanabe H, Kimata K, Line S, Strong D, Gao LY, Kozak CA, et al. Mouse cartilage matrix deficiency (cmd) caused by a 7 bp deletion in the aggrecan gene. Nat Genet 1994;7:154–157.
107. Warman ML, Abbott M, Apte SS, Hefferon T, McIntosh I, Cohn DH, et al. A type X collagen mutation causes Schmid metaphyseal chondrodysplasia. Nat Genet 1993;5:79–82.

108. Vu TH, Shipley JM, Bergers G, Berger JE, Helms JA, Hanahan D, et al. MMP-9/gelatinase B is a key regulator of growth plate angiogenesis and apoptosis of hypertrophic chondrocytes. Cell 1998;93:411–422.
109. Luo G, Ducy P, McKee MD, Pinero GJ, Loyer E, Behringer RR, et al. Spontaneous calcification of arteries and cartilage in mice lacking matrix GLA protein. Nature 1997;386:78–81.
110. Kocher MS, Shapiro F. Osteogenesis imperfecta. J Am Acad Orthop Surg 1998;6:225–236.
111. Ducy P, Desbois C, Boyce B, Pinero G, Story B, Dunstan C, et al. Increased bone formation in osteocalcin-deficient mice. Nature 1996;382:448–452.
112. Danielson KG, Baribault H, Holmes DF, Graham H, Kadler K E, Iozzo RV. Targeted disruption of decorin leads to abnormal collagen fibril morphology and skin fragility. J Cell Biol 1997;136:729–743.
113. Gilmour DT, Lyon GJ, Carlton MB, Sanes JR, Cunningham JM, Anderson JR, et al. Mice deficient for the secreted glycoprotein SPARC/osteonectin/BM40 develop normally but show severe age-onset cataract formation and disruption of the lens. EMBO J 1998;17:1860–1870.
114. Xu T, Bianco P, Fisher LW, Longenecker G, Smith E, Goldstein S, et al. Targeted disruption of the biglycan gene leads to an osteoporosis-like phenotype in mice. Nat Genet 1998;20:78–82.
115. Gelb BD, Shi GP, Chapman HA, Desnick RJ. Pycnodysostosis, a lysosomal disease caused by cathepsin K deficiency. Science 1996;273:1236–1238.
116. Saftig P, Hunziker E, Wehmeyer O, Jones S, Boyde A, Rommerskirch W, et al. Impaired osteoclastic bone resorption leads to osteopetrosis in cathepsin-K-deficient mice. Proc Natl Acad Sci USA 1997;95:13,453–13,458.
117. Hayman AR, Jones SJ, Boyde A, Foster D, Colledge WH, Carlton MB, et al. Mice lacking tartrate-resistant acid phosphatase (Acp 5) have disrupted endochondral ossification and mild osteopetrosis. Development 1996;122:3151–3162.
118. Rittling SR, Matsumoto HN, McKee MD, Nanci A, An XR, Novick KE, et al. Mice lacking osteopontin show normal development and bone structure but display altered osteoclast formation in vitro. J Bone Miner Res 1998;13:1101–1111.
119. Liaw L, Birk DE, Ballas CB, Whitsitt JS, Davidson JM, Hogan BL. Altered wound healing in mice lacking a functional osteopontin gene. J Clin Invest 1998;101(Suppl. 1):1468–1478.
120. Yoshitake H, Rittling SR, Denhardt DT, Noda M. Osteopontin-deficient mice are resistant to ovariectomy-induced bone resorption. Proc Natl Acad Sci USA 1999;96:8156–8160.
121. Soriano P, Montgomery C, Geske R, Bradley A. Targeted disruption of the c-src proto-oncogene leads to osteopetrosis in mice. Cell 1991;64:693–702.
122. Boyce BF, Yoneda T, Lowe C, Soriano P, Mundy GR. Requirement of pp60c-src expression for osteoclasts to form ruffled borders and resorb bone in mice. J Clin Invest 1992;90:1622–1627.
123. Packer SO. The eye and skeletal effects of two mutant alleles at the microphthalmia locus of Mus musculus. J Exp Zool 1967;165:21–45.
124. Walker DG. Spleen cells transmit osteopetrosis in mice. Science 1975;190:785–787.
125. Hodgkinson CA, Moore KJ, Nakayama A, Steingrimsson E, Copeland NG, Jenkins NA, et al. Mutations at the mouse microphthalmia locus are associated with defects in a gene encoding a novel basic-helix-loop-helix-zipper protein. Cell 1993;74:395–404.
126. Corral DA, Amling M, Priemel M, Loyer E, Fuchs S, Ducy P, et al. Dissociation between bone resorption and bone formation in osteopenic transgenic mice. Proc Natl Acad Sci USA 1998;95:13,835–13,840.

19 Transgenic Mouse Models of Prostate Cancer

Robert J. Matusik, PhD,
Naoya Masumori, MD, PhD,
Tania Thomas, PhD, *Thomas Case*, BS,
Manik Paul, BS, *Susan Kasper*, PhD,
and Scott B. Shappell, MD, PhD

Contents

HUMAN PROSTATE CANCER

The reproductive organs are not required for an individual's survival but are required for survival of the species. As the individual approaches adulthood, the prostate undergoes developmental changes, resulting in maturation of this gland at puberty. At this stage, the prostate becomes a differentiated gland that produces proteins and other substances fundamental for reproduction and survival of the species. By the age of 50, as many as 30% of all men will harbor microscopic foci of prostate adenocarcinoma (CaP), and the incidence increases with age. In the United States, CaP is clinically diagnosed in approx 10% of men during their lifetime (189,000/yr), where it will claim 31,900 lives each year (13% of male cancer deaths) *(1)*. The recent rising incidence of CaP *(2)* has plateaued *(3)*, but the high prevalence of this disease and the aging of the US population still makes this a cancer that demands prompt attention.

Too little is known about the etiology of CaP, limiting our attempts at nonsurgical curative therapy or effective treatments of advanced disease. The most likely precursor lesion of the majority of clinically detected invasive carcinomas is identified as high-

From: *Contemporary Endocrinology: Transgenics in Endocrinology*
Edited by: M. Matzuk, C. W. Brown, and T. R. Kumar © Humana Press Inc., Totowa, NJ

grade prostatic intra-epithelial neoplasia (PIN) *(4,5) (6)*, which appears to be androgen-dependent *(7)*. Low-grade PIN may emerge as early as the third decade of life *(8)*. High-grade PIN is characterized by partial disruption of the basal-cell layer, loss of differentiated cell function including secretory proteins, increased proliferative potential, nuclear alterations, and aneuploidy *(9–11)*. The origin of these premalignant lesions remains unknown, as do the triggers which lead to their progression to invasive disease. Although high-grade invasive adenocarcinoma are clearly aggressive, a continuing dilemma in the clinical management of the majority of CaP is the lack of adequate markers that can distinguish between latent or slow-growing tumors and potentially aggressive forms of histologically similar disease. Some potential prognostic markers in CaP include DNA ploidy, nuclear morphometry, proliferation antigens, and known oncogenes and tumor-suppressor genes *(12,13)*. To date, DNA content, the overexpression of the c-*erb*B2 protein *(14–17)*,and the linkage of familial disease to chromosome 1 and Xq27-28 show promise as predictors of cancer progression *(18–21)*.

The proliferation and differentiation of prostatic tissue is influenced by androgen action, and androgens initally play at least a permissive role in the development of CaP *(22)*. Since CaP continues to require androgens for growth, androgen-ablation therapy has been one of the methods of choice in treating patients with advanced prostatic carcinoma *(23,24)*. Initially, >80% of the patients respond to androgen ablation, but the duration of the response is usually only 12–18 mo *(24)* and tumor growth progresses to androgen-independence, often reflected by an increase in serum prostate-specific antigen (PSA) levels *(25)*. The loss of hormonal dependence is complex and poorly understood. Mutations of the androgen-receptor (AR) gene may be a mechanism in which androgen independence develops in some cases *(26–35)*. Recent reports indicate that in metastatic androgen-independent prostate cancer, as many as 50% of patients may have AR mutations, which may lead to activation of the receptor by other steroid hormones, such as estrogen and progesterone *(24)*. Tilley et al. *(36)* found mutations in the AR, some of which occurred in exon 1. Other examples of AR alterations include gene amplification in 23% of recurrent hormone-refractory prostate-cancer samples *(37)* and decreases in the CAG repeats (glutamine tracks), which may increase the risk of developing carcinoma of the prostate *(38–40)*. Antiandrogens such as flutamide are also used in combination with luteinizing hormone-releasing hormone agonists to block any residual effects of androgens produced by the adrenal glands *(22)*. Scher reported that when flutamide therapy failed, using another antiandrogen such as a bicalutamide significantly decreased serum PSA levels in some patients *(41–43)*, and likewise, patients who failed bicalutamide therapy may subsequently respond to flutamide treatment *(42)*. Also, a 21% of the patients showed both a clinical and serum PSA response upon complete withdrawal from the anti-androgen *(44)*. Studies in mouse models have reported the presence of AR mutations that permit the progression of CaP in castrated mice *(45,46)*. These data suggest that the AR-mediated response of CaP cells is not inactivated, but may be activated through an alternative pathway.

These clinical and research observations raise basic scientific questions about the molecular events that promote the formation and progression of PIN lesions, the progression from latent or low grade tumors to locally aggressive and metastatic CaP, and the switch from androgen-dependent to hormone-refractory tumor growth. Our understanding of carcinogenesis in the prostate has been hampered by a lack of suitable animal models and cell lines that cover the spectrum of this disease. Recently, questions have been raised about the prostatic origin of two available animal models, the Dunning tumor

(47) and the chemically induced Pollard models *(48,49)*. In addition to the few available human prostate cancer cell lines and xenograft transplantation into nude mice, these models correlate with late stages of disease, and are of limited value in studying early events in tumorigenesis. New animal models are needed to meet the following criteria: (1) prostate-restricted tumor development; (2) presence of putative precursor lesions resembling human PIN; (3) stochastic development of tumors; (4) initial androgen-dependence of tumors; (5) tumor progression to androgen-independence after androgen withdrawal; and 6) metastasis, including lymph nodes and bone. It is unlikely that any one model will meet all of these criteria; rather, a series of models may be required to adequately reproduce key features and clinical responses seen in human disease. In order to understand what goes wrong when CaP occurs, we need to understand the mechanisms that lead to prostatic growth and differentiation. This insight will also provide us with the tools necessary to make new animal models for CaP.

COMBINATIONAL GENE CONTROL

Combinational gene control offers a conceptual model to explain *cell determination* and *cell differentiation*. When the mammalian embryonic cell divides, a decision must be made as to which regulatory proteins will appear in the two daughter cells. In turn, the daughter cells must determine which regulatory proteins will be induced during the next cell division, and so on until the mature embryo develops. The paradigm of *combinational gene control* assumes that the regulatory protein added to each new cell must be self-perpetuating, giving that cell a memory *(50,51)*. The mechanism by which regulatory proteins induce the next wave of gene transcription was originally proposed as autoregulation, and recent examples of this type of autoregulation include chromatin structure *(52,53)*, DNA methylation *(54,55)*, and/or imprinting identified in at least 19 genes *(51,56)*. The regulatory proteins added during development do not always initially change gene expression but may await the appearance of a final regulatory protein that now dispatches the cell to its final transformation. This cascade can account for *cell determination*, where the cell is assigned a developmental fate within an organ, and subsequently *cell differentiation*, where the assigned cell now emerges with its own unique character producing a specific array of proteins. As few as 25 regulatory proteins can potentially specify 10,000 different cell types *(51)*. Thus, the regulatory proteins do not need to be unique, but they must appear in the proper combination of multiple factors to result in tissue-specific expression *(50)*. Although this model is intended to arrange events in a temporal fashion from one cell division to the next, cellular interactions also occur as the embryo is subdivided into distinct regions. These spatial interactions among cell types contribute signals to adjacent cells to divide and differentiate. Using this model, we can create a time line of events that result in prostatic organogenesis.

Rodent Prostate-Cell Determination

The rodent prostate gland consists of four different lobes defined as ventral, dorsal, lateral, and anterior (coagulating gland), where the epithelial cells are specialized and express specific protein products. Each lobe is anatomically distinct, but the dorsal and lateral lobes, because of their size and proximity, are often examined collectively as the dorsolateral prostate. The rodent dorsolateral prostate is considered to be homologous to the human prostate peripheral zone *(57)*, the most common site of prostate cancer in humans.

In the rodent, the urogenital sinus (UGS) begins to develop in the embryo day (E) 10–11 of gestation, and testicular androgen production in the mouse begins E 12.5–13, peaking at E 17–18 gestation, thereafter declining until birth *(58,59)*. 5α-reductase activity is detected in the UGS at E 14.5 *(60)*, and coincides with Wolffian-duct differentiation into the epididymis, seminal vesicles, and ductus deferens as the Müllerian duct degenerates. Normal rodent prostate development is first observed during gestation at E 18.5 in the mouse, or E 19.5 in the rat *(61,62)*. At that time, a pair of ventral prostate buds and outgrowths of the anterior prostate (coagulating gland) appear. Within the next 24 h, considerable development of the ventral, dorsal, and lateral prostate occurs, and the *in utero* hormones—including androgen and estrogen—imprint prostatic cells to respond later in a temporal and spatial manner *(62,63)*. The proximal prostate ducts appear during fetal development, and later give rise to the distal branching structures which begin to proliferate in the newborn. At birth, the prostatic lobes are distinct and the ventral ducts develop more rapidly than other prostatic lobes. Final differentiation of the prostate occurs when sexual maturation is reached. Differences in the ductal branching patterns of the prostatic lobes occur as a result of both temporal changes and spatial influences. For example, early prostatic budding in the fetus is controlled by the periurethral mesenchyme, while later growth requires the ventral mesenchymal pad. Furthermore, the production of lobe-specific secretory proteins is induced by mesenchymal cells that originate from defined areas of the UGS *(63,64)*.

Cell determination for the reproductive glands occurs in the embryo, but the final stages of *cell differentiation* take place after birth and during sexual maturation, and are dependent upon androgens. For example, androgen resistance prevents normal embryonic and pubertal development of the genotypic male. Mutations in the AR result in androgen-insensitivity syndrome (AIS, formerly termed testicular feminization males or *tfm*) where patients are genotypic males (46,XY) but the resulting phenotype is female or they may appear with ambiguous external genitalia *(65–68)*. In *tfm* mice, the Wolffian-duct degenerate, and a female-like ureter, a shortened vagina, and external female genitalia develop in the male. The *tfm* mice never develop prostates. Clearly, the AR plays a vital role in the development of the male reproductive tract *(69)*. AR roles differ in mesenchymal and epithelial cells. Since the *tfm* mouse reflects the AIS human phenotype, the mouse model permits an examination of the critical role that the mesenchymal-epithelial cell interactions play in organogenesis *(70,71)*. Cunha's seminal work showed that *tfm* mesenchymal cells, when recombined with wild-type epithelial cells from the UGS and implanted into the kidney capsule, would histologically develop into an organ that was vaginal-like, whereas wild-type UGS mesenchymal cells combined with *tfm* epithelial cells would result in a differentiated prostate *(72)*. Cunha went on to prove that female vaginal stroma could direct epithelium to form a prostate in response to androgens, but that these vaginal stromal cells increasingly lost this ability when the newborn mouse reached 20 d of age. However, vaginal epithelium could always be converted to a ductal prostate structure by UGS mesenchymal cells in the presence of androgens *(73)*. Therefore, the AR in the mesenchymal cells (UGS or newborn vaginal) is essential to induce a signal that permits both prostatic ductal growth and instructs the epithelium to differentiate *(74,75)*. This process is completed during sexual maturation (*see* review, ref. *76*). Thus, precise temporal event(s) within a mesenchymal cell can cause a spatial event(s) in an adjacent epithelial cell to switch that epithelial cell into a new pathway.

Current approaches to study the molecular mechanism of morphogenesis of the prostate have been limited by a lack of identified potential regulatory proteins that are key to the process. A large number of studies have identified growth factors that communicate to the cell surface to start a cascade of events; however, direct control of transcription by *critical gene regulatory proteins* has been limited to only a few candidates. The importance of homeobox genes for *cell determination* in the reproductive tract was demonstrated in the Pax-2 knockout mouse, where the seminal vesicles are absent but the prostate is normal *(77)*. Although Pax-2 is not required for prostate development *(77)*, other *gene regulatory proteins* that are likely to be important include the AR *(62–64,72–76)*, estrogen receptor (ER) α and β *(78–82)*, Hoxd-13 *(83,84)*, and Nkx3.1 *(85)* genes.

The AR is present in many tissues in both males and females; therefore, its presence alone is not sufficient to explain prostatic-cell determination. Nevertheless, the AR is a *critical gene regulatory protein* in the development and differentiation of the male accessory sex organs, because genotypic XY males with defective AR (AIS/*tfm*) develop phenotypically as females, where the degree of the female phenotype correlates with the level of activity of the defective AR *(65–68)*. These data indicate that the AR does not work alone, but is hardwired into the developmental program of the prostate.

Studies in which the administration of estrogen occurred during gestation in neonatal animals, or in combination with androgens to adult animals to induce prostatic tumors, all indicate a fundamental role for estrogen in the prostate (*see* review, ref. *78*). Diethylstilbestrol (DES) treatment between 1 and 5 d in neonatal mice will induce prostatic utricle tumors, a mullerian duct remnant *(79)*, and prostatic dysplasia within 1 yr *(86)*. However, exposure of the fetal mouse to low levels of estrogen will stimulate prostate development with an eventual twofold increase in the AR levels, a response which is opposite to the effect seen when high levels of estrogen are administered *(87)*. The combined treatment of estrogen and androgen will induce hyperplasia, dysplasia, and carcinoma in the dorsolateral prostatic lobes of the Noble rat *(88–91)*, suggesting a direct role for estrogen in that prostatic lobe. In rats, neonatal treatment with estrogen will decrease prostatic size and reduce AR levels in the prostate *(80)*. The recent cloning of the ER β from the rat prostate—a tissue that highly expresses this gene *(81,82)*—further suggests a unique role for estrogen in the rodent prostate, but the low level of this novel ER β in the human prostate makes its function unclear in man *(92)*. The importance of estrogen in the male accessory glands has been highlighted by the ER α gene knockout mouse (ERKO) where the males are infertile *(93–95)*.

The importance of homeobox genes in *cell determination* in the reproductive tract was demonstrated in the Pax-2 knockout mouse, in which the seminal vesicles are absent, but the prostate is normal *(77)*. The 39 Hox genes are organized into four clusters, Hox A, B, C, and D. These homeobox genes are activated in a temporal fashion starting at the 3'-end of the locus which is expressed at the cephalic end of the body and ending at the 5'-end which is expressed caudally. Hoxd-13 is located at the most 5'-end of the HoxD gene cluster and is believed to be involved in the morphogenesis of the male accessory sex organs. Expression of the Hoxd-13 gene is detected at E 15 in the embryonic mouse UGS, and gene expression in the prostate declines postnatally between d 1 and 60 *(84)*. Hoxd-13 mRNA is localized primarily to the epithelium with low levels being observed in the mesenchyme *(83)*. Hoxd-13 gene expression parallels both the budding of the embryonic mouse prostate that begins around E 18 *(62)* and the branching of prostatic ducts, where branching is exponential in the first 2 wk postnatally and reaches completion around d 60 at sexual maturation *(64)*. In Hoxd-13 gene-knockout mice *(96)*, the usual postnatal prostatic ductal branching is dramatically diminished *(84)*, where the dorsal lobes are hypoplastic and the anterior lobes are absent *(97)*. Thus, this model suggests that Hoxd-13 is involved in *cell determination*

The recently described mammalian homeobox Nkx3.1 gene is expressed in the developing male reproductive organs. The Nkx3.1 gene is the first vertebrate homolog to the *Drosophila* natural killer (NK)-3 family *(98)*. In the mouse, Nkx3.1 gene expression is detected in the primordial buds in the pelvic area of the UGS at E 14.5, in the developing ventral prostatic buds at E 17.5, and subsequently in the anterior and dorsolateral buds. Expression is highest in the proliferating epithelial cells at the distal portion of the prostatic ductal tree at a time when organ development is nonresponsive to androgens *(76)*. After birth, Nkx3.1 gene expression increases during sexual maturation when the prostatic ductal morphogenesis is at its peak. Nkx3.1 gene expression remains elevated in the adult, decreases upon castration, and is induced with androgen treatment in castrated rodents *(85,98)* and in cell culture *(99)*. Nkx3.1 gene expression remains elevated in the adult, decreases upon castration and is re-induced by androgen treatment in rodents *(85,98)* or cell culture *(99)*. In a Nkx-knockout mouse model, deletion of this gene results in defects in prostatic ductal organization, leading to prostatic epithelial-cell hyperplasia, dysplasia, and a loss of secretory proteins (100). Sciavolino et al., suggested that their data is consistent with both a mesenchymal-cell induction of epithelial-cell Nkx3.1 gene expression during early development when the tissue has little/no responsiveness to androgens, and androgen regulation of Nkx3.1 gene in the differentiated epithelial cells of the mature prostate *(85,100)*. The Nkx3.1 protein is predicted to be the first in a series of related genes that are important in prostatic *cell determination* that occurs in the embryo and in the final stages of proliferation and *cell differentiation* that occurs after birth.

Rodent Prostate-Cell Differentiation

Biochemical analysis exists on the rat prostate, but limited information is published on the secretory proteins of the mouse prostate. Also, no rodent analog for human PSA has been identified. The rat ventral lobe differs most significantly from the other lobes because its protein products are quite distinct *(101–103)*. The rat anterior lobe produces many of the androgen-regulated proteins seen in the dorsolateral prostate *(104)*. The seminal-vesicle secretion II (SVSII) is a major secretory product of both the seminal vesicle and the dorsolateral prostate *(103,105)*, and Dorsal Prostate I and II are found in the dorsal prostate *(104,106)*. Probasin (PB) is detected in all lobes of the prostate and the seminal vesicles, with lateral lobe expression the highest (100%) followed by dorsal (33%), anterior (14%), ventral (4%), and seminal vesicles (2%) *(103,107,108)*. Some of these secretory proteins are detected in the prostate between 2–3 wk of animal age (102; 103) but the major increase in synthesis occurs with *cell differentiation*, which is finalized with sexual maturation at around 6 wk of age.

The prostate-specific expression of differentiation associated protein products can be used to serve as markers to identify prostatic cells and the promoters of these genes may contain the necessary information to target transgenes to prostatic cells. In an ever-increasing number of model systems, it is apparent that genes of differentiated function that are expressed in a cell specific manner are controlled by the same regulatory proteins that dictate organ determination *(109–130)*. Therefore, the promoters of tissue-specific genes may not only target genes in a tissue-specific manner, but may also contain key information that will lead us to the factors that control organogenesis.

Promoters that Target the Prostate

Attempts have recently been made to create prostate-cancer models in transgenic mice. As both oncogenes *(131–135)* and tumor-suppressor genes *(136–139)* have been implicated in the development of human prostate cancer, altering expression of these genes specifically in the prostate gland represents a rational approach for developing new model systems. Transgenes have been expressed in the prostate of transgenic mice by promoters that are not normally functional in that organ. It this case, the site of integration of the DNA fragment may be instrumental in permitting transgene expression in the prostate. The random nature of this event prevents routine use of these promoters in a reproducible manner. The hormone-regulated mouse mammary-tumor virus long-terminal repeat (MMTV-LTR), which targets gene expression to the mammary gland in transgenic mice, is also reported to express transgenes in the prostate at low levels. Only two promoters have been reported to specifically target transgenes to the prostate of transgenic mice.

To target genes to the prostate in transgenic mice, the promoter of the human PSA gene would appear to be a logical choice, since the protein is selectively expressed by human prostate tissue. However, the initial short PSA-promoter fragment functions poorly in human prostate-cancer cells *(140)*, and this fragment will not target gene expression specifically to the mouse prostate *(141)*. New efforts have identified PSA DNA sequences that enhance expression in cell culture *(142,143)*, but this construct has not been tested in transgenic animals. However, two new reports show that at least 6,000 basepairs (bp) of the PSA promoter is required to obtain PSA expression in trangenic mice *(144,145)*. Further work has demonstrated that intergenic sequences within 12 kb that separate the PSA gene from the glandular kallikrein-1 gene serve as a strong regulatory region that enhances transgene expression in transgenic mice *(146)*. Although 5'-flanking fragments of the PSA gene are being analyzed for sequences that target prostate-specific reporter gene expression, these enhancers plus the PSA promoter have not yet been used to target transgenes that disrupt prostatic function in transgenic mice.

The promoter of the rat PB gene is androgen regulated in human prostate cancer cell lines *(147)*, and functions specifically in prostatic cells when compared to nonprostatic cells in vitro *(148)*.Patrikainen et al., report that the sequences between –278 and –240 bp are critical for both androgen and prostate-specific regulation of the PB gene *(149)*, an area in juxtaposition with the previously defined AR-binding site-1 (ARBS-1) *(135,150)*. Androgen regulation involved a second ARBS-2 that functions in a cooperative manner with ARBS-1 to define the androgen-response region (ARR) (–244 to –96 bp) the probasin promoter as defined by –426 to +28 bp *(151)*. Further, this small probasin (PB) promoter from –426 to +28 bp is sufficient to target androgen regulation of transgene specifically to the epithelial cells in the prostate of transgenic mice *(135)*. A large PB fragment from approx –11,000 to +28 bp contains the ARR and additional enhancers that will increase the level of transgene expression in prostatic-epithelial cells of the transgenic mice *(152)*. Both the small probasin (sPB) and large probasin (LPB), and a new ARR_2PB *(152a)* demonstrate hormonal and developmental regulation of transgene expression in the transgenic mouse prostate *(135,152)*. In addition to a reporter gene *(135)*, a number of transgenes have been placed under the control of the PB promoter in transgenic mice. The sPB promoter targeted the N-acetyltransferase 2 gene to the prostate in transgenic mice with no apparent phenotype *(153)*. Also, sPB was linked to the *ras*T24 oncogene, resulting in epithelial-cell hyperplasia *(154)*, whereas when it was coupled to the early region of the Simian Virus 40 genes (SV40) transgenic mice developed prostatic dysplasia and prostate cancer *(155)*. The sPB-directed *ras*T24 and SV40 early-region constructs in transgenic mice represent the first models of prostatic transformation that was not complicated by "leaky" expression of the transgene in other male tissues.

COMPARISON OF CHARACTERISTICS OF EXISTING TRANSGENIC MODELS OF PROSTATE CANCER

Promoters and Transgenes

Targeting oncogenes and tumor-suppressor genes to the prostate in transgenic mice represents a rational approach for developing new mouse models *(156)*. For example, using the prostate-specific sPB promoter, the *rasT24* gene was targeted to the prostate to create hyperplasia. However, no PIN was detected after 6 mo *(154)*. Unfortunately, many of the gene promoters used to date do not function in a prostate-specific manner and tumors develop in other primary sites. A prostatic hyperplasia model was reported using the MMTV-LTR promoter coupled to the *int-2* gene *(157)*. Analysis of this model revealed that it represented a hyperplasia of the ampullary gland *(135,158)*, a mouse gland that has no human counterpart. Using the MMTV promoter to express keratinocyte growth factor (*kgf*)in the prostate led to papillary hyperplasia after 9 mo, but also led to *kgf* expression in the seminal vesicles, vas deferens, mammary gland, salivary gland, and Harderian gland *(159)*. The C3(1) promoter linked to the polyoma middle-T gene resulted in prostatic hyperplasia and dysplasia *(160)*, whereas the same promoter linked to the *bcl-2* gene resulted in the proliferation of the epithelium and stroma of the prostate by 3 mo of age *(161)*. Disruption of Mxi1, results in prostatic dysplasia in 1-yr-old mice, suggesting that a Mxi1 deficiency results only in preneoplastic lesions *(162)*. The human Nkx3.1 gene is androgen-regulated *(163)*, and was mapped to 8p21, a chromosome region frequently deleted in human prostate cancer *(99)*. Upon further investigation, the Nkx3.1 was intact *(164)*, however, a polymorphic change was detected in a codon which resulted in an altered binding of this homeobox protein to its *cis*-DNA element *(165)*. A Nkx-knockout mouse develops defects in prostatic ductal organization, leading to prostatic epithelial-cell hyperplasia, dysplasia, and a loss of secretory proteins *(100)*. These transgenic animals and knockout mice have provide insight into the importance of specific gene in prostatic development, but have not provided animal models for prostate cancer (for a summary of transgenic models for prostate cancer, *see* Table 1 *[156]*).

SV40 Large T- and Small T-Antigen (Early Region)-Expressing Transgenic Mice

Prior to our study *(166)*, all transgenic mouse lines targeting SV40 oncogenes to the prostate used the intact early region that resulted in the expression of both large T-antigen (Tag) and small Tag gene *(156)*. Since both p53 and retinoblastoma (*RB*) have been linked to the development of human CaP *(136–138,167–170)*, it is reasonable to target expression of large Tag, which has been shown to effectively bind and inactivate these two tumor-suppressor proteins *(171,172)*. Expression of the small Tag may have numerous effects, including inhibition of protein phosphatase 2A *(173)* that activates

Table 1

Comparison of Characteristics of Existing Transgenic Models of Prostate Cancer*

Transgene	*Target issues*	*Low-grade PIN*	*High-grade PIN*	*Invasive carcinoma*	*Metastasis*	*Strain*
C3 (1)/ SV40 large T, small t	VP>DLP, urethral gland, mammary, gland, sweat gland	8–12 wk	20 k	28 k	No	FVB/N, B10D2
PB (–426 bp)/ SV40 large T, small t	DLP>VP	5–8 wk	8–12 wk	>12 wk	Lymph nodes, lung, liver, bone	B6, B6/C57B
PB (11.5 kb) /SV40 large T	VP, DLP, AP	12–20 wk	12–20 wk	>20 wk	Rarely	CD-1
FG/SV40 large	VP, DLP, adrenal cortex, brown fat, (other tissues in neonate)	PIN observed		16–20 wk	Lymph nodes, adrenal, lung, bone, thymus	C57BL/6J, X CBA/J
CR-2/SV40 large T, small t	Prostate	8–10 wk	12 wk	12 wk	Lymph nodes, liver, lung, bone	FVB/N
C3 (1)/ Py-MT	VP, DLP, AP, E, VD, lung, mammary gland	Hyperplasias, dysplasias		Yes	No	FVB/N
C3 (1)/ bcl-2	VP, testes, uterus	Epithelial and stromal proliferation after 3 mo		No	No	C57B16J, X CBA/J
PB (–426 bp) /rasT24	VP, DLP	Hyperplasia after 6 mo, no PIN		No	No	FVB/N
MMTV/kgf	VP, DLP, seminal vesicles, VD, mammary gland, SG, Harderian gland	Papillary hyperplasia after 9 mo		No	No	FVB/ NHd

[a]SG, salivary gland; VD, vas deferens; E, epididymis; VP, ventral prostate; DLP, dorsolateral prostate; AP, anterior prostate; PIN, prostatic intraepithelial neoplasia.

*Adapted from Workgroup 3: transgenic and reconstitution models of prostate cancer, JE Green, et al., eds., The Prostate, Wiley-Liss, Inc., 1998;36:59–63.

multiple intracellular signaling pathways, including mitogen-activated protein kinase and protein kinase A-mediated phosphorylation cascades. Small Tag also activates the cyclin D1 promoter, which involves both the extracellular signal-regulated kinases and the stress-activated protein-kinase pathways *(174–176)*. Transgenic mice expressing both large and small Tags have been used to create several cancer models *(177)*.

Small PB Promoter Plus MAR. Targeting of the SV40 large and small Tag genes *(156)* by using the small PB (–426/+28 bp) promoter linked to the chicken lysozyme matrix-attachment region (MAR) sequence resulted in Line 8247, which has now been renamed the TRAMP (transgenic adenocarcinoma mouse prostate) model. Although the MAR sequence may facilitate transgene expression *(135,178)*, the chicken lysozyme sequence may also alter the spatial and temporal expression of genes under control of the small PB promoter. TRAMP mice rapidly develop mild to severe hyperplasia prior to neoplasia *(155)*. Greenberg's laboratory describes these tumors as poorly differentiated prostatic carcinoma that metastasize as early as 12 wk of age *(179,180)* and rapidly develop as an androgen-independent tumor *(181)*. However, early castration (at 4 wk of age) will decrease the appearance of tumors and increase animal survival *(182)*. However, some mice will develop androgen-independent tumors even if castrated at this early age *(182)*. Advantage: The TRAMP line quickly develops advanced stages of prostate cancer that originate in the dorsolateral lobe (human analog) and rapidly metastasize. This may be a useful model to test therapies for aggressive/metastatic cancer. In addition, a spontaneous mutation in the boundary of the hinge- and ligand-binding domain of the AR has been identified in both human CaP and TRAMP, supporting the usefulness of the model because it demonstrates similar mechanism for tumor progression *(46)*. Disadvantage: Tumor progression is extremely rapid *(179)*, and occurs at variable rates *(180)*, making it difficult to crossbreed other transgenic lines with TRAMP in order to test the role of other transgenes that contribute to tumor progression. Also, androgen ablation has a variable impact on tumor progression, suggesting that androgen-independent cells already exist in intact mice between 4 and 12 wk of age *(181,182)*.

Fetal Globin Gene Promoter. Transgenic mice generated with the fetal globin gene promoter linked to the SV40 large and small Tag genes develop prostate, adrenocortical, and brown adipose tumors *(183)*. The prostate tumors are comprised of numerous neuroendocrine and epithelial-cell-like elements, and expression of the transgene is found in the fetal heart, lung, testes, thyroid, fat, and adrenals *(183)*. Advantage: Human neuroendocrine prostate cancer is rare, and when patients become resistant to hormonal therapy, neuroendocrine cells appear to increase in number in these tumors *(184)*. This mouse model is an androgen-independent neuroendocrine prostate cancer that metastasizes to lymph nodes, and to the adrenals lungs, bones, and thymus. Thus, it is a unique model for this rare prostate cancer. Disadvantage: The fetal globin gene promoter is not prostate-specific, and causes primary tumors in a number of sites. Therefore, it is unlikely that the fetal globin gene promoter can be used in any reproducible fashion to create a mouse model for CaP.

C3(1) Promoter. The intact C3(1) gene, which is specifically expressed by the rat ventral prostate, was originally reported to function only in the prostate in transgenic mice *(185)*. Yet, the C3(1) promoter fragment itself does not function in a prostate-specific manner in transgenic mice *(160,183,186,187)*. Although prostate cancers develop in mice carrying the C3(1) promoter linked to the SV40 large and small Tag genes *(156)*, the effectiveness of this model is severely limited by the high incidence of neoplasia in male mice at other sites, including the lung, epididymis, seminal vesicles, vas deferens, ampullary gland, urethra, and mammary gland *(160)*, in addition to the thyroid, salivary glands, and nasal epithelium *(186)*. In C3(1)-SV40 Tag transgenic mice, CaP is first seen at 7 mo of age, occurring in mice at frequency of 19% in the ventral lobe and 3% in the dorsolateral lobe. PIN appears to progress further to invasive carcinoma, but metastasis are rare *(188)*. Advantage: The prostate develops PIN, which evolves into invasive carcinoma after 28 wk. The long transition period to cancer development permits the study the genetic changes responsible for tumorigenesis. Disadvantage: The tumors appear to be androgen-independent, and develop at a high frequency in the ventral lobe, a lobe that is not analogous to the human prostate.

Cryptdin-2 Promoter. The cryptdin-2 (CR-2) promoter (–6500 to +34 bp) directs expression of genes to the intestine *(189)* and the prostate *(190)* in transgenic mice. Transgenic mice that express the SV40-Tag transgene develop prostate hyperplasia by 7–8 wk of age, PIN within 8–10 wk, high-grade PIN at 12 wk, and metastasis to lymph nodes, liver, lung, and bone by 6 mo *(190)*. At 8 wk of age, the primary transformed cell type appears to be a neuroendocrine cell that subsequently produces AR-negative metastatic tumors. The specificity of the CR-2 promoter to target transgenes to prostatic neuroendocrine cells has been confirmed in the CR2-growth-hormone (GH) transgenic mouse line *(45)*. Advantage: The rapid growth of the transformed neuroendocrine cells in the prostate resemble PIN, and provide evidence that PIN is a preneoplastic stage leading to highly metastatic neuroendocrine tumors. This tumor is an interesting AR-independent neuroendocrine model of CaP. Disadvantage: The specificity of the cryptdin-2 promoter to limit transgene expression to a specific organ is still not well-defined. Thus, tumors detected in tissues other than the prostate may represent new primary growths and/or prostatic metastasises. Furthermore, the CR-2 promoter is not an androgen-regulated promoter, and therefore results in AR-negative tumors that may primarily target transgene expression to neuroendocrine cells in the prostate.

SV40 Large TAG-Expressing Transgenic Mice

The TRAMP and LPB-Tag models are often confused as representing the same transgenic model for prostate cancer. However, tumor development and progression is significantly different between the TRAMP model *(155,179–181)* and the LPB-Tag model *(166)*. Several marked differences between how these models were generated exist, which may explain the dramatic differences in prostate tumor phenotype. First and likely most important, the TRAMP mice express the SV40 early region, which includes both the large and small Tags, while the LPB-Tag mice express a deletion mutation of the SV40 early region that results in expression of only the large Tag. Second, TRAMP was created by using both the small PB (–426/+28 bp) promoter and the chicken lysozyme MAR sequence to facilitate transgene expression. It remains to be seen whether the chicken lysozyme MAR alters the spatial and temporal expression of the small PB promoter as well as androgen regulation, since TRAMP tumors appear to contain androgen-independent cells by 12 wk of age *(181)*. The LPB-Tag construct uses the a LPB fragment (approx 11,500 bp) that adds upstreams enhancers to the –426/+28 PB promoter, resulting in higher transgene expression (Both the sPB and LPB promoters contain well-characterized androgen-regulatory sequences *[107,147,150,151,191,192]*). Third, TRAMP represents a single transgenic line. The site of DNA integration may influence the transgene expression and phenotype. Since seven different transgenic lines with generally similar phenotypes were established with the LPB-Tag construct, the site of DNA integration alone cannot explain its phenotype. Fourth, TRAMP uses C57BL/6 mice which, when crossbred with FVB mice, have an even higher incidence of metastasises *(179,180)*, suggesting that genetic background influences gene expression as previously reported in tissue recombinants *(193)*. All of the LPB-Tag lines are created in and maintained in the CD-1 strain.

LPB-Tag Adenocarcinoma Transgenic Models. Eleven LPB-Tag founders were generated, and the founders were numbered as 12T-*n*, where *n* = 1–11. Six founders established seven transgenic lines (line 12T-7 diverged into a 12T-7 fast [f] and a 12T-7 slow [s] line, where each received a different transgene copy number from the founder and developed tumors with different growth rates). Consistently, 100% of the males develop prostate neoplasia regardless of the line of origin. In general, similar histopathological stages occurred in the prostates, but at different time intervals. Thus, prostate growth curves ranged from very slow-growing to very rapid-growing *(166)*. First, proliferation of epithelial cells occurs in numerous separate ductal regions. Since nuclear atypia occurs, the epithelial-cell transformation is best described as low-grade dysplasia or the equivalent of low-grade prostatic intraepithelial neoplasia (LGPIN). Only the nuclei, in areas of proliferation and atypic, are immunoreactive with an antibody to the SV40 large Tag. Soon, all the epithelium becomes dysplastic, and in most lines, marked stromal-cell proliferation surrounds the ductal structures. During this process, multifocal and then essentially uniform progression of the LGPIN to high-grade PIN occurs. This lesion is characterized by nuclear stratification, nuclear enlargement, nuclear hyperchromasia, and increased mitoses and apoptosis *(166)*. Although extensive and involving essentially the entire prostate, the neoplastic cells appear to be confined to normal or pre-existing basement membrane-lined ducts/glands. The degree of epithelial proliferation is particularly striking in the 12T-7f line, where the enlarged prostate is paralleled histologically by complex branching large- and small-gland profiles (Fig. 1A), with similar nuclear atypia (Fig. 1B) and perineural invasion (Fig. 1C). The lesion generally maintains a lobular configuration and is accompanied by a markedly cellular stroma (Fig. 1A,B) *(166)*. Criteria for early invasion are still being established in mouse models, and may be somewhat different than invasive human CaP *(190)*. Micro acinar architecture at the base of high-grade PIN-containing ducts/glands are suspicious for frank adenocarcinoma *(190)*. Such lesions have been observed prior to the time-point of metastasis in the 12T-10 line. Unequivocal foci of stromal invasion have been observed, including in the 12T-7f line *(166)* and the 12T-10 line which develops a neuroendocrine prostate cancer *(190a)*. The rapidly growing, large size of most neoplastic prostate tumors in many of these lines limits are maintained these mice for any prolonged periods of times. Rarely do these lines develop metastatic cancer. However, if mice from these transgenic lines are castrated (androgen deprivation) at this late stage

Fig. 1. *(opposite page) In situ* and invasive prostate carcinoma in Tag Mouse. **(A)** Marked epithelial proliferation with complex large and small glandular profiles and associated hypercellular stroma in 22-wk-old 12T-7f mouse (original magnification, 100×). **(B)** Higher magnification, showing marked nuclear atypia, with nuclear enlargement, coarse chromatin, enlarged nucleoli, and increased mitosis (original magnification, 400×). **(C)** Perineural invasion by poorly differentiated carcinoma in 43-wk-old 12T-8 mouse. Large nerve immediately outside of prostate, with surrounding adipose tissue at top (original magnification, 400×).

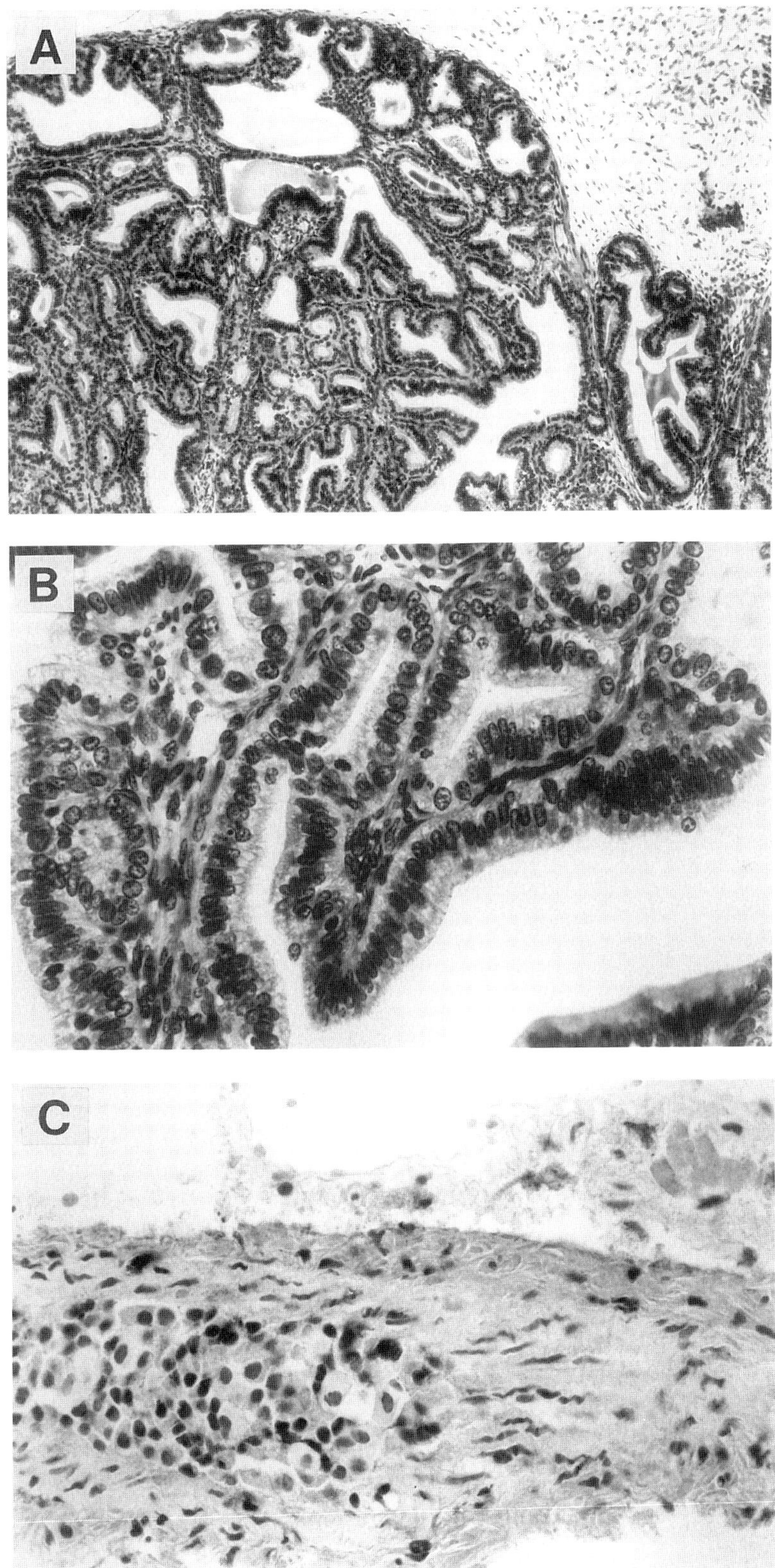
A
B
C

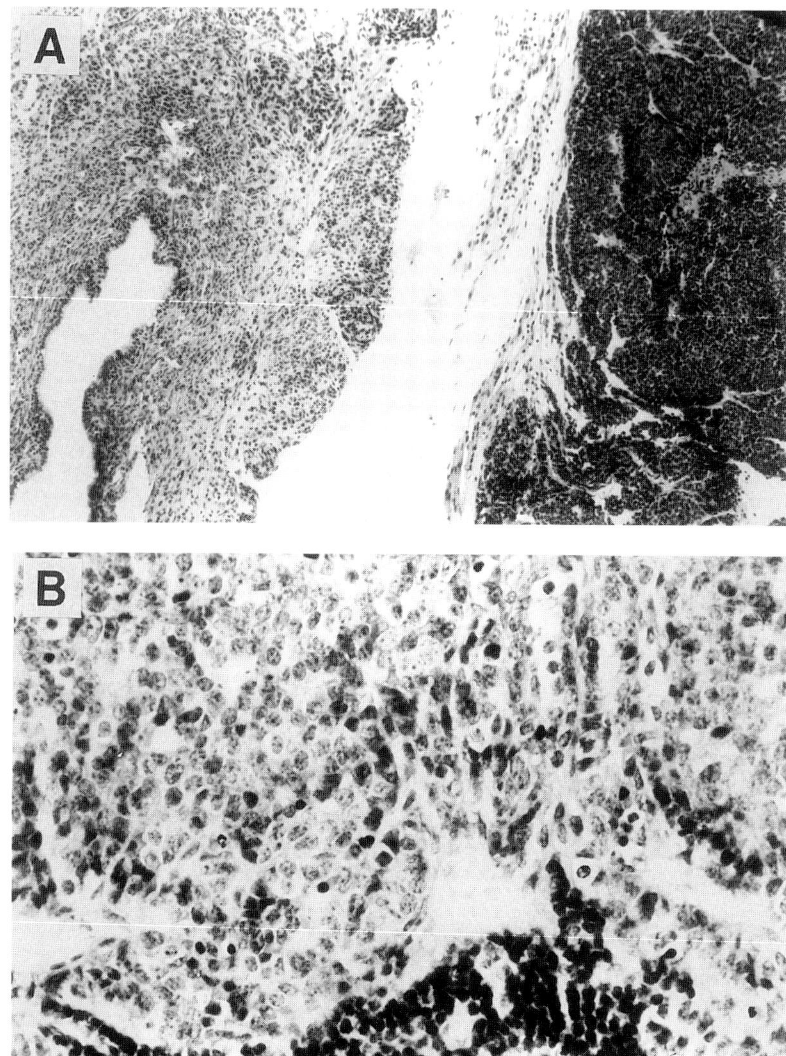

Fig. 2. Regression, regrowth, and metastasis of carcinoma in previously castrated Tag Mouse (12T-7f). **(A)** Atrophic dilated prostate gland with enlarged lumen and surrounding cellular stroma *(left)* and adjacent poorly differentiated carcinoma *(right)* in ventral prostate of 12T-7f mouse 8 wk status postcastration (original magnification, 100×). **(B)** Lymph node metastasis of poorly differentiated carcinoma in 12T-7f mouse 8 wk status postcastration. Residual lymphoid tissue (small lymphocytes) evident at top (original magnification, 400×).

of pre-invasive and minimally invasive tumor development, the primary prostate lesion will regress and begin to regrow between 2 and 6 mo after castration *(194)*. The lesion at this stage is characterized by extensive, frankly invasive, locally advanced, and metastatic poorly differentiated carcinoma (Fig. 2A,B). The events that occur during androgen deprivation that result in tumor progression are now being characterized. Advantage: The LPB-Tag models give reproducible prostate-specific tumors that are uniform in growth rate within a transgenic line and have reproducible phenotypes among different LPB-Tag founder lines. Since the large Tag is under the control of an androgen-regulated

PB promoter, these tumors are also androgen-regulated for growth. Disadvantage: The prostates develop extensive HGPIN and some local invasion, but rarely metastasize spontaneously.

LPB-Tag Transgenic Neuroendocrine Carcinoma Model. Although focal neuroendocrine differentiation is common in otherwise usual prostate adenocarcinoma, frank neuroendocrine or small-cell carcinomas occur in approx 5% of patients *(195)*. Reports on the AR status of neuroendocrine cells vary, but often these tumors are AR-negative *(196)*. Although rare, small-cell carcinomas are very aggressive, and they often have visceral metastasis, do not respond to androgen deprivation therapy, and lead to a poor patient prognosis *(197,198)*. In a recent study, repeated biopsies were performed, starting with 60 androgen-dependent prostate tumors that were followed through the course of androgen deprivation therapy. In treated men, the number of neuroendocrine cells within the biopsy increased as the tumor progressed *(184)*, suggesting that neuroendocrine cells may be involved in the emergence of AR-negative tumors.

A recently characterized LPB-Tag mouse line, 12T-10, shows tremendous promise as a model of evolving prostate carcinoma, including progression to androgen-insensitive metastatic disease, with neuroendocrine differentiation *(190a)*. Similar to other lines, the prostate in the 12T-10 mouse shows epithelial proliferation, with features compatible with low- and subsequent high-grade PIN. In contrast to some of the other lines, the tumor appearance in the 12T-10 line and prostate enlargement are slower, as previously reported *(166)* However, essentially all mice of this line develop metastatic disease by 44 wk of age. Metastasis are identified in the lymph nodes, lung, and liver (Fig. 3A). These metastatic lesions show histologic features of neuroendocrine differentiation, including scant cytoplasm, granular ("salt and pepper") chromatin, and rosette formation (Fig. 3B). Neuroendocrine differentiation has been confirmed by immunohistochemical studies (e.g., immunohistochemically positive for serotonin and chromogranin) and ultrastructurally. At earlier time-points (i.e., prior to development of metastasis), the *in situ* dysplastic proliferations show focal areas cytologically suggestive of neuroendocrine differentiation, and the prostate eventually shows foci of microinvasive and then extensively invasive carcinoma. At the time-points of tumor metastasis, the prostate shows unequivocal invasive carcinoma, also with features of neuroendocrine differentiation. Neuroendocrine differentiation of these prostate pre-invasive and invasive lesions has also been documented immunohistochemically.

The metastatic tumors in the 12T-10 mouse remain immunopositive for the presence of nuclear large Tag. The primary and metastatic neuroendocrine tumors appear AR-negative or faintly positive by immunohistochemical staining, suggesting that the LPB fragment can be regulated in neuroendocrine cells by signals other then the AR. Further, spontaneous changes likely occur during the slow growth of this tumor, resulting in an AR-negative 12T-10 tumor phenotype that shows progressive neuroendocrine differentiation and metastasises. Advantage: Under the control of an androgen-dependent PB promoter, the 12T-10 mouse develops prostate precursor lesions, which progress spontaneously to invasive carcinoma, and evolve to an AR-negative neuroendocrine cancer, which metastasises. Disadvantage: The primary tumor is a small and slow-growing lesion that first metastasize in mice that are 30 wk of age or older.

In summary, the large Tag gene is under the control of the androgen-regulated LPB promoter that targets transgene expression to prostatic-epithelial cells. However, this large fragment of DNA may have other hormonal-regulatory elements that function

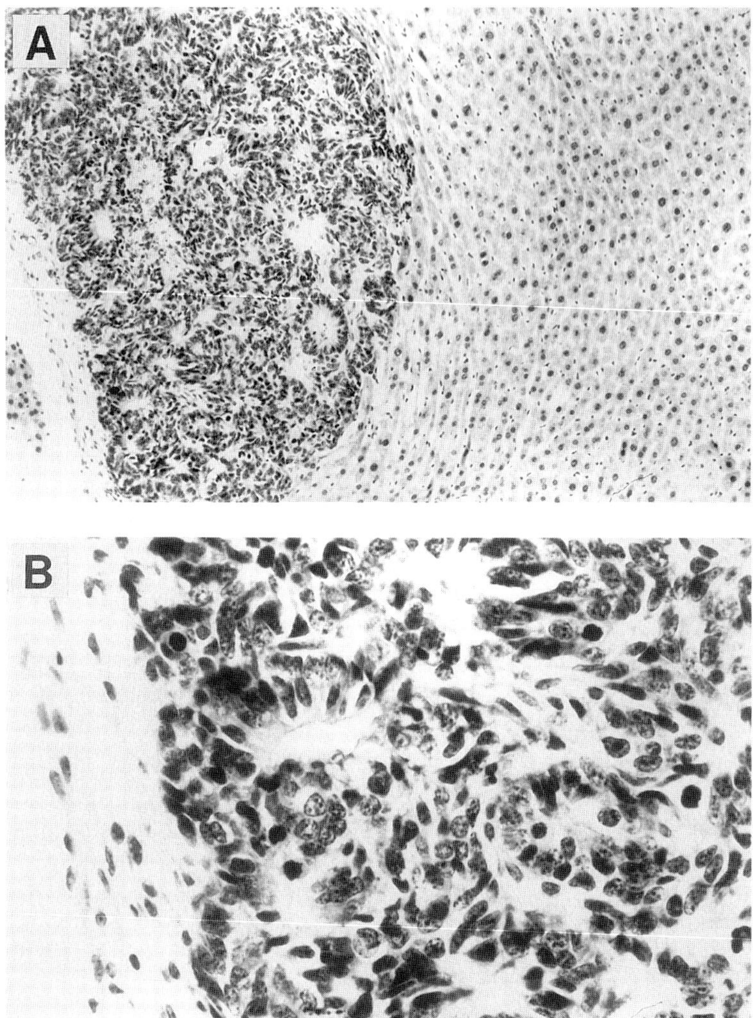

Fig. 3. Metastatic carcinoma with neuroendocrine features in 12T-10 Tag mouse. **(A)** Liver metastasis in 44-wk-old mouse. Unequivocal invasive carcinoma, with neuroendocrine differentiation, is seen at this and earlier time-points in the prostate (not shown) (original magnification, 100×). **(B)** Higher magnification of metastatic lesion, with neuroendocrine differentiation. Tumor cells with scant cytoplasm, round-to-oval nuclei, granular-to-coarse chromatin, and occasional nuclear molding. Glandular differentiation or rosette formation evident at bottom of field (original magnification, 400×).

independent of the AR. For example, transgenic mice produced with large probasin-chloramphenicolacetyl transferase (LPB-CAT) construct show CAT activity in the prostate prior to the sexual maturation of the mouse, which occurs with increased testosterone levels *(152)*. This early expression of LPB-CAT may be caused by low levels of androgens during development, or the LPB promoter may respond to other developmental signal(s). In addition, the LPB-Tag 12T-10 prostatic neuroendocrine carcinoma continues to express the large Tag gene in AR-negative cells. The transformation of both secretory epithelial and neuroendocrine cells in the LPB-Tag transgenic lines suggest that a common stem cell does differentiate into prostatic-secretory epithelial and neuroendocrine cells.

CONCLUSION

Tissue-specific promoters contain key information on how genes are regulated and expressed in specific cell populations within an organism. Although long lists of transcription factors that regulate genes have now been identified, the general mechanism by which these transcription factors work to restrict gene expression to certain populations remains an enigma. Tissue-specific promoters offer unique opportunities to target the expression of transgenes to specific organs in transgenic animals, and to target therapeutic genes to treat diseased organs in humans. Tissue-specific expression of transgenes in transgenic mice have provided opportunities to explore the mechanism involved in normal development and to create new mouse models for disease states. The specter of gene therapy invites our imagination to target new classes of drugs that may treat acquired diseases and correct genetic diseases. However, we should not forget that the DNA of these tissue-specific promoters contains the information to explain the general mechanism by which transcription factors work to restrict gene expression to certain populations of cells.

ACKNOWLEDGMENTS

This work is supported by R01-CA76142 to RJM from the National Cancer Institute, NIH; by PC970260 to SBS from the US Department of Defense; and by the Frances Williams Preston Laboratories of the TJ Martell Foundation.

REFERENCES

1. Anonymous American Cancer Society. http://www.cancer.org. 2001.
2. Potosky AL, Miller BA, Albertsen PC, Kramer BS. The role of increasing detection in the rising incidence of prostate cancer. Jama 1995;273:548–552.
3. Merrill RM, Potosky AL, Feuer EJ. Changing trends in U.S. prostate cancer incidence rates. J Natl Cancer Inst 1996;88:16–18.
4. Graham SD, Jr, Bostwick DG, Hoisaeter A, Abrahamsson P, Algaba F, di Sant'Agnese A, et al. Report of the committee on staging and pathology. Cancer 1992;70:359–361.
5. Graham SD, Jr. Critical assessment of prostate cancer staging. [review]. Cancer 1992;70 (Suppl):269–274.
6. Bostwick DG, Pacelli A, Lopez-Beltran A. Molecular biology of prostatic intraepithelial neoplasia [Review]. Prostate 1996;29:117–134.
7. Ferguson J, Zincke H, Ellison E, Bergstrahl E, Bostwick DG. Decrease of prostatic intraepithelial neoplasia following androgen deprivation therapy in patients with stage T3 carcinoma treated by radical prostatectomy. Urology 1994;44:91–95.
8. Helpap BG, Bostwick DG, Montironi R. The significance of atypical adenomatous hyperplasia and prostatic intraepithelial neoplasia for the development of prostate carcinoma. An update [Review]. Virchows Arch 1995;426:425–434.
9. Montironi R, Scarpelli M, Galluzzi CM, Diamanti L. Aneuploidy and nuclear features of prostatic intraepithelial neoplasia (PIN [Review]. J Cell Biochem Suppl 1992;16H:47–53.
10. Petein M, Michel P, van Velthoven R, Pasteels JL, Brawer MK, Davis JR, et al.. Morphonuclear relationship between prostatic intraepithelial neoplasia and cancers as assessed by digital cell image analysis. Am J Clin Pathol 1991;96:628–634.
11. Nagle RB, Brawer MK, Kittelson J, Clark V. Phenotypic relationships of prostatic intraepithelial neoplasia to invasive prostatic carcinoma. Am J Pathol 1991;138:119–128.
12. Bostwick DG, Burke HB, Wheeler TM, Chung LW, Bookstein R, Pretlow TG, et al. The most promising surrogate endpoint biomarkers for screening candidate chemopreventive compounds for prostatic adenocarcinoma in short-term phase II clinical trials. J Cell Biochem Suppl 1994;19:283–289.
13. Veltri RW, Partin AW, Epstein JE, Marley GM, Miller CM, Singer DS, et al. Quantitative nuclear morphometry, Markovian texture descriptors, and DNA content captured on a CAS-200 Image

analysis system, combined with PCNA and HER-2/neu immunohistochemistry for prediction of prostate cancer progression. J Cell Biochem 1994;19(Suppl):249–258.
14. Mellon K, Thompson S, Charlton RG, Marsh C, Robinson M, Lane DP, et al. p53, c-erbB-2 and the epidermal growth factor receptor in the benign and malignant prostate. J Urol 1992;147:496–499.
15. Grizzle WE, Myers RB, Arnold MM, Srivastava S. Evaluation of biomarkers in breast and prostate cancer. J Cell Biochem 1994;19(Suppl):259–266.
16. Myers RB, Srivastava S, Oelschlager DK, Grizzle WE. Expression of p160erbB-3 and p185erbB-2 in prostatic intraepithelial neoplasia and prostatic adenocarcinoma (see comments). J Natl Cancer Inst 1994;86:1140–1145.
17. Bostwick DG. c-erbB-2 oncogene expression in prostatic intraepithelial neoplasia: mounting evidence for a precursor role (editorial; comment). J Natl Cancer Inst 1994;86:1108–1110.
18. Bova GS, Isaacs WB. Review of allelic loss and gain in prostate cancer. World J Urol 1996;14:338–346.
19. Gronberg H, Isaacs SD, Smith JR, Carpten JD, Bova GS, Freije D, et al. Characteristics of prostate cancer in families potentially linked to the hereditary prostate cancer 1 (HPC1) locus (see comments). JAMA 1997;278:1251–1255.
20. Gibbs M, Chakrabarti L, Stanford JL, Goode EL, Kolb S, Schuster EF, et al. Analysis of chromosome 1q42.2-43 in 152 families with high risk of prostate cancer. Am J Human Genet 1999;64:1087–1095.
21. Xu J, Meyers D, Freije D, Isaacs S, Wiley K, Nusskern D, et al. Evidence for a prostate cancer susceptibility locus on the X chromosome. Nat Genet 1998;20:175–179.
22. Wilding G. The importance of steroid hormones in prostate cancer. Cancer Surv 1992;14:113–130.
23. Brewster SF, Simons JW. Gene therapy in urological oncology: principles, strategies and potential. Eur Urol 1994;25:177–182.
24. Taplin ME, Bubley GJ, Shuster TD, Frantz ME, Spooner AE, Ogata GK, et al. Mutation of the androgen-receptor gene in metastatic androgen- independent prostate cancer (see comments). N Engl J Med 1995;332:1393–1398.
25. Kelly WK, Scher HI. Prostate specific antigen decline after antiandrogen withdrawal: the flutamide withdrawal syndrome. J Urol 1993;149:607–609.
26. Newmark JR, Hardy DO, Tonb DC, Carter BS, Epstein JI, et al. Androgen receptor gene mutations in human prostate cancer. Proc Natl Acad Sci USA 1992;89:6319–6323.
27. Veldscholte J, Berrevoets CA, Ris-Stalpers C, Kuiper GG, Jenster G, Trapman J, et al. The androgen receptor in LNCaP cells contains a mutation in the ligand binding domain which affects steroid binding characteristics and response to antiandrogens. J Steroid Biochem Mol Biol 1992;41:665–669.
28. Veldscholte J, Berrevoets CA, Brinkmann AO, Grootegoed JA, Mulder E. Anti-androgens and the mutated androgen receptor of LNCaP cells: differential effects on binding affinity, heat-shock protein interaction, and transcription activation. Biochemistry 1992;31:2393–2399.
29. Gaddipati JP, McLeod DG, Heidenberg HB, Sesterhenn IA, Finger MJ, Moul JW, et al. Frequent detection of codon 877 mutation in the androgen receptor gene in advanced prostate cancers. Cancer Res 1994;54:2861–2864.
30. Schoenberg MP, Hakimi JM, Wang S, Bova GS, Epstein JI, Fischbeck KH, et al. Microsatellite mutation (CAG24—>18) in the androgen receptor gene in human prostate cancer. Biochem Biophys Res Commun 1994;198:74–80.
31. Culig Z, Klocker H, Eberle J, Kaspar F, Hobisch A, Cronauer MV, et al. DNA sequence of the androgen receptor in prostatic tumor cell lines and tissue specimens assessed by means of the polymerase chain reaction. Prostate 1993;22:11–22.
32. Klocker H, Culig Z, Hobisch A, Cato AC, Bartsch G. Androgen receptor alterations in prostatic carcinoma. Prostate, 1994;25:266–273.
33. Culig Z, Hobisch A, Cronauer MV, Cato AC, Hittmair A, Radmayr C, et al. Mutant androgen receptor detected in an advanced-stage prostatic carcinoma is activated by adrenal androgens and progesterone. Mol Endocrinol 1993;7(12):1541–1550.
34. Tan J, Sharief Y, Hamil KG, Gregory CW, Zang DY, Sar M, et al. Dehydroepiandrosterone activates mutant androgen receptors expressed in the androgen-dependent human prostate cancer xenograft CWR22 and LNCaP cells. Mol Endocrinol 1997;11:450–459.
35. Fenton MA, Shuster TD, Fertig AM, Taplin ME, Kolvenbag G, Bubley GJ, et al. Functional characterization of mutant androgen receptors from androgen-independent prostate cancer. Clin Cancer Res 1997;3:1383–1388.
36. Tilley WD, Buchanan G, Hickey TE, Bentel JM. Mutations in the androgen receptor gene are associated with progression of human prostate cancer to androgen independence. Clin Cancer Res 1996;2:277–285.

37. Koivisto P, Kononen J, Palmberg C, Tammela T, Hyytinen E, Isola J, et al. Androgen receptor gene amplification: a possible molecular mechanism for androgen deprivation therapy failure in prostate cancer. Cancer Res 1997;57:314–319.
38. Kazemi-Esfarjani P, Trifiro MA, Pinsky L. Evidence for a repressive function of the long polyglutamine tract in the human androgen receptor: possible pathogenetic relevance for the (CAG)n-expanded neuronopathies. Human Mol Genet 1995;4(4):523–527.
39. Irvine RA, Yu MC, Ross RK, Coetzee GA. The CAG and GGC microsatellites of the androgen receptor gene are in linkage disequilibrium in men with prostate cancer. Cancer Res 1995;55(9):1937–1940.
40. Coetzee GA, Ross RK. Re: Prostate cancer and the androgen receptor (letter). J Natl Cancer Inst 1994;86(11):872,873.
41. Scher HI, Kolvenbag GJ. The antiandrogen withdrawal syndrome in relapse prostate cancer. Eur Urol 1997;31:24–27.
42. Scher HI, Liebertz C, Kelly WK, Mazumdar M, Brett C, Schwartz L, et al. Bicalutamide for advanced prostate cancer: the natural versus treated history of disease. J Clin Oncol 1997;15:2928–2938.
43. Joyce R, Fenton MA, Rode P, Constantine M, Gaynes L, Kolvenbag G, et al. High dose bicalutamide for androgen independent prostate cancer: effect of prior hormonal therapy. J Urol 1998;159:149–153.
44. Kelly WK, Scher HI. Prostate specific antigen decline after antiandrogen withdrawal: the flutamide withdrawal syndrome. J Urol 1993;149:607–609.
45. Gregory CW, Hamil KG, Kim D, Hall SH, Pretlow TG, Mohler JL, et al. Androgen receptor expression in androgen-independent prostate cancer is associated with increased expression of androgen-regulated genes (In Process Citation). Cancer Res 1998;58:5718–5724.
46. Buchanan G, Yang M, Nahm HS, Han G, Bentel JM, Matusik RJ, et al. Mutations at the boundary of the hinge and ligand binding domain of the androgen receptor confer increased transactivation function. Mol Endocrinol 2001;15(1):46–56.
47. Pollard M. Commentary on the Dunning tumor. Prostate 1995;26:287–289.
48. Cohen MB, Heidger PM, Lubaroff DM. Gross and microscopic pathology of induced prostatic complex tumors arising in Lobund-Wistar rats. Cancer Res 1994;54:626–628.
49. Pollard M. The Lobund-Wistar rat model of prostate cancer. J Cell Biochem 1992;16H(Suppl):84–88.
50. Gierer A. Molecular models and combinatorial principles in cell differentiation and morphogenesis. Cold Spring Harb Symp Quant Biol 1974;38:951–961.
51. Anonymous. Control of gene expression: combinatorial gene control is the norm in eucaryotes. In: Alberts B, Bray D, Lewis J, et al., eds., Molecular Biology of The Cell, Garland Publishing, Inc, New York, NY, 1994, pp. 445–453.
52. Bodnar JW, Bradley MK. A chromatin switch. J Theor Biol 1996;183:1–7.
53. Beato M, Eisfeld K. Transcription factor access to chromatin. Nucleic Acids Res 1997;25:3559–3563.
54. Tate PH, Bird AP. Effects of DNA methylation on DNA-binding proteins and gene expression. Curr Opin Genet Dev 1993;3:226–231.
55. Graessmann M, Graessmann A. DNA methylation, chromatin structure and the regulation of gene expression. EXS 1993;64:404–424.
56. Bartolomei MS, Tilghman SM. Genomic imprinting in mammals. Annu Rev Genet 1997;31:493–525.
57. Price D. Comparative aspects of development and structure in the prostate. In: Vollmer EP, Kauffmann G, eds., Biology of the Prostate and Related Tissues, US Government Printing Office, Washington DC, 1963, pp. 1–28.
58. Huhtaniemi I. Fetal testis—a very special endocrine organ. Eur J Endocrinol 1994;130:25–31.
59. O'Shaughnessy PJ, Baker P, Sohnius U, Haavisto AM, Charlton HM, Huhtaniemi I. Fetal development of Leydig cell activity in the mouse is independent of pituitary gonadotroph function. Endocrinology 1998;139:1141–1146.
60. Tsuji M, Shima H, Terada N, Cunha GR. 5 alpha-reductase activity in developing urogenital tracts of fetal and neonatal male mice. Endocrinology 1994;134:2198–2205.
61. Cooke PS, Young P, Cunha GR. Androgen receptor expression in developing male reproductive organs. Endocrinology 1991;128:2867–2873.
62. Timms BG, Mohs TJ, Didio LJ. Ductal budding and branching patterns in the developing prostate. J Urol 1994;151:1427–1432.
63. Timms BG, Lee CW, Aumuller G, Seitz J. Instructive induction of prostate growth and differentiation by a defined urogenital sinus mesenchyme. Microsc Res Tech 1995;30:319–332.
64. Sugimura Y, Cunha GR, Donjacour AA. Morphogenesis of ductal networks in the mouse prostate. Biol Reprod 1986;34:961–971.

65. Kazemi-Esfarjani P, Beitel L, Trifiro M, Kaufman M, Rennie PS, Sheppard PC, et al. Substitution of valine 865 by methionine or leucine causes complete or partial androgen insensitivity, respectively with distinct androgen receptor phenotypes. Mol Endocrinol 1993;7:37–46.
66. Pinsky L, Trifiro M, Kaufman M, Beitel LK, Mhatre A, Kazemi-Esfarjani P, et al. Androgen resistance due to mutation of the androgen receptor. Clin Invest Med—Medecine Clinique Et Experimentale, 1992;15:456–472.
67. Brown TR, Lubahn DB, Wilson EM, French FS, Migeon CJ, Corden JL. Functional characterization of naturally occurring mutant androgen receptors from subjects with complete androgen insensitivity. Mol Endocrinol 1990;4:1759–1772.
68. Wiener JS, Teague JL, Roth DR, Gonzales ET, Jr, Lamb DJ. Molecular biology and function of the androgen receptor in genital development. J Urol 1997;157:1377–1386.
69. Sugimura Y, Cunha GR, Donjacour AA, Bigsby RM, Brody JR. Whole-mount autoradiography study of DNA synthetic activity during postnatal development and androgen-induced regeneration in the mouse prostate. Biol Reprod 1986;34:985–995.
70. Philip R, Brunette E, Kilinski L, Murugesh D, McNally MA, Ucar K, et al. Efficient and sustained gene expression in primary T lymphocytes and primary and cultured tumor cells mediated by adeno-associated virus plasmid DNA complexed to cationic liposomes. Mol Cell Biol 1994;14:2411–2418.
71. Cunha GR. Role of mesenchymal-epithelial interactions in normal and abnormal development of the mammary gland and prostate. Cancer 1994;74:1030–1044.
72. Cunha GR, Lung B. The possible influence of temporal factors in androgenic responsiveness of urogenital tissue recombinants from wild-type and androgen-insensitive (Tfm) mice. J Exp Zool 1978;205:181–193.
73. Cunha GR, Chung LW, Shannon JM, Reese BA. Stromal-epithelial interactions in sex differentiation. Biol Reprod 1980;22:19–42.
74. Hayashi N, Cunha GR, Parker M. Permissive and instructive induction of adult rodent prostatic epithelium by heterotypic urogenital sinus mesenchyme. Epithelial Cell Biol 1993;2:66–78.
75. Donjacour AA, Cunha GR. Assessment of prostatic protein secretion in tissue recombinants made of urogenital sinus mesenchyme and urothelium from normal or androgen-insensitive mice. Endocrinology 1993;132:2342–2350.
76. Cunha GR, Donjacour AA, Cooke PS, Mee S, Bigsby RM, Higgins SJ, et al. The endocrinology and developmental biology of the prostate. Endocr Rev 1987;8:338–362.
77. Torres M, Gomez-Pardo E, Dressler GR, Gruss P. Pax-2 controls multiple steps of urogenital development. Development 1995;121:4057–4065.
78. Santti R, Newbold RR, Makela S, Pylkkanen L, McLachlan JA. Developmental estrogenization and prostatic neoplasia. Prostate 1994;24:67–78.
79. Newbold RR, Bullock BC, McLachlan JA. Mullerian remnants of male mice exposed prenatally to diethylstilbestrol. Teratog Carcinog Mutagen 1987;7:377–389.
80. Prins GS. Developmental estrogenization of the prostate gland. In: Naz, R, ed., Prostate: Basic and Clinical Aspects, CRC Press, Boca Raton, FL, 1987, pp. 245–263.
81. Kuiper GG, Enmark E, Pelto-Huikko M, Nilsson S, Gustafsson JA. Cloning of a novel receptor expressed in rat prostate and ovary. Proc Natl Acad Sci USA 1996;93:5925–5930.
82. Kuiper GG, Carlsson B, Grandien K, Enmark E, Haggblad J, Nilsson S, et al. Comparison of the ligand binding specificity and transcript tissue distribution of estrogen receptors alpha and beta. Endocrinology 1997;138:863–870.
83. Oefelein M, Chin-Chance C, Bushman W. Expression of the homeotic gene Hox-d13 in the developing and adult mouse prostate. J Urol 1996;155:342–346.
84. Podlasek CA, Duboule D, Bushman W. Male accessory sex organ morphogenesis is altered by loss of function of Hoxd-13. Dev Dyn 1997;208:454–465.
85. Sciavolino PJ, Abrams EW, Yang L, Austenberg LP, Shen MM, Abate-Shen C. Tissue-specific expression of murine Nkx3.1 in the male urogenital system. Dev Dyn 1997;209:127–138.
86. Pylkkanen L, Makela S, Valve E, Harkonen P, Toikkanen S, Santti R. Prostatic dysplasia associated with increased expression of c- myc in neonatally estrogenized mice. J Urol 1993;149:1593–1601.
87. vom Saal FS, Timms BG, Montano MM, Palanza P, Thayer KA, Nagel SC, et al. Prostate enlargement in mice due to fetal exposure to low doses of estradiol or diethylstilbestrol and opposite effects at high doses. Proc Natl Acad Sci USA 1997;94:2056–2061.
88. Noble RL. Sex steroids as a cause of adenocarcinoma of the dorsal prostate in Nb rats, and their influence on the growth of transplants. Oncology 1977;34:138–141.

89. Noble RL. Production of Nb rat carcinoma of the dorsal prostate and response of estrogen-dependent transplants to sex hormones and tamoxifen. Cancer Res 1980;40:3547–3550.
90. Ho SM, Yu M. Selective increase in type II estrogen-binding sites in the dysplastic dorsolateral prostates of noble rats. Cancer Res 1993;53:528–532.
91. Ho SM, Yu M. Hormonal regulation of nuclear type II estrogen binding sites in the dorsolateral prostate of noble rats. J Steroid Biochem Mol Biol 1995;52:233–238.
92. Mosselman S, Polman J, Dijkema R. ER beta: identification and characterization of a novel human estrogen receptor. FEBS Lett 1996;392:49–53.
93. Couse JF, Lindzey J, Grandien K, Gustafsson JA, Korach KS. Tissue distribution and quantitative analysis of estrogen receptor- alpha (ERalpha) and estrogen receptor-beta (ERbeta) messenger ribonucleic acid in the wild-type and ERalpha-knockout mouse. Endocrinology 1997;138:4613–4621.
94. Hess RA, Bunick D, Lee KH, Bahr J, Taylor JA, Korach KS, Lubahn DB. A role for oestrogens in the male reproductive system (see comments). Nature 1997;390:509–512.
95. Sharpe RM. Do males rely on female hormones? (news; comment). Nature 1997;390:447,448.
96. Dolle P, Dierich A, LeMeur M, Schimmang T, Schuhbaur B, Chambon P, et al. Disruption of the Hoxd-13 gene induces localized heterochrony leading to mice with neotenic limbs. Cell 1993;75:431–441.
97. Warot X, Fromental-Ramain C, Fraulob V, Chambon P, Dolle P. Gene dosage-dependent effects of the Hoxa-13 and Hoxd-13 mutations on morphogenesis of the terminal parts of the digestive and urogenital tracts. Development 1997;124:4781–4791.
98. Bieberich CJ, Fujita K, He WW, Jay G. Prostate-specific and androgen- dependent expression of a novel homeobox gene. J Biol Chem 1996;271:31,779–31,782.
99. He WW, Sciavolino PJ, Wing J, Augustus M, Hudson P, Meissner PS, et al. A novel human prostate-specific, androgen-regulated homeobox gene (NKX3.1) that maps to 8p21, a region frequently deleted in prostate cancer. Genomics 1997;43:69–77.
100. Bhatia-Gaur R, Donjacour AA, Sciavolino PJ, Kim M, Desai N, Young P, Norton CR, et al. Roles for Nkx3.1 in prostate development and cancer. Genes Dev 1999;13:966–977.
101. Ho KC, Snoek R, Quarmby V, Viskochil DH, Rennie PS, Wilson EM, et al. Primary structure and androgen regulation of a 20- kilodalton protein specific to rat ventral prostate. Biochemistry 1989;28:6367–6373.
102. Zhang YL, Zhou ZX, Zhang YD, Parker MG. Expression of androgen receptors and prostatic steroid-binding protein during development of the rat ventral prostate. J Endocrinol 1988;117:361–366.
103. Matusik RJ, Kreis C, McNicol P, Sweetland R, Mullin C, Fleming WH, et al. Regulation of prostatic genes: role of androgens and zinc in gene expression. Biochem Cell Biol 1986;64:601–607.
104. Wilson EM, French FS. Biochemical homology between rat dorsal prostate and coagulating gland. Purification of a major androgen-induced protein. J Biol Chem, 1980;255:10,946–10,953.
105. Dodd JG, Kreis C, Sheppard PC, Hamel A, Matusik RJ. Effect of androgens on mRNA for a secretory protein of rat dorsolateral prostate and seminal vesicles. Mol Cell Endocrinol 1986;47:191–200.
106. Matusik RJ, Cattini PA, Leco KJ, Sheppard PC, Nickel BE, Neubauer BL, et al. Regulation of gene expression in the prostate. In: Karr JP, Coffey DS, Smith RG, et al., eds., Molecular and Cellular Biology of Prostate Cancer, Plenum Press, New York, NY, 1991, pp. 299–314.
107. Dodd JG, Sheppard PC, Matusik, RJ. Characterization and cloning of rat dorsal prostate mRNAs. Androgen regulation of two closely related abundant mRNAs. J Biol Chem 1983;258:10,731–10,737.
108. Spence AM, Sheppard PC, Davie JR, Matuo Y, Nishi N, McKeehan WL, et al. Regulation of a bifunctional mRNA results in synthesis of secreted and nuclear probasin. Proc Natl Acad Sci USA 1989;86:7843–7847.
109. Li S, Crenshaw EB, Rawson EJ, Simmons DM, Swanson LW, Rosenfeld MG. Dwarf locus mutants lacking three pituitary cell types result from mutations in the POU-domain gene pit-1. Nature 1990;347:528–533.
110. Gage PJ, Brinkmeier ML, Scarlett LM, Knapp LT, Camper SA, Mahon KA. The Ames dwarf gene, df, is required early in pituitary ontogeny for the extinction of Rpx transcription and initiation of lineage- specific cell proliferation. Mol Endocrinol 1996;10:1570–1581.
111. Camper SA, Saunders TL, Katz RW, Reeves RH. The Pit-1 transcription factor gene is a candidate for the murine Snell dwarf mutation. Genomics 1990;8:586–590.
112. Godfrey P, Rahal JO, Beamer WG, Copeland NG, Jenkins NA, Mayo KE. GHRH receptor of little mice contains a missense mutation in the extracellular domain that disrupts receptor function. Nat Genet 1993;4:227–232.

113. Lin SC, Lin CR, Gukovsky I, Lusis AJ, Sawchenko PE, et al. Molecular basis of the little mouse phenotype and implications for cell type-specific growth (see comments). Nature 1993;364:208–213.
114. Rhodes SJ, Chen R, DiMattia GE, Scully KM, Kalla KA, Lin SC, et al. A tissue-specific enhancer confers Pit-1-dependent morphogen inducibility and autoregulation on the pit-1 gene. Genes Dev 1993;7:913–932.
115. DiMattia GE, Rhodes SJ, Krones A, Carriere C, O'Connell S, Kalla K, et al. The Pit-1 gene is regulated by distinct early and late pituitary- specific enhancers. Dev Biol 1997;182:180–190.
116. Sornson MW, Wu W, Dasen JS, Flynn SE, Norman DJ, O'Connell SM, et al. Pituitary lineage determination by the Prophet of Pit-1 homeodomain factor defective in Ames dwarfism. Nature 1996;384:327–333.
117. Sheng HZ, Moriyama K, Yamashita T, Li H, Potter SS, Mahon KA, et al. Multistep control of pituitary organogenesis. Science 1997;278:1809–1812.
118. Sheng HZ, Zhadanov AB, Mosinger Jr B, Fujii T, Bertuzzi S, Grinberg A, et al. Specification of pituitary cell lineages by the LIM homeobox gene Lhx3. Science 1993;272:1004–1007.
119. Ingraham HA, Lala DS, Ikeda Y, Luo X, Shen WH, Nachtigal MW, et al. The nuclear receptor steroidogenic factor 1 acts at multiple levels of the reproductive axis. Genes Dev 1994;8:2302–2312.
120. Civitareale D, Lonigro R, Sinclair AJ, Di Lauro R. A thyroid-specific nuclear protein essential for tissue-specific expression of the thyroglobulin promoter. EMBO J 1989;8:2537–2542.
121. Kimura S, Hara Y, Pineau T, Fernandez-Salguero P, Fox CH, Ward JM, et al. The T/ebp null mouse: thyroid-specific enhancer-binding protein is essential for the organogenesis of the thyroid, lung, ventral forebrain, and pituitary. Genes Dev 1996;10:60–69.
122. Madsen OD, Jensen J, Petersen HV, Pedersen EE, Oster A, Andersen FG, et al. Transcription factors contributing to the pancreatic beta-cell phenotype. Horm Metab Res 1997;29:265–270.
123. Jonsson J, Carlsson L, Edlund T, Edlund H. Insulin-promoter-factor 1 is required for pancreas development in mice. Nature 1994;371:606–609.
124. Offield MF, Jetton TL, Labosky PA, Ray M, Stein RW, Magnuson MA, et al. PDX-1 is required for pancreatic outgrowth and differentiation of the rostral duodenum. Development 1996;122:983–995.
125. Ohlsson H, Karlsson K, Edlund T. IPF1, a homeodomain-containing transactivator of the insulin gene. EMBO J 1993;12:4251–4259.
126. Brunk BP, Goldhamer DJ, Emerson CP, Jr. Regulated demethylation of the myoD distal enhancer during skeletal myogenesis. Dev Biol 1996;177:490–503.
127. Faerman A, Goldhamer DJ, Puzis R, Emerson CP, Jr, Shani M. The distal human myoD enhancer sequences direct unique muscle- specific patterns of lacZ expression during mouse development. Dev Biol 1995;171:27–38.
128. Weintraub H, Genetta T, Kadesch T. Tissue-specific gene activation by MyoD: determination of specificity by cis-acting repression elements. Genes Dev 1994;8:2203–2211.
129. Goldhamer DJ, Brunk BP, Faerman A, King A, Shani M, Emerson Jr CP. Embryonic activation of the myoD gene is regulated by a highly conserved distal control element. Development 1995;121:637–649.
130. Lefebvre V, Zhou G, Mukhopadhyay K, Smith CN, Zhang Z, Eberspaecher H, et al. An 18-base-pair sequence in the mouse proalpha1(II) collagen gene is sufficient for expression in cartilage and binds nuclear proteins that are selectively expressed in chondrocytes. Mol Cell Biol 1996;16:4512–4523.
131. Fleming WH, Hamel A, MacDonald R, Ramsey E, Pettigrew NM, Johnston B, et al. Expression of the c-myc protooncogene in human prostatic carcinoma and benign prostatic hyperplasia. Cancer Res 1986;46:1535–1538.
132. Matusik RJ, Fleming WH, Hamel A, Westenbrink TG, Hrabarchuk B, MacDonald R, et al. Expression of the c-myc proto-oncogene in prostatic tissue. Prog Clin Biol Res 1987;239:91–112.
133. Buttyan R, Sawczuk IS, Benson MC, Siegal JD, Olsson CA. Enhanced expression of the c-myc protooncogene in high-grade human prostate cancers. Prostate 1987;11:327–337.
134. Dodd JG, Morris G, Miller TL, Ramsey E, Johnston B, Pettigrew NM, et al. Oncogenes and the prostate. In: Farnsworth WE, Ablin RJ, eds., The Prostate as an Endocrine Gland, CRC Press, Boca Raton, FL, 1996, pp. 49–66.
135. Greenberg NM, DeMayo FJ, Sheppard PC, Barrios R, Lebovitz M, Finegold M, et al. The rat probasin gene promoter directs hormonally and developmentally regulated expression of a heterologous gene specifically to the prostate in transgenic mice. Mol Endocrinol 1994;8:230–239.
136. Myers RB, Oelschlager D, Srivastava S, Grizzle WE. Accumulation of the p53 protein occurs more frequently in metastatic than in localized prostatic adenocarcinomas. Prostate 1994;25:243–248.

137. Bookstein R, Rio P, Madreperla SA, Hong F, Allred C, Grizzle WE, et al. Promoter deletion and loss of retinoblastoma gene expression in human prostate carcinoma. Proc Natl Acad Sci USA 1990;87:7762–7766.
138. Bookstein R, Shew JY, Chen PL, Scully P, Lee WH. Suppression of tumorigenicity of human prostate carcinoma cells by replacing a mutated RB gene. Science 1990;247:712–715.
139. Bookstein R. Tumor suppressor genes in prostatic oncogenesis. J Cell Biochem 1994;19(Suppl):217–223.
140. Riegman PH, Vlietstra RJ, van der Korput JA, Brinkmann AO, Trapman J. The promoter of the prostate-specific antigen gene contains a functional androgen responsive element. Mol Endocrinol 1991;5:1921–1930.
141. Schaffner DL, Barrios R, Shaker M, Rajagopalan S, Huang S, Tindall DJ, et al. Transgenic mice carrying a PSArasT24 hybrid gene develop salivary gland and GI tract neoplasms. Lab Invest 1995;72:283–390.
142. Cleutjens KB, van Eekelen CC, van der Korput HA, Brinkman AO, Trapman J. Two androgen response regions cooperate in steroid hormone regulated activity of the prostate-specific antigen promoter. J Biol Chem 1996;271:6379–6388.
143. Schuur ER, Henderson GA, Kmetec LA, Miller JD, Lamparski HG, Henderson DR. Prostate-specific antigen expression is regulated by an upstream enhancer. J Biol Chem 1996;271:7043–7051.
144. Wei C, Willis A, Tilton BR, Looney RJ, Lord EM, Barth RK, et al. Tissue-specific expression of the human prostate-specific antigen gene in transgenic mice: implications for tolerance and immunotherapy. Proc Natl Acad Sci USA 1997;94:6369–6374.
145. Cleutjens KB, van der Korput HA, Ehren-van Eekelen CC, Sikes RA, Fasciana C, Chung LW, et al. A 6-kb promoter fragment mimics in transgenic mice the prostate- specific and androgen-regulated expression of the endogenous prostate- specific antigen gene in humans. Mol Endocrinol 1997;11:1256–1265.
146. Wei C, Callahan BP, Turner MJ, Willis RA, Lord EM, Barth RK, et al. Regulation of human prostate-specific antigen gene expression in transgenic mice: evidence for an enhancer between the PSA and human glandular kallikrein-1 genes. Int J Mol Med 1998;2:487–496.
147. Rennie PS, Bruchovsky N, Leco KJ, Sheppard PC, McQueen SA, Cheng H, et al. Characterization of two cis-acting elements involved in the androgen regulation of the probasin gene. Mol Endocrinol 1993;7:23–36.
148. Brookes DE, Zandvliet D, Watt F, Russell PJ, Molloy PL. Relative Activity and Specificity of Promoters from Prostate-expressed Genes. Prostate 1998;35:18–26.
149. Patrikainen L, Shan J, Porvari K, Vihko P. Identification of the deoxyribonucleic acid-binding site of a regulatory protein involved in prostate-specific and androgen receptor- dependent gene expression. Endocrinology 1999;140:2063–2070.
150. Kasper S, Rennie PS, Bruchovsky N, Sheppard PC, Cheng H, Lin L, et al. Cooperative binding of androgen receptors to two DNA sequences is required for androgen induction of the probasin gene. J Biol Chem 1994;269:31,763–31,769.
151. Kasper S, Rennie PS, Bruchovsky N, Lin L, Cheng H, Snoek R, et al. Selective activation of the probasin androgen responsive region by steroid hormones. J Mol Endocrinol 1999;22:313–325.
152. Yan Y, Sheppard PC, Kasper S, Lin L, Hoare S, Kapoor A, et al. A large fragment of the probasin promoter targets high levels of transgene expression to the prostate of transgenic mice. Prostate 1997;32:129–139.
152a. Zhang J-F, Thomas TZ, Kasper S, Matusik RJJ. A small composite probasin promoter confers high levels of prostate-specific gene expression through regulation by androgens and glucocorticoids in vitro and in vivo. Endocrinology 2000;141:4698–4710.
153. Leff MA, Epstein PN, Doll MA, Fretland AJ, Devanaboyina US, Rustan TD, Hein DW. Prostate-specific human N-acetyltransferase 2 (NAT2) expression in the mouse. J Pharmacol Exp Ther 1999;290:182–187.
154. Barrios R, Lebovitz RM, Wiseman AL, Weisoly DL, Matusik RJ, DeMayo F, Lieberman MW. RasT24 driven by a probasin promoter induces prostatic hyperplasia in transgenic mice. Transgenics 1996;2:23–28.
155. Greenberg NM, DeMayo FJ, Finegold MJ, Medina D, Tilley WD, Aspinall JO, et al. Prostate cancer in a transgenic mouse. Proc Natl Acad Sci USA 1995;92:3439–3443.
156. Green JE, Greenberg NM, Ashendel CL, Barrett JC, Boone CW, Getzenberg RH, Henkin J, et al. Transgenic and reconstitution models of prostate cancer. Prostate 1998;36:59–63.
157. Tutrone RF, Jr, Ball RA, Ornitz DM, Leder P, Richie JP. Benign prostatic hyperplasia in a transgenic mouse: a new hormonally sensitive investigatory model. J Urol 1993;149:633–639.

158. Donjacour AA, Thomson AA, Cunha GR. Enlargement of the ampullary gland and seminal vesicle, but not the prostate in int-2/Fgf-3 transgenic mice. Differentiation 1998;62:227–237.
159. Kitsberg DI, Leder P. Keratinocyte growth factor induces mammary and prostatic hyperplasia and mammary adenocarcinoma in transgenic mice. Oncogene 1996;13:2507–2515.
160. Tehranian A, Morris DW, Min BH, Bird DJ, Cardiff RD, Barry PA. Neoplastic transformation of prostatic and urogenital epithelium by the polyoma virus middle T gene. Am J Pathol 1996;149:1177–1191.
161. Zhang X, Chen MW, Ng A, Ng PY, Lee C, Rubin M, et al. Abnormal Prostate Development in C3(1)-bcl-2 Transgenic Mice. Prostate 1997;32:16–26.
162. Schreiber-Agus N, Meng Y, Hoang T, Hou H, Jr, Chen K, Greenberg R, et al. Role of Mxi1 in ageing organ systems and the regulation of normal and neoplastic growth. Nature 1998;393:483–487.
163. Prescott JL, Blok L, Tindall DJ. Isolation and androgen regulation of the human homeobox cDNA, NKX3.1. Prostate 1998;35:71–80.
164. Voeller HJ, Augustus M, Madike V, Bova GS, Carter KC, Gelmann EP. Coding region of NKX3.1, a prostate-specific homeobox gene on 8p21, is not mutated in human prostate cancers. Cancer Res 1997;57:4455–4459.
165. Bhatia-Gaur R, Sciavolino PJ, Desai N, Gridley T, Abate-Shen C, Shen MM. The Nkx3.1 homeobox gene is required for normal prostate development. Biology of Prostate Growth Symposium, Bethesda, MD, 1998; (abstract).
166. Kasper S, Sheppard PC, Yan Y, Pettigrew N, Borowsky AD, Prins GS, et al. Development, progression and androgen-dependence of prostate tumors in transgenic: a model for prostate cancer. Lab Invest 1998;78:319–334.
167. Bookstein R, MacGrogan D, Hilsenbeck SG, Sharkey F, Allred DC. p53 is mutated in a subset of advanced-stage prostate cancers. Cancer Res 1993;53:3369–3373.
168. Isaacs WB, Carter BS, Ewing CM. Wild-type p53 suppresses growth of human prostate cancer cells containing mutant p53 alleles. Cancer Res 1991;51:4716–4720.
169. Brooks JD, Bova GS, Ewing CM, Piantadosi S, Carter BS, Robinson JC, et al. An uncertain role for p53 gene alterations in human prostate cancers. Cancer Res 1996;56:3814–3822.
170. Brooks JD, Bova GS, Isaacs WB. Allelic loss of the retinoblastoma gene in primary human prostatic adenocarcinomas. Prostate 1995;26:35–39.
171. Levine AJ, Momand J. Tumor suppressor genes: the p53 and retinoblastoma sensitivity genes and gene products. Biochim Biophys Acta 1990;1032:119–136.
172. Levine AJ. Tumor suppressor genes. Bioessays 1990;12:60–66.
173. Pallas DC, Shahrik LK, Martin BL, Jaspers S, Miller TB, Brautigan DL, et al. Polyoma small and middle T antigens and SV40 small t antigen form stable complexes with protein phosphatase 2A. Cell 1990;60:167–176.
174. Sontag E, Fedorov S, Kamibayashi C, Robbins D, Cobb M, Mumby M. The interaction of SV40 small tumor antigen with protein phosphatase 2A stimulates the map kinase pathway and induces cell proliferation. Cell 1993;75:887–897.
175. Wheat WH, Roesler WJ, Klemm DJ. Simian virus 40 small tumor antigen inhibits dephosphorylation of protein kinase A-phosphorylated CREB and regulates CREB transcriptional stimulation. Mol Cell Biol 1994;14:5881–5890.
176. Frost JA, Alberts AS, Sontag E, Guan K, Mumby MC, Feramisco JR. Simian virus 40 small t antigen cooperates with mitogen- activated kinases to stimulate AP-1 activity. Mol Cell Biol 1994;14:6244–6252.
177. Furth PA. SV40 rodent tumour models as paradigms of human disease: transgenic mouse models. Dev Biol Stand 1998;94:281–287.
178. McKnight RA, Shamay A, Sankaran L, Wall RJ, Hennighausen L. Matrix- attachment regions can impart position-independent regulation of a tissue-specific gene in transgenic mice. Proc Natl Acad Sci USA 1992;89:6943–6947.
179. Gingrich JR, Barrios RJ, Morton RA, Boyce BF, DeMayo FJ, Finegold MJ, et al. Metastatic prostate cancer in a transgenic mouse. Cancer Res 1996;56:4096–4102.
180. Hsu CX, Ross BD, Chrisp CE, Derrow SZ, Charles LG, Pienta KJ, et al. Longitudinal cohort analysis of lethal prostate cancer progression in transgenic mice. J Urol 1998;160:1500–1505.
181. Gingrich JR, Barrios RJ, Kattan MW, Nahm HS, Finegold MJ, Greenberg NM. Androgen Independent Prostate Cancer Progression in the TRAMP Model. Cancer Res 1997;57:4687–4691.
182. Eng MH, Charles LG, Ross BD, Chrisp CE, Pienta KJ, Greenberg NM, et al. Early castration reduces prostatic carcinogenesis in transgenic mice. Urology 1999;54:1112–1119.
183. Perez-Stable C, Altman NH, Brown J, Harbison M, Cray C, Roos BA. Prostate, Adrenocortical, and Brown adispose tumors in fetal globin T antigen transgenic mice. Lab Invest 1996;74:363–373.

184. Jiborn T, Bjartell A, Abrahamsson PA. Neuroendocrine differentiation in prostatic carcinoma during hormonal treatment. Urology 1998;51:585–589.
185. Allison J, Zhang YL, Parker MG. Tissue-specific and hormonal regulation of the gene for rat prostatic steroid-binding protein in transgenic mice. Mole Cell Biol 1989;9:2254–2257.
186. Maroulakou IG, Anver M, Garrett L, Green JE. Prostate and mammary adenocarcinoma in transgenic mice carrying a rat C3(1) simian virus 40 large tumor antigen fusion gene. Proc Natl Acad Sci USA 1994;91:11,236–11,240.
187. Yoshidome K, Shibata MA, Maroulakou IG, Liu ML, Jorcyk CL, Gold LG, et al. Genetic alterations in the development of mammary and prostate cancer in the C3(1)/Tag transgenic mouse model (Review). Int J Oncol, 1998;12:449–453.
188. Shibata MA, Maroulakou IG, Jorcyk CL, Gold LG, Ward JM, Green JE. p53-independent apoptosis during mammary tumor progression in C3(1)/SV40 large T antigen transgenic mice: suppression of apoptosis during the transition from preneoplasia to carcinoma. Cancer Res 1996;56:2998–3003.
189. Bry L, Falk P, Huttner K, Ouellette A, Midtvedt T, Gordon JI. Paneth cell differentiation in the developing intestine of normal and transgenic mice. Proc Natl Acad Sci USA 1994;91:10,335–10,339.
190. Garabedian EM, Humphrey PA, Gordon JI. A transgenic mouse model of metastatic prostate cancer originating from neuroendocrine cells. Proc Natl Acad Sci USA 1998;95:15,382–15,387.
190a. Masumori N, Thomas TZ, Case T, Paul M, et al. A probasin-large T antigen transgenic mouse line develops adeno- and neuroendocrine-carcinoma having metastatic potential. Can Res 2000; in press.
191. Snoek R, Rennie PS, Kasper S, Matusik RJ, Bruchovsky N. Induction of cell-free, in vitro transcription by recombinant androgen receptor peptides. J Steroid Biochem Mol Biol 1996;59:243–250.
192. Snoek R, Bruchovsky N, Kasper S, Matusik RJ, Gleave M, Sato N, et al. Differential transactivaiton by the androgen receptor in prostate cancer cells. Prostate 1998;36:256–263.
193. Thompson TC. Genetic predisposition and mesenchymal-epithelial interactions in ras+myc-induced carcinogenesis in reconstituted mouse prostate. Mol Carcinog 1993;7:165–179.
194. Kasper S, Vaikunth S, Thomas T, Case T, Thomas J, Paul M, et al. Androgen independent progression of prostate cancer in LPBTag transgenic mice. 8th Annual Society for Basic Urologic Research, Prouts Neck, ME, 1998; (abstract).
195. di Sant'Agnese PA, Cockett AT. Neuroendocrine differentiation in prostatic malignancy. Cancer 1996;78:357–361.
196. Bonkhoff H. Neuroendocrine cells in benign and malignant prostate tissue: morphogenesis, proliferation, and androgen receptor status. Prostate 1998;8(Suppl):18–22.
197. Tetu B, Ro JY, Ayala AG, Johnson DE, Logothetis CJ, Ordonez NG. Small cell carcinoma of the prostate. Part I. A clinicopathologic study of 20 cases. Cancer 1987;59:1803–1809.
198. Amato RJ, Logothetis CJ, Hallinan R, Ro JY, Sella A, Dexeus FH. Chemotherapy for small cell carcinoma of prostatic origin. J Urol 1992;147:935–937.

20 Neural- and Endocrine-Cell-Specific Immortalization Using Transgenic Approaches

T. Rajendra Kumar, PHD

CONTENTS

INTRODUCTION

During development, in response to a wide range of intracellular and extracellular signals, eukaryotic cells choose a spectrum of "fates." Under normal physiological conditions, cells grow, progress through a series of cell cycle events, divide, and differentiate *(1,2)*. Depending upon the cell type and the terminally differentiated function, a fraction of cells also undergo apoptosis (i.e., cell death) or cell-cycle arrest, until they are activated by specific stimuli and re-enter the cell cycle *(1,2)*. Overall, the cell phenotypes are dictated and governed by a balanced activity between various positive regulators (growth factors, oncogenes) and negative regulators (tumor suppressors) of the cell cycle *(1,2)*. Aberrant activities of either of these regulators result in excessive proliferation, leading to immortalization at the cellular level and the initiation of tumors at the organism level. Alternatively, cells may undergo accelerated senescence and eventually die because of these aberrant events *(1,2)*. Molecular analysis of these critical events has many implications in understanding the origin and nature of human cancers and possible therapeutic intervention and cure by various genetic approaches.

Neural and endocrine cells are both ectodermal and endodermal in origin, and constitute an important group within the body. Usually, these cells replicate only during a

From: *Contemporary Endocrinology: Transgenics in Endocrinology*
Edited by: M. Matzuk, C. W. Brown, and T. R. Kumar © Humana Press Inc., Totowa, NJ

limited time span and terminally differentiate, but lose the differentiated function prior to or during the process of tumorigenesis *(3)*. The incidence of clinically significant neuroendocrine tumors in humans is quite high compared to tumors of non-neuroendocrine origin, and these tumors are often highly metastatic *(3)*. Over the past several years, knowledge of cell-cycle regulators, tumor suppressors, oncogenes, various apoptotic regulators, and the developmental aspects of several neuroendocrine cells has expanded enormously. In addition, recent genetic advances in the manipulation of the mouse genome by transgenic and gene-targeting approaches have greatly facilitated an understanding of the origin, development, and maintenance of the differentiated function of many neuroendocrine cells and tumors *(4,5)*. These technologies have also been successfully utilized in immortalizing neuroendocrine cells in vivo, an event that is otherwise difficult to achieve by regular in vitro cell transfection strategies.

This chapter, explores in vitro approaches to neural- and endocrine-cell immortalization, and describes various transgenic and knockout mouse models with neuroendocrine-cell immortalization, and presents aspects related to in vitro derivation of novel cell lines from tumors that have been induced in vivo in these genetically modified mutant mice.

IN VITRO APPROACHES FOR NEURAL- AND ENDOCRINE-CELL IMMORTALIZATION

As previously mentioned, normal cells attain replicative senescence in vitro with progressive cell divisions. The molecular mechanisms that control these events are complex, and are responsible for normal-size maintenance of various tissues and organs in vivo. However, these mechanisms can be overridden in vitro by expression of oncogenes, various cell-cycle components, or mutant forms of tumor suppressors, all of which promote cell divisions leading to immortalization of the transfected cells. Some of these approaches are described briefly in the following section.

Immortalization by DNA-Tumor Virus-Encoded Proteins

At least two decades ago, specific immortalizing genes were identified for adenovirus, polyoma virus, SV40 virus, and papillomavirus *(6)*. Although each of them is capable of immortalizing many mammalian cells, SV40-virus-encoded large T-antigen (SV40 TAg) or its various mutant forms have been extensively used, either alone or by cotransfection with various other cooperative agents *(7)*. In addition, the immortalization of cells by SV40 TAg has been achieved by various strategies, including direct transfection, retroviral-mediated transduction, or expression by site-specific recombination *(7)*.

SV40 TAg-Mediated Immortalization

SV40 TAg is a 90-kDa nuclear protein that binds many proteins through distinct binding domains. Its functions range from adenosine triphosphate (ATPase) and helicase activities to initiation and maintenance of cellular transformation *(8)*. The transforming ability of SV40 TAg is its binding to tumor-suppressor proteins such as p53 and Rb, thus enabling the cells to proliferate uncontrollably *(8)*. Several types of primary cells have been directly transfected with SV40 TAg encoding plasmids and selected for foci or colony formation in vitro. These are typically subcloned further or propagated in vivo in nude mice. Although SV40 TAg is extremely efficient, the main problem is that the transformed cells fail to maintain their differentiated function, and the subsequent application of such cells for understanding physiologically relevant functions is less meaningful *(8)*. Often, a combination of more than one transforming agent in addition to SV40 TAg seems to be more potent for efficient immortalization of some types of cells. Such cooperativity between different types of transforming agents has been well-established *(2)*.

SV40 TAg and Cooperativity with Other Transforming Agents

One of the best studied secondary oncogenes in mediating cooperativity with SV40 TAg during transformation and immortalization of cells is the Ha-*ras* oncogene. The *ras* family members were originally discovered in human tumors, including medulloblastomas. These proteins function as signaling activators similar to many membrane-bound G-proteins *(9)*. Although the mechanistic details are unknown, cotransfection of Ha-*ras* oncogene into neuroendocrine cells with SV40 T-Ag results in several characteristics of the immortalized cells that are completely different from cells immortalized by SV40 TAg alone *(9)*. In addition, cooperativity between the oncogenes results in efficient immortalization of many primary cells, and in retention of the many differentiated functions. For example, ovarian granulosa cells immortalized by Ha-*ras* and SV40 TAg demonstrate many features superior to cells transformed by SV40 TAg alone-transformed cells, including better cell morphology, well-preserved structural organization of the cytoskeletal network, hormone-responsiveness, enhanced steroid production, and quick doubling times *(10)*. Further, these cells are also highly tumorigenic when injected into nude mice *(10)*.

Two other classes of oncogenes that are known to act cooperatively with SV40 TAg are human papillomavirus early proteins E6 and E7 and the adenovirus early proteins E1A and E1B. One common mechanism of cellular transformation and immortalization by all these proteins seems to be their ability to bind and inhibit the activities of many cell cycle proteins including the tumor suppressors, p53 and Rb *(2)*.

Conditional Immortalization by Temperature-Sensitive (ts) SV40 TAg Mutants

Extensive structural studies aimed at delineating the functional domains of SV40 TAg have resulted in identification of regions that are important and necessary for the DNA-helicase-binding and ATPase-binding activities of SV40 TAg. These functions are relevant for the cell-transforming and immortalizing properties of TAg. These studies have also led to the generation and characterization of various point mutants of SV40 TAg that have thermosensitive properties *(11)*. For example, at a permissive temperature of 33°C, expression of ts TAg causes cells to rapidly proliferate, and when the temperature is shifted to 39–41°C, cells stop dividing and undergo differentiation at this nonpermissive temperature *(11)*.

More than 20 ts mutants of SV40 TAg have been characterized in detail. Some of these demonstrate absolute thermosensitivity, whereas others are semisensitive. Among these, the best-characterized and extensively utilized variant is the A58 ts mutant *(12)*. This variant harbors a point mutation (with Ala-438 $\rightarrow$ Val substitution) and exhibits sensitivity to thermal shifts between 33–41°C. The ts A58 has been successfully utilized in vitro and in vivo to immortalize several types of neuroendocrine cells *(12)*. One of the advantages of conditional immortalization by ts A58 is the choice of shifting the cell fates to both replication and differentiation phases, and systematically examining the differentiated phenotypes of an immortalized cell line. These strategies will be further discussed in the following sections.

Immortalization by Retrovirus-Mediated SV40 TAg Transduction

Introduction of oncogenes into primary cells by conventional biochemical methods is straightforward, and results in immortalization. However, many neural cells subjected to this protocol do not retain differentiated functions, a common problem *(13)*. One efficient approach to bypass this limitation is the delivery of SV40 TAg oncogene via retroviral transduction. This strategy has additional advantages over the direct oncogene transformation of cells, such as stable integration of the viral genome into and successful transmission from the host-cell genome *(13)*. In addition, the direct delivery of such "packaged" recombinant viral vectors carrying oncogenes in vivo is also feasible with this approach. Several types of neural cells and neural-cell progenitors have been successfully immortalized using this approach to study the differentiated properties.

As a further variation of the retroviral transduction, a reversible immortalization scheme has also been developed based on a site-specific recombination strategy *(14)*. In this method, the cells are subjected to transfection by the polycistronic constructs (drug-selectable marker and oncogene) followed by retrovirus-mediated transfer of SV40 TAg, which is subsequently excised by site-specific recombination using the Cre/loxP system. This approach eliminates the transferred oncogene, i.e., SV40 TAg following transient transfection of Cre recombinase enzyme after the initial induction of cell proliferation *(14)*. This method offers the advantage of initiating the immortalization process in a temporal fashion (depending on the time of excision of the oncogene), thus identifying cell characteristics during the time course of development.

Immortalization by Telomere Maintenance

The repetitive DNA sequences present at the end of linear chromosomes are known as telomeres. Telomeres progressively shorten in length with successive rounds of cell replication, leading to cellular senescence and eventual cell death *(15)*. The enzyme telomerase, which prevents this process and maintains telomere integrity, is a ribonucleoprotein. Telomerase is present in undetectable levels in normal somatic cells, but is highly active in many tumor cells *(15)*. Thus, it is believed that overexpression of the enzyme may lead to immortalization of cells through stabilization of telomere length. Two types of recent evidence suggest the importance of telomerase-mediated immortalization. First, ectopic expression of the catalytic subunit of the telomerase enzyme in the neuroectodermal retinal pigmented epithelial cells results in indefinite multiplication of these cells, and both senescence and subsequent crisis are completely blocked *(16)*. Second, a catalytically inactive, dominant negative mutant form of human telomerase, when expressed in immortalized cells, eliminates the endogenous telomerase activity, and leads to shortening of telomere length and eventual death of tumor cells *(17)*. Future studies involving telomere-length maintenance processes may result in immortalization of rare cell types. Furthermore, directed expression of mutant forms of the enzyme into cells and tissues may offer a novel therapeutic approach for treating some cancers.

Immortalization by Other Oncogenes

Cancer generally results from alterations in multiple biochemical pathways. This is not surprising, considering the diversity of cell types and the existence of complex interactions between several signaling networks within the cell. Although SV40 TAg-mediated immortalization is the most widely studied aspect of in vitro tumor-cell biology, several investigators have used multiple oncogenes to achieve immortalization of cells *(2)*. These include *myc*, c-jun, mutant forms of p53, c-*ras*, and v-*src*. In many instances, the relevant gene product is overexpressed, and this is sufficient to cause immortalization of the cells. This is also true for many growth factors, which they may cause immortalization of cells when overexpressed *(2)*. The cross-communication between various biochemical pathways activated by growth factors and the interactions with downstream targets have been characterized in sufficient detail, and are beyond the scope of this chapter. The reader is referred to several textbooks and recent reviews related to this topic *(2,18–20)*.

IN VIVO APPROACHES FOR IMMORTALIZATION OF NEURAL AND ENDOCRINE CELLS

Although immortalization of neuroendocrine cells by in vitro methods has been achieved, the resulting cells/tumors do not always maintain the differentiated functions to completely understand the cellular and developmental aspects. Although retroviral transduction is widely applicable, this method is limited by viral integration only after a minimum of one round of replication of the transfected cell type, and therefore requires replication of progenitor cells *(13)*. The feasibility of mouse germline manipulation, such as introduction of site-specific mutations into mouse embryonic germ cells *(21)* and generation of transgenic mice *(4,22)*, offer infinite opportunities to immortalize a desired cell type in vivo. Using these genetic approaches, it is now possible to precisely regulate any endogenous or foreign gene-expression patterns in a controlled and spatiotemporal manner. Thus targeted tumorigenesis can theoretically be achieved during the developmental program, the cell-fate commitment phase, or even during the postmitotic phase of any given cell type. The conventional targeted tumorigenesis (transgenic) approach involves pronuclear microinjection of mouse embryos with oncogenes driven by cell-specific promoters, and subsequent transfer of such embryos into pseudopregnant foster mothers. The identification of founder progeny carrying the stably integrated transgene and subsequent transmission to further generations permits the establishment of pedigrees of various lines of mice. These transgenic mice can be monitored for the development of tumors over a period of time, in the cell/tissue of interest. In the gene-targeting approach, deletion of a tumor-suppressor gene in embryonic stem (ES) cells is first achieved at one or both loci, and subsequently germline transmission of these mutated ES cells is obtained. If the targeting event occurs at only one locus, then heterozygous mutant mice are first generated, and these are intercrossed to obtain homozygous mutant mice that completely lack the protein encoded by the deleted gene. To date, only a few mouse models lacking tumor suppressors have been generated *(23)*. However, there has been an intense search to identify and characterizing both tissue-specific and global tumor suppressor genes, using many in vivo genetic approaches. Both conventional and ES-cell-based transgenic approaches are important, and complement each other to understand the developmental aspects of immortalization, the role of various modifier factors in tumorigenesis, and for the successful isolation and establishment of novel cell lines. The following sections, examine targeted tumorigenesis in various neural and endocrine systems, and later describe neuroendocrine tumors in mouse models deficient in tumor suppressors or other genes.

Targeted Tumorigenesis of Neural and Endocrine Cells in Transgenic Mice

Targeted expression of oncogenes or growth factors driven by neuroendocrine cell-specific promoters in transgenic mice often results in the development of tumors. These tumors have been a valuable resource for the generation of many novel cell lines, or for studying analogous human neuroendocrine diseases. Major problems associated with this approach are the rapid growth of aggressive tumors leading to morbidity, mice that survive becuase of the slow growth of the resulting tumors, or mice that are sterile for known or unknown reasons. However, the specificity of the tumorigenesis and the advantage of propagating the tumors in vivo in nude mice make this approach highly useful for various studies.

Directed Expression of Oncogenes to the Nervous System

The cellular complexity of the brain and the postmitotic nature of mature differentiated neurons present a major challenge for successful immortalization of cell types within the brain. This section illustrates the use of SV40 TAg to immortalize different neuronal cell lineages in transgenic mice with a few examples.

During the initial phases of development of transgenic mouse technology, Brinster and colleagues have developed a mouse model in which SV40 TAg gene expression is regulated by the mouse metallothionein (MT) promoter (24). One of the mouse lines established using this fusion transgene developed choroid plexus tumors, with nearly 100% penetrance. Some mice also developed thymic hypertrophy and kidney pathology. Only the affected tissues demonstrated the presence of SV40 TAg mRNA and protein, but not the nontumorous tissues, suggesting that SV40 TAg gene activation may necessary for the onset of tumorigenesis *(24)*. Since this study, many research groups have studied the choroid plexus, even when SV40 TAg gene expression is driven by various cell-specific promoters using transgenic mice.

The immortalization of hypothalamic gonadotropin-releasing hormone (GnRH) neurons by targeted tumorigenesis in transgenic mice *(25)* marked an exciting new era in mammalian reproductive biology. GnRH is a decapeptide released from the neurosecretory neurons of the ventral median hypothalamus. Its primary role is to regulate the biosynthesis and secretion of the pituitary gonadotropins, luteinizing hormone (LH), and follicle-stimulating hormone (FSH) *(26)*. In the normal mouse, GnRH-secretory neurons are low in number, and are scattered in the rostral hypothalamus. Sophisticated grafting and developmental studies using dye-labeling and tracking methods have previously shown that the GnRH neurons actually originate in the olfactory placode and migrate toward the pre-optic area (POA) and hypothalamus *(27)*. When the mouse GnRH gene was appropriately expressed in the genetic background of hypogonadal *(hpg)* (a naturally occurring mouse mutant caused by a deletion in the GnRH gene), hypogonadism in these mice was corrected, and normal LH and FSH levels were restored in the serum *(28)*. This key experiment suggested that it might be possible to target expression of SV40 TAg gene to the GnRH neurons. In 1990, Mellon and colleagues, successfully immortalized GnRH neurons in transgenic mice, using a 2.3-kb of the rat GnRH gene 5' flanking region, including the mRNA start sites fused to the entire coding region of SV40 TAg *(25)*. Nine founder mice carrying the transgene were generated, and all of them were hypogonadal and sterile. Seven mice developed choroid plexus brain tumors, which were GnRH-negative and demonstrated epithelial morphology. In two of the mice, hypothalamic tumors were observed. In one mouse, a large tumor was observed, extending from the dorsal border of the optic chiasma to below the internal capsule and displacing the anterior commissure. Northern-blot analysis demonstrated high-level expression of GnRH and SV40 TAg mRNAs specific to the tumor tissue. A cell line, named GT-1, was successfully derived from this tumor, and has since been extensively used to study GnRH biology and cell signaling within the GnRH neurons *(29)*.

While GnRH neurons have been successfully immortalized in vivo, similar transgenic strategies using a 872-bp growth hormone-releasing factor (GRF) promoter-SV40 TAg fusion gene did not result in immortalization of GRF neurons in mice *(30)*. These transgenic mice had normal hypothalamic functions, but demonstrated severe thymic hyperplasia because of inappropriate production of T-cell maturation factors of the thymic epithelial cells. However, later studies using a larger 5-kb GRF promoter driving SV40 TAg gene expression resulted in tumors of exclusive hypothalamic origin *(31)*.

Targeted immortalization of catecholaminergic neurons using the regulatory region of the tyrosine hydroxylase (TH) gene driving SV40 TAg expression has been achieved *(32)*. This finding illustrates the advantage of using the transgenic approach over retroviral transduction, because TH is expressed only in postmitotic neurons. Furthermore, only a limited number of TH-producing cells in the central nervous system (CNS) are scattered in multiple sites, thus precluding isolation of a pure population of cells for in vitro experiments *(32)*. TH is the first and rate-limiting enzyme that catalyzes the conversion of L-tyrosine to dihydroxyphenylalanine, from which all the catecholomines are biosynthesized. Within the CNS, TH is expressed in the olfactory bulb, midbrain, other brainstem regions, and in the peripheral nervous system, and is highly expressed in the chromaffin cells of the adrenal medulla *(32)*. Several studies have delineated the 5' upstream regulatory region of the TH gene that confers cell-specific expression in transgenic mice and in cultured cells. Based on this information, transgenic mice were generated in which SV40 TAg expression was regulated by a 773-bp of the rat TH promoter sequences. These mice develop both brain and adrenal tumors as early as 15 wk with motor-function deficits—predominantly seizures, hunched posture, and an enlarged head. The tumors in the brain were largely restricted to the ventral side of the brain—more precisely, between the midbrain and rostral brainstem *(32)*. In other tissues, no tumor developed and TAg expression was not found. Morphologically, the tumors resembled primitive neuroectodermal tumors, and histologically, they were composed of small, densely packed, undifferentiated cells with scant cytoplasm. Both the brain and the adrenal tumors were immunopositive for TH and TAg proteins. Additionally, some non-TH-producing cells were also found immunoreactive for TAg protein. Although the 773-bp TH promoter targeted expression of SV40 TAg to appropriate regions in the brain and adrenal glands, TH-producing cells within the olfactory bulb and superior cervical ganglia were not stained for TAg protein. Three different cell lines were developed from the TH-positive tumors from the brains of the transgenic mice, and were further characterized *(32)*.

The olfactory system is a unique and dynamic neural system. The olfactory neuroepithelium consists of three main cell types—olfactory sensory neurons, sustentactular cells (support or glial-like cells), and basal or stem cells *(33)*. These cells are present in a pseudostratified columnar neuroepithelium that has synaptic projections to the olfactory bulb. The olfactory neuroepithelium undergoes continual turnover in the adult animal, and represents a neural tissue consisting of cells at various developmental stages of neurogenesis *(33)*. Less is known regarding the molecular mechanisms of development of the olfactory system. In recent years, families and subfamilies of mammalian and fish odorant receptors have been cloned and characterized *(34)*. To examine the cellular dynamics of the olfactory neuroepithelium and to establish cell lines of olfactory neuronal origin, a hybrid oncogene encoding the SV40 TAg fused to the 3.0-kb of cell-specific transcription regulatory elements of the rat olfactory marker protein (OMP) was used to generate transgenic mice *(35)*. In the normal mouse, OMP transcripts are selectively and highly expressed within mature olfactory sensory neurons. The encoded protein is a 19-kDa acidic soluble protein with no known function. Several lines of transgenic mice were generated, and one line demonstrated the olfactory tissue-specific expression of TAg mRNA similar to the endogenous OMP transcripts *(35)*. In general, tumor development was very slow, and paralleled the level of expression of the transgene, which was also low compared to the mouse OMP mRNA. In older transgenic mice at 10 mo, a dramatic hypoplasia in the neuroepithelium resulted in a reduced number of olfactory receptor neurons; however, the basal cells and sustentacular cells were normal *(35)*. The olfactory bulb, which is the synaptic target for olfactory sensory neurons, was also found to be abnormal. Although the tumors developed slowly in these transgenic mice, large tumors originating from the nasal cavity were present by 1 yr and invaded the maxillary and cranial structures through the palate *(35)*. This tumor growth also resulted in circling behavior, postural tilt, and locomotor inactivity. Clonal cell lines were subsequently derived from the dissociated tumors in primary culture *(35)*.

The pineal gland is an important neuroendocrine tissue, and is the best-studied model system for circadian regulation in mammals *(26)*. The suprachiasmatic nucleus within the brain is the center for circadian activity, and is known to regulate the expression of the key pineal genes, tryptophan hydroxylase (TPH) and N-acetyl transferase (NAT) genes. TPH catalyzes the first step of serotonin biosynthesis, both in the neuroendocrine pineal cells and in the serotonergic neurons in the raphe nuclei *(26)*. Serotonin is the precursor for the synthesis of the pineal neurohormone melatonin, whose secretion is also regulated by circadian variation. The molecular mechanisms of circadian regulation of expression of pineal TPH and NT enzymes are largely unknown. Because of the small size of the pineal gland, it is often difficult to obtain sufficient number of pinealocytes in primary culture. Initial transgenic studies demonstrated that a 6.1-kb 5' upstream region of the mouse TPH gene can direct the expression of lacZ specifically to the pineal gland and dorsal raphe nuclei *(36)*. This promoter region has been utilized as a targeted immortalization strategy to drive the expression of SV40 TAg to pinealocytes in transgenic mice *(37)*. Mice derived from two independent transgenic lines have been established from pronuclear microinjections, and both of these lines develop a characteristic cranial bulge caused by pineal tumors which develop as early as 6–8 wk and can kill the mice by 12–15 wk of age because of the invasive tumor. The tumors stain positive for TPH immunoreactivity when coronal brain sections from a transgenic mouse at 8 wk of age are analyzed *(37)*. Double transgenic mice have been generated by intercrossing the TPH-lacZ with the TPH-SV40 TAg mice to more clearly identify the tumor-positive cells. These tumors were eventually used for successful development of clonal pinealocyte-cell lines *(37)*. Thus, targeted immortalization using transgenic mice has been successfully used to generate an unlimited pure population of pineal cells from an otherwise tiny gland to further study the molecular and cellular biology of the pineal gland.

Similar transgenic strategies have been used to immortalize retinal neurons, Purkinje cells, astroglial cells, and Schwann cells *(38)*. Additionally, several lines of transgenic mice have been established that develop neuroblastomas by targeted expression of *MYCN*, *ret*, or *Ras* protooncogenes *(38)*.

TARGETED TUMORIGENESIS OF THE PITUITARY CELL LINEAGE

The pituitary gland develops through a series of inductive interactions between the diencephalon and Rathke's pouch as it invaginates from the oral ectoderm. Subsequently, several classes of transcription factors are activated and expressed in a precise spatio-temporal fashion in the proliferating cells of Rathke's pouch, leading to the commitment and establishment of distinct cell lineages *(39,40)*. In the adult, the pituitary gland is organized into three anatomically distinct lobes: the anterior, intermediate, and posterior lobe. The anterior pituitary gland consists of a heterogenous population of trophic-hormone-producing cells, including the somatotropes, lactotropes, corticotropes, thyrotropes, and gonadotropes. Each of the anterior pituitary-cell types is responsive to both hypothalamic and gonadal factors (steroid and peptide), and synthesizes, stores, and secretes distinct hormones such as growth hormone (GH, somatotropes), prolactin (lactotropes), adrenocorticotropin (ACTH, corticotropes), thyroid-stimulating hormone (TSH, thyrotropes) and the gonadotropins, LH and FSH (gonadotropes). The intermediate lobe consists primarily of melanotropes, whereas the posterior lobe is the storage site for neurohormones synthesized from hypothalamic neurosecretory neurons *(26)*.

In humans, pituitary tumors are mostly benign monoclonal adenomas arising from adenohypophyseal (anterior pituitary) cells, and often express and excessively secrete the pituitary hormones leading to endocrine syndromes. These include the most common, hyperprolactinemia, acromegaly, Cushing's disease, and the rare hyperthyroidism *(41)*. The existence of a network of autocrine, paracrine, and endocrine interactions, and the diversity of signaling pathways within the pituitary cells provide a complex scenario to understand and to develop novel mouse models for studying the pathobiology of pituitary tumors *(41)*. Both gain-of-function and loss-of-function mouse models have been invaluable in analyzing the pituitary tumorigenesis and deriving novel cell lines to explore transcriptional regulation and cell signaling within distinct anterior pituitary-cell types.

Somatotropes and Lactotropes. During embryonic pituitary development, somatomammotrope progenitor cells transiently exist in small numbers, and produce both GH and prolactin before commitment to a final differentiated cell type (i.e., somatotype-only GH-producing cell) or lactotrope (only prolactin-producing cell) *(40,41)*. A homeodomain protein named Pit-1 or GHF-1 is required for somatotrope andlactotrope commitment, differentiation, and later on expansion and maintenance of somatotrope- and thyrotrope-cell lineage. Pit-1 mRNA is detectable prior to expression of GH and prolactin in mature cells, and Pit-1 protein can autoregulate its own gene expression *(40,41)*. To study the mechanisms of initial activation of Pit-1 in the early phases of pituitary development, and to eventually isolate the somatotropic progenitor-cell population, a 15-kb 5' flanking region of the rat Pit-1 gene was used to target SV40 TAg specifically to the anterior pituitary in transgenic mice *(42)*. These mice were phenotypically small and infertile, and developed pituitary tumors. Northern blot analysis confirmed the expression of SV40 TAg and Pit-1 mRNA only in the pituitary-tumor cells, but not in other tissues of the transgenic mice. Immunohistochemistry of the tumors revealed TAg and Pit-1 staining, but no GH or prolactin staining *(42)*. These observations suggest that use of the Pit-1 promoter results in immortalization of a progenitor cell prior to the differentiation of somatotrope and lactotrope lineages.

Metallothionein (MT)-1-GHRH transgenic mice also develop pituitary adenomas. All of the mice demonstrated GH-immunoreactive tumors by 10–24 mo, and some of the tumors also secreted prolactin and TSH *(43)*. These findings provide evidence that prolonged exposure of the pituitary in vivo to a hypothalamic growth factor causes proliferation, hyperplasia, and eventually adenoma of adenohypophyseal cells *(43)*.

Prolactinomas are quite common in humans. Prolactin secretion is normally under negative regulation by dopamine, an inhibitory neurotransmitter for lactotropes. Pituitary adenomas with lactotrope hyperplasia have been observed in transgenic mice in which either NGF is ectopically produced *(44)* or TGFα is locally produced *(45)*, using rat prolactin promoters that specifically target the transgenes to lactotropes. Recently, targeted overexpression of the neuropeptide galanin in lactotropes has been achieved using a rat prolactin promoter *(46)*. These female transgenic mice develop pituitary lactotrope hyperplasia and hyperprolactinemia, but no tumors. There are several GH- and prolactin-secreting tumor-cell lines already established that have provided excellent models to study the transcriptional regulation of GH and prolactin gene expression *(39,40)*.

Gonadotropes. Gonadotropes represent the least abundant (5–7%) cell type within the anterior pituitary. A fully differentiated gonadotrope expresses three gonadotropin subunits: α-glycoprotein hormone (αGSU), LHβ, and FSHβ. These subunits undergo posttranslational glycosylation modifications, and noncovalently associate to form functional heterodimers (47). The gonadotropin subunits are synthesized and secreted in response to hypothalamic GnRH stimulus, and are both positively and negatively regulated by activins and inhibins (produced within the pituitary and the gonads) or by gonadal steroids, such as progesterone, estrogen, and testosterone *(26)*. α-GSU is first expressed during anterior pituitary development (E 10.5), whereas LHβ- and FSHβ-subunits appear much later, at E 16.5 and E 17.5 respectively. α-GSU is also expressed in thyrotropes, where it heterodimerizes with thyroid-stimulating hormone β-subunit (TSH) *(40)*.

Gonadotropin-secreting tumors have been identified in many mammalian species. However, human tumors are often considered "null-adenomas," even when they secrete one or two gonadotropin subunits. Immortalized gonadotrope tumor cell lines that are fully differentiated and express all the three subunits have not yet been identified. Thus, the detailed mechanisms of transcriptional regulation of gonadotropin subunits have lagged behind compared to the other anterior pituitary-cell-specific markers or hormones.

Transgenic targeted oncogenesis has been effectively used to immortalize pituitary gonadotrope lineage in vivo at discrete stages, and subsequently derive novel cell lines representing each developmental phase. Mellon and colleagues have utilized a series of α-GSU promoters or a LHβ promoter to drive the gonadotrope/thyrotrope-specific expression of SV40 TAg in transgenic mice. First, they generated a transgenic mouse model utilizing a 1.8-kb promoter sequence from human α-GSU gene. These mice develop pituitary gonadotrope tumors and express only α-GSU but not the hormone-specific LHβ- and FSHβ-subunits *(48)*. However, these tumors express GnRH receptor, steroidogenic factor-1 (SF-1), Lim-2, and Lim-3. In contrast, a longer 5.5-kb promoter sequence from the same human α-GSU gene directs the SV40 TAg expression to both gonadotrope and thyrotrope lineages. These tumors express only α-GSU and no GnRH receptor, SF-1, LHβ, or FSHβ *(49,50)*. Thus, the immortalization event must occur at a developmental stage prior to than that obtained with the 1.8-kb human α-GSU promoter.

As predicted from the "sequential immortalization during discrete developmental phases," SV40 TAg gene expression driven by a rat LHβ 5' region induces gonadotrope-tumor formation at a later developmental stage in transgenic mice. These tumors express α-GSU, LHβ, GnRH receptor, SF-1, Lim-2, and Lim-3, but not FSHβ-subunit. Thus, almost fully differentiated gonadotrope tumors are induced when the oncogene is expressed with LHβ-promoter sequences *(50)*. Since the FSHβ-subunit gene is the last gonadotropin subunit to be expressed, we utilized regulatory sequences of human FSHβ gene to drive the expression of SV40-tsA58 TAg to gonadotropes in transgenic mice *(51)*. Our goal was to eventually obtain a fully differentiated and immortalized gonadotrope cell line. Male, but not female, FSHβ -ts TAg transgenic mice develop slow-growing, gonadotrope-enriched, nodular tumors that express all three gonadotropin subunits (i.e., α-GSU, LHβ, and FSHβ), GnRH receptor and SF-1 *(51)*. The anterior pituitary pathology in these mice progressed from gonadotrope hyperplasia to nodular adenomas with decreasing immunoreactivity for FSHβ and LHβ, and phenocopied the ultrastructural characteristics of human null-cell adenomas *(51)*.

More recently, Chin and colleagues have utilized 1.2 kb of the 5' flanking sequences of the mouse GnRH receptor, and have targeted SV40 TAg gene expression specifically to gonadotropes *(52)*. These mice also develop fully differentiated gonadotrope tumors, and express GnRH receptor, α-GSU, LHβ, and FSHβ *(52)*. It is also important to note that only the gonadotrope-cell lineage is specifically immortalized with α-GSU-, LHβ-, FSHβ-, and GnRH-receptor promoters.

Thyrotropes. Thyrotropin, or TSH, is also a heterodimeric glycoprotein similar to the gonadotropins. The α-GSU is also expressed in thyrotropes, in addition to gonadotropes *(47)*. Thyrotrope-cell lineage is affected in some of the mouse mutants, which have defects in somatotrope/lactotrope commitment, differentiation, and maintenance *(39,40)*. Notably, pit-1 seems to be important for thyrotrope differentiation. Also, a transcription factor known as TEF is expressed in thyrotropes *(40)*. Mouse thyrotrope tumors develop in human α-GSU (5.5 kb)-SV40 TAg transgenic mice, and one subset of cell lines derived from these pituitary tumors known as TαT1 cells expresses both α-GSU- and TSHβ-subunits *(50)*. In contrast, in a murine transplantable thyrotrope tumor, MGH101A, the TSHβ gene is repressed because of the absence of potential transacting factors in these cells *(53)*. Targeted pituitary tumorigenesis in vivo using a 1.1-kb human TSHβ promoter-TAg hybrid transgene has also been achieved. These mice are small, develop aggressive pituitary tumors by 6 wk, demonstrate a wasting syndrome, and die by 9 wk of age *(54)*. The tumors express high levels of TAg protein, but do not express TSHβ. Undifferentiated, faint immunopositive tumor cells express GH, prolactin, and LH, suggesting that immortalization occurs at an early progenitor-cell stage prior to the specification of TSHβ-expressing mature thyrotrope-cell population *(54)*.

Corticotropes. Corticotropes are the central target-cell types of the pituitary for the hypothalamic peptide, corticotrophin-releasing factor (CRF) and for the adrenal cortex-derived glucocorticosteroids *(26)*. The hypothalamus-pituitary-adrenal axis is important for stress homeostasis and for neuroimmunomodulatory interactions *(26)*. Corticotropes express the multifunctional polyprotein precursor, proopiomelanocortin (POMC), which upon further site-specific proteolytic processing by subtilisn-like enzymes generates small peptides (i.e., ACTH, MSH, and β-endorphins). The POMC gene is expressed in the hypothalamic arcuate nucleus, and also in the melanotropes of the pituitary intermediate lobe *(26)*. Elevated ACTH levels (caused by a microadenoma representing corticotrope hyperplasia) or elevated glucocorticosteroids (caused by an adrenal cortical adenoma) in a chronic situationy lead to the characteristic Cushing's syndrome in humans *(3)*. Similarly, patients with CRF-producing tumors develop Cushing's syndrome, as well as corticotrope hyperplasia. The common symptoms of the disease are truncal obesity, hypertension, hyperglycemia, muscular weakness, and dystrophic skin changes. At least two transgenic mouse models phenocopy Cushing's syndrome. In the first model, transgenic mice were generated by expressing polyoma virus large TAg cDNA driven by a polyoma early-region promoter *(55)*. These mice develop progressive adenomas of the pituitary between 9–16 mo of age, have high ACTH levels, demonstrate increased body wt, and hypertrophied adrenal glands, and exhibit medullary hyperplasia *(55)*. In the second model, pituitary-directed expression of leukemia inhibitory factor (LIF) driven by a 4.6-kb mouse α-GSU promoter results in Cushing's syndrome *(56)*. These mice demonstrate high cortisol levels, and exhibit dwarfism, hypogonadism, truncal obesity, and thin skin. The pituitary gland from these transgenic mice shows corticotrope hyperplasia, and a striking reduction in somatotrope-, lactotrope-, thyrotrope-, and gonadotrope-cell populations consistent with their phenotypes *(56)*. In addition, multiple Rathke-like cysts lined by ciliated cells often seen in nasal epithelium are evident in the pituitary. Thus, an early expression (E 10.5) of LIF—a neuroimmunomodulator—in transgenic mice leads to rerouting of anterior pituitary progenitor cells to a corticotrope lineage resulting in Cushing's syndrome *(56)*.

Melanotropes. Melanotropes are a cell population that resides in the pituitary intermediate lobe. Melanotropes, like corticotropes, also express POMC gene products, which have diverse bioactivities. Intermediate-lobe tumors are rare in mice; however, pituitary adenomas of the intermediate lobe are relatively common in canine and equine species *(57)*. Low and his colleagues have delineated the melanotroph-, corticotroph-, and hypothalamus-specific targeting sequences on the POMC gene promoter, using a series of in vitro and transgenic expression systems *(58)*. They have also targeted tumorigenesis to the melanotrope lineage in transgenic mice, using a 700-bp promoter region of rat POMC gene that drives the expression of SV40 TAg transcripts *(57)*. These mice develop massive intermediate-lobe tumors that express TAg and POMC, but not other pituitary hormones. Biochemical analysis has revealed that these tumors process POMC peptides similar to the normal mouse melanotropes and produce high proportions of acetylated and carboxy-terminal-shortened β-endorphins, and amino-terminal acetylated α-MSH, but no 1-39 form of ACTH and β-lipotropin *(57)*. The mice develop enlarged adrenals, and demonstrate cortical hyperplasia in the zonae fasiculata and reticularis and high glucocorticoid levels in the serum. When dissociated pituitary-tumor cells are transplanted into nude mice, subcutaneous (sc) tumors readily develop. The transplanted mice produce large amounts of glucocorticoids, but POMC processing is different compared to that in the primary pituitary tumor *(57)*. Several novel melanotrope-cell lines, which secrete prolactin releasing factors, have eventually been developed from these tumors *(59)*. Thus, a targeted oncogenesis approach in transgenic mice results in induction of an otherwise rare melanotrope tumor, helping to delineate distinct biochemical processing of POMC peptides and to achieve derivation of novel melanotrope cell lines that complement the existing mouse corticotrope tumor-cell line AtT20, and in identifying novel prolactin-releasing factors. For a detailed discussion on POMC biology and tumors in POMC-expressing cells, refer to Chapter 15.

Intermediate-lobe tumors have also been observed infrequently in transgenic mice in which the expression of SV40 TAg gene is regulated by a rat vasopressin promoter *(60)*, or by 4.2 kb of mouse *c-kit* 5' flanking sequences *(61)*.

Targeted Tumorigenesis of the Thyroid Gland

The thyroid gland controls the basal metabolic rate by influencing the levels of thyroid hormones, thyroxine (T_4) and triiodothyronine (T_3). The thyroid gland is embryologically derived from the pharyngeal epithelium. Anatomically, the thyroid gland is divided into lobules, which are composed of dispersed follicles. The thyroid follicles are filled with thyroglobulin, the major protein that produces thyroid hormones in response to TSH *(3)*. The stromal cells interspersed between thyroid follicles are known as parafollicular cells or C cells, which produce calcitonin. Calcitonin regulates bone-calcium homeostasis. Common disorders of the thyroid in humans include multinodular goiter, Grave's disease, and hyperfunctional adenoma *(3)*. Thyroid carcinomas are rare, comprising only 1.5% of all cancers. Four main types of human thyroid carcinomas are known: papillary carcinoma (the most frequent of the four), follicular carcinoma, medullary carcinoma, and the very rare anaplastic carcinoma *(3)*. Interestingly, each of these four human cancers can be mimicked in transgenic mouse models, which develop analogous thyroid tumors.

Alterations in three important signal transduction pathways caused by somatic genetic changes have been described in human thyroid tumors. These include mutations in the TSH-receptor, mutations affecting the small G proteins Ras or Gs α-subunit, and mutations leading to aberrant activities of the tyrosine kinase receptor genes *ret* and *trk*. More recently, mutations in the *ret* locus have also been mapped, which cause multiple endocrine neoplasia (MEN) type II syndrome in humans *(3)*. Medullary thyroid carcinoma is one of the features of this syndrome in humans. These human clinical genetics studies have helped to design transgene constructs and to generate mouse models with targeted tumorigenesis in the thyroid gland.

It is well known that elevated cAMP levels promote both cell proliferation and differentiation of thyroid epithelial cells. Previously, a constitutively active mutant Gsα-subunit (Arg-201 $\rightarrow$ His) known as *gsp* has been identified in human growth-hormone-secreting pituitary adenomas (62). The *gsp* mutation inhibits the intrinsic GTPase activity of the α-subunit, and thus maintains Gsα in an active form, resulting in a constitutive activation of the adenylate cyclase *(62)*. This mutant form of *gsp* was specifically targeted to thyroid glands by a 2-kb bovine thyroglobulin promoter in transgenic mice (63). Between 8–12 mo, these mice develop hyperplastic nodules consisting of hypertrophied follicles with aberrant cells in the tumors. In many of the transgenic mice, papillary foci develop and invade the thyroid gland *(63)*. These foci contain elevated cAMP levels, actively uptake ^{125}iodine, and secrete large amounts of T_4 and T_3. Thus, this transgenic mouse model mimics autonomously functioning (insensitive to TSH) human thyroid nodules *(63)*.

Papillary and Follicular Carcinomas. Papillary carcinomas account for the majority of thyroid carcinomas in human patients who have had previous radiation exposure. The papillary thyroid carcinoma oncogene (RET/PTC) is formed as a result of genetic rearrangements on human chromosome 10 at the RET locus *(3)*. These include the loss of the ligand (GDNF)-binding extracellular domain of the RET receptor, and a chimeric fusion with at least three different constitutively active promoters of genes that are expressed in thyroid follicle cells *(3)*. Thus, the RET tyrosine kinase is constitutively active by fusion with the *H4, RIa*, and *ele1* promoters on chromosome 10, leading to the development of papillary thyroid carcinoma *(3)*. These resulting oncogene "chimeras" are designated as RET/PTC1, RET/PTC2, and RET/PTC3, respectively. Based on in vitro and in vivo studies, the oncogenic activity of one of the oncogenes, RET/PTC1, is believed to be caused by its dimerization through leucine zipper motifs at the amino-terminal end, and subsequent stimulation of tyrosine phosphorylation *(64)*.

Several transgenic models for papillary carcinoma of the thyroid have been established. In many of these models, the RET/PTC1 oncogene or a human-activated Ras gene (c-Ha-Ras) have been specifically targeted to thyroid glands, using the well-characterized bovine or rat thyroglobulin promoters *(65,66)*. In the "RET/PTC1 mouse" model, as early as E 16–18, thyroid follicles develop rapidly, leading to a distorted follicle formation, and demonstrate reduced radioiodide-concentrating activity. Bilateral thyroid tumors develop as early as postnatal d 4. Many cytological features of these locally invasive thyroid tumors accurately resemble human papillary carcinomas *(66)*. Thus, tissue-specific expression of a "suspected human oncogene" in transgenic mice confirms that RET/PTC1 is a specific genetic lesion that leads to the development of papillary carcinoma.

In the "c-Ha-Ras model," tumors are not exclusively thyroid-specific. Some of the lines of mice also develop tumors of the lung and thymus *(67)*. In addition, the thyroid papillary carcinoma in one transgenic line appears to be influenced by genetic background. For example, on a DBA/2J background, thyroid dysgenesis, and growth retardation leading to premature death are evident *(67)*.

Transgenic mice with thyroid-targeted expression of the human papillomavirus E 7 driven by a bovine thyroglobulin promoter have also been generated *(68)*. Many of these mice develop differentiated and functionally regulated goiters casued by thyroid-cell proliferation and accumulation of colloid. These cells eventually become locally invasive, dedifferentiate, and at later stages demonstrate many histological characteristics of human thyroid papillary and follicular carcinomas *(68)*.

Medullary Carcinoma. Whereas the chimeric RET/PTC1 oncogene causes papillary thyroid carcinoma, germline mutations in RET protooncogene alone can cause medullary thyroid carcinoma (MTC) in humans *(3)*. The common residues mutated in RET in patients affected with MTC include the five cysteines located in the extracellular domain and residues in the intracellular tyrosine kinase domain, including Glu768Asp and Val804Leu *(69)*. As mentioned earlier, the RET gene is mutated in 95% of families with MEN-2, a genetic disorder manifested in medullary thyroid carcinoma, pheochromocytoma, and hyperparathyroidism *(3)*. Ontogenically, MTCs are malignant neuroendocrine neoplasms derived from the thyroid parafollicular, or C cells. These tumors secrete calcitonin, and occasionally other neuropeptide hormones including somatostatin, serotonin, and vasoactive intestinal peptide. In addition to its expression in thyroid C cells, the calcitonin gene is expressed as an alternatively spliced form in populations of sensory neurons in the dorsal-root ganglia (DRG) and motor neurons of the spinal cord, where the encoded product is called calcitonin gene-related peptide (CGRP) *(70)*. Although both DRG neurons and C cells are derived from neural-crest precursor cells, unlike C cells, DRG neurons are refractory to transformation. This clinical observation is also true in murine models. When 2 kb of calcitonin/CGRP-promoter sequences directed the expression of SV40 TAg gene in transgenic mice, MTCs occurs, but no tumors within the DRG develop *(70)*. These earlier experiments with the rat calcitonin/CGRP promoter have led to the more recent studies, in which transgenic mice have been generated that express human MEN-2A mutant form of the RET protooncogene under the regulatory control of a calcitonin/CGRP promoter *(71)*. The MEN-2A mutant form of the RET protooncogene consists of a point mutation engineered by site-directed mutagenesis at cys 634 to arginine. This results in an aberrant homodimerization leading to the ligand-independent constitutive activation of the intracellular tyrosine kinase *(71)*. Four independent lines of transgenic mice harboring this mutant transgene have been produced. The transgene is expressed in the thyroid of transgenic, but not nontransgenic, litter mates by RT-PCR analysis. In addition, an inconsistent expression of the transgene in the adrenals, kidneys, and in brain, and no expression in the lung, heart, liver, skin, or gut of the transgenic mice has been found *(71)*. From 3 wk to 4 mo, these transgenic mice develop bilateral C-cell hyperplasia. Subsequently, multicentric MTC develops after 8 mo, with many cytological features resembling human MTC. The thyroid pathology in many of these mice is correlated with an abnormal increase in plasma calcitonin levels, and also with readily detectable calcitonin immunoreactivity of the parafollicular C-tumor cells *(71)*. Similar to an advanced-stage human MTC, in the livers of one transgenic line at 15 mo, metastatic nodules are present. This liver dissemination of malignant C cells is a typical feature of human MTC *(71)*. Thus, this transgenic mouse model accurately recapitulates the human MTC features, and provides genetic evidence that the MEN-2A mutant form of RET is oncogenic in parafollicular C cells.

When the expression of the v-Ha-*ras* transgene is driven by the same calcitonin/CGRP promoter described here, the transgenic mice also develop medullary-thyroid carcinomas, with high incidence from 6 mo to 1 yr of age *(72)*. The transgene is expressed in the thyroid at high levels, and there is a dramatic C-cell hyperplasia and evidence of prominent immunohistochemical staining for calcitonin within these cells. By 1 yr a majority of the mice consistently develop uni- or bilateral thyroid tumors, which are often locally invasive and hemorrhagic *(72)*. Furthermore, a small percentage of tumors co-express thyroglobulin (the follicular epithelial-cell marker), and calcitonin (a neuroendocrine marker) in a subset of cells. Thus, this transgenic model establishes that follicular and C-cells may share a common lineage during development, based on the existence and development of mixed medullary-follicular-thyroid carcinomas in these mice *(72)*.

Additionally, in two other models, transgenic mice have developed medullary-thyroid carcinomas. In one model, expression of SV40 TAg driven by a c-kit promoter leads to multiple neuroendocrine tumors, including the medullary thyroid-carcinomas *(61)*. These tumors are derived from C-cells, and stain for calcitonin immunoreactivity. In the second model, transgenic mice have been generated, in which expression of a C-terminal truncated middle-T and wild-type small-TAg is regulated by a polyoma virus early promoter itself *(73)*. These mice develop multifocal, bilateral medullary-thyroid carcinomas with 100% penetrance. The tumors express RET protein abundantly, intensely stain for calcitonin immunoreactivity, and secrete high levels of plasma calcitonin *(73)*. Although the incidence of thyroid carcinomas is rare in humans, these murine models are important because the majority of MEN-2 patients demonstrate medullary thyroid carcinomas.

Anaplastic Carcinoma. Anaplastic carcinomas of the thyroid are very rare (<5% of all thyroid cancers), but they are highly aggressive and typically represent the poorly differentiated tumors of the thyroid follicular epithelium *(3)*. It is speculated that human anaplastic carcinoma develops from more differentiated tumors as a result of other genetic alterations, such as the loss of the p53 tumor-suppressor gene.

A transgenic mouse model in which SV40 TAg gene expression is targeted to the thyroid gland using a 2-kb bovine thyroglobulin promoter shows phenotypic features of human anaplastic thyroid carcinomas *(74)*. These mice develop dramatic enlargement of the thyroid gland, demonstrate a rapid cachexia, and die prematurely (as early as 9 d postnatally). Survival of the transgenic mice can be prolonged by thyroid-hormone treatment, which usually exhibit abnormally low T_4 levels *(74)*. The lines of mice that survive longer demonstrate TAg staining specifically in the thyroid follicular cells, which are hyperplastic starting from the day of birth. Mice at later ages demonstrate poorly differentiated thyroid adenocarcinomas, with no detectable expression of thyroglobulin, thyroperoxidase, and low-level expression of TSH receptors *(74)*. In the majority of thyroids obtained from transgenic mice, the normal follicular organization is disrupted and colloid is absent, but a normal level of iodine-uptake activity is present. Thus, in this mouse model, dedifferentiation and thyroid structural organization are first initiated, followed by the rapid and aggressive development of numerous tumor nodules leading to premature death of the mice *(74)*. These features are characteristic of human anaplastic thyroid carcinoma, as mentioned earlier, and are therefore useful in understanding the molecular genetic events that lead to this human syndrome.

Targeted Tumorigenesis of the Pancreas

Pancreas is an endoderm-derived tissue, and histologically consists of two compartments, the exocrine and the endocrine glands. The exocrine pancreas constitutes the major portion of the two glands, and is comprised of many small glands called acini *(3)*. Pancreatic acini are filled with zymogen granules containing precursor forms of digestive enzymes, trypsin, chymotrypsin, aminopeptidases, elastase, amylase, lipase, phospholipases, and nucleases *(3)*. The pancreatic duct is the "draining system" for many of the channels that release the contents of each secretory acinus. The active forms of digestive enzymes are synthesized mainly by proteolytic processing at specific amino-acid residues, in response to both neural and hormonal stimuli *(3)*. The endocrine pancreas consists of many small clusters of cells—the islets of Langerhans—and elaborate four major functionally distinct hormones; glucagon, insulin, somatostatin, and pancreatic polypeptide (PP). The four major cell types that synthesize and secrete these hormones are α, β, δ, and PP, respectively *(3)*. Both glucagon and insulin are important for glucose homeostasis—somatostatin acts locally to regulate the release of glucagon and insulin, and PP stimulates secretion of gastric and intestinal enzymes and inhibits intestinal motility. Insulin-dependent and non-insulin-dependent diabetes are the major human diseases of the pancreas that affect the function of islet-β cells, leading to defects in insulin/glucose homeostasis and insulin signaling *(3)*. Carcinoma of the pancreas is the fifth most common cause of death from cancer in the United States. The precise genetic lesion responsible for this type of cancer is still unknown. Morphologically, these cancers are characterized by some regions of ductal dysplasia and intraductal tumor growth (ductal adenocarcinoma) *(3)*. Transgenic mouse models demonstrating tumors of both the exocrine and endocrine pancreas have been generated. Some of these models are described in the following sections.

Tumors of the Exocrine Pancreas. Efforts to develop transgenic mouse models for tumors of the exocrine pancreas have relied upon the use of rat elastase I promoter sequences to regulate the expression of various oncogenes. Elastase is one of the serine proteases synthesized in the exocrine pancreas acinar cells and secreted into the gut. During mouse embryogenesis, the elastase gene is expressed at E14, when the acinar cells begin to differentiate *(75)*. Subsequently, the expression peaks for a few days, and then levels off a few weeks after birth. Approximately 200 bp upstream of the transcription start site of the rat elastase I gene, including the promoter and enhancer elements, have been shown to be important for correct embryonic activation (i.e., E 14) and appropriately targeted expression of reporter genes (human GH) in pancreatic acinar cells *(75)*. Several potential transforming oncogenes have been independently targeted to acinar cells in transgenic mice using 7.2-kb 5' flanking sequences containing this enhancer. In the early studies by Brinster and Palmiter, either a wild-type or a mutant human c-Ha-ras proto-oncogene has been targeted to acinar cells in transgenic mice *(76)*. Transgenic mice that express wild-type c-Ha-ras develop subtle anomalies, but do not develop pancreas tumors, whereas mice expressing the mutant protooncogene develop aggressive acinar-cell tumors during early embryonic development. In contrast, when the SV40 TAg gene is expressed under the regulatory control of the same elastase promoter, pancreatic acinar-cell tumors develop progressively, and fully manifest at the adult stage in transgenic mice *(75)*. All of the three independent lines of transgenic mice die by approx 6 mo of age as a result of pancreatic cancer. Extensive developmental analyses of this mouse model suggest that the tumorigenesis is histologically a two-stage process. The first stage is a TAg–dependent preneoplastic state characterized by a progression from hyperplasia to dysplasia of the exocrine pancreas by an increased percentage of tetraploid cells, and by an arrest in acinar-cell differentiation *(75)*. The second stage is characterized by the formation of monoclonal tumor nodules with discrete aneuploid DNA content. The tumors express high levels of TAg mRNA and protein. Although not initially observed, subsequent studies have identified insulinomas and delta cell (D-cell) hyperplasia at later stages of tumor progression, in addition to acinar-cell tumors in these mice *(77)*. One possible reason for this unique D-cell hyperplasia (TAg-negative cells) seems to be a secondary event in these cells of abnormal growth in the exocrine pancreas. However, the occurrence of insulinomas in these mice is believed to be a result of the inclusion of SV40 early region (enhancer) sequences in the transgene construct.

Targeted expression of the *myc* oncogene using a 3-kb promoter of rat elastase I gene in transgenic mice results in development of mixed acinar/ductal pancreatic adenocarcinomas from 2–7 mo of age *(78)*. The pancreatic pathology in these mice is unique compared to the previous models, because transformed acinar-derived cells appear within islet cells. Together, these experiments with targeted tumorigenesis of the pancreas in transgenic mice suggest that the nature of initiating oncogenic stimulus can exert a profound influence upon tumor pathogenesis and progression *(75)*.

Tumors of the Endocrine Pancreas. In humans, the incidence of pancreatic islet-cell (endocrine) tumors is rare compared to that of the exocrine pancreas. Although most islet-cell tumors elaborate pancreatic hormones, some of them may be nonfunctional *(3)*. Islet-cell tumors may be single or multiple but mostly are benign solid tumors. β-cell tumors (also called insulinomas) are the most common of islet-cell tumors, and the patients are characterized by hyperinsulinemia, and consequently, hypoglycemia *(3)*. The other rare islet-cell tumors include α-cell tumors (also called glucagonomas), which secrete large amounts of glucagon and δ cell tumors (also called somatostatinomas), and the asymptomatic (despite high levels of PP secretion) PP-cell tumors.

β-Cell Tumors. Almost 15 yr ago, Hanahan developed a transgenic mouse model that reproducibly demonstrates the heritable formation of pancreatic β-cell tumors *(79)*. The oncogene (SV40 TAg) expression was exclusively specific to the islet-β cells in these mice. Injections of two independent transgenic constructs gave rise to multiple lines of mice. In one case, 660 bp of rat insulin II promoter/enhancer-containing sequences have been fused to the SV40 TAg early region consisting of the protein coding and termination sequences *(79)*. In the second case, approx 520 bp of the same regulatory region have been fused in an opposite orientation to the transcription of the SV40 TAg gene *(79)*. Transgenic mouse lines carrying either of the transgenes have been established. Most of the progeny derived from these lines develop islet-cell hyperplasia, whereas some develop solid tumors of the pancreas and die prematurely between 9 and 12 wk of age *(79)*. However, the transgenic mice survive longer when maintained on a high-sugar diet, indicating that they suffer from hypoglycemia caused by islet-cell hyperplasia. The mice that survive longer eventually develop highly vascularized solid tumors with no evidence of metastases *(79)*. Histologically, the islet-cell tumors consist predominantly of insulin-producing β-cells (immunochemical staining), but rarely glucagon-(α-) and somatostatin (δ)-producing cells. Only the β-cells within the pancreas but not other tissues intensely stain for SV40 TAg protein, confirming that islet-cell tumors arise as a result of oncogenic stimulus *(79)*. Thus, these transgenic mice develop islet-cell tumors containing pure populations of insulin-producing β-cells.

Since the initial report of this insulinoma mouse model, Hanahan and colleagues have extensively used this model, and have investigated the biology of tumorigenesis in general, the role of angiogenic factors in transitions from hyperplastic to neoplastic stages, and the effects of anti-angiogenic drugs on multistage carcinogenesis of the pancreas in these mice, as well as deriving novel insulin-producing cell lines *(80)*.

Heritable tumors of the endocrine pancreas have also been induced in transgenic mice that carry a moloney murine sarcoma virus-SV40 TAg hybrid gene. Similar to many human insulinomas, these mice develop single benign pancreatic tumors. They contain mostly insulin-producing β-cells, and the tumors also metastasize to the liver, spleen, and lymph nodes *(81)*.

α-Cell Tumors. Glucagon is the islet α-cell marker, and is synthesized and processed similar to insulin as a precursor form known as preproglucagon. Two additional glucagon-like peptides, GLP-I and GLP-II, are released when this precursor peptide is cleaved *(26)*. Glucagon and these peptides have also been detected in intestinal mucosa cells, and in various regions of brain, by immunohistochemical methods. However, *in situ* hybridization methods have confirmed that glucagon transcripts are localized only to neurons of the nucleus tractus solitarii in the brain stem *(82)*. Transgenic founder mice that harbor a rat preproglucagon promoter (~900 bp)-SV40 TAg hybrid transgene have been generated, and subsequently, stable transmission of the transgene to progeny has been achieved in independent lines of mice. The expression of TAg is initiated at E 10 during development in the α-cells of the pancreas in transgenic embryos. From E 10 until the young adult stage, distribution of the pancreatic α-cells and their histology appears normal in these transgenic mice *(82)*. By 5 mo of age, α-cell proliferation has been observed and these cells co-express TAg and glucagon proteins. By 9–12 mo of age, solid and highly vascularized α-cell tumors arise in these mice, and the cells co-express TAg, glucagon, and GLP-I. As a result of the tumorigenesis in the α-cells, very few endocrine cells of other lineages within the pancreas are obvious by immunochemical staining methods, presumably because of local effects of hyperplastic α-cells *(82)*. The majority of the transgenic mice that develop these pancreatic tumors die at approx 10 mo for unknown reasons. Interestingly, tumors of the intestinal mucosa cells (L-cells) and brain tumors (despite TAg expression in many brain regions) have not been observed in these mice. Thus, transgenic mice that express an oncogene as a reporter driven by cell-specific regulatory elements of the glucagon gene confirm the region-specific tumorigenesis in the pancreas, gut and brain *(82)*.

Other Mouse Models with Pancreatic Cancer. It is well-established that many neuroendocrine cells share common developmental lineages. Several transgenic mouse models have been produced that develop tumors in the pancreas and multiple neuroendocrine tissues. For example, a longer 5' upstream regulatory region (~2 kb) of rat preproglucagon-SV40 TAg hybrid gene induces tumors of the endocrine pancreas (α-cells), and tumors of the endocrine cells of the stomach and of the small and large intestine *(83)*. Similarly, mice transgenic for a vasopressin-SV40 TAg gene develop islet β-cell tumors and anterior pituitary tumors *(60)*. Transgenic mice in which SV40 TAg gene expression is driven by regulatory elements of gastrin (normally expressed in the gastric antrum and fetal but not adult islet cells) develop hepatobiliary-tract tumors, hyperplasia of stromal antral cells, and pancreatic ductal hyperplasia *(84)*. Tumors of the pancreas, pituitary, thyroid, and antral stomach have been characterized in transgenic mice that carry an upstream glucokinase promoter-SV40 TAg transgene *(85)*. Finally, transgenic mice that harbor a transgene comprised of 1.6 kb of 5' flanking region of the secretin-SV40 TAg gene develop insulin-producing pancreatic tumors, tumors of the small intestine, and colon tumors containing glucagon-expressing cells *(86)*.

Targeted Tumorigenesis of the Male Reproductive System

The differentiation and development of the male reproductive system occurs through a series of interactions between various intratesticular and extratesticular factors. Whereas spermatogenesis occurs in the seminiferous tubules within the testis, sperm storage and transport take place in accessory sex glands, including the epididymis and vas deferens. Other accessory glands such as the prostate, seminal vesicles, and preputial and bulbourethral glands also influence sperm function through their secretions. In humans, testicular germ-cell tumors and prostate cancer are the two major diseases of clinical significance *(3)*. In general, tumors of male accessory sex glands other than the prostate are very rare. The following sections, describe mouse models for testicular tumors and accessory-gland tumors. Targeted tumorigenesis of the prostate is described in Chpater 19 of this book.

Testicular Tumors. The specification of the testis from undifferentiated bipotential gonads during early embryogenesis is dictated by interactions between male sex-determining genes Sry, Sox9, SF-1, GATA-1, and others *(87)*. Once committed to the "male pathway," migration and colonization of the germ cells occurs in the somatic-cell milieu of the testis. The sex cords give rise to Sertoli cells and the stromal component develops into Leydig cells, the two major somatic-cell types. Sertoli cells synthesize and secrete Müllerian-inhibiting substance (MIS), which prevents development of the female internal sex organs *(87)*. Leydig cells produce testosterone, the male sex steroid that supports the development of the male reproductive structures and maintenance of the accessory glands. During the early prepubertal phase, Sertoli cells respond to FSH and Leydig cells respond to LH, and these signaling events are critical to support germ cell development leading to the formation of functional sperm. Sertoli cells also secrete important peptides, inhibins (α-β heterodimers), and activins (β-β homodimers and heterodimers), which regulate FSH secretion from the pituitary and are also produced locally within the gonadotropes. Although testicular germ-cell tumors are most common in humans, very few transgenic mouse models develop this type of cancer. Similarly, only a few mouse models have been generated that develop testicular sex cord-stromal tumors: Sertoli-, Leydig-, or mixed-cell tumors.

Sertoli Cells. MIS is one of the earliest factors secreted by the Sertoli cells during the early stages of testis development. Although the fetal role of MIS is well-established, its functions in the adult are unknown despite its continued expression in the testis. To genetically manipulate Sertoli cells in vivo, targeted tumorigenesis of testicular Sertoli cells has been achieved in transgenic mice using a human MIS (~2 kb 5' flanking promoter sequences) SV40 ts1609 TAg hybrid transgene *(88)*. Two of seven founder males that carried and expressed the oncogene exclusively in the testis developed visibly enlarged scrotal areas. Pathological analysis revealed large, bilateral, and vascularized testicular tumors, but no other tumors. Histologically, the tumors contain masses of cells separated from the inner surface of the tubules, with no evidence of ongoing spermatogenesis. The tumors in these mice were classified as malignant gonadal stromal tumors composed of both anaplastic spindle-shaped Sertoli cells and more differentiated, epithelioid Sertoli cells *(88)*. Subsequently, Sertoli-cell lines have been derived from these tumors that retain only some of the differentiated characteristics of the normal cells, but do not express detectable levels of inhibin α, MIS, or FSH receptor. More recently, Picard and colleagues have developed another transgenic mouse line in which wild-type SV40 TAg expression is directed by a longer 3.6 kb 5' flanking region of the human MIS gene *(89)*. These mice also develop heritable Sertoli-cell tumors with many features resembling human Sertoli-cell tumors. Sertoli-cell lines derived from the tumors in these mice at 6.5 d postnatally secrete MIS into the culture medium and express the type II MIS receptor *(89)*. Thus, these experiments with P6.5 and adult mice suggest that the timing of cell-line derivation plays a critical role, even when the oncogene expression is driven by a developmentally regulated promoter. In contrast to the two previous models, transgenic mice carrying a metallothionein (MT)-polyoma virus TAg transgene develop late-onset testicular tumors that initially consist of proliferating Sertoli cells. But when secondary transplantable tumors are derived in nude mice, they demonstrate a unique mixed germ-cell-sex-cord tumor phenotype *(90)*. Thus, the type of oncogene and its cellular specificity to immortalize a given cell type varies, depending upon the promoter from which it is derived and on the viral origin.

Leydig Cells. Leydig cells are steroidogenic cells producing testosterone. In humans, Leydig-cell tumors are the most common tumors derived from gonadal stroma *(3)*. The combined incidence of Leydig-cell and Sertoli-cell tumors is only 5–10% of all the testicular cancers *(3)*. Although Leydig-cell tumors have been reported in some of the non-inbred strains of mice and chronic administration of estrogens results in Leydig-cell tumor formation in mice, testicular tumors rarely develop in rodents. There are two well-characterized mouse Leydig tumor-cell lines established almost two decades ago: MA-10 and MLTC-1 *(91)*.

In at least four transgenic mouse models, Leydig-cell tumors have been observed. In the first model, transgenic mice carrying a mMT-1-polyomavirus TAg transgene develop testicular Leydig-cell adenomas and seminal-vesicle enlargement at approx 10 mo of age. The Leydig cells from these tumors are functionally active, and secrete at least 8.5-fold higher than normal levels of testosterone *(90)*. Although the expression of TAg is detected very early, the tumor pathology is evident only at later stages. In addition,

Leydig cells prepared from the early-stage testis undergo crisis in culture, suggesting that additional oncogenic stimuli must be required in conjunction with TAg gene for tumor formation *(90)*. In the second model, transgenic mice harboring a human papillomavirus type 16-E6 and E7 oncogenes develop bilateral testicular Leydig-cell tumors with 100% penetrance *(92)*. The transformation by these dual oncogenes is dominant in all the inbred genetic backgrounds tested. The tumors also express high levels of 3 beta-hydroxysteroid dehydrogenase (3-βHSD) and other enzymes required for androgen metabolism, suggesting that the Leydig cells in the tumors retain their differentiated function *(92)*. In the third model, transgenic mice have been generated in which directed expression of SV40 TAg gene is achieved under the regulatory control of a 6-kb mouse inhibin α-subunit promoter *(93)*. These transgenic mice develop aggressive testicular tumors consisting primarily of Leydig cells with 100% penetrance. Signs of hyperplasia in the testis are evident as early as 6–7 d of age, and the tumors in the testis are immunologically stained positive with TAg and inhibin-α antibodies. The tumors are gonadotropin-dependent, express 3-βHSD enzyme, and contain high-affinity receptors for LH/hCG *(93)*. Although Sertoli cells also weakly express TAg in the testis of these transgenic mice, no tumors in this cell type have been observed, confirming that within the testis, Leydig cells are more susceptible to oncogenic stimulus by SV40 TAg when it is expressed from the inhibin-α-subunit promoter *(93)*. Whereas in all the above models, oncogenes have been directed to Leydig cells to achieve tumorigenesis, in the fourth model, Leydig-cell tumors arise in transgenic mice in which aromatase cDNA is overexpressed under the regulatory control of mouse mammary-tumor virus promoter (MMTV-LTR) *(94)*. MMTV-LTR is known to be active in both mammary and male reproductive tissues. About 50% of the male transgenic mice overexpressing aromatase are infertile and/or morphologically demonstrate testicular enlargement. Histopathological analysis shows uni- or bilateral Leydig-cell tumors containing large polygonal shape cells and multiple hemorrhagic cysts *(94)*. The tumors express large amounts of aromatase, and estrogen receptor α, and secrete high levels of estrogen into serum. Thus, this transgenic mouse model confirms that altered levels of estrogen produced locally or in serum by directed expression of aromatase may lead to noninvasive Leydig-cell tumors that closely resemble many human Leydig-cell tumors (94).

Germ Cells. Germ-cell tumors of the testis develop from primordial germ cells. During embryogenesis, primordial germ cells migrate from the yolk sac to the genital ridge *(87)*. The clinical features of germ-cell tumors depend on the age of the patient and the anatomical localization and histological composition of the tumors. Three distinct classes of germ-cell tumors can be distinguished in the human testis. The first group includes the teratomas, which are yolk sac-derived and usually originate before puberty. The second group, known as testicular germ-cell tumors of the adolescents and adults, is the most frequent type and comprises the seminomas and nonseminomas. Typically, these tumors arise after puberty. The third group, which appears in elderly men is known as the spermatocytic seminoma *(87)*. There are no well-characterized transgenic mouse models which develop testicular germ-cell tumors. The inbred 129 mouse strain develops spontaneous germ-cell tumors with only 1–2% incidence. Several cytogenetic and molecular genetic studies have recently identified a potential tumor-suppressor gene, pgct1, on mouse chromosome 13 in a region that is syntenic to a region of human chromosome 5 that is implicated in human male germ-cell tumor susceptibility *(95)*. In one transgenic mouse model, in which HPV type 16 oncogenes E6 and E7 are expressed, seminomas develop in the testis *(92)*. In another recent model, SV40 TAg gene expression is directed to haploid male germ cells using a 2.3-kb 5' flanking sequences of rat proacrosin gene *(96)*. Interestingly, none of the transgenic mice develop testicular tumors, confirming that spermatids show no susceptibility to transformation by SV40 TAg *(96)*. It is anticipated that in the future, efforts to mimic the human germ-cell tumor phenotypes in mice may be made possible by a more thorough understanding of the genetics of the human disease.

Male Accessory Sex-Gland Tumors. Two transgenic mouse models have been developed that demonstrate hyperplasia/tumors of the accessory sex glands. In one model, overexpression of vascular endothelial growth factor (VEGF) is overexpressed in the epididymis of transgenic mice using MMTV-LTR promoter sequences *(97)*. These mice are infertile because of spermatogenic arrest. The epididymis epithelial cells demonstrate hyperplasia, and there is increased angiogenesis in the subepithelial region *(97)*. In the second model, targeted expression of SV40 TAg has been achieved using an androgen-responsive promoter region of rat prostatic steroid binding protein, C3(1) in the prostate, urethral, and bulbo-urethral glands of transgenic mice *(98)*. These mice develop adenocarcinoma of the urethral and bulbo-urethral epithelium (in addition to prostate carcinoma) after 7 mo of age. The tumor progression in these affected tissues correlates to the expression of TAg and p53. Elevations of intracellular and extracellular TGFβ1 and extracellular TGFβ3 are observed that may augment the tumor growth, and the tumors are also responsive to androgens *(98)*. This is the first transgenic mouse model for urethral and bulbo-urethral-gland carcinomas, and may be useful for further analysis of the normal and tumor development of these male sex-accessory glands.

Targeted Tumorigenesis of the Female Reproductive System

The differentiation and development of the female reproductive system, similar to the male reproductive system, occurs through multiple interactions through intra-ovarian and extra-ovarian factors. During the early stages of development, MIS is undetectable in the female reproductive system, and hence the female ducts and internal sex organs persist *(87)*. Because testosterone is also absent, the Wolffian-duct derivatives do not differentiate. In humans, ovarian, uterine and cervical cancers are the most common types of malignancies *(3)*. This section, describes mouse models for ovarian, uterine, and cervical tumors. Tumors of the mammary gland are described in Chapter 10.

Ovarian Tumors. During early embryogenesis, ovarian differentiation occurs by a "default" pathway. Although the absence of Sry (and Y-chromosome in an XX genotype embryo) renders this "fate," an orphan nuclear receptor—Dax-1—and unidentified factors appear to establish the ovarian pathway *(87)*. Folliculogenesis within the ovary is a highly-regulated and coordinated cyclic process dependent on multiple factors. Each round of folliculogenesis begins with the recruitment of a primordial follicle that undergoes progressive changes, leading to the growth and differentiation of the somatic, granulosa, and theca cells (the inner and outer layers of the follicle), and culminating in ovulation, or atresia *(99)*. Germ-cell cancers in females are very rare, but are more common in males. The predominant ovarian cancers in women originate from surface epithelial cells, whereas the stromal-derived cancers (granulosa cell-, thecal cell-derived) are of low incidence *(3)*. To date, there are no mouse models for human ovarian epithelial cancers. Mutations in mouse homologs of tumor-suppressors or oncogenes that have been implicated in human ovarian-epithelial cancers, do not lead to ovarian cancer *(3)*. Transgenic mouse models for human stromal-cell/granulosa-cell tumors have been generated, as described in the following section.

Granulosa Cells. Granulosa cells are important targets for gonadotropin action, and they communicate with both thecal cells and oocytes. Approximately 15–20% of pediatric ovarian tumors are granulosa-cell tumors that arise between infancy and menarche, and thus are known as juvenile-onset granulosa tumors *(3)*. Although circumstantial evidence suggests that these tumors are gonadotropin-dependent, there are no definitive data to confirm this. Transgenic mice that overexpress a human/mouse hybrid FSH (hFSHβ targeted to the pituitaries) or those that ectopically overexpress human FSH (human α-human FSHβ-subunit expressing mice) do not develop ovarian tumors, and instead develop polycystic ovaries, similar to human patients with ovarian hyperstimulation syndrome *(100,101)*. In contrast, granulosa-cell tumors and theca-interstitial-cell tumors arise beginning at 4 mo of age in transgenic female mice in which the expression of an LH analog (LHβ gene fused to carboxyl terminus of hCGβ-subunit gene) is directed to the pituitary gonadotropes using a bovine α-GSU promoter (*see* Chapter 13). Further studies on this transgenic mouse model by Nilson's group have suggested that LH induction of these tumors is genetic-strain-specific *(102)*. The granulosa-cell tumors are more prevalent on a CF-1 genetic background, but not on a hybrid genetic background when progeny mice are obtained by crosses between CD-1 male transgenic mice and C57BL/6 or SJL female mice *(102)*.

Targeted tumorigenesis of the ovarian granulosa cells has also been achieved by transgenic expression of SV40 TAg using the previously mentioned inhibin-α promoter (either 6 kb or 2 kb) *(103)*. The penetrance of ovarian tumorigenesis in these mice is 100%. The tumor histopathology reveals damaged follicular architecture containing many proliferating granulosa cells with scant cytoplasm, but no features of ovarian epithelial cells. The granulosa-cell tumors stain intensely with an antibody to TAg, and the ovarian tumors also express FSH and LH receptors and inhibin-α subunit *(103)*. These tumors have been the eventual source for derivation of granulosa-cell lines and characterization of the gonadotropin-mediated signaling within granulosa cells *(103)*.

Germ Cells. The actual number of germ cells present in the normal ovary is more than the number that are ovulated. Physiologically, germ-cell death may be direct (intrinsic to this cell type), or indirectly mediated via somatic-oocyte cell interactions, or via somatic-cell (granulosa) death (atresia). Thus, the reproductive potential of the female is tightly regulated *(99)*. A number of cell-death (apoptotic) regulators have been shown to be expressed in oocytes and granulosa cells within the human or mouse ovarian follicles *(104)*. Bcl-2 is an apoptosis-suppressing protein, and Bcl-2 has been targeted to the ovaries of transgenic mice using the inhibin-α-subunit promoter. The transgene is expressed in granulosa cells, and 25% of older transgenic female mice develop benign cystic ovarian teratomas *(105)*. The majority of these tumors contain cells resembling those from the respiratory tract and intestine, and in some, neuron-like cells are evident. Thus, preventing ovarian somatic-cell death by overexpressing the anti-apoptotic protein Bcl-2 leads to germ-cell tumorigenesis *(105)*. Surprisingly, when Bcl-2 is overexpressed directly and specifically in oocytes by using the zona pellucida (ZP) protein–3 (oocyte-specific) 5'-flanking sequences, follicular atresia and germ-cell apoptosis (natural or chemotherapy-induced in vitro) are prevented, and no germ-cell tumors have been observed *(106)*.

Uterine Tumors. The uterus consists of three distinct layers: the epithelial, endometrial (stroma), and myometrial (smooth-muscle) layers. Ovarian steroids play critical roles during the differentiation of the uterus and the cyclic changes that manifest during the menstrual/estrous cycle and pregnancy *(3)*. Uterine leiomyomas are benign smooth-muscle tumors that occur in 20–30% of women over 30 yr of age. Several cytogenetic and molecular genetic studies in humans have identified two genes coding for high-mobility group (HMG) proteins, HMGIC and HMGIY, as potential pathophysiological candidates for this disease *(107)*. Similarly, loss of heterozygosity (LOH) involving the human CUTL1 (cut like 1) locus at the 7q22 chromosomal position has been implicated in sporadic uterine leiomyomas. CUTL1 encodes a homeobox-domain containing cell-cycle regulatory protein *(107)*.

There are two transgenic mouse models in which targeted expression of viral oncogenes result in the formation of uterine smooth-muscle tumors (which mimic human leiomyomas). In one model, the SV40 TAg gene has been targeted to the myometrium using either a 1-kb or a 120-bp promoter region of the rat calbindin D-9K (CaBP9K) gene *(108)*. CaBP9K is normally expressed in myometrium, is estradiol-dependent, and is repressed by progesterone. CaBP9K is also expressed in the intestine, kidney, and lung.

These transgenic mice develop leiomyomas beginning at 2 mo of age with 100% penetrance. Lines that carry the 1-kb promoter-transgene develop lung and kidney tumors, but the 120-bp promoter transgene-carrying lines develop exclusively uterine leiomyomas. In all the leiomyomas, TAg expression is only detected in the smooth-muscle cells (SMCs) of the uterus *(108)*. Despite the presence of proliferating fusiform SMCs in the myometrial layers, no metastases or local invasion by the tumors is evident, thus truly representing the features of analogous benign human uterine leiomyomas *(108)*. Notably, leiomyomas in these transgenic mice develop only after puberty, when estradiol levels begin to rise. This dependency on estradiol for the development of tumors is mediated via an estrogen-responsive element in the CaBP9K promoter. Ovariectomy of the transgenic female mice at puberty prevents the development of the uterine tumors, and ovariectomy of tumor-bearing mice causes regression of the tumors *(108)*. Thus, these transgenic mice offer an ideal model to study estradiol-dependent and oncogene-mediated human uterine leiomyomas in an in vivo context.

Leiomyomas of the uterus also develop in another transgenic model in which polyomavirus TAg is targeted to multiple tissues (mammary gland, testis, etc.) using MMTV-LTR sequences. Similar to the previous model, these female transgenic mice frequently develop estradiol-dependent uterine SMC tumors, which express abundant levels of TAg protein *(109)*. Consistent with the human LOH studies, immunoprecipitation experiments confirm that polyomavirus TAg can efficiently sequester CUTL1 protein from the leiomyoma-tumor extracts. Thus, the development of the leiomyomas may indeed depend on the regulation of the CUTL1 protein to affect the cell cycle, leading to tumor development *(109)*.

Estrogen/estrogen-receptor signaling pathway alterations may also lead to uterine tumors. This is illustrated by transgenic experiments by Korach and colleagues *(110)*. Transgenic mice that overexpress estrogen receptor α (ERα) using the MT I promoter are neonatally (1–5 d) exposed to diethylstilbestrol (DES), and the development of tumors in these and the wild-type mice are monitored. By 4 mo of treatment, the transgenic (26%) but not the wild-type mice develop premalignant lesions leading to the uterine adenocarcinomas. The DES-treated transgenic mice develop uterine adenocarcinoma at 8 mo, with high incidence (73%) compared to DES-treated wild-type mice (46% incidence). Most of these tumors originate at the junction of the uterine and cervical epithelium. Additional preneoplastic lesions include squamous metaplasia and atypical hyperplasia of the uterus. Thus, the levels of ERα are an important factor in development of these estrogen-responsive tumors *(110)*.

Cervical Tumors. It is estimated that 500,000 women worldwide are afflicted with cervical cancer, and at least 45% of them die as a result of the metastatic spread of these tumors *(3)*. The cervix is composed of three distinct layers—the ectocervix, the endocervix, and the transformation or transition zone interspersed between these two layers. Most cancers of the cervix originate from the metaplastic stratified epithelium lining the transformation zone *(3)*. Human papillomavirus (HPV) types 16 and –18 are found in 80–90% of these invasive cervical cancers. Two oncoproteins encoded by these viruses, E6 and E7, are known to inactivate p53 and retinoblastoma (Rb) proteins and promote cell cycle. Similar to uterine cancers, cancer of the cervix is also estrogen-dependent. HPV infection is observed in pregnancy, and long-term use of estrogen-containing oral contraceptives increases the risk of HPV neoplasia and malignancy *(3)*. Because the presence and expression of the HPV virus itself is not sufficient for carcinogenesis of the cervix, estrogen seems to be an important cofactor in this pathway.

To evaluate the possible effects of estrogen on HPV-associated neoplasia, Hanahan and colleagues have generated transgenic mice expressing the HPV16 oncogenes under the control of 2 kb of the human keratin-14 promoter, and have treated these mice with time-release 17β-estradiol pellets *(111)*. This chronic estrogen exposure induces a 100% penetrant multistage neoplastic progression in the squamous epithelium of the cervix and vagina. Sixty percent of the treated—but not untreated—transgenic mice eventually develop invasive cancers of the cervix within the female reproductive tract *(111)*. Thus, this targeted tumorigenesis model identifies an important synergistic cooperation between chronic estrogen exposure and the HPV16 oncogenes in promoting cancers of the female reproductive tract. This model will now be useful for studying interactions of various other cofactors that induce cervical cancers, such as progesterone/estrogen combinations, hydrocarbons, and other individual oncogenes of HPV *(111)*. Since the invasive cancers of the cervix in these mice arise through a series of well-defined histological changes, many drugs that may interfere with actions of HPV oncogenes can be tested at the preclinical level in this model.

Targeted Tumorigenesis of the Adrenal Gland

The adrenal gland is structurally and functionally divided into two major regions, the cortex and the medulla. The steroid-producing cortex consists of three layers: mineralocorticoid (aldrosterone)-producing zona glomerulosa (the outermost layer), the glucocorticoid (cortisol)–producing zona fasciculata (the intermediate layer), and sex-steroid-producing zona reticularis (the innermost layer). The adrenal medulla is composed of neural crest-derived neuroendocrine cells, chromaffin cells, which synthesize and secrete catecholamines, (norepinephrine and epinephrine) in response to signals from preganglionic nerve fibers in the sympathetic nervous system *(3)*. As mentioned in earlier sections, the adrenal cortex is downstream in the neuroendocrine stress axis. Transgenic mice with directed expression of oncogenes specifically in the cortex or medulla have been generated. Some of these are described below.

Adrenocortical Tumors. In humans, adrenocortical hyperplasia and cortical tumors cause adrenal hyperfunction syndromes, including Cushing syndrome, which is characterized by hypercortisolism. Similarly, some forms of adrenocortical adenomas, or adrenocortical hyperplasia, are associated primarily with hyperaldosteronism *(3)*. Excess secretion of aldosterone, secondary to elevated plasma renin levels, is also seen in clinical cases. An excessive levels of aldosterone cause sodium retention and potassium excretion, leading to hypertension and hypokalemia *(3)*.

Immortalized mouse adrenocortical Y-1 cells have been the main reagents for previous studies on adrenal steroidogenesis. However, these cells are not fully differentiated because they do not express all the specific markers *(112)*. To develop alternate in vitro models for adrenocortical steroidogenesis, Mellon and colleagues have used a 2.4-kb of promoter fragment of gene-encoding human cytochrome P450 cholesterol side-chain cleavage enzyme to direct the expression of SV40 TAg in transgenic mice *(112)*. Although this promoter is active in all steroidogenic cells (kidney, gonads, and adrenals), the two founder female transgenic mice develop only adrenocortical tumors. The tumor cells resemble those from the sex steroid-producing zona reticularis layer. Subsequently, highly differentiated adrenal-cell lines have been developed from these tumors that are cAMP-responsive, express all the steroid enzymes, secrete progesterone and also express mouse renin-1 mRNA *(112)*.

Adrenocortical tumorigenesis with 100% penetrance is also observed in the mouse α-inhibin-TAg transgenic model when both male and female transgenic mice are gonadectomized *(113)*. The affected layer is the zona reticularis, with many TAg immunopositive nuclei-containing cells. The gonadectomized transgenic mice demonstrate elevated serum levels of progesterone, estradiol, and immunoreactive dimeric inhibin, but not corticosterone. Administration of recombinant inhibin suppresses these adrenal tumors *(113)*. Since mouse-α-inhibin mRNA is detectable in adrenal tumors in gonadectomized transgenic mice, it appears that the gonadal-derived inhibin (in the intact transgenic mouse) autoregulates itself by suppressing its expression in the adrenal gland. However, this hypothesis does not hold true for inhibin-α deficient mice which do not express inhibin-α in any tissue. When these inhibin-knockout mice are gonadectomized, they also develop aggressive adrenal tumors. However, the inhibin-α-TAg transgenic model has been useful for further studies on LH-induced LH-R expression in adrenocortical cells and in delineating the mechanisms of adrenocortical hyperplasia seen in the bLHβ-CTP transgenic mouse model, which have elevated LH levels *(114)*.

Adrenal Medullary Tumors. In humans, pheochromocytomas are neoplasms of the adrenal medulla. Occasionally (~15%), they may also arise in extra-adrenal sites, such as paragangliomas. Most of the pheochromocytomas (90%) are sporadic in occurrence; however, at least 10% of them occur in one of the mostly autosomal dominant familial syndromes, including MEN II and von Hippel-Lindau syndromes *(3)*. The most significant clinical feature of pheochromocytomas is catecholamine-mediated hypertension, and rarely these tumors also secrete ACTH, somatostatin, and other peptide hormones.

Two groups have generated transgenic mouse models that develop primitive adrenal medullary tumors or pheochromocytomas. In one model, neuroepithelial tumors of the adrenal medulla occur in transgenic mice that express SV40 TAg under the regulatory influence of an 872 bp of the hypothalamic GHRH promoter *(30)*. These mice do not develop hypothalamic tumors, but instead form adrenal medulla tumors with 100% penetrance between 7–16 wk as a result of medullary tumor mass. The cortex is represented as a thin compressed rim within the gland (30). The tumors contain low levels of catecholamines and cortisol compared to those in adrenal tissue extracts from normal mice. This suggests that the tumors represent an immortalization event during the primitive stages of adrenal development. In accordance with this, cultured adrenal cells from these tumors mimic many features of primitive neuroectodermal tumors, such as neuroblastomas *(30)*.

In the second model, transgenic mice that express SV40 TAg gene from the erythroid transcription factor GATA-1 promoter-enhancer region develop pheochromocytomas *(115)*. Very large bilateral tumors arise between 2 and 3 mo of age, not metastatic to kidney, but contained within the adrenal capsule. Cell lines have been obtained from these tumors, and detailed morphological, biochemical, and functional characterization studies have suggested that these cells represent both neuronal (160-kd neurofilament staining) and endocrine (chromogranin staining) lineages *(115)*.

The preceeding sections include many examples of the power of targeted tumorigenesis using transgenic mice in many neural- and endocrine-cell lineages. Mouse models that develop non-neuroendocrine tumors (not described) have also been generated and extensively studied—for example, tumors of the lung, bone, kidney, and skin. Similarly, more models will be generated in the future to understand further tumor development in the placenta, parathyroid gland, and tissues otherwise nonpermissive to immortalization.

The following section, describes endocrine tumors in gene-knockout mutant mice derived by gene targeting (ES-cell) technology.

Endocrine Tumors in Gene-Knockout Mutant Mice

Cancer is a polygenic disease. Genetic and statistical analysis of a large number of human retinoblastoma cases by Knudson has led to the "two-hit hypothesis." According to this theory, the first mutation in a tumor-suppressor locus in the germline is accompanied by a second somatic/sporadic mutation (a recessive mutation) in order to fully manifest the cancer phenotype *(116)*. Gene-knockout mouse models are useful in many ways to study human tumorigenesis on a large scale. Haplo-insufficiencies in heterozygous conditions can be studied in mice. The role of modifier factors, the role of genetic-background in tumor predisposition, and the derivation of cells bearing single- or double gene targeted loci are all feasible with this approach *(116)*. Although almost all identified human tumor-suppressor homologs in mice have been deleted, only a few of these mouse models develop similar human phenotypes, of which only a few demonstrate endocrine tumors *(117)*. Endocrine tumors also develop in some knockout mouse models other than those in which a tumor-suppressor gene is disrupted. Some of these mouse models are described in this section.

A Fraction of p53-Deficient Mice Develop Gonadal Tumors

Mutations in the *p53* gene are frequently found in many human cancers. p53 is ubiquitously expressed in multiple cell types, and plays a variety of roles in the cell cycle, during early embryogenesis, in radiation-induced DNA damage (apoptosis) and as a "global" tumor-suppressor *(118)*. Mutations in *p53* are associated with the inherited cancer susceptibility Li-Fraumeni syndrome. The p53-deficient mouse model is the first tumor-suppressor mouse model to be generated using gene targeting in ES cells. These mutant mice develop normally, but are susceptible to spontaneous tumor formation in various tissues *(118)*. Malignant lymphomas (mostly thymic T-cell lymphomas) and hemangiosarcomas predominantly arise in these mice. Although p53 mutations are observed in many gonadal cancers, only one of 30 mutant mice of a hybrid (C57Bl/6/129SvEv) genetic background analyzed over a period of more than 1 yr, developed gonadoblastoma of the testis, and only one heterozygous mouse developed embryonal carcinoma of the testis *(118)*. In addition, at 60 wk, one male chimera developed choriocarcinoma surrounded by ovarian tissue at a very late stage, and another male chimera developed a Leydig-cell tumor. Although many of the tumor types seen in p53-deficient mice are similar to those in Li-Fraumeni patients, the most frequent breast and brain tumors in these patients are infrequent in heterozygous mutant mice. In contrast, when the p53 mutation was bred onto a 129/SvEv inbred genetic background, testicular tumors arose in homozygous mice as frequently as lymphomas on a C57BL/6 background. As mentioned previously, wild-type 129/Sv-strain mice have a modest susceptibility to testicular teratomas, and absence of p53 dramatically increases this predisposition to testicular tumors on this genetic background *(119)*. p53-deficient mice have also been useful for studying both spontaneous and carcinogen-induced tumorigenesis *(120)* in genetic manipulations using mutant p53 transgenes to analyze the dominant-negative effects on accelerated tumorigenesis *(121)*, in delineating the dosage effects of the p53 alleles on tumor formation and in exploring mechanisms of cell-cycle regulation both in vivo and in vitro *(122)*.

Spontaneous Multiple Neuroendocrine Tumors in Rb-Heterozygous Mutant Mice

The Rb gene is one of the common tumor-suppressor genes and is frequently mutated in many types of human cancers, most predominantly in retinoblastomas of the eye. Similar to p53, Rb is ubiquitously expressed, and encodes a nuclear phosphoprotein *(123)*. The phosphorylation status of the Rb protein is critical during cell-cycle progression and for formation of stable complexes with various cell-cycle proteins and oncogenic proteins of several DNA tumor viruses *(123)*. Despite its ubiquitous expression in many cell types, germline mutation of the Rb gene in humans leads to predisposition to retinoblastoma, with high penetrance *(123)*.

Mutant mice deficient in Rb have been generated independently by three groups *(124–126)*. The homozygous mice die in mid-to-late gestation, with defects in the hematopoietic system and in the central and peripheral nervous system. Interestingly, Rb-heterozygous mice develop spontaneous pituitary tumors with nearly 100% penetrance between 2 and 11 mo *(127,128)*. Multiple tumor foci are present in the intermediate lobe of the pituitary gland, which at advanced stages often infiltrate into multiple regions within the brain. The tumor cells express intermediate lobe-specific POMC-derived peptides, and α-MSH is abundantly expressed by immunohistochemical analysis *(127,128)*. Tumor progression in these heterozygous mutant mice is also correlated with elevated serum levels of α-MSH. The tumors do not express anterior pituitary or neuropeptide markers, but express some of the glial and neural-cell markers with some degree of variability. The predisposition to intermediate-lobe melanotropes is further confirmed by the loss of the wild-type Rb allele and the absence of full-length Rb protein in the tumors by both Southern blot and Western blot analyses *(128)*. Further studies have suggested that early loss of the Rb gene in heterozygous mutant mice is associated with impaired growth innervation during melanotrope tumorigenesis *(129)*. Dopaminergic innervation is normally critical for cell cycle and apoptosis events in the melanotropes, and therefore these studies indicate that the Rb protein plays a key role in neuron-neuroendocrine cell interactions *(129)*. Subsequent studies have also identified multiple neuroendocrine tumors in Rb-heterozygous mice, including thyroid C-cell, parathyroid, and adrenal medullary tumors *(130)*. Rb mutant mice have also been genetically rescued by transgenic expression of a human Rb cDNA transgene driven by human Rb promoter in multiple tissues. Both developmental defects and tissue-specific tumorigenesis are completely rescued by the human transgene in these mice *(131)*. Thus, gene-targeting approaches have provided a unique Rb mouse model for addressing tissue-specific tumor predisposition by inactivation of a ubiquitously-expressed tumor-suppressor gene. The Rb mouse model also illustrates that a tumor-suppressor inactivation in mice does not always result in phenotypes often seen in human patients with mutations in the corresponding tumor suppressor gene.

Pituitary Tumors and Testicular Hyperplasia in p27-Deficient Mice

Cell-cycle progression is dependent on formation of a series of cyclin and cyclin-dependent kinase (CDK) complexes. The kinase activity of these complexes can be blocked by two families of CDK inhibitors, known as Cip/Kip (p21, p27, and p57) and Ink4 (p16, p15, p18, and p19). The ability of these two family members to inhibit CDK-cyclin complexes blocks the G1 to S phase transition of the cell cycle *(2)*. $p27^{Kip1}$ is widely expressed in multiple cell types in both humans and mice. $p27^{Kip1}$-deficient mice are larger, and display multiple-organ hyperplasia, including gonadal and adrenal hyperplasia *(132–134)*. Although mutant males are fertile, mutant females are infertile and do not undergo the normal ovarian folliculogenesis program leading to corpus luteum formation. The striking feature of p27-deficient mice, similar to Rb-heterozygous mice, is the development of pituitary intermediate-lobe adenomas at high incidence *(132–134)*. These tumors occur beginning at 12 wk of age, and contain a large number of pleomorphic and atypical cells. The tumors are also often cystic and hemorrhagic, but are noninvasive up to 7 mo of age. The tumor cells highly express three melanotrope-peptides derived from POMC—namely, α-MSH, ACTH, and β-endorphin—whereas anterior pituitary and posterior pituitary markers are unchanged *(132–134)*. The remarkable similarity in pituitary intermediate-lobe tumors between Rb-heterozygous and p27-deficient mice suggests that these two proteins may participate in the same pathway that normally limits melanotrope-cell proliferation. p27-heterozygous and homozygous mutant mice are also predisposed to tumors in multiple tissues when treated with X-rays or ethyl nitrosourea (ENU). In these studies, it has been demonstrated that the wild-type allele in tumors in p27-heterozygous mice is not mutated or silenced, in contrast to the situation observed with many tumor-suppressor genes *(135)*. Thus, gene-targeting studies have identified p27 to be haplo-insufficient for tumor suppression, and this CDK inhibitor does not fulfill Knudson's "two-hit" criterion for a tumor-suppressor gene *(135)*.

Tumorigenesis in Double-Mutant Mice Lacking General Cell-Cycle Regulators

The advantage of gene-knockout technology is the ability to systematically study tumorigenesis pathways in mice by identifying several modifier factors. It is often possible to identify synergism between factors that participate in the same pathway. This synergism can result in earlier occurrence of the tumors or development of more aggressive tumor phenotypes. Typically, intercrosses between mutant mice lacking different cell-cycle regulators or tumor suppressors have been made, and double-mutant mice of various genotype combinations have been developed *(136)*. These mice are then monitored for tumor initiation, development, survival, and occurrence of any novel phenotypes that are not normally seen in the mutant mice deficient in only one of the two genes *(136)*.

Cooperativity in tumorigenicity has been observed in mutant mice that are genotypically Rb-heterozygous and p53-nullizygous. These double-mutant mice have reduced viability, and develop novel types of tumors not typically seen in either Rb or p53-deficiency alone. These include pinealoblastomas, islet cell adenomas, bronchial-epithelial hyperplasia, and retinal dysplasia *(137,138)*. The genetic crosses between p27 and Rb mutant mice have also suggested important roles for these proteins in the pituitary intermediate lobe and thyroid C-cell. Mutant mice that are Rb+/– and p27–/– develop more aggressive thyroid C-cell carcinomas, and both thyroid and pituitary tumors develop much earlier than mice harboring the individual mutation *(139)*. Therefore, p27 and Rb cooperate to suppress tumor development involving overlapping signaling pathways. Similarly, functional interaction between p18^{Ink4c}, p19^{Ink4d}, and p27 has been discovered recently. Double-mutant mice that lack two CDK inhibitors (p18 and p27) develop a unique spectrum of neuroendocrine tumors, including pituitary, adrenal, thyroid, parathyroid, testis, pancreas, duodenum, and stomach cancers *(140)*. p19^{Ink4d} and p27 double-mutant mice demonstrate hyperproliferative neuronal-cell populations in all parts of the brain that are normally quiescent in wild-type mice *(141)*. Thus, genetic intercrosses between mutant mice lacking cell-cycle regulators will provide clues to the mechanisms of origin and development of multiple neuroendocrine tumors.

Pituitary Lactotrope Adenomas in Dopamine D2 Receptor-(D2-R)-Deficient Mice

Signaling through dopamine D2 receptors on lactotropes of the anterior pituitary is critical for prolactin gene expression and secretion. Genetic evidence in mice that lack D2R supports the hypothesis that functionally reduced dopamine inhibition of pituitary function leads to the development of pituitary tumors *(142)*. D2R-deficient female mice have persistent hyperprolactinemia associated with extensive hyperplasia of lactotropes, while mutant males do not have any lactotrope lesions up to 1 yr of age *(143)*. However, during later stages, both male and female mutant mice (at 17–20 mo of age) develop lactotrope adenomas. The tumors in females enlarge up to 50 times the normal pituitary size, and immunostaining reveals presence of monohormonal prolactin-cells in these tumors *(143)*. Mutant males exhibit multifocal, microscopic lactotrope adenomas with immunoreactivity for estrogen receptors and Pit-I transcription factor (the differentiated markers for lactotropes). Thus, these gene-targeting studies have identified two aspects of pathobiology of pituitary adenomas. First, the prolonged loss of dopamine inhibition with a concomitant long exposure to prolactin leads to pituitary adenomas, and tumorigenesis involves a sequential process of hyperplasia leading to neoplasia only in females *(143)*. Furthermore, these studies demonstrate the lactotrope adenomas consist of only prolactin-producing cells without any detectable bihormonal (both GH and prolactin) mammosomatotrope component *(143)*. More discussion on this tumor model and the uterine adenomas in this model has been presented in Chapter 15.

Gonadal and Adrenal Tumors Develop with 100% Penetrance in α-Inhibin-Knockout Mice

The generation and characterization of the inhibin-deficient mouse model illustrates the application of gene-targeting strategies to identify the novel tumor-suppressor role with restricted tissue specificity of a well-known secreted protein, inhibin *(144)*. Inhibin-deficient mice develop gonadal sex-cord-stromal tumors with 100% penetrance as early as 4 wk of age, accompanied by weight loss, cachexia, and destruction of hepatic and gastrointestinal-cell lineages, leading to death by 20 wk of age *(145,146)*. Gonadectomized mutant mice live longer, but develop adrenal tumors with 100% penetrance (*see* Chapters 13 and 14). The progression of these gonadal stromal tumors (granulosa- and

Sertoli-cell tumors) in the α-inhibin mutant mice has been further characterized by generating a series of double-transgenic mouse models, and thus identifying important modifier factors, including activin receptor IIA *(147)*, MIS *(148)*, androgens *(149)*, FSH, LH *(101,150)*, and follistatin, which modulate the gonadal stromal tumorigenesis pathway and the activin-inducd cachexia syndrome. Because inhibin-deficient mutant mice lose body wt with the progression of the gonadal tumors, accompanied by cachexia wasting syndrome, these parameters have been a useful index of monitoring the effects of various modifier factors in double-mutant mice *(136)*. Future studies with this model will involve unraveling the inhibin signaling pathways and cell-cycle events in gonadal cells to completely understand the specificity of inhibin's tumor-suppressor activity.

Prostate Epithelial Hyperplasia in Mxi 1-Deficient Mice

Mad family members are potent antagonists of Myc oncoproteins. The long arm of human chromosome 10 at region 24–26 has been implicated to contain a putative tumor suppressor, Mxi 1—a Mad family member—notably in human prostate cancer *(151)*. Mice deficient in Mxi 1 have increased susceptibility to multiple tumors. At about 1 yr of age, the mutant males develop prostate-epithelial hyperplasia *(151)*. The prostate from these mice demonstrate microscopic foci of enlarged glands containing hypercellular acini and dysplastic cells, with no evidence of neoplastic transformation *(151)*. Therefore, the absence of Mxi 1 alone is sufficient for preneoplastic lesions in the prostate. It may possible that a full-blown prostate adenocarcinoma may occur in Mxi 1-deficient mice after longer latency periods and upon mutatons in additional genetic loci.

The preceeding sections have illustrated with some examples the transgenic approaches to endocrine-cell-specific immortalization. Some of these and additional transgenic mouse models are summarized in Tables 1 and 2.

STRATEGIES FOR DERIVATION OF CELL LINES FROM TRANSGENIC MUTANT MICE WITH INDUCED IMMORTALIZATION

Both gain-of-function and loss-of-function mutant mice have been useful in deriving cell lines from the tumors developed in vivo in these mice as a result of the designed genetic changes. Some of these strategies are briefly described here.

Derivation of Cell Lines from Tumors with Directed Expression of SV40 TAg in Transgenic Mice

As described in previous sections, targeted tumorigenesis in multiple tissues has been achieved in transgenic mice by driving SV40 TAg gene expression using a variety of cell-specific promoters. The tumors from these mice have been the source for deriving cell lines from rare endocrine-cell types (Fig. 1). This strategy is the most direct and straightforward, and often results in the isolation of differentiated cell lines. Typically, the tumor tissue is minced and the dispersed cells cultured at 37°C in the presence of 10–15% fetal calf serum, often without any additional growth-factor supplementation. Usually, between 2 and 6 mo, when foci form and cells grow more rapidly, limiting dilution method is employed for subcloning cells and for eventual isolation of clonal cell lines *(48)*. This strategy is also useful for preparing cell lines from tumors at discrete stages of development, depending on the aggressiveness of the tumors in vivo *(50)*.

Table 1

Representative List of Endocrine Tumors Induced by Targeted Expression of Viral TAg Gene in Transgenic Mice

Tissue/cell type immortalized	*Promoter sequences used to drive SV40 TAg or (growth factors)*[a]	*Refs.*
Pinealocyte	Mouse TPH	*(37)*
Hypothalamic neurons		
Magnocellular	Rat GnRH	*(25)*
Catecholaminergic	Rat TH	*(32)*
Sertonergic	Mouse TPH	*(37)*
Anterior pituitary		
Gonadotropes	Human α-GSU	*(48–50)*
Gonadotropes	Rat LHβ	*(50)*
Gonadotropes	Human FSHβ	*(51)*
Gonadotropes	Mouse GnRH-R	*(52)*
Somatotropes	Rat Pit 1	*(42)*
Lactotropes	Rat prolactin (NGF)	*(44)*
Thyrotropes	Rat TSHβ	*(54)*
Corticotropes	Polyoma early region	*(55)*
Intermediate lobe		
Melanotropes	Rat POMC	*(57)*
Testis		
Sertoli cells	Human MIS	*(88,89)*
	Mouse MT I	*(90)*
Leydig cells	Mouse MT I	*(90)*
	Mouse inhibin α	*(93)*
	MMTV (aromatase)	*(94)*
Ovary		
Granulosa cells	Bovine α-GSU (LHβ-CTP)	*(102)*
	Mouse inhibin α	*(103)*
Uterus	Rat calbindin D-9K	*(108)*
	MMTV-LTR	*(109)*
	Mouse MT (ERα)	*(110)*
Adrenal		
Cortex	Human cytochrome P450	*(112)*
	Mouse inhibin α	*(113)*
Medulla	Rat GHRH	*(30)*
	Rat GATA 1	*(115)*
Pancreas		
Exocrine pancreas	Rat elastase I	*(75)*
Hepatobiliary tract	Rat gastrin	*(84)*
Endocrine pancreas		
α-islet cells	Rat preproglucagon	*(82)*
β-islet cells	Rat insulin II	*(79)*
	Rat vasopressin	*(60)*
Thyroid		
Thyroid follicular cells	Bovine thyroglobulin ($G_{S\alpha}$)	*(63)*
Medullary	Rat calcitonin/CGRP	*(70)*
	Rat c-kit	*(61)*
Anaplastic	Bovine thyroglobulin	*(74)*

[a]Promoters driving expression of genes other than TAg are indicated in parenthesis.

Table 2

Knockout Mouse Models with Endocrine-Cell Tumors

Knockout mouse model	*Tumor type*	*Refs*
p53	Testicular teratoma	*(118)*
Rb-heterozygous	Melanotrope	*(127,128)*
p27	Melanotrope	*(132–134)*
	Testicular hyperplasia	*(132–134)*
	Thyroid C-cell	*(132–134)*
	Adrenal cortex	*(132-134)*
Lats1	Ovarian sarcoma	*(163)*
Dopamine D2-R	Lactotrope adenoma	*(143)*
α-inhibin	Sertoli/granulosa-cell	*(144)*
MIS	Leydig-cell	*(148)*
Mxi1	Prostate hyperplasia	*(151)*
Brca1 conditional	Mammary gland	*(164)*
MEN-2B	Thyroid C-cell hyperplasia	*(165)*
	Pheochromocytoma	*(165)*

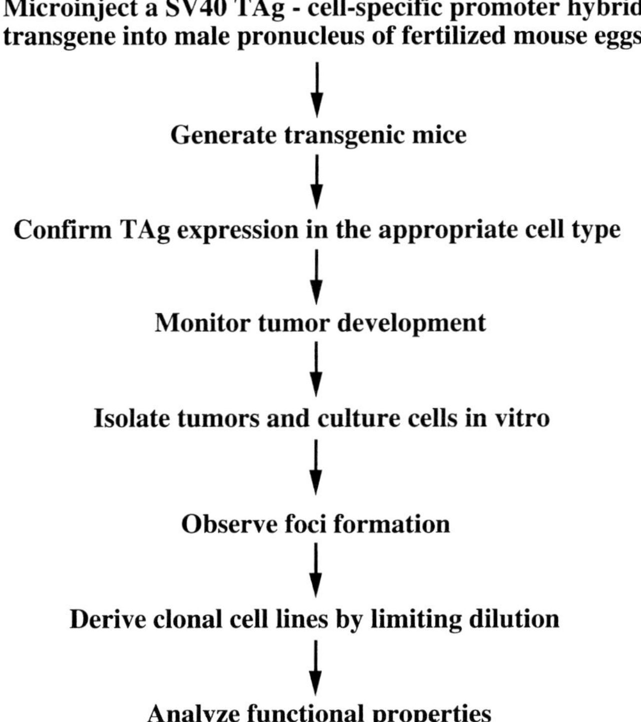

Fig. 1. Important steps involved in derivation of cell lines from tumors induced in transgenic mice by targeted expression of an oncogene.

Although some of these cell lines continue to express SV40 TAg protein, they lose their differentiated function and are not useful for many further studies.

Derivation of Conditionally Immortalized Cell Lines from Transgenic Mice with Directed Expression of ts TAg Genes

To circumvent the difficulties in maintaining the differentiated properties of many immortalized cell lines for functional studies, two transgenic approaches have been used. In the first approach, transgenic mice have been produced with cell-specific promoters directing the expression of a ts variant of TAg gene to desired tissues in vivo. Although the ts TAg protein is thermolabile, in many instances tumors do arise in mice and such tumors are usually slow-growing and do not cause early lethality of the transgenic mice that harbor the transgene *(88)*. Subsequently, cell lines have been derived that rapidly proliferate at the permissive temperature (33°C) and differentiate and express markers at the nonpermissive temperature (39°C). In the second approach, the ts variant of TAg gene has been targeted to multiple tissues under the regulatory influence of an interferon-inducible promoter that is ubiquitously active *(152)*. Depending on the level of activity of the promoter in a given tissue from this mouse model, further expression of the TAg can be achieved by induction in culture with interferon treatment of the cells derived from the tissue of interest. Both these methods have been successfully utilized to obtain clonal endocrine-cell lines *(152)*. At both the temperatures (33 and 39°C), the cells show distinct morphology, and they are functionally different. Thus, this approach results in a conditional immortalization of cells by temperature shifts in vitro, and is useful for analyzing both immortalization and differentiation events from a single-cell population.

Derivation of Conditionally Transformable Cell Lines from Bitransgenic Mice Expressing a Tetracyclin-Regulated Oncogene

In another transgenic approach of deriving cell lines that are conditionally immortalized, two separate lines of transgenic mice are generated *(153)*. One transgenic line of mice is derived by pronuclear microinjection of gene sequences encoding the bacterial tetracycline repressor (tet R) fused to the activating domain of the herpes simplex virus (HSV) protein VP16 under the regulatory control of a cell-specific promoter. A second line of transgenic mice is derived harboring the SV40 TAg gene under the control of a tandem array of *tet* operator sequences and a minimal promoter, which in itself is not sufficient for gene expression. These two lines of mice are then intercrossed to generate double-transgenic mice. Expression of the tet R fusion protein activates TAg transcription only in specific cells (depending on the promoter used), resulting in the development of tumors in the bitransgenic mice *(153)*. Cell lines prepared from such tumors can then be manipulated by the presence or absence of the drug tetracycline and its addition inhibits the cell proliferation at any given point of time. Tetracycline or its analogs can also be administered in vivo to bitransgenic mice to achieve conditional immortalization *(153)*.

Derivation of Cell Lines from Knockout Mutant Mice

Cell lines can also be derived from knockout mice that are deficient in general tumor suppressors or in important cell-cycle regulators. For example, cell lines derived from p53-deficient mice generally demonstrate increased replication times, and these p53-deficient cells can be readily immortalized *(154)*. Sometimes, loss of one p53 allele (cells obtained from p53-heterozygous mice) is sufficient enough for an effective immortalization event in vitro. It is also possible to generate, by genetic intercross, a combined mutant mouse strain that is heterozygous for a tumor suppressor (by knockout strategy) and also carries an oncogene under the regulation of a cell-specific promoter (by transgenic approach) *(155)*. Cells derived from such tumors have a complement of both a dominant oncogene and loss of a recessive tumor suppressor, and thus may have a better chance for efficient immortalization in vitro and for derivation of a novel cell line. It is also feasible to double-target both loci of a given tumor suppressor or cell-cycle regulator in ES cells and establish these mutant ES cell lines for a number of normal and cancer cell-cycle regulation studies *(156)*.

CONCLUSIONS AND FUTURE DIRECTIONS

This chapter illustrates the power of genetic manipulation in targeted tumorigenesis of various endocrine-cell types. Over the past two decades, hundreds of transgenic and knockout mutant mouse strains have been developed with tumor phenotypes in specific tissues. These mice will continue to be models for understanding the origin, progression, and molecular mechanisms of suppression of many human cancers. In addition, the fundamental issues of cell-cycle regulation can be studied using these tumor-prone mouse models.

More recent developments in the broad area of functional genomics offer enormous potential for future studies of cancer biology. Large-scale ENU mutagenesis programs coupled with retroviral insertion strategies in mice offer a feasible approach to identify many novel tumor suppressors that are relevant to human cancers *(157–159)*. Megabase-range chromosomal rearrangements often seen in human malignancies can possibly be modeled in corresponding regions in the mouse genome by cre-lox chromosome engineering technology *(160)*. It is now possible to establish "molecular profiling" of many human cancers by using high-throughput DNA microarray and protein array approaches, and thus, large-scale gene expression profiles of thousands of genes can be monitored at one time *(161)*. The cancer genome anatomy project at NIH primarily aims to catalog many such "profiles" for future studies. The recent advances in cell-cycle regulation have facilitated the identification of novel protein partners functionally important for cell-cycle progression in mice *(162)*. These may be potential targets for successful therapeutic intervention of many cancers. Hopefully, genetic approaches involving transgenic mice will soon be the molecular foundations for understanding the pathobiology of all cancers.

ACKNOWLEDGMENTS

I am grateful to Dr. Malcolm J. Low for introducing me to the fascinating field of targeted tumorigenesis in transgenic mice and for his encouragement all these years. I thank Dr. Hannes Vogel for his critical reading of the manuscript and for his comments. Finally, I thank Ms. Shirley Baker and Mr. Kelly Hart for their skillful and timely assistance in preparing this manuscript.

REFERENCES

1. Murray A, Hunt T. The Cell Cycle. An Introduction. Oxford University Press, New York, NY, 1993.
2. Cooper GM. Oncogenes. Jones and Bartlett Publishers, Boston, MA, 1995.
3. Robbins Pathologic Basis of Disease. W.B. Saunders, Phialdelphia, PA, 1999.

4. Kumar TR, Matzuk MM. Gene knockout models to study the hypothalamus-pituitary-gonadal axis. In: Shupnik MA, ed. Gene Engineering and Molecular Models in Endocrinology. Humana Press, Totowa, NJ, 2000.
5. Camper SA, Saunders TL, Kendall SK, et al. Implementing transgenic and embryonic stem cell technology to study gene expression, cell-cell interactions and gene function. Biol Reprod 1995;52:246–257.
6. Butel JS, Lednicky JA. Cell and molecular biology of simian virus 40. implications for human infections and disease. J Natl Cancer Inst 1999;91:119–134.
7. Jha KK, Banga S, Palejwala V, Ozer HL. SV40-Mediated immortalization. Exp Cell Res 1998;245:1–7.
8. Mole SE, Gannon JV, Ford MJ, Lane DP. Structure and function of SV40 large-T antigen. Philos Trans R Soc Lond B Biol Sci 1987;317:455–469.
9. White JA, Carter SG, Ozer HL, Boyd AL. Cooperativity of SV40 T antigen and ras in progressive stages of transformation of human fibroblasts. Exp Cell Res 1992;203:157–163.
10. Suh BS, Amsterdam A. Establishment of highly steroidogenic granulosa cell lines by cotransfection with SV40 and Ha-ras oncogene. induction of steroidogenesis by cyclic adenosine 3'-5'-monophosphate and its suppression by phorbol ester. Endocrinology 1990;127:2489–2500.
11. Reynisdottir I, Prives C. Two conditional tsA mutant simian virus 40 T antigens display marked differences in thermal inactivation. J Virol 1992;66:6517–6526.
12. Chou JY. Differentiated mammalian cell lines immortalized by temperature- sensitive tumor viruses. Mol Endocrinol 1989;3:1511–1514.
13. Cepko CL. Immortalization of neural cells via retrovirus-mediated oncogene transduction. Annu Rev Neurosci 1989;12:47–65.
14. Westerman KA, Leboulch P. Reversible immortalization of mammalian cells mediated by retroviral transfer and site-specific recombination. Proc Natl Acad Sci USA 1996;93:8971–8976.
15. Colgin LM, Reddel RR. Telomere maintenance mechanisms and cellular immortalization [published erratum appears in Curr Opin Genet Dev 1999;9(2):247]. Curr Opin Genet Dev 1999;9:97–103.
16. Bodnar AG, Ouellette M, Frolkis M, et al. Extension of life-span by introduction of telomerase into normal human cells. Science 1998;279:349–352.
17. Hahn WC, Stewart SA, Brooks MW, et al. Inhibition of telomerase limits the growth of human cancer cells. Nat Med 1999;5:1164–1170.
18. Werner H, Le Roith D. New concepts in regulation and function of the insulin-like growth factors. implications for understanding normal growth and neoplasia. Cell Mol Life Sci 2000;57:932–942.
19. Schroeder JA, Lee DC. Transgenic mice reveal roles for TGFalpha and EGF receptor in mammary gland development and neoplasia. J Mammary Gland Biol Neoplasia 1997;2:119–129.
20. Frame S, Balmain A. Integration of positive and negative growth signals during ras pathway activation in vivo. Curr Opin Genet Dev 2000;10:106–113.
21. Bradley A. Site-directed mutagenesis in the mouse. Recent Prog Horm Res 1993;48:237–251.
22. Palmiter RD, Brinster RL. Germ-line transformation of mice. Annu Rev Genet 1986;20:465–499.
23. Macleod KF, Jacks T. Insights into cancer from transgenic mouse models. J Pathol 1999;187:43–60.
24. Messing A, Pinkert CA, Palmiter RD, Brinster RL. Developmental study of SV40 large T antigen expression in transgenic mice with choroid plexus neoplasia. Oncogene Res 1988;3:87–97.
25. Mellon PL, Windle JJ, Goldsmith PC, et al. Immortalization of hypothalamic GnRH neurons by genetically targeted tumorigenesis. Neuron 1990;5:1–10.
26. Malven P. Mammalian Neuroendocrinology. CRC Press, Boca Raton, FL, 1993.
27. Wu TJ, Gibson MJ, Rogers MC, Silverman AJ. New observations on the development of the gonadotropin-releasing hormone system in the mouse. J Neurobiol 1997;33:983–998.
28. Mason AJ, Pitts SL, Nikolics K, et al. The hypogonadal mouse. reproductive functions restored by gene therapy. Science 1986;234:1372–1378.
29. Mellon PL, Wetsel WC, Windle JJ, et al. Immortalized hypothalamic gonadotropin-releasing hormone neurons. Ciba Found Symp 1992;168:104–117.
30. Asa SL, Kovacs K, Stefaneanu L, et al. Pituitary adenomas in mice transgenic for growth hormone-releasing hormone. Endocrinology 1992;131:2083–2089.
31. Nogues N, Magnan E, De Grandis P, et al. Expression of a fusion gene consisting of the mouse growth hormone- releasing hormone gene promoter linked to the SV40 T-antigen gene in transgenic mice. Mol Cell Endocrinol 1998;137:161–168.
32. Suri C, Fung BP, Tischler AS, Chikaraishi DM. Catecholaminergic cell lines from the brain and adrenal glands of tyrosine hydroxylase-SV40 T antigen transgenic mice. J Neurosci 1993;13: 1280–1291.

33. Mombaerts P. Molecular biology of odorant receptors in vertebrates. Annu Rev Neurosci 1999;22:487–509.
34. Buck LB. The molecular architecture of odor and pheromone sensing in mammals [comment]. Cell 2000;100:611–618.
35. Largent BL, Sosnowski RG, Reed RR. Directed expression of an oncogene to the olfactory neuronal lineage in transgenic mice. J Neurosci 1993;13:300–312.
36. Huh SO, Park DH, Cho JY, et al. A 6.1 kb 5' upstream region of the mouse tryptophan hydroxylase gene directs expression of E. coli lacZ to major serotonergic brain regions and pineal gland in transgenic mice. Brain Res Mol Brain Res 1994;24:145–152.
37. Son JH, Chung JH, Huh SO, et al. Immortalization of neuroendocrine pinealocytes from transgenic mice by targeted tumorigenesis using the tryptophan hydroxylase promoter. Brain Res Mol Brain Res 1996;37:32–40.
38. Rossant J. Manipulating the mouse genome. implications for neurobiology. Neuron 1990;4:323–334.
39. Rosenfeld MG, Bach I, Erkman L, et al. Transcriptional control of cell phenotypes in the neuroendocrine system. Recent Prog Horm Res 1996;51:217–238.
40. Dosen JS, Rosenfeld MG. Signaling mechanisms in pituitary morphogenesis and cell fate determination. Curr Opin Cell Biol 1999;11:669–677.
41. Shimon I, Melmed S. Genetic basis of endocrine disease. pituitary tumor pathogenesis. J Clin Endocrinol Metab 1997;82:1675–1681.
42. Lew D, Brady H, Klausing K, et al. GHF-1-promoter-targeted immortalization of a somatotropic progenitor cell results in dwarfism in transgenic mice. Genes Dev 1993;7:683–693.
43. Mayo KE, Hammer RE, Swanson LW, et al. Dramatic pituitary hyperplasia in transgenic mice expressing a human growth hormone-releasing factor gene. Mol Endocrinol 1988;2:606–612.
44. Borrelli E, Sawchenko PE, Evans RM. Pituitary hyperplasia induced by ectopic expression of nerve growth factor. Proc Natl Acad Sci USA 1992;89:2764–2768.
45. McAndrew J, Paterson AJ, Asa SL, et al. Targeting of transforming growth factor-alpha expression to pituitary lactotrophs in transgenic mice results in selective lactotroph proliferation and adenomas. Endocrinology 1995;136:4479–4488.
46. Cai A, Hayes JD, Patel N, Hyde JF. Targeted overexpression of galanin in lactotrophs of transgenic mice induces hyperprolactinemia and pituitary hyperplasia. Endocrinology 1999;140:4955–4964.
47. Pierce JG, Parsons TF. Glycoprotein hormones: structure and function. Annu Rev Biochem 1981;50:465–495.
48. Windle JJ, Weiner RI, Mellon PL. Cell lines of the pituitary gonadotrope lineage derived by targeted oncogenesis in transgenic mice. Mol Endocrinol 1990;4:597–603.
49. Alarid ET, Windle JJ, Whyte DB, Mellon PL. Immortalization of pituitary cells at discrete stages of development by directed oncogenesis in transgenic mice. Development 1996;122:3319–3329.
50. Alarid ET, Holley S, Hayakawa M, Mellon PL. Discrete stages of anterior pituitary differentiation recapitulated in immortalized cell lines. Mol Cell Endocrinol 1998;140:25–30.
51. Kumar TR, Graham KE, Asa SL, Low MJ. Simian virus 40 T antigen-induced gonadotroph adenomas. a model of human null cell adenomas. Endocrinology 1998;139:3342–3351.
52. Albarracin CT, Frosch MP, Chin WW. The gonadotropin-releasing hormone receptor gene promoter directs pituitary-specific oncogene expression in transgenic mice. Endocrinology 1999;140:2415–2421.
53. Yusta B, Alarid ET, Gordon DF, et al. The thyrotropin beta-subunit gene is repressed by thyroid hormone in a novel thyrotrope cell line, mouse T alphaT1 cells. Endocrinology 1998;139:4476–4482.
54. Maki K, Miyoshi I, Kon Y, et al. Targeted pituitary tumorigenesis using the human thyrotropin beta-subunit chain promoter in transgenic mice. Mol Cell Endocrinol 1994;105:147–154.
55. Helseth A, Siegal GP, Haug E, Bautch VL. Transgenic mice that develop pituitary tumors. A model for Cushing's disease. Am J Pathol 1992;140:1071–1080.
56. Yano H, Readhead C, Nakashima M, et al. Pituitary-directed leukemia inhibitory factor transgene causes Cushing's syndrome. neuro-immune-endocrine modulation of pituitary development. Mol Endocrinol 1998;12:1708–1720.
57. Low MJ, Liu B, Hammer GD, et al. Post-translational processing of proopiomelanocortin (POMC) in mouse pituitary melanotroph tumors induced by a POMC-simian virus 40 large T antigen transgene. J Biol Chem 1993;268:24,967–24,975.
58. Young JI, Otero V, Cerdan MG, et al. Authentic cell-specific and developmentally regulated expression of pro- opiomelanocortin genomic fragments in hypothalamic and hindbrain neurons of transgenic mice. J Neurosci 1998;18:6631–6640.

59. Hnasko R, Khurana S, Shackleford N, et al. Two distinct pituitary cell lines from mouse intermediate lobe tumors. a cell that produces prolactin-regulating factor and a melanotroph [see comments]. Endocrinology 1997;138:5589–5596.
60. Stefaneanu L, Rindi G, Horvath E, et al. Morphology of adenohypophysial tumors in mice transgenic for vasopressin-SV40 hybrid oncogene. Endocrinology 1992;130:1789–1795.
61. Baetscher M, Schmidt E, Shimizu A, et al. SV40 T antigen transforms calcitonin cells of the thyroid but not CGRP- containing neurons in transgenic mice. Oncogene 1991;6:1133–1138.
62. Bosse P, Bernex F, De Sepulveda P, et al. Multiple neuroendocrine tumours in transgenic mice induced by c-kit- SV40 T antigen fusion genes. Oncogene 1997;14:2661–2670.
63. Cho JY, Sagartz JE, Capen CC, et al. Early cellular abnormalities induced by RET/PTC1 oncogene in thyroid- targeted transgenic mice. Oncogene 1999;18:3659–3665.
64. Coppee F, Depoortere F, Bartek J, et al. Differential patterns of cell cycle regulatory proteins expression in transgenic models of thyroid tumours. Oncogene 1998;17:631–641.
65. Feunteun J, Michiels F, Rochefort P, et al. Targeted oncogenesis in the thyroid of transgenic mice. Horm Res 1997;47:137–139.
66. Michiels FM, Caillou B, Talbot M, et al. Oncogenic potential of guanine nucleotide stimulatory factor alpha subunit in thyroid glands of transgenic mice. Proc Natl Acad Sci USA 1994;91:10,488–10,492.
67. Michiels FM, Chappuis S, Caillou B, et al. Development of medullary thyroid carcinoma in transgenic mice expressing the RET protooncogene altered by a multiple endocrine neoplasia type 2A mutation. Proc Natl Acad Sci USA 1997;94:3330–3335.
68. Russo AF, Crenshaw EBD, Lira SA, et al. Neuronal expression of chimeric genes in transgenic mice. Neuron 1988;1:311–320.
69. Schuffenecker I, Billaud M, Calender A, et al. RET proto-oncogene mutations in French MEN 2A and FMTC families. Hum Mol Genet 1994;3:1939–1943.
70. Spada A, Lania A, Ballare E. G protein abnormalities in pituitary adenomas. Mol Cell Endocrinol 1998;142:1–14.
71. Tong Q, Xing S, Jhiang SM. Leucine zipper-mediated dimerization is essential for the PTC1 oncogenic activity. J Biol Chem 1997;272:9043–9047.
72. Johnston D, Hatzis D, Sunday ME. Expression of v-Ha-ras driven by the calcitonin/calcitonin gene-related peptide promoter. a novel transgenic murine model for medullary thyroid carcinoma. Oncogene 1998;16:167–177.
73. Felici A, Giorgio M, Krauzewicz N, et al. Medullary thyroid carcinomas in transgenic mice expressing a Polyoma carboxyl-terminal truncated middle-T and wild type small-T antigens. Oncogene 1999;18:2387–2395.
74. Ledent C, Dumont J, Vassart G, Parmentier M. Thyroid adenocarcinomas secondary to tissue-specific expression of simian virus-40 large T-antigen in transgenic mice. Endocrinology 1991;129:1391–1401.
75. Hammer RE, Swift GH, Ornitz DM, et al. The rat elastase I regulatory element is an enhancer that directs correct cell specificity and developmental onset of expression in transgenic mice. Mol Cell Biol 1987;7:2956–2967.
76. Ornitz DM, Hammer RE, Messing A, et al. Pancreatic neoplasia induced by SV40 T-antigen expression in acinar cells of transgenic mice. Science 1987;238:188–193.
77. Bell Jr RH, Memoli VA, Longnecker DS. Hyperplasia and tumors of the islets of Langerhans in mice bearing an elastase I-SV40 T-antigen fusion gene. Carcinogenesis 1990;11:1393–1398.
78. Sandgren EP, Quaife CJ, Paulovich AG, et al. Pancreatic tumor pathogenesis reflects the causative genetic lesion. Proc Natl Acad Sci USA 1991;88:93–97.
79. Hanahan D. Heritable formation of pancreatic beta-cell tumours in transgenic mice expressing recombinant insulin/simian virus 40 oncogenes. Nature 1985;315:115–122.
80. Bergers G, Javaherian K, Lo KM, et al. Effects of angiogenesis inhibitors on multistage carcinogenesis in mice. Science 1999;284:808–812.
81. Gotz W, Schucht C, Roth J, et al. Endocrine pancreatic tumors in MSV-SV40 large T transgenic mice. Am J Pathol 1993;142:1493–1503.
82. Efrat S, Teitelman G, Anwar M, et al. Glucagon gene regulatory region directs oncoprotein expression to neurons and pancreatic alpha cells. Neuron 1988;1:605–613.
83. Lee YC, Asa SL, Drucker DJ. Glucagon gene 5'-flanking sequences direct expression of simian virus 40 large T antigen to the intestine, producing carcinoma of the large bowel in transgenic mice. J Biol Chem 1992;267:10,705–10,708.

84. Montag AG, Oka T, Baek KH, et al. Tumors in hepatobiliary tract and pancreatic islet tissues of transgenic mice harboring gastrin simian virus 40 large tumor antigen fusion gene. Proc Natl Acad Sci USA 1993;90:6696–6700.
85. Jetton TL, Moates JM, Lindner J, et al. Targeted oncogenesis of hormone-negative pancreatic islet progenitor cells. Proc Natl Acad Sci USA 1998;95:8654–8659.
86. Lopez MJ, Upchurch BH, Rindi G, Leiter AB. Studies in transgenic mice reveal potential relationships between secretin-producing cells and other endocrine cell types. J Biol Chem 1995;270:885–891.
87. Parker KL, Schimmer BP, Schedl A. Genes essential for early events in gonadal development. Cell Mol Life Sci 1999;55:831–838.
88. Peschon JJ, Behringer RR, Cate RL, et al. Directed expression of an oncogene to Sertoli cells in transgenic mice using mullerian inhibiting substance regulatory sequences. Mol Endocrinol 1992;6:1403–1411.
89. Dutertre M, Rey R, Porteu A, et al. A mouse Sertoli cell line expressing anti-Mullerian hormone and its type II receptor. Mol Cell Endocrinol 1997;136:57–65.
90. Lebel M, Mes-Masson AM. Establishment and characterization of testicular cell lines from MT- PVLT-10 transgenic mice. Exp Cell Res 1994;213:12–19.
91. Ascoli M. Molecular basis of the regulation of the lutropin/choriogonadotropin receptor. Biochem Soc Trans 1997;25:1021–1026.
92. Li Q, Yoshioka N, Yutsudo M, et al. Human papillomavirus-induced carcinogenesis with p53 deficiency in mouse. novel lymphomagenesis in HPV16E6E7 transgenic mice mimicking p53 defect. Virology 1998;252:28–33.
93. Kananen K, Markkula M, el-Hefnawy T, et al. The mouse inhibin alpha-subunit promoter directs SV40 T-antigen to Leydig cells in transgenic mice [published erratum appears in Mol Cell Endocrinol 1996;122(1):109,110]. Mol Cell Endocrinol 1996;119:135–146.
94. Fowler KA, Gill K, Kirma N, et al. Overexpression of aromatase leads to development of testicular leydig cell tumors . an in vivo model for hormone-mediated TesticularCancer. Am J Pathol 2000;156:347–353.
95. Muller AJ, Teresky AK, Levine AJ. A male germ cell tumor-susceptibility-determining locus, pgct1, identified on murine chromosome 13. Proc Natl Acad Sci USA 2000;97:8421–8426.
96. Nayernia K, Samani AA, Klaproth S, Engel W. Haploid male germ cells show no susceptibility to transformation by simian virus 40 large tumour antigen in transgenic mice. Cell Biol Int 1998;22:437–443.
97. Korpelainen EI, Karkkainen MJ, Tenhunen A, et al. Overexpression of VEGF in testis and epididymis causes infertility in transgenic mice. evidence for nonendothelial targets for VEGF. J Cell Biol 1998;143:1705–1712.
98. Shibata MA, Jorcyk CL, Devor DE, et al. Altered expression of transforming growth factor betas during urethral and bulbourethral gland tumor progression in transgenic mice carrying the androgen-responsive C3(1) 5' flanking region fused to SV40 large T antigen. Carcinogenesis 1998;19:195–205.
99. Elvin JA, Matzuk MM. Mouse models of ovarian failure. Rev Reprod 1998;3:183–195.
100. Kumar TR, Fairchild-Huntress V, Low MJ. Gonadotrope-specific expression of the human follicle-stimulating hormone beta-subunit gene in pituitaries of transgenic mice. Mol Endocrinol 1992;6:81–90.
101. Kumar TR, Palapattu G, Wang P, et al. Transgenic models to study gonadotropin function: the role of follicle- stimulating hormone in gonadal growth and tumorigenesis. Mol Endocrinol 1999;13:851–865.
102. Keri RA, Lozada KL, Abdul-Karim FW, et al. Luteinizing hormone induction of ovarian tumors: oligogenic differences between mouse strains dictates tumor disposition. Proc Natl Acad Sci USA 2000;97:383–387.
103. Kananen K, Markkula M, Rainio E, et al. Gonadal tumorigenesis in transgenic mice bearing the mouse inhibin alpha-subunit promoter/simian virus T-antigen fusion gene: characterization of ovarian tumors and establishment of gonadotropin- responsive granulosa cell lines. Mol Endocrinol 1995;9:616–627.
104. Morita Y, Tilly JL. Oocyte apoptosis. like sand through an hourglass. Dev Biol 1999;213:1–17.
105. Hsu SY, Lai RJ, Finegold M, Hsueh AJ. Targeted overexpression of Bcl-2 in ovaries of transgenic mice leads to decreased follicle apoptosis, enhanced folliculogenesis, and increased germ cell tumorigenesis. Endocrinology 1996;137:4837–4843.
106. Morita Y, Perez GI, Maravei DV, et al. Targeted expression of Bcl-2 in mouse oocytes inhibits ovarian follicle atresia and prevents spontaneous and chemotherapy-induced oocyte apoptosis in vitro. Mol Endocrinol 1999;13:841–850.
107. Van de Ven WJ. Genetic basis of uterine leiomyoma: involvement of high mobility group protein genes. Eur J Obstet Gynecol Reprod Biol 1998;81:289–293.

108. Blin C, L'Horset F, Romagnolo B, et al. Functional and growth properties of a myometrial cell line derived from transgenic mice. effects of estradiol and antiestrogens. Endocrinology 1996;137:2246–2253.
109. Webster MA, Martin-Soudant N, Nepveu A, et al. The induction of uterine leiomyomas and mammary tumors in transgenic mice expressing polyomavirus (PyV) large T (LT) antigen is associated with the ability of PyV LT antigen to form specific complexes with retinoblastoma and CUTL1 family members. Oncogene 1998;16:1963–1972.
110. Couse JF, Davis VL, Hanson RB, et al. Accelerated onset of uterine tumors in transgenic mice with aberrant expression of the estrogen receptor after neonatal exposure to diethylstilbestrol. Mol Carcinog 1997;19:236–242.
111. Arbeit JM, Howley PM, Hanahan D. Chronic estrogen-induced cervical and vaginal squamous carcinogenesis in human papillomavirus type 16 transgenic mice. Proc Natl Acad Sci USA 1996;93:2930–2935.
112. Mellon SH, Miller WL, Bair SR, et al. Steroidogenic adrenocortical cell lines produced by genetically targeted tumorigenesis in transgenic mice. Mol Endocrinol 1994;8:97–108.
113. Kananen K, Markkula M, Mikola M, et al. Gonadectomy permits adrenocortical tumorigenesis in mice transgenic for the mouse inhibin alpha-subunit promoter/simian virus 40 T-antigen fusion gene: evidence for negative autoregulation of the inhibin alpha- subunit gene. Mol Endocrinol 1996;10:1667–1677.
114. Kero J, Poutanen M, Zhang FP, et al. Elevated luteinizing hormone induces expression of its receptor and promotes steroidogenesis in the adrenal cortex. J Clin Invest 2000;105:633–641.
115. Cairns LA, Crotta S, Minuzzo M, et al. Immortalization of neuro-endocrine cells from adrenal tumors arising in SV40 T-transgenic mice. Oncogene 1997;14:3093–3098.
116. Sharan SK, Bradley A. Role of transgenic mice in identification and characterization of tumour suppressor genes. Cancer Surv 1995;25:143–159.
117. Kumar TR, Matzuk MM. Transgenic mice as models of disease. In: Jameson JL, ed., Textbook of Molecular Medicine. Humana Press, Totowa, NJ, 1998, pp. 97–110.
118. Donehower LA, Harvey M, Slagle BL, et al. Mice deficient for p53 are developmentally normal but susceptible to spontaneous tumours. Nature 1992;356:215–221.
119. Donehower LA, Harvey M, Vogel H, et al. Effects of genetic background on tumorigenesis in p53-deficient mice. Mol Carcinog 1995;14:16–22.
120. Harvey M, McArthur MJ, Montgomery Jr CA, et al. Spontaneous and carcinogen-induced tumorigenesis in p53-deficient mice. Nat Genet 1993;5:225–229.
121. Harvey M, Vogel H, Morris D, et al. A mutant p53 transgene accelerates tumour development in heterozygous but not nullizygous p53-deficient mice. Nat Genet 1995;9:305–311.
122. Venkatachalam S, Shi YP, Jones SN, et al. Retention of wild-type p53 in tumors from p53 heterozygous mice. reduction of p53 dosage can promote cancer formation. EMBO J 1998;17:4657–4667.
123. Adams PD, Kaelin Jr WG. Negative control elements of the cell cycle in human tumors. Curr Opin Cell Biol 1998;10:791–797.
124. Lee EY, Chang CY, Hu N, et al. Mice deficient for Rb are nonviable and show defects in neurogenesis and haematopoiesis. Nature 1992;359:288–294.
125. Jacks T, Fazeli A, Schmitt EM, et al. Effects of an Rb mutation in the mouse. Nature 1992;359:295–300.
126. Clarke AR, Maandag ER, van Roon M, et al. Requirement for a functional Rb-1 gene in murine development. Nature 1992;359:328–330.
127. Hu N, Gutsmann A, Herbert DC, et al. Heterozygous Rb-1 delta 20/+mice are predisposed to tumors of the pituitary gland with a nearly complete penetrance. Oncogene 1994;9:1021–1027.
128. Williams BO, Schmitt EM, Remington L, et al. Extensive contribution of Rb-deficient cells to adult chimeric mice with limited histopathological consequences. EMBO J 1994;13:4251–4259.
129. Nikitin A, Lee WH. Early loss of the retinoblastoma gene is associated with impaired growth inhibitory innervation during melanotroph carcinogenesis in Rb+/- mice. Genes Dev 1996;10:1870–1879.
130. Nikitin AY, Juarez-Perez MI, Li S, et al. RB-mediated suppression of spontaneous multiple neuroendocrine neoplasia and lung metastases in Rb+/- mice. Proc Natl Acad Sci USA 1999;96:3916–3921.
131. Chang CY, Riley DJ, Lee EY, Lee WH. Quantitative effects of the retinoblastoma gene on mouse development and tissue-specific tumorigenesis. Cell Growth Differ 1993;4:1057–1064.
132. Nakayama K, Ishida N, Shirane M, et al. Mice lacking p27(Kip1) display increased body size, multiple organ hyperplasia, retinal dysplasia, and pituitary tumors. Cell 1996;85:707–720.
133. Kiyokawa H, Kineman RD, Manova-Todorova KO, et al. Enhanced growth of mice lacking the cyclin-dependent kinase inhibitor function of p27(Kip1). Cell 1996;85:721–732.

134. Fero ML, Rivkin M, Tasch M, et al. A syndrome of multiorgan hyperplasia with features of gigantism, tumorigenesis, and female sterility in p27(Kip1)-deficient mice. Cell 1996;85:733–744.
135. Fero ML, Randel E, Gurley KE, et al. The murine gene p27Kip1 is haplo-insufficient for tumour suppression. Nature 1998;396:177–180.
136. Matzuk MM, Kumar TR, Shou W, et al. Transgenic models to study the roles of inhibins and activins in reproduction, oncogenesis, and development. Recent Prog Horm Res 1996;51:123–154.
137. Williams BO, Remington L, Albert DM, et al. Cooperative tumorigenic effects of germline mutations in Rb and p53. Nat Genet 1994;7:480–484.
138. Harvey M, Vogel H, Lee EY, et al. Mice deficient in both p53 and Rb develop tumors primarily of endocrine origin. Cancer Res 1995;55:1146–1151.
139. Park MS, Rosai J, Nguyen HT, et al. p27 and Rb are on overlapping pathways suppressing tumorigenesis in mice. Proc Natl Acad Sci USA 1999;96:6382–6387.
140. Franklin DS, Godfrey VL, Lee H, et al. CDK inhibitors p18(INK4c) and p27(Kip1) mediate two separate pathways to collaboratively suppress pituitary tumorigenesis. Genes Dev 1998;12:2899–2911.
141. Zindy F, Cunningham JJ, Sherr CJ, et al. Postnatal neuronal proliferation in mice lacking Ink4d and Kip1 inhibitors of cyclin-dependent kinases. Proc Natl Acad Sci USA 1999;96:13,462–13,467.
142. Kelly MA, Rubinstein M, Asa SL, et al. Pituitary lactotroph hyperplasia and chronic hyperprolactinemia in dopamine D2 receptor-deficient mice. Neuron 1997;19:103–113.
143. Asa SL, Kelly MA, Grandy DK, Low MJ. Pituitary lactotroph adenomas develop after prolonged lactotroph hyperplasia in dopamine D2 receptor-deficient mice. Endocrinology 1999;140:5348–5355.
144. Matzuk MM, Finegold MJ, Su JG, et al. Alpha-inhibin is a tumour-suppressor gene with gonadal specificity in mice. Nature 1992;360:313–319.
145. Matzuk MM, Finegold MJ, Mather JP, et al. Development of cancer cachexia-like syndrome and adrenal tumors in inhibin-deficient mice. Proc Natl Acad Sci USA 1994;91:8817–8821.
146. Li Q, Karam SM, Coerver KA, et al. Stimulation of activin receptor II signaling pathways inhibits differentiation of multiple gastric epithelial lineages. Mol Endocrinol 1998;12:181–192.
147. Coerver KA, Woodruff TK, Finegold MJ, et al. Activin signaling through activin receptor type II causes the cachexia- like symptoms in inhibin-deficient mice. Mol Endocrinol 1996;10:534–543.
148. Matzuk MM, Finegold MJ, Mishina Y, et al. Synergistic effects of inhibins and mullerian-inhibiting substance on testicular tumorigenesis. Mol Endocrinol 1995;9:1337–1345.
149. Shou W, Woodruff TK, Matzuk MM. Role of androgens in testicular tumor development in inhibin-deficient mice. Endocrinology 1997;138:5000–5005.
150. Kumar TR, Wang Y, Matzuk MM. Gonadotropins are essential modifier factors for gonadal tumor development in inhibin-deficient mice. Endocrinology 1996;137:4210–4216.
151. Schreiber-Agus N, Meng Y, Hoang T, et al. Role of Mxi1 in ageing organ systems and the regulation of normal and neoplastic growth. Nature 1998;393:483–487.
152. Jat PS, Noble MD, Ataliotis P, et al. Direct derivation of conditionally immortal cell lines from an H-2Kb- tsA58 transgenic mouse. Proc Natl Acad Sci USA 1991;88:5096–5100.
153. Efrat S, Fusco-DeMane D, Lemberg H, et al. Conditional transformation of a pancreatic beta-cell line derived from transgenic mice expressing a tetracycline-regulated oncogene. Proc Natl Acad Sci USA 1995;92:3576–3580.
154. Harvey M, Sands AT, Weiss RS, et al. In vitro growth characteristics of embryo fibroblasts isolated from p53- deficient mice. Oncogene 1993;8:2457–2467.
155. Colucci-D'Amato GL, Santelli G, D'Alessio A, et al. Dbl expression driven by the neuron specific enolase promoter induces tumor formation in transgenic mice with a p53(+/-) genetic background. Biochem Biophys Res Commun 1995;216:762–770.
156. Abuin A, Bradley A. Recycling selectable markers in mouse embryonic stem cells. Mol Cell Biol 1996;16:1851–1856.
157. Woychik RP, Klebig ML, Justice MJ, et al. Functional genomics in the post-genome era [published erratum appears in Mutat Res 19983;422(2):367]. Mutat Res 1998;400:3–14.
158. Justice MJ, Noveroske JK, Weber JS, et al. Mouse ENU mutagenesis. Hum Mol Genet 1999;8:1955–1963.
159. Su H, Wang X, Bradley A. Nested chromosomal deletions induced with retroviral vectors in mice. Nat Genet 2000;24:92–95.
160. Ramirez-Solis R, Liu P, Bradley A. Chromosome engineering in mice. Nature 1995;378:720–724.
161. Hughes TR, Marton MJ, Jones AR, et al. Functional discovery via a compendium of expression profiles. Cell 2000;102:109–126.
162. Liu Q, Guntuku S, Cui XS, et al. Chk1 is an essential kinase that is regulated by Atr and required for the G(2)/M DNA damage checkpoint. Genes Dev 2000;14:1448–1459.

163. St. John MA, Tao W, Fei X, et al. Mice deficient of Lats1 develop soft-tissue sarcomas, ovarian tumours and pituitary dysfunction. Nat Genet 1999;21:182–186.
164. Xu X, Wagner KU, Larson D, et al. Conditional mutation of Brca1 in mammary epithelial cells results in blunted ductal morphogenesis and tumour formation. Nat Genet 1999;22:37–43.
165. Smith-Hicks CL, Sizer KC, Powers JF, et al. C-cell hyperplasia, pheochromocytoma and sympathoadrenal malformation in a mouse model of multiple endocrine neoplasia type 2B. EMBO J 2000;19:612–622.

Index

D

F

G

M